P9-ASJ-785

The American Railroad Passenger Car

THE JOHNS HOPKINS UNIVERSITY PRESS   BALTIMORE AND LONDON

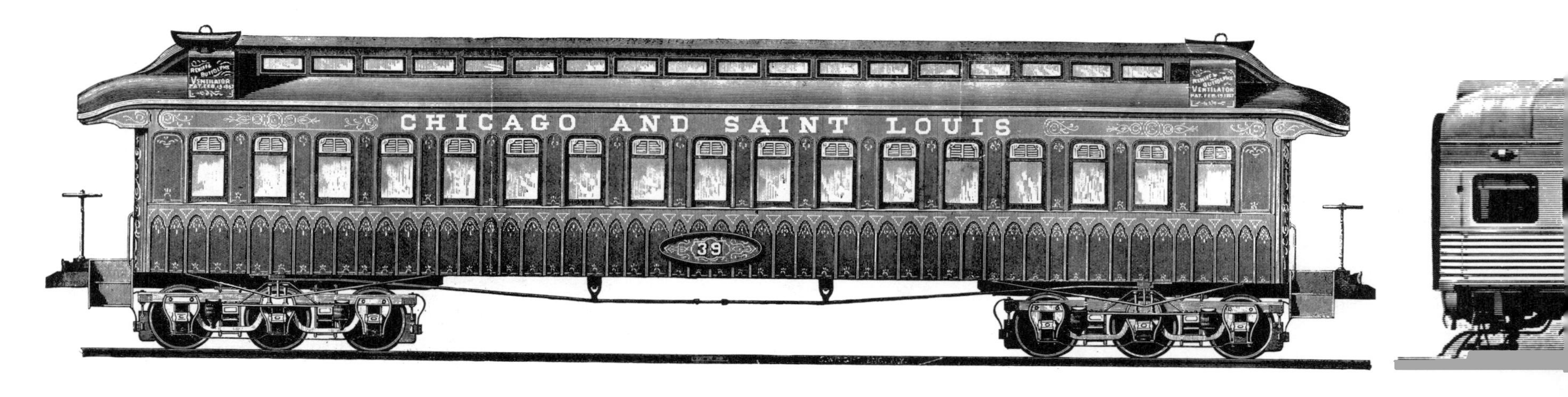

The American Railroad Passenger Car

JOHN H. WHITE, JR.

Part 1

Printed in the United States of America

Originally published in one hardcover volume, 1978
Third printing, 1979
Johns Hopkins Paperbacks edition, published in two softcover volumes, 1985
Fourth printing, 1991

The Johns Hopkins University Press
701 West 40th Street
Baltimore, Maryland 21211
The Johns Hopkins Press Ltd., London

Library of Congress Cataloging in Publication Data

White, John H.
The American railroad passenger car.

"Originally published in one hardcover volume, 1978"—
Vol. 1, t.p. verso.
Bibliography: v. 2, p. 677
Includes index.
1. Railroads—United States—Passenger-cars—History.
I. Title.
[TF455.W45 1985] 625.2'3'0973 84-26161
ISBN 0-8018-2743-4 (pbk.: set: alk. paper)
ISBN 0-8018-2722-1 (pbk.: v. 1: alk. paper)
ISBN 0-8018-2747-7 (pbk.: v. 2: alk. paper)

For Christine

CONTENTS

PREFACE

AMERICANS once spent a considerable portion of their lives in railroad cars. From approximately 1860 to 1930 railroads were the most common means of inland transportation, and at their peak just before the First World War, 98 percent of all intercity travel was by rail. No other means of transportation, not even the automobile, has achieved such a monopoly. Each day millions of passengers mounted the end platform steps of steam cars, some to travel a few miles to a suburban home, others to live within the swaying body of a cinder-laden Pullman for a week as it crossed the continent.

This vehicle that was once intimately intertwined with the lives of all Americans has rapidly receded from view. What was commonplace has grown obscure, even unknown. An abundance of surviving artifacts and documents makes it possible to reconstruct everyday life in the American home of that period in great detail, but information about the railroad car, particularly its early years, is difficult to recover. Early cars were junked when they were condemned as obsolete. Engineering offices systematically discarded old drawings, photographs, and specifications (the historian's enemy is always the fastidious housekeeper). The data that have survived are scattered, incomplete, and occasionally contradictory.

It is the purpose of this book to provide a picture of American railroad passenger cars as they were and as they are now. It will try to answer such basic questions as what they looked like at various periods, how they were made, what materials went into their construction, who made them, how the interiors were arranged, how many of each type were in service, and what they cost. The subject is a complicated one, for the passenger car is the product of many crafts and technologies. Even the day coach evolved into a complex conveyance with a maze of auxiliary machinery and systems. Long before steam railroads were envisioned, cabinetmakers, wheelwrights, blacksmiths, upholsterers, foundry workers, drapers, painters, and even silversmiths combined their skills to produce carriages. Some were elegant and light, others plain and rugged. Many were custom-made to suit the buyer's taste or purse. All reflected man's ingenuity in adapting and reshaping natural elements into useful products for human convenience.

Thus the railroad car was born of the established coach maker's tradition. Carpenters and bridge builders soon introduced structural designs that permitted larger cars. In more recent times steelmakers and electrical and aeronautical engineers have contributed their technologies. The evolution of the passenger car has always been dependent on known methods and materials; it has never inspired a major technical breakthrough. Although its progress might be criticized as overly conservative, it has moved forward in a slow but constant series of changes during its 150-year development.

In any technical history it is necessary to deal with the origins of important inventions. Establishing who did what becomes a somewhat arbitrary exercise, but there seems no way to avoid it. Obviously there is little genuine novelty in mechanics, and every inventor has predecessors whose work was close to his. Alexander von Humboldt claimed that an invention goes through three stages: first its existence is doubted, next its importance is denied, and finally the credit is given to someone else.

A more important question about every major invention is when it was accepted into actual practice. The naïve assumption of many popular histories is that once an invention is born, it is immediately put into everyday use. Far more often, a sound idea is initially rejected and then reintroduced at a later date. Determining when an innovation became standard practice is extremely difficult because the event was rarely newsworthy; naturally the technical press tends to emphasize new or unusual ideas. It is the plan of this book, however, to trace the popular acceptance of inventions and to concentrate on ordinary standard designs. Novelties and experiments are included, but they are not emphasized.

My main purpose here is to describe the passenger car as a physical object. I have included some information about the economics and history of travel in order to explain the appearance and disappearance of certain types of car—the railway mail car, for example. Also, the general evolution of the passenger car is tied to the growth of the railroad system, since longer trips in turn called for more comfortable cars. And growth brought increasing competition between parallel lines, which courted passengers by introducing such costly luxuries as dining cars and air conditioning. But the reader should not expect a complete discussion of rail travel history, or a description of each class in every railroad's passenger rolling stock.

Some readers may ask why another book on railroad cars should be added to an extensive bibliography which already includes the work of Lucius M. Beebe, Arthur D. Dubin, Robert J. Wayner, and other authorities. Yet the existing books concentrate on name trains and luxury cars, detailing the joys and splendors of first-class travel. It is easy for a reader to forget that the majority of passengers occupied coach seats, ate box lunches, and sat up all night. The more nostalgic literature deals in velvet-smooth rides with sunset vistas of mountain lakes through picture windows. There is no mention of the incessant rattle and clatter, or of the dust and cinders in pre-air-conditioned days. In fact travel by any mode, even by rail, is often a miserable experience, and I have attempted to cover this side of the story as well.

The existing books do not attempt to describe ordinary cars. In addition, they say little about construction, and they ignore design except for the topic of decor. As early as 1894 *The Railroad Car Journal* pointed out: "Perhaps no branch of the mechanical arts is so poorly supplied with literature as the art of car construction—particularly such as relates to its past history and development." Sixty years would pass before August Mencken's book *The Railroad Passenger Car* (1957) ventured into these uncharted waters. At that time, railroad historical literature was still confined to the evolution of locomotives and the corporate growth of the railroad companies.

This book emphasizes nineteenth-century developments rather than more recent ones, because the nineteenth century was the formative period when the basic shape, arrangement, and types of car came into being. And whereas little meaningful work is being done on the early period, the trade literature and the increasing number of publications for railroad enthusiasts are abundantly documenting twentieth-century developments.

When I began work on the subject in 1968, the railroad passenger car seemed to be fading into the ranks of antique modes of transportation. Few new cars had been ordered since 1955, and the complete abandonment of long-distance trains seemed inevitable. I felt that I might be writing the preface to the end of passenger car history. But Amtrak was created, and a generous public purse has kept the trains running and has even funded a new generation of cars. The new ones do not materially differ from the established lightweight designs, however, and for this reason I saw no need to go beyond the cutoff date of 1970 which I had originally set.

Most travelers fail to appreciate the passenger train for the miracle it is. In essence it forms a small city on wheels, with lighting, heating, air conditioning, food services, toilets, washrooms, a water supply, and sleeping, seating, and lounging facilities for perhaps a thousand people. All these systems must be fitted into the cramped spaces available between the side panels, under the floor, and in utility closets. Space is at a premium, making miniaturization a necessity. Dependability, weight, and cost are also crucial factors. Perhaps even more remarkable, this 1,400-ton city on wheels crosses the countryside at 80 miles an hour through all extremes of weather. It is the perfection of this movable city, a triumph of American technology, that we will survey in the following pages.

Acknowledgments

Many individuals and institutions assisted me in the preparation of this volume, and it is a pleasure to have an opportunity to express my gratitude. For eight years the staff of the Smithsonian Library showed great energy and patience in borrowing the technical journals and books needed for my research. Jack Goodwin, Charles Berger, Frank Pietropaoli, and Lucien R. Rossignol of the library staff were particularly helpful. The manuscript was typed and meticulously proofread by Mary E. Braunagel—to quote George Hilton, her typing efficiency "never ceased to impress me." Dorothy Young, Jean Stensland, Luwan Brown, and Elizabeth James were also helpful in assembling lists, photographs, and tables.

The advice, criticism, and encouragement of Arthur D. Dubin and William D. Edson were invaluable. Their knowledgeable comments did much to improve the text. I am also indebted to Dr. Louis Marre for his critical reading of the final chapter, on rail cars. Many errors and omissions were brought to my attention by these experts.

The following persons provided information or illustrations needed to complete this study: A. Andrew Merrilees, Douglas Wornom, Robert Wayner, Everett DeGolyer, Jr., G. M. Best, Lee Rogers, the late L. W. Sagle, R. C. Reed, C. M. Clegg, Cornelius Hauck, Nora Wilson, John Keller, Lester S. Levy, Ralph Greenhill, George Hart, George Krambles, Lawrence S. Williams, E. P. Alexander, Thomas T. Taber, Herbert H. Harwood, Jr., and Peter B. Bell. Institutions that provided information or illustrations are the National Archives, the Library of Congress, the Historical Society of Pennsylvania, the Lancaster County Historical Society, the Hagley Foundation Library, the Harvard School of Business' Baker Library, the Newberry Library, the Deutches Museum, the Peale Museum, the Baltimore and Ohio Transportation Museum, Pullman, Inc., the Budd Company, the Southern Railway, General Motors Corporation, The Railroadians, and the Hall of Records-Delaware Archives. The special attention over many years of John McLoed and Helen Roland, librarians at the Association of American Railroads, must also be mentioned.

The Simmons-Boardman Company, through its long-time publisher, Mr. Robert G. Lewis, kindly permitted us to reproduce many line drawings from past issues of the *Railway Age* and the *Car Builders' Cyclopedia.*

Direct financial support for the production of this book came from the Allegheny Foundation—Richard M. Scaife, Chairman; the National Museum of History and Technology—Brooke Hindle, Director, Silvio A. Bedini, Deputy Director, and Robert G. Tillotson, Assistant Director; Mr. W. Graham Claytor, Jr.; Morgan Guaranty Trust Company of New York; Mr. John F. Ruffle; and the Smithsonian Institution.

The staff of The Johns Hopkins University Press showed more than the expected professional cooperation in producing the book, and I wish to thank the following persons: Henry Y. K. Tom, Nancy Essig, Patrick Turner, Susan Bishop, Mary Lou Kenney, and finally Mary Barnett, whose skillful copy editing did so much to tighten and smooth the final text.

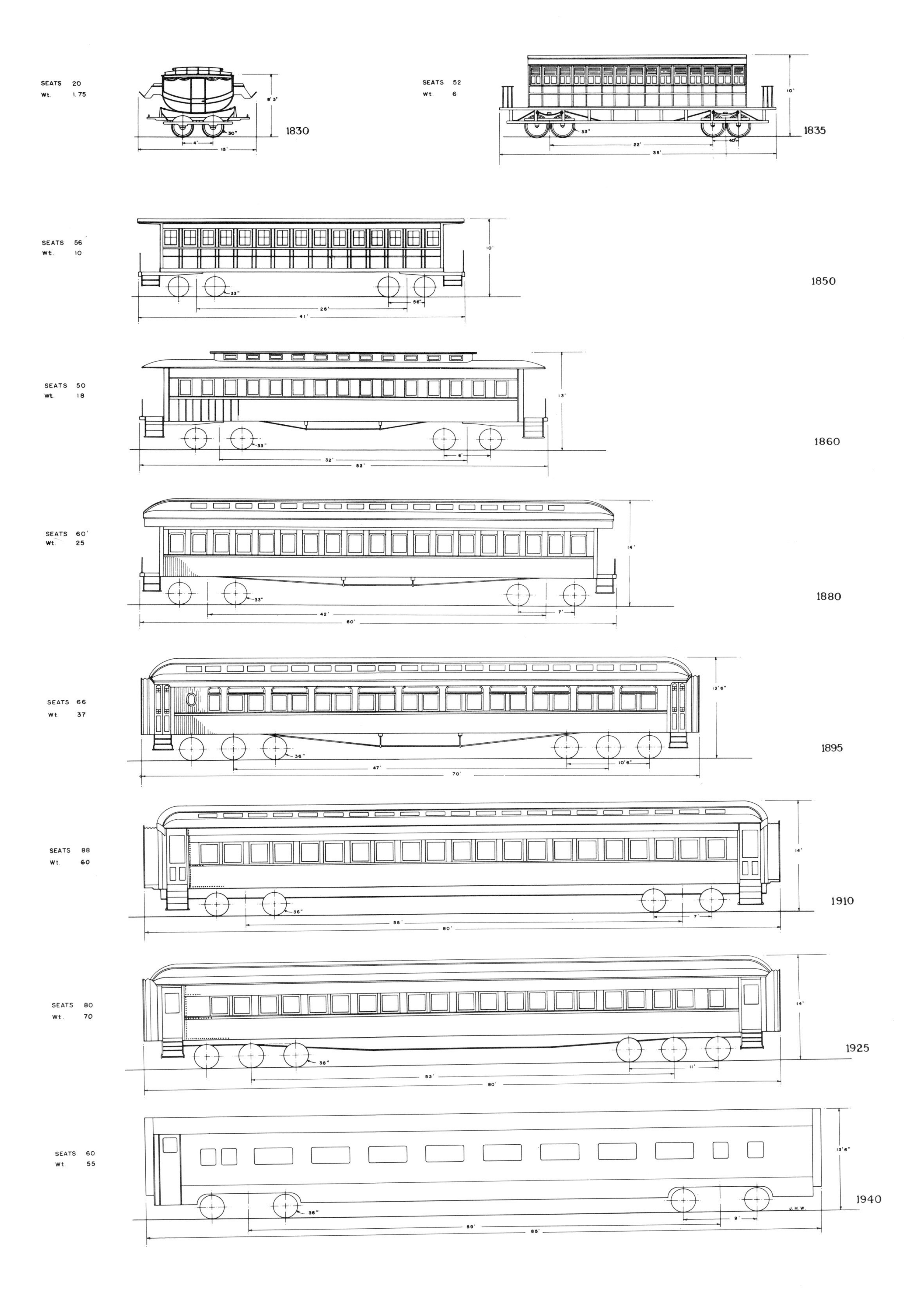
SEATS 20
Wt. 1.75
1830
SEATS 52
Wt. 6
1835
SEATS 56
Wt. 10
1850
SEATS 50
Wt. 18
1860
SEATS 60
Wt. 25
1880
SEATS 66
Wt. 37
1895
SEATS 88
Wt. 60
1910
SEATS 80
Wt. 70
1925
SEATS 60
Wt. 55
1940
J.H.W.

The American Railroad Passenger Car

CHAPTER ONE

The Day Coach in the Wooden Era

1830–1910

THROUGHOUT THE HISTORY of railroad passenger travel, the common day coach has been the largest and most important single class of car. It was not only the mainstay of the passenger fleet, but the archetype of all passenger cars. Yet it has not received the attention that has been given to the splendid cars built for first-class travel. The trade press featured parlor and sleeping cars as attractions far more newsworthy than the humdrum day coach. Railroad managers concentrated on promoting first-class travel as the keystone of prestige and profit. But as one manager, Daniel Willard, president of the Baltimore and Ohio Railroad, said in the 1920s: "We are spending too much time on the problems of the Pullman passenger. Eighty per cent of our patrons ride in day coaches, yet we barely give them twenty per cent of our attention."[1] Because of its central position in the history of passenger cars, the day coach will receive first attention in this book.

Four-Wheel Pioneers

The first American passenger cars were simple four-wheel carriages that can be divided into three basic types: the traditional curve-sided stagecoach, the compartment style with three tandem bodies built as a unit, and the minor subclass of so-called Gothic cars.

A favorite story about these early cars (and one with no basis in fact) is that old stagecoach bodies were hastily mounted on flanged wheels for railway service. At one time or another this anecdote has been pinned on every pioneer car in the country. It is true that most of the earliest cars were built by established carriage makers, who naturally constructed bodies identical with those used for road coaches.

An American pattern of swell-sided road coach was developed in Troy, New York, and Concord, New Hampshire, in the mid-1820s. As the predecessor of the more familiar stagecoach of the mid-nineteenth century, it had a more rounded body profile. By the time railroads came to America, this style of coach was widely used in the United States. Thus the melon-shaped stagecoach body, familiar to many readers as a product of the Wild West, was developed in the Northeastern section of the country long before the West was settled. The swelled sides conformed to the human body, giving maximum space at the middle where seated passengers need it. Cutting away the lower portion of the body not only reduced the weight but also provided better clearance for the high wheels that were necessary for travel on unimproved roads.

The lengthwise curve of the body was designed to accommodate the leather-strap suspension. Steel springs were so expensive and unreliable that carriage builders devised the leather strap or thoroughbrace suspension to avoid them. Heavy strips of leather were sewn and buckled together to form great elastic bands that would give and stretch with the motion of the coach. In addition, the round body could rock to and fro on the limber leather springs, dissipating the jolts from the roadway.

When he was approached on short notice for car bodies, the typical carriage builder would produce what he already knew. Richard Imlay, who began carriage building in Baltimore around 1828, was asked to build the first cars for the Baltimore and Ohio Railroad. The cheapest and simplest approach was, of course, to turn out bodies duplicating the standard mail coach of the period. The design was proved, the patterns were available, and the men in the shop were familiar with the work. Imlay built only the bodies; the railroad provided the running gears, complete with wheels. The first car, one of the very first railroad passenger coaches built in this country, was finished in May 1830 and appropriately named *Pioneer*.[2] Five others made from the same pattern were completed by August. A contemporary newspaper account praising one of the cars spoke of its "beauty, comfortable arrangement and excellence of construction."[3] It mentioned rooftop seats and a canvas awning but gave few other details. A more precise picture of the construction of these cars is found in the mechanical drawing of the car *Ohio* from the B & O's 1831 annual report (see Figure 1.58 in the Representative Cars section at the end of this chapter).

Imlay's production was not limited to local Baltimore railroads. He built two cars for the Little Schuylkill Railroad, delivering them in November 1831. He built six for the Newcastle and Frenchtown Railroad, completing one of them, *Red Rover*, in February 1832.[4] As shown by the two Newcastle and Frenchtown cars in a broadside issued in 1833 (Figure 1.1), the similarity between them and the *Ohio* was unmistakable. Imlay also produced cars for the Paterson and New York (later the Paterson and Hudson River) and the B & O railroads.[5] In 1832 he constructed twenty cars similar to the Newcastle and Frenchtown lot for the Philadelphia, Germantown, and Norristown Railroad.[6] The Baltimore daily press gave the names and color schemes of three of them: the *President* was blue and gold, the *Robert Morris* green and gold, and the *Philadelphia* sage and gold. The practice of naming cars and painting each in different colors was limited to the first years of the railroad car in America.

Another early builder of stagecoach railroad cars was James Goold (1790–1879) of Albany, New York. Goold opened a carriage shop in 1813, and he had produced scores of road coaches when the Mohawk and Hudson Railroad came to him in 1831 for six car bodies. Like Imlay, he built on the existing pattern of road coach. The original drawing furnished the contractor by the railroad's chief engineer, John B. Jervis (Figure 1.2), is close to the cars that Imlay was building in Baltimore.[7] It had larger wheels and a longer wheelbase to improve tracking, but since it, like Imlay's, depended on thoroughbrace suspension, it offered little improvement in passenger comfort. Goold continued building railroad cars until about 1870. While something is known of his later production, no specific instances of other thoroughbrace cars can be found. It is probable, of course, that he made similar cars for other local railroads.

Stagecoach cars were being built in New England as well. The Boston and Providence car of about 1834, described in detail in the Representative Cars section (Figure 1.62), is a surviving

Figure 1.1 A broadside dated June 1, 1833, showing some cars produced for the Newcastle and Frenchtown Railroad by Richard Imlay. (Bureau of Railway Economics)

Figure 1.2 James Goold of West Albany built the first cars for the Mohawk and Hudson Railroad in 1831. This drawing is contemporary. (The American Railway, *1892, p. 139*)

example. It is in fact the only original car of this type in existence, and it represents the final development of the swell-sided car for railway service. The body is larger in all dimensions than the Imlay or Goold cars, but it is far from an ideal design. No other specific examples of New England stagecoach-bodied cars can be offered, though it is probable that the first cars of Osgood Bradley, a carriage builder of Worcester, Massachusetts, who turned car builder in 1833–1834, were of the road coach style. He had produced road coaches for several years before he made his first cars for the Boston and Worcester Railroad.

The shortcomings of stagecoach-body cars were evident from the beginning. They were regarded as a temporary expedient until a more suitable body style could be found. Some lines avoided them entirely, and it might be safely estimated that fewer than one hundred saw service on American railways. The swell-sided body had been developed expressly for horse-drawn road coaches, where size and weight were more important than they were on the railroads. Larger, more commodious cars were not only desirable but possible for railway service. The stagecoach body did not lend itself to infinite expansion. A major reason for the curve-sided body was to provide clearance for the high spoke wheels of the road coach. It was superfluous in railway cars, since wheels larger than 36 inches in diameter were rarely used. Furthermore, it was expensive and difficult to build because of the many compound curves. The curving sides would not accommodate glass windows, and the leather curtains that were used instead proved unsatisfactory with steam locomotives because of the speed and the smoke. The seats were fixed so that some passengers were obliged to ride backwards—a position which many people complain about and which for some reason seems particularly annoying in railway travel. The thoroughbrace suspension only reinforced the galloping motion of the four-wheel car, accentuating the already poor ride of the single-truck carriage. The running gear was entirely unsprung and thus free to clatter and pound over the lightly built track.

COMPOSITE OR TANDEM-BODY CARS

A more rational form of car body for railway service was discovered by uniting several compartments as a single structure. Such bodies might be infinitely extended on a single frame, but for four-wheel cars they were invariably limited to three compartments. Tandem bodies were introduced at the same time as the swell-sided cars and were soon considered the more suitable body style. They appear to have entirely superseded the swell-sided cars by the mid-1830s and were undoubtedly the most numerous four-wheel passenger cars in the United States.[8] The tandem body originated with the European city omnibus of the

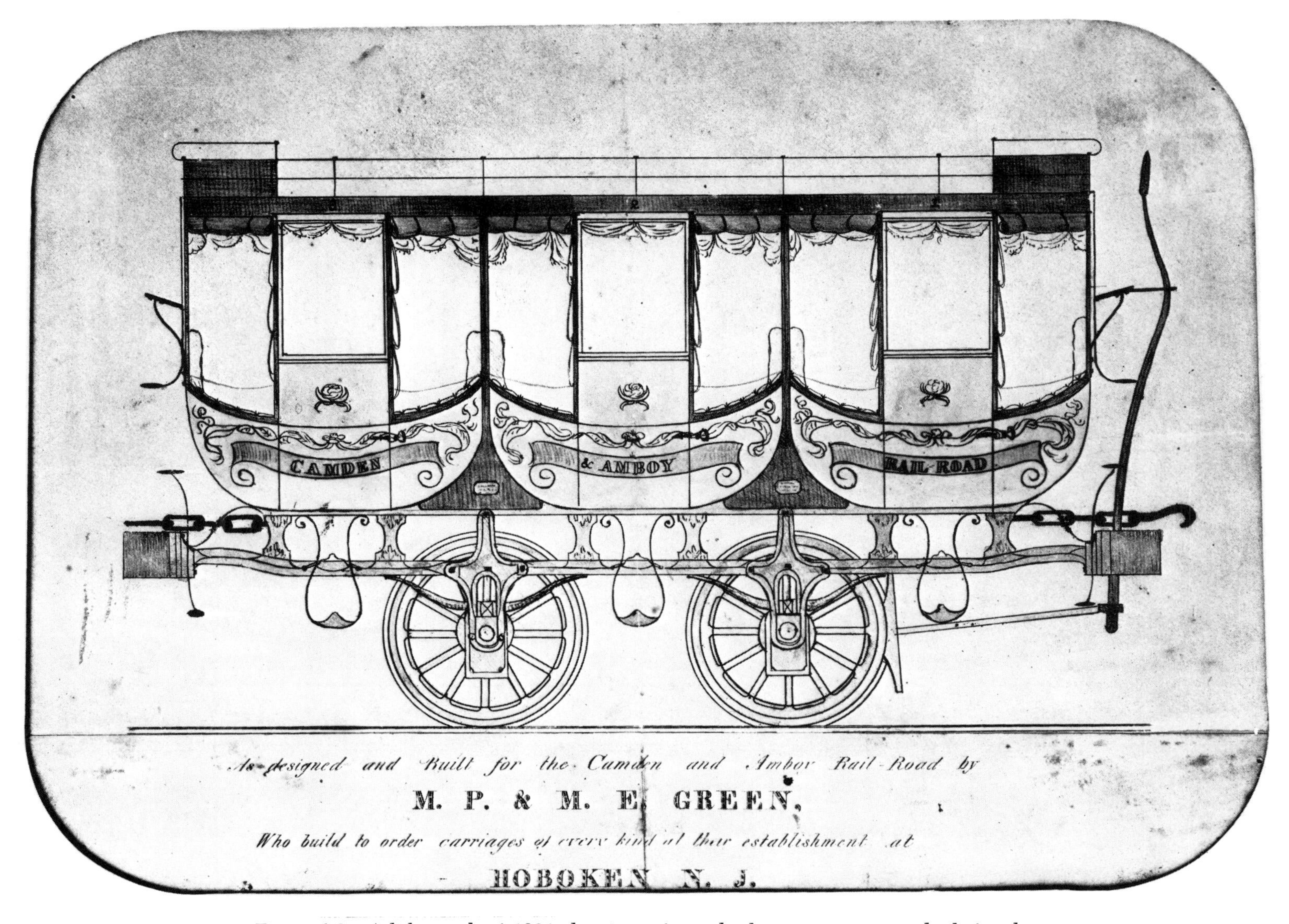

Figure 1.3 A lithograph of 1831 showing a four-wheel compartment car built for the Camden and Amboy Railroad. The influence of the swell-sided road coach is still evident. (Smithsonian Neg. 25012-E)

period. The design was copied by early British railways, as shown in the famous Ackermann lithographs of the first trains on the Liverpool and Manchester Railway. These prints were widely circulated, and some of the copies that went overseas were undoubtedly scrutinized by American mechanics. It is also probable that city omnibuses on this pattern were operating in the larger American cities. The early inspection visits of American engineers to British railways form another obvious avenue for the importation of the tandem design.

The earliest American examples of a composite-body car were those built for the Camden and Amboy Railroad late in 1831 by M. P. and M. E. Green of Hoboken, New Jersey (Figure 1.3). A Philadelphia newspaper of the day reported that the cars seated 36 passengers and had wooden-spoked wheels, which the paper described as being safer than imported iron wheels.[9] The lithograph issued by the builder gives a more precise idea of the arrangement. It is not an altogether praiseworthy design. The extreme bowed side panels of the individual compartments retained the defects of the swell-sided body. Window sashes could not be used; side curtains were again necessary. The body was placed high over the large wheels. The short wheelbase encouraged a rocking gait, and the chain coupling was unsatisfactory for a train of cars. The car had leaf springs, however—a decided advance over thoroughbraces. The body ran the full length of the frame, which was an impossibility with thoroughbrace cars because the body had to be inside the jacks. The body was wider than the usual road coach, so that five to six passengers could sit abreast. The Green shop built other cars of this style for the New Orleans and Carrolton Railroad.[10] In its February 27, 1836, issue the *American Railroad Journal* said that more cars by the same maker were bound for the Petersburg and Roanoke Railroad.

Figure 1.4 This engraving is often labeled "the first streetcar," but in fact it represents a railroad coach produced by Stephenson in 1832. (Smithsonian Neg. 10861-B)

Just a year after the first Green cars were delivered, John Stephenson, a young coach and omnibus builder of New York

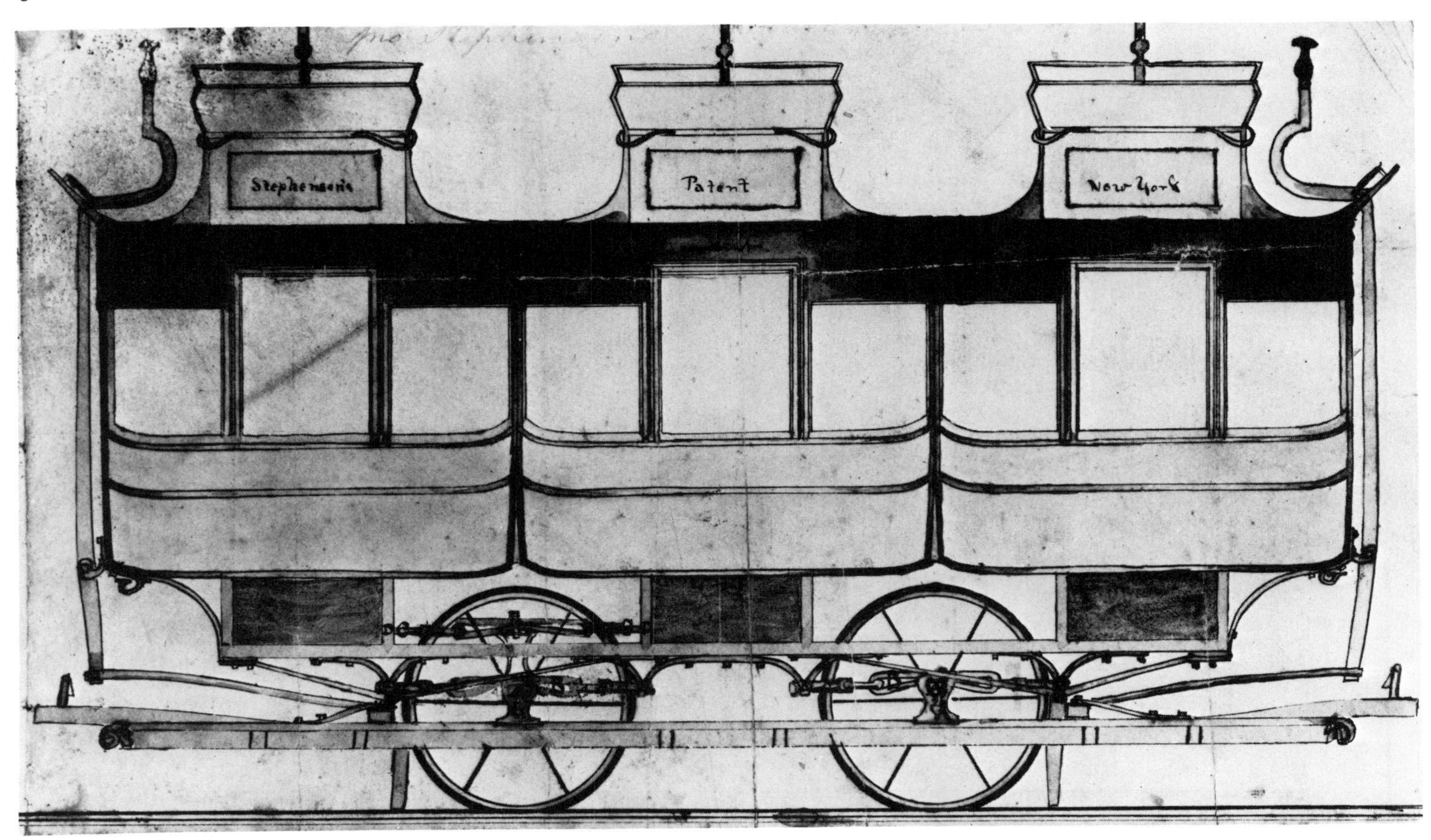

Figure 1.5 John Stephenson's patent drawing of 1833. The name President *is faintly inscribed over the center door. A car of this name was made for the New York and Harlem Railroad. (New York University Engineering Library)*

City, produced an improved style of composite-body car for the New York and Harlem Railroad. In later years Stephenson made his mark as a streetcar builder, and when referring to his original car, the *John Mason*, he called it the first street car (Figure 1.4). This claim can be argued both ways; surely the car was no more a streetcar than the earliest horse-drawn cars used by the other pioneer American railroads. The Harlem railroad did begin streetcar-like operations in lower Manhattan a few years later, but the *Mason* and its counterparts were intended for ordinary railroad service.

A contemporary account spoke of the *John Mason* as resembling an omnibus "or rather several omnibuses attached to each other, padded with fine cloth and with handsome glass windows, each [car] capable of containing outside and inside, fully forty passengers."[11] This description failed to take note of the improvements in Stephenson's design compared with its predecessors. By building the body around the wheels, it was possible to lower it by nearly a foot. The floors became drop wells between the wheels, with the seats over the wheels. Stephenson was issued a patent for this idea in April 1833. He treasured the papers, signed by Andrew Jackson, all his life. The original patent is now in the New York University Engineering Library in New York City. The drawing attached to the papers does not agree with the familiar line cut of the *John Mason*; it is marked *President* in a faint handwritten inscription over the center doorway (Figure 1.5). The *President*, the *Mentor*, and the *For-Get-Me-Not* are among the known names of cars delivered to the New York and Harlem by Stephenson. A comparison of the two drawings shows several differences: the roof seats on the *President* are the most obvious. Vertical posts indicate that there was a canopy over the roof seats, but the upper part of the drawing is missing. The absence of side doors is a serious and puzzling draftsman's error. The second leather-strap "spring" over the left wheel indicates a supplementary or alternate method of suspension. The general plan is the same in both drawings and represents considerable progress over Green's idea for a compartment coach. Stephenson's body was simpler and more squared off; the body panels were nearly flat. The dropped floor wells provided a lower center of gravity and greater headroom. Except for the leather-strap springs, Stephenson's design was one of the best available to American railroads of the period. He built cars of this type for the Paterson and Hudson River Railroad, the Brooklyn and Jamaica Railroad, and lines in Cuba and Florida.

One of the first engineers to recognize that a slavish adherence to established road coach designs was not necessary or even desirable for railway travel was John B. Jervis. In 1832 Jervis prepared the simplest possible design for compartment cars for the Mohawk and Hudson Railroad. The body was a plain box. The only concession to orthodox styling was the curved panel moldings at the bottom of each car. (A more detailed description of these cars is given in Figure 1.61, in the Representative Cars section at the end of the chapter.) It is probable that Jervis's design is a fair approximation of most of the composite-body, four-wheel cars built in the United States. It is surely the logical simplification of a basic design. Evidence is thin, and it is risky to be dogmatic on any point concerning the earliest cars. But from what information is available, it seems safe to state that of the American four-wheel passenger cars, the three-tandem compart-

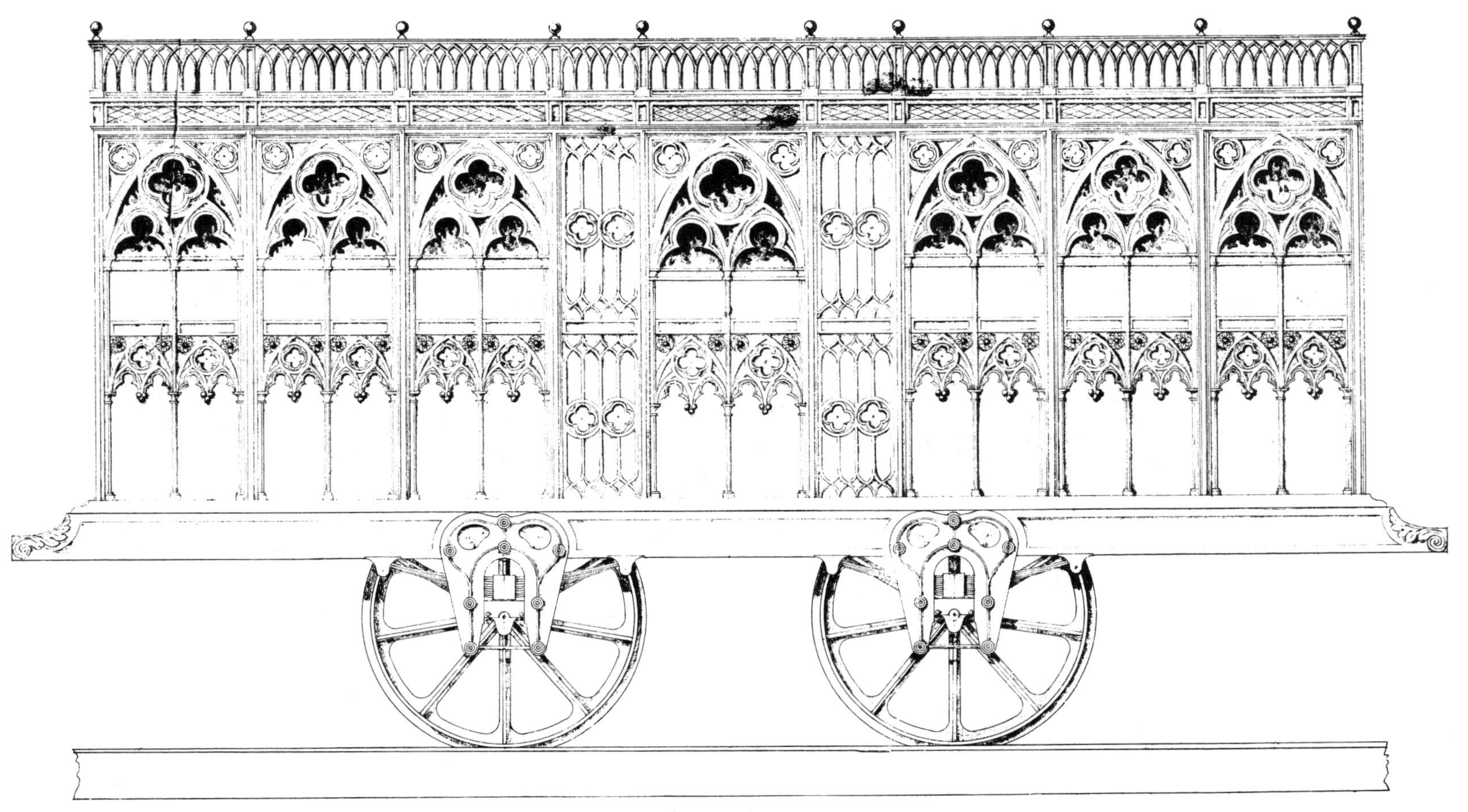

Figure 1.6 A Gothic car designed by Asa Whitney in 1834 for the Mohawk and Hudson Railroad. Presumably this is a study drawing.

ment with side doors was the most popular form. Some four-wheelers were built with end doors and platforms and center aisles, but only the sketchiest information exists for cars on this plan. By the time the center-aisle plan was adopted, the four-wheel passenger car was made obsolete by double-truck eight-wheel cars.

GOTHIC CARS

At best a minor subclass of the compartment car, the Gothic car is known only through a few examples. The most familiar of these began operation in 1837 on the Erie and Kalamazoo Railroad running between Toledo, Ohio, and Adrian, Michigan. A drawing prepared in 1881 in consultation with one or more old officers of the road has been widely reproduced.[12] It shows a three-compartment car with an elevated central compartment. The floor of this central area is dropped sufficiently for a baggage space beneath the passenger compartment. The roof and body ornamentation suggests the Gothic style. A second drawing, now in the University of Michigan's Transportation Library (Ann Arbor), is also identified as the Erie and Kalamazoo Gothic car but appears to be of more modern origin. It shows a boxy four-wheel car indistinguishable from an ordinary four-wheel caboose of, say, the 1880s. The validity of this drawing can be seriously questioned. Unfortunately, little concrete information exists on the car, its builder, or its final history.

The Gothic car, or at least the term Gothic cars, originated some five years before the Erie and Kalamazoo coach: Gothic cars were reportedly in service on the Mohawk and Hudson Railroad in 1832.[13] A drawing by Asa Whitney, master mechanic of the road in its early years, clearly shows a car that can only be described as Gothic (Figure 1.6). If the wheels were removed, it would be a perfect plan for an elaborate transept screen in a Gothic cathedral. But it is difficult to believe that so elaborate a vehicle was constructed for use outside the Vatican railway; the probability is that it was a draftsman's exercise. Gothic cars of some form, however, were in use on the Mohawk and Hudson. The 1840 inventory lists twenty-four "Gothic Coaches" as purchased in 1835 and 1836.[14] The neighboring Tonawanda Railroad built six Gothic cars in its own workshops.[15] F. A. Ritter Von Gerstner, the Austrian engineer who studied American railroads in the period 1838–1840, reported that these seated 24, weighed 2¼ to 2½ tons each, and cost $800 each. The cars were known locally as camelbacks, yet because of their classical Grecian exterior they were endowed with a certain architectural grace. But the interiors, with their awkward raised center sections, "afforded all the inconvenience that the most disobliging railroad manager could desire."[16]

The defects of the various four-wheel cars used in this country were not only obvious but numerous. Both the traveling public and the railway managers were loud in their criticism. The wheelbase or spread of the axles was necessarily limited so that sharp curves could be negotiated. This restricted the body length to no more than 20 feet. In 1832 the New York and Harlem attempted to build some long four-wheelers. The journals were apparently made with enough side play for the curves, but the support of only four wheels was not enough, and the body sagged at the middle. A small-wheeled four-wheel truck had to be placed at the center of the car for support.[17] After this discouraging episode no more long single-truck cars were built.

The short body cut up into three sections resulted in dwarfish compartments, which became uncomfortably crowded when they were filled. Roof seats were added to increase capacity, but they were formidable to climb into and provided a disagreeable and dangerous ride. The cars swayed and rolled, low bridges were hazardous to rooftop passengers, and smoke and sparks poured from the locomotive. Seated on high among a jumble of baggage, a traveler often regretted his willingness to "Make room for ladies! Come gentlemen, jump up on the top—plenty of room there."[18] In the opinion of many early railroad travelers, these cars were ugly boxes devoid of even the most meager comfort.

Four-wheel railway vehicles are plagued with a characteristic galloping motion. When the forward set of wheels strikes the rail joint, the body noses down; the springs recoil, and so the pattern is repeated at each joint. This motion, noticeable at low speeds, becomes intolerable and even dangerous as velocity increases. It can be somewhat lessened by spreading the wheelbase, but a rigid wheelbase of much over 7 feet was not possible when relatively sharp curves were to be traversed. The long-wheelbase cars passed through curves with more friction, thus requiring more powerful locomotives because of the additional drag.

The worst fault of the four-wheel car was its inherently unsafe structure. A broken axle or wheel meant certain derailment. In a time of unreliable axles, wheels, and rails, failures were common. A sound-appearing piece of iron might be riddled with fissures or blowholes.

During its first year of operation the Camden and Amboy had a terrible wreck because of a broken axle: the four-wheel cars derailed, killing two people and injuring many others. As late as 1850 the Western Railroad in Massachusetts continued to operate four-wheel cars. In that year a broken axle dropped a car body to the tracks, killing three.[19]

After only a few years' experience, railroad managers realized that something better than the common four-wheel car was needed. It was obvious that the four-wheeler did not lend itself to further improvement; the car was only a makeshift hastily assembled from existing highway vehicle technology. New thinking was needed to develop a conveyance specifically suited to railway requirements.

A car with six-wheel running gears might appear to be the logical successor to the single-truck car. Six-wheelers became popular on European railroads at an early date, and a large number continue in service today. The center pair of wheels stabilized the ride and permitted construction of relatively long car bodies. But this presumably natural transition did not come about in America. The only six-wheel passenger cars observed by Von Gerstner in his exhaustive tour of American railroads were four coaches on the obscure New Orleans and Nashville Railroad. In these the center pair of wheels had no flanges to assist in rounding curves. The only other American railroad known to possess a six-wheel passenger car was the Baltimore and Susquehanna. In 1843 the road had one six-wheeler for passenger service as well as a fairly large number of six-wheel freight cars. From the evidence available it seems clear that the six-wheel car was a rare, almost nonexistent species in this country.

Eight-Wheel Cars—The Beginning of the American Coach

The most stable railway vehicle is that mounted on two independent wheel assemblies known as trucks. These carriages are attached to either end of the car body by center pins and are free to turn or swivel. They provide a flexible arrangement that enables the car to traverse rough track and sharp curves with remarkable ease. The double-truck car has been the ideal running gear for railway vehicles since its introduction. To the present day it has maintained a leading position because of its smooth, easy ride. It was, of course, particularly well suited to the uneven and lightly built tracks of the early American railway.

The eight-wheel car offered more than a flexible running gear, for the car body could be made as long as desired. Double-truck cars provided seating equal to three four-wheeled cars. The saving in cost and dead weight was substantial. Three single-truck cars would have six end bulkheads, twelve wheels, and six couplers. The single double-truck car of equal capacity would have only two bulkheads, eight wheels, and two couplers. If the compartment style of car had continued in use, as it did in Europe, the additional weight of the side doors and interior bulkheads would have been added, making the dead-weight factor of the four-wheel car even more disproportionate to the eight-wheel car. In America the car body was constructed as a single, open compartment. It was the simplest, lightest, and cheapest plan for a railroad coach. The great common compartment was proclaimed as more "republican" in character, and thus more agreeable to the democratic ideals of the American traveling public, than the exclusive compartment car. In the common compartment passengers mingled freely without distinction of class or station. (This sentiment was true enough in ordinary coach travel, but it became something of a myth after the early introduction of first-class sleeping and parlor cars.) By using end doors, the side panels below the windows could be made as a continuous truss. With side doors, extremely heavy floor sills were necessary to support the body. It was argued that multiple side doors permitted quicker loading and discharge of passengers. But in this country, where station stops were less frequent than in Europe, end doors were found adequate.

The double-truck idea was projected in the dawn of the railway era long before passenger carrying lines existed. William Chapman (1749–1832), a British civil engineer, obtained a patent in 1812 for single and double "bogies" for railway service.[20] Chapman's patent specified trucks to spread the weight and improve the trackings of locomotives. He later built several engines with this improvement, and he envisioned the application of the scheme to passenger cars as well. In 1825 he forecast a commodious passenger carriage, complete with dining facilities, rolling grandly across the British Isles. On a more practical level in the same year, Thomas Tredgold, the author of many widely read engineering treatises, suggested that eight-wheel cars would provide more even distribution of weight upon the track. The idea apparently languished in England; the first instance of an eight-

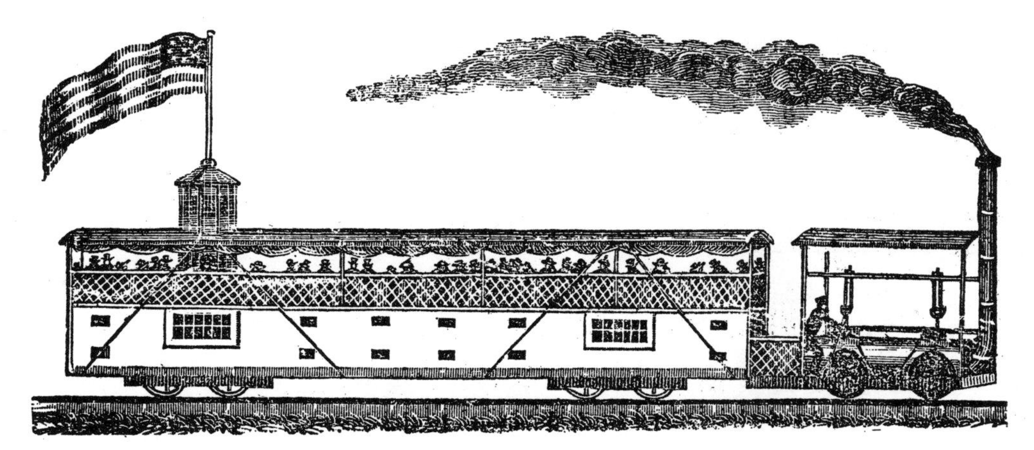

Figure 1.7 R. P. Morgan's double-deck eight-wheel car, patented in July 1827, was illustrated in a Boston newspaper, The American Traveler, *April 14, 1829. (Railroadians of America)*

wheel car was the ceremonial carriage made for the Duke of Wellington on the occasion of the Liverpool and Manchester Railway's opening in September 1830.[21] An open car with a red canopy, it was 32 feet long. Apparently it was intended only as a special barge for the celebration, since it was rarely used afterward. Certainly it set no pattern, for British railways remained content with four-wheel cars for many years.

In America a plan for an eight-wheel car was put forward just before the railway age opened. In 1829 Richard Morgan of Stockbridge, Massachusetts, proposed the construction of a huge two-story land barge for railway travel (Figure 1.7).[22] The car was to have sleeping "births" on the first floor and a gay fresh-air promenade on the second. The captain (conductor) was provided with a cupola lookout, which in turn was surmounted by a giant flag. Morgan may have projected his scheme as early as July 27, 1827, when he was granted a patent for improvements in railroad cars. Unfortunately, the papers were lost in the 1836 Patent Office fire and were not among those restored in later years. Morgan did promote a friction bearing, and this may have been all that was covered by the patent. However, newspaper notices with a line drawing appeared in the Boston *American Traveler* in April 1829 and were considered important enough to be copied by a New York paper.

In the same year at Quincy, Massachusetts, primitive eight-wheel cars were being used to carry blocks of granite. Mechanics on the Baltimore and Ohio inadvertently stumbled on the eight-wheel arrangement by loading long timbers on two four-wheel cars. The plan worked so well that double-truck cars were built to haul cordwood and other products. Several were in service late in 1830.

It was not long before the same idea was tried out for passengers. In 1828 or 1829 Joseph Smith of Philadelphia visited the B & O offices with a model of a long, rectangular passenger car.[23] Smith argued that this body style was more suitable for railway service than the ordinary stagecoach. The model, which lacked wheels, apparently made no immediate impression. Construction of the railroad was just getting underway, and rolling-stock considerations were necessarily postponed. Richard Imlay claimed that he suggested long-bodied cars on eight wheels in 1829 to the Newcastle and Frenchtown Railroad, but that the disinterest of railroad officials and the timidity of a partner prevented the construction of a sample car.[24]

These proposals were not entirely lost, for George Brown, treasurer of the B & O, is credited with reviving the idea several years later after passenger service was established. The line's experience with eight-wheel freight cars led to the decision to build a similar running gear for a new long-bodied passenger car. The car would be called the *Columbus*, which seems appropriate for so adventurous an experiment.[25] At a time when the stagecoach style of car was in vogue, the *Columbus* represented a radical departure. The work was performed under the direction of the road's master carpenter, Conduce Gatch, who later insisted that he was also the designer. Although Ross Winans claimed the design and has received popular credit for the eight-wheel car in general, Gatch's assertion was supported by half a dozen old shopmen who were actually engaged in building the *Columbus*. These conflicting claims were aired at length in the famous eight-wheel car case (discussed in the following section).

No contemporary drawings or descriptions of the *Columbus* have survived, but around 1853 Winans prepared a reconstruction sketch which is reproduced here as Figure 1.8. Gatch attacked the sketch, insisting that it had many errors; a second sketch has been prepared to show Gatch's version of the car (Figure 1.9). Winans's drawing portrays a squat, octagonal-ended car with doors at each corner. A wooden railing backed by wire mesh surrounds the flat roof. The drawing is badly proportioned: the window openings are too low and the headroom is insufficient for even a very short passenger.

Gatch, on the other hand, declared that the body measured about 7 feet by 24 feet and was painted yellow and drab. The window openings were 2 feet square. He agreed that the ends were octagonal but insisted that there was only a single door at each end and that iron stairways occupied the opposite triangular spaces. Gatch explained the peculiar octagonal-end entrance of the car by saying that it was simply the only plan that occurred to him at the time. Since no end-entrance or platform-style cars had yet been built, there was nothing to copy. The *Columbus* was an experiment, and many features of the design proved unsatisfactory. According to Gatch, an ornamental iron railing circled the roof—which was arched and not flat, as Winans's sketch indicated. The trucks had roller side bearings and were originally fitted with plain journals. Winans's friction wheels were not placed on the trucks until the following year. Four longitudinal benches provided seating for about 40 passengers. In 1832 roof

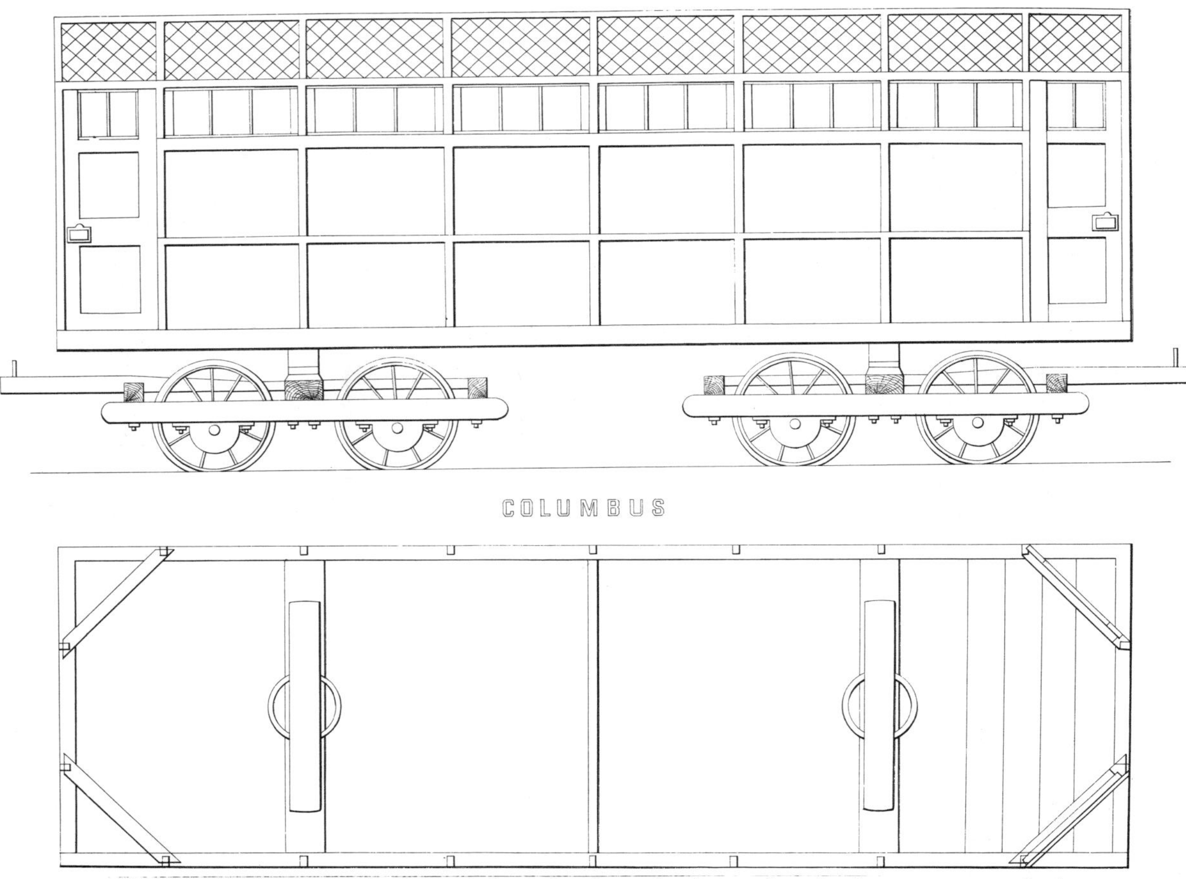

Figure 1.8 Winans's drawing of the Columbus, *the first eight-wheel passenger car built by the B & O shops in 1831. The drawing, dating from about 1853, is a reconstruction. (Peale Museum)*

seats, an extended iron railing, an awning, and an interior stairway were added to increase the car's capacity. Other minor differences can be seen by comparing the two drawings.

Work began on the *Columbus* in the spring of 1831. In late June the *Baltimore Patriot* said that "a spacious new carriage running upon eight wheels" would soon be in service. Its capacity was given, with journalistic exaggeration, as 150 passengers. According to another Baltimore paper, the *Columbus* entered service on July 4. Its first trip was horse-drawn, but on July 13 it was pulled by steam to the Ellicott City end of the line. Surprisingly, the car did not take well to the sharp curves on the upper end of the line and went off the track. Gatch blamed the mishap on the unsprung trucks. He placed india-rubber pads between the side frames and bolsters of the trucks, and this elementary suspension improved the car's tracking enough for further operation. It was seen in service by one old employee as late as 1837.

The *Columbus* was successful enough to encourage the construction of more eight-wheel cars. Between 1832 and 1834 several experimental coaches were built at the B & O's Mt. Clare shops. Descriptions of three have survived. They were produced in the following order: the *Winchester*, the *Dromedary*, and the *Comet*.[26] But instead of refining the basically sound square-body plan, the B & O wandered into a blind-alley detour of exotic drop-frame compartment cars with side entrances. The *Winchester* represents a giant step backwards (Figure 1.10). Three swell-side stage bodies with side doors were mounted on an eight-wheel running gear. It was again necessary to use side curtains instead of window glass. Like the *Columbus*, the trucks were unsprung. The second car in the series, the *Dromedary*, had a drop frame that lowered the passenger compartment to about 10 inches from the rails. This arrangement permitted easier boarding. The main reason for using it, however, was that of safety: in the event of a broken axle, the underslung car was considered less likely to upset.

Winans's version of the *Dromedary* shows a mechanically unsound scheme (Figure 1.11). The truck mounting is particularly weak; the overhead truss support is as ungainly as it is ineffectual. A rack for cordwood is shown at the right end. The trucks are sprung, which represents some advance over the previous cars. Gatch's sketch also portrays the depressed-center plan, but it shows a long, continuous compartment with a single side entrance (Figure 1.12). Both Winans and Gatch agreed that small glass sashes were placed over the window openings. In Gatch's drawing the wooden frames are clad with sheet iron, and this detail is confirmed in the 1834 B & O annual report, which said that the passenger car frames were "plated with iron."

The *Comet*, the last of the ill-conceived trio, had five narrow bodies on a single drop frame and was mounted on live-spring trucks. It is possible that other experimental eight-wheel passenger cars were built at Mt. Clare. The 1834 annual report mentioned that additional new cars were under construction and that four old single-truck cars were being rebuilt on the eight-wheel pattern. In 1840 Von Gerstner saw abundant evidence of past follies in the Mt. Clare yard. He reported: "At no other railroad station in North America does one find . . . so many old and new cars of every size and shape, so many broken wheels and axles and demolished car bodies."[27] It was obvious, he said, that every conceivable plan for rolling stock had been tried and that the B & O had paid high for the experience gained. As early as 1834 the road's chief officers were convinced that the eight-wheel passenger car was the superior plan. The president's message in the annual report contained a strong endorsement:[28]

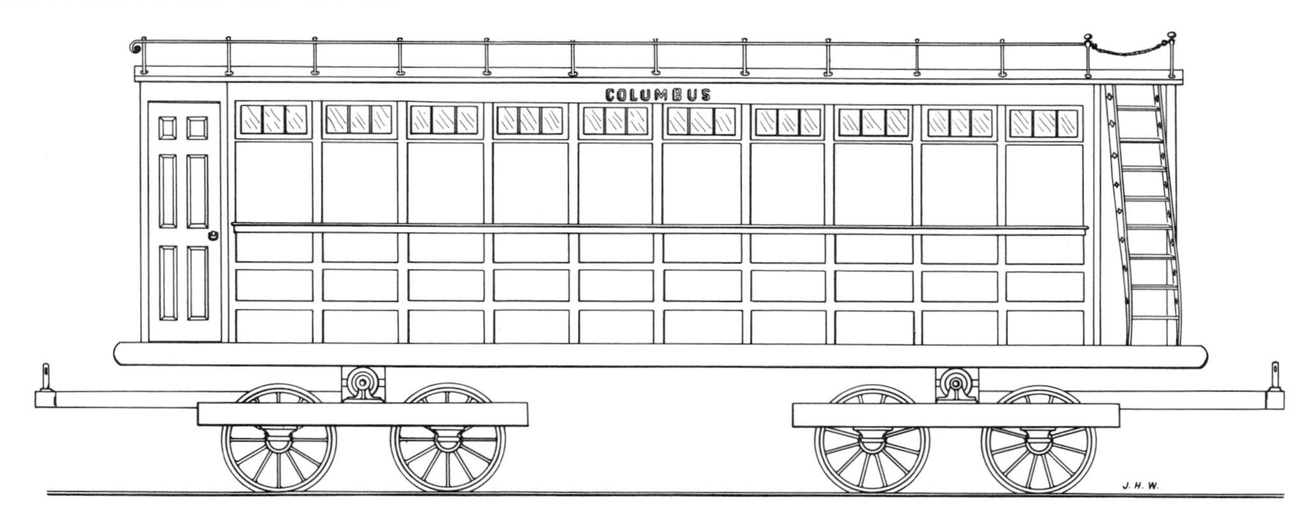

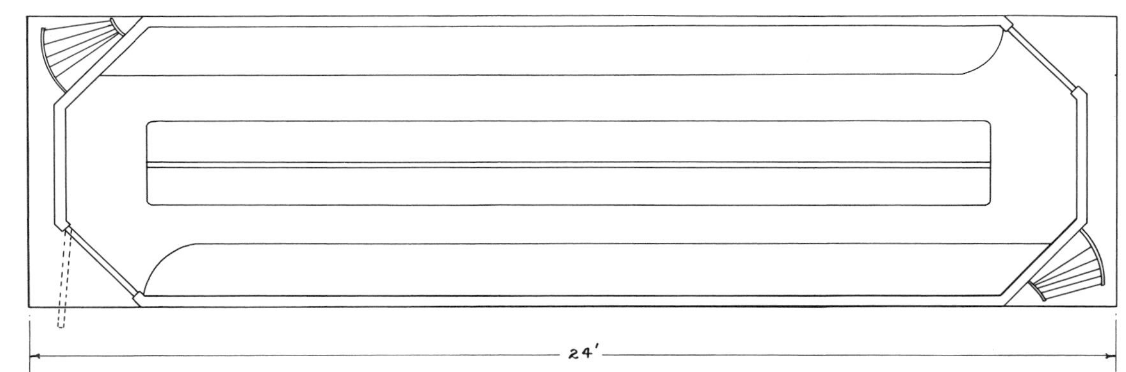

Figure 1.9 A drawing of the Columbus *based on the description by Conduce Gatch, who contested Winans's claim to be the car's designer.* (*Drawing by John H. White, Jr.*)

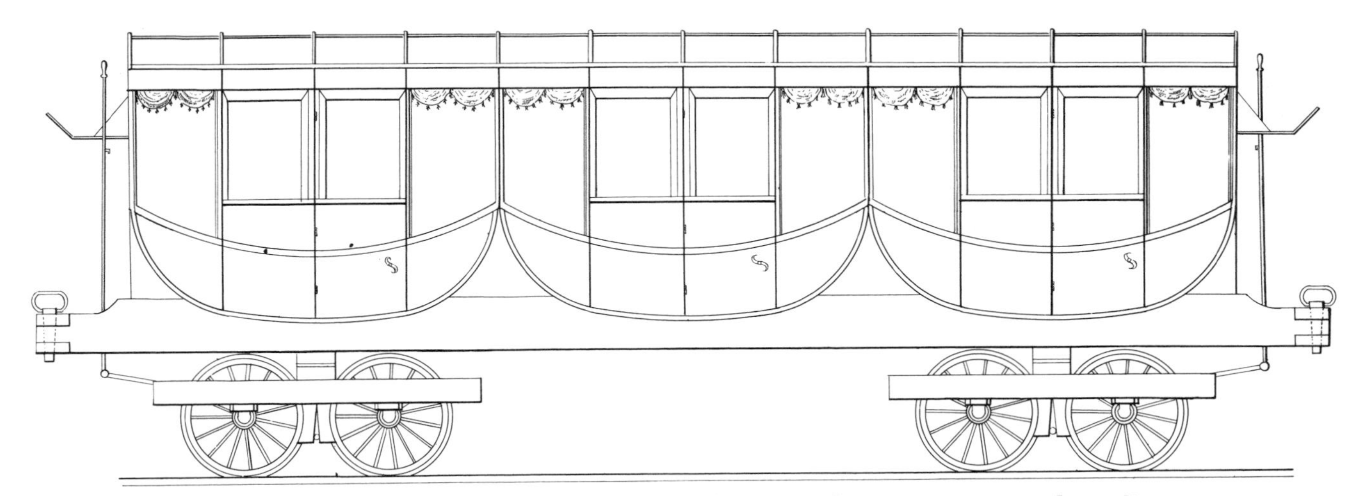

Figure 1.10 The Winchester, *built by the B & O around 1832, was among the earliest eight-wheel cars constructed. It shows the adaptation of swell-sided coach bodies for railroad use.* (*Peale Museum*)

After the experience of several years, the Board have come to the determination of employing an eight wheel car invented by Ross Winans for the transportation of passengers. This consists of two sets of ordinary running gear with steel springs, each set having what may be called a rose bolt, equidistant from the centre of motion of each wheel. The two sets are placed at the desired distance apart, and connected by a frame, stiffened by a thin iron plate on the sides, on which frame the body of the car rests. The great advantages of this mode of construction consist in the steadiness of the car, when moving on a curved road, or on one whose surfâce is uneven or slightly out of repair; for the two sets of wheels accommodate themselves to the inequalities of the surface, without affecting the car, resting as it does on the centre of each set. Another important object, which is attained, is safety: for while, in the ordinary four wheeled cars, the breaking of a wheel or an axle, might be productive of the most fatal consequences, such could not be the case, where, with eight wheels, there would be enough left to support it on the track until the train could be stopped. Again, upon a curved road it is necessary to place the pairs of wheels of a four wheeled car as near as possible, so as to diminish friction, and this makes it necessary to build the body of the car very short, to prevent a disagreeable, and at times, dangerous vibration from side to side while the car is in rapid motion. In the eight wheel cars, on the contrary, the pairs of wheels of each set are placed as near together as can be desirable, under any circumstances,

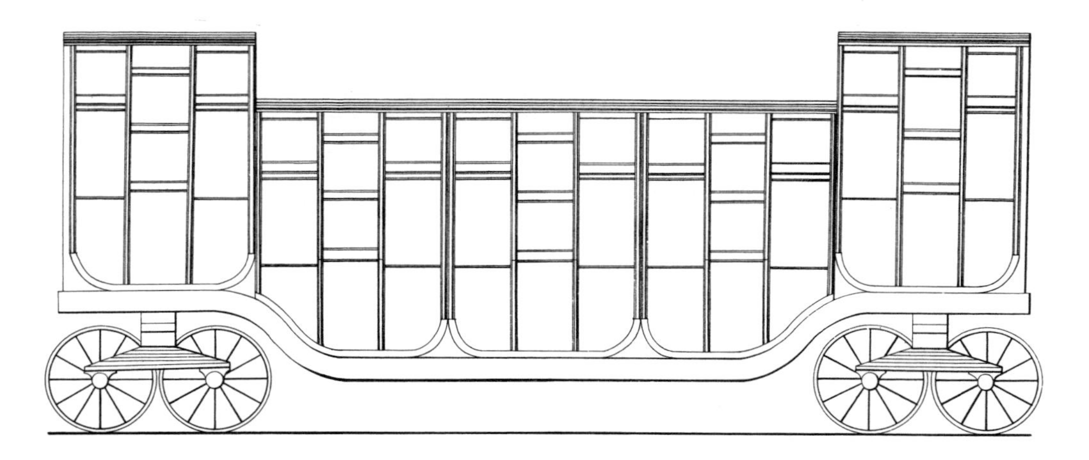

Figure 1.11 Two other early eight-wheel cars, the Dromedary *and the* Comet, *were built by the B & O in 1834. (Peale Museum)*

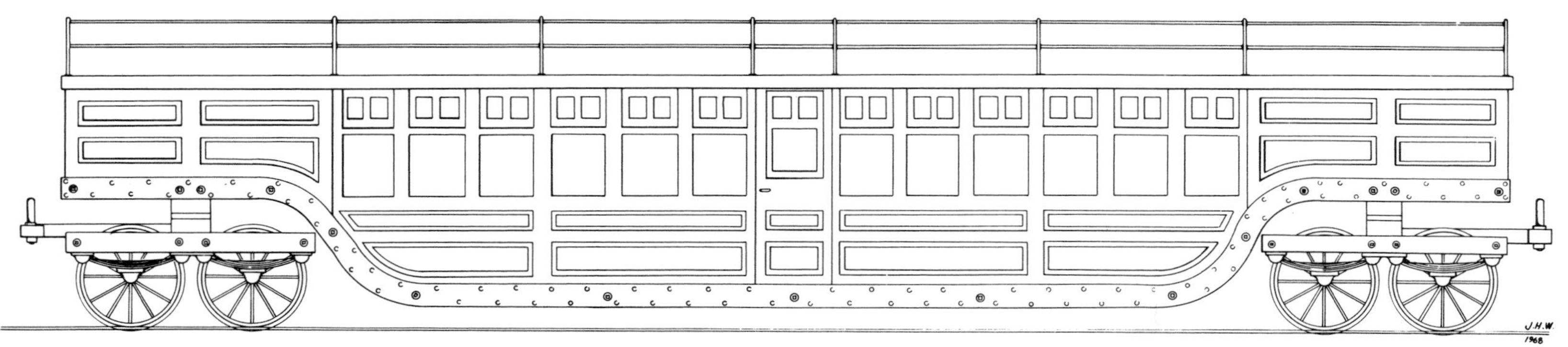

Figure 1.12 This drawing of the Dromedary *is based on a sketch made by Conduce Gatch around 1855. (Traced by John H. White, Jr.)*

and the sets themselves may be as far apart as may be necessary to accommodate a car of any length, without increasing the friction on the rails when passing curves of the least radius, or producing the lateral motion complained of. In point of economy too, this mode of construction is much preferable to the one hitherto employed.

The B & O decided to order ten eight-wheel cars for the Washington branch and wisely resolved to return to the rectangular body. Important improvements had accrued since the building of the *Columbus*, however. The octagonal end door gave way to the now familiar end door, with platform entrances. A center aisle was flanked by double cross seats. The classic American day coach was born.

The first Washington branch cars were ready in early July 1835. By October of that year the B & O had twenty-five eight-wheel passenger cars in service (see Figures 1.69 and 1.70 in the Representative Cars section at the end of the chapter).

The B & O deserves full credit for introducing and developing the eight-wheel passenger car years before most roads thought beyond the simplest style of four-wheelers. In this early period, however, there were also a few other builders who showed a strong interest in experimenting with eight-wheel designs. The first builder outside of the B & O was Richard Imlay, who constructed a number of four-wheel cars at his Baltimore coach factory. By 1833 he had opened a new car works next to the Baldwin and Norris locomotive shops in the Bush Hill section of Philadelphia. Once there, Imlay's interest in eight-wheel cars was renewed. In 1832 he had built a sample car for Jonas P. Fairlamb, who obtained a patent on January 19, 1833, that shows an eight-wheel running gear.[29] The next year Imlay decided to construct an eight-wheel car on his own plan. The body design has been credited to a model built as early as 1829 by either George Fultz or Laban Proctor.[30] It was on a drop-center plan similar to Gatch's reconstruction of the *Dromedary*. Whatever the source of the design, Imlay began work on his first eight-wheel car, the

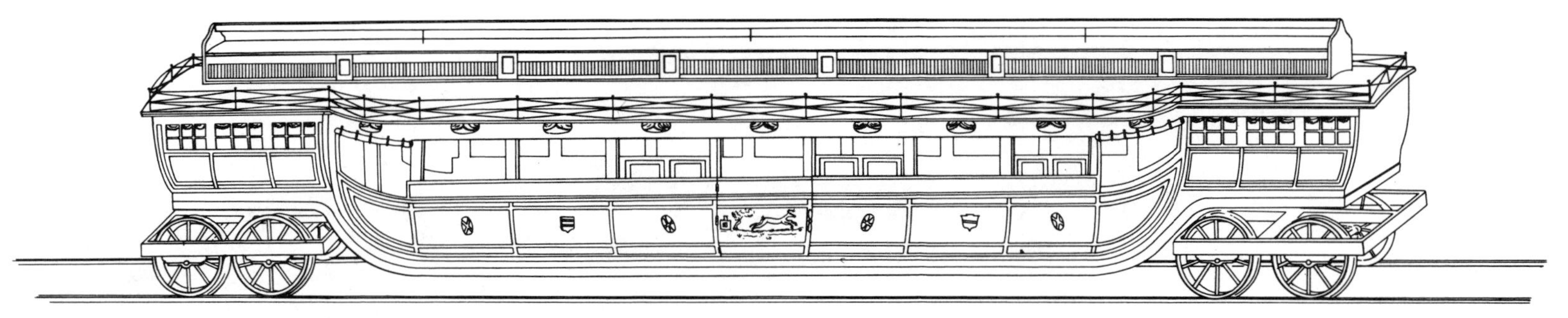

Figure 1.13 A possum-belly, double-truck car designed by Richard Imlay of Philadelphia. The drawing is based on a lithograph produced around 1836. (Traced by John H. White, Jr.)

Victory, in August 1834. It was described as ready for service in the *Philadelphian* of July 3, 1835:

It is divided into three apartments, two being designed for private parties of ladies and gentlemen, while the third [the central compartment] is a kind of ordinary. The car will be drawn by four matched horses, and run from the corner of Chestnut and Broad to the viaduct or bridge on the Schuylkill, or to any other place on the railroad when chartered by private parties for that purpose.

The model of the *Victory* is illustrated in Figures 1.67 and 1.68 in the Representative Cars section. A lithograph showing a full-sized car of the same design was issued around 1836 by the well-known Philadelphia printmakers Lehman and Duval, but only a poor photocopy of the print has survived. A modern tracing made from it shows the car's construction (Figure 1.13). After a trial on the Philadelphia and Columbia, the 36-foot-long car was sold to the Philadelphia, Germantown, and Norristown Railroad. The line was satisfied with the *Victory* and, according to the directors' minutes, ordered three more eight-wheel cars from Imlay in 1835 at a cost of $1,600 each. Continued approval of eight-wheel cars was indicated in 1837, when the minutes record that two four-wheel cars were rebuilt as the new double-truck *Norristown*.[31]

Imlay's design was defective in many respects. He followed the drop-center or "possum-belly" plan that had already been rejected by the B & O because of the complicated and weak floor frame. The side-door entrance, swell sides, and open windows were obvious defects. The end compartments over the truck could be reached only by mounting two or three steps. The lack of end doors or platforms prevented passage from one car to the next. The leaf-spring truck bolster, patented on September 21, 1837, by Imlay (Patent No. 389), proved unstable and was little used. One notable improvement made by Imlay was the raised or clerestory roof; he elevated the central portion of the roof to admit additional air and light. This idea seemed to have been forgotten by the industry for a while after Imlay was forced to quit the car-building trade, but it was revived with considerable success in the early 1860s.

Despite their shortcomings, Imlay's original cars had many purchasers. W. Milnor Roberts recommended them to the private car operators of the Philadelphia and Columbia after a personal inspection of the *Victory*. A car was promptly placed on the line and named for Roberts.[32] Within a few years the road was largely stocked with Imlay cars, all of them eight-wheelers.[33] The largest carried 60 passengers and cost $2,000. All had water closets, and many were fitted with ladies' compartments—separate rooms set aside for women and their escorts. Years later, an old employee of the Philadelphia and Columbia still remembered the early Imlay cars and praised their easy ride.[34]

Imlay's business flourished during the next years. He not only built cars for the nearby roads but also had customers referred directly by the prominent locomotive builder M. W. Baldwin. For example, in a typical letter dated February 17, 1837, Baldwin stated:[35]

Mr. Imlay of our city will make passenger car No. 1 on 8 wheels 37 ft. long, projecting roof [clerestory], side seated or shifting backs on cast iron wheels, with axles 3¼ diameter finished in handsome style, and capable of carrying comfortably 60 passengers all seated, for the sum of 1750 Dolls. No. 2 similar to No. 1 but not so well finished, 1600 Dolls. No. 3 1550 Dolls. Passenger on 4 wheels, . . . with shifting backs, cross seats, capable of carrying seated 24 passengers, to be well finished for 800 Dolls. No. 2 also on 4 wheels 725 Dolls.

In addition to day cars, Imlay was among the first to offer sleeping cars (see Chapter 3). These cars, first offered in 1837 or 1838, were not built on the depressed-center plan; they were ordinary square-body cars with end doors and platforms. It is probable that Imlay had recognized the folly of the drop-center cars by this time and had switched over to the square-body plan for day cars as well.

Although Imlay was a pioneer in the field and seemed destined to become a major car builder, his business foundered in the financial depression following the 1837 panic. The shop closed around 1840, and Imlay drifted into other lines of employment.

Through the B & O and Imlay, the gospel of the eight-wheel car quickly spread among the mid-Atlantic railroads. The idea was carried to the Newcastle and Frenchtown in 1834 by a former B & O employee, James B. Dorsey.[36] His new employers were skeptical at first when Dorsey spoke of the superior performance of the eight-wheel cars, but he persisted. To prove his point he built a crude double-truck flatcar from two four-wheel freight cars. The success of this test car led to construction of eight freight cars and finally, in the spring of 1835, to a passenger coach.

Next to fall into line was the Philadelphia, Wilmington, and Baltimore. The southern end of the road was built by Benjamin H. Latrobe between 1835 and 1837.[37] Because of his experiences with the B & O, he faithfully recommended the eight-wheel car to the road's directors. But the directors sided with William Strickland, the engineer for the northern end of the line, in opposing "the large cars." He maintained that they were heavier, more expensive, and more likely to go out of order. Strickland also claimed, incorrectly, that no other Eastern road had adopted them. In the spring of 1835, just before the final pattern of the eight-wheel car was perfected, the Wilmington line decided to continue buying four-wheel cars. With some reluctance the directors agreed to try a double-truck baggage car, and from the beginning its performance was good enough to convince them of their error. Before the year was out their shops were busy splicing short four-wheel bodies together for remounting on eight wheels.

Immediately after the P W & B capitulation came the Camden

Figure 1.14 Camden and Amboy Railroad passenger cars built about 1836, shown at the time of their retirement thirty years later. (Chaney Neg. 14489)

and Amboy Railroad. Construction of eight-wheel cars began in its repair shops in 1836, and by 1840 fifteen double-truck cars were in service. They were patterned very closely on the Washington branch cars, as will be apparent from a comparison of Figures 1.69 and 1.70. The underhung truss remained popular with the C & A for some time; cars of this construction were in service until the early 1870s. Those pictured in Figure 1.14 may be cars of the first lot that were later modernized with extended platform roofs and roof ventilators. Another scene of the 1870s showing a car on a swing bridge illustrates the need for retaining the underhung truss (Figure 1.15). The center side door would have cut an inside-the-panel truss in half.

A second New Jersey line to adopt the eight-wheel car was the Morris and Essex Railroad. The first eight-wheeler was placed on the road in February 1838. Von Gerstner reported that it was a giant vehicle capable of seating 90 passengers and that it had a baggage compartment under the car.

Southern railroads, often dismissed as slow to accept technical innovation, were quick to recognize the merits of the American-style coach. One reason is that many of these roads were founded in the mid- or late 1830s, just as the reputation of the eight-wheel car was being established. In 1837 the Monroe Railroad and Banking Company had two eight-wheel cars under construction in Philadelphia, possibly at Imlay's shop.[38] In the same year the South Carolina Railroad began acquiring double-truck coaches. This pioneering Southern line was already stocked with four-wheelers, but it went ahead with a costly program of substitution after being convinced of the new style's superiority. In the road's 1840 annual report the four-wheelers were contemptuously described as "old and nearly valueless."

Not content with simply copying the invention of the Yankee car shops, the South Carolina proceeded to develop its own style of car body. The hogsheads that were used to transport tobacco throughout the South suggested this form of construction to one of the road's directors, Colonel James Gadsden. These mammoth barrels, cheap and hardy containers, seemed well suited to the strenuous labor of railway service.[39] On Gadsden's suggestion a freight car was built in September 1837 by the master carpenter, George S. Hacker. The fact that this style of construction was familiar to the local mechanic undoubtedly kindled enthusiasm for the bizarre design. The road produced more barrel-shaped freight cars and then, in 1840, a passenger car in the same style (Figure 1.16). The body was 30 feet long, with 30-inch platforms and twenty 15- by 30-inch windows. Heavy staves 1¼ by 5 inches were held together with iron bands.

The unprepossessing cars were "frequently condemned by strangers at first sight," the 1843 annual report admitted. The directors claimed, however, that once the initial shock was over, the discerning passenger would find that the cars had many advantages. In a masterful series of rationalizations, the merits of the barrel car were summarized in the *American Railroad Journal*.[40] According to this article the cars were lighter, simpler to build, and cheaper than square cars. The round body was nearly fireproof, for it shed live cinders as fast as they fell; lateral wind pressure was dissipated; and in a derailment the car rolled down the embankment without injury. Hacker patented the design on January 21, 1841 (No. 1937), and the South Carolina reports were full of praise for the hogshead car. Happily, the design was confined to the home road; the prospect of barrel cars "flying with the speed of a locomotive" held little appeal elsewhere.

Other Southern roads, while avoiding the South Carolina's misguided digression, showed a strong attraction to the eight-wheel car. Nearly all the Southern lines visited by Von Gerstner in his travels of 1838–1840 were well stocked with American coaches. He reported that the Wilmington-to-Raleigh line in North Carolina had eight very elegant double-truck coaches that were provided with ladies' compartments and water closets.[41] Many of the lines contained half-fare compartments in the baggage car for

Figure 1.15 Camden and Amboy Railroad coach of about 1840, shown in a scene from the 1870s. (E. P. Alexander Collection)

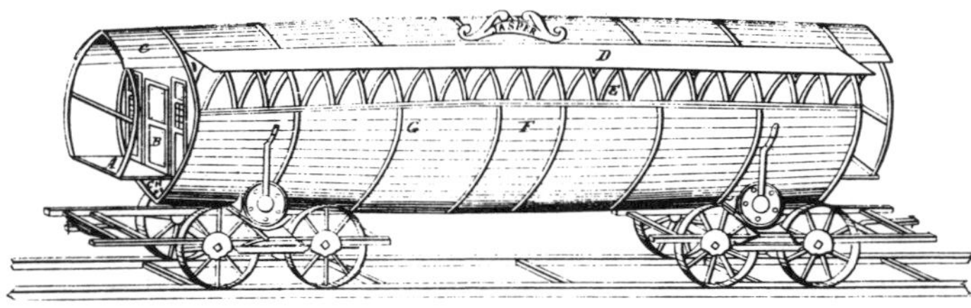

Figure 1.16 Hacker's hogs head car, built in the shape of a gigantic barrel, was produced for service on the South Carolina Railroad in 1840. (Southern Railway)

Negroes, but Negroes could also ride in the coaches at full fare. Von Gerstner pointed out that this was in fact better than the treatment given Northern Negroes, who had to ride in separate cars no matter what the fare.

By the late 1830s the entire car-building industry had begun to approve of the eight-wheel car. In 1837 Stephenson stopped pushing cars with omnibus bodies and developed a distinctive, if not beautiful, design for double-truck passenger coaches.[42] The whole side of the body was made as a lattice truss. This plan, which was among the first to employ bridge-construction methods for car building, created a very strong car. The truss assisted materially in supporting the body and lessened the need for heavy frame sills. Long self-supporting car bodies thus became possible. The cars were popularly known as X- or Diamond frame cars.

One X-frame car built for the New York and Harlem proved too high for the overhead clearances and it was sold to the Auburn and Syracuse Railroad in August 1839. A contemporary newspaper described it as a "traveling palace . . . gorgeously flowered and varnished."[43] The car, renamed *Auburn*, was reserved for through passengers. It was fitted with thirty double reversible seats and had a stovepipe running the length of the car for more even heating. The glass windows were stationary; adjacent wooden panels were opened for air. Cars of the same pattern were built for the Mohawk and Hudson, the New Jersey Railroad and Transportation Company, and the New York and Erie. One of them is pictured in the lower right corner of Figure 1.17, a lithograph showing Stephenson's factory. (Another contemporary, somewhat foreshortened, sketch is shown in Dunbar's *A History of Travel in America.*[44]) Stephenson failed in 1843, however, and that ended the production of Diamond frame cars. He was able to reopen, but he blamed the "bad paper" of the steam roads for his failure and said that henceforth he would build omnibuses and horsecars only when payment was to be in cash.

In Wilmington, Delaware, the car builders Betts, Pusey, and Harlan (later reorganized as the mighty Harlan and Hollingsworth Company) also came to approve of the eight-wheel car. In a printed prospectus of 1839 the firm stated that during the short period of two years since their general adoption in this country, eight-wheel cars had given "universal satisfaction."[45] Thirty-nine passenger and twenty-eight mail, baggage, and freight cars on

Figure 1.17 Stephenson's lattice frame car, shown in the lower right of this lithograph, was produced for several railroads, including the New York and Erie. The print dates from about 1840.

this plan had already been delivered. The prospectus listed seven major Middle Atlantic and Southern railroads receiving these cars and described the various classes of passenger car:

1st Class Car. For 60 passengers, body 32 feet in length and 8 feet 6 inches in width, spring seats crosswise, with shifting backs, trimmed in best style, with silk or worsted, venetian blinds to the windows, the platforms protected by an ornamental iron railing, with iron steps to ascend the same, draw springs and buffers, zinc roof, wheels of cast iron, (chilled, 3 feet in diameter, effective double acting brakes, the body secured in the strongest manner by iron bolts and brace rods; painting in best style, with gilt border and lettering, and the entire finished in the most workmanlike manner

for . . . . 493 Liv. St. or 2,400 dollars.
Wrought iron tires on the wheels
additional . . . . 200 ″
Private apartment for ladies, completely trimmed and
carpeted . . . . 100 ″

2d Class Car. For 60 passengers, body as above, seats crosswise, with shifting backs, hair or cloth seating, worsted head lining, festoons, and sun curtains, iron railing and steps to the platforms, spiral draw springs and buffers, zinc roof, wheels as above, brakes and brace rods, neatly painted and varnished

for . . . . 411 Liv. St. or 2,000 dollars.
Wrought iron tires on the wheels
additional . . . . 200 ″
Private apartment for ladies, neatly fitted
up . . . . 60 ″

3d Class Car. For 60 passengers, body as above, seats crosswise, with shifting backs, cloth or leather seating, plain iron railing, canvass roof painted and sanded, wheels and brakes as above, trimmed and painted in a plain neat style,

for . . . . 339 Liv. St. or 1,650 dollars.
Wrought iron tires on the wheels
additional . . . . 200 ″

4th Class Car. For 60 passengers, body as above, plain wooden seats, with shifting backs painted, common iron railing on platforms, wheels, brakes and canvass roof as above, and plainly painted on the outside,

for . . . . 287½ Liv. St. or 1,400 dollars.
Wrought iron tires on the wheels
additional . . . . 200 ″

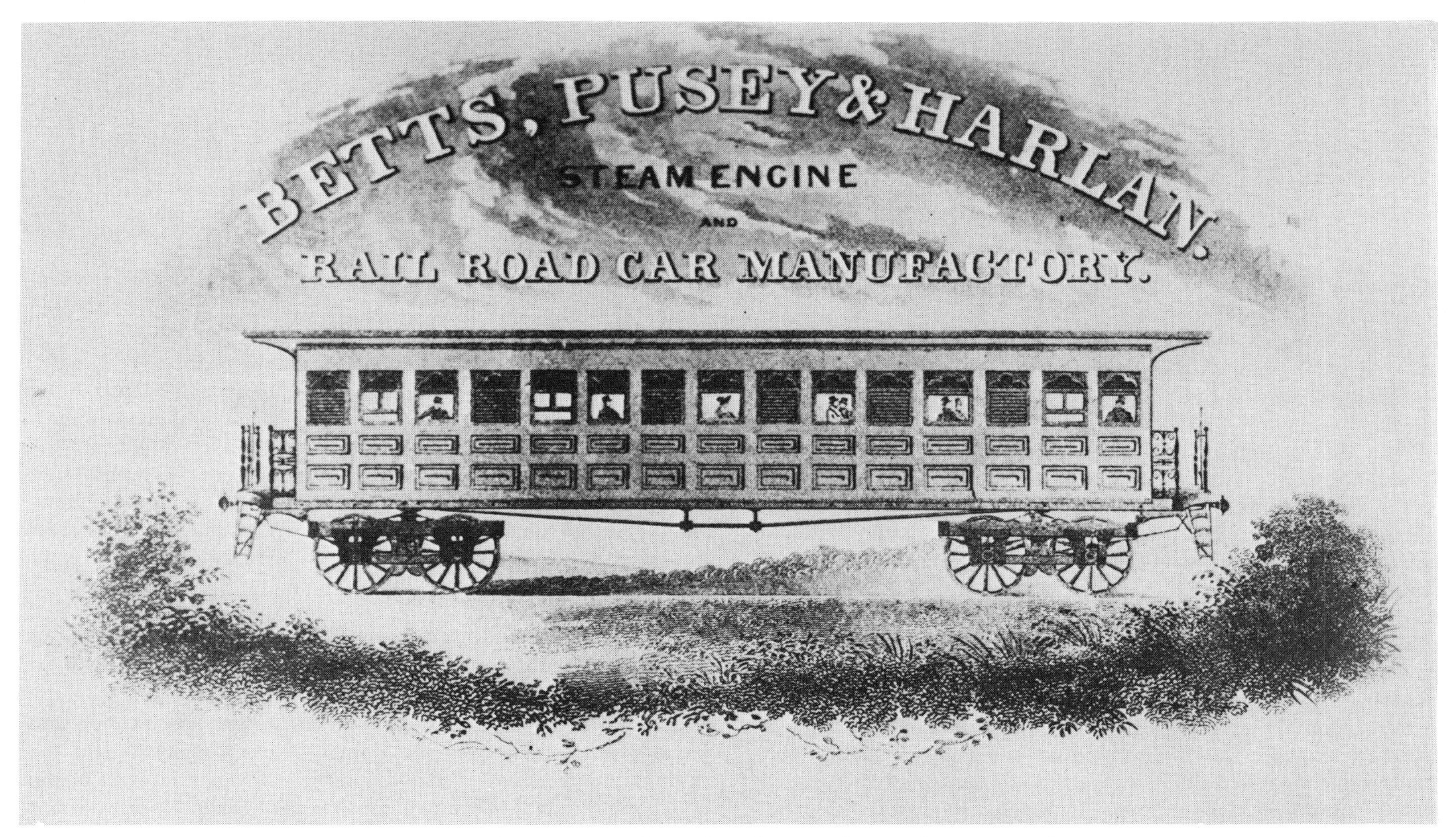

Figure 1.18 A predecessor of Harlan and Hollingsworth of Wilmington, Delaware, included this engraving in an advertising brochure of 1839. (Association of American Railroads)

Printed in English and French, the prospectus was addressed to the railroad companies of Europe. Whether any foreign orders were secured is a question, but it is known that American cars were supplied to European roads by other U.S. builders at an early date. The front page of the prospectus was graced with a fine lithograph of a passenger coach (Figure 1.18). The drawing is well proportioned and mechanical in nature, although the human figures shown at the windows appear to be only about one-half life size.

Acceptance of the eight-wheel car was well advanced in some parts of the country by late 1838, but as the *American Railroad Journal* stated:[46]

> Much as this description of Railroad car is in repute in the Southern and a portion of the Middle States—yet in this State [New York] and East of us, there is a singular prejudice in favor of the four-wheel cars, and against the new form.

In general New England was slower to adopt the American car than other settled regions of the United States. At this time the Northeastern states were showing the same conservatism in locomotives: a major locomotive builder of the area was content with the rigid-frame English style of engine some years after all other domestic manufacturers had turned to the truck locomotive. It should also be recognized that the New England roads had already invested heavily in four-wheel cars, and many simply could not afford to discard them after only a few seasons.

Although resistance to the new form was marked in the Northeast, it was not universal there. As early as 1833 the Boston and Worcester was considering buying trucks from Winans with bodies to be furnished by Stephenson. The road decided against it, however, and did not place eight-wheel cars in service until 1837 or 1838.[47] Even so these cars, built by Davenport and Bridges, were among the very first of their type in New England. A year or so earlier the Hartford and New Haven Railroad had asked Winans the price of eight-wheel cars. He sent the following quotation; it cannot be determined if the road actually placed an order.[48]

> The 8 wheels, the axles made of swedes iron faggoted under a trip hammer, the 4 iron bolsters made same as axles, the 8 light & 4 heavy English steel springs, the two small carriage frames or running gear made mostly of wrought iron, the large frame for the body to rest on with iron suspension braces, the four friction rollers & various castings for supporting the springs boxes &c. the iron railing surrounding the platform at either end of the car, the two brakes one at each end of the car so arranged as to brake or apply friction to all of the wheels when necessary for stopping or retarding the motion of the car, and the fitting up of everything complete ready to receive the body cost $1200. the body cost $850. making the sum of $2050.

The Boston and Providence's first eight-wheel car was completed at the company workshops in September 1838 (see Figure 1.75 in the Representative Cars section). The car was so successful that the line rebuilt some four-wheel cars on the double-truck pattern. Two or even three bodies were united, and the old running gears were remodeled as trucks and placed beneath them.[49]

In December of the same year the Nashua and Lowell received several American-style coaches from the shop of Jeremiah Meyers in Attleboro, Massachusetts.[50] (Meyers secured the order by underbidding Davenport and Bridges, but it was a profitless contract that put him out of business.) The Boston and Lowell built its first eight-wheel car in April 1840 in the company repair shops, under the direction and to the design of the master mechanic. The drawing made from the original car which was still in service in 1853, was offered in testimony at the eight-wheel car case.[51] It was explained that the left truck shows the first mode of suspension and the right truck that which was in use when the drawing was made.

While these examples demonstrate that some of the New England roads were only a few years behind other regions in adopting the eight-wheel car, there is evidence of continuing resistance to the new form in New England. As late as 1853 the Eastern Railroad defended four-wheel passenger cars, claiming that they required less power to pull them and passed through curves with less friction and hence less wheel wear (arguments that, ironically, were usually reserved for the double-truck car). The Western Railroad, later the Boston and Albany, was purchasing new four-wheel passenger cars some years after they were considered obsolete. One of these, built by the Springfield Car and Engine Company in 1848, was involved in an accident which brought about some pointed criticism of the road for clinging to outmoded equipment.[52] In fairness it should be noted that the Western was also running eight-wheel cars as early as 1842. An engraving made after a daguerreotype taken the same year shows cars of this type (Figure 1.19).

In spite of regional variations, it is fair to say that in only five years the double-truck car displaced the four-wheel carriage. By 1840 it was the unchallenged standard on American railroads. Its rapid spread across the nation must be regarded as one of the most complete and revolutionary innovations in American railroad technology. It is also something of an American phenomenon, for eight-wheel cars were not widely used elsewhere in the world until rather late in the nineteenth century.

THE EIGHT-WHEEL CAR CASE

There is a disturbing tendency in some histories of technology to credit basic inventions to the genius of one man. Among the most fallacious, yet best-established reputations are James Watt's as the inventor of the steam engine, Robert Fulton's as the inventor of the steamboat, and George Stephenson's as the inventor of the locomotive. To this list might be added Ross Winans's as the inventor of the eight-wheel car. Many such claims originate with enthusiastic biographers who puff the work of their subjects. Other men simply become folk heroes and are popularly crowned with greatness. Still others actively sought recognition in their own lifetime for personal gain or the need to satisfy a voracious ego.

Most standard references, including the major encyclopedias, unquestioningly assign the eight-wheel car to Winans. Yet the eight-wheel car, like most such devices, was a conglomerate invention that evolved from the labor of several resourceful mechanics. One man rarely envisions a whole machine at once. A basic idea may occur to any number of artisans when a new need arises, and each studies and refines it in his own way. Winans was certainly involved in the early development of the eight-wheel car. He deserves partial, but not exclusive, credit for its perfection.[53]

Winans documented his claim in a patent of October 1, 1834, for improvements in railroad cars. The long specification covered many points in the construction of eight-wheel cars. According to his critics, it exhibited as much or more genius than the invention itself. The principal claim specified an extremely short-wheelbase truck with live-spring side frame. In the final paragraph, however, Winans specifically disclaimed inventing the eight-wheel car, referring to "the running of cars or carriages upon eight wheels, this having been previously done."

Winans's patent was not in fact the first in the country that broadly touched upon such cars; Jonas P. Fairlamb (1833) and Ephraim Morris (1829) had previously secured papers on eight-wheelers. The patent law was very liberal at the time that Winans's patent was issued. A properly drafted specification and payment of fees secured the patent. No experts examined the draft for conflicts with previous claims, nor was the "state of the art" considered. It is surprising that the wheel itself was not patented. Winans appended no drawing to the patent until after the first lawsuit in 1838, and then he included an illustration of a boxcar rather than a passenger car. Another major discrepancy was the fact that the drawing showed a wooden-beam truck rather than the live-spring design called for in the specification.

If Winans believed himself the creator of the eight-wheel car, he suffered in silence for more than four years before pressing the claim. If he invented and built the *Columbus*, as he claimed, why didn't he apply for a patent in 1831? And why did the text of his 1834 patent deny that he had originated the eight-wheel car? As a wheel manufacturer Winans was in regular contact with major railroads and car builders across the land, yet nearly all stated that Winans had never hinted at his right to a royalty on the eight-wheel cars for which he supplied the wheels. It was uncharacteristic behavior for a man who was renowned for his outspoken independence.

Winans' critics claimed that the patent could not cover an idea which had been abandoned to the public for nearly four years. They also suggested that Winans did not apply for his patent until his chief rival, Conduce Gatch, retired from railroad service in May 1834. In Winans's defense it should be said that an inventor is entitled to a grace period in which to improve and assess the importance of his discovery. It should also be noted that the president's message in the 1834 B & O annual report acknowledges Winans as the inventor of the eight-wheel car. As an employee of the road, Winans reportedly waived his claim against the B & O early in 1834. The document was not produced, however, during the long years of litigation that followed; either it never existed or it was lost.

The first suit that Winans filed was against the Newcastle and Frenchtown Railroad in July 1838 for violating his patent by operating eight-wheel cars without paying him a fee. It was the beginning of twenty years of legal proceedings. The jury could not agree, but before a new trial was ordered the defendant

EXPRESS TRAIN ON WESTERN RAILROAD.

(From a Daguerreotype, made in 1842, under the direction of Charles Van Benthuysen.)

AFTERNOON TRAIN BETWEEN ALBANY AND SPRINGFIELD.

STILLMAN WITT, Superintendent at Albany.

THOMAS W. ALLEN, Master Mechanic.
D. S. WOOD, Engineer.

JOHN B. ADAMS, Conductor.
HORACE H. BABCOCK, Ticket Agent.

Figure 1.19 Early examples of eight-wheel passenger cars on the Western Railroad, later the Boston and Albany. This engraving was based on a daguerreotype made in 1842.

voluntarily paid Winans a judgment of $500. This half victory was enough to encourage him to start action against others. He approached the Baltimore and Susquehanna next, asking for the modest sum of $500 late in 1839. The railroad countered with an offer of $350, which Winans refused. He waited until the road's president was in Philadelphia and then had that official served with a summons. Faced with an inconvenient suit away from the home office, the Baltimore and Susquehanna quickly agreed to Winans's demands. His claims now established in two states, Winans turned to the Portsmouth and Roanoke, which meekly paid him a license fee. The Philadelphia and Reading also agreed to a friendly, out-of-court settlement for the minimum fee of $500.

The strategy was simple: establish the patent's validity by a series of small settlements. Even if the payments were trivial, the precedents confirmed the soundness of Winans's claim. Curiously, at this juncture Winans did not press his advantage. It was not until seven years later that he unexpectedly renewed legal action against the Troy and Schenectady Railroad. A small, financially weak line, it was in no position to fight off so wealthy and vigorous an opponent as Ross Winans of Baltimore. Apparently the suit was prompted by the pending expiration of the 1834 patent. Winans sued the road for $100 a year for every eight-wheel car in operation, plus a settlement for the past use of such cars. While the suit was underway, a special bill was hurried through Congress in 1848 to extend the potentially valuable patent for another seven years. In June 1850 the U.S. Cicuit Court ruled in Winans's favor, but settlement was delayed by appeals. Eventually, a new trial was denied, and Winans accepted a paltry settlement of $100.

The real prize in the Troy and Schenectady suit was legal confirmation of Winans's claim, for the entire industry was now open to a severe levy. Before initiating the expected barrage of lawsuits, he agreed to share the spoils with his attorneys, C. D. Gould and J. A. Spenser, if they would finance and direct the proceedings. Receipts from his past legal battles had been disappointing, and Winans did not want to proceed alone. Between 1852 and 1853 Gould and Spenser started action against eight New York railroads as well as against Eaton and Gilbert, car builders of Troy. Separate suits were entered against A. & W. Denmead, car and locomotive builders in Baltimore, and the Eastern Railroad.

This massive legal assault was the climax of the Twenty Years' War against the Railroads, as Winans's series of patent suits came to be called. It was now a matter of grave importance to the railroad industry. Ruin lay ahead if Winans's extraordinary claim was not nullified. Winans was characterized as the arch-villain of the railway supply trade, a pariah and an "imposter." He was called the "inventor of everything." These attacks were based on more than the car case, for Winans was also conducting court actions on his alleged patent rights to six-wheel connected locomotives and to the variable exhaust. Moreover, he was engaged in a vitriolic public debate with the directors of the B & O over the merits of his Camel locomotives. It is not surprising that his railway machinery business declined rapidly after the mid-1850s.

Recognizing the seriousness of the Winans suits, the recently formed New York Central Railroad took on the defense of smaller New York railroads and assisted Eaton and Gilbert. It hired two first-rate lawyers, William Whiting and William E. Hubbell. On its part the Eastern Railroad spent $20,000 in legal preparation and published a 1299-page transcript of its defense. "Neither time, nor labor nor money was spared. The history of railroads was thoroughly investigated from their origin both in the United States and in England. Every old workshop was searched for plans and drawings of early railroad cars—every venerable engineer and machinist was questioned of his experience relative to railroad carriages. No geologist ever examined more carefully for antediluvian remains."[54]

Chief among the witnesses for the defense in the series of suits was Conduce Gatch, who constructed the first double-truck cars on the B & O. Gatch was steadfast in his claim that he designed and built these cars, including the prototype *Columbus*. Several old workmen supported his contentions. Only one, Oliver Cromwell, sided with Winans, but even he would not credit Winans with more than the body design. Evidence given by Gridley Bryant, builder of eight-wheel freight cars for the Granite Railway in 1829, and the abundance of printed and patent literature on eight-wheel cars previous to Winans's 1834 patent, made a shambles of the Winans case. The testimony was long and repetitious; one trial ran for 75 days. Winans lost one action after another, but as a railroad pamphlet of the time described it, he kept pressing "with a pertinacity of purpose rarely equalled." At last the New York and Erie case was heard by the U.S. Supreme Court in its December 1858 term. Judgment was awarded the defendants on January 10, 1859. So ended Winans's incredible legal adventure—the Twenty Years' War was over. The weight of the evidence was against him, but the publicity, as misinterpreted by later writers, seems to have ensured Winans's enthronement in history as the father of the eight-wheel car.

Age of the Arched-Roof Car: 1840–1860

After the late 1830s, when the eight-wheel car became firmly established, there were twenty quiet years of contentment with the existing day coach pattern. The same design was repeated again and again with little or no change. Why tamper with a simple, inexpensive plan that served every requirement so well? It seemed unnecessary if not imprudent to experiment. A typical car of 1840 had a rectangular body rarely over 30 feet long by 8½ feet wide. Before 1845 the glass windows were generally stationary, with sliding panels for ventilation. The ceiling was low, providing headroom of just over 6 feet. The seats were closely spaced and had narrow cushions and low backs. The trucks were on a short wheelbase. The total weight was about 8 tons.

An example of such a car was the *Morris Run*, built in 1840 for the Tioga Railroad by Harlan and Hollingsworth (Figure 1.20). When it was exhibited some forty years later at the Railway Appliance Exposition, it was described as being 36 feet long by 8 feet, 4 inches wide, with a 6-foot, 4-inch headroom clearance.[55] Inside-bearing trucks were seldom used on American cars. The truss rods shown were also rare at this early period, although they became a standard fixture on later wooden cars. The seats had cast-iron frames, and the cushions were covered in leather. Single candle lamps were placed at each end of the car. The original cost was given as $2,000. In 1883 the car had run more than a million miles. It continued in service for another ten years, when it was destroyed by fire.[56] Similar cars are shown in the Representative Cars section (see Figures 1.70 to 1.81) at the chapter end.

By the mid 1840s a few modest improvements were evident. Some cars had 35-foot bodies. Movable window sashes replaced stationary glazed frames, and more and larger windows replaced the upper panels. Roof canopies extended over the end platforms to protect boarding passengers. Roof-mounted ventilators were introduced. But the headroom remained restricted, the seats continued to be narrow and closely spaced, lighting was inadequate, and many cars did not have water closets or ice-water dispensers. The interior finish of these cars was generally plain. Common woods were typically used and painted over. Head linings (ceiling coverings) were somtimes omitted, and the exposed wooden roof structure was simply painted. Seat cushions were of leather or broadcloth, and the cheaper cars had wooden-slat seats.

More luxurious cars were occasionally built, however. In 1845 Davenport and Bridges constructed some large coaches seating 70 persons for the Eastern Railroad.[57] They were described as 9½ feet wide, or a foot wider than the usual day coach. The seats were mounted on pivots, upholstered like fine armchairs, and covered in a rich red-velvet plush. The interior paneling was mahogany, and the floor was laid in a diamond pattern of cherry and black walnut.

The early eight-wheel cars were very lightly built. The floor frame, the basic support of the entire car, was the simplest post-and-lintel form. Two side sills 6 by 7 inches, supplemented only by some cross and short end timbers, composed the frame. Center sills, truss rods, or body-panel truss were not generally used, nor were they needed for the short car bodies of the period. The body bolster was built up from three timbers; two to four short beams parallel to the side sills ran between the body bolster and the end sill. Ten or more cross timbers were placed between the bolsters. An example of such a frame is given in the 1845 Eaton and Gilbert drawing (Figure 1.79). John Kirby (1823–1915), a prominent master car builder, remembered these frames on the first cars he helped build in the late 1840s.[58] Another pioneer car builder said that the Boston and Albany was using this simple but obviously weak style of framing in 1870.[59] According to *The National Car Builder*, the old form was still found on some roads five years later.

Figure 1.20 Harlan and Hollingsworth built the Morris Run *in 1840 for the Tioga Railroad. The car was preserved as a relic until its loss by fire in 1893.* (Semi Centennial Memoir of Harlan and Hollingsworth)

The body structure was even more fragile than the main frame. It was composed of thin vertical and horizontal pieces very much like the balloon house frame of the day. Thin wooden panels covered the frame. No truss or cross braces were used. The whole framework was like an egg crate resting on two matchsticks—an exaggeration, of course, but not far off. The pre-1855 passenger cars was a generally feeble structure.

Why did the industry fancy this obviously weak construction? Wood-building technology was an ancient art that offered centuries of experience in elaborate bridge and roof trussing. Good timber was abundant and cheap. The explanation appears to lie in an overriding concern to hold down size and weight. Lack of capital had led to the construction of exceedingly light, cheap railways in this country. Flimsy track and bridges could support only the lightest rolling stock. Locomotive power was accordingly limited; hence the cars that were being pulled had to be underweight. Low ceilings, closely spaced seats, and skimpy framing produced cars of maximum capacity and minimum weight. The absence of auxiliary apparatus, water closets, ventilating equipment, vestibules, and other devices that characterized later cars held down both weight and cost. Certainly for short daylight trips such cars were satisfactory.

By the late 1840s, however, somewhat larger and better-furnished cars were in demand. The Louisa Railroad of Virginia submitted the following specification to Harlan and Hollingsworth in 1848:[60]

Length of Car Out and Out 43 feet 4 inches, with two drop platforms of best yellow heart pine 2 ft. 8 × 4 feet.

Body of Coach from out to out 38 ft. Long × 8 ft. 8 inches wide.

Bottom Rail of best heart yellow pine 38 ft. long 10×5 inches.

Top Rail 42 feet long 8×3 inches best heart pine or oak, to be confined to the pillars with two rivets to each pillar.

Bolsters to be 12×13 best white oak, 4 knee plates of wrought iron 5/8 thick by 10 inches wide, each knee plate confined to the bolster with 3 bolts running through the bolster 3/4 in. diameter and 3 running through the bottom railing of 5/8 diameter.

(This a substitute for the circular bar of iron extension bolts.) An inside plank riveted to each pillar with 3 rivets and bolted to the bottom rail between every other window with a bolt of 5/8 diameter, with a nut at the bottom, the plank to be 2½ inches thick by 13 deep.

The Ladies Apartment 6 ft. 4 in. long with a lamp box each side of the door with glass doors inside and out. Seats in the ladies apartment to be red plush sofas, back of the same.

Gentlemen's Apartment to have a passage 16 inches wide, the floor laid with the best yellow pine dressed to 1¼ inches. Seats put crosswise, made of Mahogany wide backs stuffed with best black hair cloth well stuffed.

Seats 2 feet 7 inches from center to center with a foot piece underneath, water cooler to hold 4 gallons, lamp box on each side of the door in the Ladies Apartment—with a stove fixed with hand railings and brass knobs.

Drop platform of best heart yellow pine for the floor 2 steps on each side 9 in. wide and 1 foot high supports on 3 pieces of iron 2½×½ confined to the platform, upright rod from the bottom to the platform to the top of the coach 1⅛ inches in diameter, balance of rod for hand railing 1 inch diameter 2 ft. high.

Door to be 5 feet 10 inches high 2 inches thick paneled of Mahogany good yellow heart pine or curled maple.

Windows to hoist fastened in from the outside with a wooden moulding instead of putty.

Figure 1.21 The cars shown here in 1883 were built around 1850 for the Cleveland, Columbus, Cincinnati, and Indianapolis Railway. They were remodeled with clerestory roofs in the 1860s. (Thomas T. Taber Collection)

Eves from one end of the car to the other 9 inches wide immediately over the window to throw off rain, etc.

Curtains of strong drab twilled worsted.

The car to be 2 inches higher on the inside than the first one made [for] the Louisa Co.

Painted handsomely on the inside and in a plain substantial manner on the outside.

To have a bar of iron on each side 3½ in. wide by ⅜ in. thick running from the front bolster to the back bolster, bending with a regular bow, so that the Top of the bow of iron just comes to the bottom of the window, the said bar to extend from the bolster to each corner in the manner of a brace, and to extend up the corner posts as high as the bottom of the window and the said bar to be made fast to each pillar with rivets, there is to be an extension bolt of ⅝ diameter between every Pillar with a hook head to hook on the 3½ inch bar and to go through the bottom rail with a nut at the bottom. Every pillar is to be confined to the bottom rail with a pilot point bolt.

Very Respectfully, E. Fontaine, Pr. L. RRCO.

In another letter the exterior color was specified as dark red. The car was built for $1,350.

More substantial passenger cars were also constructed in Cleveland for the Cleveland, Columbus, and Cincinnati Railroad (Figure 1.21). These cars, completed in late 1849 and early 1850, were made on the plan of William F. Smith (1826–1878), who was the line's master car builder for nearly thirty years. The cars were 41 feet long by 6½ feet wide and seated 60 passengers.[61] Ash-frame sills were cut and squared by hand in a nearby forest, as no sawmill was available. No center sills were used, but the body was made sag-resistant by a slight camber or bow in the sills. Continuous siding rather than individual panels, together with "plenty of glue," stiffened the body. The cars were still in service in 1883, and they were reported to be holding up better than many new cars. Some had derailed several times and a few had rolled down embankments, but all survived these mishaps. The truss rods and clerestory roofs that can be seen in Figure 1.21 were added in later years.

During this period a peculiar style of car was evolved by E. A. Stevens for the Camden and Amboy Railroad. In 1848 Stevens revised the old underhung-truss frame used on the road's first eight-wheel cars by boxing in the truss.[62] This improved the appearance of the cars (though they still could not be called attractive), and the line claimed that it provided two giant sled runners that would support the car in case of a broken axle. The cars, known as Rockersills, bore a strong resemblance to a canal-boat and were said to be among the lightest and strongest for their size in service.[63] They were built in large numbers through the late 1850s; a few were even constructed as late as 1864. A train of Rockersill cars on one of the United Railroads of New Jersey is shown in Figure 1.22.

By the early 1850s increased attention was being given to passenger comfort. Larger, more commodious cars were coming into service, though the simple arched roof style remained in fashion. The growth of the railroad network made longer trips possible. The first east-west trunk lines were opening, and night travel became more common in journeys of 200, 300, and even 500 miles. Car bodies were stretched to 40 feet; headroom rose to 7 feet; car weights increased to 10 or even 12 tons. Longer, heavier cars called for stronger framing. The addition of center sills was the most obvious remedy. Two to four timbers running the full length of the body did much to strengthen the car. These were usually somewhat lighter than the side sills.

A more startling change in construction was the introduction of the side-panel truss. This wooden truss built inside the lower

Figure 1.22 A scene on the United Railroad of New Jersey about 1870 pictures Rocker-sill cars constructed between 1848 and 1864. (E. P. Alexander Collection)

body panel greatly stiffened the side sills, permitting long but relatively light car bodies. Thus the side truss, made up of many light members, took the place of excessively large floor timbers that would otherwise be necessary for long bodies. The idea was, of course, taken directly from wooden bridge technology, but it was an imaginative solution and must be regarded as among the most important American contributions to wooden car construction. It was suitable only for the American style of coach; side-door cars did not lend themselves to the side-truss design.

The origins of the side-panel truss are obscure. They might be traced back to the underhung frames on the B & O Washington branch cars. Stephenson's X-frame design was another predecessor. But who first thought of the neat, simple side-panel scheme is unknown.

Specific examples of American coaches for this important transitional period are unfortunately rare. Several pictures from the early 1850s (Figures 1.23 to 1.26) show nothing of the cars' construction. In 1853 the Michigan Central built a number of 60-foot cars, giants for their day, which would require strong framing, and it can be speculated that side trusses were employed.[64] The first concrete example of this construction so far uncovered is a group of 65-foot cars built for the New York and Erie Railway in 1855 and 1856. Curiously, despite its great track gauge of 6 feet, the Erie had been content with short cars in its early years. Some were as much as 11 feet wide, it is true, but their length was held to 36 feet. The new long cars were built to the design of Calvin A. Smith, master car builder at the company repair shops. The top cord of the truss was arched at the center. The diagonal braces were iron rods. The Erie used this simple form of trussing for many years; after 1855, however, the top cord and window sill were made as one rather than as two independent members, and the diagonal braces were wooden instead of iron rods. Drawings and more information are given in the Representative Cars section (Figures 1.83 to 1.85).

The trend toward big cars was well underway by the mid-1850s. The weight of some cars was over 15 tons. Most roads accommodated the load by increasing journal bearing sizes and wheelbases, but in general they found the ordinary four-wheel truck adequate. Others contended that more wheels and axles were needed to spread the load and ensure safe movement of heavy cars over the existing roadbeds. Improved riding quality was another justification offered. The adherents of multiple-axle trucks created something of a vogue for twelve- and sixteen-wheel cars between about 1855 and 1865.

Isolated examples of twelve-wheel cars can be found a decade earlier in the mid-1840s, when the Camden and Amboy began fitting cars with six-wheel trucks. The main reason was safety: spreading the load over extra axles lessened the possibility of a fracture or derailment. The South Carolina Railroad also operated twelve-wheel cars, according to a note in the *American Railroad Journal* of April 1846. And the Michigan Central's annual reports show that it began building twelve-wheel cars a year or two later. Like the Camden and Amboy, it became largely stocked with these cars; by 1855 forty were in service, with six-

Figure 1.23 *The* Illustrated London News, *April 10, 1852, depicted the interior of an American railway car of that period.*

Figure 1.24 *The Number 164, dating from about 1855, is shown in this photograph taken some forty years later. Note the small ventilators in the letter board. (Library of Congress, Van Name Collection)*

Figure 1.25 Kimball and Gorton produced this coach for the Pacific Railroad (of Missouri) around 1852. The illustration is a contemporary lithograph. (Historical Society of Pennsylvania)

teen more under construction. The Illinois Central was another Western line that adopted twelve-wheelers (drawings for one of them are in Figures 1.86 to 1.87).

If twelve wheels were good, it seemed to some builders that sixteen would be better. They attached two four-wheel trucks to a secondary pivoting framework at each end of the car. It is reported that Davenport and Bridges built such a car in 1839, though it cannot be confirmed. Osgood Bradley built cars of this type for the Norwich and Worcester in 1857 and for the Eastern Railroad.[65] A German visitor described these or identical cars running on one of the New York–Boston roads as having bodies 52 feet long, truck centers of 37 feet, and truck wheelbases of 9 feet.[66] The arrangement was particularly favored for sleeping cars; both Woodruff and Pullman had sixteen-wheelers. But this style was never widely used, and no sixteen-wheel cars are known to have been built for passenger service after the late 1860s. Among their many defects were the extra cost, the dead weight of the complicated running gear, the added friction in passing through curves, and the complicated brake rig. The Old Colony purchased some sixteen-wheel cars and regretted it, complaining that they did not take curves well and even occasionally turned over.[67]

While the trend to larger cars was unmistakable, not all major trunk roads could adopt them. Some Eastern lines that double-tracked at an early date placed the tracks too close together, with clearance for narrow cars only. The Philadelphia and Columbia, which laid its second track in 1834, was forced to operate cars nearly 2 feet narrower than those on other standard-gauge roads. The Reading built its second track on too narrow a plan in the mid-1840s.[68] The resulting cars were too narrow for ordinary double seats; like the narrow-gauge cars of the 1870s, they had double seats on one side of the aisle and a single on the opposite side. Between 1856 and 1860 the line was rebuilt at considerable expense to allow another 2 feet of clearance. Even this was not enough, however, and the Reading was noted for its narrow cars until it was rebuilt again in later years. The Philadelphia, Baltimore, and Wilmington was another road cursed with a restricted loading gauge. New cars that it received from Harlan and

Figure 1.26 The Central Railroad of New Jersey's Number 159, dating from about 1855, is shown in this scene at the time of its retirement around 1895. (Library of Congress, Van Name Collection)

Figure 1.27 Jackson and Sharp produced this handsome clerestory roof car around 1867. It illustrates the earliest style of clerestory with monitor ends. (Hall of Records, Delaware Archives)

Hollingsworth as late as 1865 were only 8 feet wide (Figure 1.92). But most American roads were built late enough or wisely enough to avoid loading-gauge problems, and car widths of 9 to 10 feet were common by the late 1850s.

Other noticeable changes came in the same period. There were improvements in seating, decoration, lighting, and heating (these are treated in Chapter 5). The outside appearance of the cars also changed. Day cars had usually been finished with horizontal rectangular panels, but now batten board (narrow vertical wooden strips covering each plank joint on the side of the car) came into favor. In a few cases plain, narrow vertical sheathing was used, although it was not widely adopted until after the 1870s.

By 1860 the arched-roof car reached its ultimate development. The overall size had materially increased, as had the facilities for passengers' comfort. The oppressively cramped, lightly built eight-wheel cars of the 1840s had given way to a solid, roomy conveyance of reasonable if not extraordinary accommodation. The discomforts of rail travel by this time were largely the result of defects in the roadbed rather than in the cars.

The Clerestory Roof

To the casual observer an American railroad car of 1840 differs little from a car of 1860. Experts could point out many changes, but in general the layman's view is correct: in outer appearance, cars made during this twenty-year span looked much alike. However, a major change in their external configuration began in the early 1860s, when the roof profile was radically altered. The central portion of the roof was raised about 18 inches, and its sides were provided with openings for light and air. This centuries-old architectural device is properly known as a clerestory, but it was referred to at the time under many names—among them the lantern, monitor, raised, deck, clear-story, elevated, and steamboat roof.

Figure 1.28 More monitor roof passenger cars outside the builder's shop in 1867. (*Hall of Records*)

Richard Imlay's clerestory roof cars of the 1830s were in the vanguard of raised roofs, of course, but they had no lasting influence. In England the Great Western Railway tried a rudimentary raised roof for a few first-class parlor cars in 1838, but again the scheme failed to catch on.[69] The clerestory plan went into dormancy for over twenty years and was then revived by Webster Wagner, who was trying to improve the ventilation of his sleeping cars. Passengers complained about the bad air in Wagner's first sleepers of 1858. When the central portion of the roof was raised, the stale air, which rose to the top, could be exhausted out of the side vents. Wagner's first clerestory roof was reportedly built in 1859. In the spring of the following year his plan, now fully developed, was praised in the *American Railway Review*: "The roof is raised in the centre, and in the sides of the elevated part are inserted twenty-eight ventilators, which preserve the air pure and sweet, and keep up a steady and healthful circulation—a desideratum which has long been desired but which until now has never been supplied."[70]

Wagner's cars, which he called Monitor roof cars, were operated on the New York Central and its associated lines. One of these, the Hudson River Railroad, was quick to notice the merits of Wagner's roof for ordinary day cars. Four coaches with the "double deck" roofs were under construction at the shops of Eaton and Gilbert in August 1860.[71] The scheme spread quickly across the country. The Michigan Central built some raised-roof cars the following spring. Several years later the company decided that to "sustain the reputation of the road" all its day cars must be rebuilt with the new, fashionable roof. The Burlington shops were likewise busy between 1862 and 1865 fitting clerestories to their cars. The Cleveland, Columbus, and Cincinnati Railroad updated some of its original cars during the same years. And in 1862 the Pennsylvania Railroad adopted its first standard-class car, the PA, which was designed in the current style with a clerestory roof.

Wagner was almost too late in recognizing the value of his innovation. Together with an able mechanic and owner of several important patents, Alba F. Smith, he hastily prepared a patent application, which was granted on September 23, 1862 (No. 36536). However, the specifications spoke only of the ventilating aspects of the clerestory roof. No windows were shown in the accompanying drawings. Yet one of the chief values of the raised roof for day cars was the additional natural light provided by the side windows. Wagner and Smith tried to establish their ownership of the design through lawsuits, but with little success. The idea had been in the public domain too long to be claimed by any one inventor.[72]

The appeal of the clerestory went beyond its practical benefits of additional air and light; it agreed with the expansive mood of the times. As one journal said, "Loftiness and width are more and more in demand in railroad cars just as they are in steamboats, churches, hotels and theatres."[73] An opportunity for architectural grandeur never before possible had entered the realm of railroad car design. The cramped, tunnel-like appearance of the arched roof car gave way to the open dignity of a more spacious edifice. No more bending or crouching for low ceilings; at last a properly attired Victorian gentleman could stand erect, top hat and all. The means were at hand for more stylish deck lights, moldings, and ornamental art-glass windows. Elaborate center lamps were now possible, and these added greatly to the decor as well as to the lighting of the interior.

During the early years of the clerestory, no one could have anticipated the variety of design treatments it would receive in the course of time. At first Wagner's Monitor design prevailed: the raised roof running nearly the full length of the body terminated in half circles at each end. Several illustrations of Monitor roof cars have already been shown (among the better views are Figures 1.27 and 1.28). A variation on this design is the Buffalo, Corry, and Pittsburgh car Number 7, which has prow-shaped clerestory ends (Figure 1.30). The deck or Monitor style continued in favor until the late 1860s and was then succeeded by the

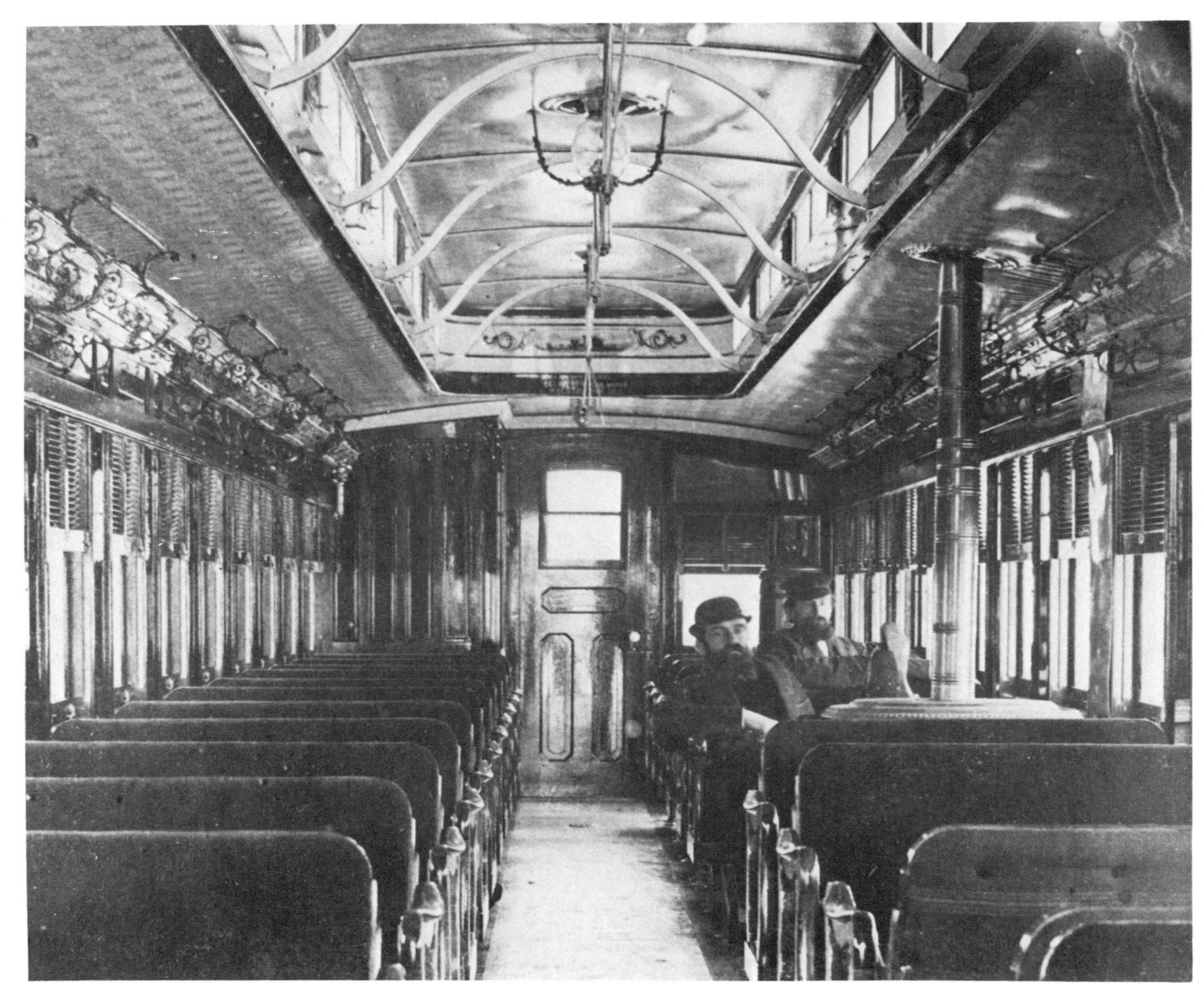

Figure 1.29 Interior photograph of a Jackson and Sharp monitor roof car. Note the decorative painting in the clerestory. (Hall of Records)

Figure 1.30 The New Haven Car Company built this car around 1866 with an unusual prow-shaped clerestory roof end. (Railway and Locomotive Historical Society)

Figure 1.31 This heavy coach dating around 1870 exhibited the broken bull-nosed roof end once popular with Pullman. (Railway and Locomotive Historical Society)

so-called duckbill roof (for examples see Figures 1.32 and 1.33). Here was an attempt to blend the clerestory ends with the main roof line. The first experiments, made with straight lines, were not too successful; later, graceful curved lines more smoothly integrated the two roof profiles. To achieve a greater decorative effect Pullman and others adopted the broken duckbill pattern, which employed several compound curves (Figure 1.31). The duckbill roof was popular in the seventies, and some roads, notably the Pennsylvania, continued to use it for almost twenty years. It was succeeded by the crescent or bullnose end, known in street-railway circles simply as the railroad roof. The crescent end was the final form of the clerestory. It prevailed from 1880 until the end of the raised roof some fifty years later (Figures 1.32 and 1.33). It was the most logical and to many, the most pleasing style of clerestory roof.

Although the clerestory roof was wholeheartedly accepted from 1860 forward, its value, beyond that of style, was challenged from the beginning. It had hardly come into use when Henry Ruttan, the promoter of a ventilating system for arched-roof cars, bitterly attacked it as a foolish, ostentatious device that served no purpose.[74] Ruttan's self-interest was apparent, but he correctly pointed out that the car builders' mania for the new design was fed by fashion and not utility. His chief complaints were the added first cost of $600 per car, the increased heating space, and the top-heavy bodies of cars with clerestories. The top-heaviness, however, was a matter of appearance rather than weight, for the roof structure was very light.

More serious criticisms could be leveled at the clerestory. Its many joints and corners offered more places for leaks and dirt-catching, so that it certainly added to maintenance costs. It never really solved the ventilating problem; complaints continued on this subject until the advent of mechanical air conditioning in modern times. But by far its greatest defect was its structural weakness. Each carline (roof rafter) was broken by four right-angle joints. Builders did not use stiffening angle brackets because they did not want to break the clean sweep of the car's interior. Metal reinforcements cut to the exact outline of carlines provided some support for the structure, but the roof remained the weakest part of these cars. It had no lateral strength and did little to support the sides of the body. In fact, the clerestory roof was said to spread the sides, for the weight of the roof caused the carlines to be pressed down and out.

The structural deficiencies of the clerestory prompted periodic clamorings for a return to the simple and strong arched roof. In the early 1880s the *National Car Builder* said that the clerestory had prevailed long enough—that a return to correct mechanical construction was overdue. The Boston and Albany began building arched-roof cars again. They were made with an extremely high arch, however, so that there was ample headroom and space for center lamps. By 1887 this design was being copied by the Boston and Lowell and the Chicago and Eastern Illinois.[75] With their introduction into the United States in 1883, William D. Mann's European-style sleeping cars were also built with plain high arched roofs. The Philadelphia and Reading adapted the old X-frame brace, which had been recommended since the 1870s as a strengthener for the flimsy clerestory.[76] The X frame ran from the side of the car to the underside of the clerestory without a break, thus avoiding the right-angle joints of the conventional construction. This plan was very strong, but it was also ugly. The Reading modified the idea by eliminating the upper arms of the X brace and forming an inverted V truss. This improved the interior appearance somewhat, but not enough to attract imitators.

For the majority of car builders and railroads, however, the

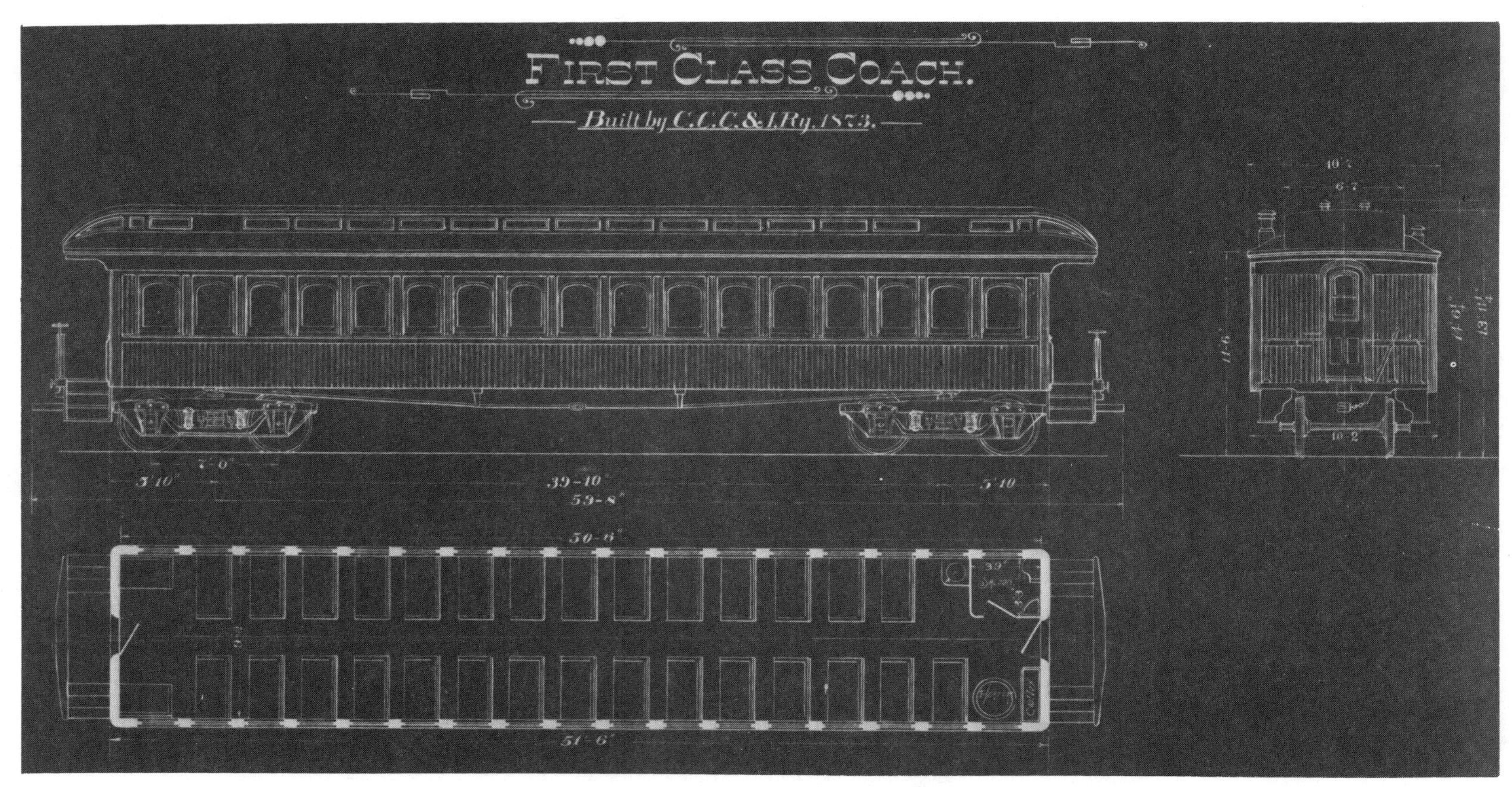

Figure 1.32 Conventional bull-nose clerestory roof end shown in a drawing of 1873. The design for these cars was prepared by the Cleveland, Columbus, Cincinnati, and Indianapolis Railway. (Railway and Locomotive Historical Society)

Figure 1.33 Coach Number 119, dating from about 1880, is another example of the bull-nose clerestory roof end. (Railway and Locomotive Historical Society)

esthetic appeal of the clerestory overruled all mechanical considerations. The raised roof was to remain a fixed part of American passenger car design until the 1930s.

The Tranquil Years: Design in the Seventies and Eighties

With the clerestory's acceptance in the early 1860s came a period of relative calm in passenger car manufacture. No radical changes in construction or arrangement took place during the next twenty-five years. It was a time for acceptance of things as they were, a time for regrouping, a time for refining the ideas of previous decades. The overall size, weight, and capacity of cars were remarkably stable throughout the period. Typical day cars were 50 to 60 feet long, weighed 20 to 25 tons, and seated 50 to 60 passengers. New appliances and auxiliary apparatus were introduced, of course. Existing auxiliaries such as the air brake and the Miller platform came into wider use, steam heating experiments began, and decorative treatment developed in ever changing directions. But no fundamental alterations in structure or radical advances in size occurred.

With the clerestory came a greater awareness of the passenger car's architectural potential. Arched windows and doors in Roman or Tudor styles became popular. Small, square windows were out of fashion, although they were still used for cheaper cars. Arched molding over the windows ended in delicate finials. Oval picture-frame moldings, called "name panels," were placed

Figure 1.34 The handsomely decorated coach Number 31 was produced by Jackson and Sharp around 1875. (Hall of Records)

in the center of the body panel. Side panels were invariably done in batten-board style. Interiors of first-class cars were finished in rare cabinet woods such as black walnut, cherry, and mahogany. Only the very cheap coaches, such as emigrant cars and excursion cars, had painted interiors. Gaudy oilcloth ceiling covers or head linings prevailed. Large, showy brass or nickel-plated center lamps were now possible because of the extra headroom provided by the clerestory. Examples of these features are shown in Figures 1.34 to 1.36.

The tranquillity of the seventies was mildly ruffled by the narrow-gauge movement, which seemed to promise a revolution in railway construction and operations.[77] The idea started in Great Britain and spread to this country through the writings of its zealous proponents. It was seized upon here by equally enthusiastic apostles, most of whom were innocent of any engineering knowledge or else chose to ignore what they did know. The major argument for narrow-gauge lines was their low first cost. Presumably the revolution would extend itself to rolling-stock design, where mystical improvements were promised. The laws of physics would be confounded; cars running on rails set at a 2- or 3-foot gauge would weigh less and carry more than standard-gauge equipment. Cars of 8 tons could be built that would carry a dead weight of only 450 pounds per passenger, compared with a standard-gauge car of 25 tons with a dead weight of 700 pounds per passenger. What the narrow-gauge partisans failed to consider was the great difference in overall size and the cramped seating and legroom. In these respects the narrow-gauge car represented a throwback to the diminutive primitives of an earlier time. The standard-gauge car was 60 feet long, 10 feet wide, and 13 feet high, with 24-inch aisles and seats 42 inches wide. The narrow-gauge car was 35 feet long, 8 feet wide, and 10½ feet high, with 17-inch aisles and seats 35 inches wide. A contemporary account attempting to gloss over the skimpy accommodations of the narrow-gauge coaches said that two persons of ordinary size could sit together comfortably; "fat men, however, are generally allowed . . . to occupy a whole seat in peace."[78] To many people who have ridden in narrow-gauge cars, it is questionable that passengers of normal dimensions can share these seats in comfort for any extended trip.

As the narrow-gauge movement developed, it was restricted to secondary lines; few trunk lines were built. Most of these secondary lines operated short local trains at very slow speeds. Consequently the great overhang of the cars presented no substantial danger, although the cars surely swayed more than their standard-gauge counterparts. Again, because these lines did not offer express service, the danger of high-speed collision was improbable and the cars could be lightly framed. Thus they could eliminate even more dead weight.

The first narrow-gauge passenger car built in this country was completed in July 1871 by Jackson and Sharp of Wilmington, Delaware, for the Denver and Rio Grande Railway (Figures 1.37 and 1.38). Named the *Denver*, it was 35 feet long and 7 feet wide.[79] It was divided into two compartments (one for smokers) and seated 34. The narrow width would not permit double seating throughout. Each side had double and single seats for half its length. The arrangement was reversed at the center of the car to maintain balance. The exterior was painted chocolate brown, with gold and yellow letters and ornamentation. The interior was black walnut. The seats were covered in scarlet plush, and the hardware and lamps were silver-plated. The total weight was 7½ tons. The road's first idea had been to order four-wheel cars on the British plan, but Jackson and Sharp insisted that a practical eight-wheel car could be built. In the following years the Wilmington builder supplied large numbers of coaches to the D & R G. By 1876 the firm had produced over 400 narrow-guage cars.

Other contract builders were quick to sight this fresh market unsullied by a supply of second-hand cars. Billmeyer and Small proclaimed themselves especially devoted to the narrow gauge and built some 2,000 cars by 1878. An example of their work is the *Eureka*, constructed in 1875 for the Eureka and Palisades Railroad in Nevada (Figure 1.39). It was 41 feet long, weighed 8½ tons, and seated 36. The richly furnished interior was said to give it "the appearance of some fairy boudoir rather than a temporary convenience for the traveling public."[80]

Harlan and Hollingsworth, Jackson and Sharp's rival in Wilmington, produced a first-class passenger car for the narrow gauges (Figure 1.40). It was similar to those turned out by other makers, but the outside width was pushed from 7 feet to 7 feet 11 inches. By cutting down the double-seat size and narrowing the aisle, the builder installed double seats the full length of the car. The capacity rose from 36 to 46.

In the Middle West, Ohio Falls and Barney and Smith sought a share of the new market. These builders offered first-class cars and some very cheap coaches as well. Ohio Falls built light excursion cars for a Kentucky road; they seated 64, weighed just under 7 tons, and could carry 125, including standees (Figure 1.41).[81] The seats were wooden slabs. For short commuter lines Barney and Smith built cheap, plainly finished cars with longitudinal benches and candle lighting for as little as $2,500. Far Western narrow-gauge roads ordered from Carter Brothers of Newark, California, who specialized in narrow-gauge rolling stock. Although the firm was never a major builder, it supplied cars for many Western, Mexican, and Central American lines.[82]

The narrow-gauge idea was pushed to its extreme in this country by the construction of 2-foot-gauge trackage. The cars for these roads, almost all located in Maine, were extremely small.

Figure 1.35 Interior photograph of the Providence and Worcester car shown in the preceding illustration. (*Hall of Records*)

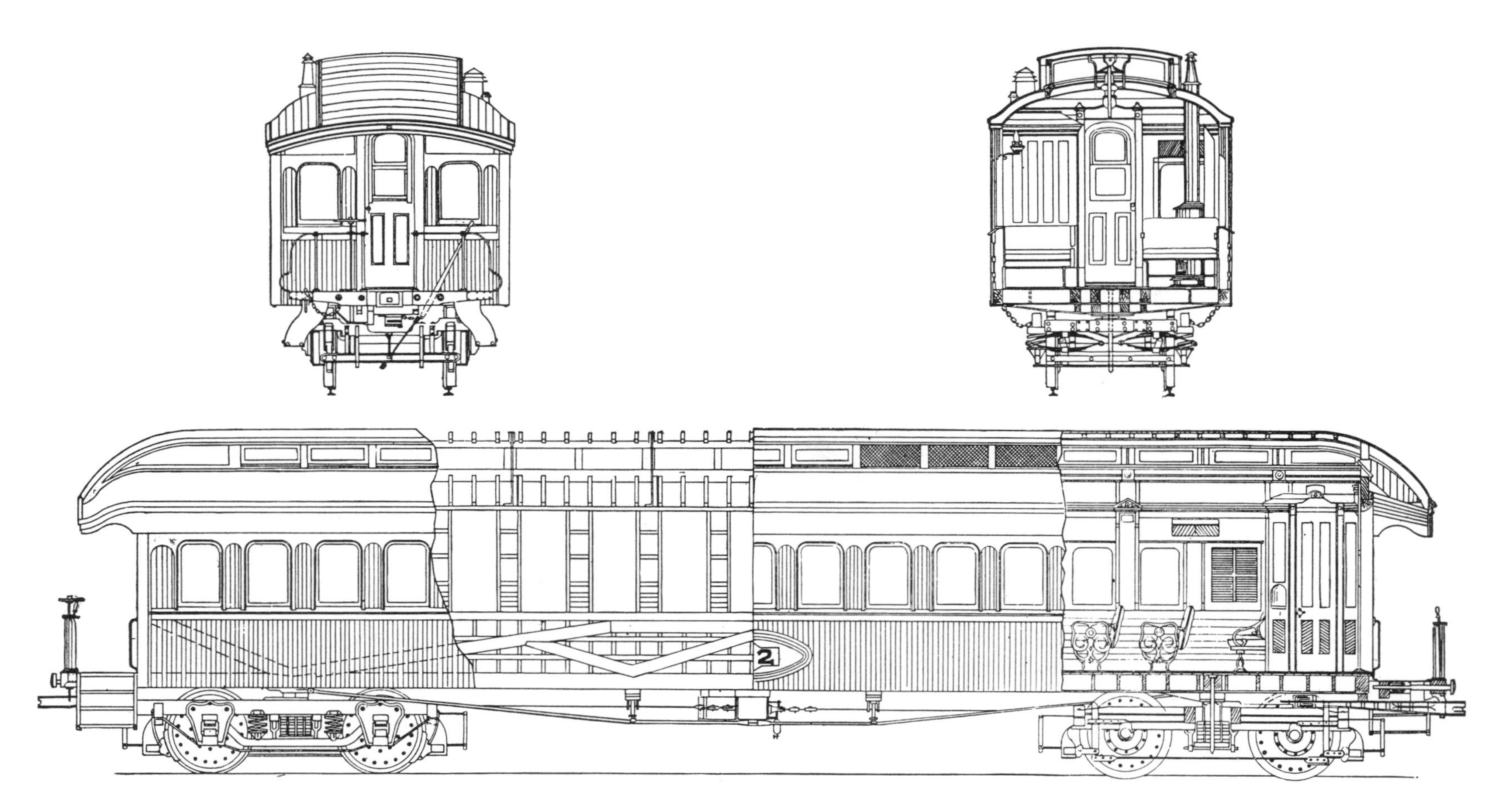

Length of Body* 50 *ft.* 10½ *in.

Figure 1.36 A standard passenger car designed by George Hackney, car superintendent of the Santa Fe Railroad. The car design dates from 1880. (National Car Builder, *January 1881*)

Figure 1.37 A narrow-gauge car produced by Jackson and Sharp in 1871. (Hall of Records)

They were built on the conventional eight-wheel pattern and were often as well finished as any standard-gauge cars, but in size they were much on the order of today's zoo railways for children. The first 2-foot-gauge car, the *Sylvan*, was built at Laconia, New Hampshire, for the Billerica and Bedford Railway in Massachusetts.[83] It measured 40 feet long by 6 feet, 2 inches wide, weighed 4½ tons, and had wheels 18 inches in diameter. The cost was $2,000. It offered seating for 30; single seats on either side of the aisle were all that the very narrow body could accommodate. In later years the width of these cars was increased to 6 feet, 7 inches, but double seating was still impossible.

The history of narrow-gauge rolling stock is short. After the initial enthusiasm, the defects of the cars, notably their limited carrying capacity and the impossibility of interchange, ended their construction for all practical purposes by the mid- and late 1880s. Narrow-gauge cars contributed little to general developments in car building. The designs were copied directly from existing technology and in essence were scaled-down versions of standard-gauge practice. It appears that little more than 10,000 miles of narrow-gauge line were ever built in this country. Thus even at its peak it accounted for only a small percentage of total mileage. Abandonment or conversion to standard gauge began swiftly; orders for new equipment evaporated, and the market was glutted with second-hand cars.

Most narrow-gauge lines surviving into this century made do with existing stock. Old cars were upgraded and repair parts were taken from surplus equipment. The D & R G kept its wooden cars in service for eighty years and more by periodic rebuilding. Steel frames, new window sashes, wheels, draft gears, and brakes, together with normal maintenance repairs, did their part in extending the service life of these cars well beyond that of their contemporaries. In the course of upgrading, their weight was raised to 16 tons, or nearly twice that of the original cars. Continued tourist traffic on the Silverton branch did force the construction of several new steel coaches in 1963 and 1964. But these were built in the old style, with scribed steel side plates to simulate wooden cars, and must be thought of as replicas rather than as attempts at modern coach design.

Figure 1.38 Interior of the narrow-gauge car in the preceding illustration. (Hall of Records)

If the narrow-gauge movement can be credited with little else, it did rekindle the debate on the dead weight of passenger cars. Designers were admonished to develop new patterns, and the iron car was perennially put forward as a solution. Car builders, however, faced conflicting demands for coaches that were light, yet strong. Heavy cars were expensive to haul, required heavier locomotives and bridges, and damaged the roadbed. Railroad operating departments called for lighter construction. The daily press, aggravated by the "horrors of travel," clamored for stronger cars. The newspaper crusade was more than yellow journalism; because of increased traffic and speeds, accidents were rising. It was not possible to serve both demands, and designers were inexorably pushed toward the side of safety. The trend since the beginning of the American railroad car had always been toward increased size and weight. Within practical cost considerations, no way was found to lessen dead weight without sacrificing strength. The paradox was never resolved; the talk of lightweight construction continued, but the practical men of the industry were resigned to ever heavier cars.

The steady increase in weight is shown in Table 1.1, which reveals that weights doubled within a decade. Before about 1870

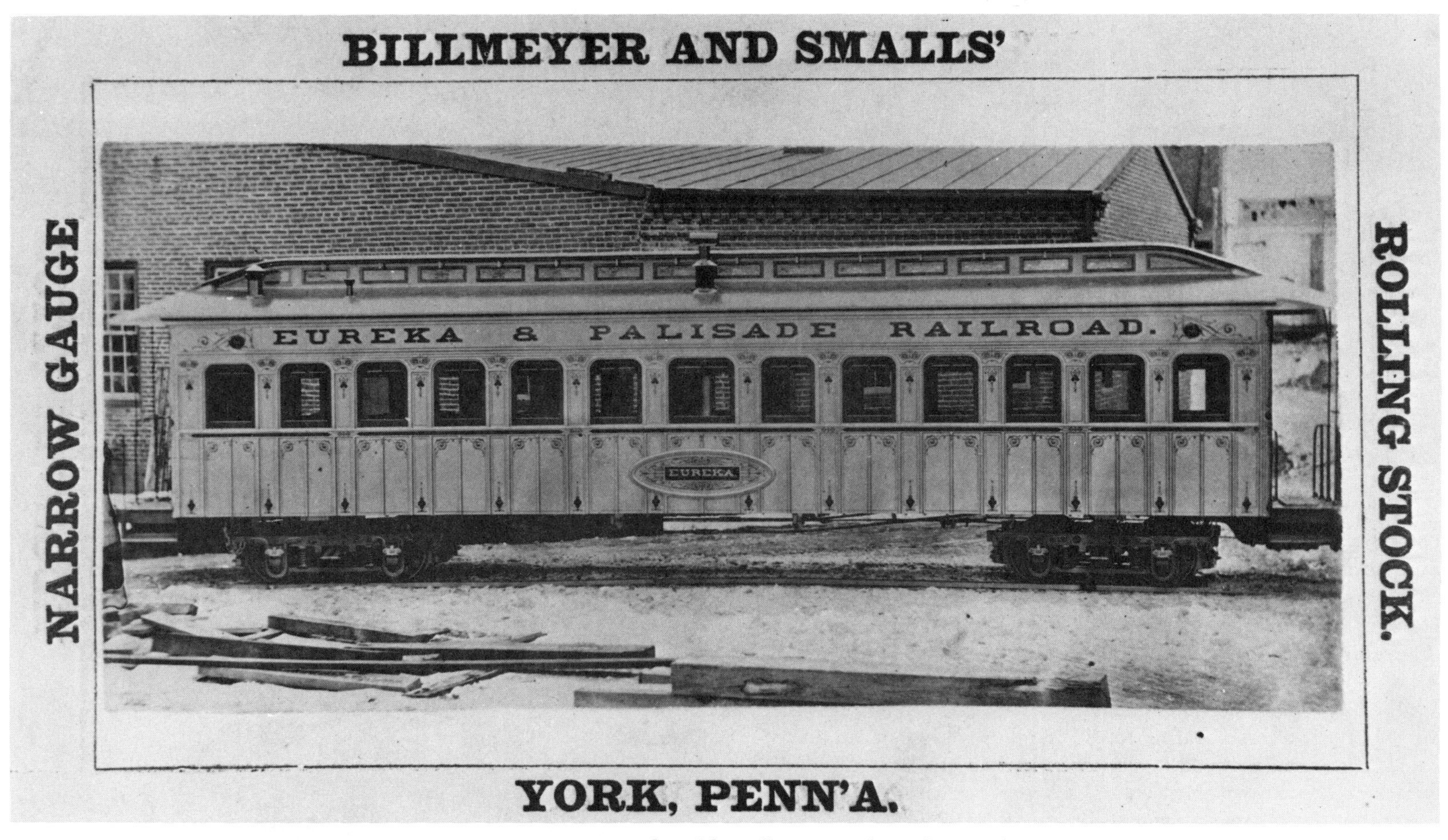

Figure 1.39 Narrow-gauge car produced by Billmeyer and Small of York, Pennsylvania, in 1875. The car seated 36 and weighed only 8½ tons. (*Smithsonian Neg. 46860-A*)

Figure 1.40 A narrow-gauge car design by Harlan and Hollingsworth, 1875. (N.C.B., *August 1875*)

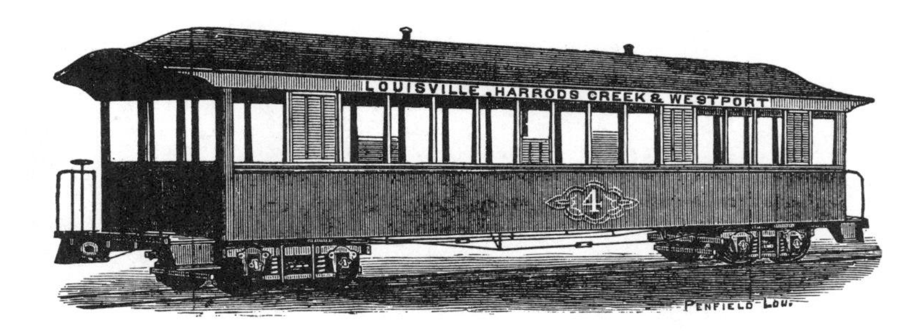

Figure 1.41 This diminutive narrow-gauge car was produced in 1875 by the Ohio Falls Car Company for summertime or excursion service. (N.C.B., *November 1875*)

the increase can be explained by simple growth in the size of cars, but after that date it was the result of adding new auxiliary equipment. Considerable extra weight came from hot-water or steam heating systems, air brakes and the stronger foundation gears they required, gas lighting with its storage tanks and piping, safety end platforms, and couplers. Seats with cast-iron frames, more glamorous interior furnishings, and larger, more fully equipped water closets also increased weight. The Boston and Albany, noted for light and rather sparsely fitted out day cars, found that half the weight of a complete body was in the stove, seats, and miscellaneous hardware.[84] In a car weighing 45,000 pounds, the body—floor frame and all—accounted for only 13,600 pounds. The hardware and trucks made up the remaining weight. In later years steel-reinforced bodies and platforms, electric lighting (batteries and generators), and six-wheel trucks added new burdens. Considering all this auxiliary apparatus, car builders were remarkably skillful in holding the total weight within reasonable bounds.

Table 1.1 Wooden Passenger Car Weights

YEAR	RAILROAD	BODY LENGTH, FT.	SEATING	WEIGHT, LBS.	WEIGHT/PASSENGER, LBS.
	COACHES				
1835	Baltimore & Ohio	30	50	12,000	240
1850	Many U.S. roads	35	56	20,000	357
1860	Many U.S. roads	48	50	36,000	720
1870	Many U.S. roads	50	50	40,000	800
1880	Boston & Albany	57	68	45,000	661
1885	Chicago & North Western	55	64	60,000	936
1890	Southern Pacific	54	62	56,300	908
1890	Fitchburg	57	73	46,000	630
1895	Northern Pacific	52	56	64,900	1,133
1900	Southern	57	60	79,500	1,325
1905	Great Northern	72	86	111,200	1,293
	SUBURBAN AND ELEVATED CARS				
1870	Pittsburgh, Ft. Wayne, & Chicago	(unknown)	30	21,500	716
1872	New York Elevated	30	48	9,400	235
1880	Illinois Central	(unknown)	40	29,000	725
1895	New York Elevated	40	48	29,300	614
	PULLMAN SLEEPERS				
1870	All Pullman clients	60	32	52,000	1,625
1880	All Pullman clients	66	27	75,000	2,777
1890	All Pullman clients	69	27	90,000	3,333
1895	All Pullman clients	70	27	111,900	4,144
1905	All Pullman clients	74	30	120,000	4,000

Concern for safety and comfort led to the construction of heavy cars for main-line service. Here the strongest cars were needed because speeds and traffic density were greatest. Most trunk lines justified their heavy coaches by pointing to the necessity for strong framing. But a few roads managed to produce lighter cars: the 6-foot-gauge Erie, where the heaviest cars might be expected, built a surprisingly light 60-foot car that weighed nearly 7 tons less than an equivalent car on the Pennsylvania. How this was achieved is not obvious from comparing the drawings. Late in the century the New Haven produced a lightweight car in which about 2½ tons of dead weight was saved by substituting Oregon fir for the usual oak frame, and aluminum for the usual cast-iron seat frames.[85] The car body framing was lightened as well (see Figure 1.42).

For suburban, emigrant, or branch-line service, light, cheap cars were considered acceptable. Speeds and traffic density were low enough to justify more lightly built rolling stock. Undersized framing, furnishings, and plain finish reduced the weight and cost. Closely spaced seats raised the seating capacity. For its branch lines the Pittsburgh, Fort Wayne, and Chicago used extremely light cars that weighed just over 10 tons.[86] They were not run faster than 15 miles an hour; at higher speeds they jumped so badly that there was danger of their leaving the track. The road's coaches for main-line service weighed 23 tons. The Illinois Central's suburban service out of Chicago grew so rapidly after 1880 that the line developed a special car for it patterned after those used on the New York Elevated. Here again, very light framing and reduced overall size held the dead weight under 15 tons (Figure 1.43).

In fact, the New York Elevated cars were the established model for lightweight construction. No writer on the dead-weight issue failed to mention them. The earliest cars weighed only 9,400 pounds, but they were so undersized and so unorthodox in design (representing the old depressed-center idea revisited) that no application to steam railroad practice was possible.[87] They were in essence little more than eight-wheel horsecars. But even the Elevated began to build heavier cars, and by the mid-1890s weights were up to 15 tons.

Special lightweight cars were constructed to meet temporary needs as well. For the U.S. Centennial of 1876, the Reading and Pennsylvania railroads built a number of suburban cars to accommodate visitors traveling from Philadelphia to the exhibition grounds. The Pennsylvania produced 100 cars of the same overall size as its standard PB coach, but the seats were crowded together so that the cars carried twelve more passengers (Figure 1.44). Their exterior and interior finish was very simple. After the Centennial the cars were used as emigrant and excursion trains. Even cheaper and more temporary cars, little more than boxcars fitted with windows and wood-slat seats, were manufactured for the Columbian Exhibition in 1893. Built for easy conversion to freight service, they were forerunners of the boxcar-like troop carriers of World War II.

The trend toward heavier construction was evident even in suburban cars. Thus in the one area where it was generally agreed that light cars were possible, builders gradually drifted toward the heavyweights. By 1904 the average weight of America's 3,919 suburban cars was given as 30½ tons each.[88] While this was considerably below the tonnage of a day coach of the period, it represents a marked increase over earlier suburban cars.

Lip service was paid to lightweight construction, but in practice, strength was the designer's goal. It is wrong to dismiss the

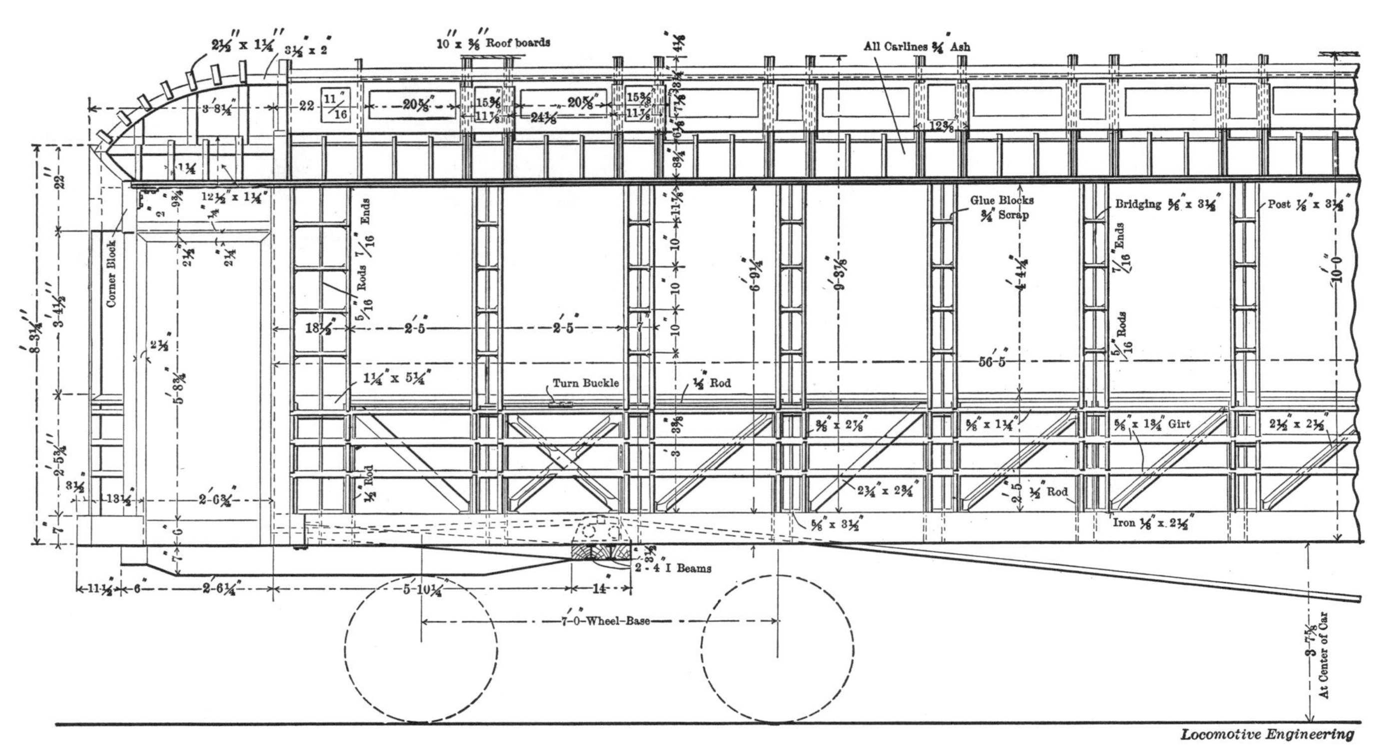

Figure 1.42 A lightweight car with an extremely sparse body and floor frame. The design was prepared by the New York, New Haven, and Hartford Railroad in 1896. (Railway and Locomotive Engineering, *June 1898*)

Figure 1.43 The Illinois Central suburban fleet running out of Chicago was equipped with lightweight cars on the same general plan used for New York Elevated coaches. This example dates from about 1880 and was produced by Jackson and Sharp. (Hall of Records)

wooden car as a flimsy matchbox or egg crate—a popular flippancy so often expressed that there seems little hope of refuting it. It would be equally wrong to pretend that the wooden car was faultless; it had many defects, but the case against it has been overstated. In the more spectacular wrecks, it did fail as its worst critics predicted. Wooden cars could not withstand the driving force of a locomotive—but neither could all-steel cars, much to the dismay of those who predicted that metal cars would usher in an age of absolute safety for railway travel.

Derailments or collisions of wooden cars, however, did not mean certain death among shattered timbers and splintered wood. In ordinary mishaps wooden cars did well, though occasions like these were not as newsworthy as the Revere, Angola, or Spuyten Duyvil disasters. The monthly accident reports in *Railroad Gazette* during the years of the wooden car are not a continuous chronicle of death and destruction. Apparently the cars offered reasonable protection to the traveler. A freak event that showed the surprising strength of wooden cars occurred in Philadelphia in the fall of 1878.[89] The train shed at the Pennsylvania Railroad's West Station was blown down by a 75-mile-an-hour gale. The cars not only supported the giant 70- by 800-foot roof, but withstood the battering effect of the 6-ton cast-iron supporting columns. One column broke in half when it struck the side of a car. None of the cars were seriously damaged.

Two years before that, a passenger train derailed on the Richmond and Danville. Several cars rolled down an embankment;

Figure 1.44 The Pennsylvania Railroad built a special class of light car for the Philadelphia Centennial Exhibition in 1876. The cars were produced in the railroad's own shops in 1875 and 1876. (Pennsylvania Railroad)

Figure 1.45 Interior of a New York Central and Hudson River Railroad coach of 1886. (Hall of Records)

Figure 1.46 Western and Atlantic coach of about 1885. Note that the interior wooden window blinds are all closed. (Hall of Records)

Figure 1.47 A Norfolk and Western coach of 1888 shows the strong influence of its parent line, the Pennsylvania Railroad. (Hall of Records)

one landed 150 feet from the tracks. Yet no one was killed, only 20 injuries were reported, and the cars remained intact. The trucks were lost, as might be expected, but no major damage occurred. Late in the wooden car era, a coach on the Lake Shore and Michigan Southern was thrown from the rails. It rolled over and straddled an opening in a washed-out embankment. The underside of the car was stripped, including the truss rods, yet repairmen discovered that the body was straight and sound despite rough handling in its removal from the accident site. Four hundred dollars put the car back in service.

To build sturdy wooden cars required the skill and experience of the master carpenter. As already suggested, he borrowed directly from the traditional technologies of the architect, the bridge builder, and the millwright. It is doubtful if any unique methods of fabrication can be credited to car builders. But the solutions that they developed to practical design problems were often subtle and ingenious. The various styles of floor and body framing are perhaps the best examples of this inventiveness.

It is true that a civil engineer might criticize the wooden railroad car as a hopelessly inept design. Certainly as a bridge it was terrible. A wooden bridge should support ten times it own weight; the passenger car was rarely required to carry half its weight. Instead the car had to absorb the draft and buffering impacts of train service. Vertically the framing needed to support only its own weight and the relatively light load of the passengers. The most severe stresses on the structure were longitudinal, from the push and pull of the train. Accordingly the designer aimed to build strength into the floor framing.

After 1870 the standard floor frame consisted of six sills. The outside members forming the main support for the body were the heaviest cross section and often measured 5 by 8 inches. Round iron truss rods an inch or more in diameter were attached to the underside of the outside sills. These rods greatly stiffened the body. If a sag developed, a turnbuckle at the center of the rod could be adjusted to draw the body straight again. The floor frame was in fact often built with a camber of 1½ inches to avoid the optical illusion of a sag. Generally the truss rod terminated in an anchor block bolted to the side sills near the body bolster. A stronger scheme nearly always used for freight cars but apparently not favored for passengers was to run the truss rod over the bolster and through the end sill. The rod extended the full length of the body and tied the frame together as well as jacking it up. It also served to support the ends of the car body which overhung the trucks.

The intermediate and center sills were slightly smaller in size, with the center sills placed close together to make an easy attachment for the drawbar. Because the frame was subject to few transverse stresses, massive cross bracing was not necessary; it was made only strong enough to space the sills and resist end shocks. End pieces were mortised to the sills. A body bolster or transom, as it was sometimes called, was bolted to the underside of the sills at each end of the car. Center plates for receiving the truck were fitted to the bolster. The bolsters were generally made of two or three timbers. After 1870 iron bolsters became more common, but the wooden bolster seems to have prevailed almost to the end of wooden construction.

Closer to the center of the body were the needle beams, which were fitted to the underside of the sills and formed a place of attachment for the queen posts. Short bridging joists were placed between the sills more as convenient nailing pieces for the floorboards than as structural members. Cross iron rods tied the floor together. Builders concerned with maintaining the squareness of body (the Pennsylvania Railroad, for example) used diagonal braces in an X pattern. Planks 1½ by 4 inches were attached to the top of the frame. An objection to this practice was that the sills had to be notched so that the floor could be laid over them. Critics said that the notching weakened the frame. The underside of the frame was covered with a wooden board ceiling, which served to deaden the sound of the running train, exclude dust, and insulate the floor. For many years the air space between the sills was filled with wood shavings for better insulation. The practice was condemned as a fire hazard in an age of oil lighting. Rock wool was sometimes substituted, but it proved too expensive for many roads.

The ordinary floor frame, simple and strong as it was, suffered from one grievous defect. The platforms of the car were made of short timbers bolted beneath the floor frame, and this weak cantilevered structure had to support the coupler, draft gear, end rails, and steps. It invariably sagged. A special committee on the subject created by the Master Car Builders in 1884 admitted that a first-hand inspection had discovered few cars without drooping platforms. Worse still, the dropped platform was vulnerable to the most serious of all possible hazards: telescoping of cars. In the event of a collision, the dropped platform allowed the car behind to climb up on it and slide directly into the first car's body.

How did such an unmechanical arrangement come about, and why was it tolerated? This form of construction existed from the beginning of the eight-wheel car. At first it was adopted to facilitate entrance to the car: the lowered platform was easier to mount than a platform at the level of the car floor. One possible reason for its perpetuation is that as new, larger cars were introduced, the dropped platform was necessary to keep drawbar heights level with the old cars still in service.[90] Another theory is that the larger wheels which became popular for passenger cars raised the level of the body; and the dropped platform was used to maintain a standard coupler height. None of these arguments is entirely convincing. It seems more likely that the dropped platform continued not because of any mechanical merit but because it had become an entrenched traditional design.

It remained for a designer outside the profession to correct this basic and long-standing defect in railroad car construction. Ezra Miller, a civil engineer only casually associated with the railroad industry, recognized the need to raise the platform to the same level as the main floor frame of the car. Miller is usually remembered as the inventor of a popular car coupler antedating Janney's design (see Chapter 7). But he should also receive great credit for the elevated platform—a relatively simple reform that did much to promote safe railway travel. Examples of level platforms, both in actual construction and in the patent record, can be found before Miller's 1865 patent. However, he became the advocate of the level platform combined with a buffered coupler, and it was adopted largely through his efforts. Miller retained the suspended platform timbers, but he inserted short spacers between the end sills and the buffer beam (the end cross timber of the platform) and raised the level to agree with that of the main frame. The entire structure was tied directly to the body bolster by iron rods. By the mid-1870s Miller's platform was widely used, and in another ten years it was considered standard equipment.

There was general agreement among builders on floor framing, but no such consensus can be found for the body frame. Though everyone agreed that there must be some form of panel side trussing, "there is such a wide difference in the ideas of the officers of different roads with regard to the amount of material . . . and the form in which it should be placed that it seems almost impossible to bring their ideas together."[91] So wrote Thomas Bissell, chairman of the Master Car Builders' committee on passenger car framing, in 1884. Since the mid-1850s car builders had delighted in perfecting new and variant forms of side-panel trussing. The history of this elusive art is difficult to trace, because few drawings exist for the early years. Some idea of the variety of forms can be gained from Figures 1.48 and 1.49.

Before 1870 car builders favored the bastard Pratt and Howe trusses, which were simple and did the job for short bodies. The Pennsylvania remained loyal to this form through the 1880s, mainly because it was adjustable during the service life of the car. But it appears that most other roads were attracted to a modified queen-post truss, generally called a combination or compression truss. It may have been suggested to some observant car builder when passing through a covered bridge, where the structure was exposed on the inside. Or possibly the external hog bracing of a river steamer inspired the anonymous inventor of the side-panel truss. The first known example of a compression-truss car is the 1870 Chicago and Alton coach described in the Representative Cars section (Figure 1.103). It is not claimed that this is actually the earliest example; the compression truss may have been used for some years, but no evidence is available. The car follows Pullman's overall design and is unusually large for a coach of that day. Pullman or Wagner may have originated the compression truss in building their heavy sleepers during the 1860s. It seems to have caught on rapidly after 1870, and a survey of existing drawings indicates that it was the most popular style by the end of wooden car construction.

The compression truss grew out of the distortion of the simple

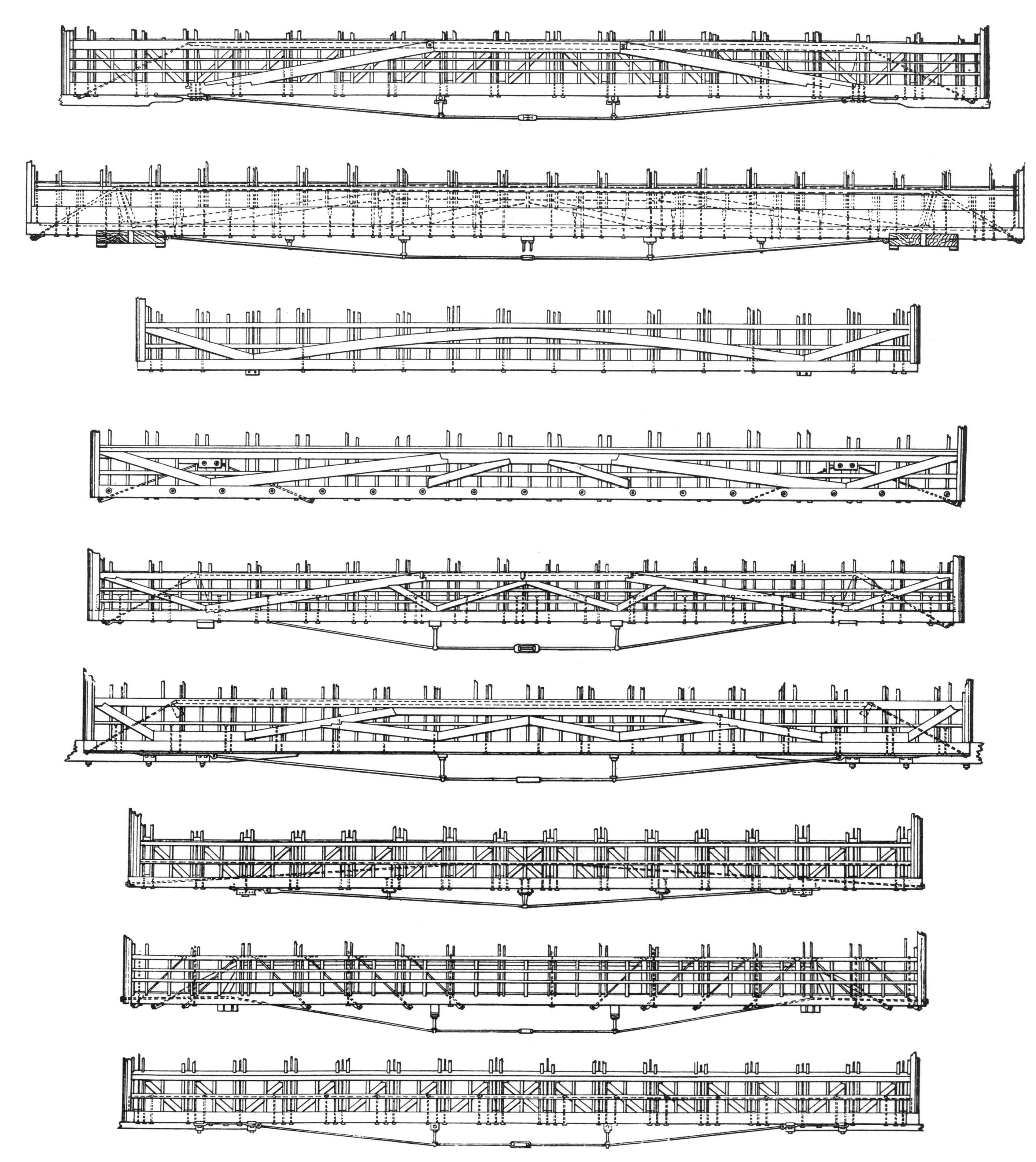

Figure 1.48 Some idea of the variety of side-frame trussing in wooden cars can be gained from this and the following illustration. (Railroad Car Journal, *September 1894*)

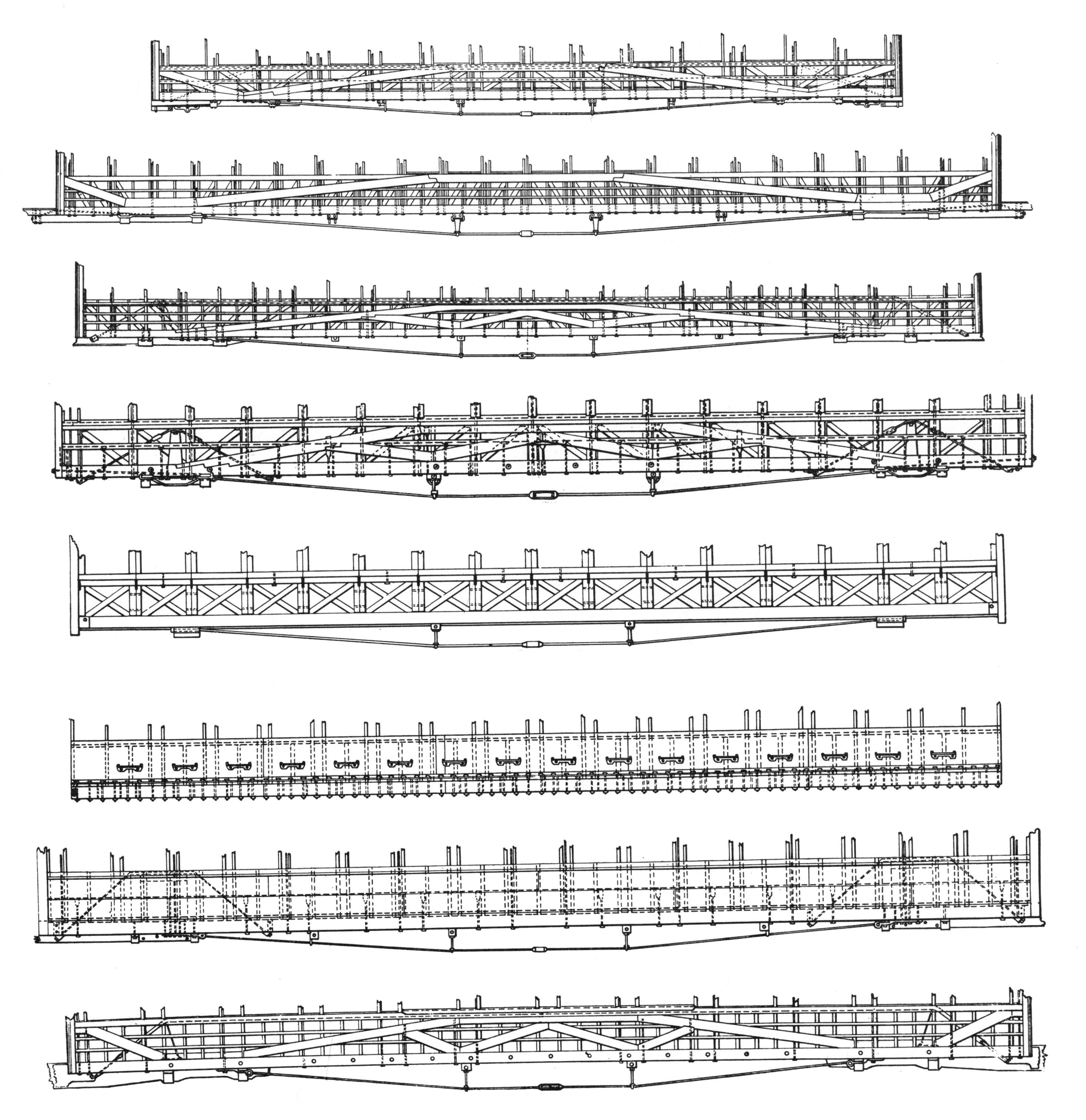

Figure 1.49 More examples of wooden passenger car side trusses. Detail third from bottom shows Chalender's iron plate truss. (Railroad Car Journal, *September 1894*)

Figure 1.50 The intricate woodwork of passenger car framing is exposed in a photograph taken in 1907. (Pullman Neg. 10281)

queen-post truss when it was unevenly loaded. The short auxiliary counterbraces were inserted to stabilize the truss. From frame drawings shown throughout this chapter, it might be supposed that the side truss was made of heavy square timbers. This is not the case; the structure must be seen "in the round" to be understood. The photographs in Figures 1.50 to 1.52, and a more careful examination of the drawings, reveal that the truss members are deep but relatively thin planks. The usual size was 1½ by 6 inches. The vertical posts were notched so that the truss planks could be set flush with the car side.

A great quantity of glue was used in car fabrication—50 to 100 pounds was not unusual. Glue was needed not only for frame assembly but also for applying outside sheathing. The glue stiffened the structure and became one of the hidden sources of strength for the lightweight wooden frame. Once the car sagged, however, it could not be straightened without breaking the glue joints. Wooden screws and nails were used as well. Iron rods also helped tie the body and floor frame together. The ends of the body overhanging the body bolsters were supported by a short rod (shaped like an inverted V) or a long iron hog brace. The hog brace, properly known as an overhang brace rod, extended the length of the body just under the belt rail (window sill). It ran on a 30-degree angle from one corner of the side sill to just under the belt rail. At this point the rod was flattened so that it would fit more easily inside the body panel. The overhang brace rod does not appear to have been used much before the long-bodied cars of the late 1880s.

The compression truss, combined with the six-sill floor frame and iron truss rods, formed a strong, relatively light wooden structure. The designer's ingenuity was taxed to devise a wooden truss stiff enough to support a 70-foot body within a 30-inch panel, but it was done. All elements of this form of construction were available by 1870, and they continued in use until the advent of the steel passenger car some forty years later.

Wood was the obvious material for car building because of America's great timber resources. Virgin stands of oak, pine, ash, and walnut offered the choicest woods, and the supply was bountiful. Iron, on the other hand, was scarce and expensive. The fledgling American mills could not compete with those of Britain, and much iron and steel were imported. The cost of the imported metals was artificially raised by tariffs meant to protect home suppliers.

Native woods seem to have been used by local builders in the early years. Locust was the principal wood in the *Columbus* (1831), local ash in the first cars made at Cleveland. The Eastern hardwood forests provided a good supply of white oak, which was long favored for car frames. With the voracious demand for hardwood, New England's woodlands soon vanished. The loggers

Figure 1.51 The timbers used in floor framing were massive, as shown in a photograph of about 1890. (Pullman Neg. 3327)

then began on the oak of New York, Ohio, and Tennessee. Wherever the white man moved, the timber disappeared. Standards declined, and timber that would earlier have been cast on the firewood pile was later accepted as staple lumber.

By the early 1870s builders were turning to yellow pine as the choice stands of hardwood disappeared. In 1872 Gilbert, Bush, and Company was reported to be importing yellow pine from Georgia and Florida.[92] Its high resin content was considered an effective insect deterrent. Yellow pine proved an excellent wood for floor sills; it was as strong as oak and was available in long, nearly flawless sections. Norway pine was another favorite for car sills. Pine eventually became the basic wood in car construction. Pullman estimated that half of the wood used in its cars was pine.[93] By the middle 1890s, however, all kinds of prime timber for long sills were becoming scarce. Builders turned to the great Pacific Northwest fir forests, where a plentiful supply of best-quality sills was available at half the cost of Eastern pine. The Northern Pacific opened a car shop in Tacoma; the Southern Pacific imported Oregon fir for its big Sacramento shops. But fir did not take over the car sill market, for it could not match the strength of oak or the better pines.

Securing long, perfect sills had been difficult since the 1880s.[94] Good-quality timber in 40-foot lengths was easy to come by, but longer pieces were generally defective at one end. That end was from the top or small part of the tree and was commonly riddled with sap pockets, shakes, and knots. Even when very tall, solid trees were available, it was not always possible to harvest them in one piece. As the timberlands receded into remote or rough terrain, the logs had to be cut into shorter lengths for haulage to the mill. Thus as longer cars were coming into fashion, the supply of prime framing timber was declining. Splicing was the only alternative. The preferred form was the 24-inch lock or ship splice, in which the ends of the timber were cut at an angle of about 20 degrees. A notch was made near the center to prevent slipping. A 6-foot timber was bolted on the spliced sill overlapping the joint. Iron plates were sometimes used as well, but most car builders felt that they were too rigid. The wooden splice bar would bend and give with the sill and thus not loosen the connecting bolts. The splice was best made near the center of the car, preferably over the needle beams but never over the bolsters. In 1901, to the surprise of many car builders, the traditional lock splice was proved inferior to the plain right-angle splice. In a test conducted for the Master Car Builders Association that year, the lock-splice breaking load was 82,000 pounds, while the plain splice broke at 140,000 pounds. But this revelation was beyond the belief of the majority of the members, who chose to stay with the old form.

The best building timber is naturally seasoned by one to two years of air drying. Pullman maintained a vast 60-acre yard for the purpose, but not all contract builders could afford to tie up so much capital. Sometimes lumber was even sawn from logs that had been in the mill pond when the order was received.[95] Such an outrageous practice may have been limited to a few cases, but the use of green lumber by unscrupulous builders was a frequent complaint. One thousand feet of unseasoned lumber might contain as much as a ton of water. Cars built from it seemed tight enough at first, but as the wood dried out and shrank, mortise joints loosened and the whole structure sagged or fell to pieces. An alternative to the expensive process of natural seasoning was to dry the lumber in a kiln. Artificial drying in heated ovens produced lumber of less strength and with more susceptibility to rot than the naturally dried product, but it was certainly preferable to green wood. As early as 1881 Jackson and Sharp placed two kilns in its Wilmington yard.[96] An unexpected upsurge in new car orders had cleared the market of seasoned lumber. When this happened, cynical builders turned to freshly sawn logs, while the honest shops erected kilns.

The blame for using inferior wood cannot be laid entirely upon the contract shops; the railroads themselves were lax in insisting

Figure 1.52 In this Chicago and West Michigan coach, under construction sometime around 1888, the side paneling has already been applied on the right.

upon the best material. Some master car builders made a first-hand selection, but too many delegated this important duty to the purchasing agent.[97] Engineering departments issued vague specifications, often only a few words, which named only the variety and grade of wood. Competing lumber dealers were tempted to supply inferior stock to win the order. The wood they furnished might be "yellow pine, grade B," or it might be sapwood or some poor variety of the species. The density, strength, and decay-resistance properties varied enormously within the subvarieties of a species. Thus the most detailed and careful specifications were necessary in order to obtain the best timber.

Rough guidelines for grading lumber were established through public law in some states by the 1840s. But it was not until the early part of this century—at the close of the wooden car era—that scientific standards were developed for use by the railroads.[98]

Another criticism leveled at car builders was their narrow-minded insistence on a few species of wood. Corner posts and window sills were made of tough, rot-resistant black walnut. But in general the principal framing timber was limited exclusively to oak or pine. Although beech, elm, cypress, and even sycamore were suggested as cheaper alternatives in some locales, they never won general acceptance.

Outside paneling or sheathing presented no unusual supply problems. It was light, conventional lumber, always available in good supply. The earliest references speak vaguely of white wood, which was presumably clear white pine or poplar. Before about 1855, rather large, rectangular panels were the rule. They tended to split, however, and for this reason, as well as from changes in fashion, were superseded by vertical batten-board or narrow tongue-and-groove siding. Batten-board paneling was favored through the 1880s. The boards were 12 to 15 inches wide; narrow strips covered each joint. But even with a carefully built body frame, it was difficult to lay on a perfectly smooth side panel. Minor defects were bound to exist in a 50- or 60-foot panel, and sighting down it at a close angle revealed many irregularities. Some leveling was possible with furring strips or blocks. The Pennsylvania Railroad shops developed a method all their own, in which thick panel boards were securely fastened to the car sides, and skillful cabinetmakers planed and scraped a quarter inch off the entire side of the car to make it true and level.[99] By the early 1890s narrow tongue-and-groove boards, 1½ to 2 inches wide, came into favor. One edge was beaded with a V-shaped chamfer. The boards were glued and blind-nailed to the car frame. They were always made of yellow poplar, a clear, close-grained wood that offered a smooth surface for painting. (The rare decorative woods used for interiors will be treated in Chapter 5).

Perfecting the Art of Wooden Car Construction

By the end of the century car building entered its Augustan age. As the *Railroad Car Journal* said, "The days of building cars by the rule of thumb have passed and men of high scientific training are rounding out the experience of the untutored carpenter."[100] The wooden car reached its high point of development in the middle 1890s, when a masterful system of spidery wood trussing produced sleek coaches. The 80-foot cars manufactured at this time exhausted the potential of wooden construction; the ultimate practical limit had been reached.

The truss plank frame, at first considered practical only for bodies up to 50 feet, proved capable of appreciable expansion. Demand for longer, more luxurious cars in the late eighties added another 10 feet, so that by the middle of the next decade, bodies of 60 feet were common for first-class cars. The New York Central created a sensation in 1893 by building the first 80-foot day coaches; the bodies were just an inch under 72 feet.[101] These were the largest day coaches constructed to that time and marked in fact the final limit for wooden car sizes. The New York Central cars were the fulfillment of a new trend in day cars that might best be described as palace day coaches. A year or two earlier the larger trunk lines had begun to purchase unusually luxurious coaches for express-train service that were modeled on the most expensive parlor and sleeping-car designs. As competition for passengers grew, these lines bought ever more extravagant cars. In 1898 the *Railroad Gazette* protested that the rivalry had gone beyond reason: "The builders are the only parties who can possibly benefit by costly trains."

The longer, heavier cars created problems that wooden construction could no longer safely handle. Car builders turned to more extensive use of iron reinforcement. Composite construction was the basic development of the 1890s. It was a compromise scheme that prolonged the wooden car age into the early twentieth century. Iron reinforcing had been used for years; iron truss rods, carlines, and corner brackets were established materials. The Illinois Central was installing iron body bolsters in the early 1850s, and their use spread in the 1870s. But builders were now working heavy steel and iron plates into the very fabric of the structure.

The earliest important application of composite construction was the Chalender plate truss, a sheet-iron plate that was substituted for the ordinary side-panel wooden truss.[102] Made of boiler iron about ¼ inch thick, it was fastened inside the body frame below the windowsill and ran the full length of the car. An angle iron riveted to the bottom of the plate was screwed to the side floor sill. Possessing the rigidity of a plate girder bridge, the Chalender truss was said to be the strongest possible style of trussing. Installed, the plates weighed 1,600 pounds and cost $90. This was somewhat more expensive than a wooden truss and, being an unfamiliar style of construction, was not favored by wooden car builders. It was also criticized because the vibration and twisting of the car body eventually loosened the screws, rendering the truss ineffectual. The idea was patented October 7, 1873 (No. 143,498), by George F. Chalender, who was then master mechanic for the Burlington and Missouri River Railroad. As a blacksmith and locomotive builder, Chalender was familiar with working boiler iron. He succeeded in promoting his idea when he became superintendent of the Chicago, Burlington, and Quincy's repair shops. The Burlington remained loyal to the Chalender truss for years after the inventor's departure, building cars on this plan through the 1890s. Few other roads appeared interested, however.

The Pullman and Wagner superintendents, Henry H. Sessions and Thomas A. Bissell, began work on rival schemes for composite construction at the same time. Sessions appears to have completed his plans first and applied for a patent in 1889 (see Figure 1.53). His major concern was to strengthen the ends of the car, an objective he achieved by bolting a 9⁄16-by 20-inch plate under the end sill in order to form a solid connection between the floor and end sill. A 3- by 4-inch angle-iron framework formed in the shape of the car's end profile was attached to the steel plate. His method was reported in use by the trade press in February 1890, but he did not receive his patents until the following year.[103] The first patent was issued June 2, 1891 (No. 453,403); the second, more general patent was granted on July 21, 1891 (No. 456,291). Pullman claimed that cars built on this plan could withstand a collision at 40 miles per hour without serious injury.[104] No test of the claim has been recorded, but it seems safe to conclude that it was little more than bravado on the part of Pullman's press agent. Sessions's scheme became standard at Pullman and was recommended by the Master Car Builders Association. It is known to have been adopted by the Boston and Maine to strengthen its day coaches, and it may well have been used by other roads.

In 1894 Sessions turned his attention to strengthening the side-panel trussing. A cantilever plate girder made of flat ⅝- by 3-inch iron bars was placed at each end of the panel truss to prevent the ends of the body from drooping. Two years later Sessions expanded his ideas for composite construction in a patented scheme which advocated an X-brace angle-iron side truss.[105] The iron web plate described in the 1891 patent was much enlarged and ran nearly the full length of the platform from the end body sill. It is not known if Sessions's side truss was ever used.

Bissell's plan for composite framing was completed late in 1890.[106] Like Sessions, he concentrated on strengthening the end bulkheads, but he also effectively tied the floor frame into the end structure (Figures 1.54 and 1.55). Six ¾- by 3½-inch flat iron bars were bolted to the end frame of the body. They were sandwiched between the wooden upright posts, and the ends were twisted at right angles for easy fastening to the frame's end sill and deck sill. A horseshoe-shaped iron plate ⅜ by 6 inches ran across the top of the bulkhead (the end deck sill). The clerestory and roof were reinforced with iron carlines which had lugs welded on for more secure attachment to the wooden framing of the car. The best features of Bissell's plan were the heavy iron plates that ran from the end sills along the side sills just past the bolsters—usually 18 feet. This plate was sandwiched between the 5- by 8-inch side sill and a 3- by 8-inch supplementary timber. In some cases the reinforcing plate ran the full length of the frame. Bissell's stated purpose in the patent specification was a judicious strengthening of the conventional wooden frame. He warned that

Figure 1.53 Reinforced end framing is shown in this construction view of 1907. (Pullman Neg. 10280)

care must be taken not to overdo it. Too much iron plating would add needless weight and cost; just the right amount of reinforcing was needed.

Bissell received a patent on August 11, 1891 (No. 457,486), and saw his idea widely used for Wagner sleeping and parlor cars. Moreover, because of the Wagner Company's close association with the Vanderbilt interests, Bissell's plan was adopted by the New York Central and its associated lines. After trying the composite frame on some baggage cars, the Lake Shore and Michigan Southern was so pleased with the results that it adopted the scheme for coaches in 1892. Within four years the road had 150 cars on this plan in service. Bissell's scheme seems to have been popular, and as late as 1909 it was used for some new cars by the Central Railroad of New Jersey.

In 1896 composite construction had become respectable enough to be recommended by the Master Car Builders Association.[107] In fact, it seemed a necessity; the Post Office Department was insisting on safer mail cars to protect its employees. Short of going into all-steel construction (a distasteful conversion seemingly to be put off as long as possible), composite construction was an attractive alternative. The association admitted that it could produce no designs which entirely avoided the Sessions and Bissell patents, but a British patent of 1884 seemed to cover the same general features and would undoubtedly void the American claims if they were taken to court. Old-line car builders like F. D. Adams, however, objected strenuously to composite fabrication because it added useless ironwork to cars already overbuilt. According to the association's report, Adams contended that the Bissell and Sessions schemes increased the weight by 5 tons. He also stated that conventional wooden construction was satisfactory on the Boston and Albany. Light, all-wooden cars built twenty-five years ago were turning in good service, Adams said, and would not be improved by the addition of so much scrap metal.

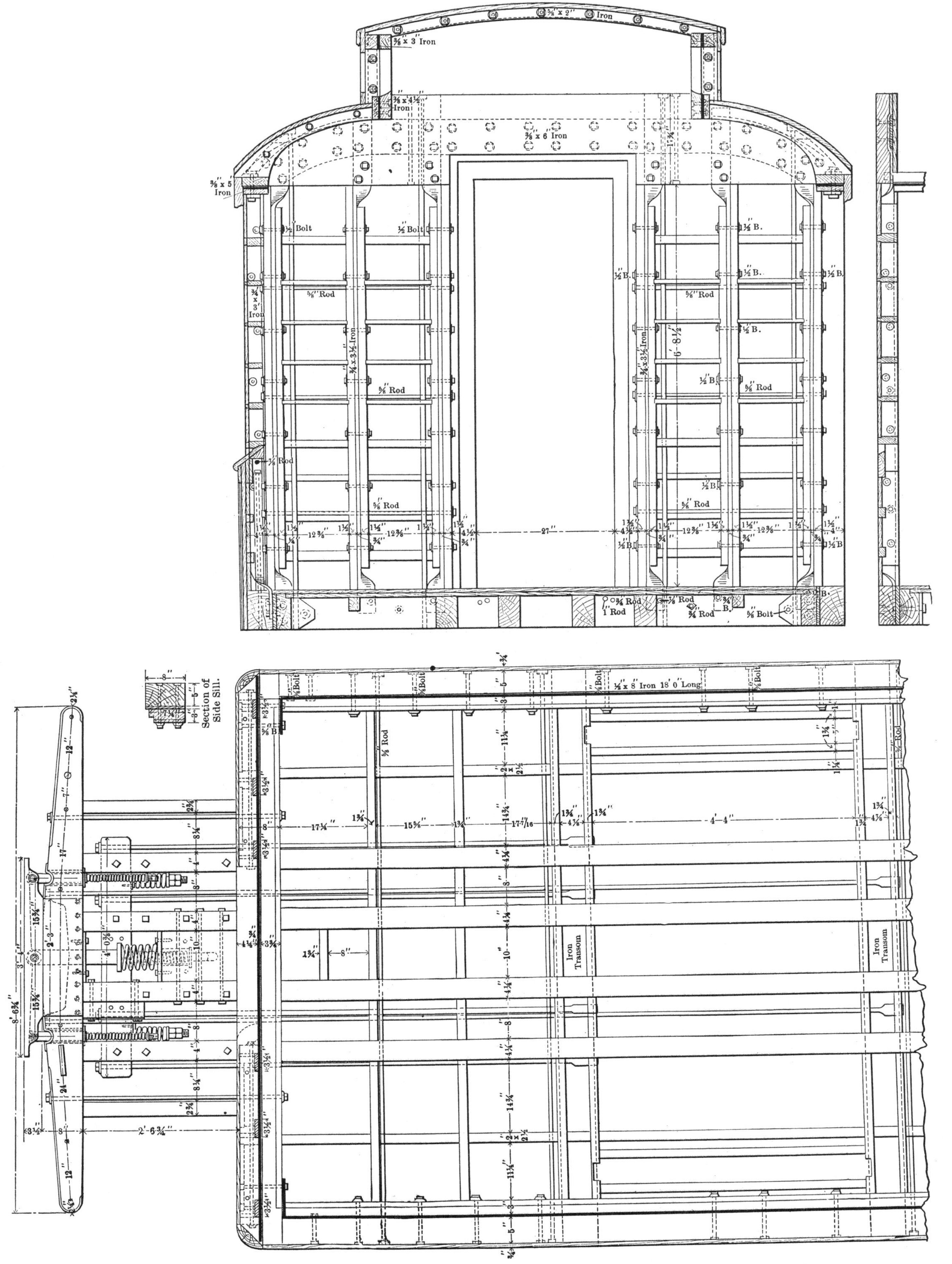

Figure 1.54 Thomas Bissell's reinforced end framing, devised in 1890, was used in many Wagner palace cars. (Railway Review, *December 6, 1890*)

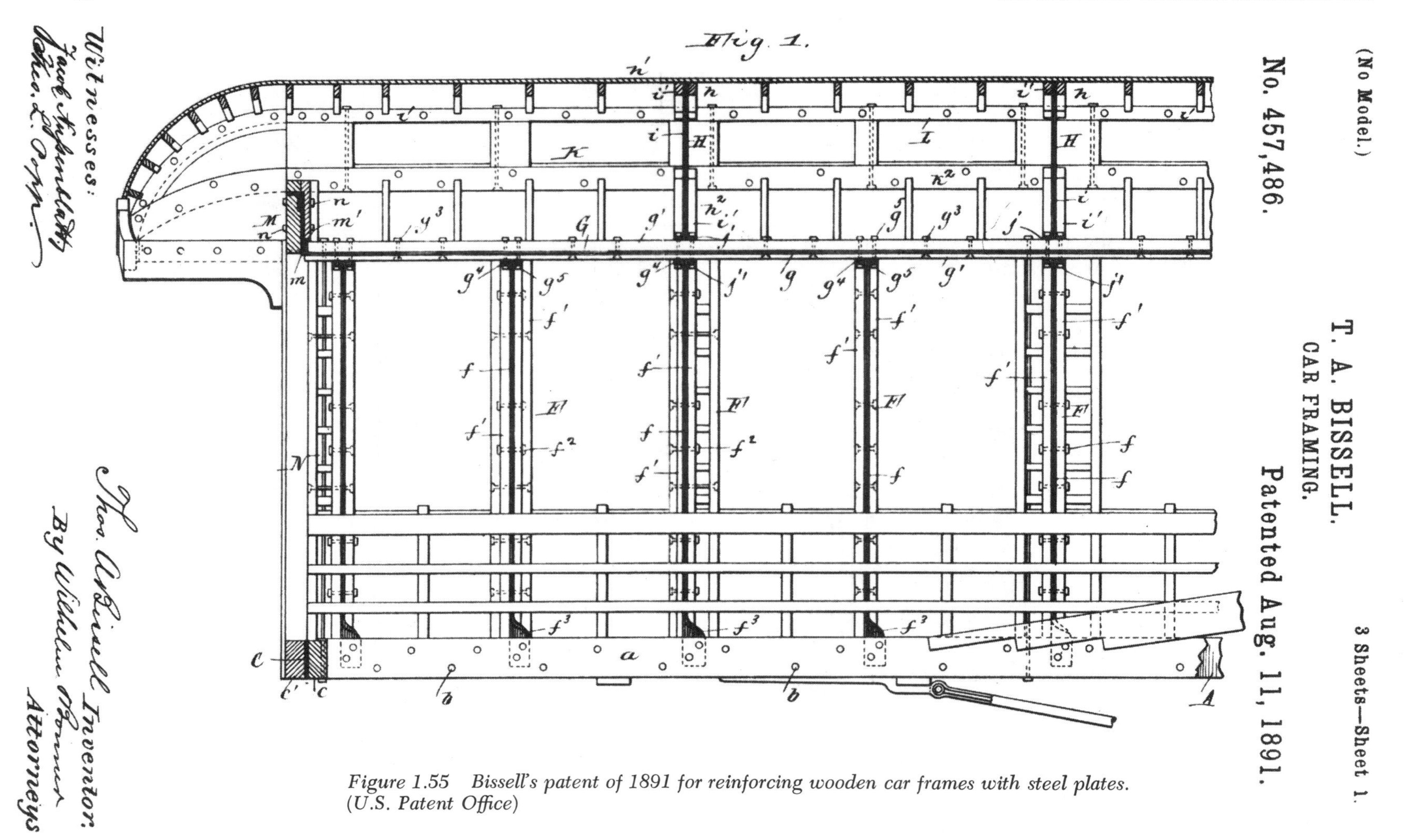

Figure 1.55 Bissell's patent of 1891 for reinforcing wooden car frames with steel plates. (U.S. Patent Office)

After the introduction of composite construction, the next notable design reform was the steel platform. The vestibules, automatic couplers, and elaborate draft gears that were coming into use by the early 1890s resulted in a greater load than wooden platforms could handle. Platform timbers were stiffened with iron plates, but the advent of wide vestibules and ever heavier draft stresses called for stronger construction. Pullman resorted to 6-inch I beams in 1895; steel construction was coming to the passenger car in gradual steps.[108] Within two years about fifty Pullmans were fitted with steel platforms. Pullman's Henry H. Sessions developed the design and secured two patents (No. 562,343 of June 16, 1896, and No. 575,994 of January 26, 1897). Just weeks after the second patent was granted, the Standard Coupler Company put Sessions's steel platform on the market. The response was good; by 1900 eighty railroads were using it. The Rock Island thought so well of the steel platform that the road installed it in some 400 cars. Thomas A. Bissell also designed a steel platform for Wagner that used Z rather than I beams. After 1899 it was manufactured by the Gould Coupler Company.[109]

End of Wooden Construction

Although composite construction extended the wooden car age into the early years of the twentieth century, pressures were mounting for all-steel fabrication. Acrimonious editorials against wooden cars had been standard fare in the daily press for years. Wooden cars were anathematized as tinderboxes, rolling death traps, and coffins-on-wheels. A popular notion developed that iron or steel cars would miraculously end the hazards of railway travel. And while it is difficult to present a single clear-cut reason why wooden cars were abandoned, the argument for safety appears to have been the most potent.

Another strong reason was that the industry had already accepted steel freight cars. By the late 1890s they had proved clearly superior to wooden cars in capacity and strength. Few wooden-frame freight cars were built after 1905, and manufacturing and repair shops accordingly converted to steel fabrication. Throughout the nineteenth century, car builders had thought of themselves strictly as a wood working industry. Proposals for iron passenger cars were dismissed as visionary and impractical. The wooden car was perfected to a high point; it was cheap, easy to build, serviceable, and most important, familiar. But the success of steel freight cars could not be denied. Once the trauma of conversion was over, traditional car shops saw little reason to resist the introduction of steel for passenger cars. The equipment was on hand and their employees were practiced in the new construction. What had formerly been a major obstacle was now an academic consideration.

The advantage of economy, long the main justification for the wooden car, was fading. Prime framing lumber was becoming both scarce and expensive. By pushing car lengths to 80 feet, builders had arrived at heavy construction that increased the weight and cost nearly to that of an equivalent steel car. Meanwhile steel was steadily declining in cost as American industry grew. Steel-working tools were not only well developed but becoming common. In addition, the railroads were ripe for the conversion. As the carriers that dominated land transportation in a wealthy manufacturing nation, they could afford and were expected to adopt the latest improvements. Whenever one trunk line purchased steel cars, competition forced other lines to follow suit.

Figure 1.56 A handsome arch-window coach was produced by Pullman in 1907. So-called art glass, a wavy, opaque green or tan material, was used in the arch sashes. (Pullman Neg. 10020)

Figure 1.57 When the Number 274 left the Pullman shop in 1909, wooden passenger cars were in their final years of production in this country. The last were produced in 1913. (Pullman Neg. 11470)

The Pennsylvania Railroad's commitment to steel cars in 1907 was the beginning of the end of wooden construction. New orders for wooden passenger equipment plummeted. In 1909 just over one-half of new orders were for wooden cars.[110] By the next year, orders were down to 29 percent of the total. In 1912 only 276 all-wooden cars were built, and many of these were for Canadian lines. The following year the last all-wooden passenger cars were produced for domestic service. After this time the only cars with wooden bodies were built for export, and most of them had steel frames.

The abrupt suspension of production did not mean that wooden cars were taken out of service. Steel and steel-frame cars rapidly preempted name-train assignments while wooden cars were downgraded to less prestigious trains, but there was no wholesale scrapping. The conversion was in fact remarkably languid. In 1912 over 90 percent of passenger cars were wooden; three years later the figure was 77 percent—comprising some 40,000 cars that made up the backbone of the fleet. In 1920 60 percent of cars were still wooden. Age took its toll during the next ten years, when more than half of the wooden cars were retired. Cutbacks in service during the early years of the Depression resulted in a dramatic reduction, so that by 1935 less than 6 percent of the fleet was wooden. It should be noted, however, that in the preceding years many wooden cars had been steel-framed, and just over 20 percent of the passenger cars in service in 1935 were of this type. By 1950 a few persistent holdouts, notably the Boston and Maine Railroad, had 276 all-wooden cars. Just three years earlier, the B & M was still assigning wooden cars to through trains. The wooden car has since disappeared—a good fifty years after its predicted demise.

Representative Cars

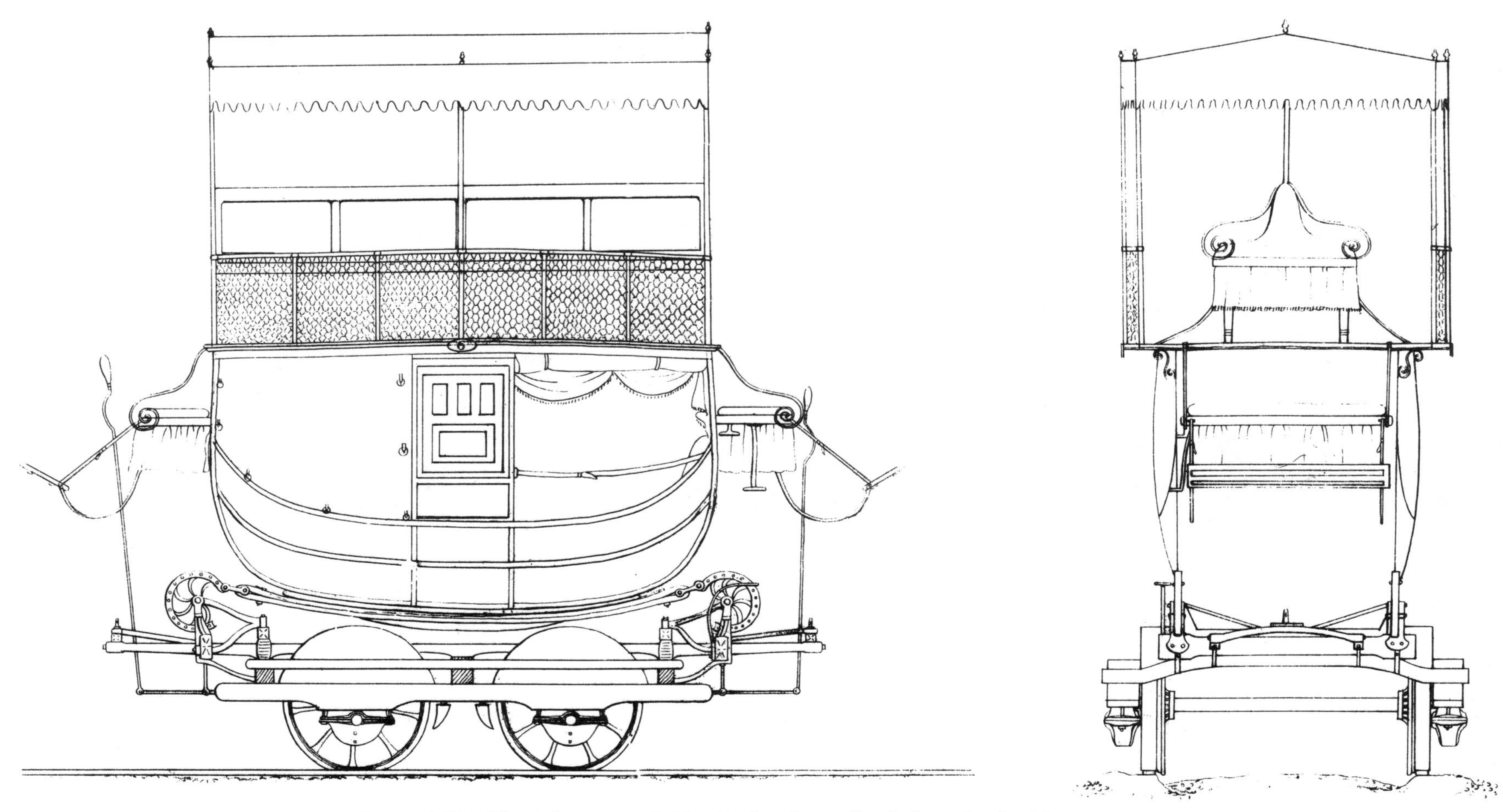

Figure 1.58 The Baltimore and Ohio Railroad's swell-sided coach, the Ohio, *was completed in 1830 by Richard Imlay. The drawing is from an original lithograph appended to the road's 1831 annual report.*

IMLAY CAR, 1830

The *Ohio* was the last of six four-wheel passenger cars completed by Richard Imlay for the Baltimore and Ohio Railroad between May and August 1830. The first car was named the *Pioneer*; others in the group were the *Frederick*, the *Maryland*, and the *Virginia*. They were the first cars built for regular passenger service on the Baltimore and Ohio and were, of course, among the very first rail passenger cars in the country. The B & O purchased another twenty-five or so four-wheel cars before adopting eight-wheel cars in 1834–1835.

This working drawing of the *Ohio* was included in the 1831 annual report of the Baltimore and Ohio. The influence of the road coach is obvious in the swell-sided body style, the suspension, and the frame. Two transverse stationary seats were fitted opposite one another inside the body. Collapsible jump seats were placed astride the doors, to be opened up once the other passengers were seated. Leather straps were buckled in place for seat backs. This was an established arrangement in road coaches of the day. While it was not the most comfortable seating possible, it did increase the inside capacity to 12 seats. Six more passengers could be accommodated on the outside end seats. The roof bench held another 12, making a total seating of 30 passengers—an extraordinary load for a 13-foot car. The cars were intended for slow-speed horse operation, but once steam locomotives were in use, the increased speed and the smoke and sparks caused passengers to abandon the roof section. The B & O annual report for 1836 gave the capacity of the four-wheel cars as 17. A brakeman stationed on one of the outside end seats took the eighteenth seating place.

The seats were upholstered, but there were few other concessions to passenger comfort—no lighting, heating, or toilet facilities. In bad weather, heavy leather curtains were dropped to cover the window openings. One of these is shown in place on the left side of the car. Cloth curtains could be raised or lowered to block out the sun. This simple arrangement was tolerable at very slow speeds, but when the car was pulled by a locomotive at even 25 miles per hour, the open windows must have let in a stream of smoke and cinders. Or when the side curtains were closed against the wind and rain, the interior of the little car as it lurched over the line must have been gloomy and claustrophobic.

The car's sole suspension was the leather thoroughbraces. Heavy leather straps (about 2 inches wide by 7 feet long), built up of several thicknesses of leather sewn together, formed each thoroughbrace. The ends were attached to jacks firmly bolted to the undercarriage frame. Crescent-shaped adjustment wheels made it possible to take up the slack as the leather stretched. Since the wheels and axles were not equipped with springs, the leather straps alone absorbed the inequalities of the track. The wheels were chilled cast iron, 30 inches in diameter. Mudguards were placed over the tops of the wheels. Winans's friction-wheel journal bearings were employed. The brakes were operated by a lever at either end of the car (the brake shoes are visible between the wheels). A perch pole drawbar not unlike that found on a road coach of the period can be seen in the drawing.

The overall length of the car was 13 feet, 2 inches; the wheelbase, 3 feet, 10 inches; extreme body width, 4 feet, 10½ inches. The weight was 1¾ tons when the coach was empty. (See Figure 1.58)

GOOLD CAR, 1831

In the spring of 1831 James Goold of Albany, New York, contracted to build six stagecoach bodies for the Mohawk and Hudson Railroad. Goold had been a carriage and coach builder since 1813. It is obvious from the specification that he was asked to furnish ordinary bodies without special modifications for railway service:[111]

To the Commissioners of the Mohawk and Hudson Railroad Company:

SIRS—I propose and agree to furnish for said railroad company six coach tops; that is to furnish jacks, and jack bolts, and braces, with thorough braces, and put them on the frames of the company's railroad carriages to support the coach tops; the coach tops to be finished and hung in the style of workmanship generally adopted in Albany and Troy, for post coaches; the materials and workmanship to be first quality; a baggage-rack and boot to be hung at each end; the length of coach body to be seven feet and four inches; five feet wide in the center, and three feet eight inches between the jacks. The general plan of the coach to conform to the plan and explanation given by the engineer of the company; to have three inside seats, the back of the end seats to be stuffed with moss, and all the seats to be stuffed with hair; to have a door on each side; to have an outside seat on each end across the top of the coach with suitable foot-board; also a seat at each end for driver or brakeman, to drop below to a suitable heighth to make the rack his foot board. An oil cloth to be rigged to the center rod on coach top, to cover baggage, and one at each end rolled to the back of the seat to protect it from rain. The whole completed, and to be hung on the carriage frames at some point on the line of said railroad, as follows:

Two coaches to be hung by the first day of July next, and the remaining four by the first day of August next; the work to be subject to the inspection of the engineer of the said railroad company. The whole to be completed as aforesaid for the sum of three hundred and ten dollars each.

It is understood that the above coaches are not to be provided with lamps or mud leathers.

JAMES GOOLD.

The written proposition is adopted on the part of the Mohawk and Hudson Railroad Company, by order of the commissioner.

JOHN B. JERVIS,
Engineer Mohawk and Hudson Railroad Company.

Albany, 23d April, 1831.

A contemporary side elevation reproduced in Figure 1.2 was acknowledged by Goold a few years before his death to be a true representation of the original design. A reconstruction drawing, shown here, was prepared in 1892 so that full-sized replicas might be built to accompany the *De Witt Clinton* locomotive at the Columbian Exhibition. The locomotive and cars are presently exhibited at the Ford Museum in Dearborn, Michigan.

These cars are very similar in construction and size to the Imlay car in Figure 1.58. Notice, however, the longer wheelbase, larger wheels, and inside axle bearings. The leather thoroughbraces are attached to stationary jacks. The seating capacity was smaller than the Imlay cars because steam operation was contemplated from the beginning, so that the roof seating was limited. A Mohawk and Hudson inventory of 1840 lists nine 9-passenger and two 6-passenger thoroughbrace cars.[112]

The overall length was 13 feet, 9 inches; the wheelbase, 5 feet; wheel diameter, 36 inches; extreme body width, 5 feet. (See Figures 1.59 and 1.60.)

JERVIS CAR, 1832

While he was chief engineer of the Mohawk and Hudson Railroad, John B. Jervis took a direct hand in designing the rolling stock. His improvements in locomotive design, notably the leading truck, brought him national attention. His approach to car construction appears more conservative than his locomotive work, as shown by this drawing. A three-compartment composite body with side doors was mounted on a simple four-wheel running gear. The body was simplified and more box-like, but a gesture toward the orthodox road coach body was made in the curved moldings on the lower panels. The driver's seat and long brake lever were also vestiges of road coach construction. The drum brake is a distinct departure, however, and possibly the

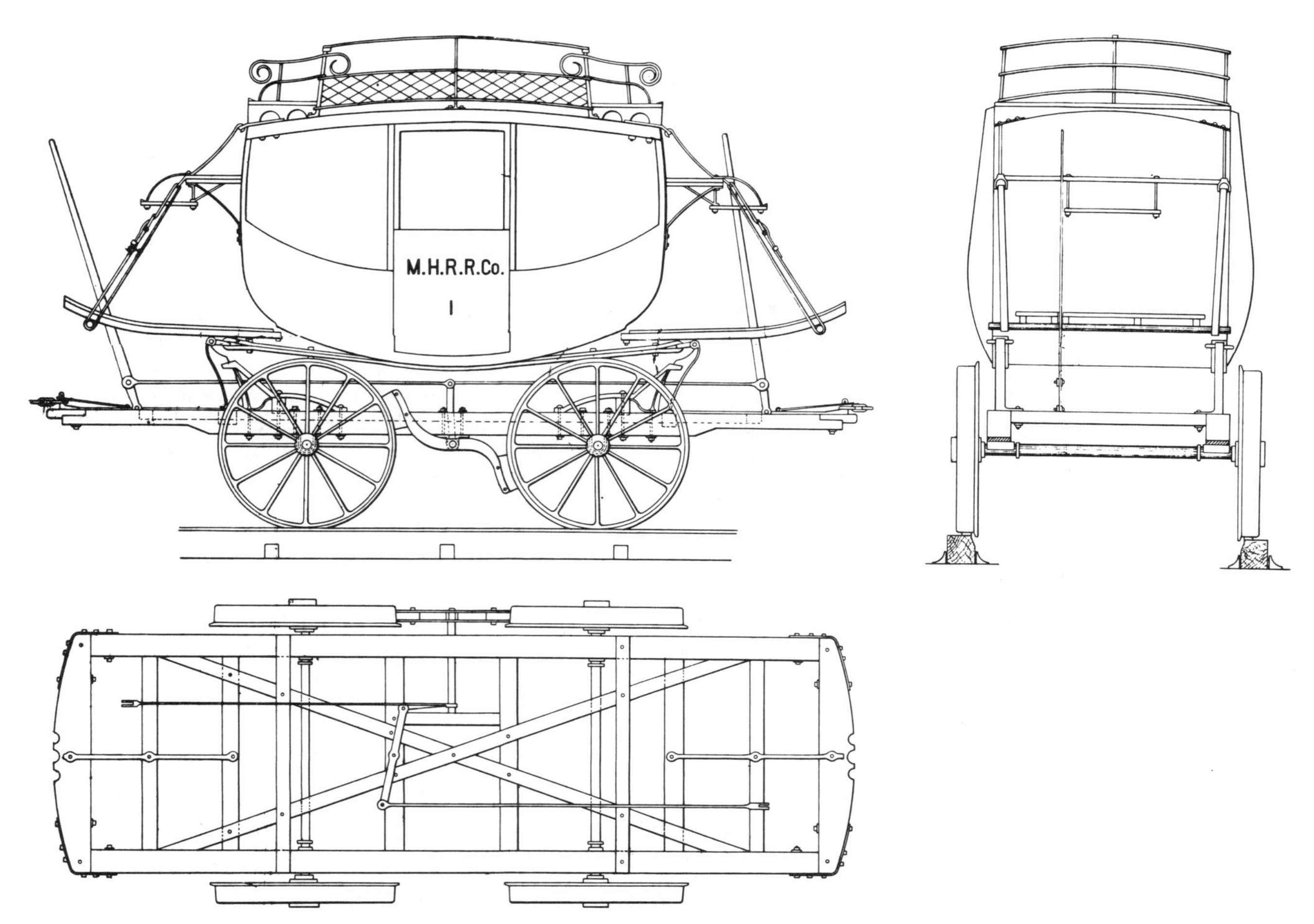

Figure 1.59 This Mohawk and Hudson Railroad four-wheel coach was built by James Goold in 1831. (New York Central Railroad)

most interesting detail of the car. The leaf springs are also worthy of notice as an alternative to the leather thoroughbrace suspension. The wrought-iron journal pedestals are, incidentally, very similar to those used on locomotives of the period.

The original drawing from which this tracing was made is dated 1832 and initialed by Jervis. It was apparently the plan followed for several cars built in Schenectady for the Mohawk and Hudson. The *Buffalo Journal* of April 25, 1832, described the cars as being made in a new and convenient "square form" with separate compartments. They were 15 feet long and seated 18; one of the cars was named *Utica.* The exact number that were built is not known, but Von Gerstner's report states that 48 four-wheel passenger cars were in service at the time of his visit in 1840.[113] He described them as having three compartments with seating for 20 first-class passengers. Second-class ticket holders were carried in freight cars.

The basic dimensions are clearly noted on the drawings. Cars of this general style weighed 2 to 2½ tons. (See Figure 1.61.)

BOSTON AND PROVIDENCE CAR, 1834–1836

The only surviving original stagecoach-style passenger car was built at the Boston and Providence Railroad repair shops in Roxbury, Massachusetts, between 1834 and 1836. It was designed and constructed by John Lightner (1811–1896), a Baltimore carriage builder who was master car builder for the railroad from 1834 to 1888.

The survival of an original swell-sided railroad passenger car of the 1830s is a fortuitous incident made even more improbable by the career of the car itself. It became obsolete just a few years after construction, for it could not operate with larger eight-wheel cars and was downgraded to branch-line service. From there it declined to hauling track laborers on work trains. Most swell-sided, thoroughbrace cars were found unsuitable for any service and were scrapped, possibly excepting the running gear, which could be salvaged for a track-construction car. But by some curious fluke, this car survived until the body was smashed in a wreck. According to one story, the pieces were gathered up and stored for many years at the Roxbury repair shops of the Boston and Providence.[114] Whether from thrift or sentiment, the car was preserved. In 1893 it was reassembled with a new running gear under the direction of its builder, John Lightner, for exhibit at the Columbian Exhibition. Around 1905 the car was given to Purdue University for its collection of historic rolling stock. It was returned to the New Haven Railroad in 1939 for exhibit at the New York World's Fair.

For years the New Haven displayed the car at various fairs and exhibitions as a prize publicity tool, but hard times forced its sale in 1951 to the Danbury Fair in Connecticut. The car is presently on exhibit there—under cover, fortunately.

The car differs from the two previous examples of swell-sided cars in that the body is wider and longer and it is hung high to clear the wheels. In general it represents a more positive departure from existing road coach design. If the modifications seem slight, consider that the railroad age had only begun, and that within three or four years builders such as John Lightner recognized that coaches were subject to fewer size limitations on the railway and should be designed differently. This car illustrates the conservative, straight-line development of an existing design for a new technology.

Figure 1.60 A reproduction of an 1831 Mohawk and Hudson coach built around 1892 for the Columbian Exposition. (*New York Central Railroad*)

The car's construction is evident from the drawings. The continuous draft sill or perch pole is similar to the Imlay car. The brake mechanism shown in section EE could brake only one set of wheels unless brakemen were stationed at both ends of the car. The wheels, it should be noted, are cast iron of an exceedingly fragile pattern.

The overall length is 15 feet, 2 inches; the wheelbase, 5 feet, 9 inches; wheel diameter, 36 inches; extreme width of body, 6 feet. The weight is estimated at 2 to 3 tons. (See Figures 1.62 to 1.64.)

NOVA SCOTIA CAR, ABOUT 1838

The history of this small four-wheel coach, now on display at the Baltimore and Ohio Transportation Museum in Baltimore, is very uncertain. According to one story, it was produced in England for the General Mining Association of Nova Scotia. But the late Robert R. Brown, a well-informed Canadian railroad historian, declared that it was turned out by the carpenters of the General Mining Association in Nova Scotia. The date of construction appears to be 1838. A railway was built to connect the Albion coal mines of the General Mining Association with Pictou Harbor, and the car was used by officials and distinguished guests of the mining company.

By 1883 it was considered enough of a relic to show at the Chicago Railway Appliance Exhibition. Ten years later it was returned to Chicago for the Columbian Exhibition. Apparently abandoned in Chicago, the car was acquired by the Baltimore and Ohio Railroad when the exhibit closed. During the B & O's Centennial in 1927, the fanciful "bridal car" story was hatched: according to this piece of puffery, the car originally carried the governor of Nova Scotia's bride to her wedding, and any young girl seeking a husband need only seat herself in the car. The story has been so often repeated that it is now a hallowed part of the car's history.

The car, with its Bombay-shaped body, seats 6. Two upholstered seats are placed facing one another across the car. The cast-iron wheels are individually suspended by leaf springs. Since there is so little information, and no illustrations of the car dating before 1883, it can be questioned whether the running gear is original. It is possible that the body was originally mounted on thoroughbraces.

The car's dimensions are given in the drawing. (See Figures 1.65 and 1.66.)

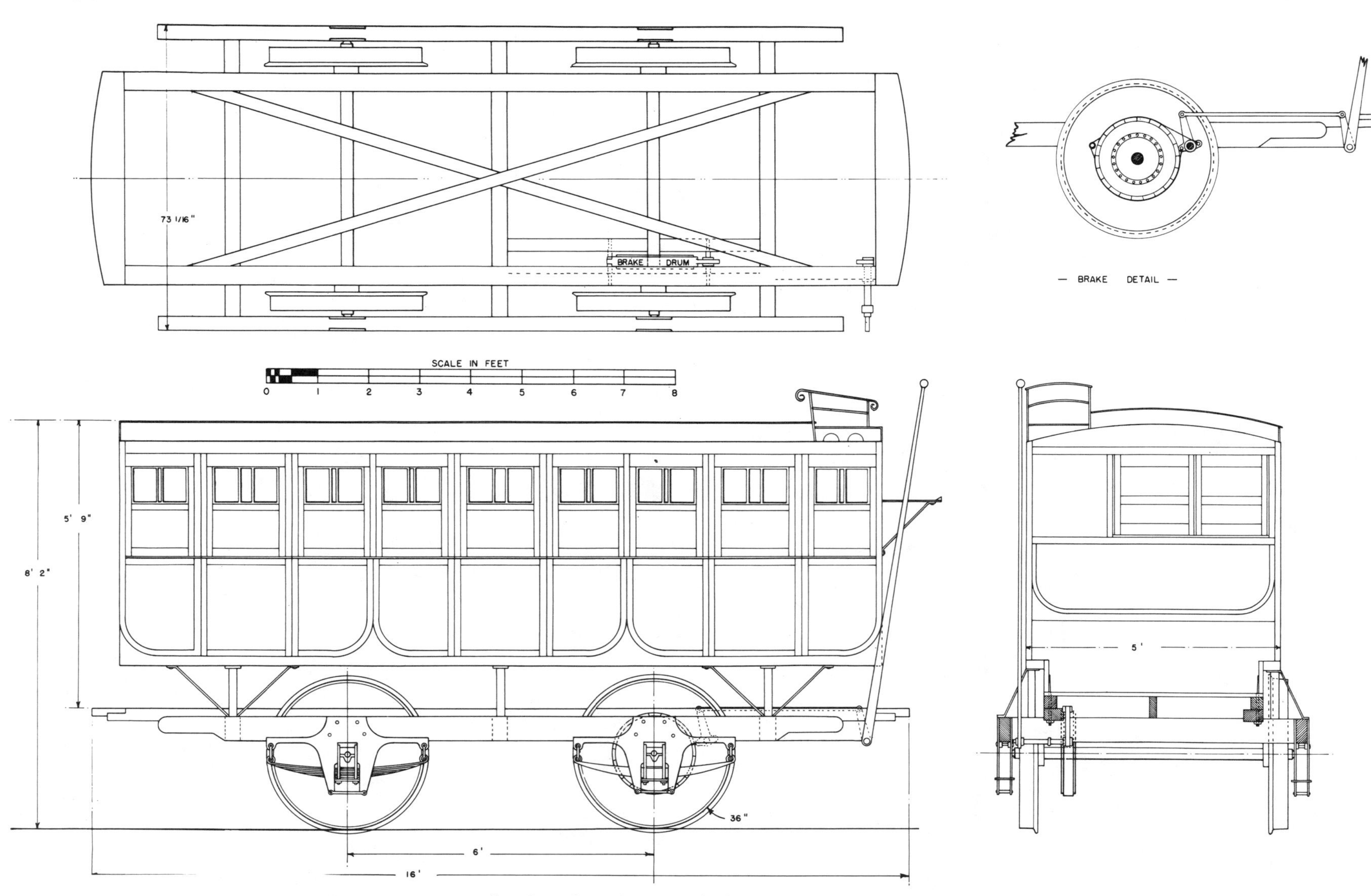

Figure 1.61 A Mohawk and Hudson coach designed in 1832 by John B. Jervis. The tracing made from an original drawing now in the Jervis Library, Rome, N.Y.

VICTORY MODEL, 1835

Before Richard Imlay constructed the *Victory*, an eight-wheel car that he completed in July 1835 for the Philadelphia, Norristown, and Germantown Railroad, he built this model in order to check out the design. The model was constructed to a large scale, roughly 1½ inches to the foot, and is over 4 feet long. It was expertly made and finished as a showpiece; Imlay probably displayed it to prospective customers. He kept it for years after his car-building career was over, and on his death in 1867 it was purchased by the Eastern Railroad Association for exhibit at its offices. The model was sent to Chicago for the Columbian Exhibition, and through circumstances not entirely clear, it came into the possession of the Baltimore and Ohio Railroad. It is now exhibited at the B & O Transportation Museum. The drawing and the following description were prepared after first-hand inspection of the model.

The body is divided into three compartments. The center underslung section is the passenger compartment. One end compartment has a water closet, but the reason for that section's division into three spaces is not clear. The other end compartment is fitted out as a pantry. Its door has an opening and a serving shelf; on one side is a desk, and at the rear are several shelves for provisions. The passenger saloon is ringed with a continuous side bench broken only at the door openings. Each side of the car has a single door. The benches are not uphol-

Figure 1.62 A Boston and Providence coach built in the railroad's shops around 1834.

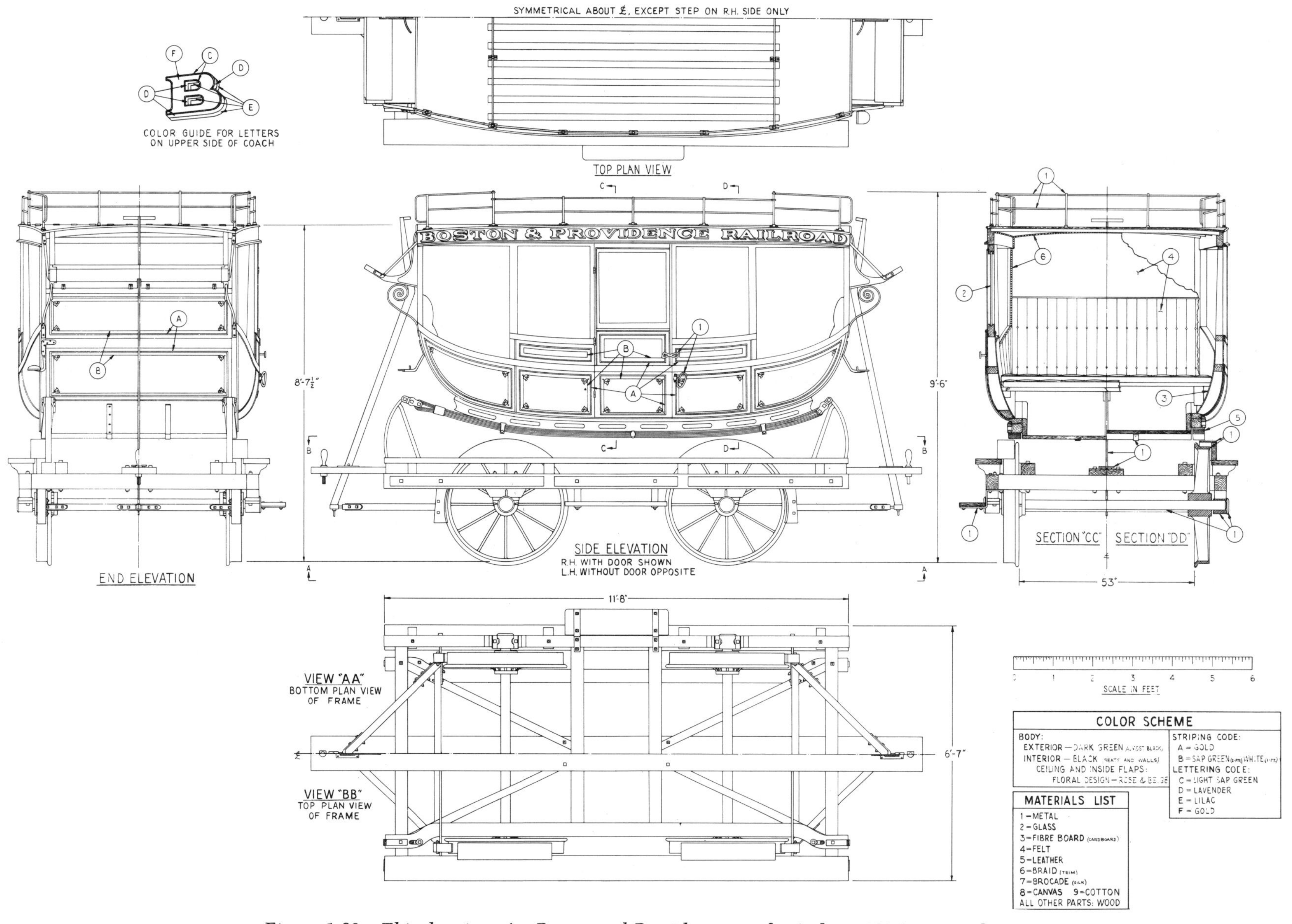

Figure 1.63 This drawing of a Boston and Providence coach of about 1834 was made from the original, which is now exhibited in Danbury, Connecticut. (*Clark Williams*)

stered, but it is possible that cushions were used. The interior of this area is covered in leather. The windows are entirely open, but it is probable that there were at least leather side curtains for protection. The lithograph in Figure 1.13 shows window sashes for the straight portions of the body, with curtains at the curved-end windows. The lack of windows in end compartments is another obvious defect of the model, and one that was corrected in the full-sized car.

The rudimentary clerestory runs the full length of the car and is open except for the light iron bars. Some provision (such as curtains) must have been made against the weather. It should be noted that the carlines run completely across the roof and do not break at the clerestory opening. The railing above the clerestory roof has been explained as a seat back, but the roof does not appear to be constructed for such a load. Lightly fabricated from thin wooden strips covered with canvas, it may have been a handrail for trainmen passing over the top of the car.

The trucks follow Imlay's leaf-spring bolster, which he patented in 1837. The instability of this arrangement is evident, but for very slow speeds it was probably satisfactory. Because the bolster was equipped with springs whereas the axles were not, the entire dead weight of the truck was free to damage the track. At some point in the model's history the spoked wheels were lost from one truck; the plate wheels are a later repair.

As a record of the exterior finish, this model is without parallel. The striping and decorative painting are finely executed, and despite successive coats of dark varnish and the scars of a long life, there is evidence that bright colors were used. The panels of the passenger saloon are bright yellow. Broad gold leaf and black bands outline the body shape. Fine brown and black lines outline the gilt bands and the window posts. The heraldry in the body panels is executed in gilt, red, and green, and finely picked out in black. The panels of the end compartments are a rich brown-red. They are striped in black and gold leaf. The figure of the phoenix is in gilt shaded in a brown wash. The car frame is naturally finished mahogany, striped in gilt. The roof is now black, but there is evidence that it was originally painted dark green. The trucks and wheels are black.

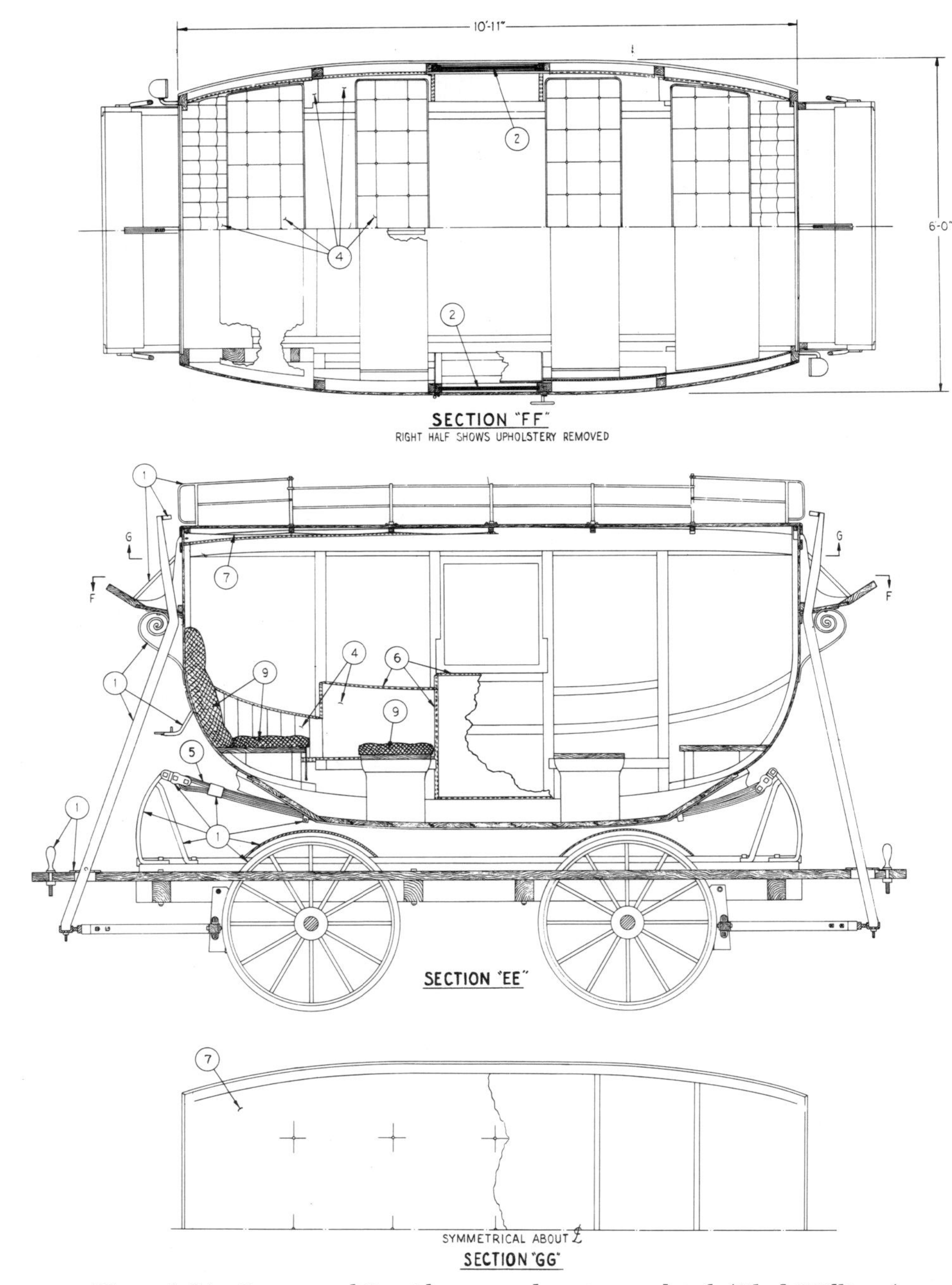

Figure 1.64 Boston and Providence coach: cutaway detail. (Clark Williams)

The general dimensions of the model are clearly shown in the drawing. It should be noted that in certain proportions the model appears out of scale with the track gauge of 7¼ inches. This is particularly true for the end compartments, where insufficient headroom is provided. (See Figures 1.67 and 1.68.)

WASHINGTON BRANCH CARS, 1835

By 1834 the Baltimore and Ohio Railroad had become convinced of the superiority of eight-wheel passenger cars. The experiments were finished, and they had made it clear that a simple box-like body carried on two four-wheel trucks was the safest and most economical passenger car possible. The road therefore decided to build cars of this pattern for the new Baltimore-Washington branch. Construction began at the company shops late in 1834. Ten cars were completed in time for the Washington branch's opening in August 1835.

The cars incorporated all the basic characteristics of the American style of passenger car and may be regarded as the prototype of that enduring pattern. Two drawings exist of these historic cars. The first (Figure 1.69) was prepared around 1853 for the series of lawsuits known as the eight-wheel car case. It was apparently taken from original drawings or from an original car then in existence but now vanished. And a contemporary drawing (Figure 1.70) of incredible detail was found at the Deutsches Museum in Munich. The origins of the drawing and the reason why it was in Germany cannot be determined, but it might be traced back to one of the numerous European engineers who came to study American railways early in the nineteenth century. In addition to the splendid engineering information, the drawing portrays the subtle elegance of American passenger cars of the period. If the 1853 drawing had been the only one to survive, it would be possible to dismiss the Washington branch cars as simple, even crude, boxes on wheels. But the contemporary drawing from Munich shows finely cut moldings, elaborate scrolls, panels, and louvers.

The car was supported by a simple box frame placed entirely under the body. The frame was stiffened by an iron truss rod. Cross sills were used. A continuous iron drawbar running the full length of the car took the place of the center sills. The body was

Figure 1.65 A coach of about 1838 operated by the General Mining Association, Nova Scotia. The photograph was taken around 1930 at Bailey's roundhouse, Baltimore.

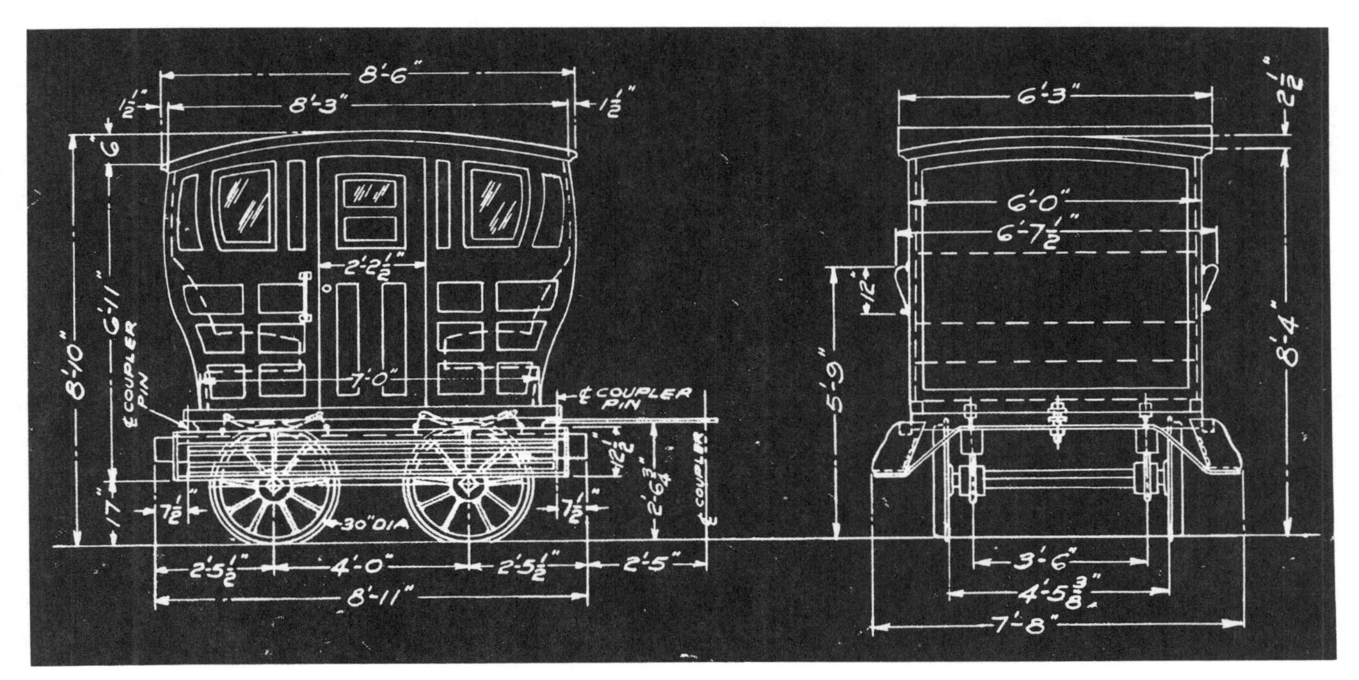

Figure 1.66 Diagram drawing of a Nova Scotia coach, 1838c.

lightly framed and was not self-supporting. Solid shutters and glass panels alternated at the window openings. Presumably the glass was stationary and the shutters could be raised to admit air. Unaccountably, the seats were made with stationary backs even though reversible seats were in use on the B & O before this time. They were well upholstered, and an early newspaper account correctly described them as high-backed and well inclined. The wooden ends were treated in the bold Empire style of the period.

The trucks were of the "live" or leaf-spring side-frame type, with wrought-iron truck bolsters. A secondary leaf spring attached to the frame of the car bore on the ends of the body bolster (item X). This complex double suspension appears unique to the B & O. The 31-inch-diameter wheels were cast iron made on Winans's "drumstick" pattern. Although not shown in the drawing, they would have had split hubs banded with wrought-iron rings. The axles were tapered, measuring 3¾ inches at the heaviest point. The journals were small, measuring 2 by 4 inches. The truck's wheelbase was 40 inches. The safety frames, shown in detail 25 of Figure 1.70, were a variation on Kite's safety beams. If the side-frame spring should break, the bracket Z′ would fall on the wheel treads and support the car until it could be stopped. The bar frame shown in the same detail was meant to support a broken axle and hold it in line.

The brake is the earliest mechanical braking mechanism that

Figure 1.67 Richard Imlay's Victory, *constructed about 1835. The model is now in the B & O Museum, Baltimore. (Smithsonian Neg. 68359)*

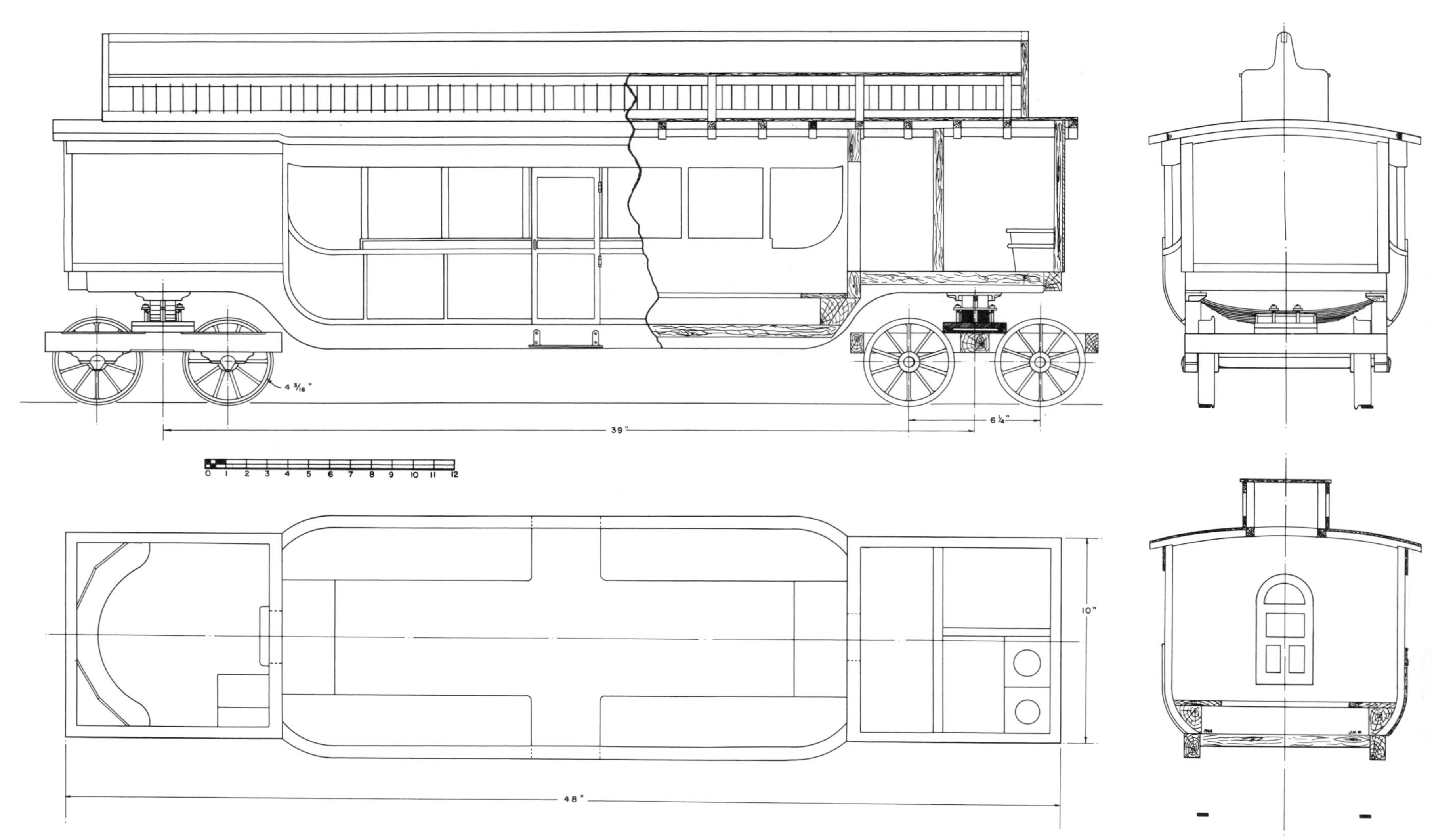

Figure 1.68 Drawing of the eight-wheel car model Victory.

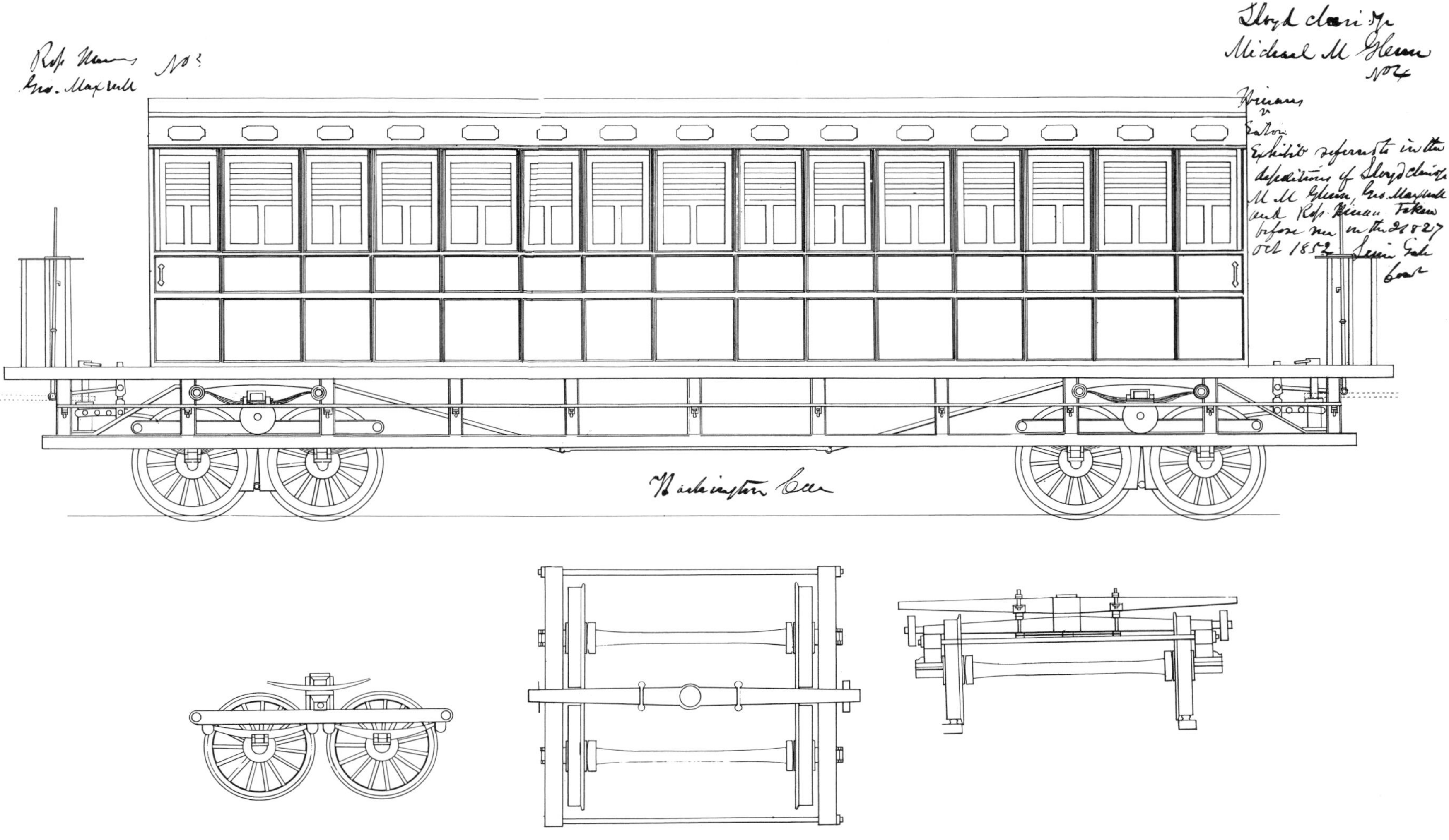

Figure 1.69 Washington branch car of the Baltimore and Ohio Railroad, 1834–1835. The drawing was made around 1853. (Peale Museum)

has been discovered. A light chain attached to the top of the upright staff (O′) at the right end of the car was stretched across the top of the train from one car to the next. Pulling the chain moved the staff, freeing a heavy counterweight lever which in turn set the brakes. The idea for this brake was credited to the brother of the president of the railroad, Evan Thomas, in the B & O's 1835 annual report.

The overall length was 37 feet, 2½ inches; the truck centers, 24 feet; the body, 32 feet, 2½ inches long by 9 feet wide by about 7 feet high. The car seated 44 and weighed an estimated 6 tons. (See Figures 1.69 and 1.70.)

CAMDEN AND AMBOY RAILROAD, CAR NUMBER 3, ABOUT 1836

This is the oldest existing eight-wheel passenger car. It was built around 1836 at the Camden and Amboy shops by Edwin Lockwood (1810–1881), the road's master car builder for over forty years. In 1840 the Austrian engineer Von Gerstner reported fifteen eight-wheel passenger cars in service on the Camden and Amboy.[115] The dimensions given in his report agree almost exactly with those of this car.

Some thirty years later, when the Pennsylvania Railroad took over operation of the New Jersey railroads, much of the obsolete equipment of the Camden and Amboy was sold. Car Number 3 was purchased for a chicken roost by Samuel Newton of South Amboy, who discarded the trucks and set the body on posts. It remained on his farm until 1892, when the Pennsylvania Railroad retrieved it for exhibit at the Columbian Exposition. The remains of an old truck were uncovered in a marl pit; the wooden members were gone, but the metal parts were salvaged to construct a partial replica. A second matching car was found hidden away in the New Jersey Meadows repair shops. It had been discovered, derelict and half buried under rubbish in a lumberyard, some years earlier by an official of the railroad, who had farsightedly removed it for safekeeping.

Many parts were missing from both cars. The interiors were stripped; much of the hardware was gone; some wooden parts undoubtedly had rotted and been replaced. Probably there had been some rebuilding over the years as well. Thus the existing car must be viewed critically, at least in its details, and not accepted without qualification as a representative example of antique car construction. After restoration both cars were sent across the country in train with the venerable *John Bull* locomotive—a magnificent publicity stunt to advertise the Pennsylvania Railroad exhibit at the 1893 Chicago fair. The history of the cars after the fair is cloudy. Both were in existence as late as 1907, when it was suggested that they be shown with the *John Bull* at the Smithsonian Institution. By 1927, however, one of the cars was gone; the Pennsylvania exhibited only the Number 3 at the Fair of the Iron Horse in Baltimore. The fate of the other car cannot be discovered. In more recent years the Number 3 was given to the Smithsonian, where it has been on display in Railroad Hall since 1963.

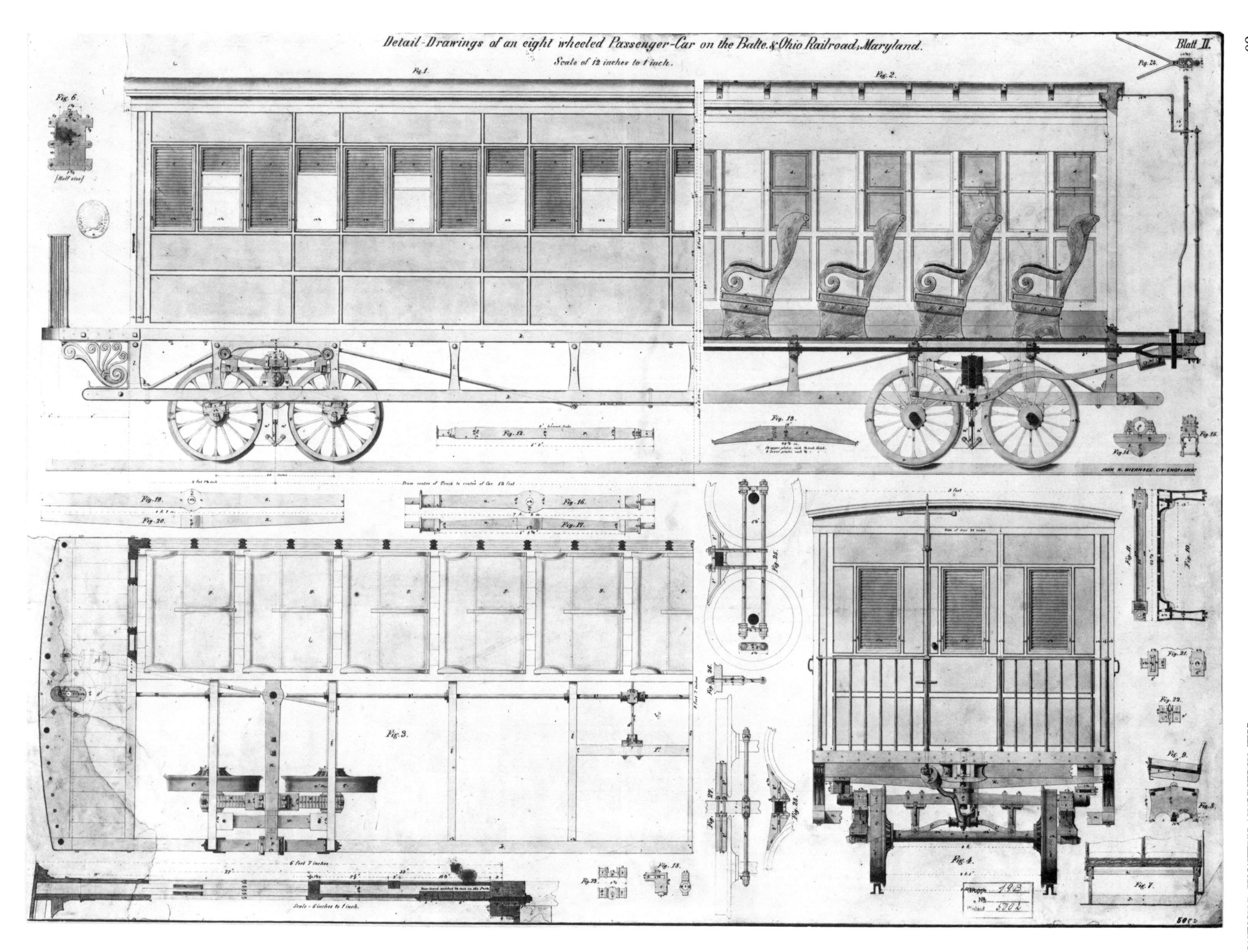

Figure 1.70 B & O Washington branch coach of 1835, from an original drawing in the Deutsches Museum, Munich, Germany.

Figure 1.71 Camden and Amboy Railroad car Number 3, built in the road's own shops around 1836. Now in the Smithsonian Institution. (Pennsylvania Railroad)

The overall proportions of the car are small. A modern passenger, even one accustomed to the barbaric poverty of tourist air travel, would feel uncomfortably cramped in this low, narrow car. The ceiling is a bare 6 feet, 4 inches high. The seats are 1 foot deep; the center aisle is 18 inches wide. But it should be remembered that the cars were meant only for short trips. A way passenger could endure a short ride down the line, and a through passenger going the full 60-mile length would surely find the ride of the eight-wheel car more agreeable than that of a pitching road coach. The seats, if narrow and low-backed, were upholstered, and the backs were reversible.

Ventilation received some attention. The small-paned glass windows were stationary, but the larger wooden panels could be opened and dropped down between the sides of the car body. The row of horizontal shutters at the top of the body (where the letter board would normally be) were hinged at their centers and could be tilted open. Lighting, however, received less attention than ventilation. A candle lamp at each end of the car, built into one of the end windows, provided the only illumination. It served to light the doorway of the car and doubled as a running light, since it was made to shine in both directions. It is possible that there once were individual side candle lamps, but no evidence of them survives in the present car.

The decorative treatment is plain and somber, possibly because of the 1893 restoration rather than the original finish. The exterior is painted olive drab outlined in black. The truck frames are cream yellow, while the wheels and hardware are black. The interior woodwork is painted a dull tan. The headlining is a rich, figured grey-green brocade. An ornamental fringe runs around the ceiling. The seats are covered in a heavy light-grey cloth, as are the armrests. Lighter, livelier colors would actually seem to be a more authentic finish for this car.

The most striking structural feature is the underslung wooden truss, which provides the car's main support. The center sill runs only between the body bolsters (which partly accounts for the sag at either end of the body). Light cross pieces and end sills form the remainder of the floor frame. The platforms are supported by separate timbers. A continuous iron plate ½ inch thick is bolted to the underside of the center sill. A clevis attached to the body bolsters receives the end of the long drawbar. The platform end of the drawbar is split to make the coupler pocket. Thus a continuous metal draft rig was created to relieve the wooden frame from this duty.

The body framing is exceedingly light, composed of thin vertical and horizontal members. The side panels are of clear white wood. The roof is fabricated in the usual manner, with arched carlines and narrow wooden sheeting covered over with painted canvas.

The trucks are wooden beams with stationary bolsters that are supported by iron truss rods, one on either side. These rods pass through the side frames, holding the assembly firmly together. A nest of seven light coil springs is fitted over each journal box,

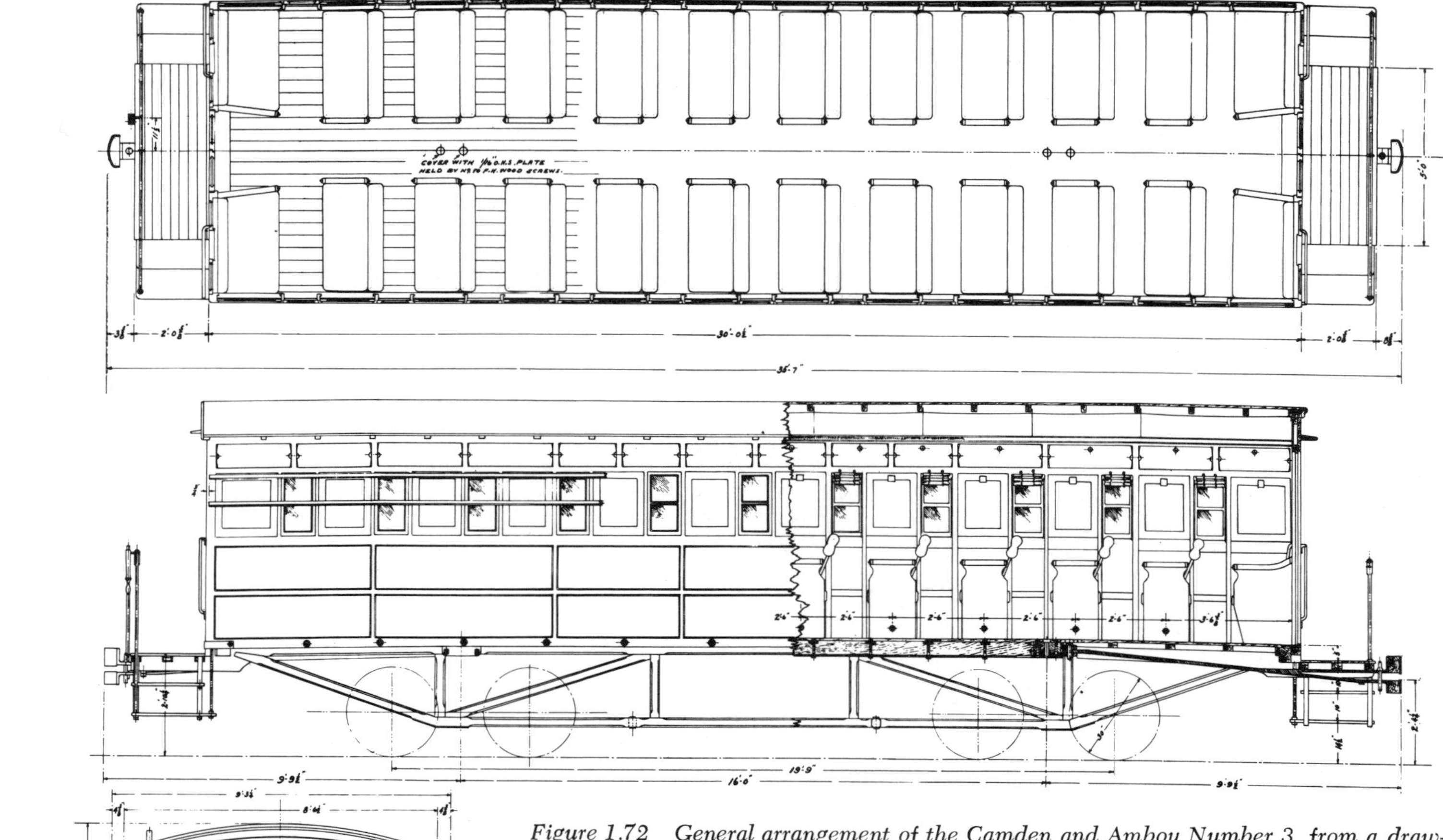

Figure 1.72 General arrangement of the Camden and Amboy Number 3, from a drawing prepared by the Pennsylvania Railroad. (Pennsylvania Railroad)

Figure 1.73 Interior of the Camden and Amboy Railroad coach. (Pennsylvania Railroad)

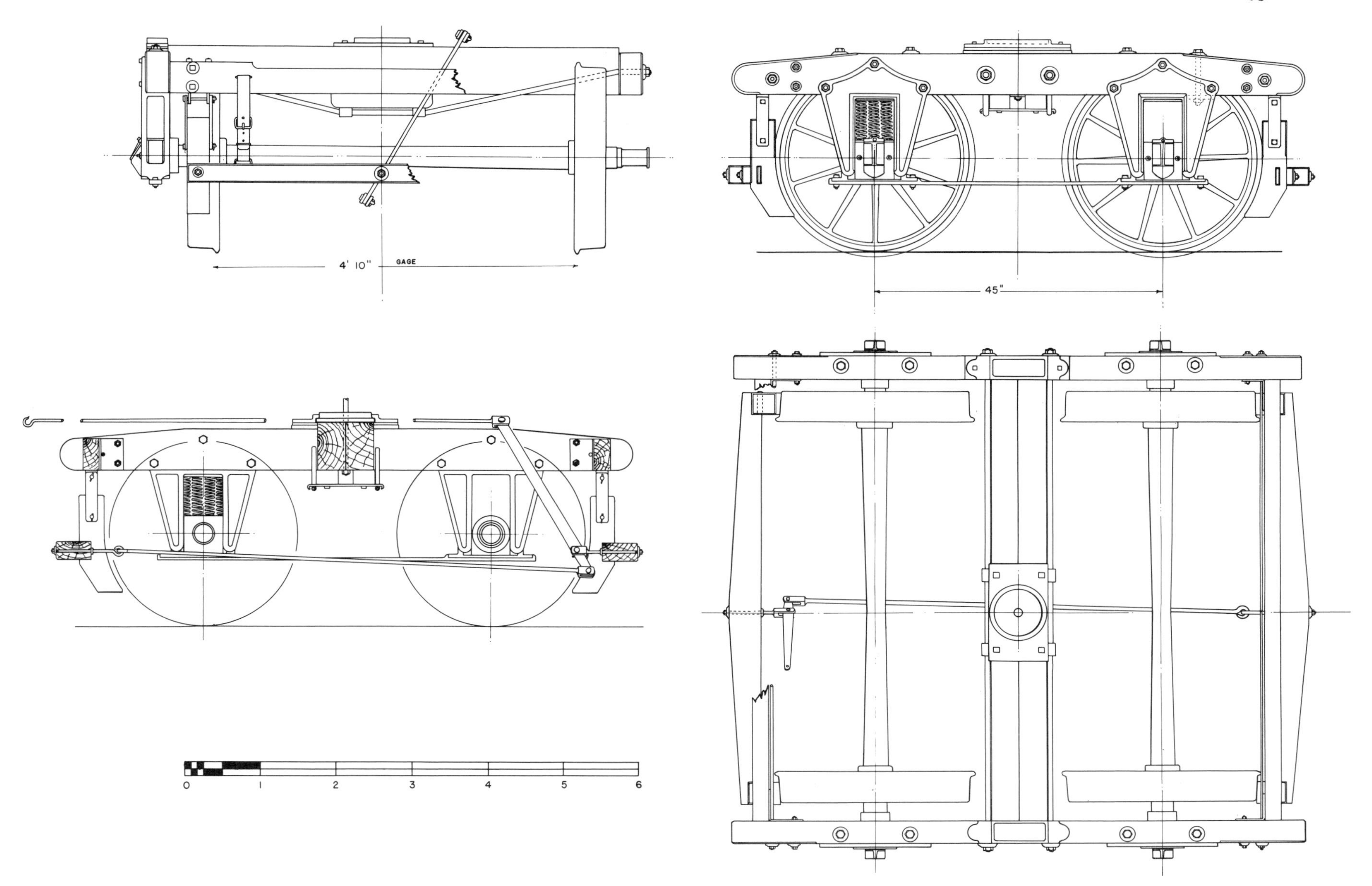

Figure 1.74 Truck detail of the Camden & Amboy Number 3 showing the original iron spoked wheels. (Drawing by John H. White, Jr.)

which is made following John Lightner's 1848 design. The journal bearings measure 2¾ by 4¼ inches. The present wheels are a curious mixture of single- and double-plate styles; no two are duplicates, and all date from the 1870s. What the original wheels were like is impossible to say, because the Camden and Amboy tried every conceivable style during its early years. In 1840 Von Gerstner said that cast-iron chilled wheels were used under the road's eight-wheel passenger cars.[116] These were probably spoked wheels.

Brakes are fitted to only one truck, and wooden brake shoes are hung outside of the wheels. The brake beams are supported by leather straps provided with buckles to adjust the position of the shoes. The brake staff has a hinged handle which folds down out of the way when not in use.

The overall length is 35 feet, 7 inches; body width, 8 feet, 6½ inches; height, 10 feet; seating, 48; weight, 14,250 pounds. (See Figures 1.71 to 1.74.)

BOSTON AND PROVIDENCE RAILROAD, EIGHT-WHEEL CAR, 1838

The Boston and Providence was among the first New England railroads to adopt eight-wheel cars. George S. Griggs, master mechanic of the line (best known as a locomotive designer), claimed credit for the road's first eight-wheel passenger car. In testimony given in the 1850s at the eight-wheel car trial, Griggs said that he began work on the car in March 1838.[117] He claims that he had not seen any other eight-wheel railroad cars except possibly those used on the Granite Railway to haul stone blocks. He completed the car in September of that year, and it proved so satisfactory that the road rebuilt a number of four-wheel cars as eight-wheelers and constructed all new passenger cars on the double-truck plan.

Together with his testimony in the lawsuit, Griggs offered this drawing of his first car, which was reproduced in the published court transcript. The drawing was made from measurements of the existing car several days before Griggs testified. It shows a narrow, low-ceilinged car in sketchy detail, though there is enough to convey the general arrangement. The posts and window sash were omitted. The extremely narrow body of 7 feet might be explained by insufficient passing clearance on the railroad tracks. (Many early roads placed their tracks close together, never anticipating the need for wider cars.) The end platforms are mounted on extended beams of the truck's side frame. The load of the draft was then placed on the truck's kingpins. The platform would swing, of course, when the car passed around a

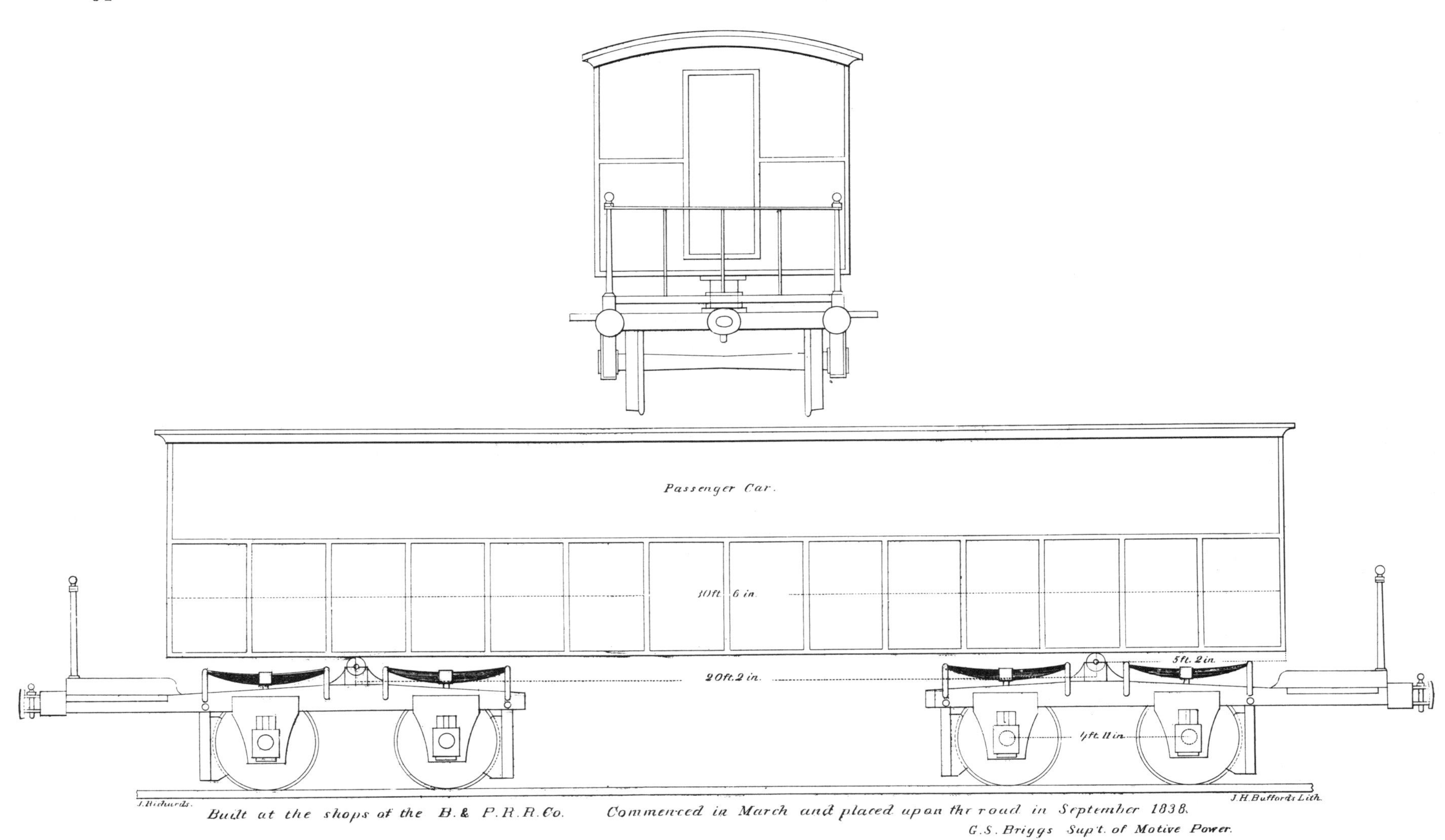

Figure 1.75 Boston and Providence eight-wheel coach, 1838. The window sash and window posts have been omitted from the drawing. (Winans vs. Eastern R.R., *1854*)

curve. Evidence that at least one other American car used this arrangement is shown in an 1845 English drawing (Figure 1.78). The plan was not satisfactory, for Griggs said that the platforms were later attached to the car body in the conventional manner. He also testified that the original 36-inch-diameter wheels were replaced by 33-inch wheels around 1845, and that the car roof had since been extended over the platforms. This car was still in service in 1853.

The car length over the couplers was 38 feet, 6 inches; the body measured 30 feet, 6 inches by 7 feet. The truck had a rather long wheelbase for the period: 4 feet, 11 inches. The seating was an estimated 45 to 50, the weight an estimated 6 to 7 tons. (See Figure 1.75.)

BALTIMORE AND OHIO RAILROAD, 48-PASSENGER CAR, 1835–1840

The drawings for this car are from a report by Michel Chevalier on internal communications in the United States, published in Paris in 1841. The design comes from the period 1835–1840. From what little information the text offers and from the measurements scaled from the drawing, it appears that this was one of the lightweight eight-wheel cars built for the old main line to Frederick. After a visit to the B & O yards in 1840, Von Gerstner said that fourteen slightly smaller cars were used on the main line, which could not accommodate the heavier Washington branch cars.[118] He reported that the main-line cars weighed 4.85 tons and seated 48. The body of the Chevalier car measured nearly 5 feet shorter than the Washington branch car in Figure 1.70.

It differed from the Washington branch cars in several minor particulars, although both followed the same general plan. The Chevalier car had a side door in addition to the end doors, and it had narrow running boards the full length of the car. The ceiling height was even more cramped; judging from the drawing, a 6-foot man could not stand erect. Two center sills were used in place of the continuous drawbar. The truck's wheelbase was only 36 inches. The counterweighted train brake was replaced by a simpler lever brake, which was actuated by stepping down on a foot treadle attached to the vertical staff (see detail 6 of Figure 1.70).

The overall length was 33 feet, 6 inches; the length of the body, 27 feet, 6 inches; width, 8 feet, 6 inches; truck centers, 22 feet. (See Figure 1.76.)

BALTIMORE AND OHIO RAILROAD, 72-PASSENGER CAR, 1840

Karl Von Ghega, another European engineer who visited the United States to inspect America's railway system, published his report in 1844.[119] The only car pictured in it was this large passenger car built in 1840. It represents a startling growth in

Figure 1.76 Baltimore and Ohio Railroad coach of about 1840, from a contemporary French text on American railroads. (M. Chevalier, Voies de Communication des Etats Unis, 1841, plate V)

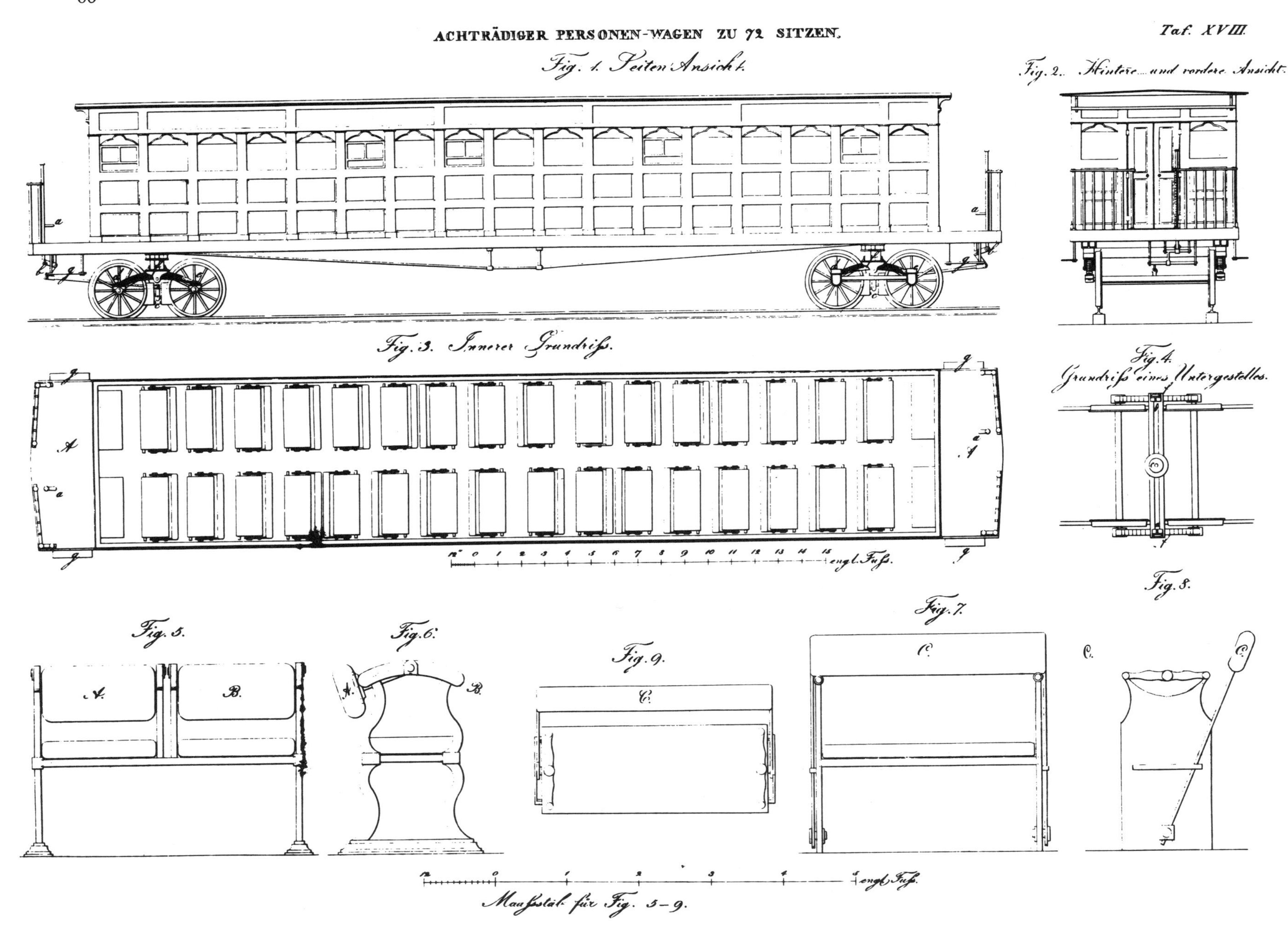

Figure 1.77 A Baltimore and Ohio Railroad coach of about 1840, from a contemporary German text. (Karl Ghega, Die Baltimore-Ohio eisenbahn, *Vienna, 1844, plate 18)*

capacity, if not in size, over the first eight-wheel cars. That 72 passengers could be crammed into a 36- by 7-foot body seems impossible, but it was achieved with extremely narrow seats and a 14-inch center aisle. The seats measured only 14 by 33 inches and were placed on approximately 2-foot centers. A new style of individual seat with an adjustable back is shown in details 5 and 6, although the mechanism for adjustment is not illustrated. Another indication of the car's meager dimensions is the inside ceiling clearance of 5 feet 11 inches.

Structurally the car shows some improvement over the earliest eight-wheel cars. The under-mounted outside frame truss has given way to ordinary side sills braced by an iron truss rod. The spring-framed truck has been retained, but in a simpler form—the secondary bolster and spring were eliminated. Although a metal bolster is used, it is made up of two light bars in place of a heavy single piece. The braking mechanism is still the foot-treadle brake, which is independent for each truck. Wood-block brake shoes can be seen between the wheels.

The builder of the car is not known; possibly it was constructed at the railroad's Mt. Clare shops. However, the Gothic arch above the windows and the general style of the body bear a strong resemblance to the Harlan and Hollingsworth drawing of 1839 (Figure 1.18).

The overall length was 42 feet; the width, 7 feet, 6 inches; truck centers, 30 feet, 9 inches; truck wheelbase, 39 inches; wheel diameter, 30 inches. (See Figure 1.77.)

AMERICAN PASSENGER CAR, 1845

The British trade journal *The Practical Mechanic and Engineer's Magazine* published this drawing in its January 1845 issue. Unfortunately, only the most general commentary accompanied the drawing; neither the builder nor the railroad was mentioned. The car was described simply as representative of "the best of these commodious vehicles," meaning American double-truck passenger cars with ladies' compartments. The only clue to its builder is the double leaf-spring draft gear, which follows the specifications of Charles Davenport's 1835 patent. However,

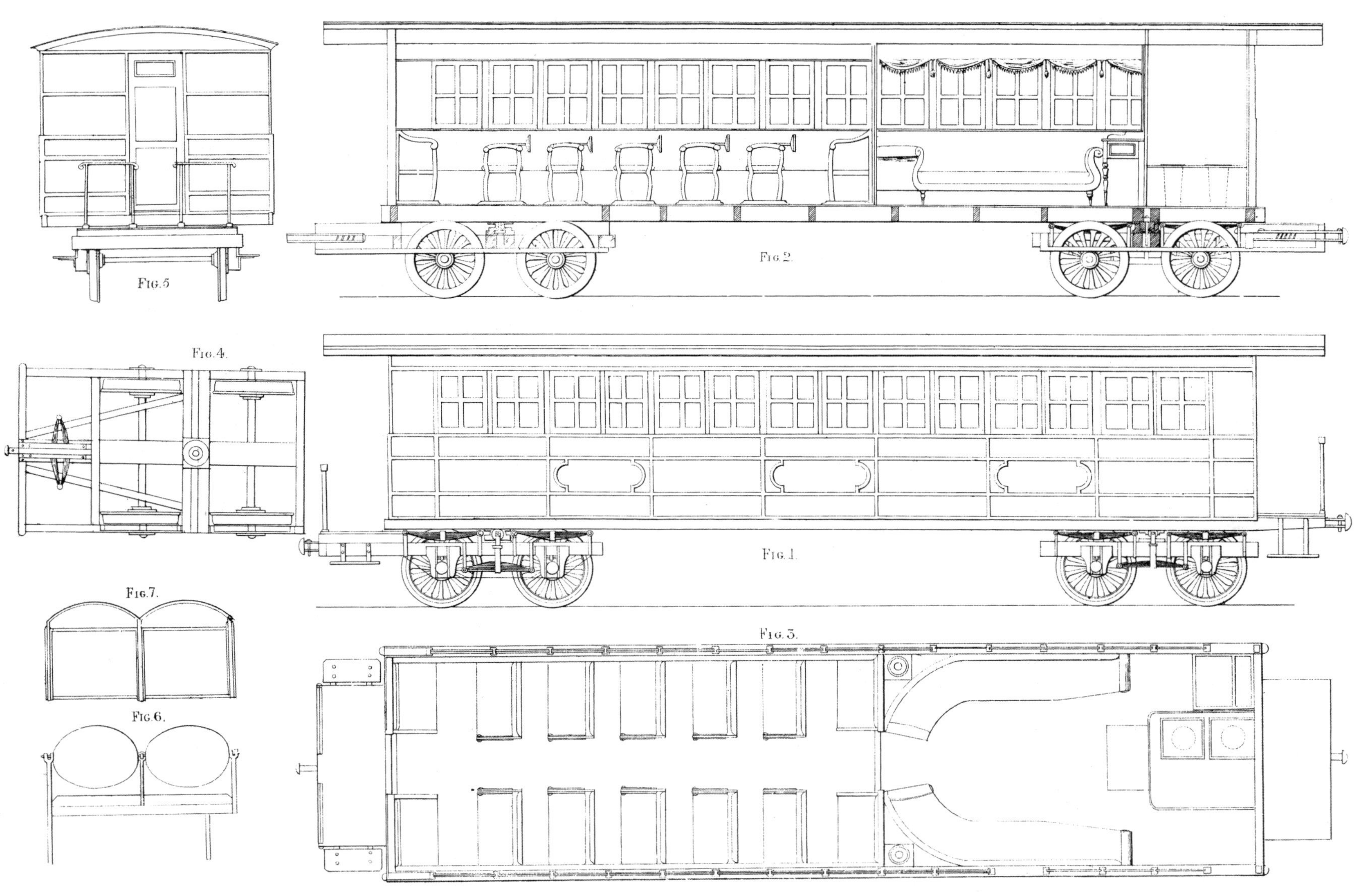

Figure 1.78 This very complete drawing of an American passenger car of 1845 was published in a British journal. Unfortunately, the builder and railroad were not identified. (Practical Mechanics Journal, *January 1845*)

there is no assurance that the car was made by Davenport and Bridges.

The car is longer and slightly wider than the Davenport and Bridges car in Figure 1.81. The ladies' compartment is more generously arranged, with elegant, curving Empire sofas, a dressing table and mirror, and a larger water closet. The truck under the passenger end of the car has extended side beams to support the end platform. At the other end of the car a more conventional truck is shown. The platform at the ladies' end of the car is carried by the sills of the car frame. A roller side bearing is cushioned by an underslung leaf spring. The axles are individually sprung—no equalizing was used. The wheels appear to be on Winans's drumstick plan.

Assuming that the wheels are 33 inches in diameter, the following dimensions can be scaled from the drawing: the length over end sills, 37 feet, 6 inches; body, 32 feet by 8 feet, 6 inches wide; truck wheelbase, 4 feet; overall height, 10 feet. The passenger compartment seated 28, the ladies' room about 10. (See Figure 1.78.)

EATON AND GILBERT CAR, 1845

This one-tenth-sized model of a second-class passenger car on the Württemberg State Railway is exhibited in the Deutches Museum, Munich, Germany. The model is expertly made, showing the construction of the original car in full detail. It is thought to be a contemporary model, possibly created by Eaton and Gilbert as an exhibition piece for prospective customers. A portion of one side is open, and the roof is removable to show the interior arrangement.

The initial rolling stock of the Württemberg Railway was purchased from American contractors. The Baldwin and Norris companies both supplied locomotives to the line in 1845, and Eaton and Gilbert was called upon to provide cars. No description of these cars has been found, but here is an earlier account prepared by the builder that agrees with the Württemberg car model:[120]

Cars to seat 52 persons are 33 feet long × 8½ wide. Roof and platform each projecting 2½ feet at each end. These [are] handsomely painted outside and well finished inside with mahogany seats. Backs to turn over. Trimmed with hair cloth, Morocco or braid cloth. The sides either cushioned with the same or finished with mahogany panels. On the floor a sutable oil cloth carpet. The inside of the roof either painted or lined with damask as may be desired. We have on each side 12 windows, the upper half of the sash stationary. The lower half to slide up with 4 steel springs on each, frames of mahogany well glazed. With a curtain to cover the windows inside. The spaces between the windows trimmed and sluffed inside and an imitation blind on the

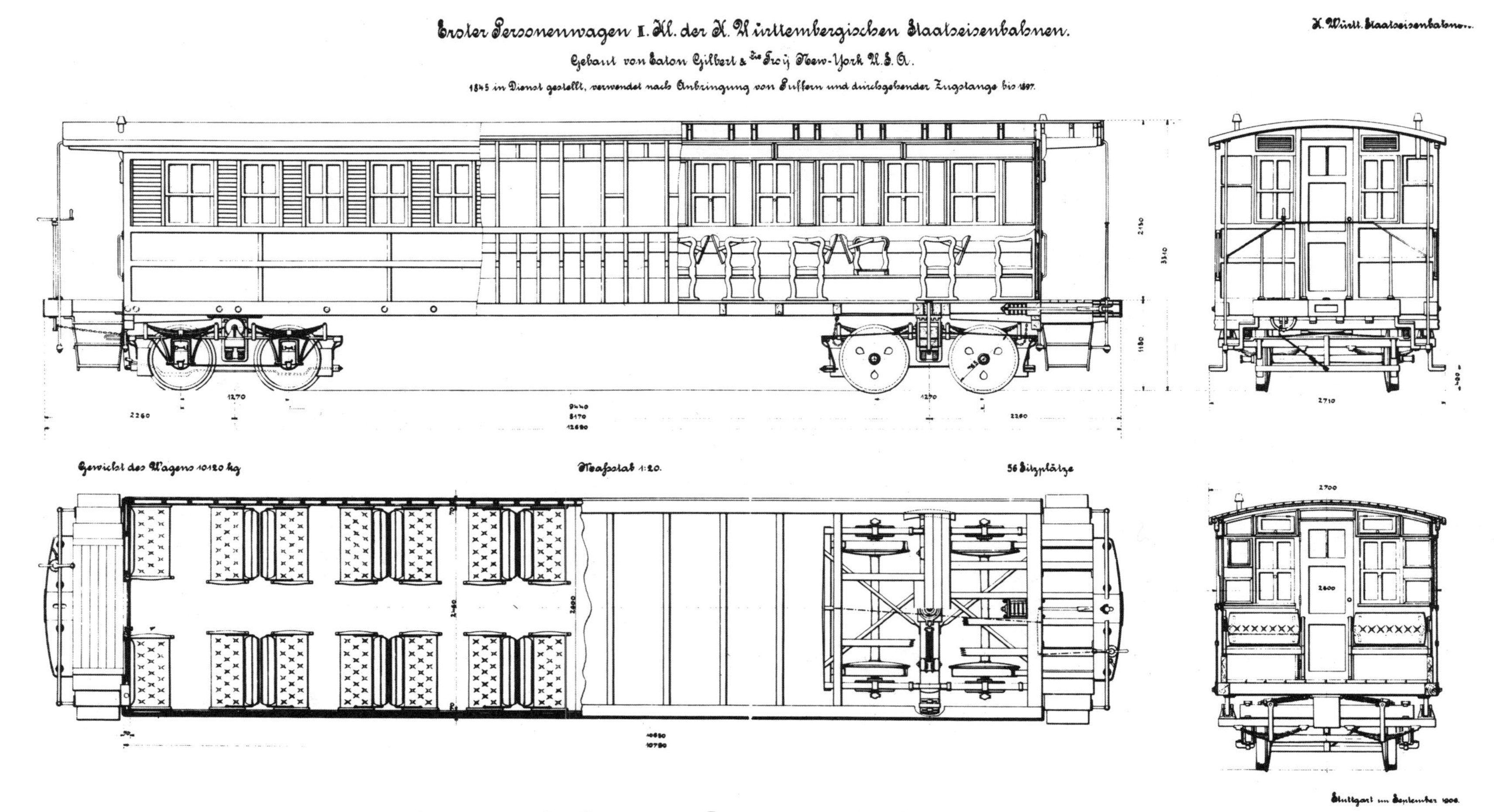

Figure 1.79 Eaton and Gilbert of Troy, New York, produced this eight-wheel passenger car for the Würtemburg State Railway in 1845. (Deutches Museum)

outside. A door in each end will furnish a passage through the center. We make a variety of trucks but the most approved is Williams Patent Iron Truck which are the only kind we can dispose of at present. There is a very good and permanent fixture for center bearing which is so hung on a stirrup as to give a side motion or rather to take away the effect of the side motion. A car of this kind we can furnish delivered on a boat at Troy for $1700. One of the same capacity for second class may be finished for from $1000 to 1200 depending entirely on the kind of finish. We have made but one of this kind and that a plain one for $1000. Those that will seat 52 are the most used. 4 wheel cars to carry half the number of passengers finished in corresponding styles may be built for $900 the first class and $600 the second class although no 4 wheel cars are called for any more here.

The iron truck and imitation outside shutter panels agree exactly with this description. From the model it can be seen that the side sills alone support the car; no center sills or iron drawbars were used. The draft gear was attached to end sills and appears to have been cushioned with a rubber spring. The body framing was equally simple, composed of light vertical members not unlike the balloon frame of a cheap wooden house. No trussing is evident. Small corner lamps with chimney vents through the roof lighted the interior and served as marker lights. The same general plan was used on the 1836 Camden and Amboy car in Figure 1.73.

The iron-frame trucks are of a particularly advanced design and refute the idea that all such assemblies before the Civil War were hard-riding, wooden-frame primitives. The neat, light, wrought-iron frame was cross-braced to hold it square. The swing bolster is after Davenport and Bridges's 1841 patent. A cast-iron pad with a side-bearing roller extends the bolster ends to the extreme width of the body in order to further stabilize the car's ride. The wheels are double-plate chilled cast iron and have cast raised letters reading "Bush and Lobdells patent March 17, 1838."

Eaton and Gilbert's design apparently suited its South German customers, for an elaborate drawing of nearly identical cars appeared in *Organ Fortschritte Eissenbahn-Wesens*, 1847, Volume 2. The description noted that the American plan of car was being followed by the Württemberg Railway for its new combination first- and second-class cars.

This drawing, prepared in 1906 at the time the model was placed in the Deutsches Museum, was made in metric scale. In English feet, the general dimensions are overall length, 41 feet; width, 9 feet; height, 10 feet, 9 inches. The truck wheelbase is 4 feet. Seating was provided for 56 passengers. (See Figure 1.79.)

DAVENPORT AND BRIDGES CAR, 1845

Davenport and Bridges of Cambridgeport, Massachusetts, was among the first railroad car builders in New England and became one of the largest car makers in the country. Before closing in the mid-1850s, it had supplied cars to railroads throughout New England, New York, and Pennsylvania. In the 1840s the firm mounted an ambitious advertising campaign with regular half- and full-page notices in the *American Railroad Journal*. One engraving that showed a working drawing of a typical car, with a longitudinal cross section and plan view, is among the best illustrations available for a car of the period (Figure 1.80). This view is complemented by an exterior perspective drawing (Figure 1.81).[121] A description of some Davenport and Bridges cars

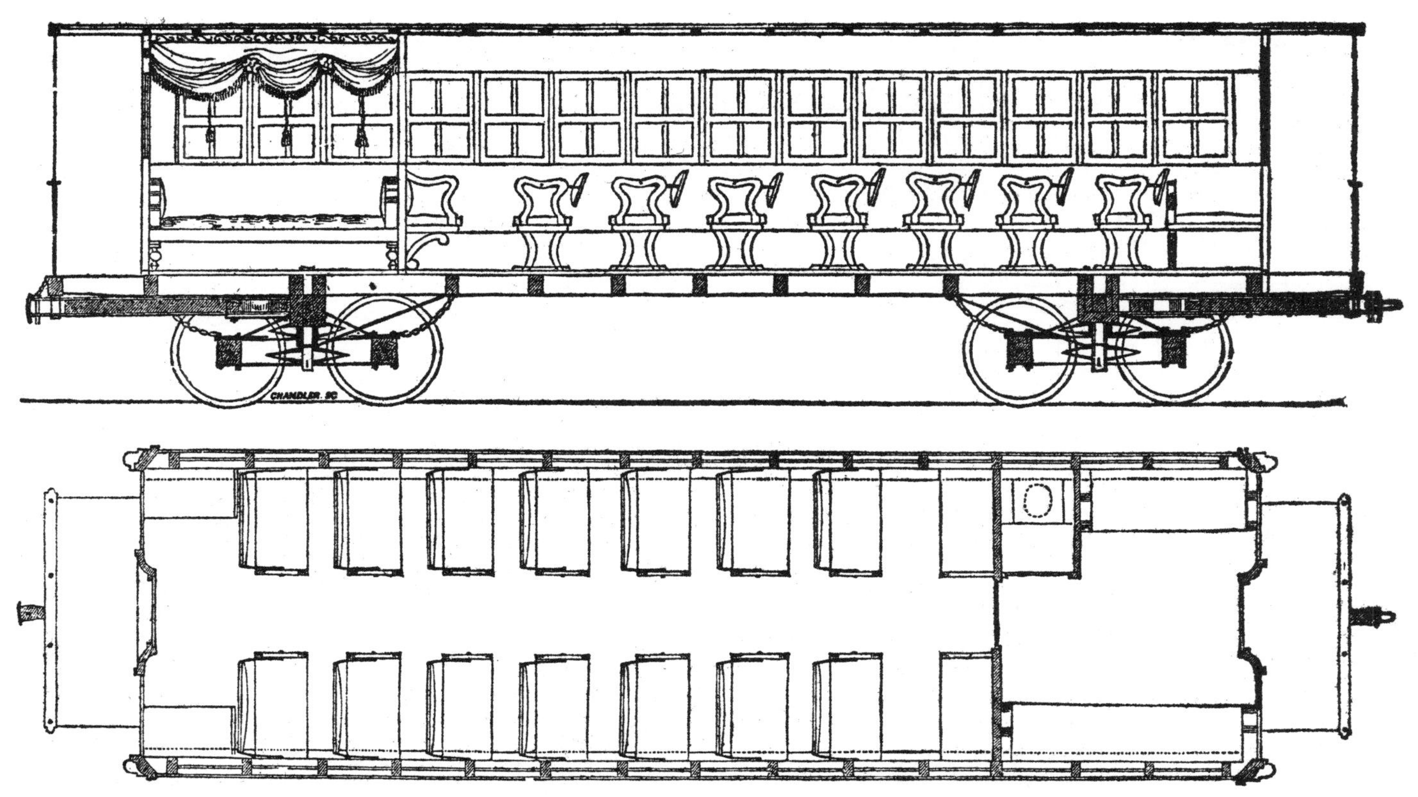

Figure 1.80 Davenport and Bridges was one of the largest builders of passenger cars in the United States in 1845 when this advertising drawing appeared. (American Railroad Journal, *November 27, 1845*)

DAVENPORT & BRIDGES CONTINUE TO MANUFACTURE TO ORDER, AT THEIR WORKS, IN CAMBRIDGEPORT, MASS. Passenger and Freight Cars of every description, and of the most improved pattern. They also furnish Snow Ploughs and Chilled Wheels of any pattern and size. Forged Axles, Springs, Boxes and Bolts for Cars at the lowest prices. All orders punctually executed and forwarded to any part of the country Our Works are within fifteen minutes ride from State street, Boston—coaches pass every fifteen minues. 1y1

Figure 1.81 A perspective illustration of a Davenport and Bridges car of 1845. (American Railroad Journal, *August 7, 1845*)

made for the Auburn and Rochester Railroad a few years earlier agrees closely with the illustrations:[122]

There are six cars, designed to form two trains. The cars are each 28 feet long and 8 feet wide. The seats are well stuffed and admirably arranged—with arms for each chair, and changeable backs that will allow the passenger to change "front to rear" by a manœuvre unknown in military tactics. The size of the cars forms a pleasant room, handsomely painted, with floor matting, with windows secured from jarring, and with curtains to shield from the blazing sun. We should have said *rooms*; for in four out of six cars, (the other two being designed only for way passengers,) there is a ladies' apartment, with luxurious sofas for seats, and in recesses may be found a washstand and other conveniences. The arrangement of the apartment for ladies, we consider the greatest improvement; and it will remedy some serious objections that have hitherto existed against railroad travelling on the part of families, especially where any of the members are in delicate health. The ladies can now have their choice either of a sofa in their own apartment, or a seat in the main saloon of the cars, as their health and inclination may require.

These cars are so hung on springs, and are of such large size, that they are freed from most of the jar, and especially from the swinging motion so disagreeable to most railroads.

The lamp of each car is so placed as to light inside and out; and

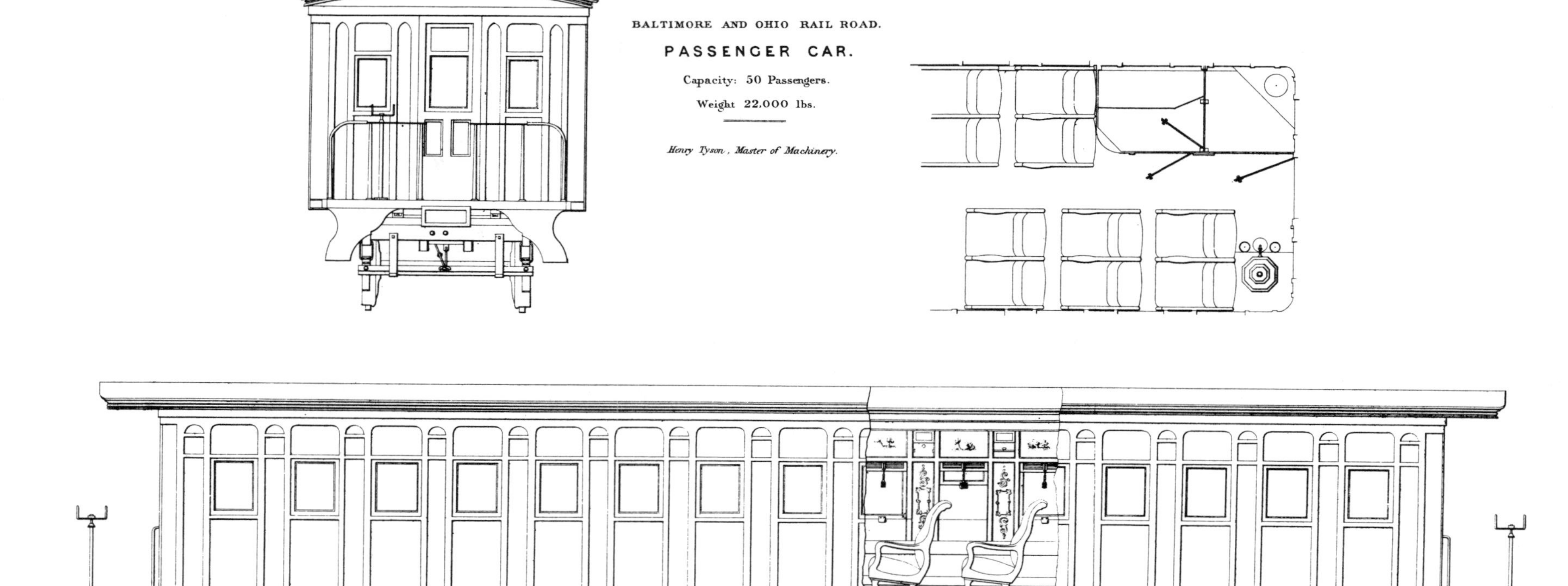

Figure 1.82 An arched-roof, eight-wheel passenger car built for the B & O to the designs of its master mechanic, Henry Tyson, in 1856. (*Douglas Galton,* Report on the Railways of the United States, *London, 1857–1858*)

last though not least, the breakers are so arranged as to be applied readily and with great power—thus guarding against the danger of collisions, etc.

We almost forgot to mention that these beautiful cars were made by Davenport and Bridges, of Cambridge, Massachusetts, and cost at low prices about $1700 each—or $10,000 for the six.

Although this account stresses the luxury of the cars, it can be seen from the drawing that the cramped ceilings and low-backed seats of ten years earlier still prevailed. The ladies' compartment, a water closet, roof ventilators, and a more generous supply of windows (with movable sash) were new concessions to passenger comfort. Extending the roof to cover the platforms was another improvement over the earlier eight-wheel cars.

Continuous sills supported the car and platforms. Short timbers attached beneath the main floor sills were used to support the platforms in later years. The drawbars, which were carried back to the body bolsters, were fabricated from wrought-iron plates riveted with a wood filler piece. The trucks were inside-bearing wrought iron made after a patent issued to Charles Davenport on August 10, 1844 (No. 3697). The trucks were of all-iron construction except for the bolster. Outside rocker side bearings stabilized the car body. Davenport did not claim credit for the iron side frames or the side truss frames with diagonal braces (the predecessor of arch-bar trucks); he conceded that these improvements had been introduced before his patent. He did maintain that his truck was lighter by 1,200 pounds and about $100 cheaper to make than the existing styles of car trucks.

By assuming a wheel diameter of 33 inches, the following dimensions can be scaled from the drawings: overall length, 34 feet; height, 10 feet; truck wheel cast, 4 feet. Seating was for 43. (See Figures 1.80 and 1.81.)

BALTIMORE AND OHIO RAILROAD, 50-PASSENGER CAR, 1856

Improvements in the B & O passenger rolling stock since the 1830s and 1840s are apparent from the new style of arched-roof car in these drawings. The car was designed in 1856 by Henry Tyson (1820–1877), master of machinery for the B & O from 1856 to 1859. The car shows no extraordinary increase in size, nor was the general construction plan altered. However, more thought was given to passenger comfort: the ceiling was raised to 7 feet, and seat spacing was more generous. Deep cushions, high padded backs, and footrests were added. A water closet, dressing room, and water cooler are shown in the plan view detail of Figure 1.82. Tiny ventilators were placed high on the car body in panels between the windows (best seen in the section detail of the side elevation), so that air could be admitted near the ceiling without opening the regular windows. Adjustable wooden blinds were used for sunshades. No information on lighting or heating apparatus is available.

The first group of Tyson cars were completed at the Mt. Clare shops late in 1856. The *Wheeling Times* wrote: "They are beautiful, substantial and safe. There is an air of comfort observable in the interior of the cars which is only surpassed by the extreme beauty and masterful skill displayed in the outside execution as to painting, carving, etc. The interiors of the cars are supplied with beautiful landscapes and light paintings in a profuse manner."[123] A few indications of the interior decoration can be seen in the drawing.

The mechanical drawing of Tyson's car is from Douglas Galton's report on American railroads to the British Parliament (1857). Truss rods were not used, and it is probable that there was no body-panel wooden truss. However, the 40-foot body was about as long as floor sills of reasonable size could support without the assistance of one or both truss reinforcement methods.

The exterior of the car was fashioned in batten-board trim, forming a pleasing series of panels. The platforms were supported by timbers fastened beneath the floor sills. This remained the typical method of construction until the end of the wooden car era.

The trucks are set well back from the platforms to shorten the total wheelbase; the truck wheelbase was only 51 inches—rather short for the period. Most other lines used the so-called "square" trucks, in which the wheelbase equaled the track gauge. The B & O's adherence to close-coupled wheels may be due to the sharp curves on the tracks of its mountain division.

The suspension of the car was sufficiently elaborate to bring joy to any spring manufacturer. Each truck had eight leaf springs (Figure 7.2). The journal boxes were individually sprung; the swing bolster was cushioned by two heavy, double elliptic springs; a pair of half elliptic springs acted to center the swing bolster. Yet with all the complexity of the suspension, no equalizing was attempted.

The iron bolster and live-spring side frame were at last abandoned; the B & O had turned to the conventional wooden-beam truck. A rather light pattern of cast-iron pedestal was used, which can be better understood from the identical trucks pictured under the camp or office car in Figure 4.95. The plate wheels shown in the photograph here appear to be of the corrugated pattern developed by Asa Whitney's company.

F. A. Stevens's brake gear was used, so that all eight wheels could be braked together from either brake staff. A single rod, visible just below the frame of the car, connected the levers on the trucks. In this car wooden blocks had given way to wrought-iron brake shoes.

The overall length was 45 feet, 6 inches; the width over body, 8 feet, 9 inches; height, 10 feet, 7 inches; truck centers, 26 feet; truck wheelbase, 4 feet, 3 inches. (See Figure 1.82.)

NEW YORK AND ERIE RAILWAY, BROAD-GAUGE CAR, 1856

This car was not typical of the period, but it is representative of the huge vehicles running on the 6-foot-gauge lines after 1855. The drawings here are the earliest concrete examples of truss framing—which, next to the eight-wheel car itself, was the single most important contribution of American designers to the art of car building.[124]

At the urging of Master Car Builder Calvin A. Smith, construction was begun on six giant passenger cars in mid-1855 at the Piermont, New York, repair shops.[125] Completion of the first lot of cars, which represented a radical departure from the short, wide cars formerly built, was reported the following year, together with a wonderfully complete description:[126]

Splendid Passenger Cars.

THE completion of a portion of the number of new passenger cars built at the Piermont shops for the Erie road, renders a notice of them appropriate and interesting at this time.

These cars are not equalled in size, accommodation, or elegance of finish, by any other public railroad conveyances in the world. Whatever may be the opinion as to the policy of running such heavy cars, universal admiration must be yielded to their comfort and luxurious finish.

The Master Car Builder at the Piermont Shops, urged the construction of some large sized passenger cars, more than a year since, and suggested the general design of their finish and furniture. MR. MCCALLUM, the superintendent, adopted the suggestion and ordered the construction of the cars. He dictated the plan of the frame, employing a bridge truss between the windows and sills, and making the upper portion of the frame independent of the lower.

The body of the car is 60 feet 3 inches long outside, and 10 feet 9 inches wide. The posts are 7 feet high and the height of the cars at the center is about 11 inches greater than of any others now on the road. There are 20 windows on each side of the car, with a single plate of double thick French plate glass, 17 by 21 inches in each. There are two windows also in each end. The car seats 74 passengers, including two in the saloon.

The outside finish of the car is singular and neat. There are three plain vertical panels under each window, and one long panel reaching from the sills to the cap, between every two windows.

The seats, throughout, are C. P. Bailey's patent reclining, day or night seats. They are very large and easy and splendidly upholstered, the covering being a rich velvet plush, costing five dollars a yard. Over one hundred yards of plush were used. This was furnished from Doremus & Nixon's, of this city.

The cars are ventilated on Foote & Hayes' patent plan, and an improvement has been made in working the pump, used in the ventilator, by a friction wheel on the axle. This will obviate the main difficulty before had with this ventilator, apart from which it is the most comfortable means of purifying, cooling, and equalizing the temperature of the air in passenger cars.

The side "pedestals," as they are called, in which the ventilating water fountains play are elegantly finished off, and a large plate glass in each gives a view of the cooling spray when the car is running. Through these "pedestals" the air is conducted from the air cap, on the roof, to the registers in the floor.

There are four pivot single seats, on each corner of each pedestal, moulded and upholstered in a most luxurious style. They are decidedly a new feature in passenger cars, although a very few are already in use in one or two other new cars on the same road.

The head-lining is an expensive and rich pattern, although not quite in good taste with the rest. In some of the other cars, a new pattern will, perhaps, be adopted. The gilt cornice around the inside is 4 inches deep and very rich.

The trucks and running gear, built by HARVEY RICE, Esq., are well got up, and although the same kind are already in use in other cars on the road, they are a novelty off of the road. The axles are spread six feet apart centers, and have both outside and inside journals. The inside journals have grooved collars on each end, and are fitted with

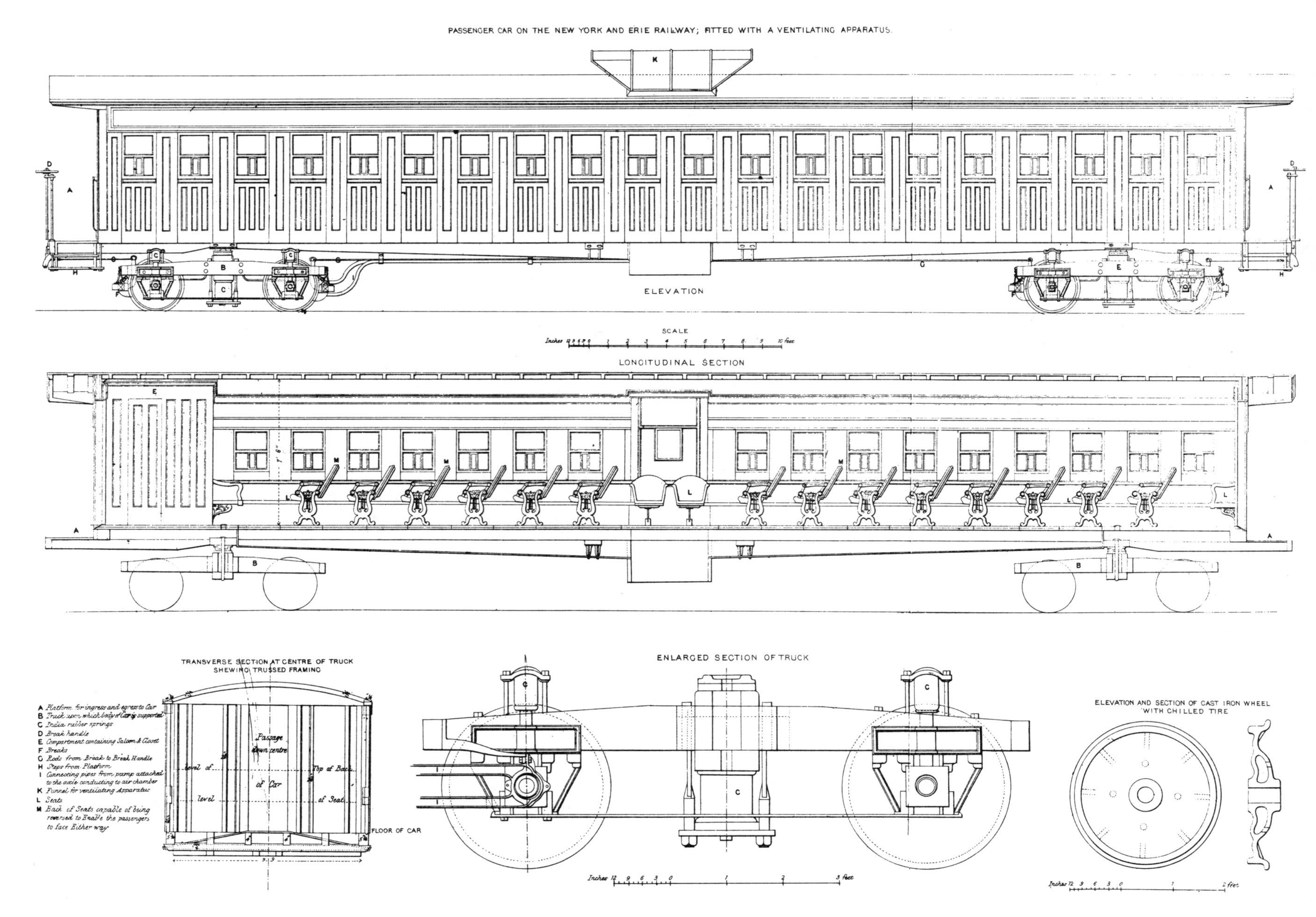

Figure 1.83 A broad-gauge passenger car produced about 1856 for the New York and Erie. (*Galton's* Report on U.S. Railways)

Harvey Rice's improved box. The outside journals have the Lightner box. The axles are from the Brunswick works in England, and the wheels, 33 inches in diameter, are Bush & Lobdell's pattern. The worst fault with the trucks is in using Rubber Springs, which, if they stand the great weight at all, will become rigid or freeze.

The cars have Hopkins' patent coupling, an improvement beyond value as respects safety, and measurably cheaper and far more convenient in use than the old coupling.

The entire weight of these cars is not far from eighteen tons, empty. It will cost more to draw passengers in these cars than in the light cars. The very sacrifice made in this respect, for the comfort and pleasure of its patrons, should commend the Erie road to the traveling public.

Some of the cars will very soon be running regularly on the night express trains.

The top cord of the truss frame described above was 3- by 4½-inch timber, slightly arched toward the center of the car. The bottom cord was the outside frame sill, composed of a 5¾- by 7½-inch timber. The wooden vertical posts, about 2¼ inches square, were set on 18-inch centers and mortised into the top and bottom cords. The truss was stiffened and drawn together by ⅜-inch iron rods placed diagonally between the posts. It was necessary, of course, to run a light parallel stringer over the arched top cord for a level window setting. Additional support to the floor sills was provided by truss rods. The upper car framing was tied together by vertical rods placed directly against the window posts. In all, it was a remarkably sophisticated framing plan, and one that continued into the 1870s.

The trucks were not equal in design competence to the general plan of the car. The suspension seems particularly defective: individual rubber springs were used for each bearing, and the stationary bolsters were carried on rubber springs nearly a foot in diameter. Inside and outside bearings were used as a safety measure in the event of a broken axle. This was a far more expensive plan than safety beams, and it entailed the added disadvantage of wear and friction should one or more of the journals be misaligned.

Dimensions taken from the drawing: overall length, 65 feet; width, 10 feet, 9 inches; height, 12 feet, 8 inches. The weight, empty, was 18 tons; the seating was for 72 if the saloon seats were included. (See Figures 1.83 to 1.85.)

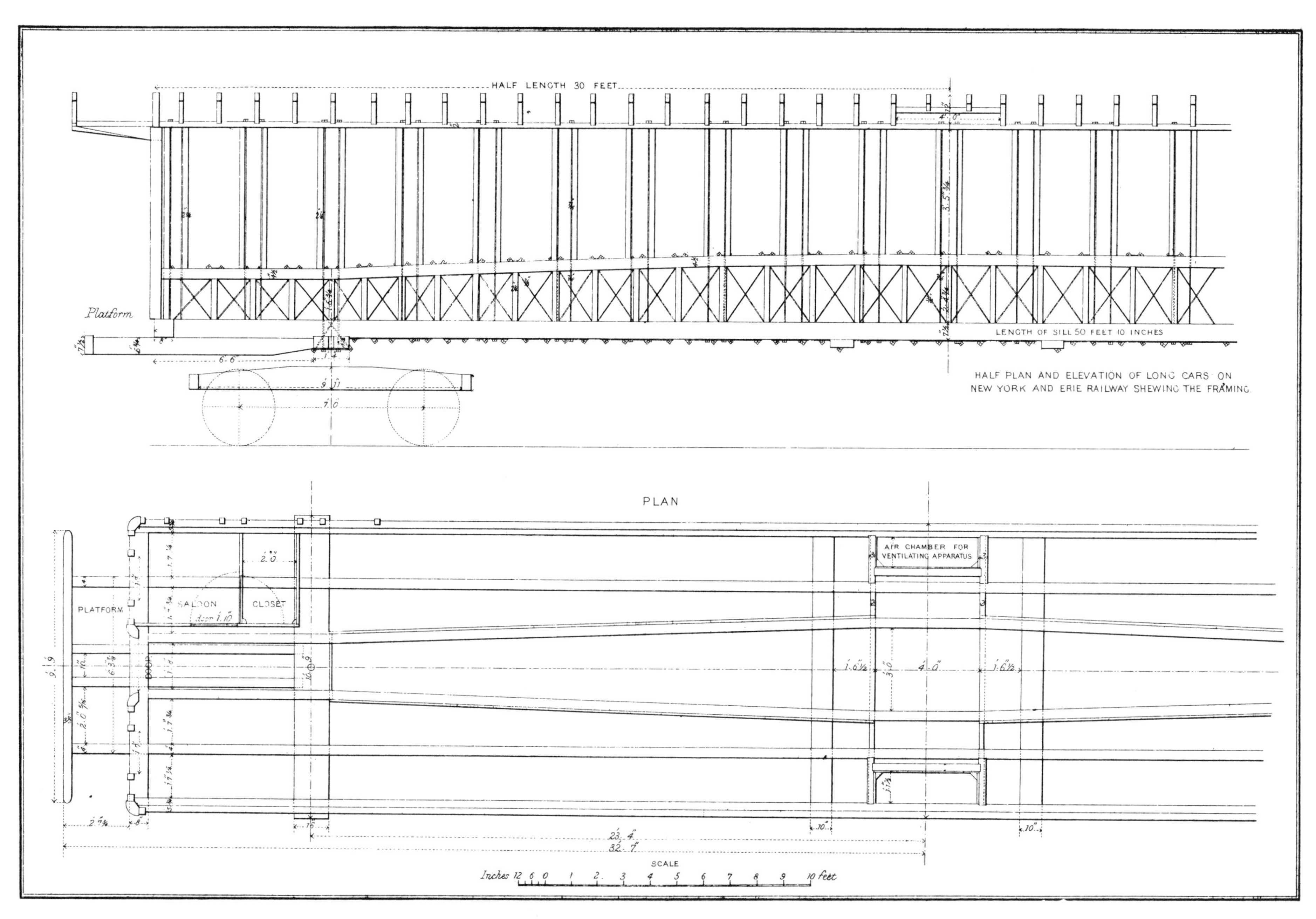

Figure 1.84 Framing details of a New York and Erie car. This illustration is the first concrete record of a side-frame truss, although the idea may have been used earlier. (Galton's Report on U.S. Railways)

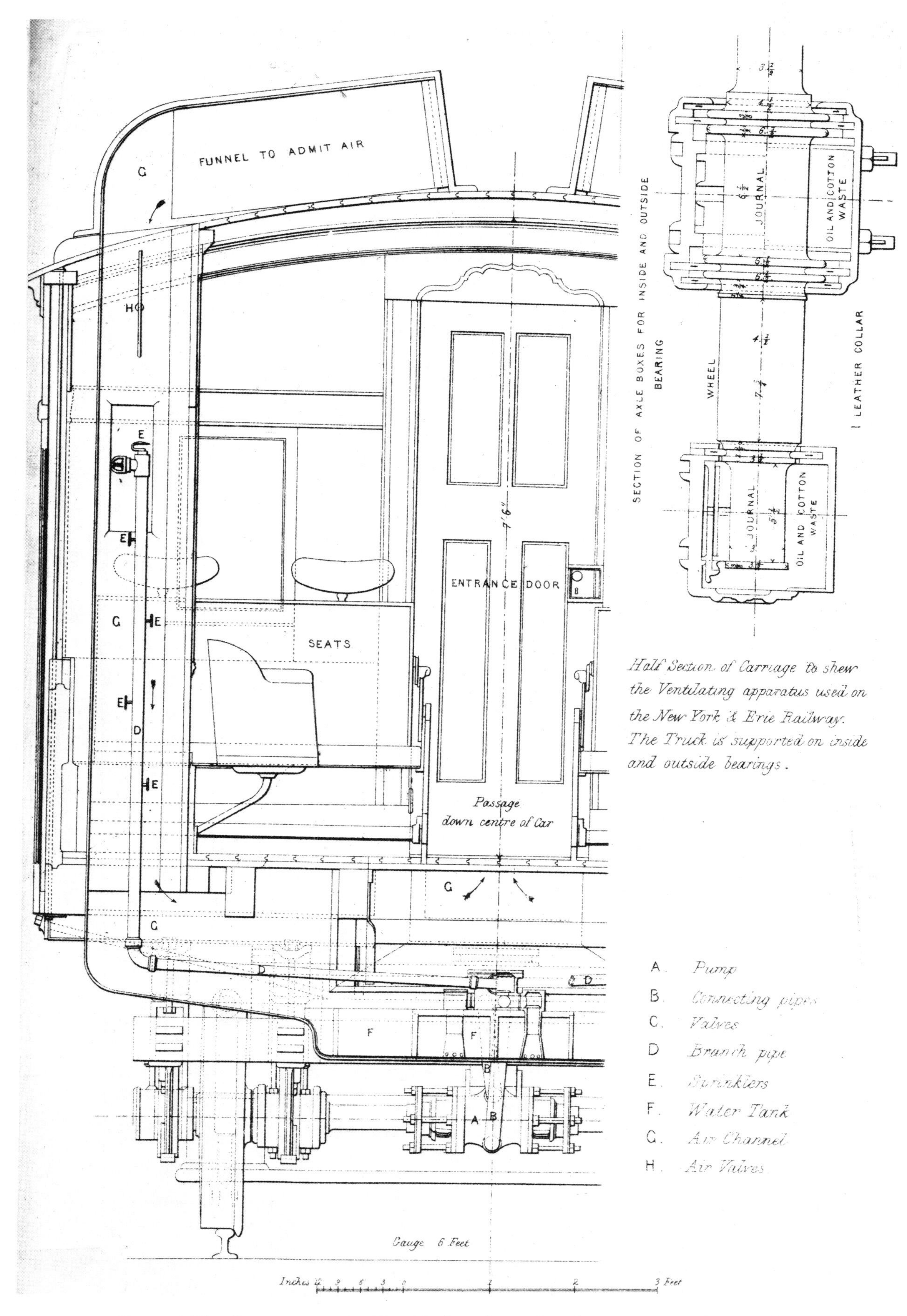

Figure 1.85 Cross section of a New York and Erie broad-gauge car of about 1856. It shows details of the Foote ventilating system. (*Galton's* Report on U.S. Railways)

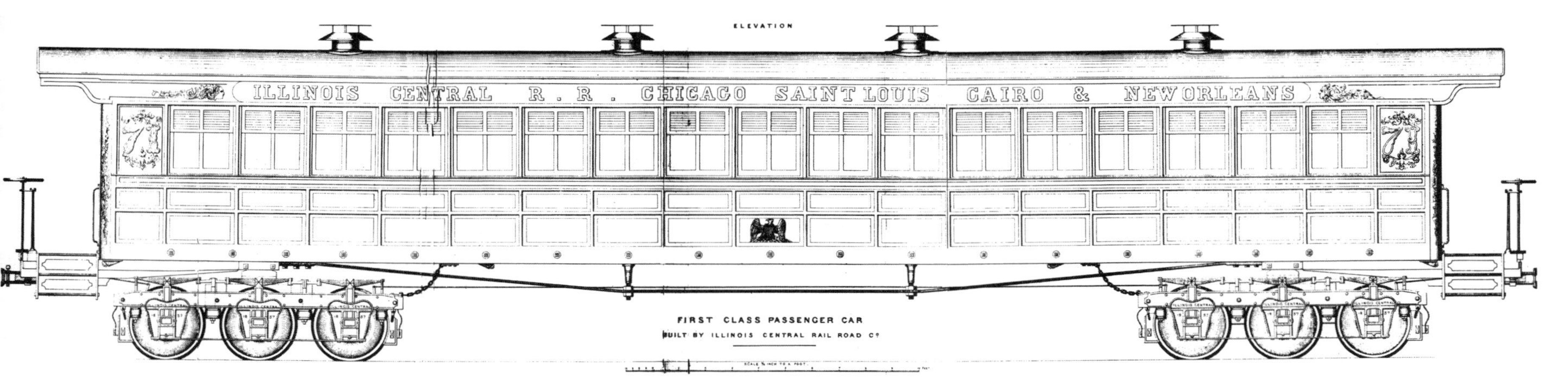

Figure 1.86 An Illinois Central twelve-wheel car of about 1857, built in the railroad shops. (*Galton's* Report on U.S. Railways)

ILLINOIS CENTRAL RAILROAD, TWELVE-WHEEL CAR, 1857

This heavy, first-class car represents the finest passenger equipment of the period. It was a large car for the time, with generous overall proportions. In contrast to the 6-foot ceilings of American cars a decade earlier, it provided 7 feet, 3 inches of headroom. The big windows and wide body are signs of generous seat spacing. However, the arched roof, paneled sides, and fancy exterior ornamentation clearly mark it as a car of the 1850s. It was probably built at the Illinois Central's Chicago car shops, which were started in 1852 by the American Car Company and sold to the Illinois Central four years later for a car repair and construction depot.

The six-wheel trucks are the most distinctive feature of the car, but as Chapter 1 mentioned, some twelve- and even a few sixteen-wheel passenger cars had been operating on American railroads ten to fifteen years earlier. However, these drawings for the Illinois Central's Number 71 are the earliest construction plans available for such a car. They are in fact the oldest authenticated illustrations of a twelve-wheel car that could be discovered. The drawings are from Douglas Galton's 1857 report on American railroads.

The body was so long (about 50 feet) that truss rods were required. It is probable that a wooden truss was built into the car sides as well. Another feature that looked forward to later cars was the letter board. Sheet-iron roof ventilators and wooden blinds are clearly shown in the drawing.

The trucks were wooden beams with cast-iron pedestals and light wrought-iron tie bars. The axles were on 42-inch centers. Each axle was sprung by half elliptic springs, and the spring hangers were cushioned by india-rubber cylindrical snubbers mounted under side-frame beams. The most significant feature of the truck's suspension was the equalizing levers connecting the springs—early examples of this important arrangement. Cast-iron plate wheels are shown. The brake gear was on Stevens's plan, and wooden-block brake shoes were used.

The overall length was 57 feet; the width, 9 feet, 6 inches; height over the roof, 10 feet, 6 inches. Seating was for about 65. (See Figures 1.86 and 1.87.)

CHARLESTON AND SAVANNAH RAILROAD, ARCHED-ROOF CAR, 1860

This drawing is one of some twenty original passenger car drawings from the Harlan and Hollingsworth Car Company of Wilmington, Delaware. (Several other drawings from the collection are reproduced later in the book.) All date from the 1860s. They are exquisitely colored in pale wash tones of yellow, grey, blue, and red. It is assumed that they were prepared for perusal by prospective customers. They were exhibited at the Columbian Exhibition in 1893 and afterwards came into the possession of the Baltimore and Ohio Railroad. In 1942 they were transferred to the United States National Museum in Washington, D.C.

The example shown here appears to be a standard-gauge car that was large for its time. The high ceiling would clear about 8 feet, offering generous headroom for an arched-roof car. The interior width was nearly 9 feet, again roomy for the period. The body was short enough so that it did not require a side-frame truss or truss rods, although a number of iron tie rods were used to strengthen the cross framing. Small, square ventilators with louvers hinged at the centers are visible in the letter board. The narrow vertical sheeting in place of the customary side body panels or batten boards was another advanced feature of the car. The trucks were ordinary wooden-beam style, with india-rubber springs over each journal and full elliptic springs for the bolsters.

The overall length was 49 feet; the height, 11 feet, 9 inches; width, 9 feet, 8 inches; truck centers, 31 feet, 6 inches; truck wheelbase, 5 feet; seating capacity, 52. (See Figure 1.88.)

LONG ISLAND RAILROAD, ARCHED-ROOF CAR, 1861

Although it was built eight months after the car in Figure 1.88, this product of Harlan and Hollingsworth was in many respects more antique. The paneled side, low ceiling, and small windows with fretwork details at the tops of the window openings are typical of earlier practice. The end lamps and roof chimneys are reminiscent of the Camden and Amboy car of 1836 (Figure 1.73).

The use of lighter sills apparently called for reinforcing truss rods. The trucks were sprung with india-rubber and elliptic springs as before, but a fully developed equalizer is shown.

The overall length was 49 feet; the height, 10 feet, 10 inches; width, 9 feet, 5 inches; truck centers, 31 feet, 3 inches; truck wheelbase, 4 feet, 8 inches. Seating capacity was 57, counting the saloon sofa. (See Figure 1.89.)

PENNSYLVANIA RAILROAD, CLERESTORY ROOF CAR, 1862

The Pennsylvania Railroad began a program of standard passenger car design in 1862 with its class PA. This style of car was

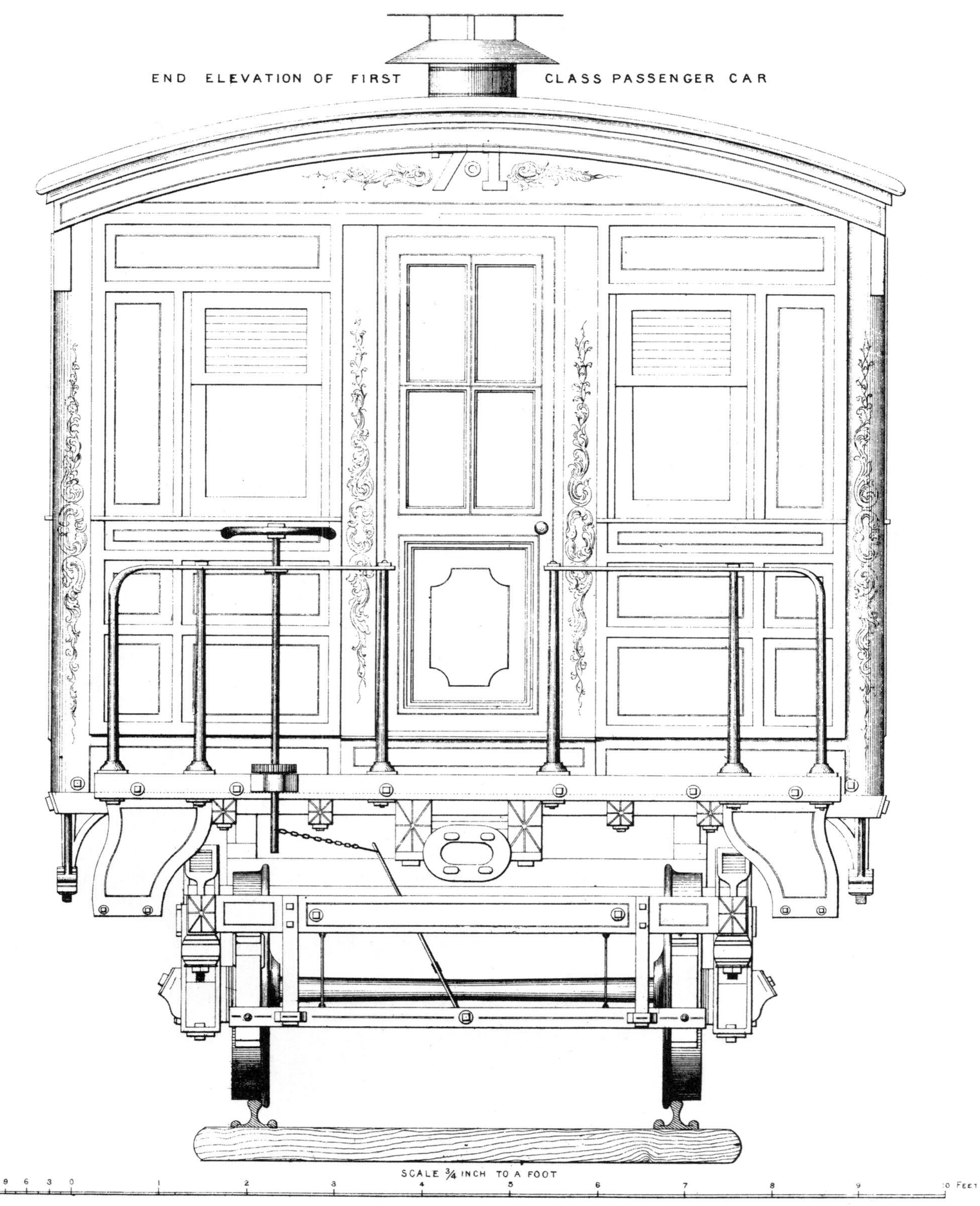

Figure 1.87 End elevation of the Illinois Central passenger car Number 71. (*Galton's* Report on U.S. Railways)

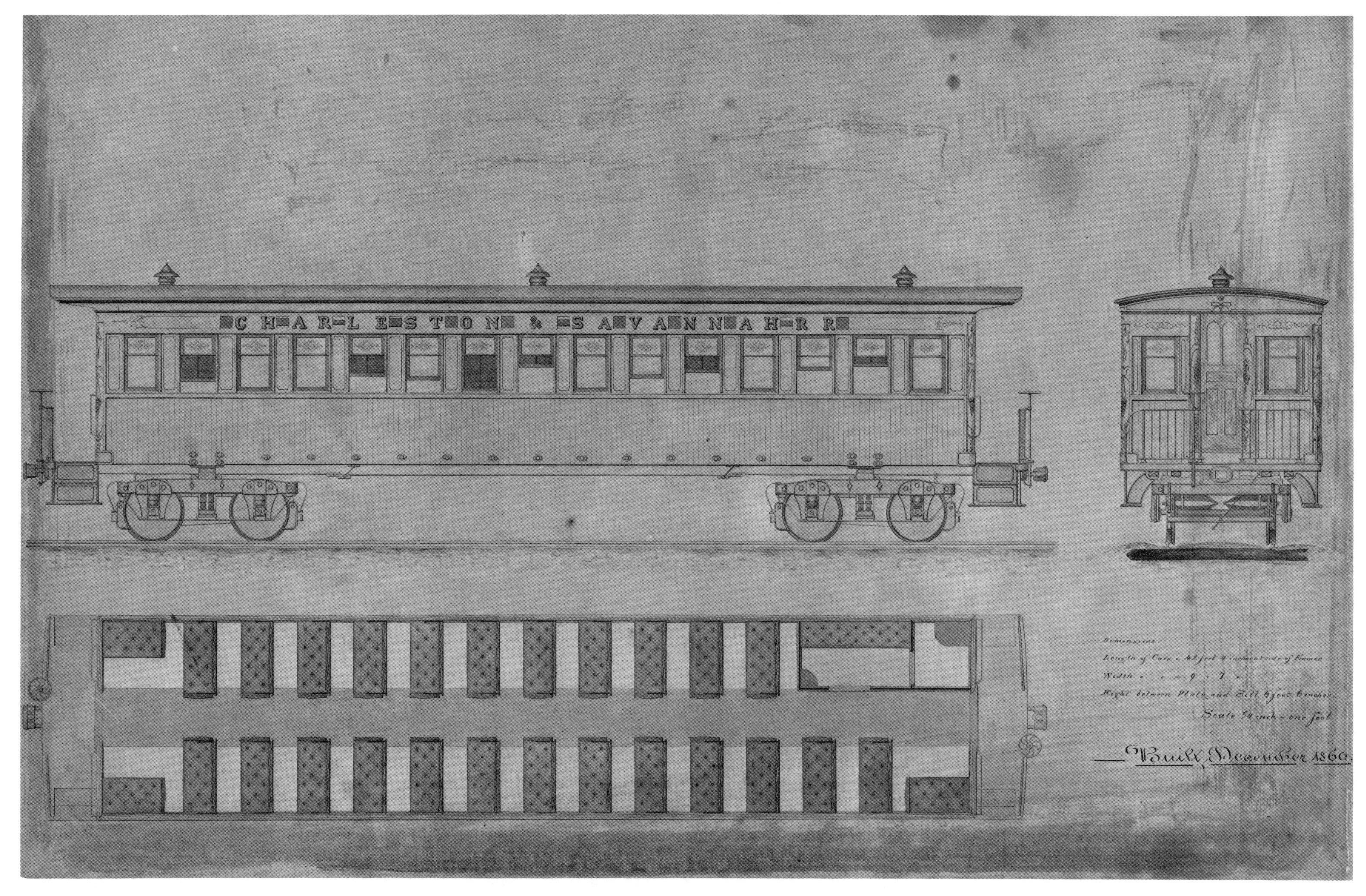

Figure 1.88 An arched-roof car, completed in December 1860 by Harlan and Hollingsworth. The car measures 49 feet over the couplers.

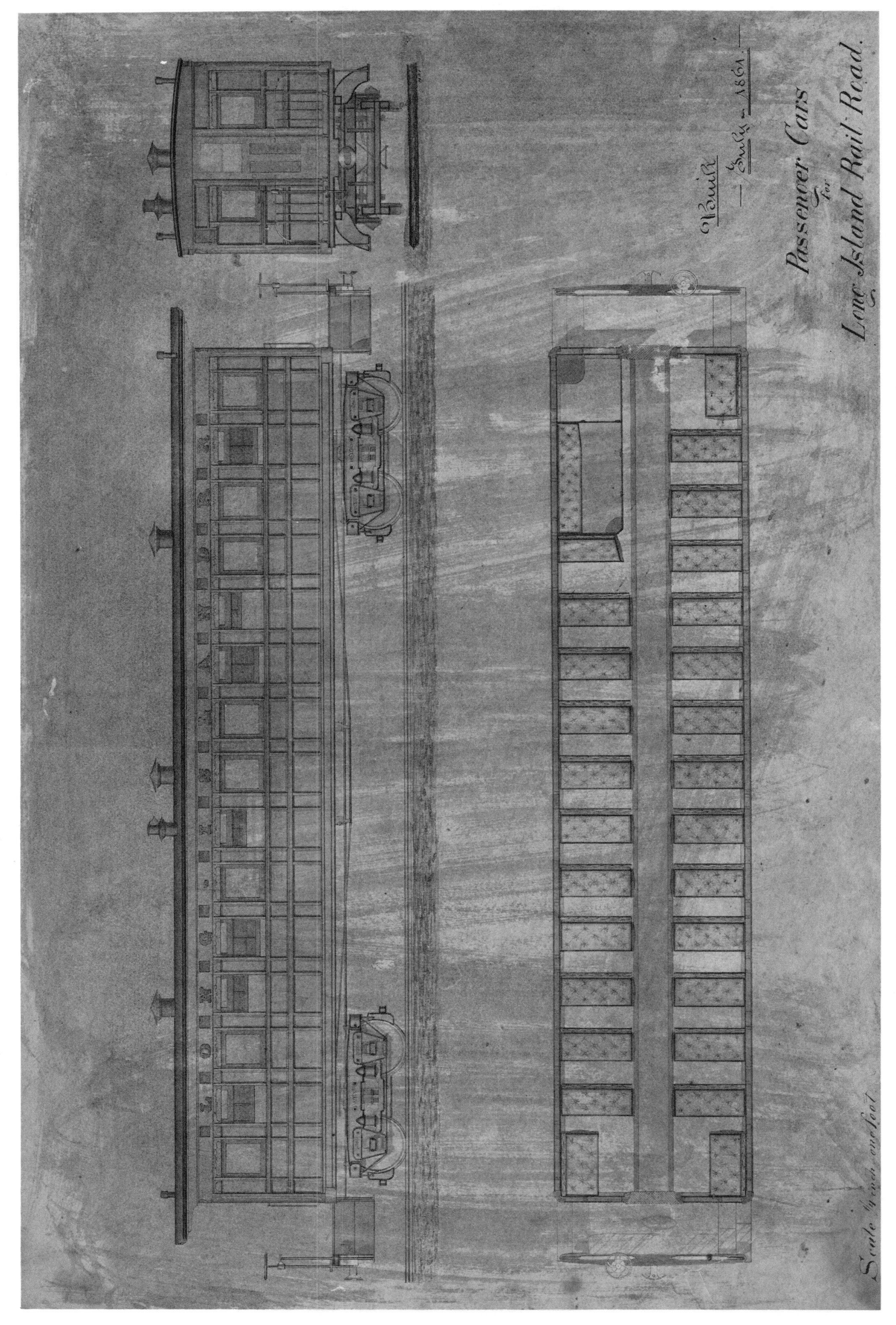

Figure 1.89 An arched-roof car completed in 1861 by Harlan and Hollingsworth.

SEATING CAPACITY, 52 PERSONS
WEIGHT, 39,300 LBS.

PASSENGER CAR, CLASS PA.

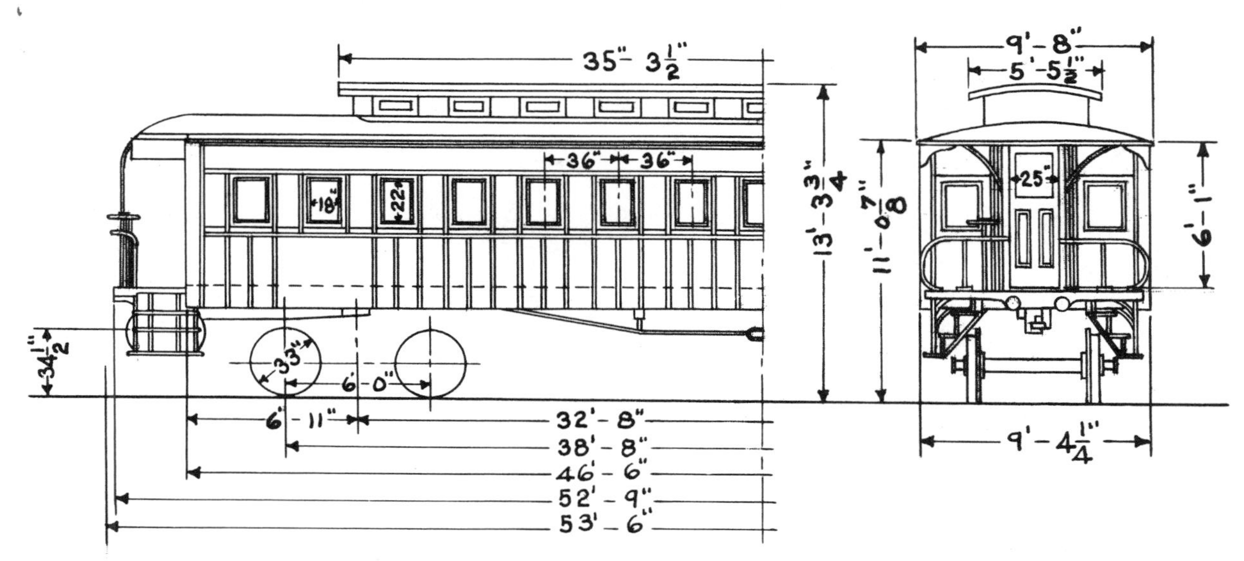

TRACED BY DOROTHY BRIGGS

Figure 1.90 A Pennsylvania Railroad class PA coach. The design was introduced in 1862.

Figure 1.91 A Pennsylvania Railroad class PA coach. The Monitor roof car design was adopted in 1862.

to be used system-wide and built to an exact common standard, whether it was constructed in the company shops or in a contract shop. The exact number produced has not been determined, but photographic evidence indicates that there were a great many PAs and that they were in service on nearly every portion of the system.

The design incorporated the most advanced styling elements of the period, including the clerestory roof. The short-lived Monitor plan was used, however. The body had a wide letter board, small, square windows, and batten-board sides. The platform canopies or hoods were made as separate structures and were not integrated with the roof—a peculiarity of construction common on the Pennsylvania until the end of wooden car construction. (See Figures 1.90 and 1.91.)

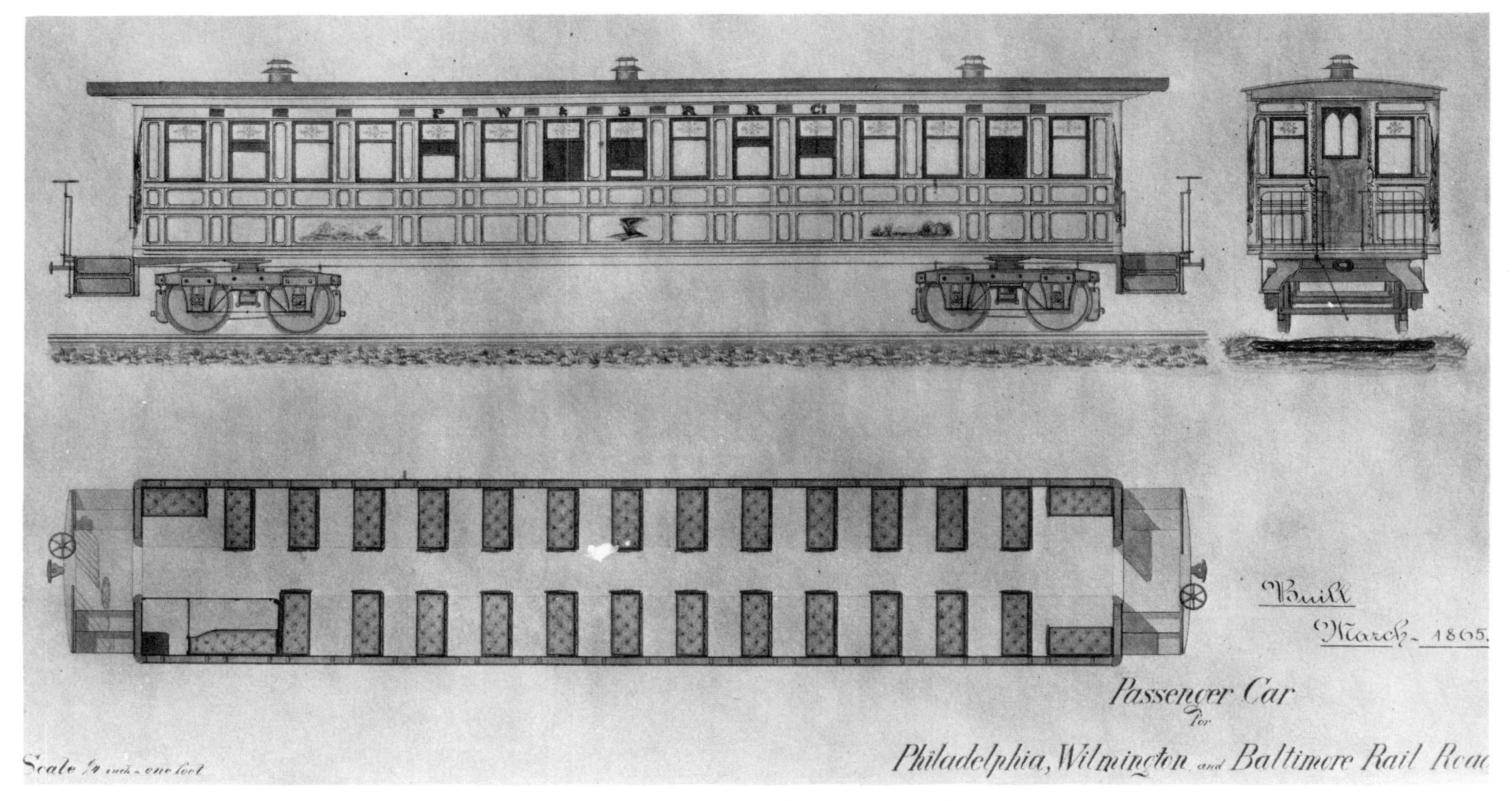

Figure 1.92 A 59-passenger coach constructed by Harlan and Hollingsworth in 1865. (Historical Society of Pennsylvania)

PHILADELPHIA, WILMINGTON, AND BALTIMORE RAILROAD, ARCHED-ROOF CAR, 1865

Here is a late example of an arched-roof passenger car, for the clerestory roof was rapidly displacing the arched roof when this car was built by Harlan and Hollingsworth. Some roads and builders were content with the old-style roof, but it was not used by any self-respecting trunk line very long after this date. The body panels, decorative painting, and short-wheelbase trucks were other indications of the anachronistic design. The total dependence on india-rubber springs (even for the bolster mounting) again showed the conservative bent of the P W & B's management. The very narrow body can be explained by restricted clearances. The Wilmington line double-tracked early, and like some other Eastern roads, built the tracks close together on a narrow roadbed. This initial economy not only necessitated major reconstruction in later years but required the use of unusually narrow cars in the interim.

The overall length was 49 feet; the width, 8 feet; the height, 11 feet (over the roof); the truck centers, 32 feet, 3 inches; the truck wheelbase, 4 feet, 3 inches. Seating capacity was 59, including the saloon. (See Figure 1.92.)

LONG ISLAND RAILROAD, CLERESTORY ROOF CAR, 1867

This car, built by Harlan and Hollingsworth for the Long Island, was an early departure from the Monitor style of clerestory. The straight-line sloping ends of the clerestory lacked the grace of the smooth-flowing duckbill ends that came into use at about this time. The small ventilator side openings show that Harlan and Hollingsworth were not going beyond Webster Wagner's original conception of the raised roof as simply a ventilating apparatus.

The plain, straight lines, square windows, and general lack of architectural ornament were in the simple plan of the old arched-roof car. The raised roof and narrow side sheeting alone classify it as a car of the 1860s.

The length over the couplers was 51 feet; the body length, 44 feet, 6 inches; width, 9 feet, 6 inches; overall height, 13 feet; truck centers, 33 feet, 9 inches; wheelbase, 66 inches; seating, 58. (See Figure 1.93.)

PENNSYLVANIA RAILROAD, CLASS PB COACH, 1867

In 1867 the Pennsylvania's class PA was succeeded by a new common standard design, the PB, which was built until 1878. It was the exact overall size of the PA, but it had a round-end clerestory roof, large arched windows, an oval number panel, and other more decorative features characteristic of the period. The truck wheelbase was extended to 6 feet. The weight was increased by 3,200 pounds, owing to the larger trucks, heavier framing, and new air brake apparatus.

The common six-sill frame was used, with the side sills measuring 5 by 8 inches and the intermediates 4½ by 7 inches. The simple side-panel truss and other framing features of the car are clearly shown in the drawing. The clerestory roof was stiffened by six wrought-iron carlines. The roof was sheathed in thin tin-plated iron sheets with soldered joints. The end hoods were made of heavy sheet iron bolted to the ends of the car. The interior was richly paneled in cherry, ash, and maple. Sliding louvered wooden sashes set inside the windows were used for sunshades. The two gaslight fixtures had two burners each. Gas tanks were attached under the car. A detailed list of costs printed in James

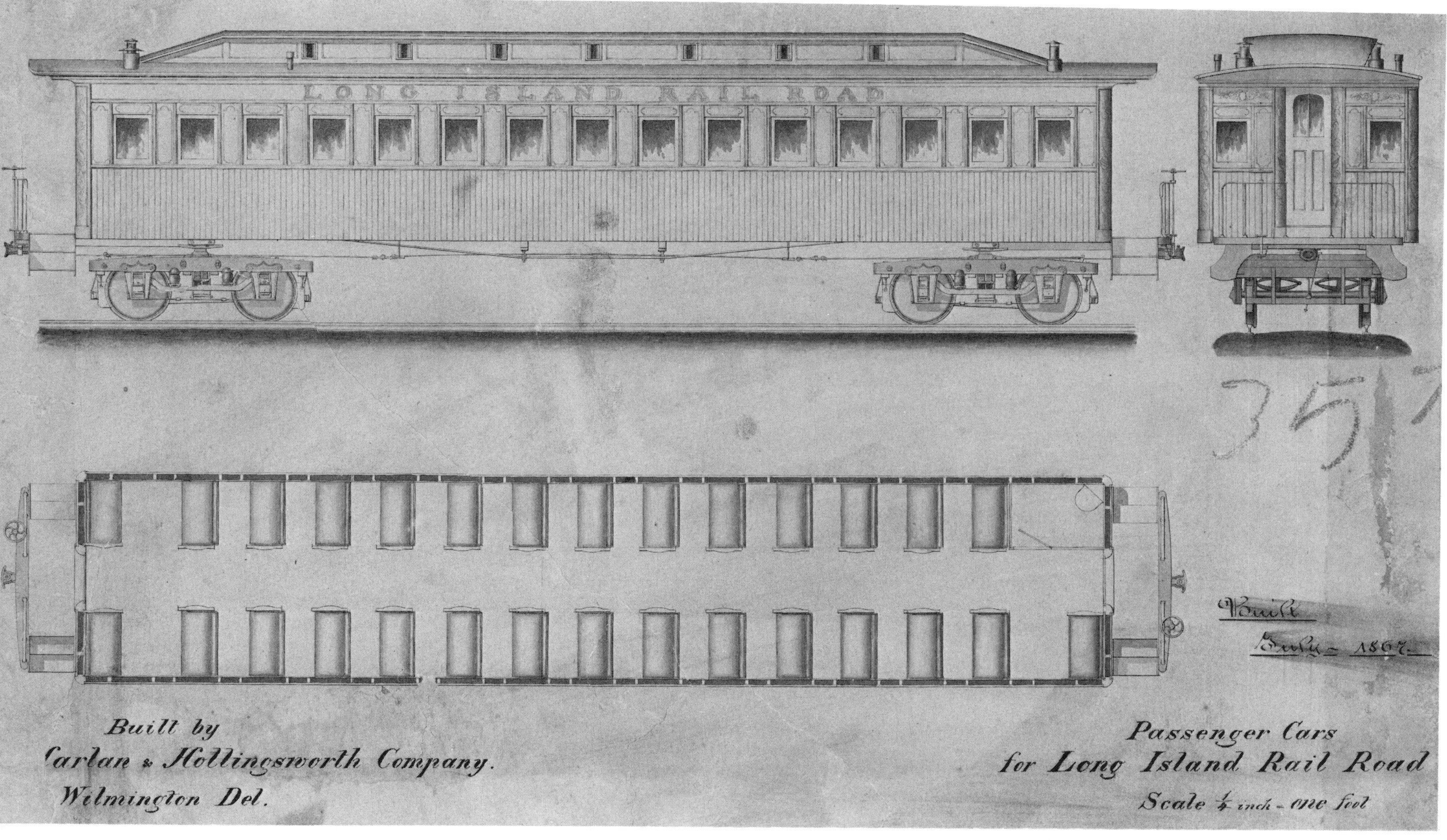

Figure 1.93 A Long Island Railroad coach with an early form of duckbill roof. The car was built in 1867.

Dredge's *The Pennsylvania Railroad* (London, 1879) gives the total cost as $4,423.75. Of this figure the labor charges were about $1,290.

With minor modifications, the PB class continued to serve as the Pennsylvania's standard coach until 1891. The intervening classes followed the size and weight of the PB. The roof profile, sash, side panels, and interior treatments changed with fashion, but the structure was faithfully built to the established pattern of 1867. No other American railroad exhibited such discipline or such steadfast devotion to a common standard.

The PBs were 53 feet, 6 inches over the couplers; 14 feet, 1⅜ inches high overall; and 9 feet, 4 inches wide. The truck centers were 33 feet; truck wheelbase, 6 feet; weight, 42,500 pounds; seating, 54. (See Figures 1.94 to 1.97.)

UNION PACIFIC RAILROAD COACH, 1868

The prospect of through passenger travel to California led Western roads to acquire additional rolling stock as the iron rails pushed westward. The Union Pacific's earliest cars had been rather primitive affairs put together in the Omaha repair shops, which could produce vehicles adequate for the provisional traffic going to the end of the track. Now, however, the Union Pacific began to buy more finished cars from the professional Eastern car shops. Among the suppliers were Harlan and Hollingsworth, whose work is shown in Figure 1.98.

Figure 1.94 The interior of a class PB coach around 1867. (Pennsylvania Railroad Guide Book, *1876)*

Figure 1.95 A Pennsylvania Railroad class PB coach built between 1867 and 1878. (Chaney Neg. 25524)

Figure 1.96 A Pennsylvania Railroad class PB coach, designed and introduced in 1867.

At that time the Monitor roof was being replaced by the duckbill or bullnose clerestory. This car represents the last of the old-style clerestory roofs built in this country. The stained-glass windows between the ventilator openings showed some appreciation of the dual light and ventilation function of the raised roof. The arched window was also more in sympathy with design trends of the late sixties.

The length over the couplers was 51 feet, 2 inches; the body width, 44 feet, 6 inches; width, 9 feet, 6 inches; height, 13 feet, 5 inches; truck centers, 34 feet, 3 inches; wheelbase, 5 feet, 6 inches; seating, 56. (See Figure 1.98.)

VIRGINIA AND TRUCKEE RAILROAD, COACH NUMBER 3, 1869

If it were not for this Nevada short line whose conservative operating policies preserved much of its original rolling stock, we would be lacking several prime examples of Western American railroad equipment. Fortunately, the ancient cars and locomotives of the Virginia and Truckee survived until their unique historic value was recognized in the late 1930s. At that time the main line to Virginia City was abandoned, and most of the older pieces became surplus property. Nearly every car of value was acquired by the movie industry or private collectors. In 1937 Paramount Studios purchased the line's two original coaches, Numbers 3 and 4, together with the cars in Figures 1.106 and 6.20.

Numbers 3 and 4 were the first coaches purchased by the Virginia and Truckee. Moreover, they were built by the pioneer West Coast car-building firm, the Kimball Manufacturing Company of San Francisco. Car building was at best a minor industry in the Western states; relatively few cars were manufactured in that part of the country at any period. It is extraordinary that these obscure relics have been shown to millions of moviegoers, few of whom realize how rare they are. It is hoped that one day the cars will be placed in a museum.

As can be seen from the photographs and drawings, car Number 3 is short and has a plain finish and design. The diagram drawing was based on measurements and photographs supplied by Gerald M. Best of Beverly Hills, California. (See Figures 1.99 to 1.101.)

CHICAGO AND ALTON RAILROAD, COACH NUMBER 39, ABOUT 1870

These drawings, published in the *Railroad Gazette* for June 10, 1871, illustrate one of ten new day coaches built in the Bloomington shops of the Chicago and Alton Railroad. For the first time they show the compression side-panel truss, one of the most important American contributions to railroad car construction.

The drawings also furnish detailed information on an unusually heavy, well-finished day coach of the period. The Chicago

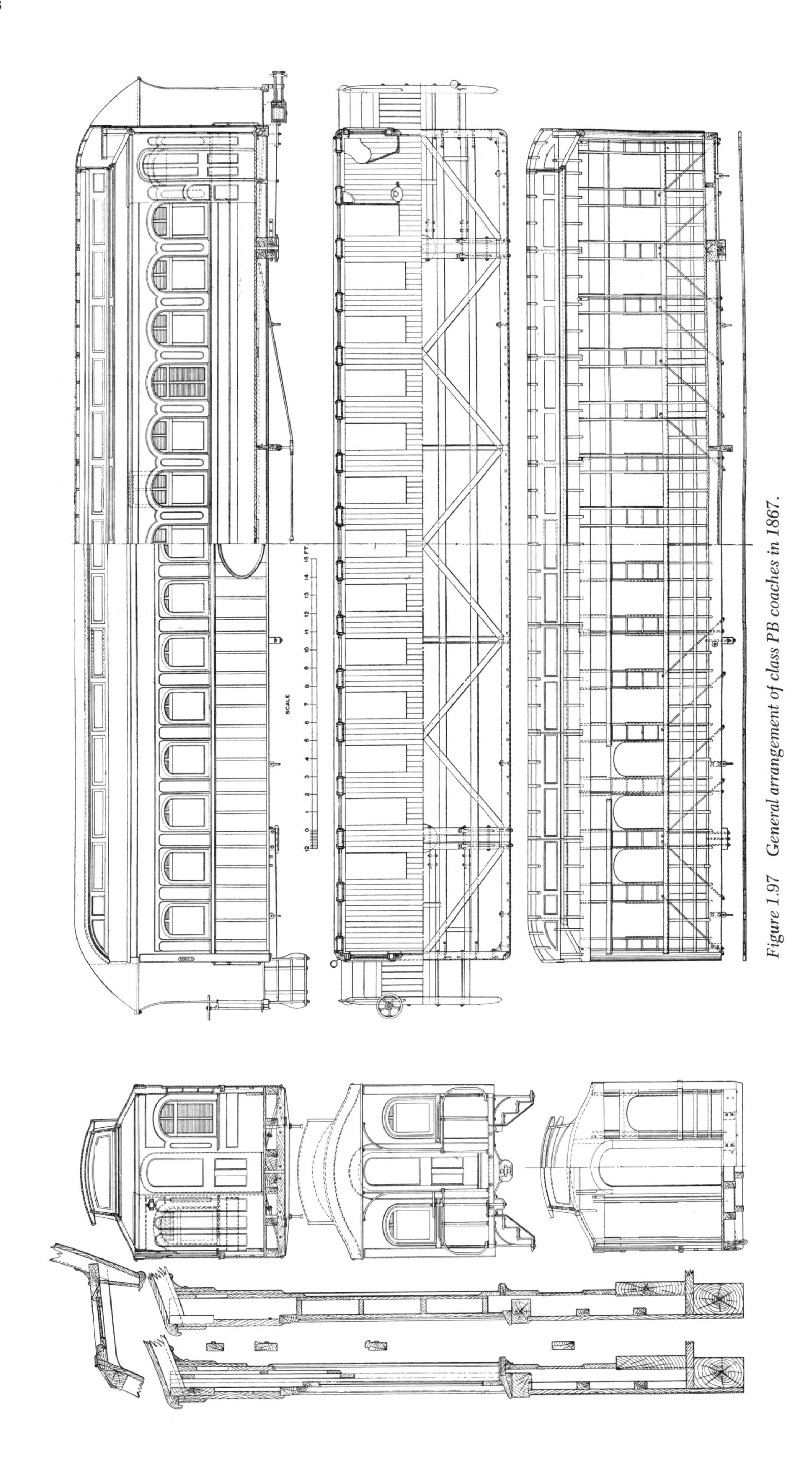

Figure 1.97 General arrangement of class PB coaches in 1867.

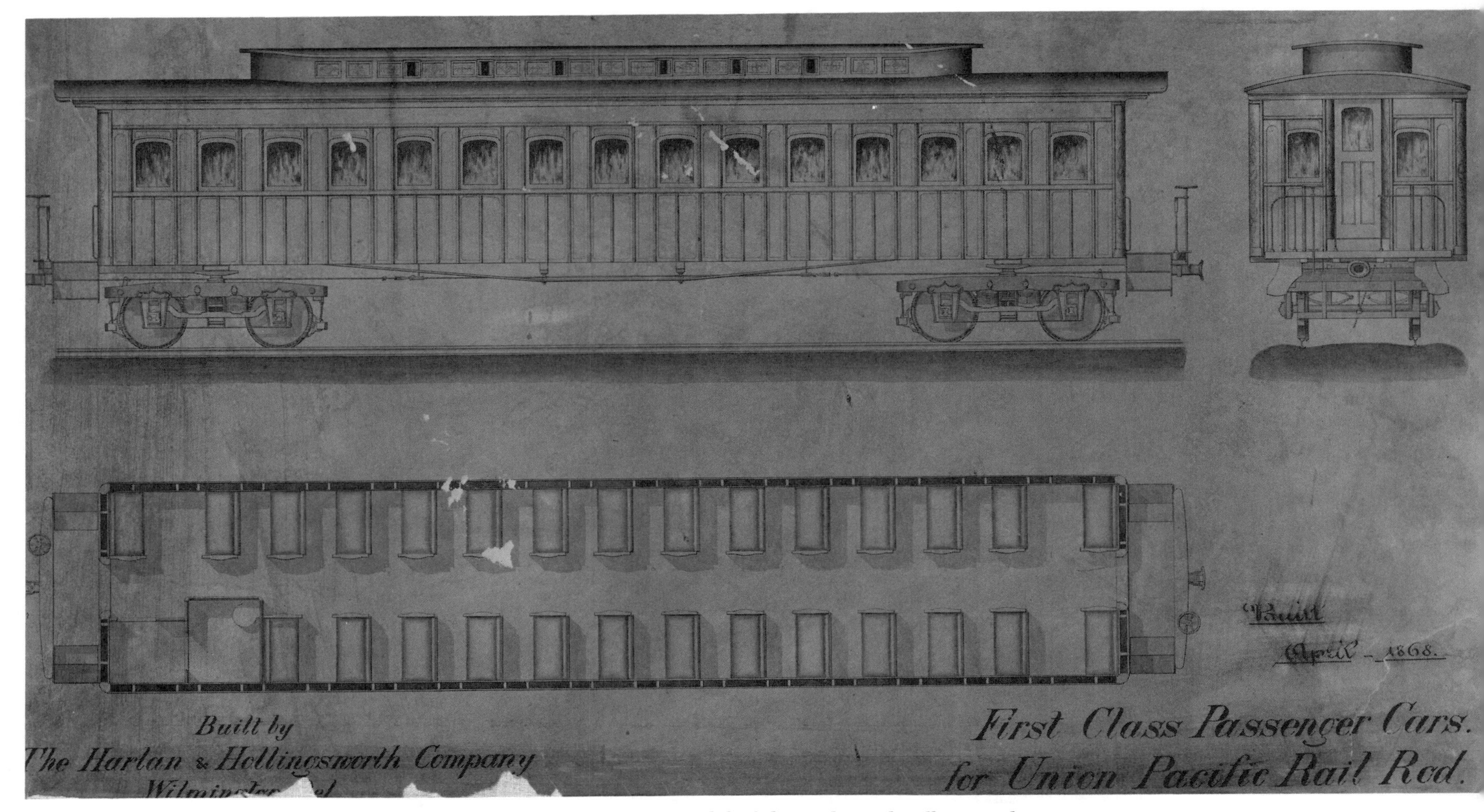

Figure 1.98 A Monitor roof coach built by Harlan and Hollingsworth in 1868.

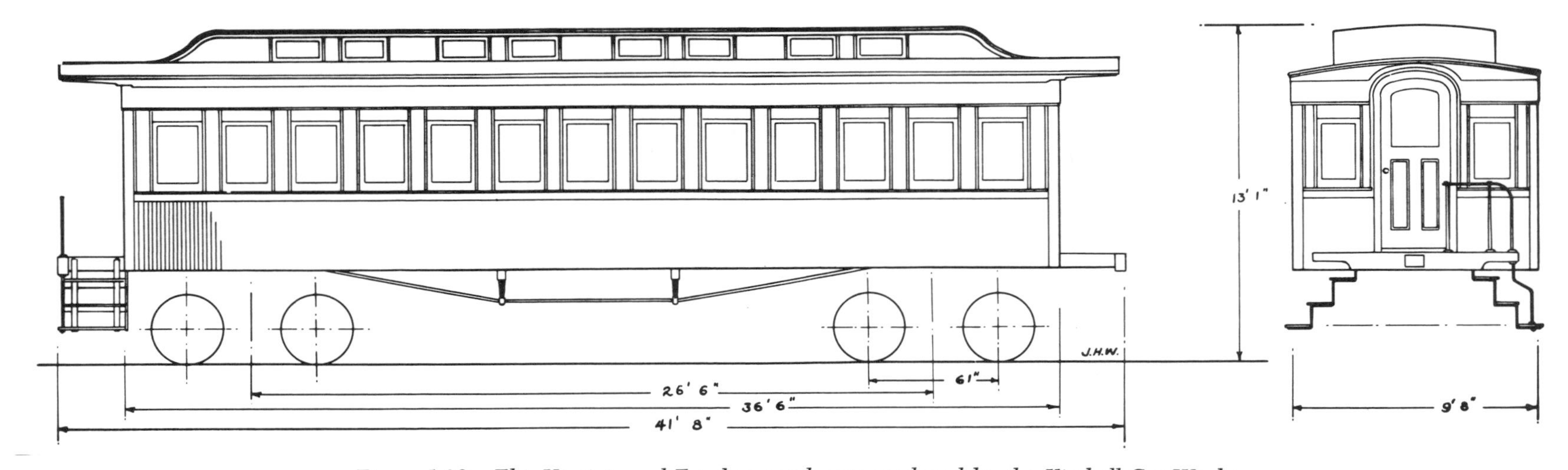

Figure 1.99 This Virginia and Truckee coach was produced by the Kimball Car Works of San Francisco in 1869.

and Alton had a tradition of providing luxurious cars; George M. Pullman refitted his first cars at the road's Bloomington shops. The road extended this policy to the day traveler, and Number 39 is obviously built much on the lines of Pullman's sleepers. It was very large for the times and was outfitted as luxuriously as any first-class car in service, except possibly for the most elaborate drawing-room cars.

In addition to its spacious interior, it was equipped with a patented ventilating system that pulled air in through screened roof hoods. The air was drawn across an open tank of water to cool it and remove dust. The clerestory, built in the old Monitor style but curiously finished with duckbill ends, served largely as an auxiliary light source. Artificial lighting was provided by five center lamps (the gas storage tanks are not shown). As a special safety feature, the car was fitted with a lock-tight antitelescoping coupler and platform invented by the railroad's president, T. B. Blackstone. Two heavy ironclad beams that projected from under the platforms and rode against corresponding timbers from the following car were designed to steady the car's motion and prevent telescoping. The couplers, fitted with adjustable

Figure 1.100 Virginia and Truckee coach Number 3 by the Kimball Car Works, 1869. (Gerald M. Best)

screws, could be drawn tight, and this eliminated the slack usually associated with link-and-pin couplers. The entire arrangement, an obvious attempt to avoid Ezra Miller's patents, appears to have been original with the C & A.

The overall length was 58 feet over the safety timbers; the body length, 50 feet; width, 10 feet, 3 inches; truck wheelbase, 8 feet, 3 inches; seating, about 60. (See Figures 1.102 and 1.103.)

BALTIMORE AND OHIO RAILROAD, CLASS A COACH, 1870

Many railroads divided their acquisition of new cars between their own shops and contract builders. The B & O depended heavily on its Mt. Clare shops in Baltimore for much of its new rolling stock. In 1870 the road began to assign serial letters to its passenger equipment. The coaches were designated class A, but unlike the Pennsylvania, the B & O made no strenuous effort to adhere to a single standard design. Small lots of cars were purchased, so that within a few years there were thirteen subdivisions of class A coaches. Cars obtained from outside builders largely followed the designs of the contractors; only the Mt. Clare cars revealed much attempt at standardization. A description of one of these coaches was given in the *National Car Builder*:[127]

The passenger work is in charge of Mr. Jacob S. Schryack. There are two erecting-shops; in one, the floor-framing of a number of passenger and postal cars are in progress, and consist exclusively of Southern pine. The length of these cars is 51 feet 7 inches by 10 feet wide outside of body. The side-sills are 5½ × 7½, the four intermediates 3¾ × 7½, the end sills 6 × 7½, and the truss-planks 2¾ × 12. Wrought-iron plates in the form of an L are placed in the corners and strongly bolted to the end and side sills. Every thing about the framing is designed to secure the greatest strength compatible with the bulk and weight of material. Our attention was specially called to the iron body-transom—as near as we can name it while waiting for the forthcoming "Dictionry of Terms"—which is attached to all the passenger-coaches built at these shops. Instead of the plain piece of timber ordinarily used, a wrought-iron plate 12 inches wide and ¾ thick is bolted to the sills, with a lip turned over at each end to hold more firmly an angular plate underneath of the same width and thickness, the central part of the lower plate resting on the truck bolster, where it is held by the king-bolt, as shown in the cut. This iron frame-work has great strength, and is not liable to shrink, warp, or split, as is often the case with wood.

In the adjoining shop were a number of cars unfinished and showing the character of the side and roof framing, the sills and plates being connected by 44 ⅝ iron rods, and the sides further strengthened by 18 ash spur-braces held by iron bolts.

Figure 1.101 End detail of Kimball coach Number 3, 1869. (Gerald M. Best)

In the paint-shop, which is a model structure of its kind, with room for thirty cars at a time, were a number completely finished and ready for the track. They have 6-wheel trucks with strong check-chains attached. Each car has two plain wood burning stoves inclosed in iron screens. The doors are locked, so the fire can not escape in case of overturn. This method of heating has proved very satisfactory. Cobb's elliptic spring seats are used, upholstered with crimson plush and provided with back-iron locks. The roofs have Creamer's exhaust ventilators, the air being admitted by supply ventilators underneath, on a line with the racks. The windows, doors, and panels have semi-circular tops, and there are sliding sashes in the doors for ventilation. The ornamental finishing is very rich and tasteful, and consists of the usual variety of choice woods. The raised panels are French walnut, and the doors are mahogany. The head-linings are made by Howard, of Hartford, and exhibit in variety, tone, and adaptation a progressive improvement in this important feature of car decoration. In one of the cars we specially noticed the admirable effect of a head-lining, so perfectly harmonizing in its subdued tints, graceful lines, and artistic blending and juxtapositon of color, with the upholstery and cabinet-work of the rest of the interior as to form in the combination a genuine work of art, not merely accidental, we are inclined to think, but the result of a right perception of what is required. Each car is supplied with water-coolers, enameled inside, and manufactured by John A. Goewey, of Albany. About three passenger-cars per month are built in this department of the works.

As can be seen from the two illustrations, the class A cars had duckbill roofs and Roman arch windows. Some were in service until the 1920s. (See Figures 1.104 and 1.105.)

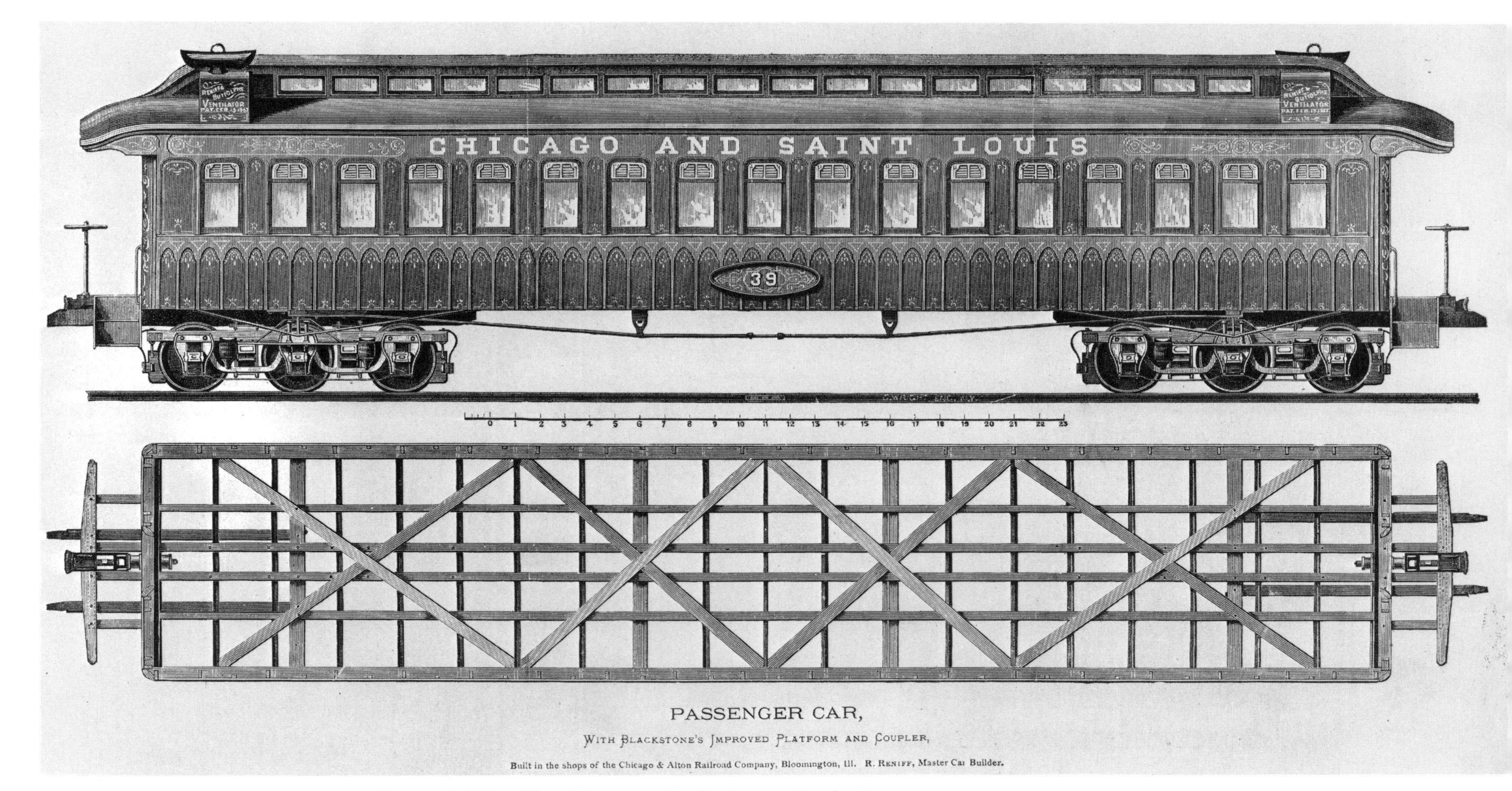

Figure 1.102 The Chicago and Alton's coach Number 39, built in the Bloomington shops around 1870. (Railroad Gazette, *June 10, 1871*)

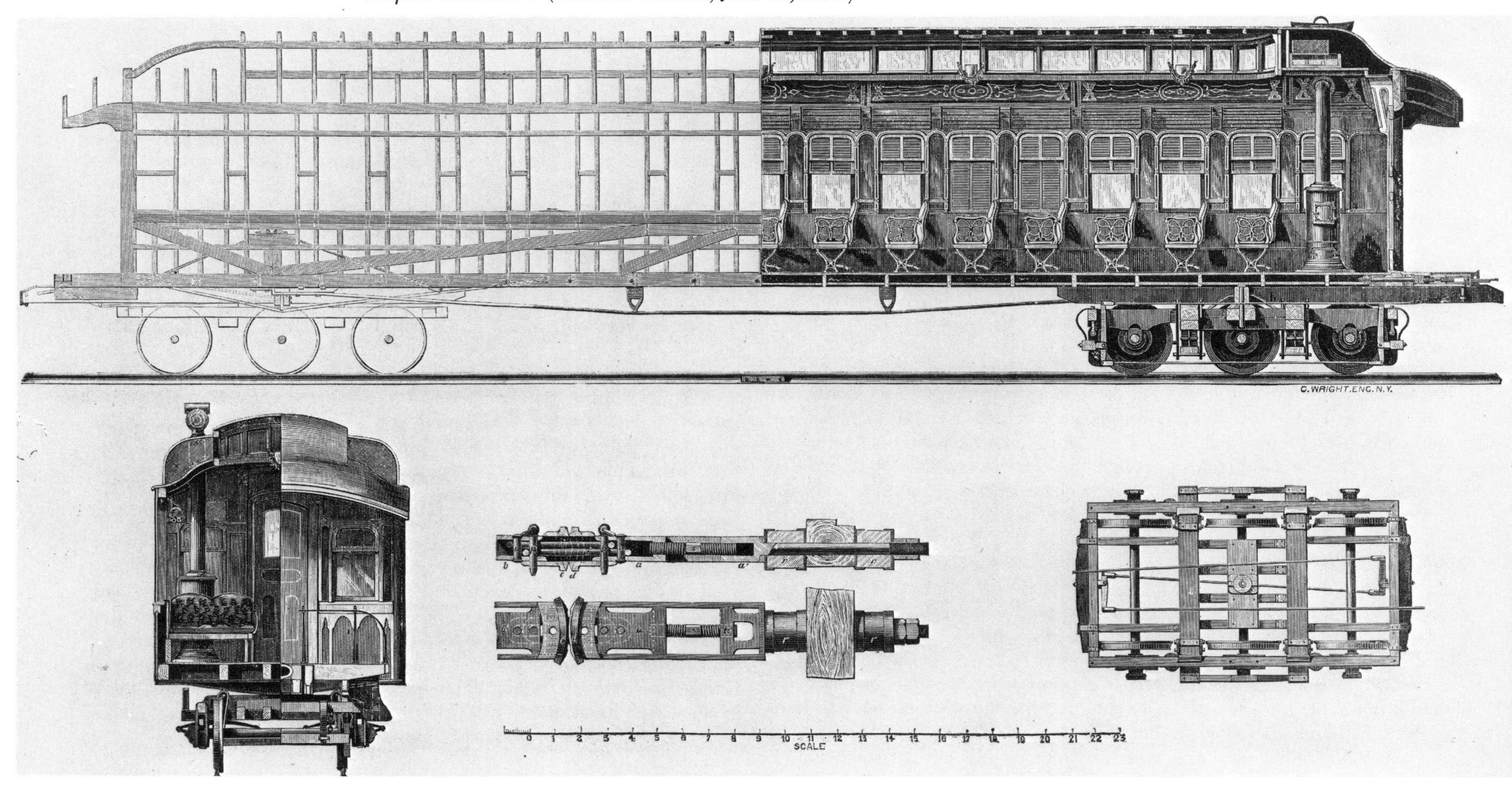

Figure 1.103 Chicago and Alton coach, longitudinal cross section, about 1870. (Railroad Gazette, *June 10, 1871*)

Figure 1.104 B & O class A coach, constructed about 1870, shown some fifty years later when it was in work service. (*Smithsonian Neg. 50742*)

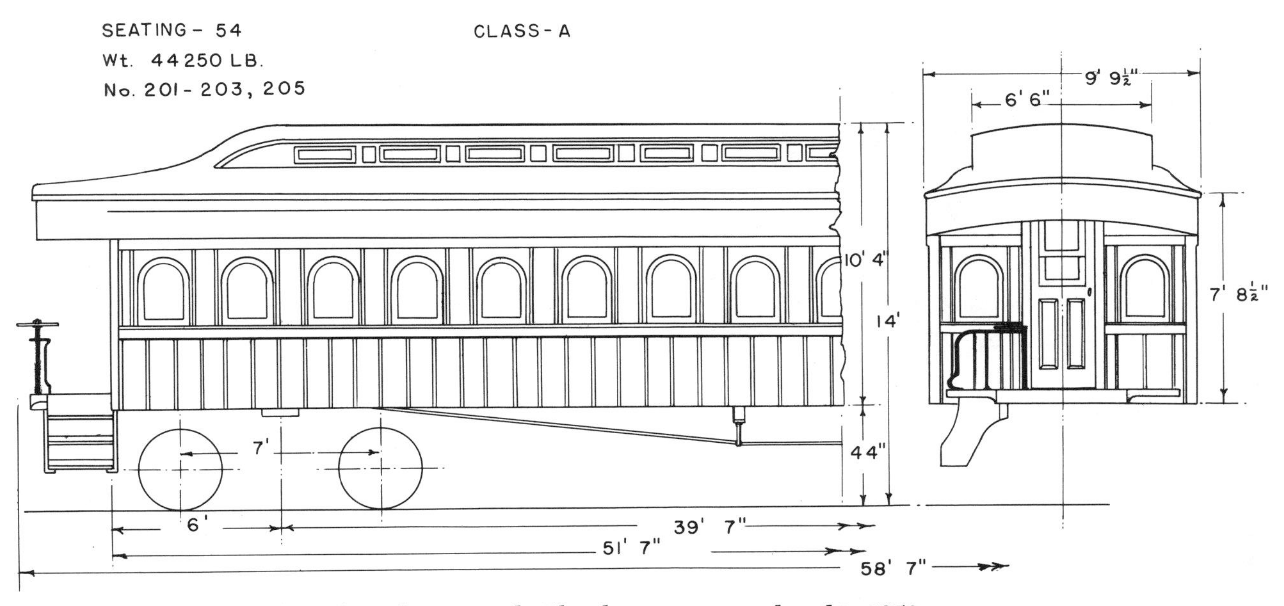

Figure 1.105 B & O class A coach. This design was introduced in 1870.

VIRGINIA AND TRUCKEE RAILROAD COACH, 1874

J. G. Brill of Philadelphia is usually thought of exclusively as a streetcar builder, but the firm was active in steam railroad car construction as well. The coach shown here was one of two that Brill constructed for the Nevada short line. They survived because of the motion picture industry's desire for historic props; both cars were purchased by Paramount Studios in 1938.

The peculiar bow-top clerestory is the most distinctive feature of these cars. It may be a unique roof style; no other examples can be located. The square windows appear small for the middle 1870s, and the large-radius end corners are another peculiarity. A partition divides the car into two compartments at the center. One section may have been reserved as either a ladies' or a smokers' compartment. Total seating was for 56 passengers. The photographs and measurements needed for preparation of the drawing were supplied by Gerald M. Best. (See Figures 1.106 to 1.108.)

ERIE RAILWAY, BROAD-GAUGE COACH, ABOUT 1874

This car, built in the railroad's own Elizabeth, New Jersey, shops, represents the final development of 6-foot-gauge construction in America. A few years after its completion the Erie suspended broad-gauge operations. Because conversion to standard gauge was underway at the time, the body was deliberately made similar to that of an ordinary day coach. Its 10-foot, 4-inch girth was in accordance with the standard-gauge lines that could handle wide cars.

Figure 1.106 Virginia and Truckee coach Number 11, built by J. G. Brill of Philadelphia in 1874. (Gerald M. Best)

Figure 1.107 Interior of the Virginia and Truckee Brill coach, 1874. (R. B. Jackson)

The car was designed by Calvin A. Smith (who also built the 1856 New York and Erie car in Figures 1.83 to 1.85). Smith did not use a truss plank but continued with the bastard Howe. The framing, arranged with economy, produced a remarkably light car for its size: only 34,600 pounds. According to a contemporary description every ounce of material was carefully calculated, and wrought iron was used in place of heavier castings to further reduce dead weight.[128]

Admirable as Smith's design was in avoiding overly heavy construction, he continued to use the long-since-discredited rubber blocks as snubbers in place of coil springs. The equalizing bar was fitted with ¾-inch rubber blocks at each end. The truck bolster was carried by the usual elliptical springs. Miller platforms and couplers were installed.

The car seated 68 in the winter; with the two stoves removed in the summer, the capacity increased to 72. (See Figure 1.109.)

CENTRAL RAILROAD OF NEW JERSEY COACH, ABOUT 1875

In 1927 the Baltimore and Ohio Railroad acquired two aging wooden coaches from the Central Railroad of New Jersey to display in its centennial pageant, the Fair of the Iron Horse.[129] Some light wooden cars of an antique pattern were needed to play a Civil War sequence presented at the fair. Ever since, these cars have been labeled as cars of the early 1860s and are popularly held to be of authentic Civil War vintage. But the arched windows, richly paneled sides, and duckbill roof in Figure 1.112 are more characteristic of the 1870s.

The Wason car works was clearly producing vehicles of a very advanced design when it began offering coaches of this pattern in the mid- and late 1860s (see Figure 1.113). A line engraving of a similar car built for the Central Pacific Railroad appeared in Poor's *Manual of Railroads* for 1869. A photograph of a sister car is reproduced in Gerald M. Best's *Iron Horses to Promontory.* According to records uncovered by Warren B. Crater, vice president of the Railroadians of America, Wason constructed about 200 cars on this design for the C R R N J between 1865 and 1882. Crater contends that two of these cars, originally numbered 255 and 261 but renumbered 1092 and 1095 in 1905, are the present B & O Museum cars 20 and 21. They were built in 1868.

The cars are strikingly similar to the C R R N J coach drawings

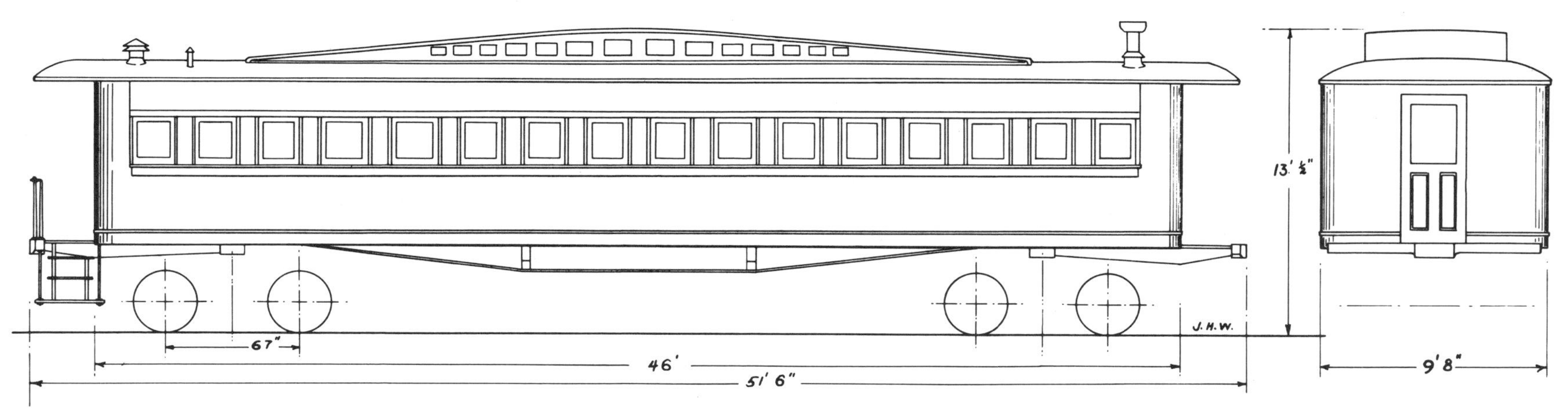

Figure 1.108 Diagram of the Virginia and Truckee Brill coach, 1874.

in the 1879 *Car Builders' Dictionary*, shown here as Figures 1.110 and 1.111. The cars now at the B & O Transportation Museum do differ from the drawings in the following particulars. The clerestory stops short of the body ends, and while the number of side windows (sixteen) agrees with that in the published engravings, the length of the bodies differs by 30 inches. The differences can be explained as design alterations to the same plan.

The drawings show small ventilator openings with screen covers placed between every second window through the sides of the car at the letter board. These permitted a flow of air when it was necessary to close the windows. A leather apron at each end of the car, on a level with the platform, helped to prevent dust from blowing up between the cars. Long rods, fitted with coil springs that reached back under the platform, held the aprons together. The platforms were of the old drop pattern that invited telescoping. The platform timbers were secured to the main frame only by stirrup bolts; no tie rods on the Miller plan are in evidence. The coupler attachment, bolted to the platform timbers with no other fastening to the main body frame, is also noticeably weak.

The overall length of the cars was 51 feet; the width, 9 feet, 10 inches; height, 13 feet, 8 inches; truck wheelbase, 6 feet; seating, 52; estimated weight, 22½ tons. (See Figures 1.110 to 1.113.)

PENNSYLVANIA RAILROAD COACH, CLASS PD, 1878

Not long after the U.S. Centennial, the Pennsylvania Railroad decided that it was time to prepare a new standard coach design. The design had to be suitable for service on the Keystone system's vast property, which extended from the United Railroads of New Jersey to the far Middle Western reaches of the Vandalia. The Pennsylvania operated over six hundred coaches on some 1,700 miles of line. The old class PB had been the company standard since 1867, but it was now considered obsolete, at least in appearance. The overall size and plan of construction was retained. The new class PD, adopted in 1878, had fifteen windows on a side like its predecessor, but it seated two less passengers than the PB. Stylistically the car was modified to reflect the purity of design favored by the great English art critic John Ruskin. The arched windows gave way to a square sash with a subtle rounded upper corner. The boldly curved sheet-metal canopy of the PB was replaced with a more straightforward wooden-frame platform roof.

The purity of design was best exhibited by the interior, which the August 19, 1881, *Railroad Gazette* described as "solid, square and honest," and without the jugglery, wood puzzles, and veneers of the old-fashioned cars. Quartered oak or ash panels engraved with the simple patterns advocated by the critic Charles Eastlake replaced the baroque mixture of ash, cherry, maple, and pine used in the PBs. The seats had wooden rather than cast-iron frames and were covered in golden brown or cherry red material. The head linings were muslin painted in deep red, grayish-brown, or light-drab tones. The windows were glazed, with polished plate glass imported from France. The hardware was polished bronze, again decorated in the simple, straight-line Eastlake style. Two stoves with hot-air registers running under the seats heated the car, and gas lighting was provided. The exterior was painted Tuscan red and striped in gold and black. This finish required sixteen coats of filler, color, and varnish. The Altoona shops could produce twelve coaches a month of this design; each car took forty days to assemble.

The length over the couplers was 53 feet, 6 inches; the width, 9 feet, 4¼ inches; height, 14 feet, 1⅜ inches; trucks, 33 feet on the center; wheelbase, 7 feet; seating, 52; weight, 21 tons. (See Figures 1.114 and 1.115.)

CINCINNATI SOUTHERN RAILWAY, COACH NUMBER 40, 1879

The diagram drawing shows one of the original coaches of the Cincinnati Southern Railway. It was constructed by Barney and Smith of Dayton, Ohio; unfortunately there are few hard engineering data on this important builder. The drawing is of interest as one of the few surviving examples of a Dayton-built car, even though the sketch exhibits no unusual mechanical features. (See Figure 1.116.)

BALTIMORE AND OHIO RAILROAD, CLASS A6 COACH, 1881

Five series and eleven years removed from the original class A coach, the A6's general dimensions and construction agreed closely with those of its predecessors. But several stylistic changes can be seen: the Roman arch windows gave way to square sashes; the duckbill roof was replaced by a bullnose shape; the window posts were made narrower and the exterior finish more subdued and unified. The first A6 was built in November 1881. Over thirty cars were made on this plan before it became obsolete in 1884. The drawings here are from the *National Car Builder* of June 1883.

The supplementary plank truss just below the belt rail was used by some builders who wished to reinforce the regular side-panel trussing. The car also had iron body bolsters and Janney

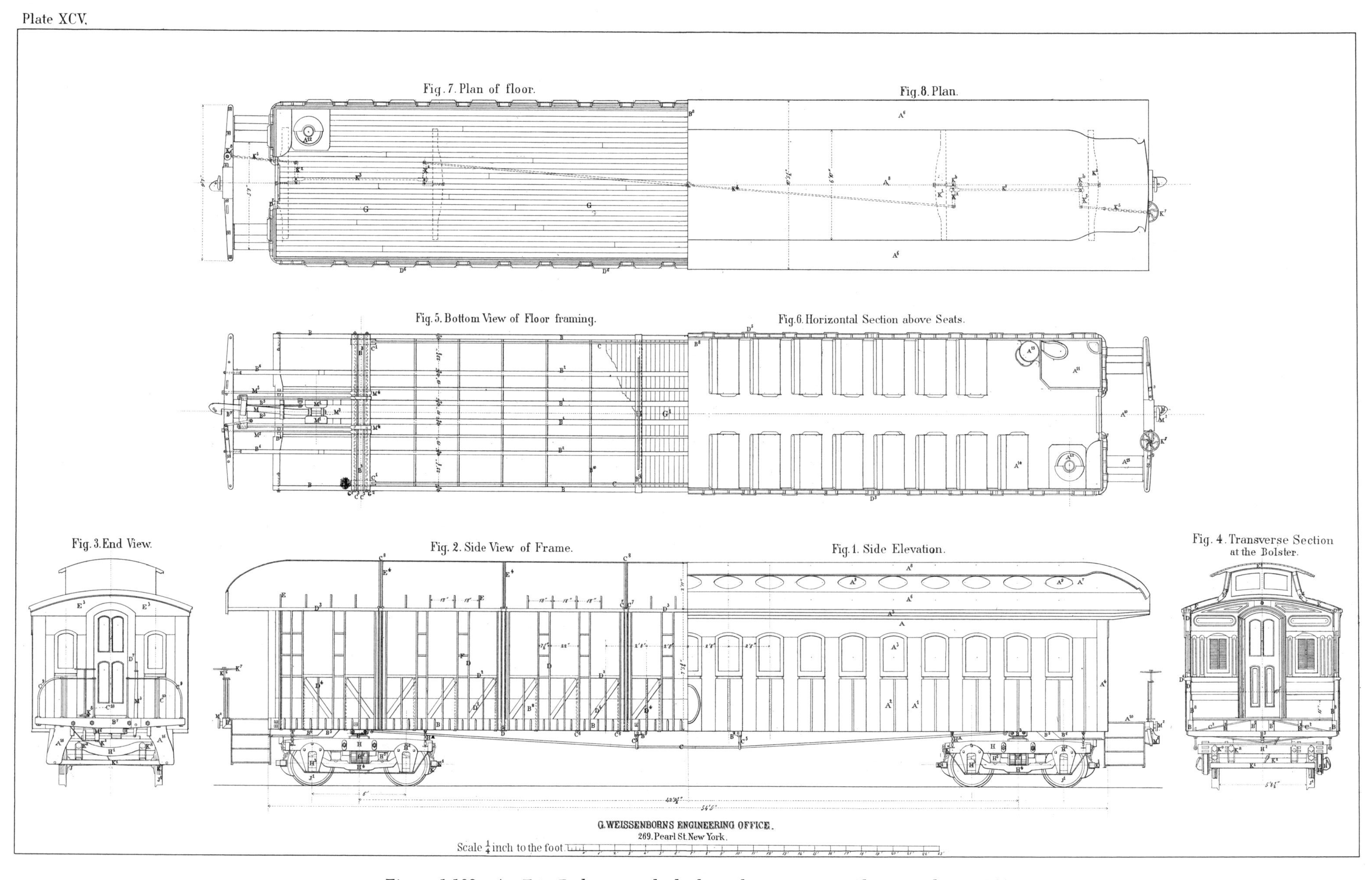

Figure 1.109 An Erie Railway coach, built at the Jersey City Shops in about 1874. (*Gustavus Weissenborn*, American Locomotive Engineering and Railway Mechanism, *New York, 1871*)

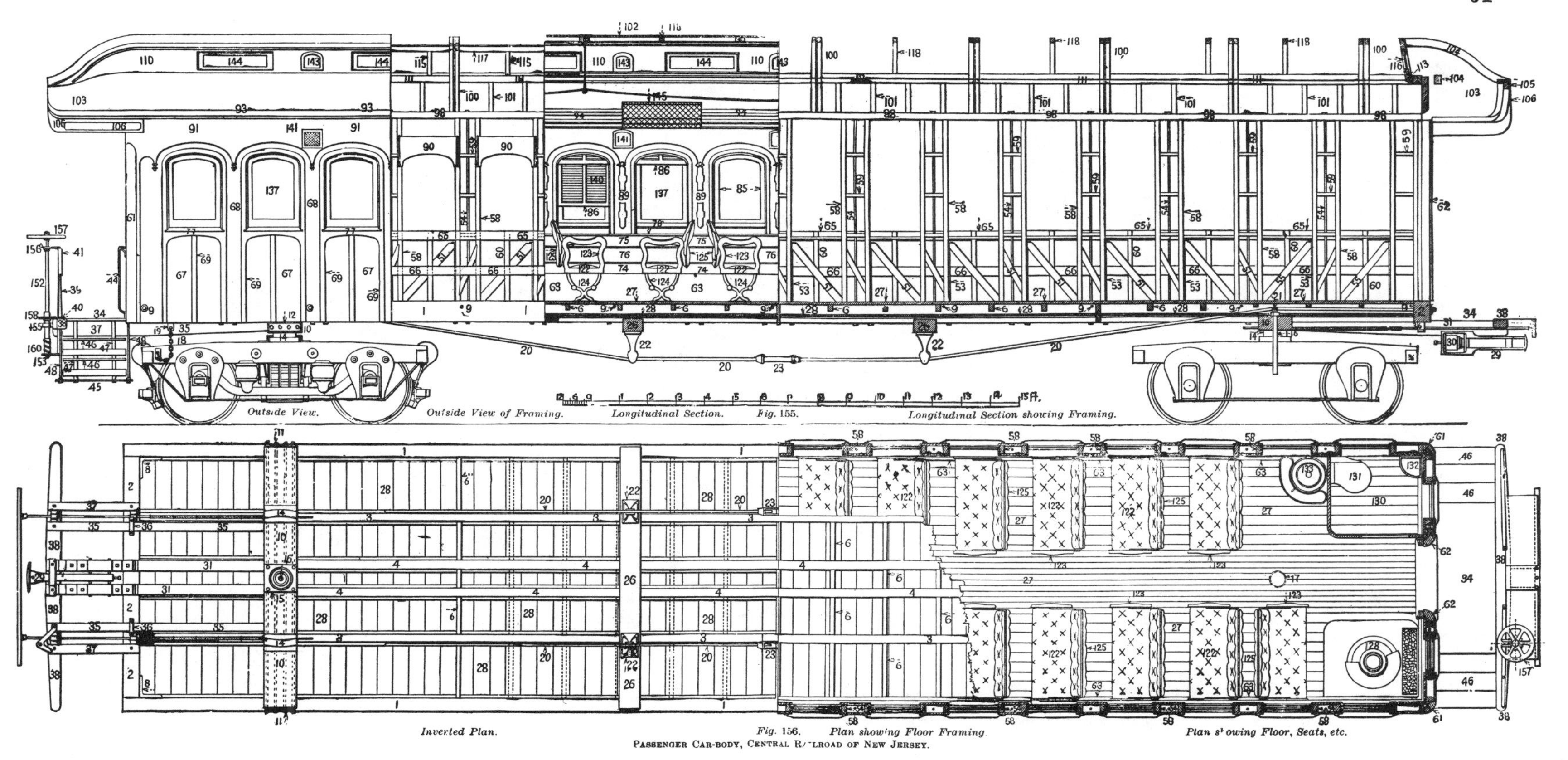

Figure 1.110 A Central Railroad of New Jersey standard passenger car, constructed about 1875. (Car Builders' Dictionary, *1879*)

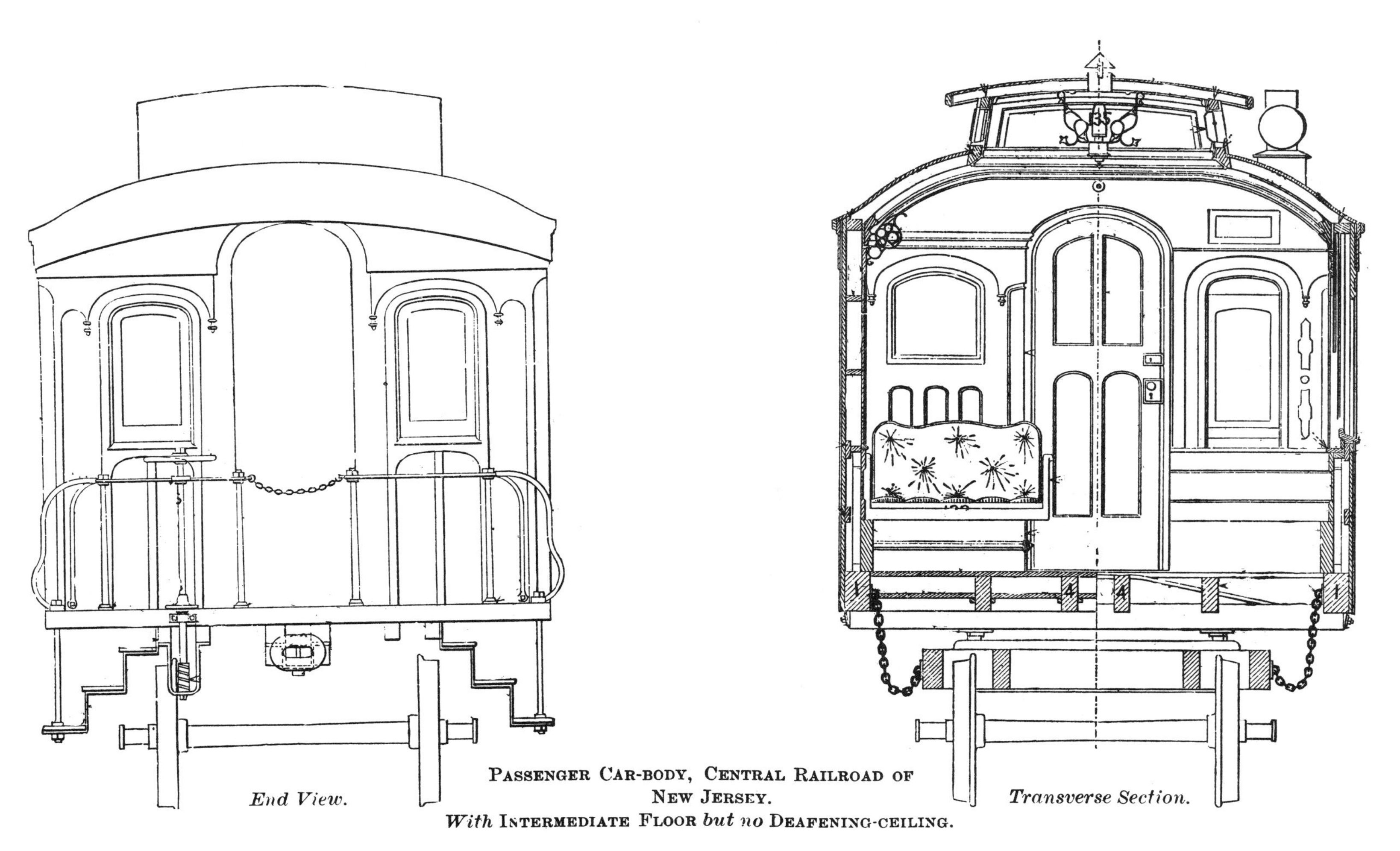

Figure 1.111 End and transverse section drawing of a Central Railroad of New Jersey passenger car of about 1875. (Car Builders' Dictionary, *1879*)

Figure 1.112 A Central Railroad of New Jersey passenger car, relettered as Baltimore and Ohio Number 21 for the 1927 Centennial Pageant of the B & O.

WASON

MANUFACTURING COMPANY,

ESTABLISHED 1845.

Builders of all descriptions of

RAILWAY CARS,

And Manufacturers of

Car-Wheels and Railway Castings.

Particular attention paid to the shipment of Cars, and Materials for Cars, to all parts of the world.

GEORGE C. FISK, President.
H. S. HYDE, Treasurer.
C. T. WASON, Secretary.

SPRINGFIELD, MASS.

couplers. The sills were yellow pine; the other framing pieces were yellow pine or ash.

The interior was paneled in mahogany. The burl walnut panels and gilt highlights of the early cars were omitted, and the painted oilcloth ceilings were replaced by oak-veneer panels. The only extravagance was in the nickel-plated and gilt luggage racks. The restrained decorative treatment was meant to avoid needless display and suggest "good, honest intelligent work," as the *National Car Builder* said.

The car seated 58 passengers and weighed 51,150 pounds. (See Figures 1.117 and 1.118.)

CHICAGO & ATLANTIC RAILWAY, COACH NUMBER 68, 1883

This car formed part of Jackson and Sharp's exhibit at the 1883 Railway Appliance Exposition in Chicago. Structurally it followed the usual pattern of American day coaches except that its finish was more dazzling, as befits a trade-fair display piece. The exterior was painted yellow, with a lake-blue border and gilt decoration. The hand railings were silver-plated. The mahogany interior was carved in the shallow Eastlake style and depicted leaves, flowers, and other "graceful and artistic specimens of hand carving," according to a description in *The Engineer.* One panel showed a spider web with a fly trapped in its fine-chiseled lines.

The best auxiliaries were installed: Foster gaslights, a Searles hot-water heating system, and Sax and Kear steel-tired wheels.

Figure 1.113 Advertisement from Poor's Manual of Railroads, *1869–1870. Note the similarity of this car with the preceding illustrations.*

Figure 1.114 A Pennsylvania Railroad passenger coach class PD, 1879. (*Smithsonian Neg. 71261*)

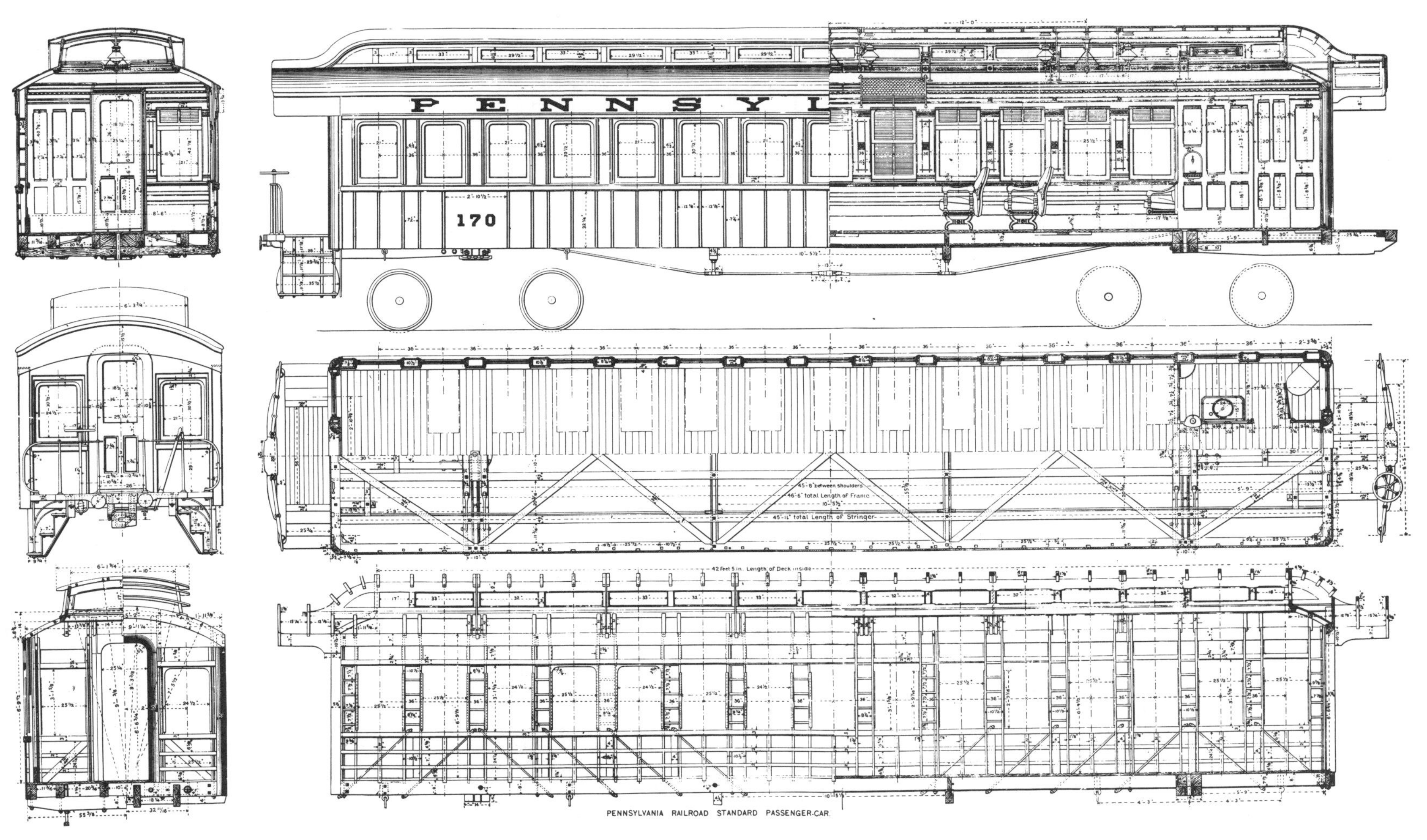

Figure 1.115 Assembly drawing of a Pennsylvania passenger coach class PD, 1880c. (Railroad Gazette, *August 5, 1881*)

The car seated 58 passengers and cost about $8,000. The drawing is from *The Engineer* of August 29, 1884 (p. 164). (See Figures 1.119 and 1.120.)

CHICAGO AND NORTH WESTERN RAILWAY, TWELVE-WHEEL COACH, 1882

The drawings and description of this heavy twelve-wheel coach are from the January 1883 *National Car Builder*. The framing was yellow pine except for the ash door and end posts, carlines, and end plates. The interior was mahogany, and the ceiling was bird's-eye maple veneer. The body bolsters were iron. Six oil lamps provided the lighting. Heat from two coal stoves was distributed by hot-air registers running the full length of the car. Miller couplers and platforms were used, and the 33-inch wheels were cast iron. One unusual feature was the absence of end windows.

The body was lemon yellow, the letter board was in ultramarine blue, and the lettering and striping were gold, picked out in red. (See Figures 1.121 and 1.122.)

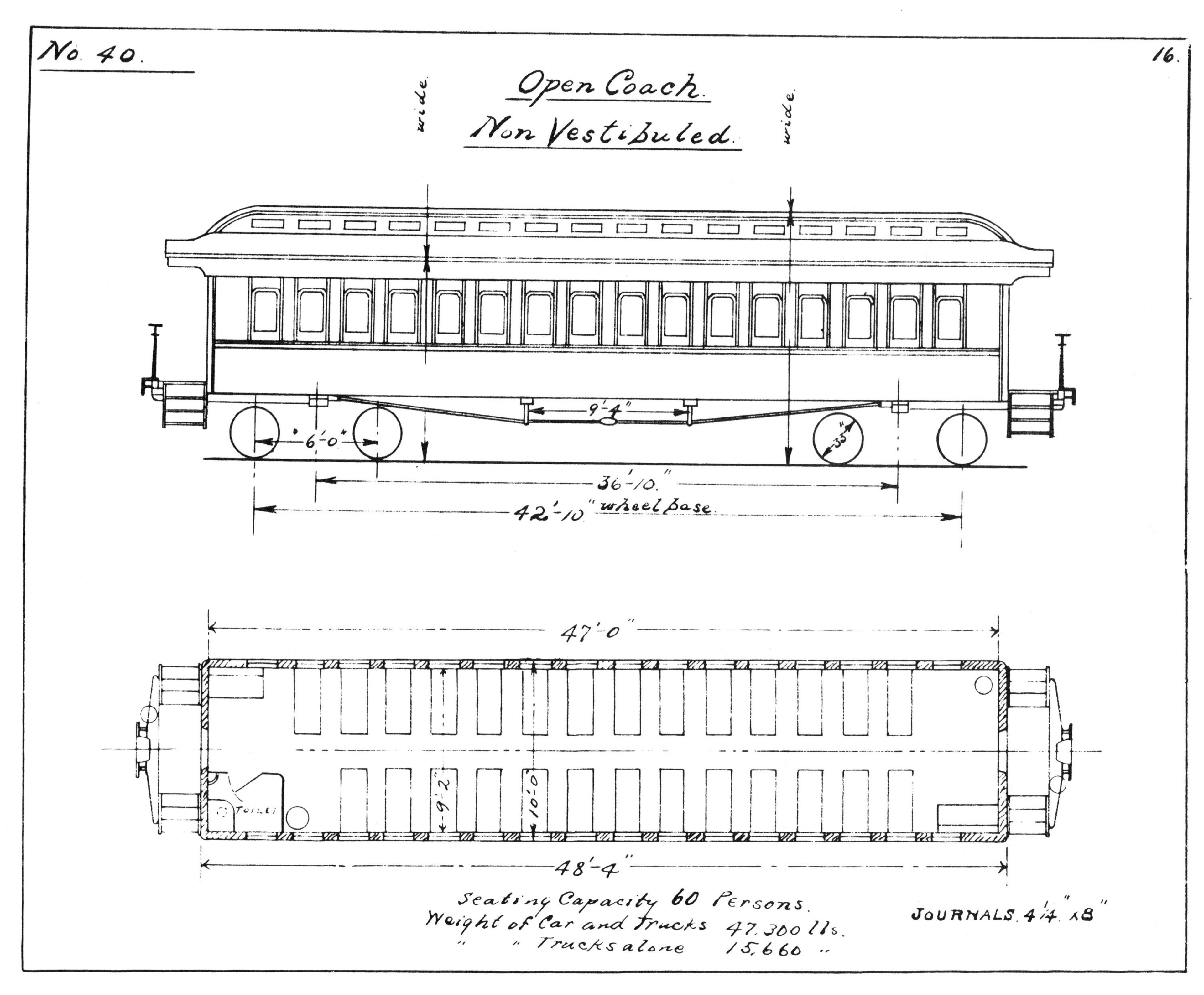

Figure 1.116 A Cincinnati Southern coach built by Barney and Smith of Dayton, Ohio, in 1879. (Southern Railway)

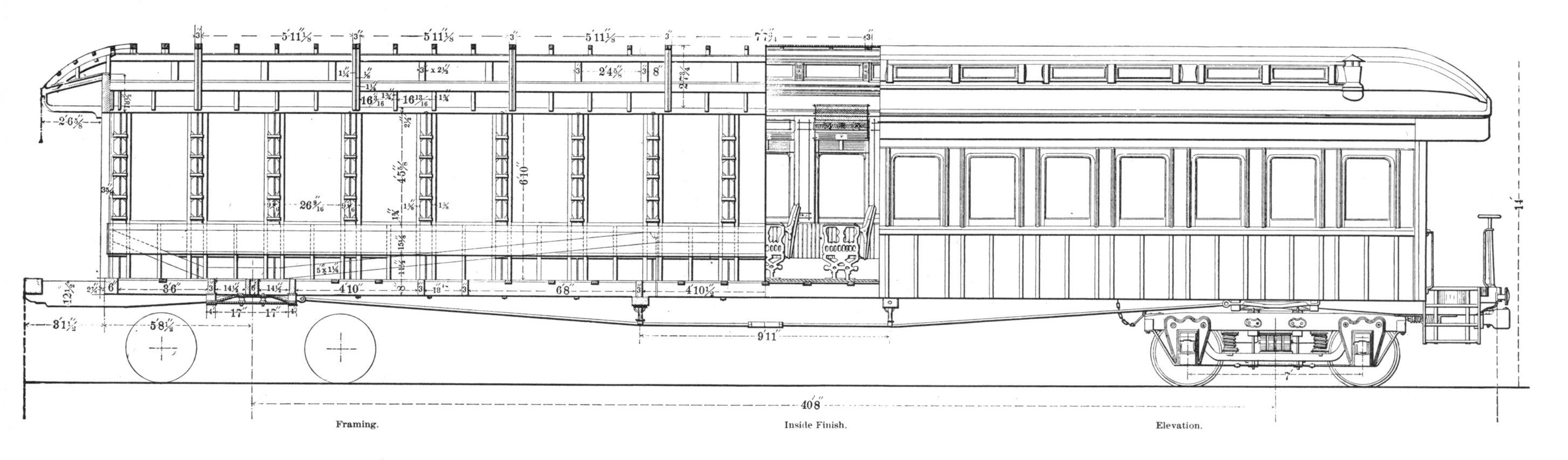

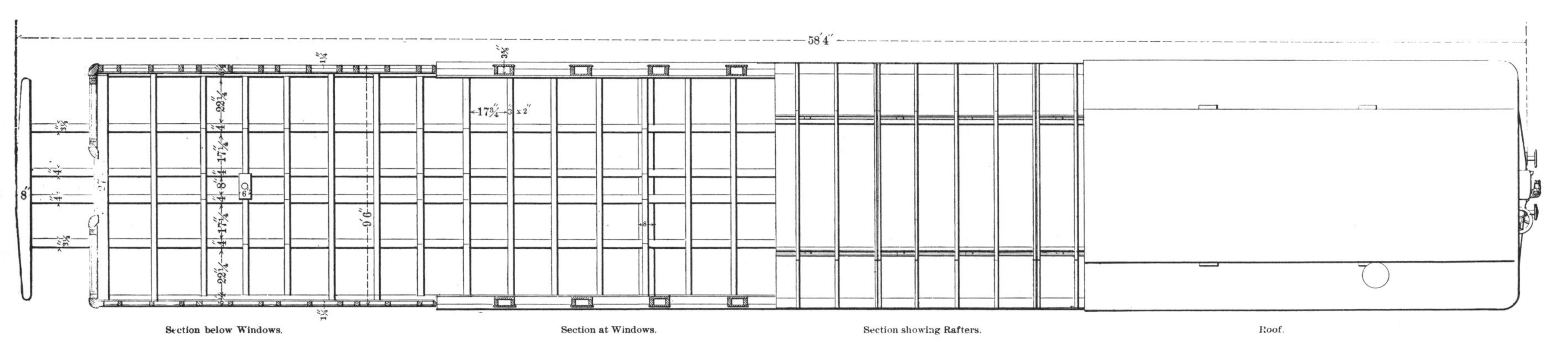

Figure 1.117 A Baltimore and Ohio coach class A-6. The design dates from 1881. (National Car Builder, *June 1881*)

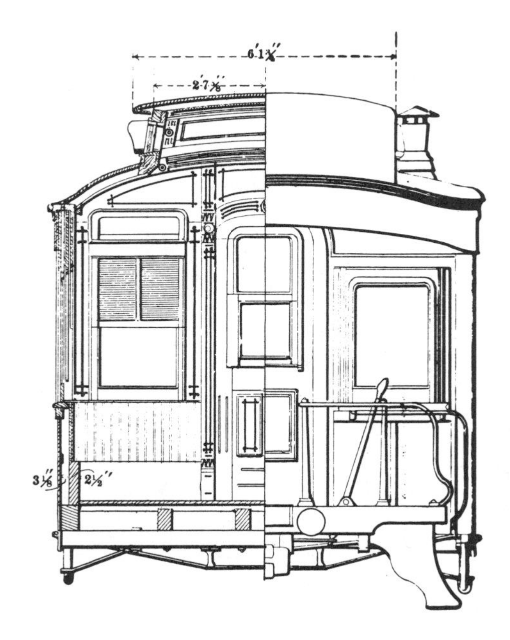

ERIE RAILWAY COACH, 1884

The Erie had suspended broad-gauge operations by the time this car entered service, but the construction of new passenger equipment continued as before at the road's Jersey City shops. The drawings from the *National Car Builder* of September 1884 show the road's loyalty to a proved design. The side-panel truss system, laid out some thirty years before by Calvin A. Smith, was still followed. Owing to the strength and light weight of Smith's plan, cars of this pattern were reported to stand well, with their sides straight and true, after years of service.

The outside panel was covered by narrow 2-inch sheeting rather than the batten board typical of the period. The square window sash and modest belt rail created an austere outside appearance. The interior was equally plain, according to the *National Car Builder*:

Plainness and durability have not been lost sight of in a desire for showy elaboration, and the result is a pleasant, cheerful and attractive interior. Birch is used in the ceiling and white ash and mahogany in the sides, the ash being obtained along the line of the road of an excellent quality, light color, and free from blemishes. Upon these ash panels, which are absolutely free from streaks and hearts, is placed a belt of mahogany. Over the windows the mahogany is ornamented with a sunken rosette, and between the windows with a rosette and engraved lines, as shown. The birch ceiling is put in with mahogany moldings stained to represent old wood, thus affording a better contrast with the ash than could be obtained from the new wood. The window moldings are ¼ round of about an inch radius, presenting no sharp corners. The seat-arms are of ash, and are strong, shapely and handsome. The seats slide upon the frames so as to be lower at the backs than at the front edges. The roof curves have been carefully studied with a view to harmony between those of the lower and upper roof. The two cross sectional elevations show the points from which the curves of the carlines are struck, and also the radii. The effect obtained is that of an elliptical head, although the curves are struck from centres. The raised roof, in consequence of the thickness of its sides being only 1½ in., is much lighter than usual, without any sacrifice of strength. The ash panels, only about ¼ in. thick, are glued directly upon the rails and posts.

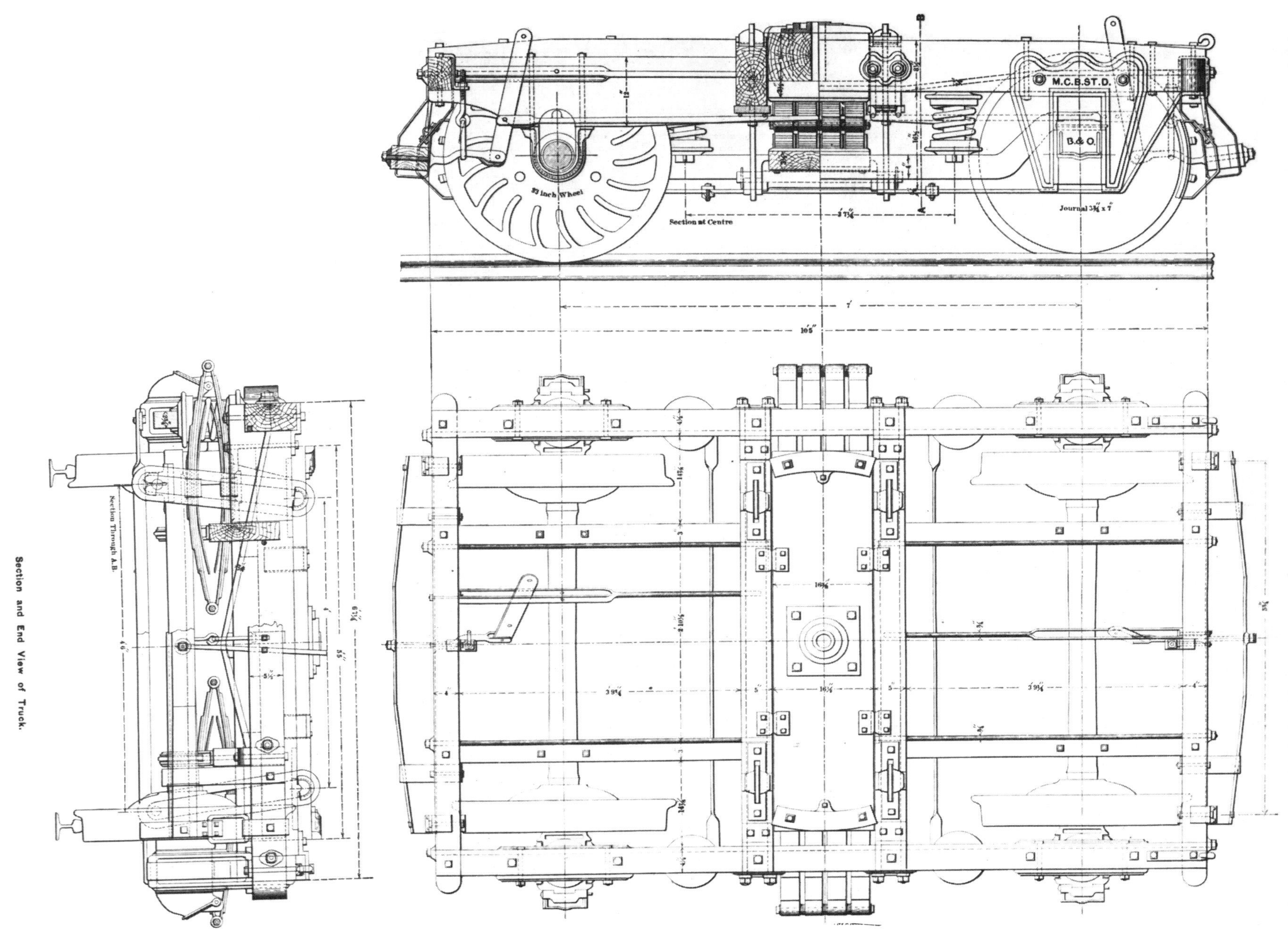

Figure 1.118 A Baltimore and Ohio Railroad passenger car truck, 1881. (National Car Builder, *June 1881*)

Figure 1.119 A Chicago and Atlantic coach built in 1883 for display at the Chicago Exposition of Railway Appliances. (*Smithsonian Neg. 71260*)

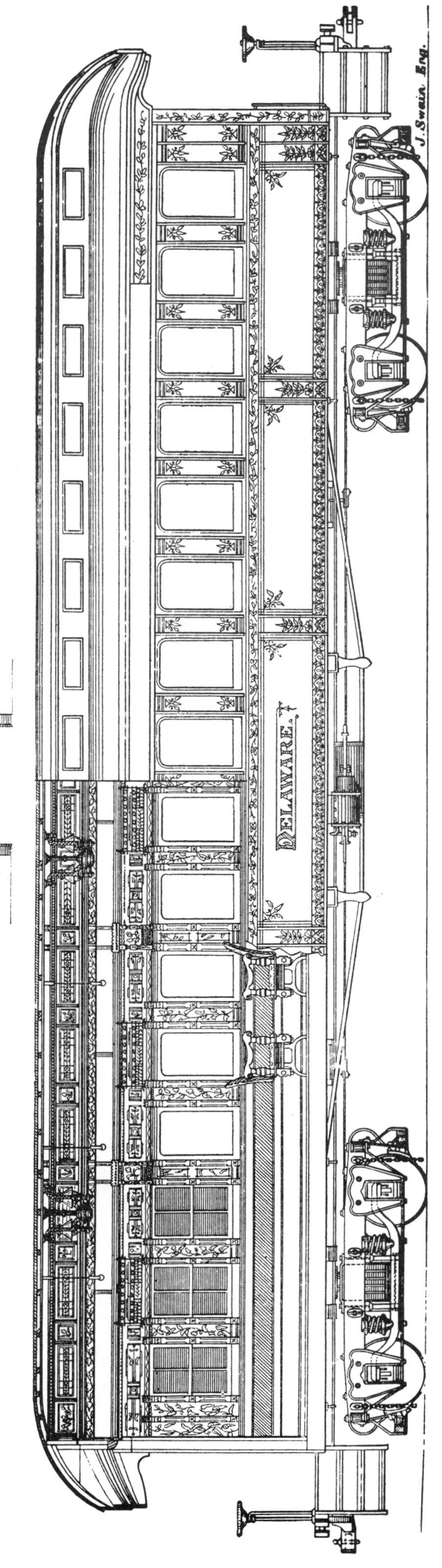

Figure 1.120 *A Chicago and Atlantic passenger car, 1883.* (Engineer, August 29, 1884)

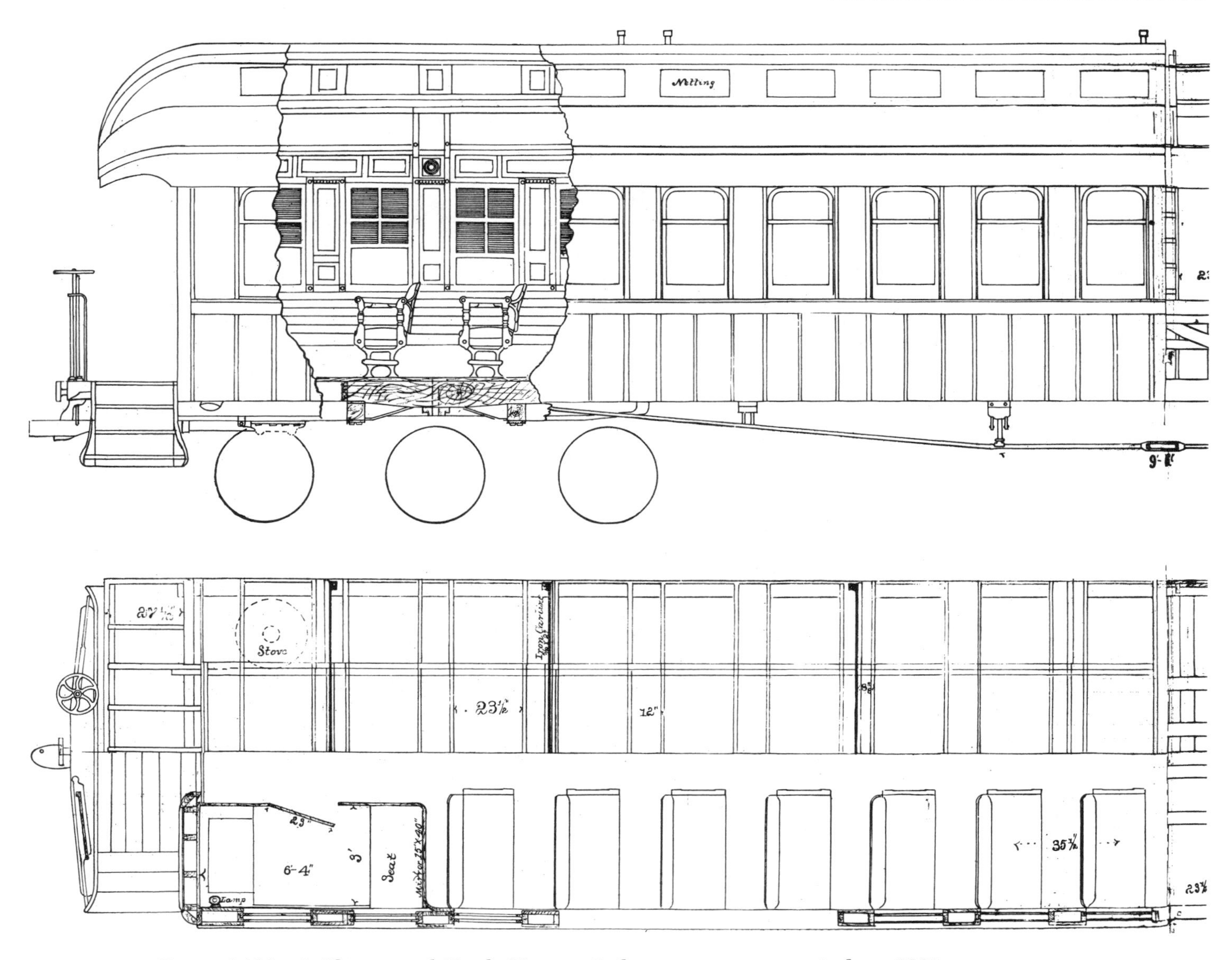

Figure 1.121 A Chicago and North Western Railway passenger car of about 1882. (National Car Builder, *January 1883*)

Pintsch gas was used, with oil side lamps for emergency lighting. The hot-water heating was of an undetermined make. The car measured 58 feet, 4 inches long over the platforms. (See Figures 1.123 to 1.125.)

PHILADELPHIA AND READING RAILROAD, ROUND-END COACH, 1884

It is not surprising that the Reading, long famous for its distinctive style of locomotives, should develop a singular day coach. These cars differed from conventional designs in two particulars: the round-corner ends and A-frame roof, which were used by few other American roads. They were first built around 1882 and continued to be produced until the end of the decade. The design attracted much comment, but no other roads seem to have copied it.

The round corners, with a radius of 2 feet, required a complicated framing built up of several heavy timbers joined by ship splices, glue, screws, and bolts. While the design had no practical justification, the road contended that the round corners would divert the truant car in a wreck. In one accident, another car glanced off the corner and only scraped the panel. It was said that the same blow would have caved in the corner of a regular car.[130] The rounded sash and glass of the corner windows was expensive, though architecturally interesting.

The A-frame roof was stronger than the conventional design but not esthetically pleasing, as can be seen from Figure 1.127. The additional moldings and the basket-like gas fixtures created a busy, cluttered ceiling. The design was seldom used by other roads.

Extra seating was provided by the corner seats and the underbody mounting of the stove. The body width was held within the undersized clearances of the Reading. The car weighed 47,360 pounds. (See Figures 1.126 to 1.129.)

CLEVELAND, COLUMBUS, CINCINNATI, AND INDIANAPOLIS RAILROAD COACH, 1884

This car had no obvious peculiarities of design such as those in the Reading car of Figure 1.126. Outwardly it was a conventional design of the period, but it did employ an unusual method of body support. In place of the customary lower panel truss was a 3- by 12-inch plank that was fastened inside the car to the posts and rested on top of the side sills. It was analogous to a plate girder although, of course, far less stiff than a metal plate. Generally such planks were used as supplementary supports for the

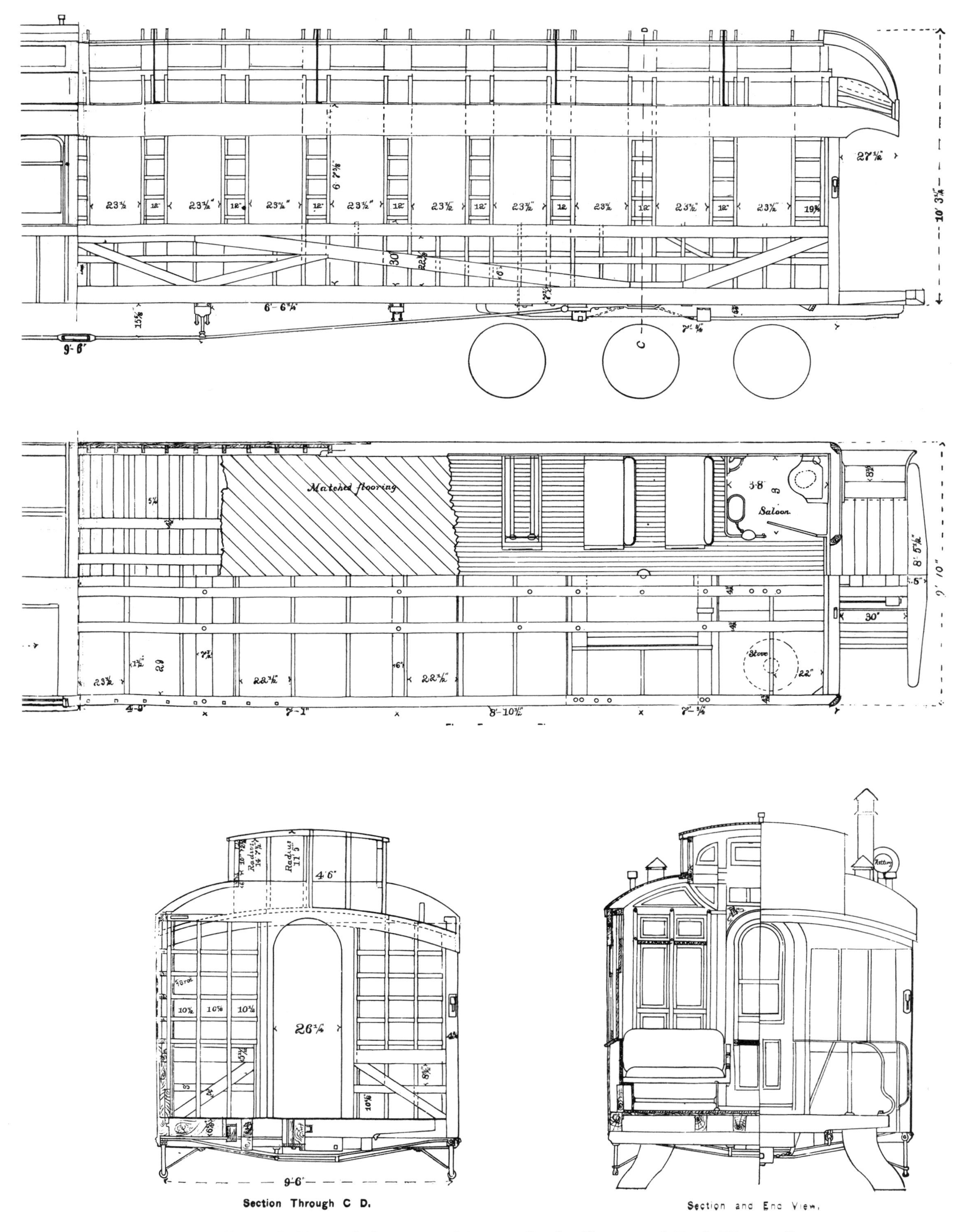

Figure 1.122 End elevation and framing details, Chicago and North Western Passenger car. (N.C.B., *January 1883*)

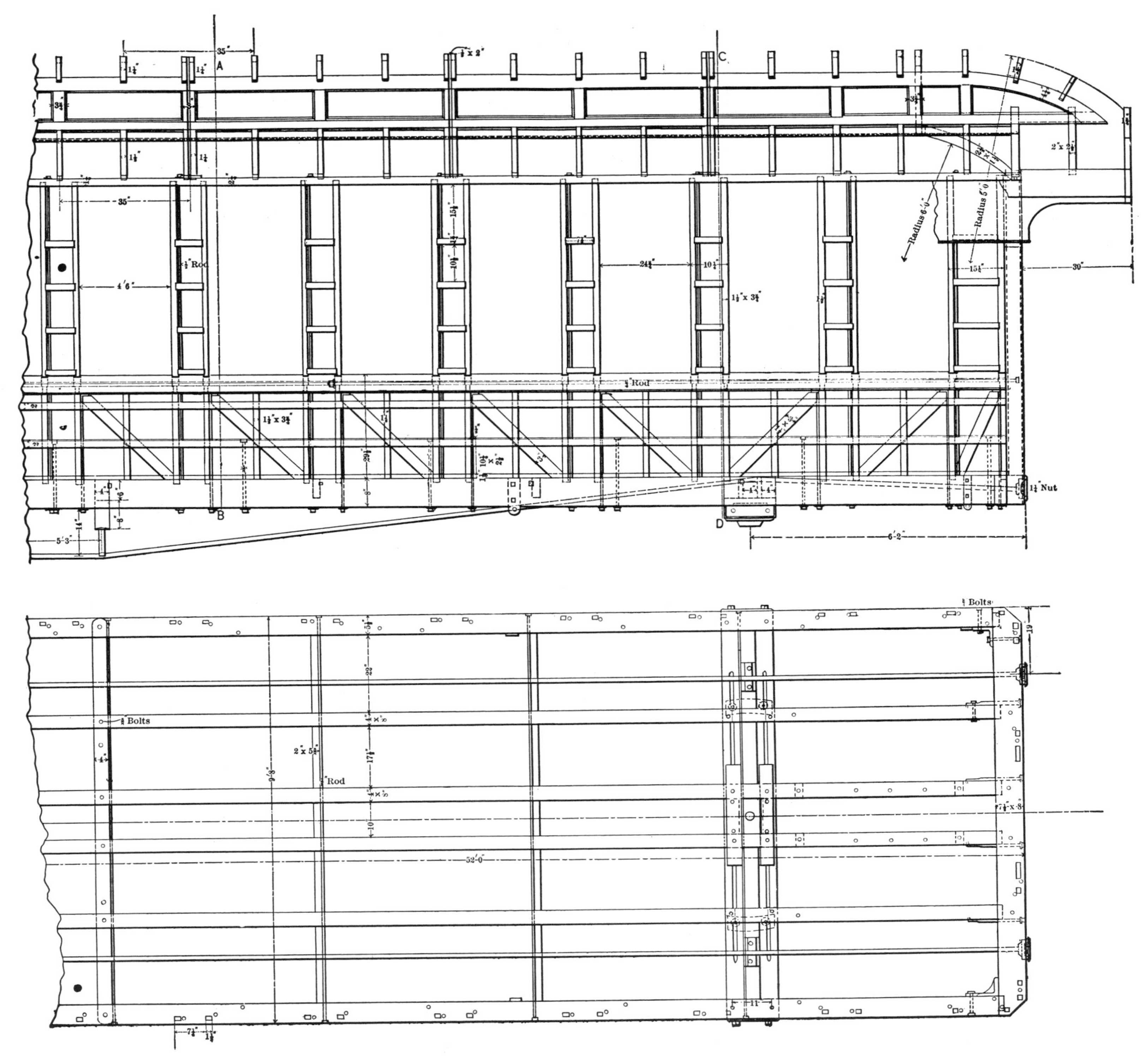

Figure 1.123 Erie Railway passenger coach framing details, 1884. (N.C.B., *September 1884*)

truss, as in the B & O class A6 coach of Figure 1.117. In the B & O coach, however, the plank was placed at the top of the panel just under the window rail.

The interior of the car was paneled in oak, cherry, and mahogany. The outside was praised as being in good taste "and notably devoid of the excessive fire-works style of decoration which becomes an eye-sore as soon as cinders and smoke have destroyed the original luster of the varnish and gilding."[131] The exterior was painted Tuscan red, with gilt striping and lettering. The car was fitted with Allen paper wheels and Miller couplers and platforms. (See Figures 1.130 to 1.132.)

BALTIMORE AND OHIO RAILROAD, CLASS A9, 1890

The Royal Blue was a plush day train intended to attract first-class travelers between New York and Washington. From 1890 to 1893 it was outfitted with luxurious new Pullman-built cars, among them Numbers 1051 to 1063. These were some of the largest and handsomest day cars operating anywhere in the United States. The majestic double windows, scroll platform railings, and enclosed vestibules reflected the latest thinking in first-class coach design.

The separate smoking room and large toilets at each end were

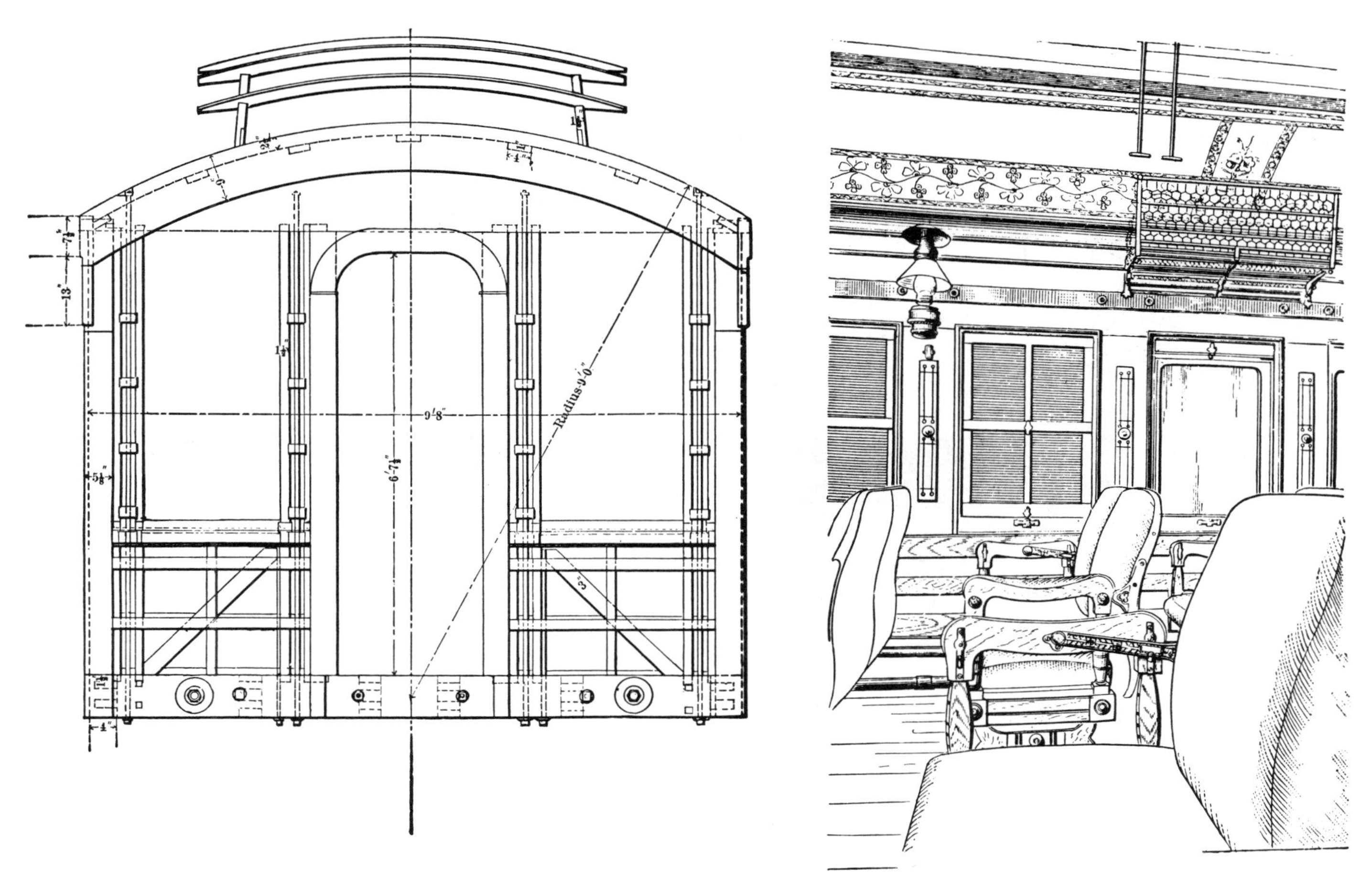

Figure 1.124 Erie Railway passenger car end and interior details, 1884. (N.C.B., September 1884)

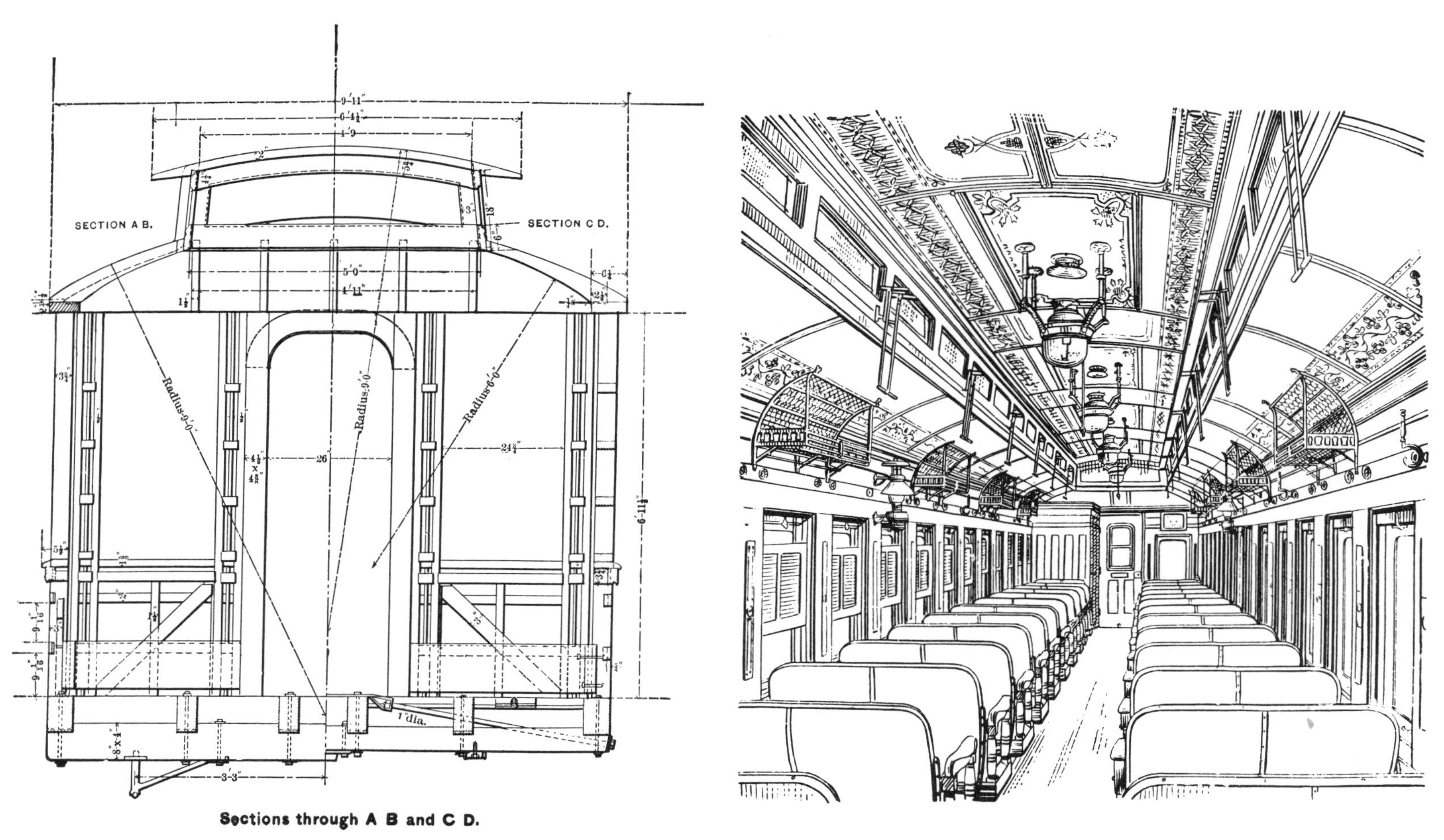

Figure 1.125 Interior details of an Erie Railway passenger car of about 1884. (N.C.B., September 1884)

Figure 1.126 A Philadelphia and Reading round-corner coach of about 1884. (N.C.B., *April 1884*)

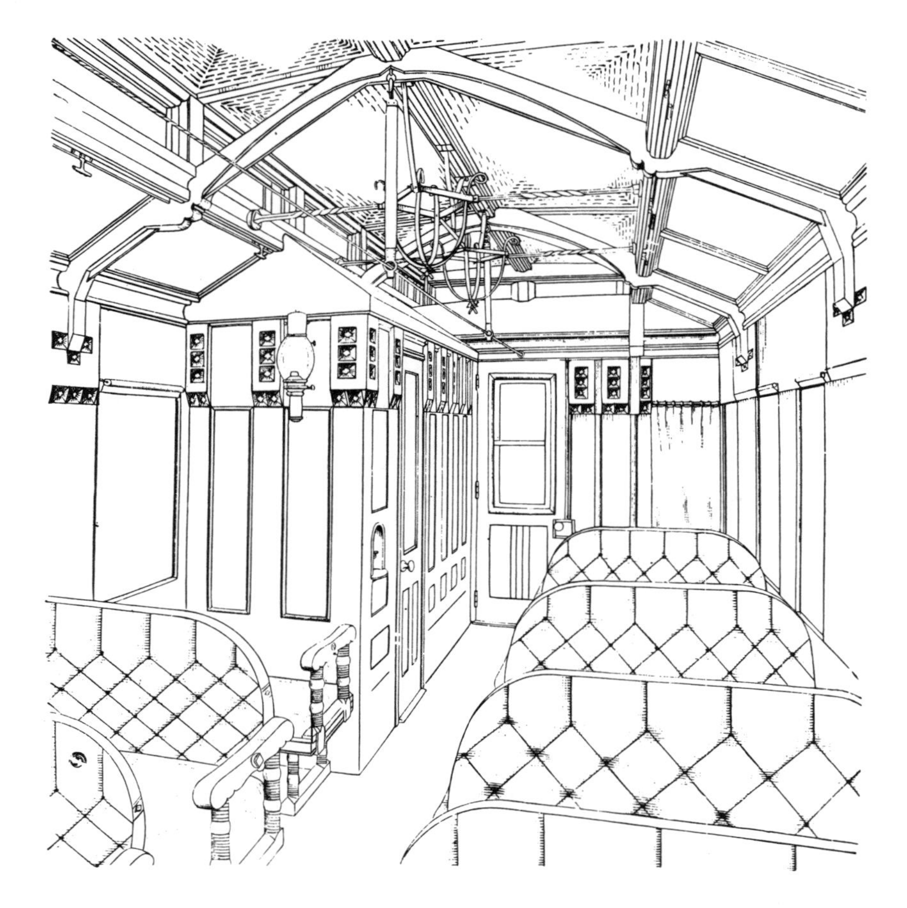

Figure 1.127 Interior of the Philadelphia and Reading round-corner coach, about 1884. (N.C.B., *April 1884*)

an extravagant use of space permissible only in first-class cars. The interiors were paneled in light mahogany; the upholstery was old-gold plush. A dazzling dark-blue exterior finish was highlighted by silver striping, with the Maryland coat of arms in full color. Vestibule ends, steam heat, and gas lighting completed the first-class furnishings of these cars.

A car approximately equal in size and vintage is displayed today in the B & O Transportation Museum, Baltimore. It is painted and lettered in the style of the Royal Blue cars, but here the similarities end. The car does not show the quality of the A9 class; it is finished as an ordinary day coach of the period. It has small rectangular windows in place of the large double openings of the original series. A photograph of the museum car taken in 1927 shows its number as 445. No further identification can be

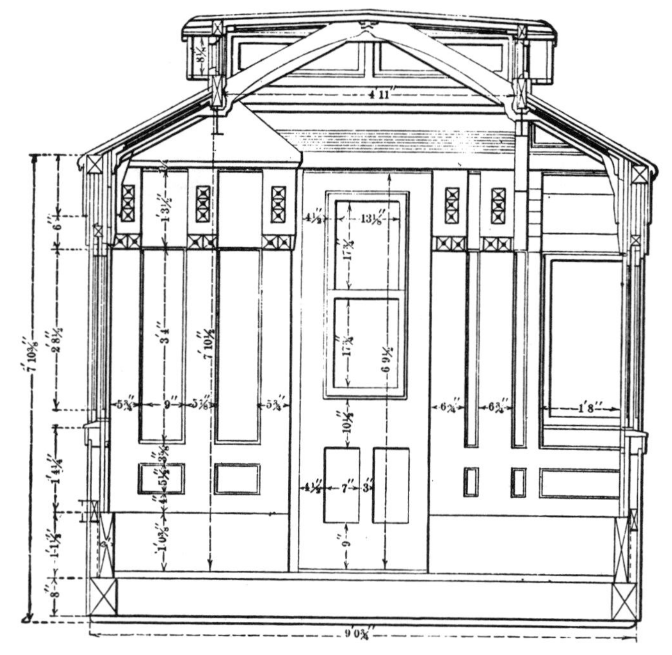
Cross Section Showing End, of Saloon.

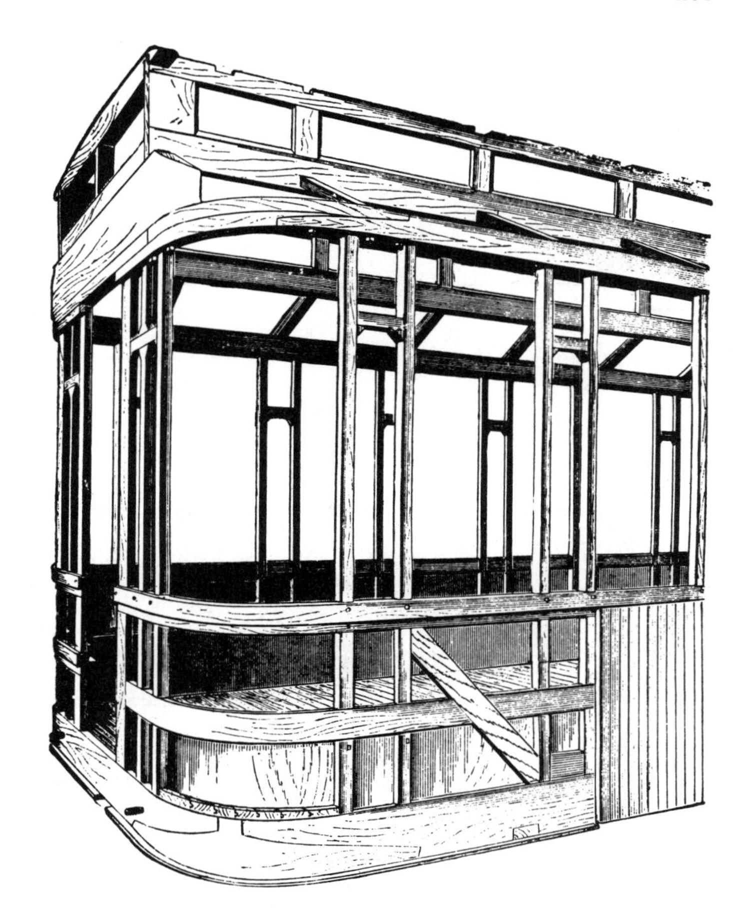
Fig. 1.—Corner Framing.

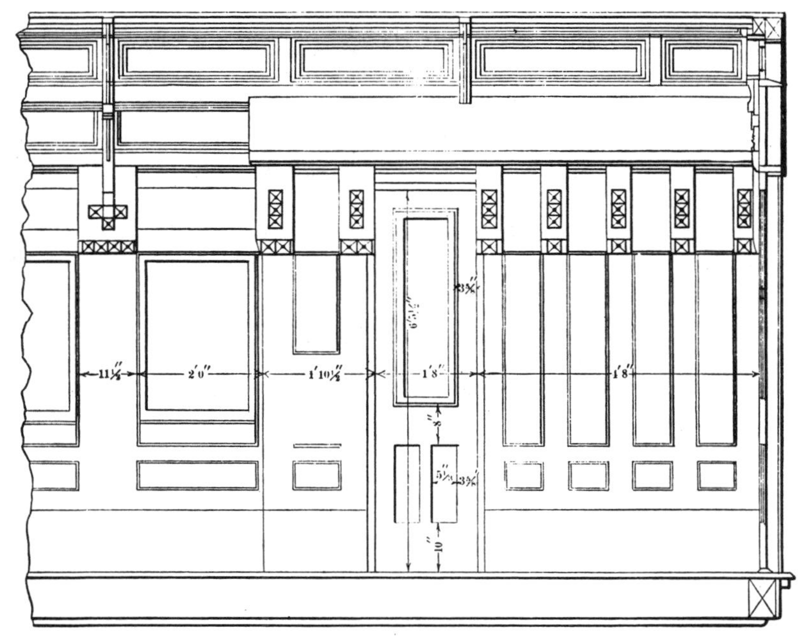
Side View of Saloon.

Figure 1.128 Philadelphia and Reading round-corner coach details, about 1884. (N.C.B., *April 1884*)

made; it does not agree with any of the B & O standard cars and may, like the "Civil War" cars in the same collection (Figures 1.110 to 1.113) have been created from available second-hand equipment to play a specific role in the 1927 Centennial celebration. Its classification as a Royal Blue car was established over the years by repeated exhibitions, until now its true history has been lost. (See Figures 1.133 and 1.134.)

BOSTON AND ALBANY RAILROAD COACH, 1890

In a departure from the usual American car of the period, this coach's severe, plain lines, single windows, thin-beaded sheathing, and arched roof give it a deceptively modern appearance. These features and several other unusual characteristics were developed by the road's master car builder, Fitch D. Adams. Compared with locomotive builders, most car builders remain shadowy figures. A few received acclaim in their lifetime, but almost none achieved the legendary status accorded to scores of locomotive men. Adams, an influential and outspoken leader in his field, deserves more than the token notice given him here.

Adams assumed direction of the B & A shops, just west of Boston at Allston, Massachusetts, in 1870. He found that the cars being built were both poorly framed and grossly overweight. He immediately adopted the six-sill floor frame and in the following year began using an arched truss plank. Adams was alone in his choice of this form of side-panel trussing, though it clearly demonstrated its superiority over the regular system. According to Adams, the arched truss weighed only one-eighth as much as a

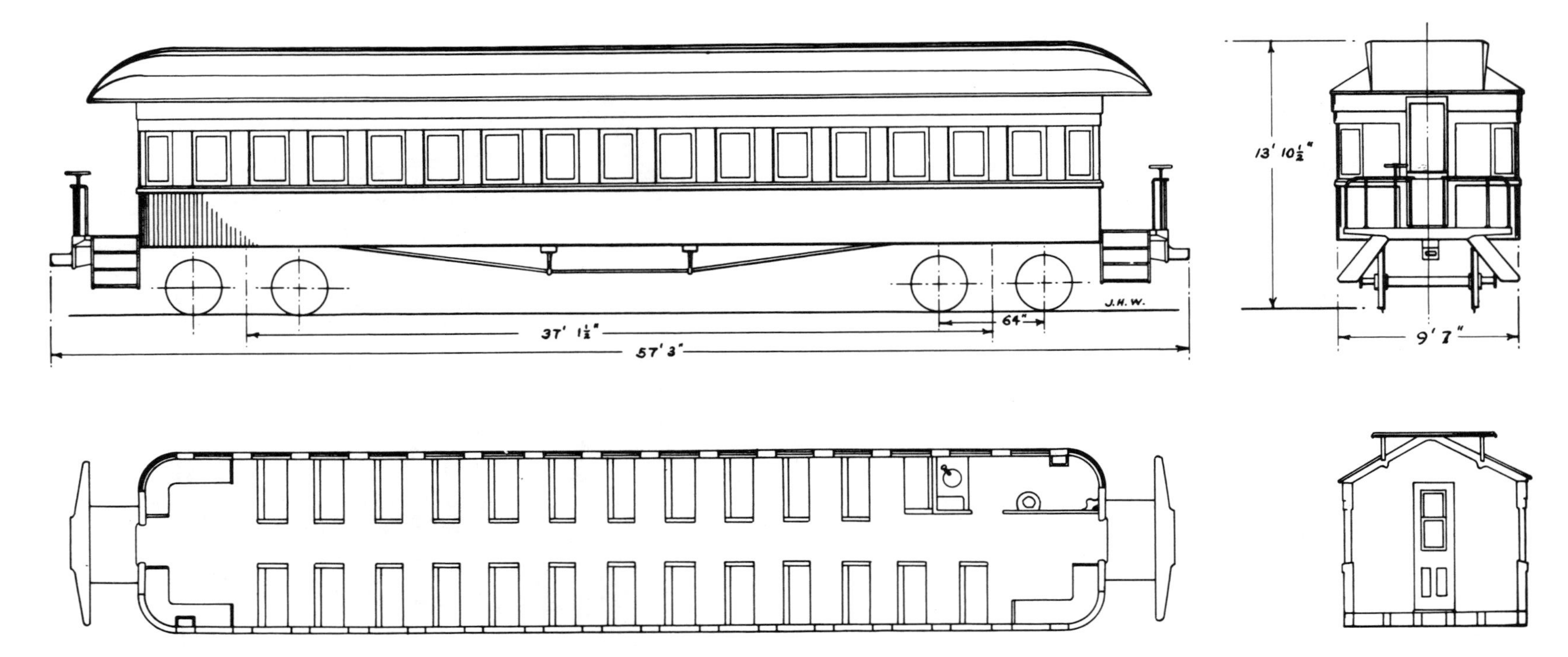

Figure 1.129 Philadelphia and Reading round-corner coach diagram drawing, about 1884. (Drawing by John H. White, Jr.)

truss plank. Old cars maintained their stiffness after twenty years of service, and the arched truss continued in use on the Boston line under Adams and his successor at least until 1899. It may well have survived until the steel car era.[132]

In the competition for lightweight construction, Adams won top honors. The trade press regularly featured his successes in this area.[133] In 1879 he fabricated a 68-passenger coach that weighed only 18 tons. Adams built drawing-room cars fully equal to Wagner's, but saved 10 tons by careful construction. The car pictured here was produced late in Adams's career and in effect summarizes his final ideas for passenger car construction. William Voss thought it important enough to include in his book, *Railway Car Construction*. Its description first appeared in the March 1891 *National Car Builder*.

The floor frame was a conventional six-sill variety made of yellow pine, except for the needle beams. The bolsters were iron. The spaces between the sills were packed with mineral wool for insulation.

The arched truss plank is clearly illustrated in the drawing. Like the posts and many other parts of the body, the trusses were made of ash. The car had four ⅞- by 5-inch truss planks, of which the outside plank was set into the posts. The double planks were apparently considered necessary for a body measuring 58 feet, 6 inches.

The arched roof is the most notable single feature, though the turtleback roof was not original with Adams. Mann sleeping cars with such roofs began operating on the Boston and Albany in the early 1880s. Adams adopted the plan in 1886, although it found little favor elsewhere in the country. The turtleback shape of the roof was designed to avoid the defects of the old arched roof by providing a high ceiling with ample headroom. The roof rose 2 feet over the sides of the car. The wooden carlines were three pieces of ½- by 1-inch ash glued together and steam-bent to the profile of the roof. Nine iron carlines were used for added strength. The absence of a clerestory made it necessary to install thirty globe ventilators.

The exterior sheathing was white wood cut into narrow strips and faced with a half-round front, creating an effect which very much resembled corrugated iron. This decorative scheme was suggested by H. H. Richardson, the architect for some suburban cars built by Adams in 1881. The window sills were capped with a thin, neatly sculptured iron casting—another example of Adams's willingness to develop new solutions for every detail of car construction.

The car was equipped with Miller couplers, Westinghouse brakes, steam heating, and 42-inch steel-tired wheels. The interior was finished in a plain Eastlake style. The head lining was Mexican mahogany. The seats were covered in golden-brown plush. The overall length was 65 feet, 6 inches; width, 10 feet, 3 inches; height over the ventilators, 13 feet, 10½ inches. (See Figure 1.135.)

CHICAGO, PEORIA, AND ST. LOUIS COACH, 1891

This first-class coach or chair car was built for through service between Chicago and St. Louis on the Chicago, Peoria, and St. Louis (Jacksonville Southeastern Line) and the Santa Fe Railroads. It and its companion cars were constructed by the St. Charles Car Company of Missouri. The plant did not begin passenger car production until 1886, but in later years it became the main passenger car works of the American Car and Foundry Company.

The general arrangement is clearly shown in the drawing.[134] The interior was paneled in mahogany with a quartered-oak headlining. The seats (in a Scarritt-Forney pattern) were covered with plush, while the smoking-room sofas were upholstered in leather. The washstands were of Tennessee marble. Heating was provided by a Baker heater. Oil lamps were used, but wiring was installed in case electric lighting should be desired at some future date. Miller couplers and Boyden air brakes completed the special equipment. The smoking room, water closets, and utility

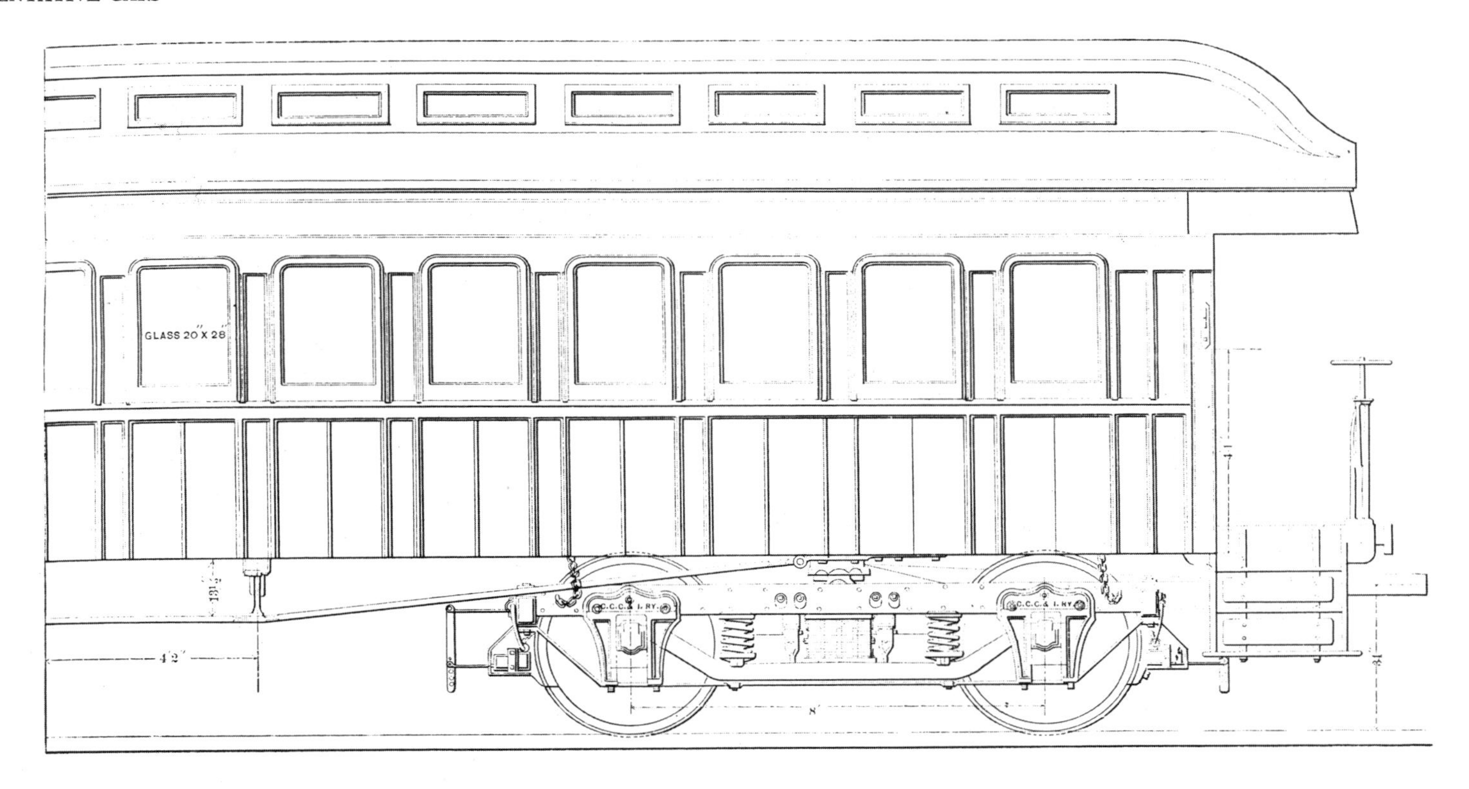

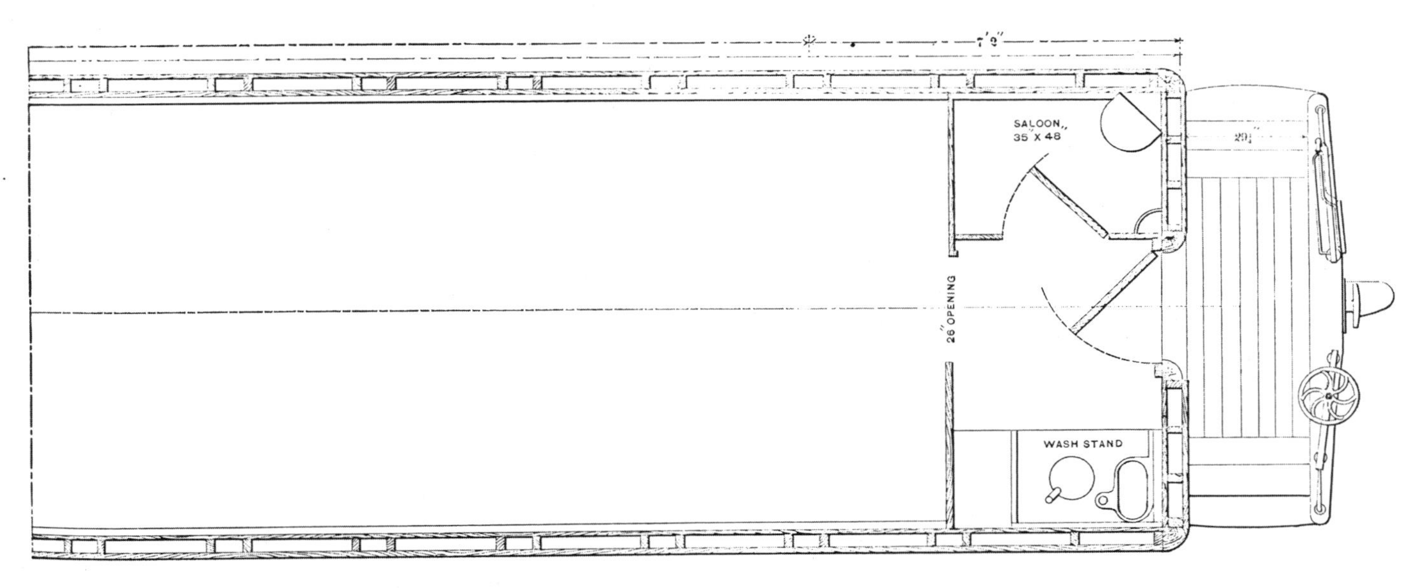

Figure 1.130 A Cleveland, Columbus, Cincinnati, and Indianapolis coach of about 1884. (N.C.B., *July 1884*)

closets restricted seating to only 42 passengers. (See Figure 1.136.)

PENNSYLVANIA RAILROAD, CLASS PH, 1892

The Pennsylvania broke away from its traditional use of the duckbill roof in 1892 to adopt a more modern clerestory profile. Vestibules and advanced appliances were used as well, but here the road's design progress seemed to halt. The body was kept short; batten-board sides and small windows reflected the conservative outlook of the Pennsylvania management.

The interior decoration was somewhat more flamboyant than the usual P R R treatment, perhaps partly because of the approaching Columbian Exposition and the anticipated rivalry for that traffic among Eastern lines. The road's sensible oak paneling was retained, but the ceiling was covered in light-blue silk tapestry. The seats were upholstered in red plush, and the floor was carpeted. The extra-wide clerestory gave the interior an agreeable feeling of space.[135] Frost gas lamps and steam heat provided an extra luxury for the 64 passengers.

The car appears to be overtrucked, yet its weight (81,100 pounds) and the desire for a smooth ride and limited axle load-

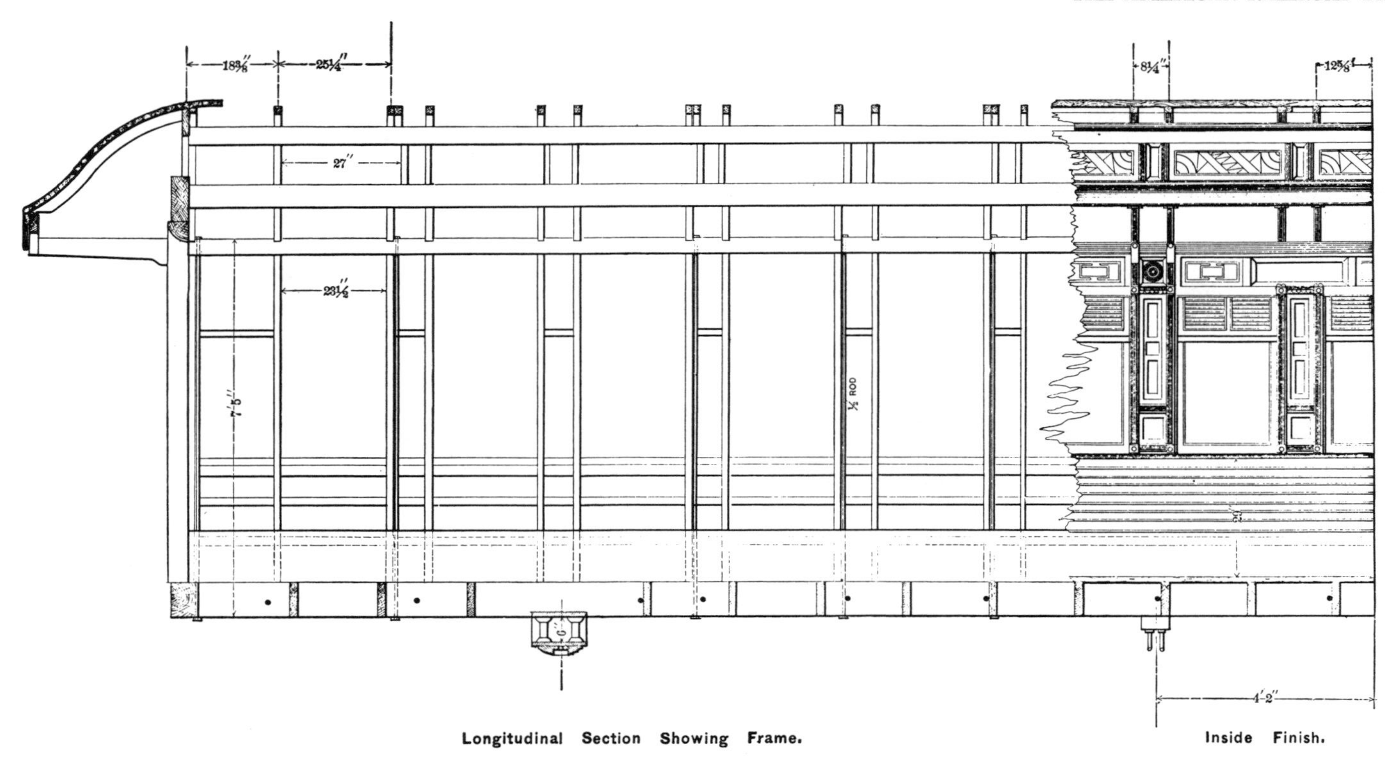

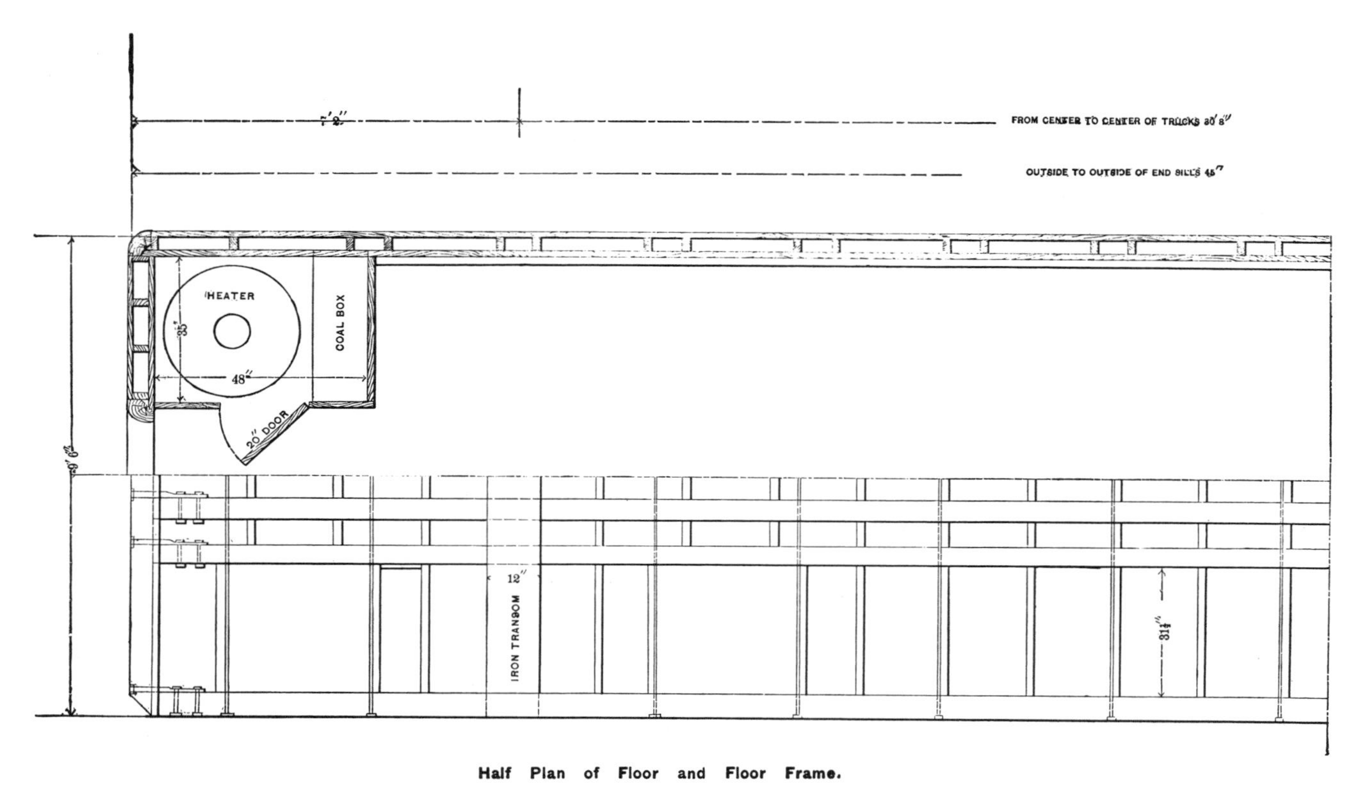

Figure 1.131 A Cleveland, Columbus, Cincinnati, and Indianapolis coach of about 1884, detail. (N.C.B., *July 1884*)

ings account for the decision to install twelve wheels on so short a car. (See Figures 1.137 and 1.138.)

ILLINOIS CENTRAL SUBURBAN CAR, 1893

Many railroads depend on old cars for suburban service, but some lines with extensive suburban operations, such as the Illinois Central's in the Chicago area, built special lightweight cars for this traffic. The general design was very similar to that of a New York Elevated car. Space and luxury were held to a minimum, because rides were short and fares low. Loading and unloading were facilitated by the wide aisles provided by longitudinal seats at either end.

Ironically, on these cars designed for utility more than appearance, featuring rattan seats and offering no toilets, the builders felt it necessary to install a solid-mahogany interior with carved panels over the end doors. A Baker stove furnished heat; Miller couplers and Westinghouse brakes rounded out the equipment. The form of lighting was not given. The car seated 48. The drawings and data are from the May 1893 *Railroad Car Journal.* See Figure 1.43 for an earlier Illinois Central suburban car of similiar design. (See Figure 1.139.)

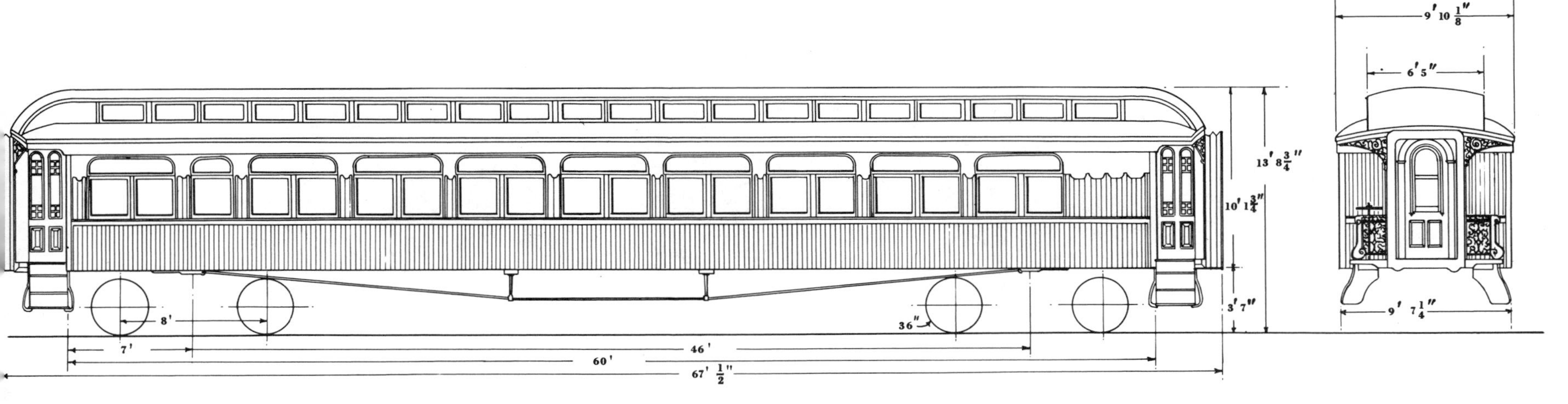

Figure 1.133 Baltimore and Ohio Railroad passenger car class A-9; a narrow-vestibule car built by Pullman in 1890. (Traced by John H. White, Jr.)

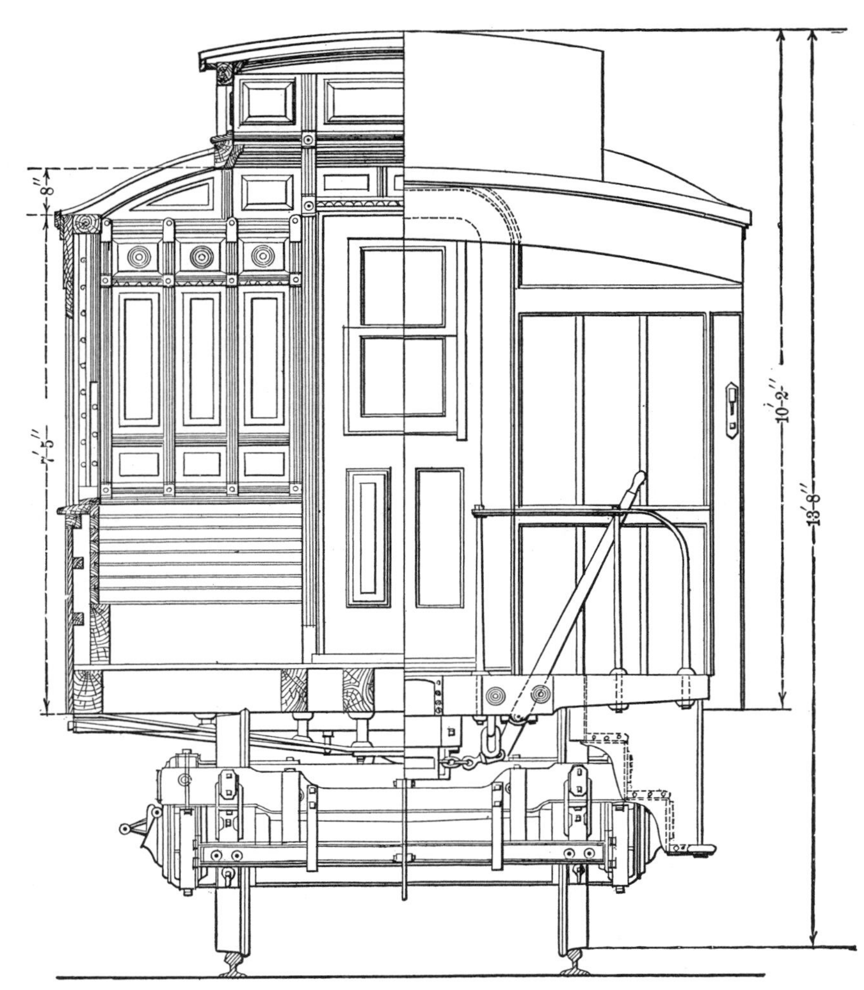

Figure 1.132 End elevation of a Cleveland, Columbus, Cincinnati, and Indianapolis passenger coach of about 1884. (N.C.B., *July 1884*)

NEW YORK CENTRAL RAILROAD, 80-FOOT COACHES, 1892–1893

In the competition for passenger traffic the New York Central attempted to outdo its rivals by the creation of supercoaches. Surely the largest all-wood coaches ever built, they were inspired more by a desire for a spectacular exhibit at the Columbian Exposition than by any practical operating considerations. The West Albany car shops built two 80-foot cars for the Empire State Express. These cars were shown at the 1893 Chicago Fair together with the celebrated 999 Locomotive. To viewers familiar with car construction, they were a sensation, as was the 112-mile-per-hour run made several months earlier with the same equipment.

The framing followed the iron-plate reinforced plan developed several years earlier by Thomas A. Bissell for Wagner sleeping cars. The floor framing was particularly heavy, as can be seen in the drawings.[136] Massive as the wood structure seems, it was apparently not sufficient to maintain the body's stiffness. Supplementary truss rods were required at either end of the body (Figure 1.141).

The exterior was painted Tuscan red, with gold and black lettering. The trucks were finished in dark brown. The interior was paneled in mahogany, and carving was held to a minimum. The seats were covered in copper-red plush. The head lining was painted white and gold. Major equipment included Gould vestibules, platforms, and couplers; Pintsch gaslights; and steam heating. Seating was provided for 84; the approximate cost was given as $8,720. The body weighed 62,200 pounds, the trucks 33,200 pounds, and the total weight was 95,400 pounds. (See Figures 1.140 and 1.141.)

NORFOLK AND WESTERN RAILWAY COACH, 1894

This drawing, one of the most complete graphic records ever prepared for an American passenger car, shows a fully equipped coach with all appliances in detail.[137] The interior was finished in oak paneling, the hardware was nickel-plated, and the aisles were covered with Brussels carpeting. The exterior was painted Tuscan red, with gold trimming and letters. The car closely followed the standard Pennsylvania Railroad design, which is not remarkable in view of the Keystone system's financial involvement in the N & W.

Frost gasoline lamps, Spear hot-air heaters, and Janney-Buhoup couplers were among the important auxiliaries used. The complete car weighed 69,350 pounds. (See Figures 1.142 to 1.144.)

NEW HAVEN RAILROAD, COACH NUMBER 1203, 1901

Serving the populous area between New York and Boston, the New Haven has always been one of the nation's largest passenger carriers. Its present service reputation hardly reflects the time when the road took pains to maintain a model fleet of cars. Representative of that more fastidious past was coach 1203, one

Figure 1.134 Baltimore and Ohio Railroad passenger car class A-9, Pullman, 1890. (Pullman Neg. 1506)

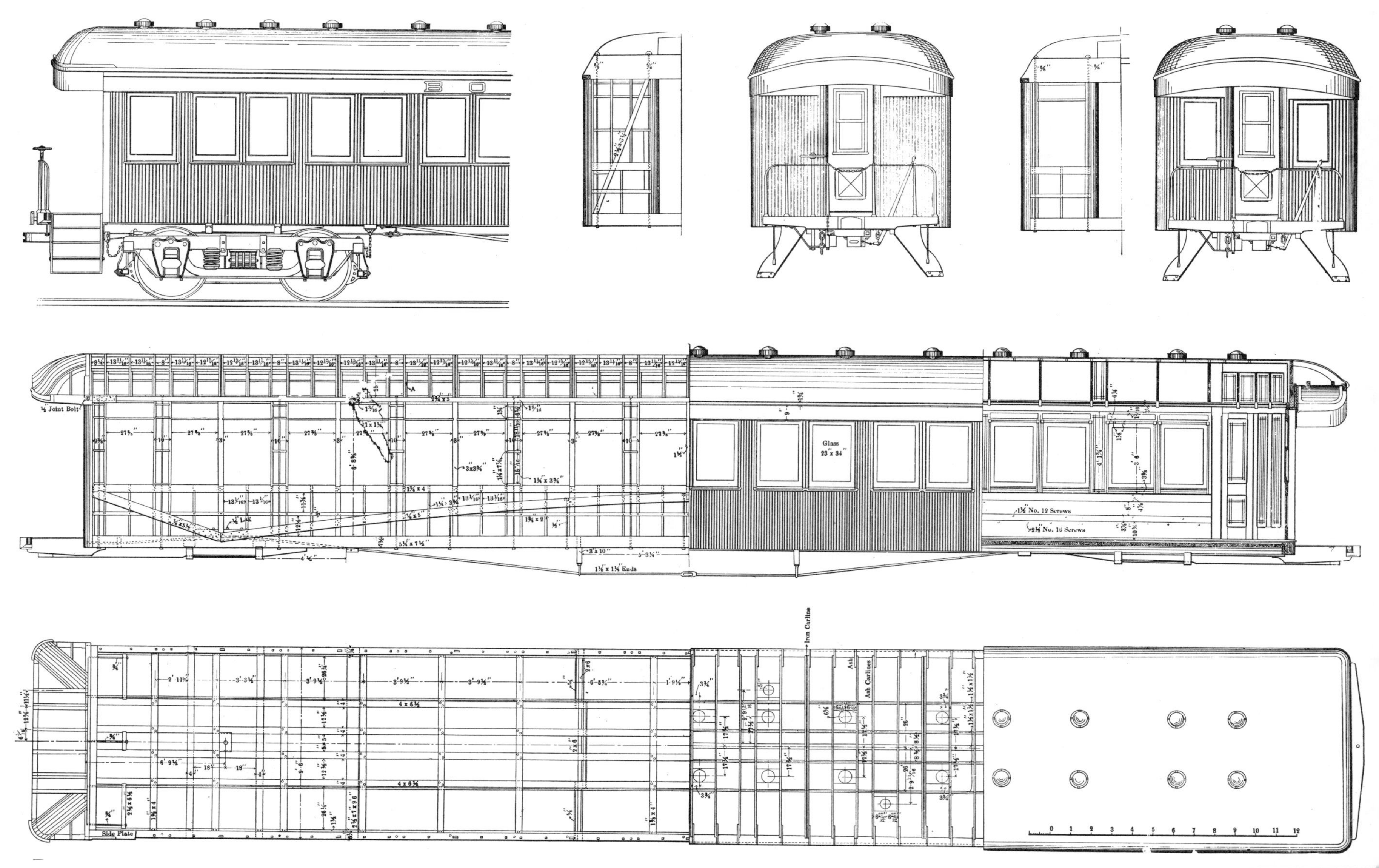

Figure 1.135 Boston and Albany arched-roof car of about 1890. (N.C.B., *March 1891*)

of twenty built in 1901 by Osgood Bradley. Like other New Haven cars of the period, it was sheathed in copper. This unusual form of exterior finish was favored by the road's master car builder, W. P. Appleyard (see Chapter 5). The car weighed 71,750 pounds.[138] (See Figures 1.145 and 1.146.)

NICKEL PLATE, COACH NUMBER 43, 1907

The New York, Chicago, and St. Louis Railroad, better known as the Nickel Plate, acquired six wooden coaches in 1907 from the American Car and Foundry Company's Jeffersonville, Indiana shops (the former Ohio Falls Car Company). The next coaches, purchased in 1910, were the last wooden cars acquired by the Nickel Plate. The road turned to steel with its subsequent order.

The Number 43's service career is typical of many late wooden cars. Bought almost at the end of the wooden era, it was a mistaken investment that the road tried to salvage by periodic rebuilding. Electric lighting and steam heating were added, and between 1923 and 1925 steel underframes were applied. In later

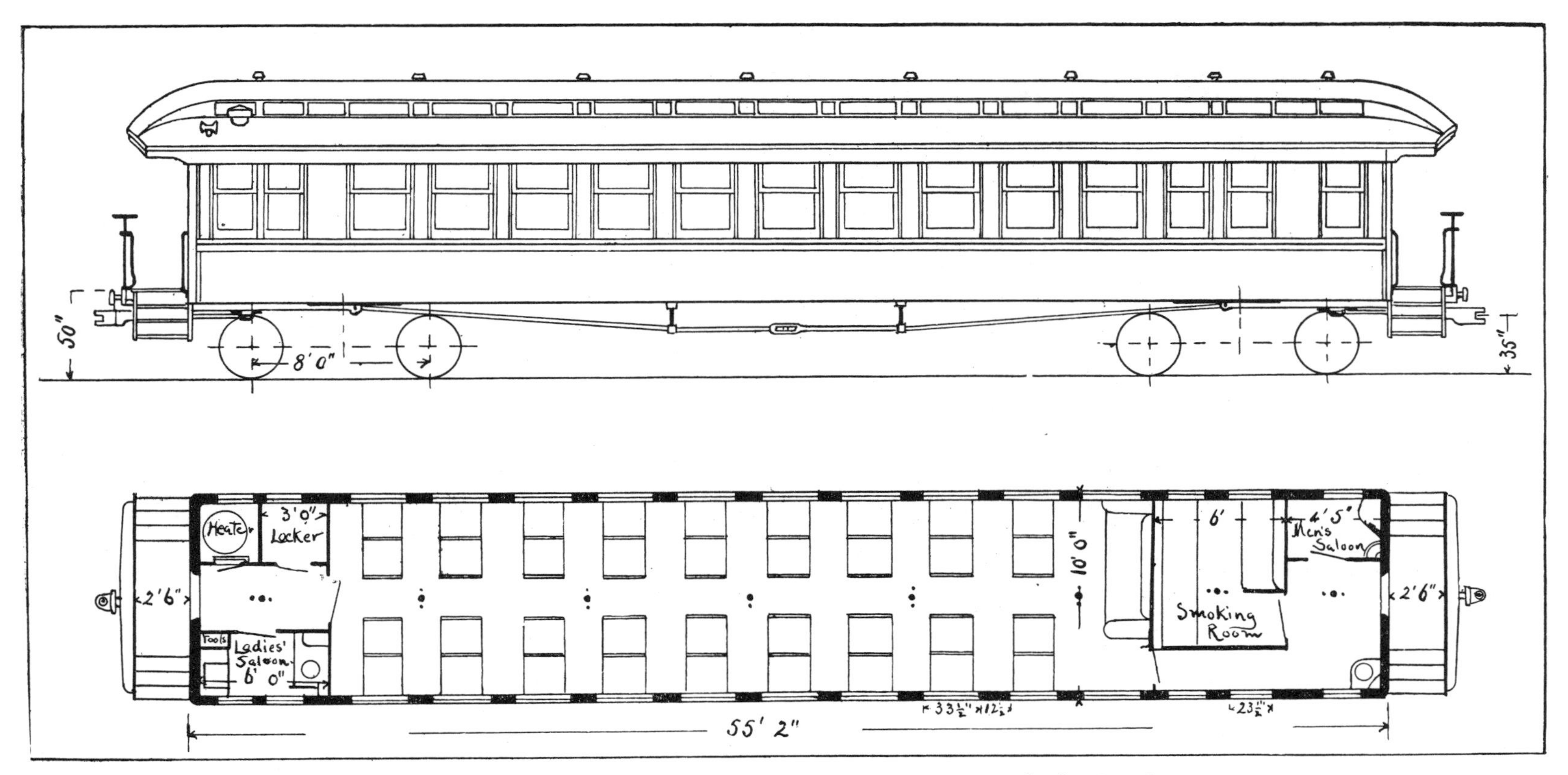

Figure 1.136 A Jacksonville Southeastern coach built by the St. Charles Car Company of St. Charles, Missouri, about 1891. (American Railroad Journal, *October 1891*)

Figure 1.137 The Pennsylvania Railroad coach Number 950, class PH, designed and introduced in 1892.

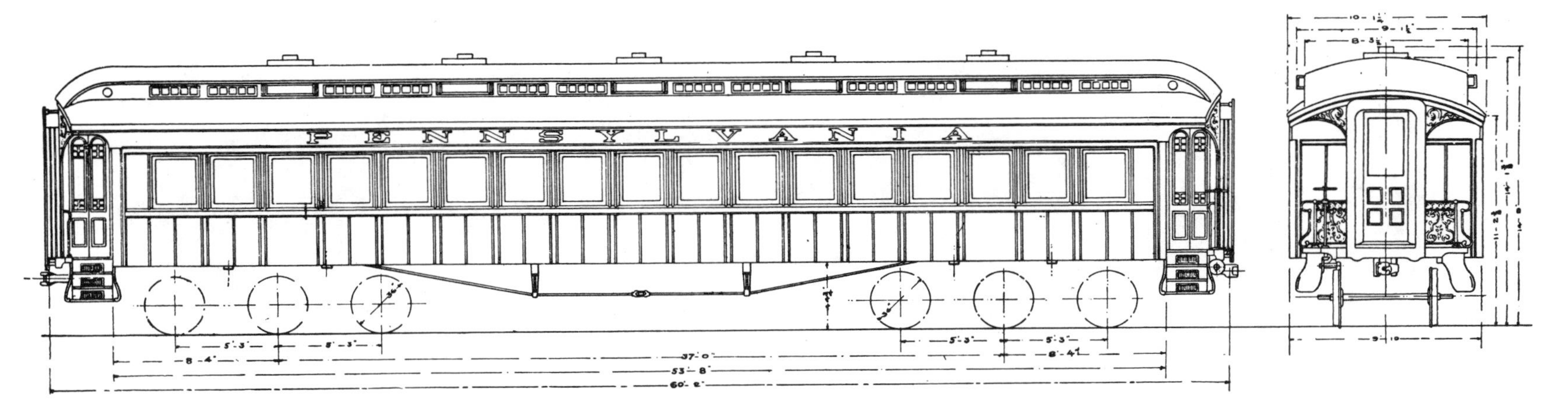

Figure 1.138 Pennsylvania Railroad passenger coach class PH, 1892.

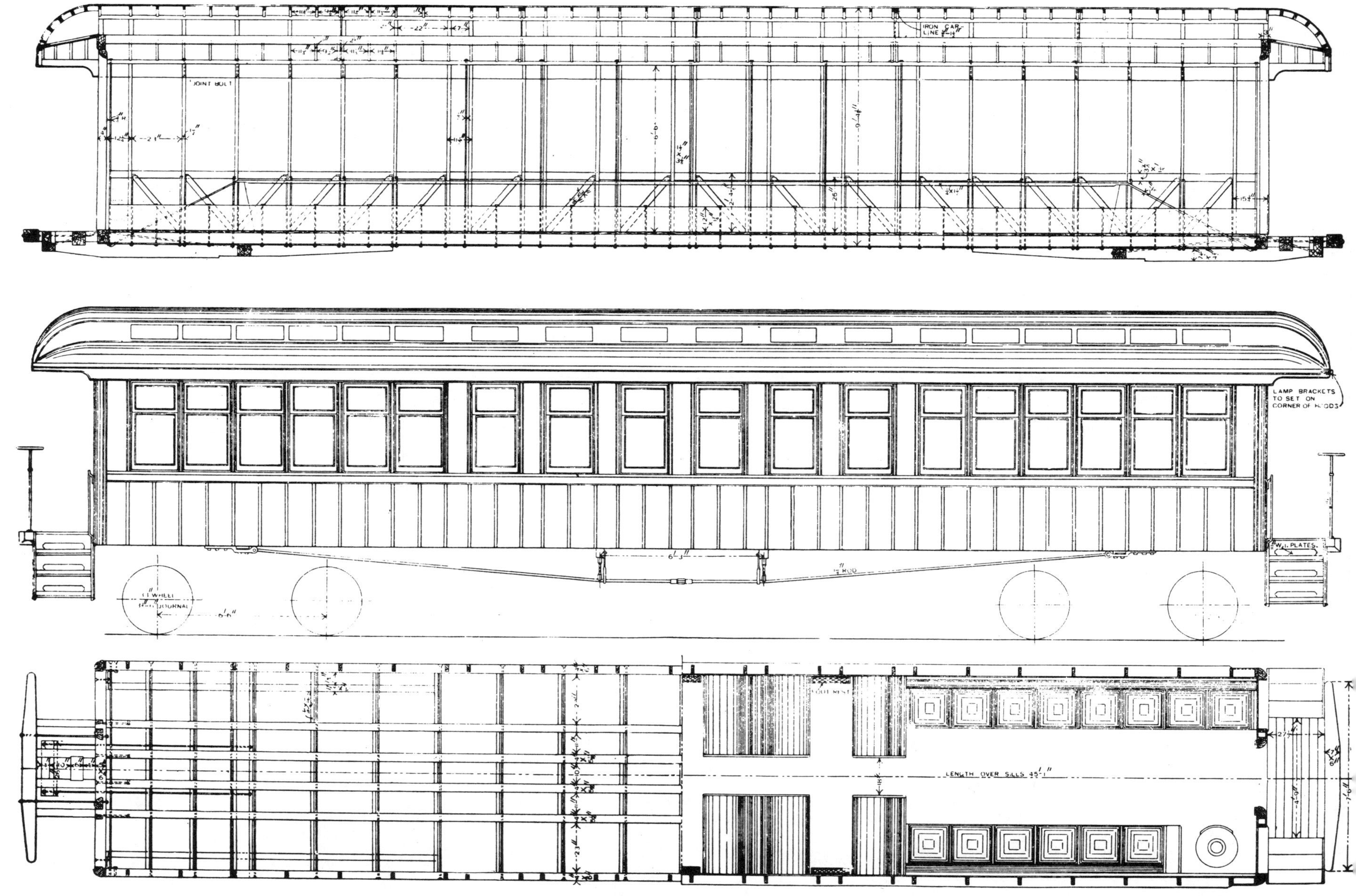

Figure 1.139 An Illinois Central Railroad suburban car built about 1893. (Railroad Car Journal, *May 1893*)

Figure 1.140 The New York Central and Hudson River Railroad coach Number 999. This 80-foot car was one of the longest wooden passenger cars constructed. (*New York Central Neg. 9493*)

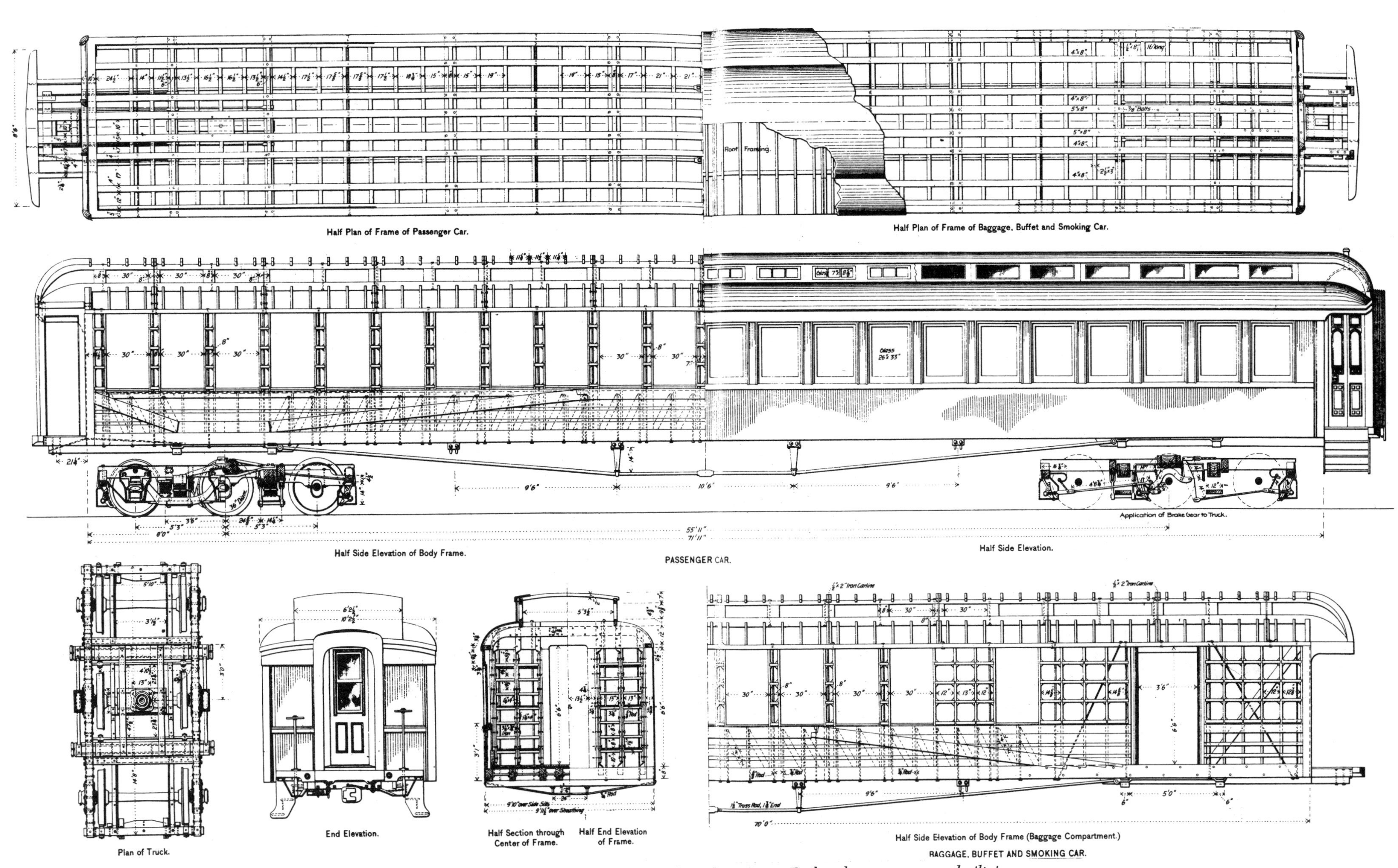

Figure 1.141 A New York Central and Hudson River Railroad passenger car built in 1893. (Engineering News, *December 14, 1893*)

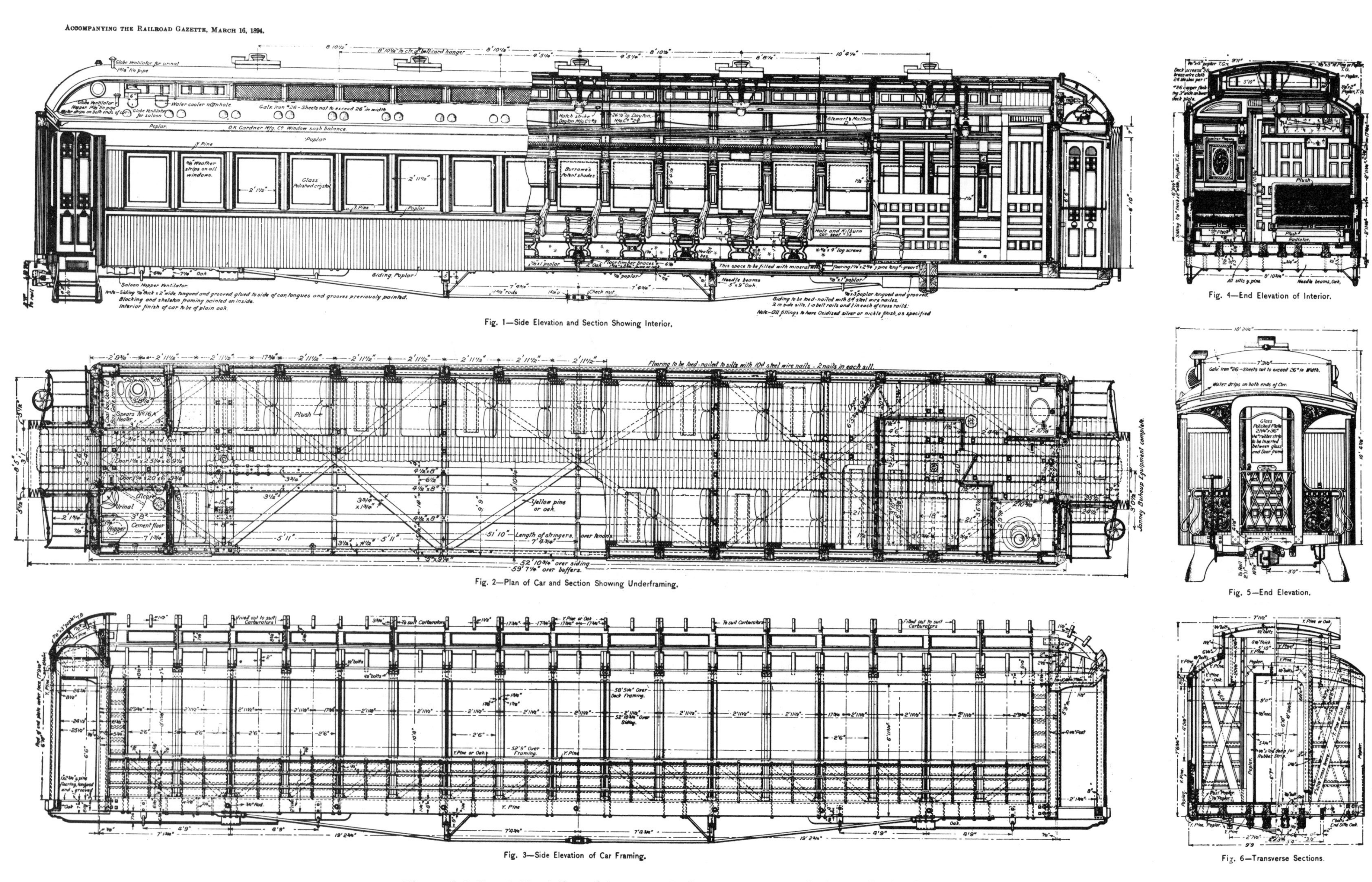

Figure 1.142 A Norfolk and Western Railway narrow-vestibule car built about 1894. (Railroad Gazette, *August 16, 1894*)

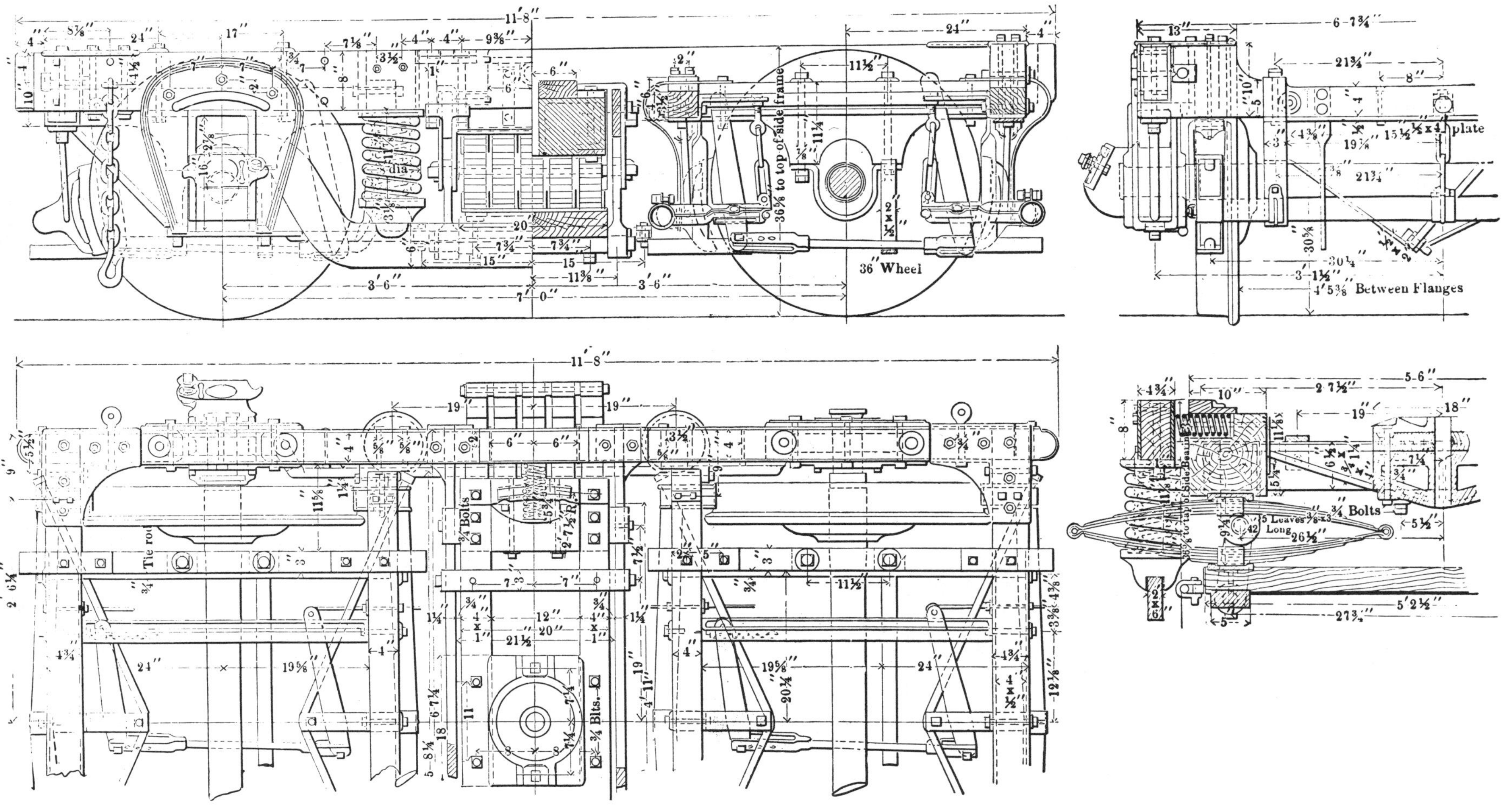

Figure 1.143 A Norfolk and Western Railway passenger car truck of about 1894. (Railway Master Mechanics' Magazine, *March 1894*)

Figure 1.144 This Norfolk and Western Railway passenger car interior of 1905 is very similar to the cars illustrated in Figures 1.142 and 1.143. (Pullman Neg. 7780)

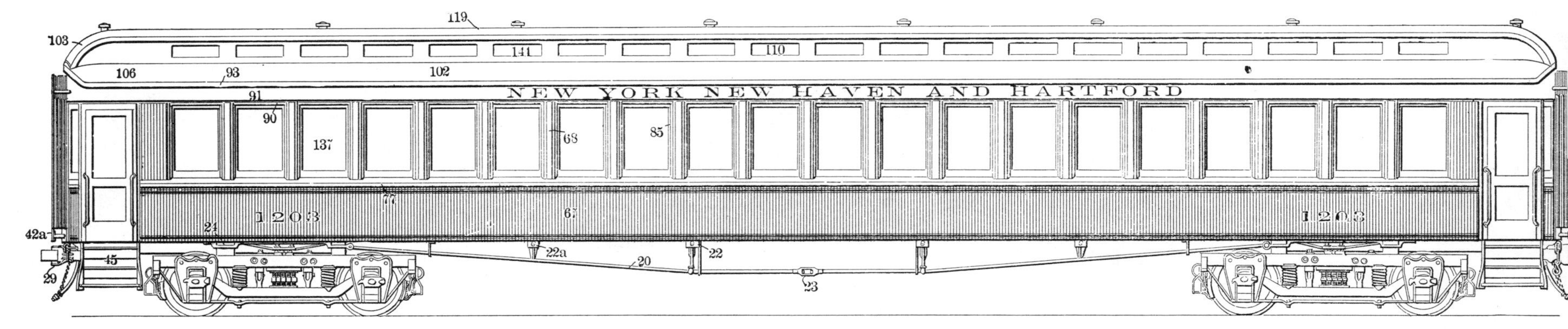

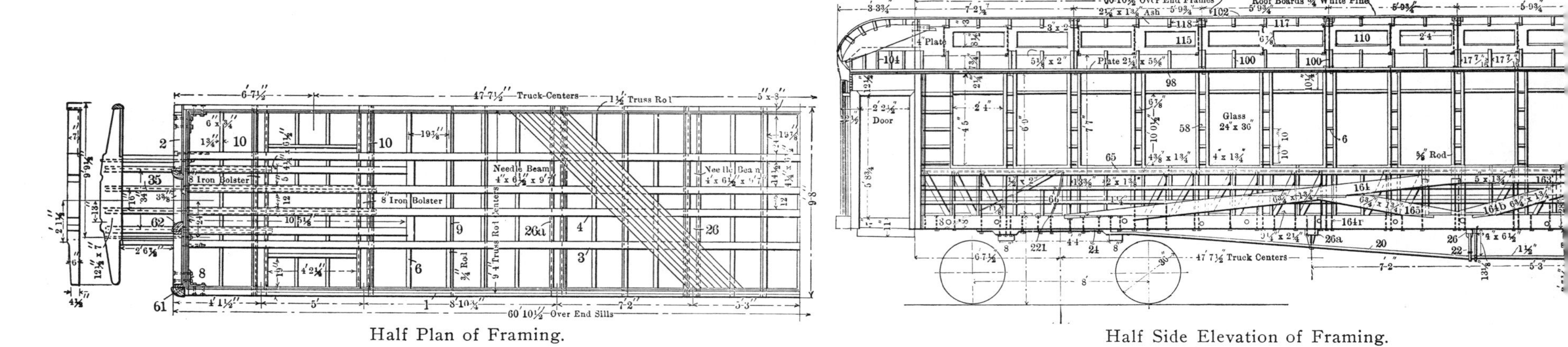

Figure 1.145 The New Haven coach Number 1203, built by Osgood Bradley in 1901. (Car Builders' Dictionary, *1906*)

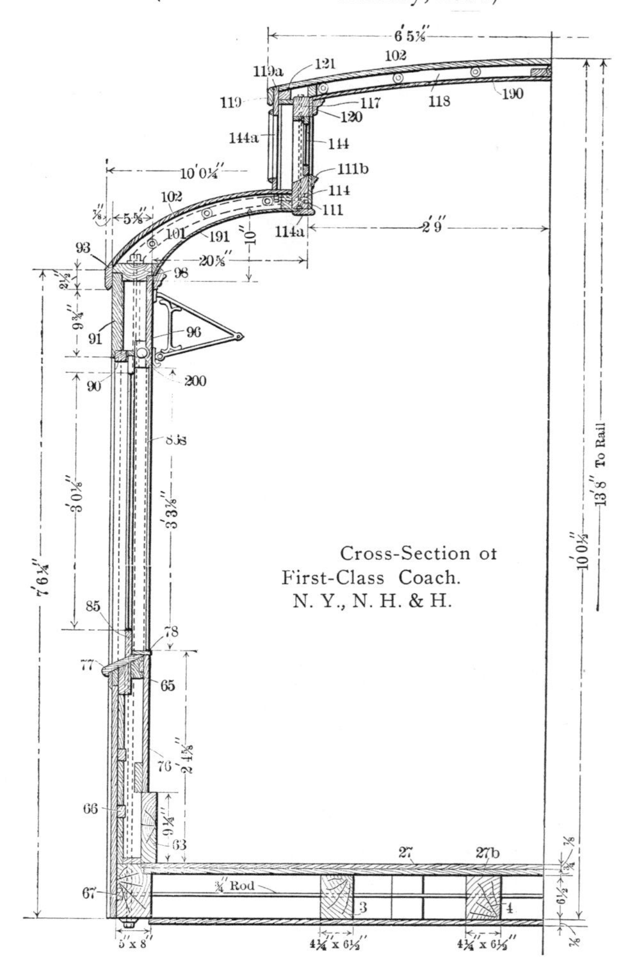

years the 43 was downgraded to work service. Many of its companion cars remained in the active passenger fleet until after World War II.

Seating was provided for 72 passengers. The weight was 46.2 tons.[139] (See Figures 1.147 and 1.148.)

NEW YORK CENTRAL RAILROAD, 1700 SERIES COACH, 1909–1911

Time was fast running out for the wooden passenger car when the American Car and Foundry Company produced the last coach in this order sometime in 1911. Steel cars had been commercially manufactured for several years. To maintain its position with rival lines, the New York Central began to take delivery of steel cars during the following year. The arch-windowed 1799 was the last new wooden car acquired by the Central, and was among the last purchased by a major domestic railroad.

The drawing here is from a diagram book published by the New York Central Railroad in 1913. By this date the car had been fitted with a steel underframe and electric lighting. Seating was for 84 passengers, and the total weight was 130,300 pounds. (See Figure 1.149.)

Figure 1.146 Cross-section drawing of the New Haven's Number 1203. (Car Builders' Dictionary, *1906*)

Figure 1.147 The New York, Chicago, and St. Louis coach Number 43, built by the Ohio Falls Car Company in 1907. (John Keller)

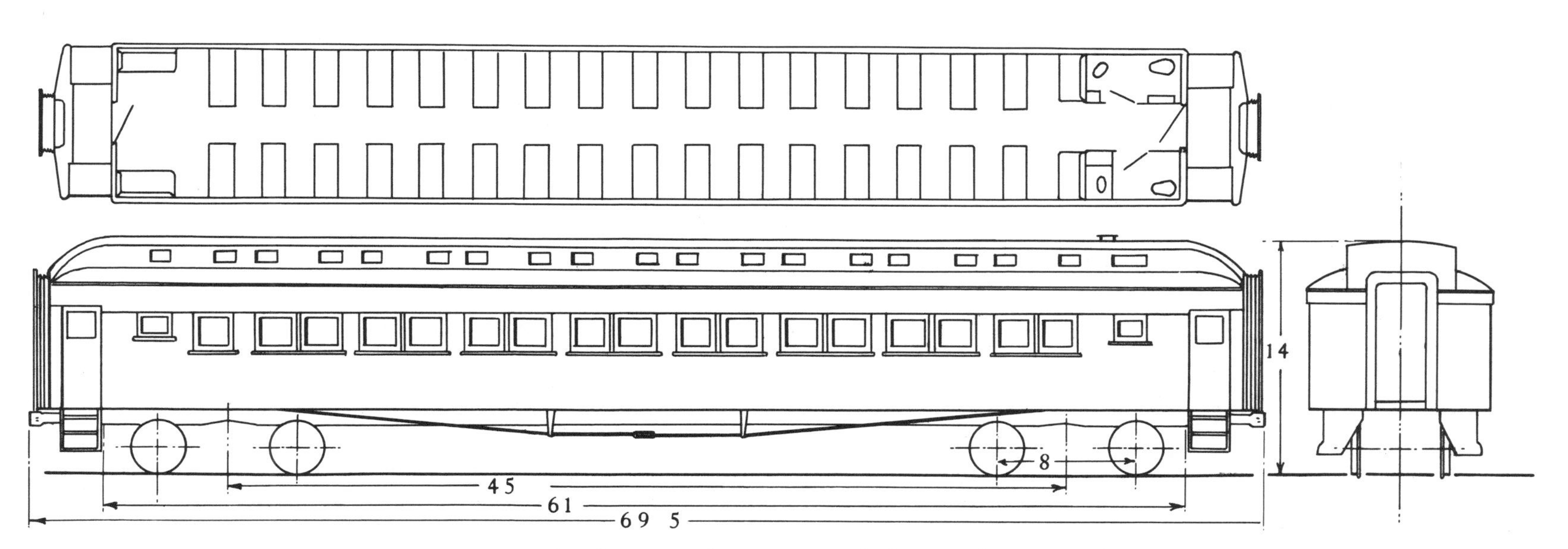

Figure 1.148 New York, Chicago, and St. Louis coach diagram drawing. (John Keller)

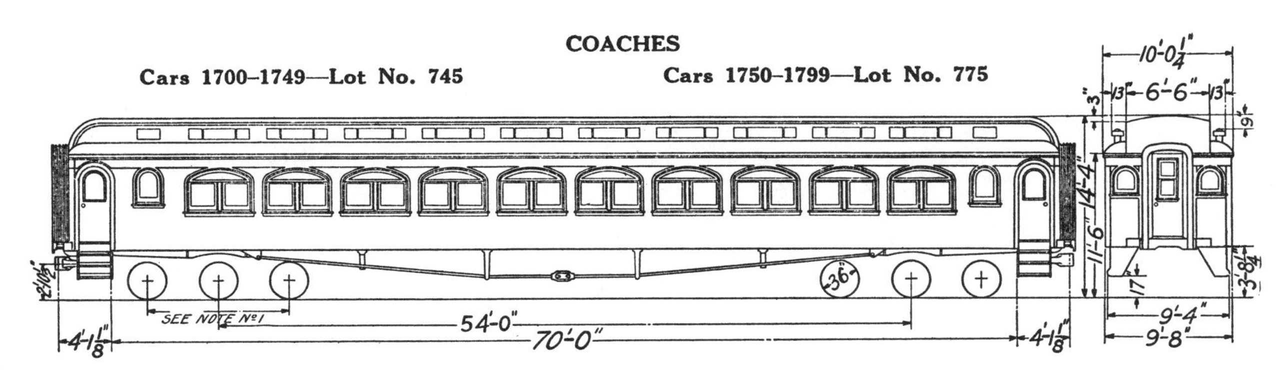

Figure 1.149 New York Central coach built in 1911 by the American Car and Foundry Company.

CHAPTER TWO

The Day Coach

ERA OF THE METAL CAR

THE HISTORY OF METALLIC PASSENGER CARS can be divided into four major periods: experimentation, introduction, full production, and streamlining. During the extended period of experimentation (1845–1902), individual inventors worked with little support or interest from the railroads or commercial car builders. In the end these inventors failed to persuade the industry to accept iron or steel cars. In the second phase, a short, intensive period of introduction (1902–1910), the industry itself came to support the movement. With the establishment of an acceptable design, which was largely an imitation of the general form of existing wooden cars, the period of full production began. It was to continue through the so-called standard or heavyweight era of American railroading until the early 1930s, when more attention was given to lightweight construction. This last phase, the streamlined era, took hold in the mid-1930s and continues to the present time. After 1955 the decline of railroad passenger traffic, except for suburban or other specialized services, reduced the market for new equipment and thus retarded further developments.

Few improvements have been longer heralded or slower in coming than the metallic passenger car. Steel cars gained acceptance at a measured pace; half a century separated the first experiments and the initial production models. Every major railway accident evoked a flurry of newspaper and magazine articles advocating collision- and fire-resistant cars. The defects of the wooden car were laboriously repeated: it was weak, likely to splinter, and highly combustible. Iron construction would end the hazards of railway travel—from the mid-1840s on, this was a standard editorial theme.

Naturally the publicity stimulated experimentation. Between 1850 and 1900 over thirty patents were granted for iron or steel passenger car construction. Some twenty sample cars were actually built, and countless others were projected. But nearly all this work was done by amateurs, whose efforts drew indifference or ridicule from professional railroad circles. These inexperienced and imprudent inventors generally produced bizarre designs that did not appeal to the practical mechanics who governed the nation's railways. And the complexity and impracticality of most designs was only one obstacle to the development of the iron car, for there were also the obvious economic disadvantages already mentioned in Chapter 1. Wooden cars were cheap, light, and for the most part, adequate for the service of that day. Even the best iron cars proved far heavier and more costly than ordinary wooden coaches.

Before American railroads had ventured more than a hundred miles from the Atlantic coast, iron passenger cars were being built in Great Britain. They were few in number, however, and most were used for the poorest class of travel. The Great Western Railway, one of the largest investors in iron, built ten cars with metal bodies in 1845 and acquired another ten three years later.[1] In the following decade, iron-frame passenger cars were introduced in both England and Germany and within a few years became relatively common in Europe.[2]

By the mid-1850s iron-bodied coal cars in considerable number were running on the Baltimore and Ohio and the Reading railroads. Many had been in use for ten years. Although durable and workaday, they appear to have attracted little attention beyond their home districts. America was far slower than Europe to accept iron for passenger conveyances.

The earliest proposal in this country for a passenger car of iron was offered in 1846 by H. L. B. Lewis of New York City.[3] Lewis envisioned an iron car body suspended between a pair of 6-foot-diameter wheels, with a smaller pair of wheels to guide the truck. A handbill distributed by the inventor described the projected car as 45 feet long and 8 feet square, with a capacity for 80 passengers. The weight less the truck was grossly underestimated at only 4 tons. The idea did not merit a patent, much less a prototype. In fact, its absurdity set the movement back a good five years.

A more practical design came from the iron-making center of Troy, New York, where in 1851 Thomas E. Warren devised a sheet-iron car following a plate girder truss.[4] The side sheets were stiffened by pressed sheet-iron columns set at regular intervals. Iron rods tied the structure together. It is possible that Warren's design was inspired by architectural ironwork fabricated in the Troy mills. Certainly the classical and Venetian mountings gave the design a strong architectural flavor. A patent (No. 10142) was granted on October 18, 1853, but as far as is known, no test cars were built.

The most enduring champion of the metallic passenger car was Bernard Joachim La Mothe, a physician who settled in New York City sometime before 1853.[5] This persistent Frenchman argued for the adoption of safe passenger cars for nearly forty years. In that time he obtained nine patents and built four passenger cars as well as countless freight cars of iron. At first La Mothe concerned himself with designs for a fireproof building. There was considerable interest at the time in such structures—particularly those of cast iron. La Mothe envisioned a latticework or basket-weave framing of iron bands.

In 1853 La Mothe adapted his scheme to railway cars and displayed a model at the New York Crystal Palace Exhibit.[6] A patent (No. 10721) issued on April 4, 1854, outlined his plan, which was an early attempt at unitized fabrication. Light iron bands were interwoven to form a strong basket-like cage. Each crossing was riveted. The cage was fastened to a foundation ring, also made of band iron, that encircled the car and formed the end and side sills. Sheet-iron panels covered the roof and sides, with windows set in at the appropriate openings. Since the center sills, floor, seats, and interior paneling were of wood, La Mothe could not claim that the car was fireproof.

La Mothe engaged a civil engineer, Alfred Sears, to put the design in a form more acceptable to the industry and to help with the promotion, and they prepared a series of pamphlets and advertisements.[7] The trade press was generous, praising the design and uncritically accepting La Mothe's exaggerated claims. It became known as the Life Preserving Car that would safeguard

LA MOTHE'S PATENT IRON RAILROAD CAR.

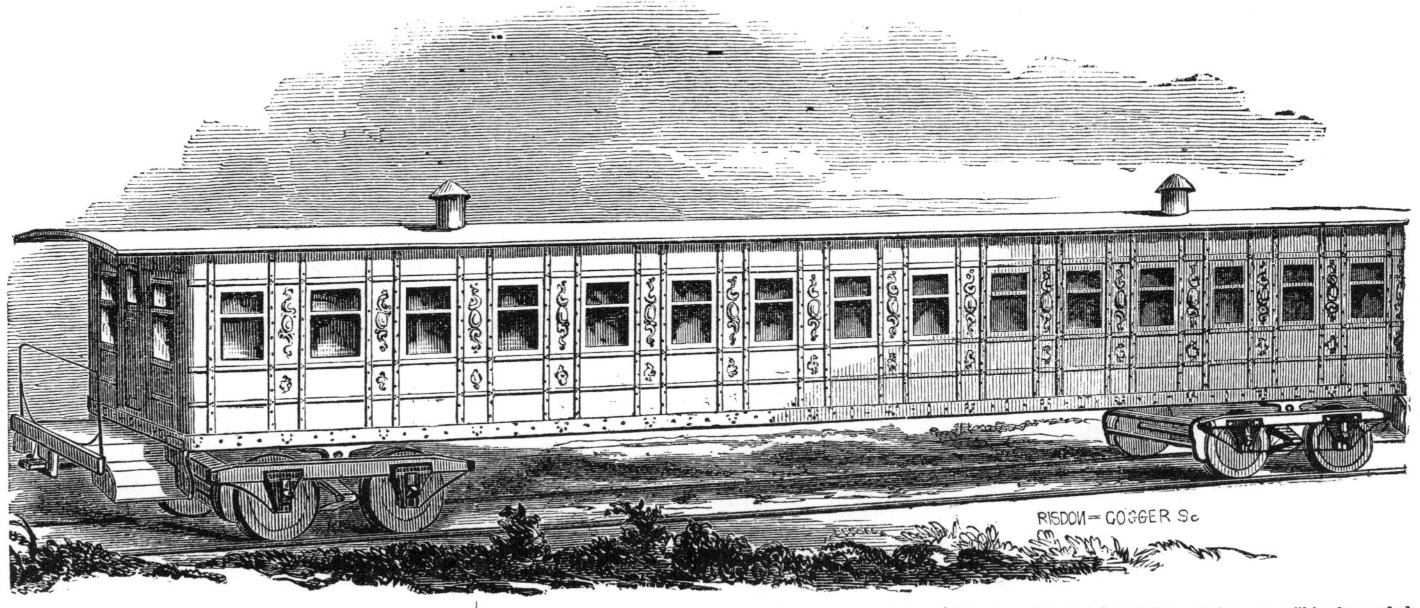

WE are now prepared to furnish this Car to railroad companies at short notice and reasonable rates.

Notwithstanding its extraordinary advantages, the prices will be arranged wholly with reference to the cost of construction—without regard to patent rights.

We are now building passenger and freight cars for several companies; and it is desirable that parties ordering give early notice of their wants.

The striking features of this principle are:—simplicity—cheapness—durability—superior safety in cases of accident—facility of repairing when damaged—and less weight compared with the wooden cars of the same capacity; these cars for 60 passengers are more than two tons lighter than the ordinary cars, while the strength is immeasurably greater.

We guarantee these points in the acceptance of orders.

The advantages may be tested by personal observation in this city. Detailed descriptions of the cars will be forwarded to parties wishing them.

ALFRED SEARS,
Civil Engineer and Architect,
Agent.
OFFICE—9 SPRUCE ST., NEW YORK.

Figure 2.1 La Mothe's patented iron car, as advertised in the American Railroad Journal *in 1855.*

passengers from any calamity. An engraving prepared for the *American Railroad Journal* showed a coach of the most ordinary appearance, indistinguishable from a wooden car of that day except for the tiny rivet heads (see Figure 2.1).

La Mothe's leaflets offered a strong, no-nonsense economic argument. For example, *How to Make Railroads Pay a Good Dividend, and at the Same Time Give Security to Life in Railroad Travel* (published in 1856) declared that iron cars would drastically reduce maintenance, renewal, and damage costs. According to the leaflet, iron cars would never wear out, burn, or permit serious injuries to passengers. Moreover, they could be built so that they were lighter than wooden cars, thereby reducing operating costs and saving wear on roadbeds and bridges. These sensible-sounding claims, together with the sensible-looking design, were buttressed by a number of models. The 1854 patent specification mentions a 4-pound model able to support 1,000 pounds. The leaflet *How to Make Railroads Pay* speaks of a model 32 inches long, 7½ inches wide, and 6½ inches high that could support 1,400 pounds. A third model 42 inches long carried a load 196 times its own weight, and although the side bulged somewhat, it sprang back to its original shape when the weight was removed.

Still no orders came, and La Mothe decided to build a test car. For the sake of economy it was a small streetcar constructed by the New York Locomotive Works of Jersey City in the spring of 1855.[8] It weighed 3,300 pounds and followed the general form of a city omnibus. Experience gained in its production suggested that this weight could be cut by 300 pounds; indeed, La Mothe insisted that he could pare the weight to a single ton with a more careful arrangement of the material.

During the next fall, Passavant and Archer, iron fabricators of New York City, were busy producing six iron horsecars on La Mothe's plan for the city of Boston.[9] Even more important, work was underway in Paterson, New Jersey, on a steam railroad coach designed by La Mothe. The job was taken not by a regular car builder but by a tinsmith, William Cundell (1804–1879), who constructed a shed on one side of his shop to accommodate the project. The Boston and Worcester Railroad was the main purchaser, but several other Boston lines held an interest for test purposes. The work went slowly, then stopped altogether with the panic of 1857; but Cundell survived the panic and had resumed construction by the spring of 1859. The specimen car was finished in July 1859. A drawing from La Mothe's 1856 pamphlet is thought to be a reasonable likeness (Figure 2.2). Press accounts offered the following information: The overall length was 51 feet, 6 inches; the body was 46 feet by 8 feet, 4 inches. Each side had fifteen windows; seating was provided for 60 passengers. The car contained 6,200 feet of band iron and over 9,200 rivets. The roof was of galvanized iron. The window sashes were brass; dark glass was used in the transom windows and end doors. The interior was handsomely outfitted. The curtains were of a rich English rep, the moldings were black walnut, and gilt mirrors occupied the corners. Scenic and patriotic views by a local artist ornamented the papiêr-mâché panels between the windows. Among them were paintings of John Smith's rescue, Evangeline, Niagara Falls, and George Washington's tomb. The exterior of the car had been coated with fish oil and white oxide of zinc to prevent rusting, then finished with yellow and brown paint.

In all this detail the actual weight of the car was never given. During construction La Mothe had claimed that it would weigh 5 tons less than an equivalent wooden coach, and before construction he had claimed that he could build a 60-passenger car of 9

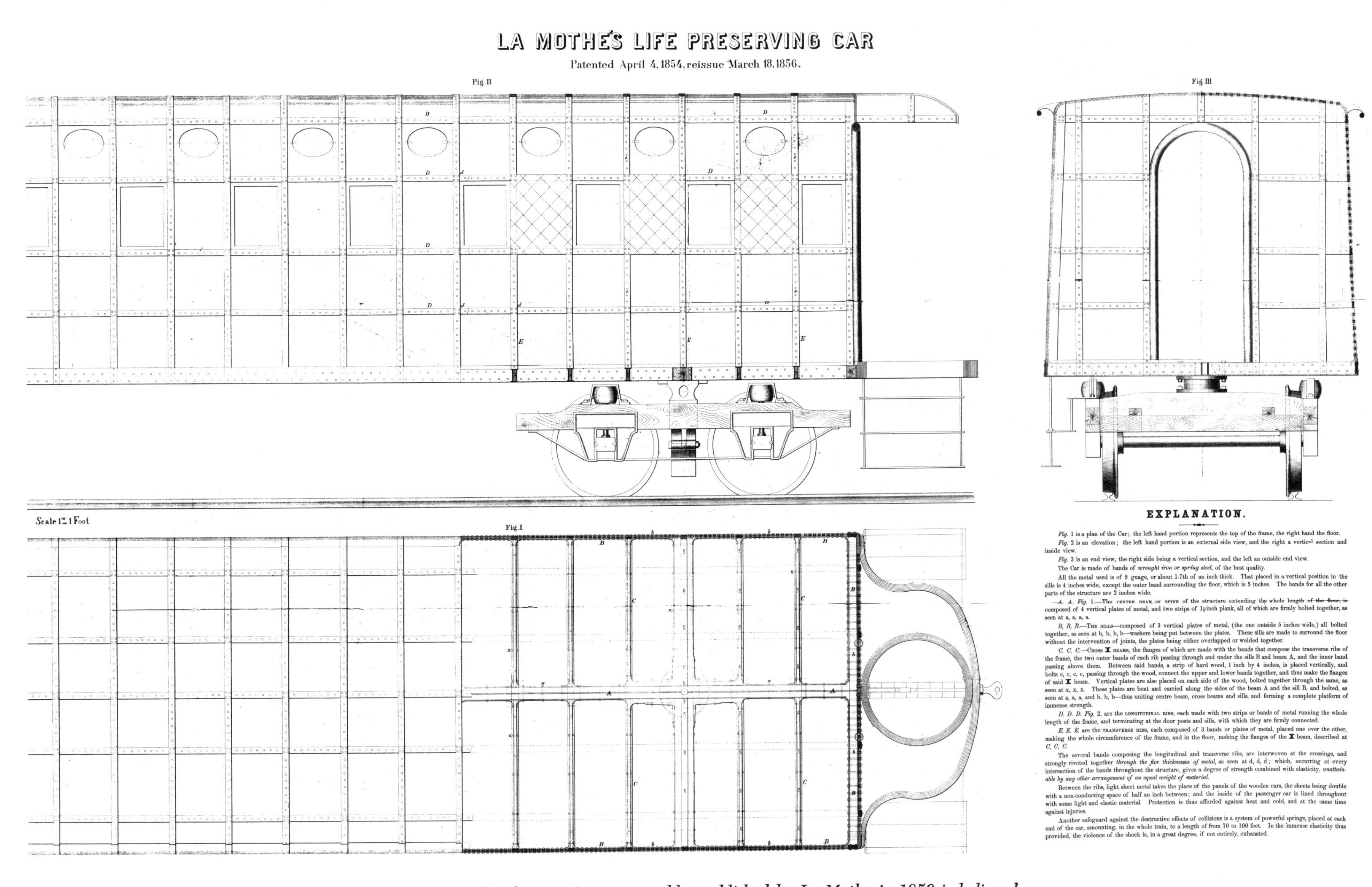

Figure 2.2 This drawing from a pamphlet published by La Mothe in 1856 is believed to resemble an iron car actually built between 1857 and 1859.

Figure 2.3 La Mothe's fourth iron car, built by William Cundell of Paterson, New Jersey, in 1861. The car ran briefly on the Hackensack and New York Railroad. (Walter Lucas)

tons. The body of that hypothetical car was to weigh 4½ tons, but a rough calculation of the weight of the strap iron and rivets used in the Cundell car nearly equals this figure. When the weights of the sheet-iron roof, side panels, floor, and furniture are added, the total equals or exceeds the weight of a wooden car. After delivery, La Mothe reduced his claim to the statement that the car was 2 to 4 tons lighter than a wooden coach, while a New York newspaper said that it was "at least" 1 ton lighter. In truth La Mothe was probably not eager to divulge the true weight, even if he knew it. Several years later the *American Railway Times* stated that the experimental iron passenger cars that had been built so far weighed more, not less, than wooden cars.[10] The journal claimed that they exceeded their wooden contemporaries by 6 tons.

Following its completion in the summer of 1859, La Mothe's specimen car ran on the New York and Erie for six months. During this time it was closely scrutinized by the engineering world. It was greeted with enthusiasm—one admirer predicted that it was "destined to work a revolution on our railroads."[11] It is true that there were complaints about the dwarfish windows and doors, but critics admitted that perfection should not be expected in a test model. In March 1860 the car was sent away to Boston, and its final disposition is unknown.[12] However, an iron car of a "crude and ill-digested pattern" was running on local trains of the Eastern Railroad out of Boston in 1870.[13] It had come into the Eastern's possession after trials on several other lines, and La Mothe himself told one of Cundell's protégés that it was his car.[14]

Obviously this car was not impressive enough to draw an avalanche of orders. Cundell's car shop remained idle until a local textile manufacturer, Robert Rennie, ordered two small cars for his switching line. These combination freight-passenger cars were delivered early in February 1861. At the same time Rennie used his influence as a director of the Hackensack and New York Railroad to arrange for purchase of a full-sized La Mothe car. This car was more conventional in appearance and more generous in overall proportions, including windows of standard size (Figure 2.3). It seated 56 and had a smoking compartment at one end. The exterior was bright red with gold lettering; the interior was plain and neat. Instead of the lavish paintings of the first car, there was wooden paneling of grained oak, relieved by black walnut and gilt. The roof and sides were insulated with felt.

This new car seemed to ensure a fair trial for La Mothe's invention. But it had been in service for only a month when the car, together with the locomotive *Bergen*, plunged through an open drawbridge into the Hackensack River with a roar heard 2 miles away. The damage was slight to the car and its passengers, and it was sent to Cundell's shops for repairs. On the night of April 14, 1861, however, the final disaster struck when a fire reduced the shop to ashes. The car did not survive. Nor did Cundell's patience; after five years in the iron car business, he had built only four cars.

While La Mothe was seeking a new manufacturing agent, he revised the general plan of construction. He abandoned band iron for tube framing, which is used today for lightweight racing cars. In La Mothe's time, however, gas or electric welding was not available. Looking for an effective fastening, he settled first on a riveted clamp, then in later designs turned to U-bolt fastenings. He rejected the obvious choice, threaded couplings, because of their cost and because cutting the threads weakens the structure. He chose common gas pipe for economy. A patent (No. 33,350), issued September 24, 1861, specified flat iron beams and some wooden cross members in the frame. It was reissued late the next year with notable improvements, including more use of large-diameter pipes and the elimination of wooden members.

A sample freight car was built on the new plan for the U.S. Military Railroad.[15] The specifications were the same as those for a pipe-frame passenger car. Inside, the car was to be 44 feet long, 9 feet wide, and 7 feet high. The projected weight, with trucks, was 15,900 pounds, but again it seems that La Mothe was grossly underestimating. When the sample freight car was completed in January 1862, it weighed only 1,500 pounds less than an armored car built of angle iron and boiler plate. La Mothe hoped that the sample car would convince General D. C. McCallum to use iron

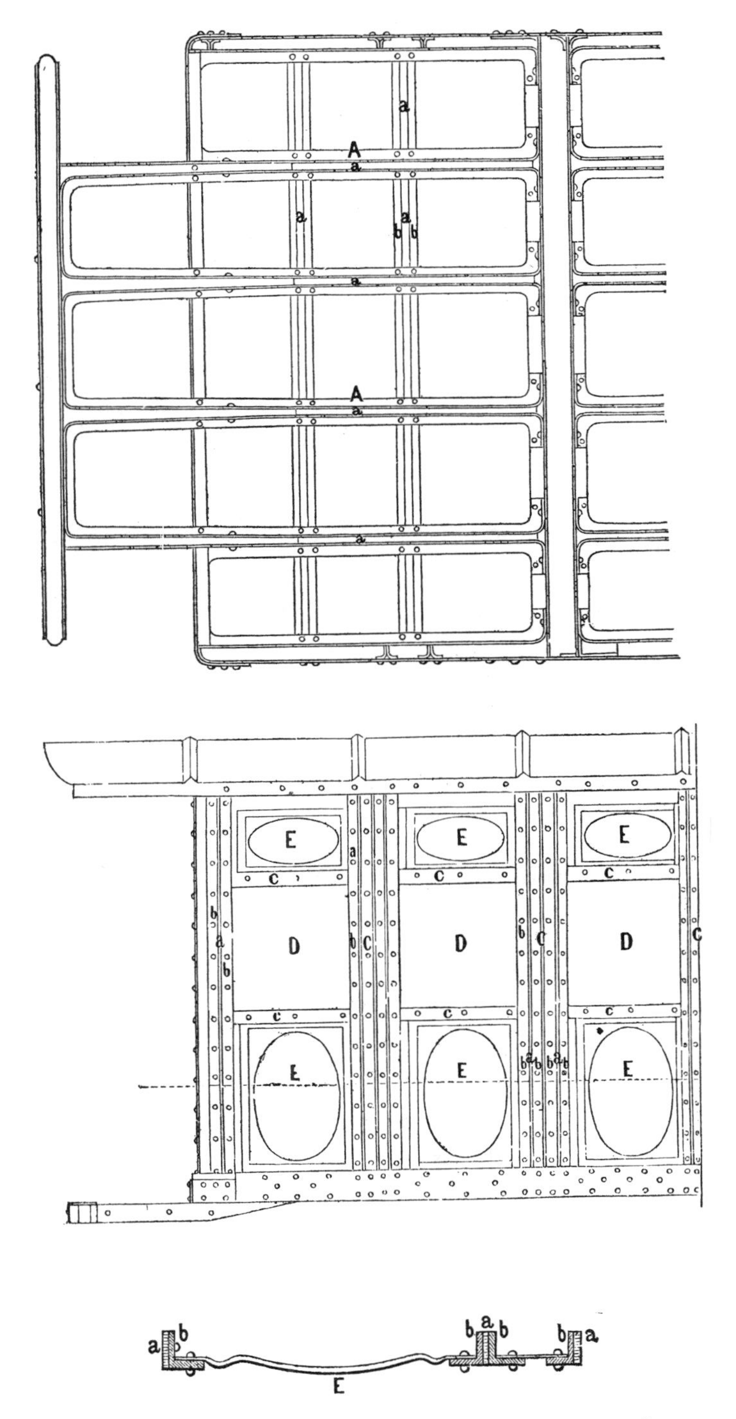

Figure 2.4 *Merrick, Hanna, and Company of New Brighton, Pennsylvania, built several iron passenger cars between 1860 and about 1863 based on this 1859 patent.* (*U.S. Patent Office*)

cars on the military lines. A second prototype freight car was also built, but the Army was not sufficiently impressed to carry the experiments further.

The following year the New York Central Railroad agreed to test La Mothe's idea. The road had recently purchased a large number of iron boxcars, which were performing well enough to stimulate interest in a metallic passenger coach. John M. Davidson, an iron safe manufacturer of Albany, was given the job because he had experience in constructing iron freight cars. The passenger car, built according to the gas-pipe plan, was ready by July 1863.[16] La Mothe declared that it was as stout as a man-of-war, yet 5 tons lighter than a wooden car. In an effort to make good on this claim, however, the car had been too lightly framed. After a special excursion to Niagara Falls the sides were notice-

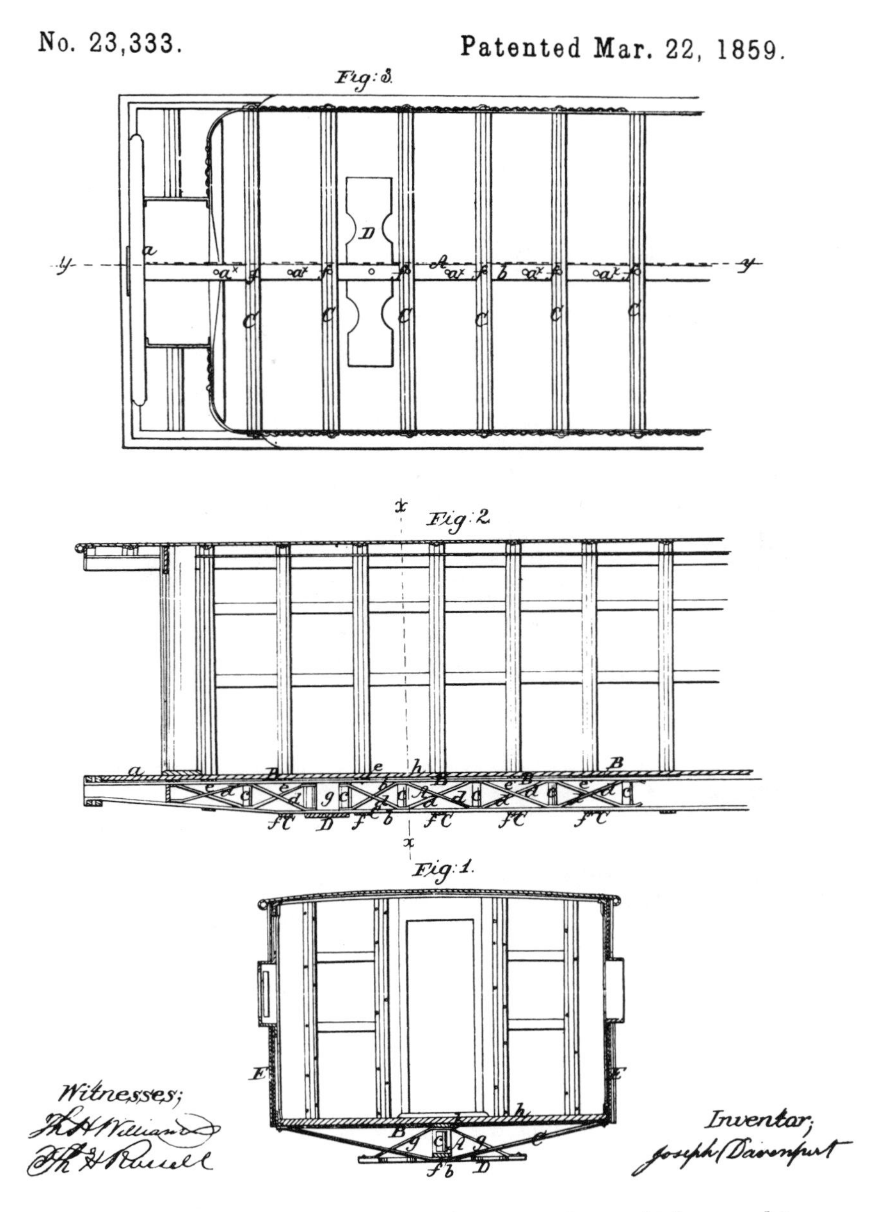

Figure 2.5 *Joseph Davenport's 1859 patent, from which several iron-bodied steam rail cars were produced in 1860.* (*U.S. Patent Office*)

ably dented; the body had twisted out of shape. Davidson went on to build more iron freight and express cars, but presumably on his own pattern. La Mothe retired, at least for the moment.

While La Mothe's first car was being finished, other mechanics were busy with similar plans. Experimentation in the Pittsburgh area was supported largely by the Pittsburgh, Fort Wayne, and Chicago Railroad. John Miner and Silas Merrick of the New Brighton, Pennsylvania, firm of Merrick, Hanna, and Company obtained a patent (No. 26,282) on November 19, 1859, for improvements in railroad cars. The drawing showed a straightforward design with an angle-iron frame riveted in a cellular form and with embossed sheet-iron side panels (Figure 2.4). The pressed panels afforded a reasonably stiff member made from lightweight sheet iron. They also provided additional embellishment in an age when any plain surface was regarded as neglected. A few months later Miner and Merrick obtained a second patent elaborating on the embossed-panel idea, and in the fall of 1860 they built a car on this plan at their New Brighton shops.[17] The body, which accommodated 56 passengers, was 40 feet, 6 inches long and 9 feet wide. The panels were made of no. 20

Figure 2.6 One of the Merrick, Hanna, and Company iron coaches is shown many years afterwards when it was in service on the South Carolina Railroad.

English stovepipe iron. Assuming that this was some form of planished iron, it is possible that the car was not painted. The seats and interior paneling were of wood. The weight was given as 26,350 pounds, or roughly the same as that of an ordinary wooden car. The builders stated that the weight could be reduced by 1 ton, but they wisely refrained from more exaggerated claims. The cost was given as $3,000, which again was about equal to that of a first-class wooden car.

The Pittsburgh, Fort Wayne, and Chicago Railroad was sufficiently pleased with the Miner and Merrick car to order four more, and all were in service by 1862.[18] Meanwhile the railroad turned its attention to reducing the operating costs of accommodation trains (short-haul trains run for local patrons and employees). Self-propelled steam cars were thought to be the answer. A car on this plan was built in 1860 by Russell and Company, car builders of Massillon, Ohio.[19] Its iron body was made after the plan of a partner in the firm, Joseph Davenport (1815c.–1912), who had received a patent (No. 23,333) on March 22, 1859, for a system of iron stanchions and bars (see Figure 2.5). The center sill was an iron latticework girder. The resulting 77-foot car was considerably longer than any coach then in service. It was divided into three compartments: the engine and boiler at the front, a baggage section next, and at the end a space for 94 passengers.

Within a year three of these cars were in service, and in 1863 the railroad ordered six more coaches from Merrick, Hanna, and Company at a cost of $3,600 each.[20] Some reports say that twenty-four cars were built, but only ten can be accounted for from contemporary sources.[21] The railroad's president used one for a private car, and it continued to serve as Number 100, the road's official car, through the 1870s. The sides were dented and scarred from countless derailments and unscheduled runs over embankments that would have wrecked a wooden car. But the regular passengers showed less liking for the iron cars than management did. Insufficient insulation made them roaring ovens in the summer and freezing boxes in the winter, and complaints were so frequent that the cars were downgraded to emigrant service. The emigrants would have none of them, however, and they were downgraded again to cabooses for local freights.

During the Civil War the government appropriated the three iron steam cars and at least one of the coaches. Evidence that the cars were in service on the military railroads is found among the wet-plate negatives often reproduced in picture histories.[22] One of the coaches was reportedly used by General George Henry Thomas as an office car. After the war it was sold to the South Carolina Railroad as surplus property (Figure 2.6). In 1893 it was hauling laborers to phosphate beds outside of Charleston. By 1925 it served as a toolhouse near the tracks of the Southern Railway (successor to the South Carolina Railroad) in Richmond, Virginia. Its final disposition is unknown. Another of the Merrick, Hanna cars had a career almost as long, for it was used as late as 1898 on the Sharpsburg and Oakland Railroad, a Pennsylvania short line.

Another experiment with iron car construction took place during the early 1860s on the Louisville and Lexington Railroad.[23] It is probable that there were earlier attempts to construct composite iron-wood cars; tin-clad mail cars have been reported as early as 1847. However, Theodore G. Shaw, master car builder for the Louisville and Lexington, took the idea to its outer limits in three coaches produced at the company's shops. The basic

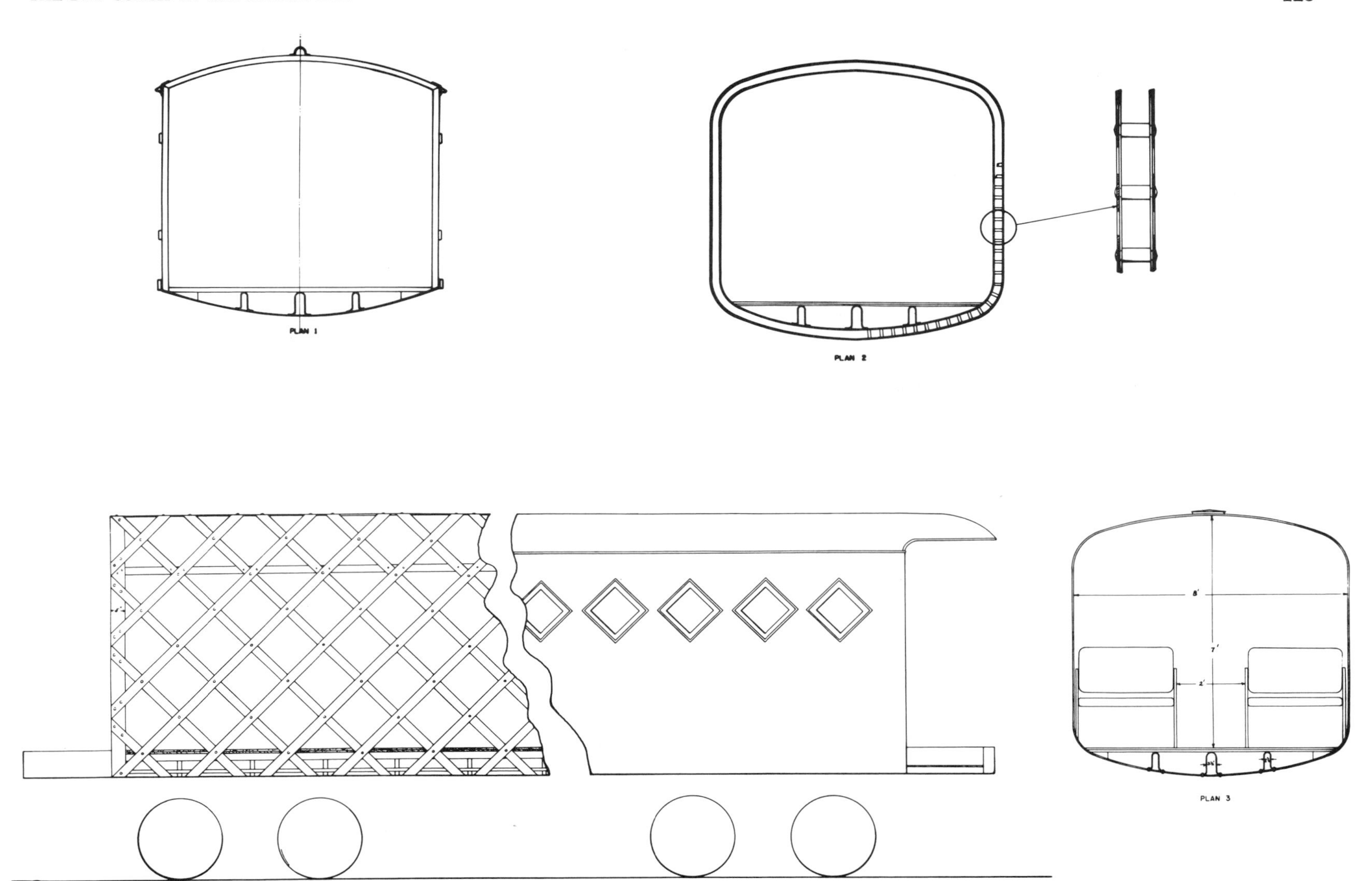

Figure 2.7 John A. Roebling's ideas for iron car construction are shown in this composite drawing made from several of his sketches of 1859. (*Traced by John H. White, Jr.*)

structure was wooden, but the exteriors were covered with sheet iron bound with heavy iron moldings (1¼- by ¾-inch bars). The base trim was a ¼- by 8-inch iron plate that completely circled the body. The Civil War appears to have ended Shaw's experiments.

The only engineering celebrity who was attracted to the problem of iron passenger cars at an early date was John A. Roebling, who had prestige, practical experience as a civil engineer, and the capital and industrial plant to fabricate sample cars. His energies were never fully devoted to the idea, however; it appears to have been a fleeting interest. Among his papers are notes and sketches dated 1859 for a tubular lattice-frame car.[24] One plan calls for a double-lattice tubular framework, in which one frame would be inside the other and the connection would be made by stay bolts, as in a locomotive firebox (Figure 2.7). Additional stiffness was provided by U-shaped channel irons; some were to be 10 inches deep. The straps for the lattice were to be ⅛ inch by 3 inches. The window frames, like those in Stephenson's Diamond cars of twenty years before, were set at a 45-degree angle. Roebling calculated the weight of the complete car, with trucks, to be just over 11 tons. The drawing developed from the inventor's sketches shows a design of surprisingly modern outlines. The fabric of the structure, however, was not much more than an elaboration of La Mothe's band-iron frame. Roebling did propose a pipe frame, and thus as far as the written record reveals, he was ahead of any other mechanic.

Roebling obtained a patent (No. 30,426) on October 16, 1860, though it did not cover the lattice-frame scheme. It specified a double-wall construction, with a bowing outer sheet for strength and added air space for insulation. A letter from Roebling's son, Washington Roebling, dated June 17, 1860, indicates that some of his father's patented ideas may have been incorporated in the Miner and Merrick cars previously described. The letter, now in the Roebling Collection at Rutgers University, briefly alludes to a visit made by the Roeblings to the New Brighton plant. The phrase "*our* car Factory" ("our" was underlined in the original) suggests more than a casual interest in the matter, but the Roeblings' main efforts were concentrated on wire rope and bridges. The evidence suggests that if Roebling had applied himself, the iron car problem might well have been solved long before it was.

Charles S. Munn of Chicago was also experimenting with iron car manufacture. It does not appear that Munn followed any patented design, but he is known to have delivered a 60-passenger iron car to the Chicago and St. Louis Railroad in the spring of 1865.[25] The car was all iron except for the doors, window sash, and interior paneling. It was said to be 3 tons lighter by actual test than an equivalent wooden car. A second car was in the

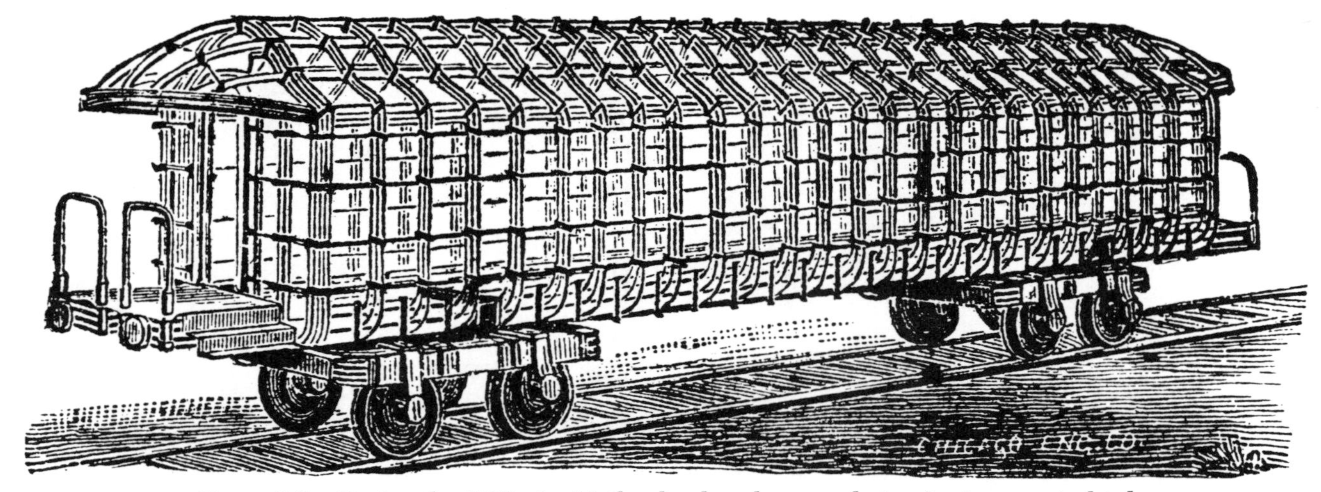

Figure 2.8 During the 1870s La Mothe developed a new design for iron or steel tube framed passenger cars. This drawing dates from 1876. (Scientific American Supplement, *September 16, 1876*)

process of manufacture in Munn's shop, which had already produced iron freight and express cars for the Chicago, Burlington, and Quincy and the Michigan Central. It appears that Munn's efforts were not favored with notable success, for there is no record of subsequent experiments.

The iron passenger car seemed thoroughly discredited. No other specimen cars were built for many years; the industry appeared content with wooden construction. La Mothe concentrated on developing a plan for iron freight cars, which had a moderate success. In 1870 he did obtain a patent (No. 105,699) for a bent-rod and -tube frame car. The design incorporated the clerestory roof and the inventor's latest ideas for uniting a framework made of round shapes. In 1877 his plan, with steel rods specified for greater security, was presented to the Master Mechanics Association (Figure 2.8), but no cars were built. La Mothe's interest in the subject continued throughout his life. In 1892 he was still active in promoting metal cars, and it is possible that he lived long enough to see the first commercial steel cars.[26]

A few freakish metal cars were produced during this time of general stagnation. In 1877, for example, Billmeyer and Small built a bullet-proof car for use in Cuba.[27] Its sides were armored with ⅜-inch iron and $\frac{3}{16}$-inch steel; iron curtains could be drawn over the windows of this 31-foot, 12-ton tank. Other scattered examples might be uncovered, but there was little substantive movement in metal car engineering until the eighties.

Coming of the Steel Car

In their 1859 patent Miner and Merrick had suggested that steel be used, but they acknowledged that wrought iron was much cheaper and possessed ample strength for all practical purposes. La Mothe's 1870 specification called for soft steel rods as the preferable material. By 1880, with the declining production costs that followed the introduction of the Bessemer process, steel had become competitive and in some cases was actually cheaper than wrought iron. Now the emphasis was on perfecting a steel car, and a new wave of experiments began.

By the 1880s European lines had turned largely to steel underframes for both freight and passenger cars, and some of these cars were exported to South America and Mexico. The trade press was filled with comment by American railway men who had seen the cars while abroad and thought that the general idea of metallic construction was worth reexamination. At least one American car builder produced steel underframe cars for export: in 1884 Harlan and Hollingsworth turned out some cars for Argentina with frames built up from I beams, channels, and tubes. One was furnished for the private use of the Argentine president. At this time Harlan and Hollingsworth was converted largely to the manufacture of iron and steel ships. As a combination ship and car manufacturer it seemed to occupy a logical position of leadership for the introduction of steel cars, but it developed no such program. The promotion of metallic cars was to remain the obsession of amateurs, cranks, and poor mechanics.

The first steel passenger car actually constructed originated with Edward Y. Robbins of Cincinnati, Ohio. Robbins was a sheet-metal fabricator who specialized in stove making. It is possible that the fearful reputation of the railroad car stove attracted his attention to the general subject of safety in railway travel. His first thoughts were probably limited to the perfection of a safe car heater, but it was plain to anyone who began studying the issue that the entire concept of railway car design called for reform. Past efforts to develop an indestructible, fireproof car were patterned after a traditional plan: a box-like form whose framework was covered with a thin wrapper. Robbins envisioned a radical departure: the basic structure would be a self-supporting tube, and the frame and basic body fabric would be a single sheet-iron unit (Figure 2.9).

Robbins expanded the idea in his patent of November 3, 1868 (No. 83,731). He favored the truly cylindrical cross section as the strongest possible form, although he conceded that a more elliptical form might be preferable. The tubular body was to be stiffened by additional attachments: U irons riveted along the length of the body just below the centerline, a sheet-iron floor with additional support from V-shaped plates or keelsons riveted under the deck, similar plates set transverse to the body to form the bolsters, and hollow ribs, to serve as structural members and air ducts, riveted inside the body at spaced intervals around its full circumference. Additional security was offered by cushioned platforms that were intended to absorb the force of most collisions. The platforms would be made of iron plates set on end, with layers of springs, cork, or rubber between them. Should the force of a collision be greater than the capacity of the end platforms to sustain it, passengers might still escape injury, for the end walls were to be made of double plates, and the interior would be thickly padded so that travelers could be thrown about like projectiles, yet come out unharmed. The stuffing might be wool, shoddy, sponge, hemp, caoutchouc, oakum, or any other

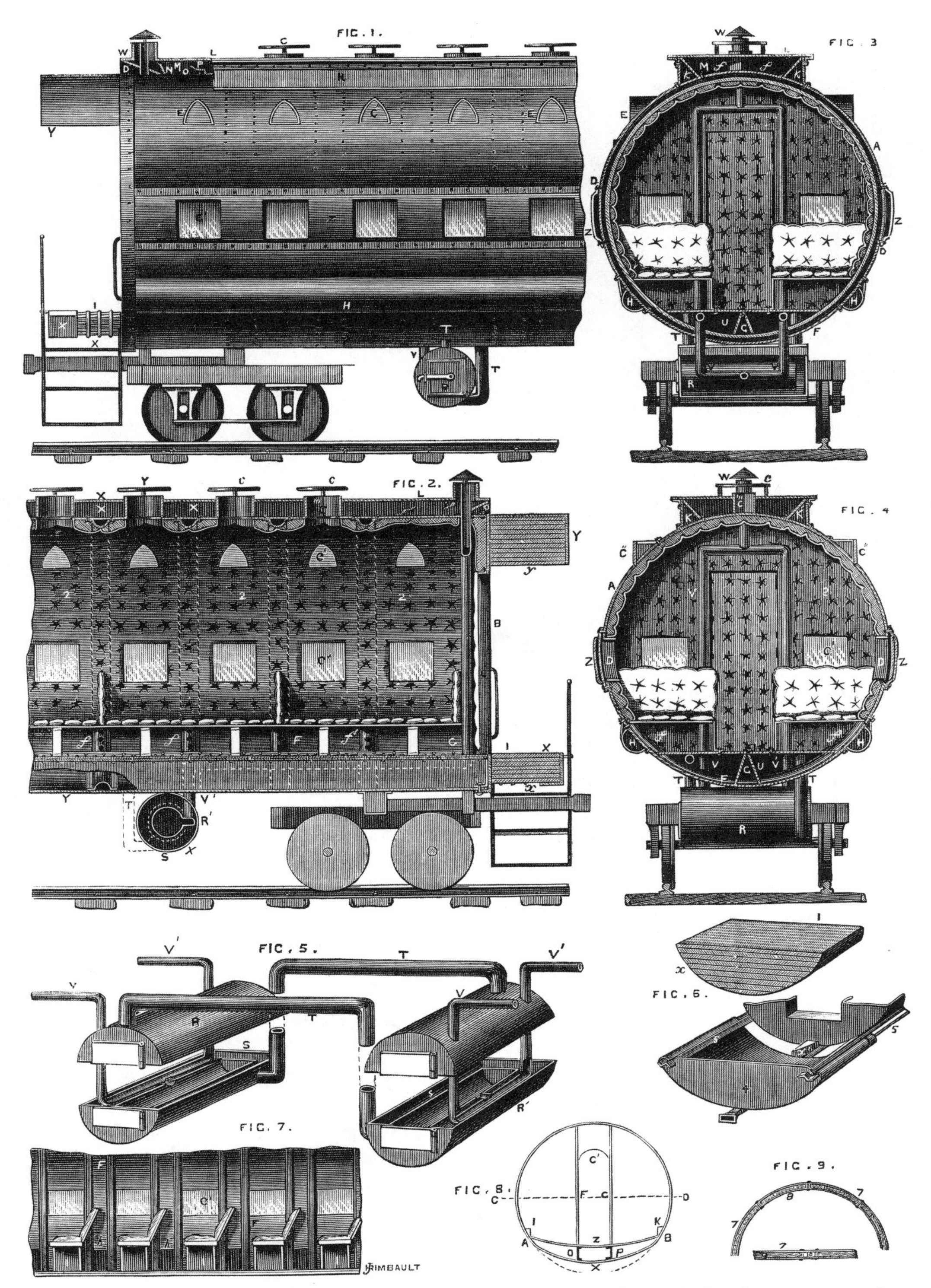

Figure 2.9 Edward Y. Robbins of Cincinnati patented a radical tubular design for metal passenger cars in 1868. (Engineering, *April 22, 1870*)

elastic material. If the stuffing was covered in fine cloth, Robbins declared, the interior would be as handsome as any car running, and certainly safer.

Robbins specified a dual hot-air heating plant suspended under the car, with ducts running under the floor. Fresh air would be fed to the furnace from the roof through the hollow, angular ribs of the body. As with all under-the-car heaters, the fire hazard was reduced because the fire was outside of the passenger compartment.

Although this idea for heating was not original, Robbins's general plan certainly was. The design was strong and reasonably practical, but incredibly ugly. The bulbous, top-heavy body, the receding roof, and the awkward union of the platform and the round end bulkhead produced a clumsy silhouette. The dwarfish side windows—not more than 15 inches square, according to the patent sketch—were among the worst defects.

Robbins's efforts to produce this cylindrical car continued for over twenty years. They exhausted as many promoters and involved as many heartaches as the misadventures of La Mothe. The first talk of producing a prototype appeared in the Cincinnati newspapers in July 1869. A large-scale model was put on exhibition, and it was announced that the full-sized car would be made

Figure 2.10 This model of Robbins's iron or steel car was exhibited to attract support for a full-sized prototype. (Engineering, *April 22, 1870*)

of $\frac{1}{8}$- to $\frac{3}{16}$-inch sheets with interior padding 4 inches thick. The promotion came to nothing, however; possibly it was a casualty of the gold panic later that year.

Robbins then tried to introduce the car in Great Britain, taking out a patent in the name of a straw party, W. Newton Mac Cartney of Glasgow.[28] A model (possibly the one exhibited in Cincinnati the year before) was shown to the Board of Trade, a national body charged with maintaining safe operations on British railways (Figure 2.10). Although the model represented a 36-foot body, for British lines Robbins proposed to construct 26-foot cars on four wheels. The usual arguments for safety, long life, and comfort were offered and quietly accepted, but the Board questioned Robbins's claim that the cars would be light in weight. It was obvious that they would exceed the estimated $4\frac{1}{2}$ to 5 tons; like La Mothe, Robbins refused to recognize that weight would never be a strong point of the metallic car. One American trade paper said that a full-sized prototype was built in Glasgow, but this report appears to have confused the model with an actual car.[29]

Nothing is known of Robbins's activities during the next fourteen years. In 1884 he reappeared in Boston as founder of the Robbins Cylindrical Steel Car Company, which had somehow raised capital to construct a sample car. The existing car builders were unable or unwilling to produce such a maverick, and Robbins turned to an iron ship builder, the Atlantic Works in east Boston. This yard was well equipped to fabricate the tubular car, for it had rolls, bulldozers, punches, and skilled metalworkers. By July 1884 the car was half finished.[30] Although the 1868 patent had specified either iron or steel, this was now the age of cheap steel, and that metal was used. The 54-foot body was made with $\frac{3}{8}$-inch bottom plates. The sides were $\frac{3}{16}$ inch thick, while the roof and clerestory monitor were made of $\frac{1}{8}$-inch and thinner sheets. The floor was $\frac{1}{8}$-inch plate. The general fabrication followed the specifications of the patent. The interior padding, to be 6 inches thick, would consist of a layer of felt placed against the shell of the car, a layer of woven wire, a layer of tufted hair, and a cloth cover. The padding would be made in sections so that they could be removed for repair or cleaning. Robbins claimed that the car would be 4 to 8 tons lighter than a wooden one. He estimated the total weight of the finished car at 22 tons. But the car was not finished; Robbins could not raise the money for completion. Some parts went to scrap, and the unfinished hulk eventually found storage in the old repair shops of the Eastern Railroad.[31]

The assets of Robbins's company were acquired by Byron A. Atkinson, a rich Boston furniture dealer whose success in the furniture business was attributed to free delivery, generous credits, and volume selling.[32] He selected Henry D. Perky to promote the enterprise. Perky, a health crank, an organizer of trade fairs, and a general speculator who was said to profit during the panics that ruined other men, became the driving force of the Steel Car Company.[33] A respectable front man, General Adna Anderson, was appointed president. Anderson had had a long career in railroad engineering. He had become a war hero as an officer with the U.S. Military Railroad. Later he served as general manager of the Wabash, engineer of the St. Louis bridge, and finally, chief engineer of the Northern Pacific. By 1885 he was exhausted, his mind was wandering, and he was retired early. He committed suicide not long after his appointment to the Steel Car Company, and it seems doubtful that he did much to promote the venture.[34]

Work resumed on the car body, now in the Eastern Railroad repair shops, in the spring of 1887.[35] Charles M. Smith was engaged to improve Robbins's designs. His most basic change was the substitution of large, dormer-like windows for the small, flush openings of the original design. A sheet-metal framework with an arched top permitted the use of large plate-glass windows. A patent (No. 366,519) for tubular railway cars that were remarkably similar to Robbins's plan was granted to Smith on July 12, 1887. The work went slowly, however. By midsummer of the next year it was reported that the car would be running in a few days,

Figure 2.11 The City of St. Joseph, *built in Chicago in 1889 after a modified version of Robbins's 1868 patent. Like one of La Mothe's cars, it was destroyed by fire shortly after delivery. (Burlington Northern)*

but nothing happened. By 1889 it was obvious that the job would never be finished at the Eastern shops. The car was sent to the Laconia Car Company in New Hampshire, where professional car fitters could complete it.[36]

While construction of the Boston car languished, Perky was talking of establishing a great steel car plant. At first he spoke of Chicago; a few months later in August 1888, he mentioned Lincoln, Nebraska, as the site and said that the Boston car would be called the *City of Lincoln*.[37] This scheme failed, and Perky shifted the proposed site to St. Joseph, Missouri.[38] With local backing he secured a tract of land east of the city and put up a building 80 by 900 feet. It was completed by the beginning of 1889, and to celebrate the new enterprise Perky organized an industrial exposition to be held in the autumn at the plant.

Sometime before the fair opened, a second specimen car was built for exhibit in St. Joseph. It was constructed in Chicago, apparently at a bridge works whose identity has not been determined.[39] The Chicago car, named the *City of St. Joseph* (Figure 2.11), was superior to the Boston car in that the roof and windows were more graceful; the awkwardness of the Boston car's projecting clerestory and window dormers was avoided. This was accomplished by modifying the shape of the body and setting the sash well inside the outer walls of the car. The *City of St. Joseph* was large; its overall length was 73 feet, 8 inches (some 12 feet longer than the Boston car), and it was mounted on six-wheel trucks. The interior was arranged as a private car, with observation, dining, and sleeping rooms.

At St. Joseph's "New Era" Exposition, the steel car was a major attraction. Ten more cars were being built in the adjacent shops to form an exhibition train that would be sent on a national tour. But during the evening of September 15, 1889, fire leveled the plant, and the *City of St. Joseph*, the cars under construction, and the prospects of the Steel Car Company vanished in the flames.

After the St. Joseph debacle the affairs of the cylindrical car company were allowed to rest for a year or two. Perky was not easily disheartened, however; he meant to give the scheme one last push. The old Boston car, which had since been finished for Atkinson, the furniture dealer, to use as a private car, was borrowed for a transcontinental tour (Figure 2.12). Perky took not only his family but also a reporter from the *New York World* to churn out releases. For some time the car stopped over in Chicago, where it was reviewed by several trade papers. The *Railway Review* called it "not nearly so business like" as the *City of St. Joseph*; certainly it had many shortcomings. Apparently it was next shown at the Columbian Exposition of 1893 in Chicago, where it attracted no orders either. More than $40,000 had been invested in the car, but it was abandoned on the fairgrounds. It was taken over by the wreckers who dismantled the exposition, and they sold it to a showman for a traveling repertory company.[40] Around 1907 the car was purchased by another showman, Leo Blondin of Oklahoma City, who traveled extensively in it over the next twenty-two years. During Blondin's ownership several fanciful stories appeared about the car. He moved the date of construction back to 1879, claimed that the car was first used under the name *Mildred* by President Garfield, and said that it was Garfield's funeral car. The validity of these tales is extremely doubtful. After Blondin's retirement in 1929, the car continued to work the show circuit. According to John Roberts of the National Museum of Transport, the car was in existence until sometime in the 1940s, when it was destroyed by fire in Hannibal, Missouri.

The Robbins car was not the only attempt to introduce steel passenger cars during the last twenty years of the nineteenth century. A dozen or more promoters were active, although few contributed more than patents, and many of their designs were derived from Robbins's cylindrical pattern. The lack of novelty indicates a certain laxity on the part of the Patent Office, which granted one license after another based on the same idea. The universal appeal of the tubular design was explained by one of the inventors, George J. Porter, in his specification of October 13, 1891 (No. 461,173): "It has been found that a car-body of sub-

Figure 2.12 The first Robbins demonstrater was constructed by fits and starts between 1884 and 1890. It is reported to have survived until the 1940s. (Railway and Locomotive Historical Society)

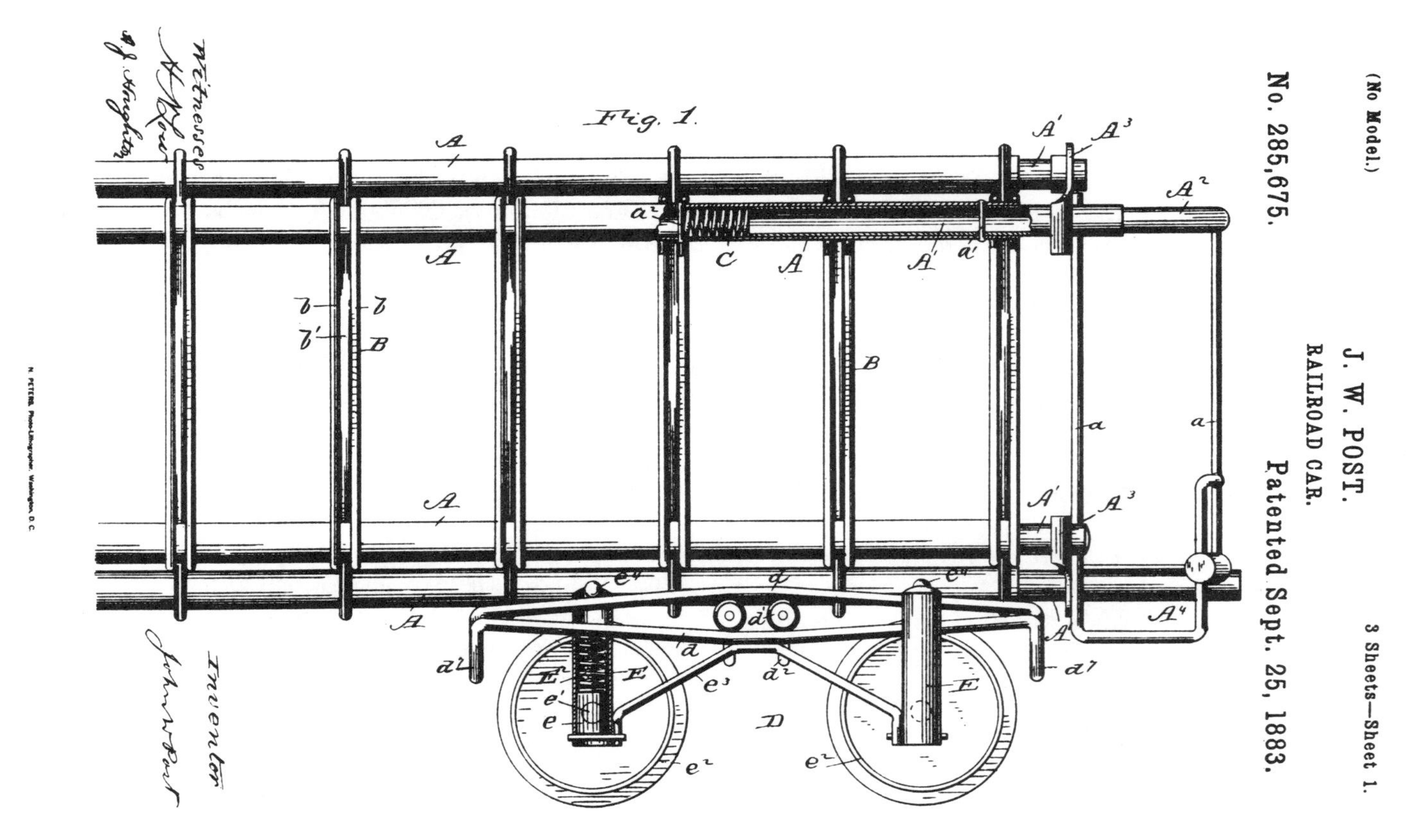

Figure 2.13 John W. Post combined La Mothe's and Robbins's ideas in this 1883 patent. (U.S. Patent Office)

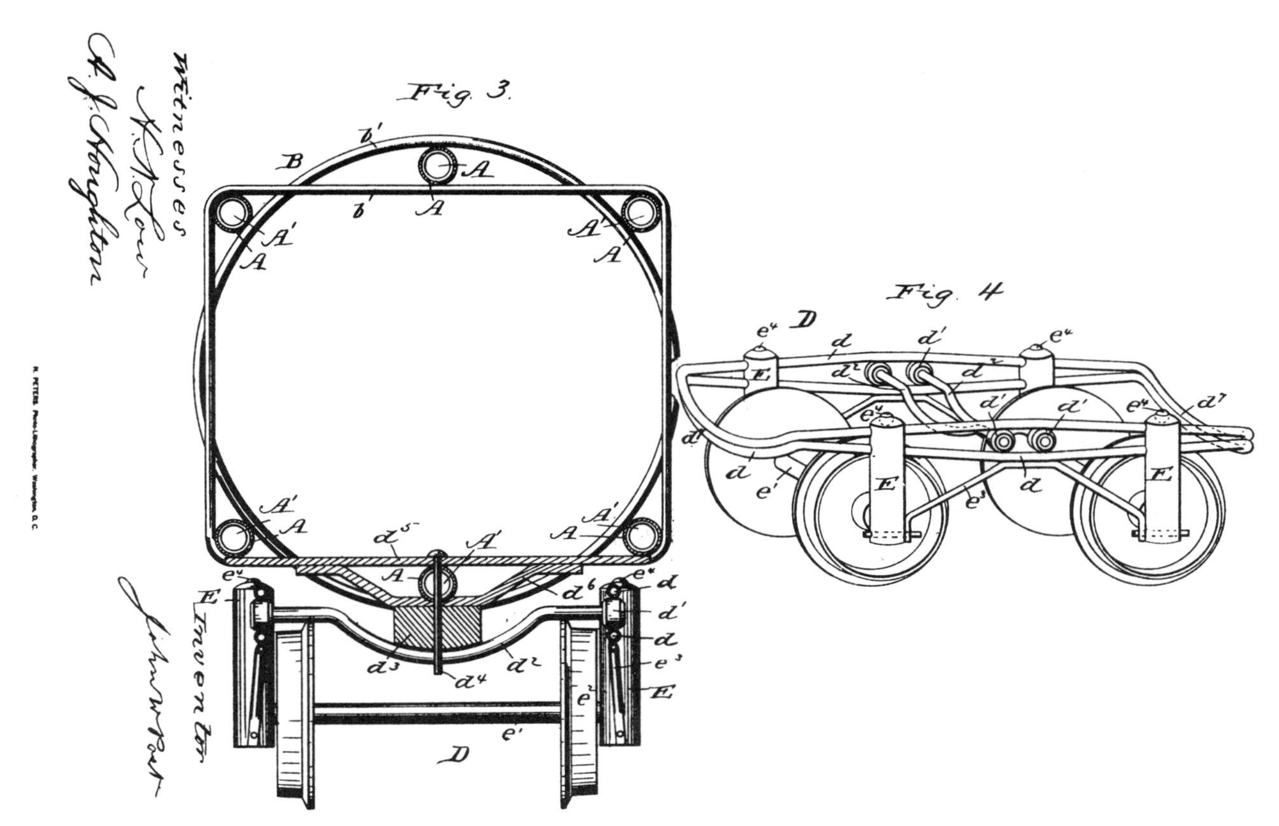

Figure 2.14 Post's 1883 patent attempted to combine the round and rectangular cross-section shape into a workable plan. (U.S. Patent Office)

stantially rectangular form in cross-section, when composed of metal is unduly heavy, and, besides this, cannot readily be made to withstand the torsional strain to which car-bodies in service are continually subjected. These facts have led inventors and artisans in devising and constructing metallic car-bodies to adapt the cylindrical form." In Porter's patent the difficulty of cylindrical fabrication was lessened by making the tube in a polygonal form of eleven straight panels.

Colonel John W. Post of New York City offered another variety of tubular body construction in his patent of September 25, 1883 (No. 285,765) by combining both a rectangular and a cylindrical shape in a single profile (Figures 2.13 and 2.14). This form of construction is best shown in a second patent (No. 366,249) granted Post in 1887. In this and several of his other patents, steel tubes were to be used for the framework. In the summer of 1890 it was announced that the Steel Tubular Car Company had been formed to exploit Post's design.[41] A plant as large as the Pullman factory was to be built near Bradford, Pennsylvania, on a 5,500-acre tract, which would contain a model town for 5,000 employees. The grandiose plans were never executed, however.

Another zealous advocate of the cylindrical car body was Casper Zimmerman of Chicago. His 1895 patent (No. 542,746) calls for nothing less than a perfectly round body with spherical ends. Like Robbins's, the interior was to be heavily padded. Zimmerman planned to add mattress springs to the interior walls and to mount a false floor. The cushioned interior, like Robbins's design nearly thirty years before, was for passenger safety, but the air space provided by the layer of springs would also contribute to the car's insulation.

More conventional-minded inventors rejected the tubular body concept and offered orthodox rectangular designs. Some of their plans were surely conscious efforts to provide a scheme that railroad managements would not immediately discount because of its bizarre appearance. Between 1887 and 1896 Chester W. M. Smith, a West Coast mechanic, prepared several patents for steel passenger cars with conventional outlines. Smith felt that the requisite strength could be gained by the use of very large one-piece sheets combined with a box girder built into the side panels near the floor. During the same period two separate inventors, Carleton B. Hutchens of Detroit and George M. Bird of Boston, offered patent designs for steel cars that were conventional in appearance (Figure 2.15).[42] None of the designs was given an actual test, however.

Two Middle Westerners, William W. Green and James Murison, reasoned that since most railroad men had shown hostility to the idea of steel cars over many years, someone should offer them a compromise scheme. Concluding that such a design should look exactly like a wooden car, Green and Murison took out a patent (No. 372,615) on November 1, 1887, for a composite wood-metal vehicle. Murison had seen these materials being combined during his youth in a Scottish shipyard at a time when, with prime timber long since exhausted, British ship builders turned to iron-frame ships.[43] A model of Green and Murison's car was exhibited in Chicago. The plan appeared so practical that the Louisville, New Albany, and Chicago Railway agreed to build a prototype mail car at its New Albany, Indiana, shops in the summer of 1888.[44] The road's master car builder, Charles Coller, worked out the details and supervised construction, and the car was finished in January of the following year.

Except for a row of rivets, it could not be distinguished from an ordinary wooden car (Figure 2.16). It was a faithful reproduction in iron and steel of the most common postal car of the day. Actually this cunning reproduction contained a fair amount of wood—so much, in fact, that it cannot properly be called fireproof, as its promoters boldly claimed. The frame was oak clad with steel plates. The window sash, doors, and flooring were wooden. The remainder of the fabric was steel. The body frame was assembled from flat plates, gas pipe, and rods. The vertical

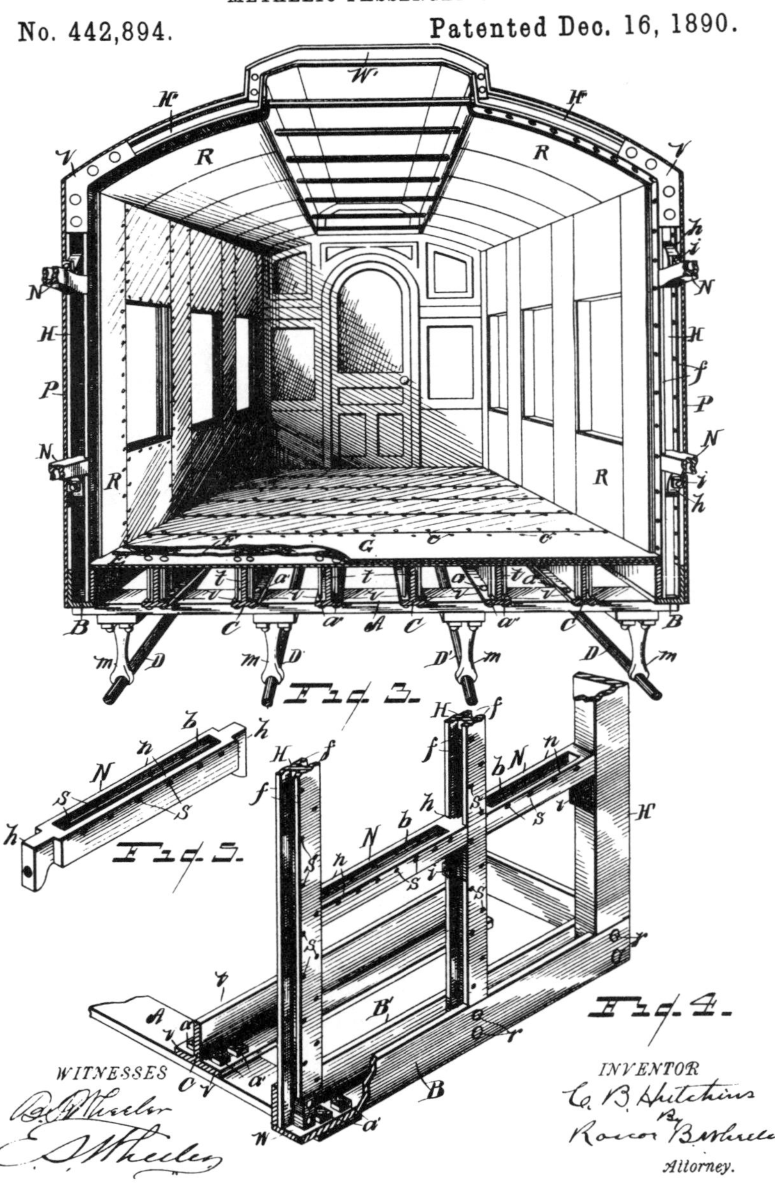

Figure 2.15 C. B. Hutchens of Detroit was one of several early inventors who hoped to perfect a practical plan for steel passenger cars. His 1890 patent was one of the more sensible schemes offered, but no test car was built. (U.S. Patent Office)

posts were 3⁄16- by 3½-inch steel plates that had been bent edgewise when cold to conform to the profile of the car's cross section. The bottoms of these posts were bent over to form feet. Bolts passing through the feet and the wooden sills tied the body to the frame. Gas pipe fitted with flanges served as spacers between frames. The exterior sides were covered in corrugated sheet metal shaped in a convincing imitation of battens. This did much to relieve the uneven appearance so common to hand-fitted sheet-metal work. The inside was covered with a ¼-inch layer of asbestos. The interior panels, also sheet metal, were lined with a second layer of the same material.

After nearly fifteen months of service over some 135,000 miles, no repairs were necessary except for replacement of a worn set of wheels and revarnishing of the exterior. Contrary to the usual complaints about excessive noise and temperature extremes, a clerk who worked in the car during the entire period said that it rode quietly and easily and was perfectly comfortable both summer and winter.

The American Fire Proof Steel Car Company was organized not long after Green and Murison's sample car entered service. The firm's plan was to manufacture passenger cars. A plant, financed by a Philadelphia syndicate, was to be erected in the coal and iron region of Alabama at Bridgeport. These prospects faded, but Green and Murison's hopes were rekindled by a wealthy lumber dealer, T. W. Harvey, and a plant was built near Chicago, 3 miles south of the Pullman factory.[45] Large orders for mail cars on the pattern of the Louisville, New Albany, and Chicago prototype were expected. The inventors claimed that their car was burglar-proof as well as fire-resistant and could be sent across the country in perfect safety. The Burlington reportedly placed an order for a steel coach. But market prospects for steel passenger cars were an illusion in 1890, and Harvey, a practical businessman, recognized the futility of the promotion and ordered the company to concentrate on developing a steel freight car. Green and Murison were dropped, and the project was entrusted to George L. Harvey.

The second concentrated effort to introduce metallic passenger cars can be said to have ended by the early 1890s. Again the attempt seems to have been frustrated by too much emphasis on novel designs. More conventional execution might have led the industry to take the steel car seriously. And as in earlier periods, the work was not carried out by established railroad car builders and so was classified out-of-hand as the foolish pursuit of mechanical cranks. Metallic passenger cars were not to receive serious consideration for another ten years.

Introduction and Acceptance of Steel Cars

In the first years of the twentieth century the railroad industry came to accept steel as the preferable material for passenger car construction. With this commitment, steel cars were on their way. The amateur inventors whose experiments with steel had long been ignored were again bypassed; the railroads turned to their own mechanical staffs or to engineering consultants. It was thus a purposeful, well-financed effort toward developing a commercially feasible metallic passenger car. The experiments of the past proved futile; the professionals saw nothing of merit in the old patents. The new steel cars were modeled on wooden cars, and orthodoxy prevailed.

For years the industry had been censured by the trade press for its conservative position on metal passenger cars. This criticism grew sharper after the introduction of steel freight cars in the late 1890s, for they made it obvious that steel framing provided a wonderful increase in capacity. Fewer cars could carry the same traffic, and unusually heavy shipments were more readily handled. The benefits were immediate and obvious.

A steel coach, on the other hand, could carry no more passengers. Being heavier and more costly, it offered only the prospect of greater safety. This advantage could be realized only in the event of a serious accident, which few officials cared to admit as a possibility on their roads. Yet it was safety that became the main argument for the adoption of steel cars. In addition, the railroad industry displayed a growing disenchantment with the complexity of wooden car construction. *Railway Age* remarked, "A more

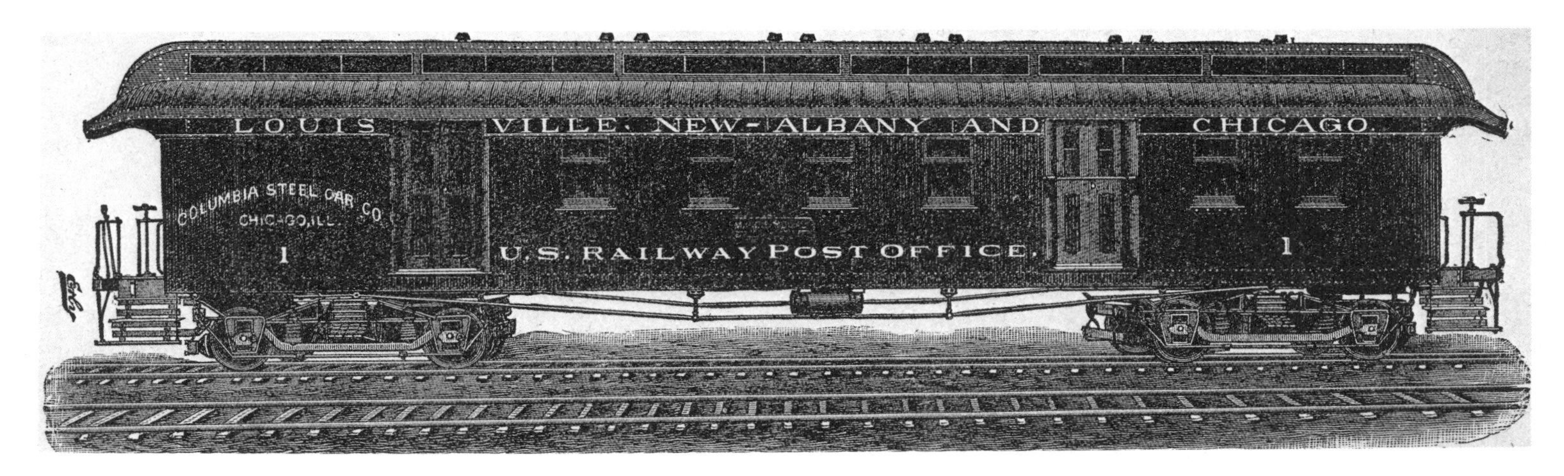

Figure 2.16 Murison and Green's composite wood and steel mail car, built in the Monon's shops in 1889, was intended to be the forerunner of a fleet of passenger cars. (Poor's Manual of Railroads, *1891*)

Figure 2.17 The first production-model steel cars were built in 1904 and 1905 for New York's Interborough subway line and the Long Island Railroad by ACF.

heterogeneous collection of trussing and bracing would be difficult to find . . . would it not be equally cheap and more satisfactory to do away with this arrangement at once and put in all-metal construction?"[46] In 1905 the *American Railroad Journal* said that passenger car design had been stagnant for twenty years.[47] It blamed the contract builders as self-serving bastions of reaction who opposed all progress, and criticized the railroads for placing so many of their orders with private concerns. To depend on what was essentially a woodworking industry for a new design was folly, the journal said; the railroads should perfect a steel passenger car in their own shops. The journal failed to note that there were some new contract builders, such as Pressed Steel Car Company, who were determined to capture the market by introducing a successful steel car.

As it happened, the action began off the main line. By 1901 construction was underway on New York's first subway, the Interborough Rapid Transit. George Gibbs was hired to prepare designs for the rolling stock.[48] A graduate engineer, Gibbs had acquired considerable railroad experience in the Milwaukee Road's mechanical department. He had done pioneering work in electric traction and railway signaling, had held several important consulting positions, and was an associate of prominent men in the engineering field. Thus when Gibbs suggested steel for the IRT subway cars, the idea was not lightly dismissed. His recommendation was not initially accepted, however, because time was short and cars were needed for the opening. The IRT ordered 500 steel-frame cars, made fire-resistant by sheathing the wooden exterior in copper and lining the floors with asbestos sheets.

The need for fireproof cars was underscored by a tragic accident in the Paris subway, in which a burning car was inaccessible to firemen and heavy smoke added to the injuries of the people who were trapped underground.[49] The IRT then decided that the composite steel-frame cars were not safe enough, and with the support of George Westinghouse and A. J. Cassatt, president of the Pennsylvania Railroad, Gibbs was asked to prepare designs for an all-steel fireproof car. The plans were finished in October 1902. The Pennsylvania Railroad began tunneling under the Hudson River to Manhattan Island in June 1903. Although electric locomotion was planned, Cassatt felt that the Pennsylvania must have fireproof cars for the tunnel operation. Toward the end of developing such cars he offered to build a sample at cost for the IRT in his Altoona shops. Completed in December 1903, the car, Number 3342, was clumsy-looking and heavier than desired. The design used only standard steel shapes, because it would have delayed delivery to create the special shapes and details necessary for a lighter, more handsome car. As it was, the car was in production for fourteen months. After tests on the Second Avenue Elevated, the IRT's president, August Belmont, was convinced of the merits of an all-steel design.

Gibbs then perfected his plan, shaving off pounds by the use of pressed shapes and aluminum panels and fittings for the interior. Only a small amount of wood was used; the floor was a poured composition material called Monolith cement. The Berwick, Pennsylvania, plant of the American Car and Foundry Company built 300 of these cars, most of them in time for the subway's opening in the fall of 1904 (Figure 2.17).[50] The worst fears about the composite cars were realized the following year, when five were totally burned following an accident.[51] After this incident, only steel cars were permitted in the subway. A number of the original Gibbs cars were in service for sixty years; one of them is preserved in the Seashore Electric Railway Museum of Kennebunkport, Maine.

The success of the IRT cars encouraged the Pennsylvania Railroad to develop a design for steam road operations. Other firms were also working on this project. As Chapter 1 mentioned, by the end of the nineteenth century the practice of reinforcing cars with steel was widespread, and it naturally suggested that cars could also be given steel underframes. In 1903 the North Western Elevated Railway of Chicago took delivery of thirty-five steel-underframe cars from the Pressed Steel Car Company. In June of the same year the Jeffersonville, Indiana, shops of the American Car and Foundry Company fabricated a 75-foot combine car for the Monon line.[52] The floor frame was built up from 8-inch I beams with timbers bolted to either side of the steel members. The body was of the usual wooden construction. Charles Coller, master car builder of the Louisville, New Albany, and Chicago, who built Green and Murison's mail car in 1889, was identified with this project as well. A more widely publicized undertaking was the steel-framed, side-door suburban cars of the Illinois Central.[53] The design was patented by two officials of the line, A. W. Sullivan and William Renshaw, on November 19, 1901 (No. 686,959). The floor frame was built up from 9-inch I beams with truss rods for additional stiffness. The ¼-inch steel-plate floor was overlaid with wooden planking. The body framing was made of 3-inch channels. Except for the usual hardware, the remainder of the car was wooden. Its weight was just over 42 tons; the road claimed a saving of 3 to 5 tons compared with an equivalent wooden car. The first car entered service in September 1903, and another sixteen were built at the Illinois Central shops (Figure 2.18). The novel design attracted considerable attention, but its ultimate effect on steel car construction was negligible. A more significant development that year was the introduction of cast-steel body bolsters by the American Steel Foundry. Meant for wooden cars, these bolsters became an important feature of the steel passenger car.

During the next three years the pace quickened. Delivery of the New York subway cars was the biggest news of 1904. Other significant developments included an order for 134 steel suburban cars from the Long Island to A.C.F.'s Berwick plant. Most of these cars, which were similar to the IRT design, were placed in service during 1905. Of greater consequence, steam roads began to experiment with full-sized cars. The Erie acquired an all-steel baggage car (Number 599) from the Standard Steel Car Company in 1904.[54] The 60-foot car was very heavy: the deep fish-belly underframe (probably the first use of this style of passenger car framing) and the 3/16-inch side plates and ⅛-inch roof plates brought the total weight to 53 tons. The same builder delivered a steel mail car of similar outline to the Erie in March 1905 (Figure 6.55). The Erie acquired a third all-steel car for express service in November. A large number of steel cars were built for the London subway by A.C.F. The Santa Fe acquired thirty-nine steel-underframe mail cars.

A paper calling for the adoption of all-steel passenger cars was read at a meeting of the Master Car Builders, who had not heard a report on this subject for twenty years.[55] In that interval, dead weight in passenger cars had advanced 100 percent, while seating capacity had risen only 30 percent. The most important part of the paper was the description of a plan for a 70-foot steel passenger car designed by George I. King of the Middletown Car Works. The center sills were 12-inch channels, and the general conformation was that of a standard car. The paper also noted that the Pressed Steel Car Company had been working on a similar plan for some time.

But not all designers were thinking along conventional lines. Around 1905 William R. McKeen, Jr., mechanical superintendent for the Union Pacific, began making steel-bodied gasoline rail cars. These distinctive vehicles, with their circular windows, side doors, and knife-edge fronts, are unlike the creations of any other designer. In 1907 McKeen modified his rail car design for use as an ordinary coach 68 feet long, which retained all the usual features of his earlier cars, including an arched roof.[56] The ends were slightly rounded and fitted with end doors and vestibules. Only one car was manufactured from this extraordinary design.

Several trunk lines had now become convinced that steel passenger coaches were a desirable investment, and in 1905 and 1906 three major railroads decided to buy some. The Southern Railway turned to a major builder of metal freight cars, the Pressed Steel Car Company, which was eager to move into the market for passenger equipment. After several years of planning, the firm erected a special shop for passenger car work. It delivered the first steel coach, the Southern's Number 1364, in time for exhibit at the June 1906 convention of the Master Car Builders (Figures 2.19 to 2.21).[57] Two more cars on the same plan followed. In this design the 66-foot body was supported by two fishbelly center sills fabricated from ⅜-inch plate. While the floor and body framing and side sheathing were of steel, considerable wood was used in the construction. The entire roof was wooden, as was the interior. Although it was not fireproof, the structure was considerably stronger than a wooden-frame coach. The mixed construction held the overall weight to 55 tons, or just 2 percent more than an equivalent wooden car. These cars were considered satisfactory at first, but in later years they were criticized for insuffi-

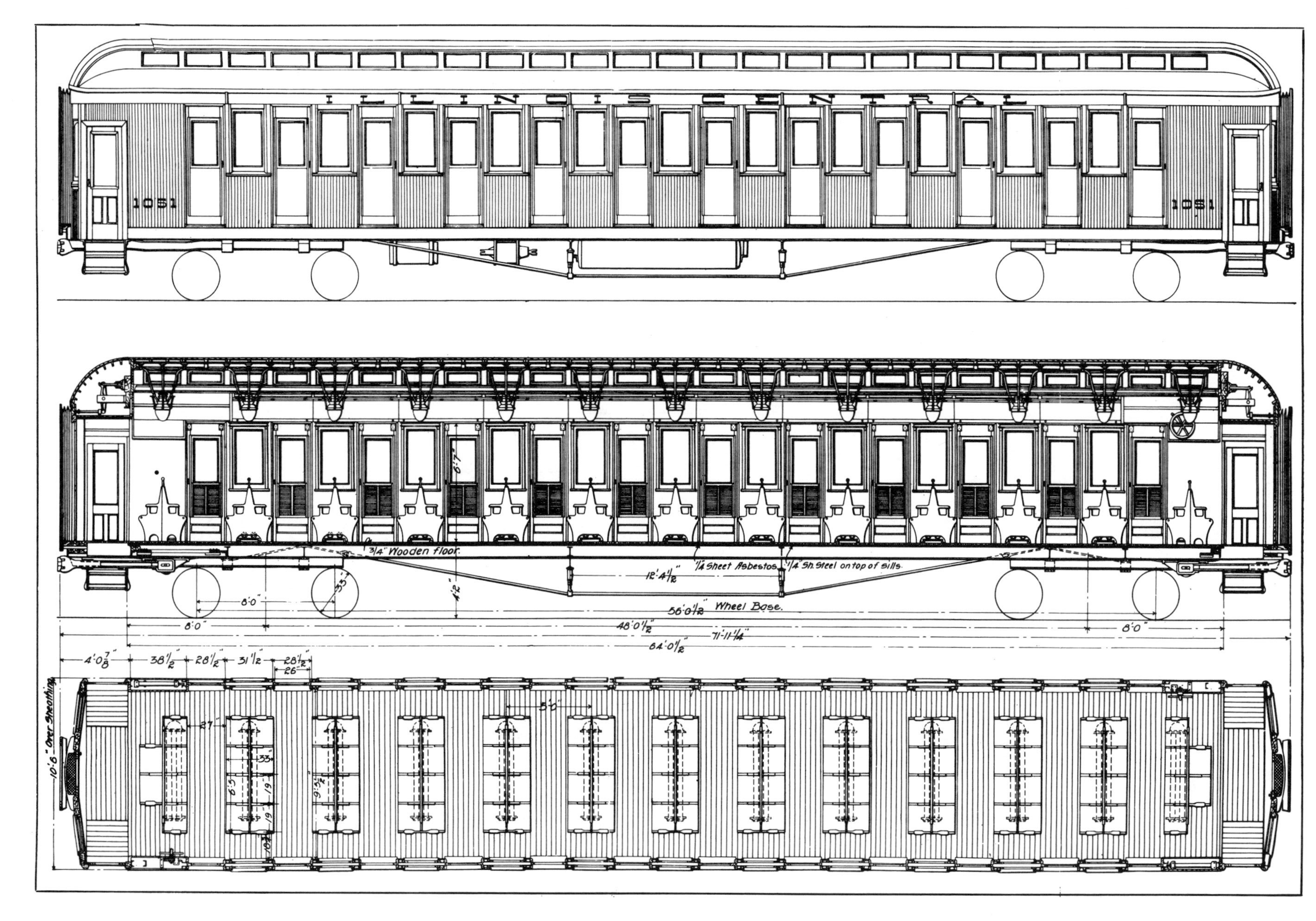

Figure 2.18 The Illinois Central built a number of steel-frame, side-door cars for its Chicago suburban service in 1903. (Railway Age, *September 4, 1903*)

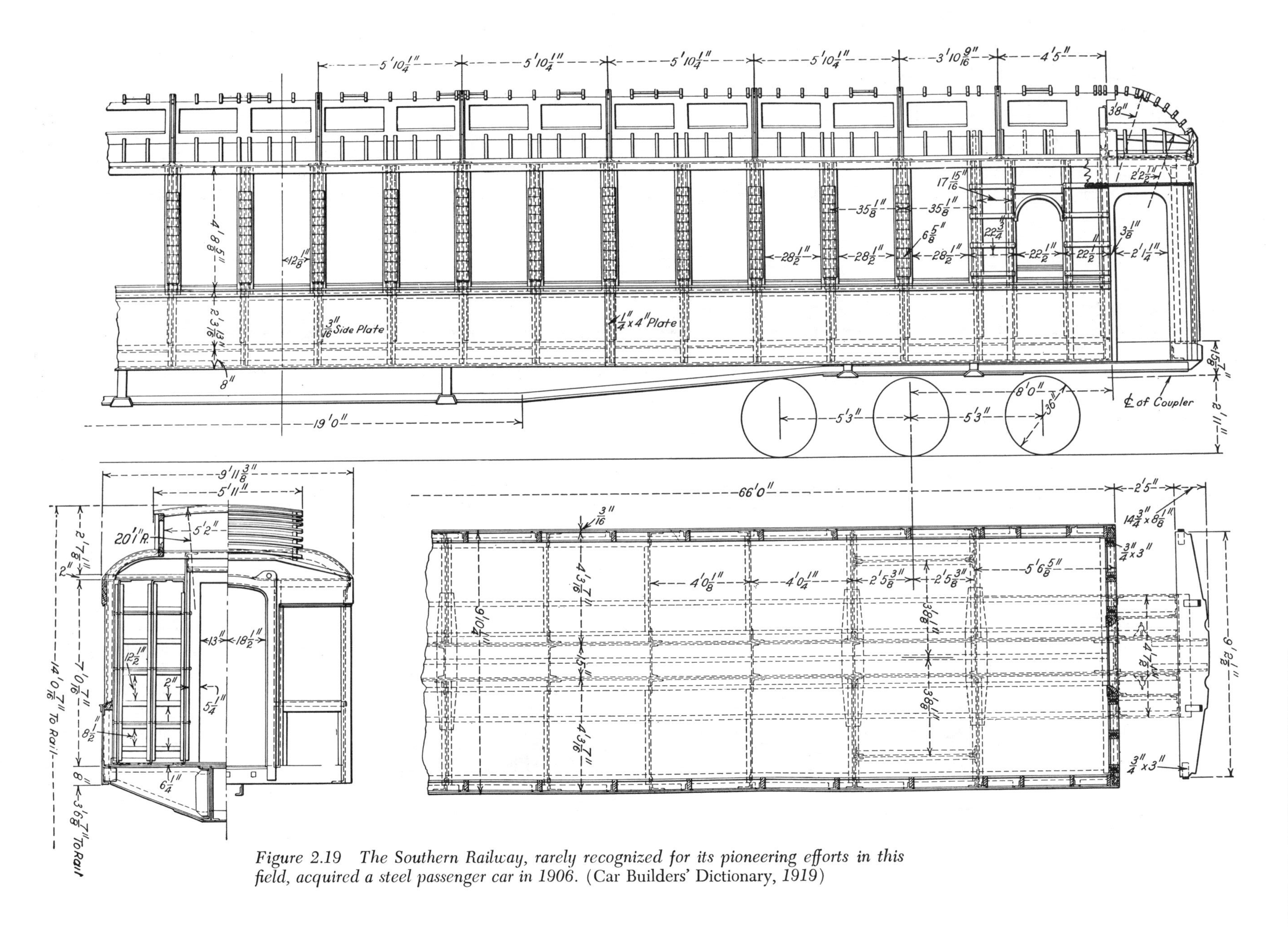

Figure 2.19 The Southern Railway, rarely recognized for its pioneering efforts in this field, acquired a steel passenger car in 1906. (Car Builders' Dictionary, 1919)

Figure 2.20 The Southern Railway's steel coach Number 1364 was built by the Pressed Steel Car Company of Pittsburgh. (Car Builders' Dictionary, *1919*)

cient cross bracing of the underframe.[58] Apparently the defect was minor, for the first car remained in service until 1953; the two others were retired in 1951.

In the Far West the Southern Pacific pioneered in introducing the steel passenger car.[59] A 60-foot coach was ordered from the road's Sacramento repair shops late in 1905 and was completed the following summer. This car, Number 1806, measured 59 feet, 10 inches long and weighed 53½ tons (Figure 2.22). The center sills were 12-inch I beams stiffened by truss rods; the side sills were 3½- by 7-inch angles. As with most experimental steel cars, it was fabricated largely from stock shapes. In this instance the fabrication was carried out in the boiler shop; the car shop was responsible for the assembly and the wooden interior. The arched or elliptical roof was the most distinctive design feature, and one that was to remain the hallmark of Southern Pacific passenger equipment for many years.

Because of the wooden interior, of course, the car was not fireproof. This was a defect that the industry eventually came to accept, as will be shown later. The other principal defects were the floor frame and the overall weight. The framing was at once heavy and weak, and the weight per passenger was the heaviest of any experimental car. This was the result of a complete dependence on the floor frame for support: the side panels were not part of the structure and thus did nothing to take up the static forces. In addition, complicated and massive cast-steel body bolsters added unnecessary weight, although these were later replaced in an effort to lighten the car. Despite its defects, Number 1806 performed well enough to stay in service until 1957.

In 1908 the Sacramento shops completed a second steel coach that showed considerable improvement, with several important modifications that had grown out of a conference of mechanical officers of the Harriman Lines.[60] Over 5 tons of dead weight had been trimmed from the new design. A total of 3,390 pounds was saved by using smaller bolster castings; the center sills were reduced to 10-inch I beams; and more care in the design of other small components made other reductions possible. The wooden interior was replaced by metal, but in overall size and appearance the car was identical to Number 1806. In service the underframing proved inadequate, and in 1911 it was replaced by the fishbelly design. The plan was generally satisfactory, however, and the Harriman Lines began a program of acquiring steel cars on a large scale. These companies became major champions of the steel passenger car, second only to the Pennsylvania Railroad in

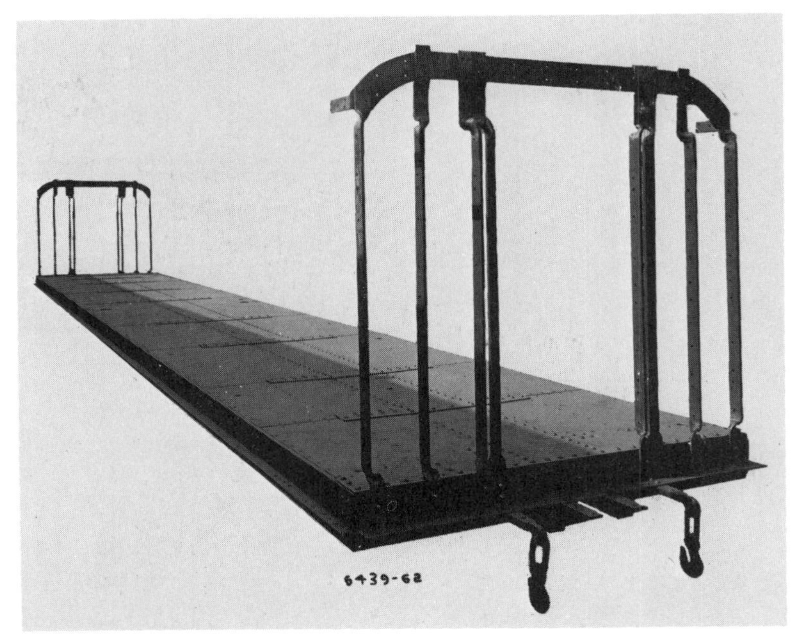

Figure 2.21 A few years after acquiring its first metal coach, the Southern began to buy steel-underframe coaches with cast-steel end frames from the Jackson and Sharp Car Company of Wilmington, Delaware.

their commitment to the elimination of wooden rolling stock (Figure 2.23). By 1910 the Harriman Lines had some 650 steel passenger cars in service or on order. During the same year they received their last deliveries of wooden cars. Once a sufficient stock of steel cars was on hand for first-class trains, however, the Southern Pacific's enthusiasm appears to have waned. In the early 1920s nearly half of the line's 1,700 passenger cars were still wooden, and it was some years before the last was retired.

The history of the steel passenger car is usually told as if it belongs exclusively to the Pennsylvania Railroad, though as this chapter has shown, many other individuals and firms also participated. Nevertheless, the Pennsylvania's role was a major one. A coach design was prepared in 1904 by the road's engineers, notably William F. Kiesel, Jr., and Charles E. Barba, in collaboration with George Gibbs, designer of the subway car. In June 1906 a sample car, Number 1651, was produced at the Altoona workshops and tested between Philadelphia and Lancaster (Figure

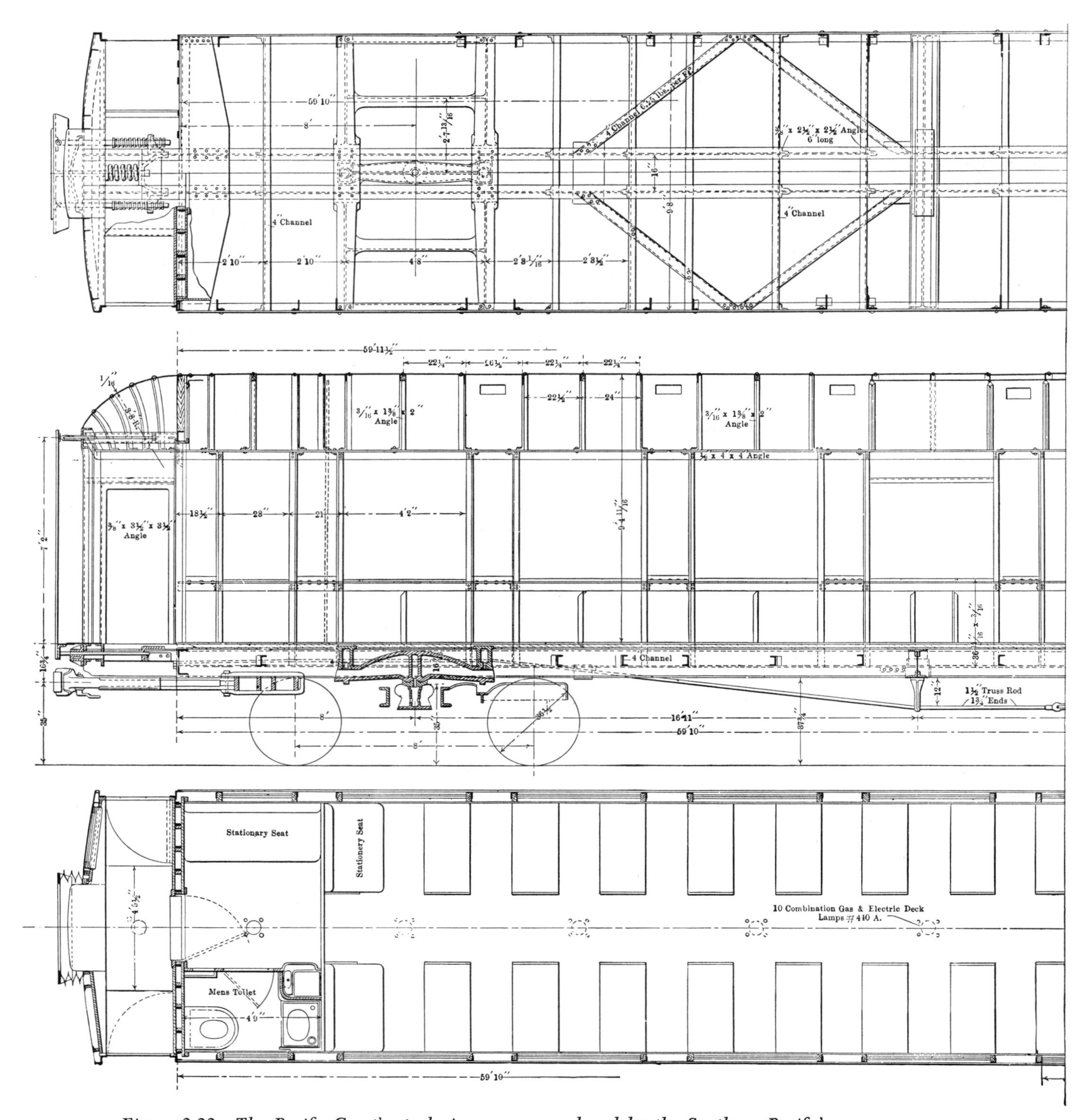

Figure 2.22 The Pacific Coast's steel pioneer was produced by the Southern Pacific's Sacramento shops in 1906. (American Railroad Journal, *January 1907*)

2.24). The 67-foot car's main support was a large box girder fabricated from two 18-inch channels and two 24-inch-wide steel plates. The interior paneling was made of composition board or treated lumber. The roof was wood with copper sheathing. In all, some 1,540 pounds of wood were used. The floor was concrete, the doors pressed steel. Yet the quantity of combustibles used, even though they were treated to resist flame, qualified the car as slow-burning but not fireproof.

Later in the same year the Long Island Railroad, a subsidiary of the Pennsylvania, took delivery of a steel car from American Car and Foundry Company. Like Number 1651, it seated 72 passengers and was just over 67 feet long. Its framing was based on a principle more like the traditional side-panel wooden car truss. Here, however, a plate girder (formed by the plates of the side panels) supported the car. The center sill, composed of 10-inch I beams, was intended to handle the draft and buffering stresses. The Long Island car contained far less combustible material than the 1651. The roof sheeting and headlining were made from a noncombustible sheet with the trade name Durite. This car was borrowed for tests on the parent company's main line so that it could be compared with the home-built car under actual road conditions. In the end, neither car was found satisfactory. They were too small to accommodate the luxury fittings required for long-distance travel. They were not fireproof, and the framing

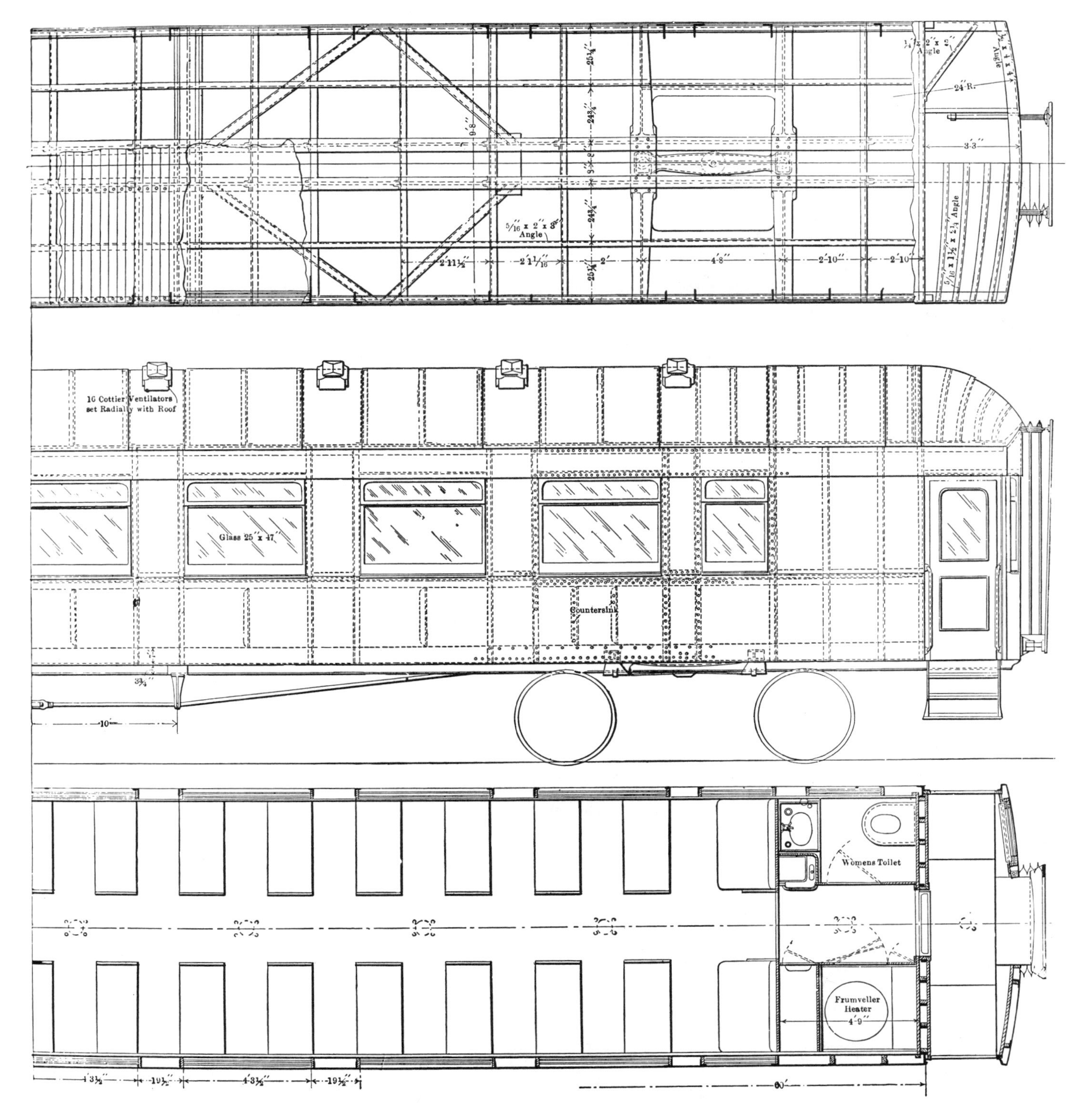

was not considered substantial enough for passenger safety in case of a wreck. The company decided that a fireproof, 80-foot passenger car seating 88 and capable of withstanding a minimum end force of 200 tons was necessary for trunk-line service. For suburban service, designs were prepared for a lighter car capable of conversion to electric traction if desired.

By early 1907 the Pennsylvania had completed plans for an 80-foot, 88-passenger car destined to be the prototype of the most famous single group of passenger coaches in this country: the class P-70 (Figure 2.25). A box-girder frame similar to that in the Number 1651 was used to support the car. The 56.75-ton car was to be carried on eight wheels. All wood was eliminated except for the seat armrests and a few other incidental parts. The interior paneling was painted sheet steel with extruded aluminum moldings. The general outline was conventional, sparse, and unashamedly mechanical. Though they had obviously made an effort to create a neat and workmanlike structure, the P-70 designers were not overly concerned with aesthetics. Nor did they show Pullman's preoccupation with disguising the method of fabrication; in the P-70 rivets, bolts, and welt plates were in plain view.

More important, the P-70 was not another experimental car but a production model. Car builders were asked to bid on an order for 500, but none quoted a price that the railroad thought fair. Negotiations continued through the spring, and in June the momentous announcement was made: 200 steel passenger cars were to be purchased.[61] This, the first large-scale commitment, truly opened the age of the steel passenger car. The historic order was divided among American Car and Foundry (which received the bulk of the work), Pressed Steel, and the Altoona shops. It was estimated that 1,000 coaches, diners, and baggage cars, together

Figure 2.23 The basic arched-roof, heavyweight coach of 1906 was duplicated in subsequent Harriman Lines orders. The Number 1866 was completed by Pullman in 1911. (Pullman Neg. 13677)

Figure 2.24 The Pennsylvania Railroad's first effort at steel coach building for main-line service was the Number 1651, completed in 1906. It was not duplicated.

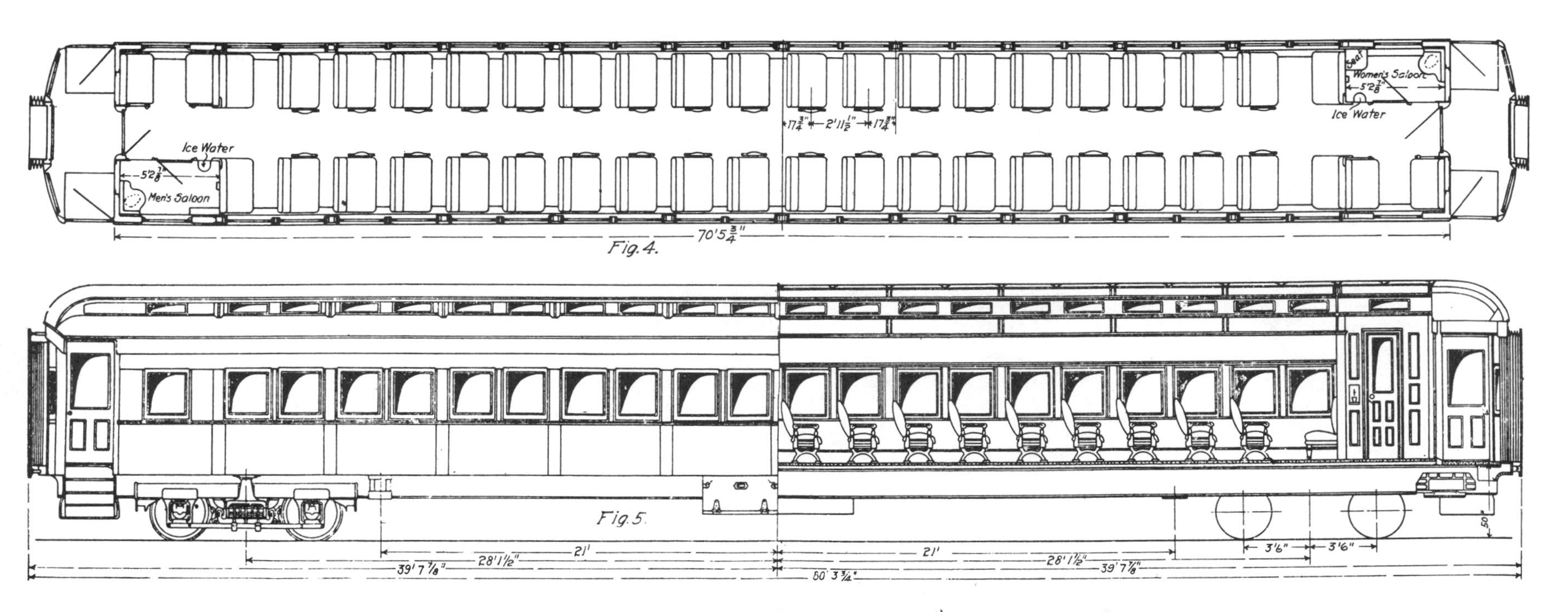

Figure 2.25 In 1907 the first of the famous class P-70s entered service. Cars on this general plan continued in production for over twenty years. (Railway Review, *June 8, 1907*)

Figure 2.26 Steel coaches such as the Number 1721 were considered essential for the safe operation of trains through the long tunnels leading to Pennsylvania Station in Manhattan.

with 500 parlor and sleeping cars, would be needed to reequip the main-line trains expected to operate in and out of the New York terminal. The projected investment was staggering, even for the giant Pennsy. The initial order would cost over 20 million dollars. Five years after the first order, twice this amount had been spent and the conversion was still incomplete.

The first P-70, Number 1652, was finished at the Altoona shops in December 1907.[62] (Unfortunately this historic car was sold for scrap in 1966, while the experimental and relatively unimportant 1651 was preserved.) The contract shops began deliveries during the next year. More orders followed; by February 1910, 324 steel cars were running and that many more were under construction or on the books.[63] Late in 1912 a third of the Pennsylvania cars were steel; some 2,800 were in service, with another 394 on order (Figures 2.26 to 2.28). Wooden passenger cars were banished from regular service on the Pennsylvania in 1928.[64] The great steel fleet, numbering 5,501, represented an investment of 100 million dollars.

The Pennsylvania's example encouraged other railroads to accept the steel car. Contract builders now felt confident enough to invest in a new plant. With some reluctance, the Pullman Company allowed itself to be pushed into the steel era by the Pennsylvania. All cars, including sleepers, running into the New York terminal had to be of the new form. The age of experimentation was over; the period of full production had begun.

Figure 2.27 A class P-70 coach interior dating from about 1910.

Figure 2.28 A late-model class P-70 produced by the Standard Steel Car Company of Butler, Pennsylvania, around 1928.

Other trunk roads were forced to follow the Pennsylvania's example if they wished to remain competitive. Their name trains, at least, had to be converted at once. As already mentioned, the Southern Pacific was close behind the Pennsylvania. By 1910 another ten major roads had steel cars on order or in service. Among these were some relatively small lines. The Lehigh Valley, for example, ordered forty steel cars late in 1910 and publicly committed itself to the purchase of steel cars only after this date (Figure 2.29).[65] The Rock Island placed a small order but would make no statement about its future plans. The Chicago and North Western was ready to convert to steel, and after receiving its first steel cars in 1910, moved rapidly toward retiring its wooden stock.[66] Within three years nearly a third of its passenger equipment was steel or steel-framed. Of its 2,085 cars, 391 were all-steel and 255 were steel-framed. At this point management's zeal waned. The road had established its image as a modernized operation, since enough steel cars were on hand for the fast name trains, and it continued to use wooden cars on the secondary, commuter, and branch-line trains. Other roads also adopted this strategy. The Southern Pacific, so quick to accept the new form,

Figure 2.29 Even relatively small lines such as the Lehigh Valley began to buy steel cars exclusively after 1910. The Number 974 was built by Pullman in 1914. (Pullman Neg. 17736)

Figure 2.30 For a few years early in the steel era, the Santa Fe and some other lines purchased steel-frame, wooden-body cars. The Number 948 was built by Pullman in 1910. (Pullman Neg. 12534)

was criticized in later years for its reluctance to complete the conversion.[67]

Some lines that wished to enter the steel era even more cautiously were content to use only steel underframes. One was the Santa Fe, which had been a pioneer in developing steel-frame baggage cars.[68] In 1909 and 1910 a large number of composite cars of all types were built for the Santa Fe by Pullman (Figures 2.30 and 2.31). The framing followed an original plan prepared by the railroad's mechanical department. The center sill was an enormous fishbelly steel-plate fabrication 32 inches deep at its heaviest point. The side sills were double 10 ¾-inch channels fabricated from ¼-inch plates and angles riveted to one side. The lower channel was visible under the car's side; the upper channel was buried inside the wooden car body. More attention to cross bracing was given in this design than in plans for other early steel cars, as can be seen from the drawings reproduced elsewhere in this volume (for example, Figure 2.19). No steel castings or special shapes were used; only a light angle-iron framework tied the upper wooden structure to the frame. By 1913 the road had about 550 steel-frame cars of this construction in service, which constituted one-third of its entire fleet. Among them were the finest cars on the line, including those used on the luxury train, the Santa Fe De Luxe.[69] They were criticized by all-steel advocates, but the Santa Fe pointed out that it had a special need for well-insulated cars to withstand the heat of desert routes. Its cars regularly encountered temperatures of more than 100 degrees, and conventional wooden-bodied cars gave the best protection. The Santa Fe contended, with considerable justification, that no effective method of insulation had been devised for steel cars. In addition, the road had already minimized the fire hazard by adopting

Figure 2.31 This composite car under construction for the Santa Fe in 1914 featured a steel upper frame. (Pullman Neg. 17867)

steam heat and electric lighting. But public criticism persisted, and with obvious misgivings the road began to introduce all-steel cars in 1913.

The Chicago, Milwaukee, and St. Paul was another company that first decided to invest in composite cars. The last of the great transcontinental lines, it needed new cars to inaugurate through service to the Pacific coast. Late in 1909 Barney and Smith delivered twenty-seven steel-framed cars, which were similar to the Santa Fe's except that they had an even more substantial underframe.[70] Like the Santa Fe cars, they combined an extremely deep fishbelly center sill with plate side sills. In this case the side plates were built up inside the car body so that they were entirely hidden. Several heavy cast-steel cross members were used. A more complete angle-iron upper framework, combined with the extraordinary underframe, made them the heaviest composite passenger cars built. At 76 tons they outweighed most all-steel cars and for this reason were considered a failure.

In 1911 the Milwaukee turned to all-steel construction, placing an order for 243 cars which, according to new designs, would each weigh only 67½ tons (Figure 2.32).[71] The composite car was dead; those that appeared in the future were conversions from wooden cars. (See Table 2.1.) Railroads were naturally eager to salvage their investment in wooden equipment, which could not be safely run with steel cars. Some lines worked out their own design of underframing, but many purchased ready-made conversion units from the Commonwealth Steel Company (Figures 2.33 and 2.34).

For example, Indiana's struggling Monon line, with modest financial resources and minor passenger traffic, still wanted to offer "as good equipment as our neighbors are running."[72] In 1918 the line began to fit Commonwealth underframes on its relatively modern fleet of thirty-five wooden cars. All-steel trucks

Table 2.1 Wood and Steel Passenger Cars in Service

	WOOD	STEEL FRAME	STEEL	SUB-TOTAL	PULLMAN	TOTAL
1910	44,900	1,100	1,100	47,100	5,200	52,300
1915	40,000	5,200	10,800	56,000	7,300*	63,300
1920	34,500	6,500	15,100	56,100	7,800	63,100
1925	26,200	9,400	21,200	56,800	8,800	65,600
1930	15,100	10,400	29,000	54,500	9,900†	63,900
1935	5,000	8,400	29,000	42,400	8,000‡	50,400
1940	2,000	6,400	29,900	38,300	6,900	45,200
1945	1,400	5,400	31,800	38,600	8,600	47,200
1950	300	3,500	33,500	37,300	6,200	43,500

* One-third of Pullman cars were all steel by this date.
† 80 percent of Pullman cars were all steel by this date.
‡ All cars were steel or steel underframe; 98.2 percent were all steel.
Source: I.C.C. reports.

Figure 2.32 *A.C.F.'s Ohio Falls plant in Jefferson, Indiana, built this coach in 1910 for the Milwaukee Road. (Railroadians of America)*

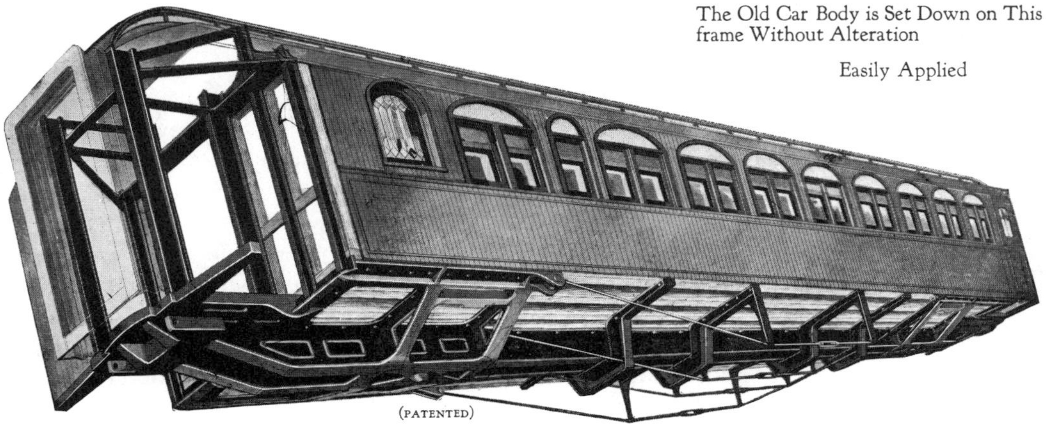

Protects Life and Property

The Old Car Body is Set Down on This Under frame Without Alteration

Easily Applied

Can Also be Arranged for Fishbelly Sills

It Fits Wooden Cars for Equal Service with Steel Cars

(PATENTED)

Figure 2.33 *Thousands of late-model wooden cars were salvaged for extra years of service by the application of steel underframes.*

Figure 2.34 *Jackson and Sharp added this steel underframe to a car being produced for the New York Central in 1909.*

and clasp brakes were installed as well. By the late 1920s the truss rods and wooden bodies seemed too ancient for the times, and a second major modernization was undertaken. A new fishbelly underframe eliminated the truss rods and provided a stronger foundation. But many changes were only cosmetic, such as sheathing the exterior of the body with no. 12 steel sheets attached by large, roundheaded wooden screws. The screw slots were filled with putty and the screws were placed exactly to imitate rivets. A deep letter board, again of sheet iron, hid the old-fashioned upper sash windows. The clerestory windows were covered over. A modern interior completed the masquerade, which few travelers could distinguish from the modern steel car. Many of these coaches remained in service until after World War II, when the Monon again improvised a new passenger fleet by remodeling surplus Army hospital cars.

The Monon's efforts to salvage wooden cars may have been extreme, but to a lesser degree some other roads followed its example. Most resorted to steel underframing only as an expedient for secondary commuter trains, and the majority of these were discarded as soon as possible. The Kansas City Southern, like the Monon, steel-sheathed some cars, but the main purpose was to disguise their antiquity—in this case by covering the arched window transoms. The Delaware and Hudson did major surgery on some of its wooden cars, not only installing steel underframes but also replacing the clerestory roof with one of arched construction. Light steel framing was introduced into the body, and the interior was completely redone. Most of the work was on commuter cars, but the line rebuilt at least one parlor-café car in this fashion.[73]

After 1912 thousands of steel cars were in service, orders for wooden cars had ceased, builders' shops were producing record numbers of new cars, and trunk lines were volubly determined to convert to steel in the interests of safer travel. Yet the pace of conversion did not satisfy certain members of Congress, for some lines displayed a noticeable leisureliness about the changeover. The New York Central and the New Haven had no steel cars until 1912, the Jersey Central waited until 1914, the Northern Pacific until 1915, the Boston and Maine until 1916. Some roads pleaded poverty; some argued that the steel car was unsuitable for their operations; some were simply too conservative. Was a push needed? Congress decided to investigate the matter in 1913. Several bills had already been introduced at the urging of the Interstate Commerce Commission. A decade earlier, the state of New York had considered but rejected a bill requiring fireproof, steel-sided cars.[74] Around 1910 a bill sponsored by the reform-minded Congressman John J. Esch of Wisconsin required rapid conversion to all-steel construction and specified the exact nature of the construction and materials. This somewhat premature action was followed by other bills. In 1912 six similar proposals were pending that allowed periods of three to six years for the conversion.[75] One of the most popular (H.R. 6142) stipulated four years. A year earlier, legislation had been enacted requiring steel-frame mail cars, thus establishing a precedent for a more general law. However, the 1911 law in effect covered government employees (mail clerks). In addition, only 1,500 cars were involved. The railroads could live with this law because it was so limited; H.R. 6142, however, would encompass all railroad passenger cars. It was visionary, impractical, and unnecessary, and it would cost over 650 million dollars. The railroad industry was determined to stop it.

The roads first argued that the adoption of steel cars, even if they completely ensured passenger safety, would have little effect on overall railroad fatalities because passengers accounted for only 3 percent of such deaths. Most of the people killed in accidents were trainmen, tramps, and pedestrians. The enormous capital that Congress would force the roads to invest in steel cars could be more effectively spent to improve rails, signals, and grade crossings. A representative of the railway supply industry put particular emphasis on this point in the Congressional hearings. He declared that a myopic concentration on rolling stock alone would prevent other needed safety improvements and possibly result in lower standards of railroad service and passenger security. Furthermore, industry management argued, the railroads could not possibly meet the four-year deadline even if they could raise the cash, because the shops could not produce cars that quickly. As it was, they were hard-pressed to build 3,200 cars a year. Conceivably 6,000 a year could be constructed, but it was unlikely that many new investors would care to enter the field for a short-run market; the capital costs were too great. According to industry estimates, ten years would be the shortest conversion schedule possible. The railroads admitted, however, that at the present rate of replacement a complete conversion would require thirty-seven years. As it happened, this seemingly outrageous estimate was a correct prediction of the actual course of events.

Several railroad officials presented their arguments against the pending legislation. Nearly all agreed that steel cars were a desirable improvement and that they should be adopted for through express trains. To some, the matter of first cost was important. A good wooden coach cost $5,500 to $7,000, whereas one of steel cost $12,500 to $14,000. The Santa Fe maintained that steel underframing was sufficient and made an issue of the discomforts of all-steel cars in hot climates. The Northern Pacific admitted that it did not have a single steel car and explained that it was overstocked with passenger equipment. The road had bought heavily in preparation for the 1909 Yukon Exposition and, now that traffic had returned to normal, owned more cars than it needed. To restock at that point seemed a foolish waste of capital.

The case of the short lines was presented by C. W. Pidcock, proprietor of numerous small roads in the South. He said that the proposed law would work an incalculable hardship on the small railroads, for they had no money for such luxuries. Anyhow, steel cars were totally unnecessary for slow-moving trains, and short lines should be exempted from any such legislation. Pidcock admitted to a personal prejudice against the newfangled "battleship cars;" they were ice-cream freezers in winter and bake ovens in summer.

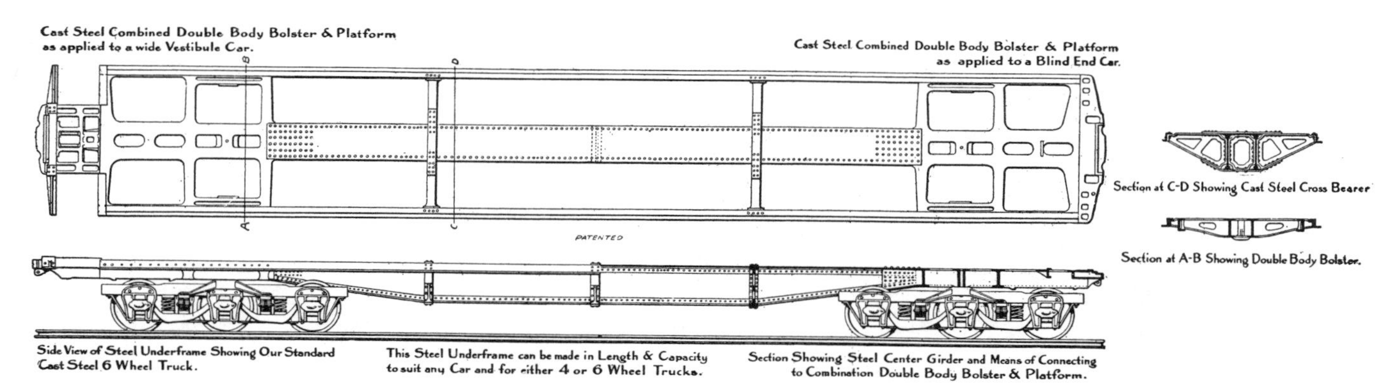

Figure 2.35 The fishbelly underframe was the most common form of passenger car framing in the heavyweight era, which roughly spanned the years 1910 to 1930.

Figure 2.36 The entire foundation of the car was its underframe. The body contributed almost nothing to the car's strength; hence the massive dimensions of the underframe. (Pullman Neg. 12484)

The industry won the battle. The law was not passed, although the I.C.C. continued to sponsor similar bills as late as 1918. The arguments against the law were generally reasonable; forced conversion at the proposed rate was unjustified. As one witness acknowledged, however, there were "some men who need urging to get up to date."

Age of the Heavyweights—Refinements in Steel Passenger Cars

So far, the steel car had brought about a revolution in materials and construction but not in form or style. Efforts to break away from existing patterns had for the most part died quietly in the storage cabinets of the Patent Office. After turning out a few experimental models, car builders settled on a general design which was seldom seriously challenged from about 1910 to 1930. Thousands were built during this period, the so-called standard era. Their main features were a clerestory roof, a riveted steel body, a pair of four- or six-wheel trucks, and a fishbelly underframe. Solid and extremely heavy, they proved wonderfully durable. Many saw more than fifty years of main-line service.

A massive steel underframe formed the backbone of the car; it was in effect a bridge on which the car body rested (Figures 2.35 to 2.37). Unlike the wooden car, the floor frame in steel cars composed almost the entire structural foundation. The side panels were generally not expected to absorb much of the static (weight) or dynamic (buffering) forces. Of several designs tried

Figure 2.37 Fabricated from heavy steel plate, castings, and bent shapes, most underframes weighed 15 to 20 tons.

out in the early steel car period, most roads preferred the fishbelly, a huge center sill made of two trapezoidal steel plates. The plates, usually 5/16 inch thick by 26 inches deep, were set about 18 inches apart. They were stiffened by angles riveted at the upper and lower edges. Sometimes a top and bottom plate were used as well. The outer ends were tapered to about 12 inches to clear the axles and truck bolster. Longer cars, 70 feet and over, almost always had fishbelly underframes. (Short or very light suburban cars were given a variety of underframes—sometimes the fishbelly, but often much simpler I or channel beam frames. Channel beams were particularly noticeable in suburban cars after the mid-1920s.)

The fishbelly design appears to have been first used in 1904 by Standard Steel. It was then adopted by Pullman, the largest passenger car builder in the world. Because of Pullman's prestige and huge production, the fishbelly was soon to be found under the majority of passenger cars operating in North America. For all practical purposes it was the standard underframe. And there was much to be said in its favor. It was enormously strong and proved effective in spanning the great lengths required for long double-truck cars. It was relatively simple and cheap to fabricate. Its principal defect was its weight of 13 to 15 tons, which was 15 to 20 percent of the car's total weight. Yet no competing design could support an 80-foot car without employing side-panel truss that nearly equaled the fishbelly's weight and far exceeded its complexity of construction. A less important disadvantage was that the fishbelly cut the underside of the car in two and thus precluded any easy arrangement of the brake rigging, piping, wiring, and other underbody gear. The low-hanging frame also prevented complete inspection of the running gear from one side of the car. Maintenance and operating departments somehow came to live with these handicaps.

The chief exception to this general plan of framing was the Pennsylvania Railroad, which stayed with the box girder it had developed for its first series of 80-foot cars in 1907 (Figure 2.38). That too was a center-sill plan, although it did not thicken at the center but remained uniform in size for the full length of the car. To ensure the strength of this member, special 80-foot plates were produced by the Cambria Steel Works. In addition, a 1-inch camber was built in. After several years' service the design was pronounced a failure by the authoritative Gutbrod Report because the cars sagged in the middle and the cross framing was

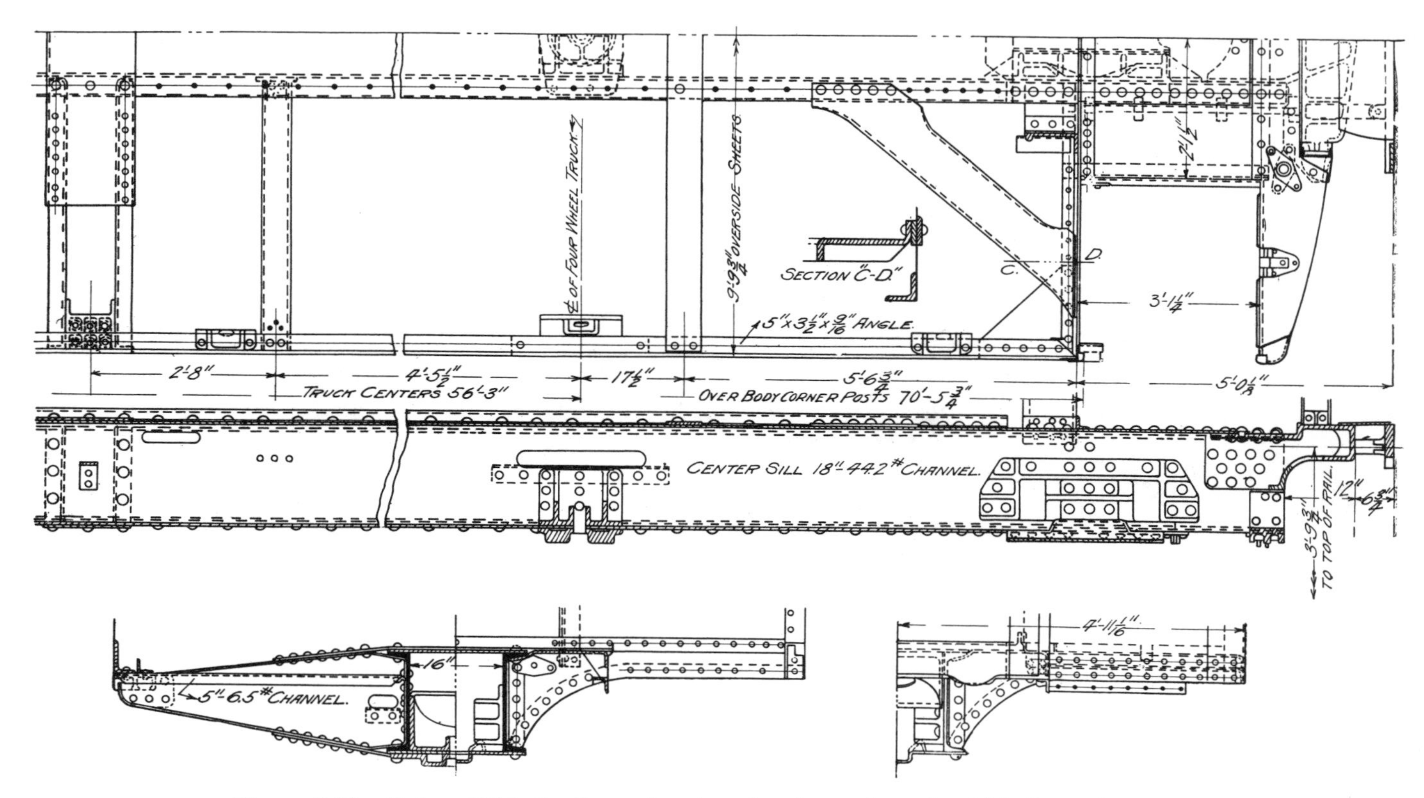

Figure 2.38 Before 1913 the Pennsylvania refused to adopt the fishbelly underframe and followed the plan shown in this drawing. (Railway Age, *June 7, 1907*)

Figure 2.39 Lewis B. Stillwell patented and oversaw the construction of about 900 passenger cars—mostly for suburban or subway service—using his pressed-steel designs. Note the side-panel truss.

inadequate.[76] In 1913 the road was reported ready to adopt the fishbelly plan, but apparently the report was untrue, for all subsequent P-70s were built with the box-girder frame.

As might be expected, several alternatives to center-sill framing were proposed. A.C.F.'s combined floor and side-panel framing has already been described. In addition to the experimental Long Island car of 1906, A.C.F. built several hundred cars, a number of them for the Rock Island, with this style of framing.[77] A more radical scheme was offered in 1906 by Lewis B. Stillwell, then consulting engineer with the Manhattan Tubes.[78] Stillwell envisioned a lightweight car that depended on pressed-steel trusses built into—and as deep as—the car body itself. The light and shallow floor frame was a secondary support. The side trusses were formed from ¼-inch pressed-steel panels. These units were riveted or butt-welded and could form a car of any desired length. The upper portion of the panels was open for windows. The roof members, also of pressed steel, were active parts of the structure and not merely a light, independent framing (Fig-

Figure 2.40 The Standard Steel Car Company built this Stillwell car for the Erie around 1928.

Figure 2.41 An interior photograph of the Stillwell car shown in Figure 2.40.

ure 2.39). Here, then, was the old wooden car's side-panel truss, now adapted to modern steel fabrication.

Fifty cars on this plan were built for the Manhattan Tubes in 1907 by Pressed Steel and A.C.F.[79] A much larger group of subway cars was completed several years later. The line seemed pleased with the design and continued to add Stillwell cars to its fleet as late as 1928. The inventor also succeeded in selling his design to several suburban operations. The Erie, for instance, began buying it in 1915 and filled out its New York suburban service largely with Stillwell cars, many of which were still in use in 1970. A late example is shown in Figures 2.40 and 2.41. In 1917 the Erie purchased a small group of Stillwells for main-line service.[80] The design was somewhat modified to replace the arched Gothic windows and peculiar ogee roof with more conventional forms (Figure 2.40), but the cars were barely full length: only 78 feet. Their impressively light weight of 59 tons was admired, but it did not lead to wide acceptance of Stillwell's design. The inventor lived to see some 900 cars, mostly short suburban ones, built on his plan; the last order arrived in 1934.

The body of the standard steel passenger car was in effect a light steel box resting on a plate-girder bridge. Its light frame of rolled shapes was covered with ⅛-inch steel plates. The side panel was usually divided into six plates. The joints were covered by a welt plate. A heavy pressed-steel belt rail was riveted below the windows. In all it was a coarse-looking, rivet-studded structure. Compared with the elegant wooden cars, it was as heavy and brutish as a battleship. Pullman, which in particular considered the rivets unsightly, attempted to produce a smooth-sided car by countersinking them, but this proved more expensive and less secure than roundheaded rivets. Next Pullman tried a hollow, pressed-metal siding made in exact imitation of wooden car siding. This plan effectively hid the rivets, and the company used it in a fair number of cars. By about 1912, however, even Pullman gave up the fight and came to accept organic steel car fabrication. Thus little concession to style or fineness was made in the heavyweights; they were plain and utilitarian.

The heavyweights' weakness of body was a more serious disadvantage. Since all of the basic structure was in the floor frame, the lightly built bodies provided little better security for passengers than wooden cars with reinforced ends. Because of the great weight of the floor frame and trucks, designers were tempted to skimp on other parts of the structure to hold down overall weight. A series of disastrous wrecks between 1911 and 1917 drew attention to this inexcusable oversight in the first generation of steel cars (Figure 2.42). Ten people were killed on the Milwaukee Road in a rear-end collision that forced a diner one-third of the way through a sleeper.[81] The wreck was described as a high-level telescope, for the diner invaded the upper portion of the sleeper's body, crashing through the end bulkhead and rolling back the roof. The frames of both cars were undamaged—a clear indication that this portion of the structure was overconstructed. Six years later even more people were killed when a Pennsylvania

Figure 2.42 Steel cars were not indestructible, as shown by this unidentified wreck photographed at the Bradley Car Works around 1920.

Figure 2.43 The steel car's vulnerability was tragically demonstrated by this telescope disaster on the Pennsylvania Railroad at Mt. Union, Pennsylvania, in 1917. (Robert C. Reed)

Commonwealth Upright End Frame in One Piece, and Commonwealth Combined Platform and Double Body Bolster for Vestibuled Cars.

Figure 2.44 Cast-steel end frames provided one solution to the problem of telescoping accidents. (Car Builders' Cyclopedia, *1928*)

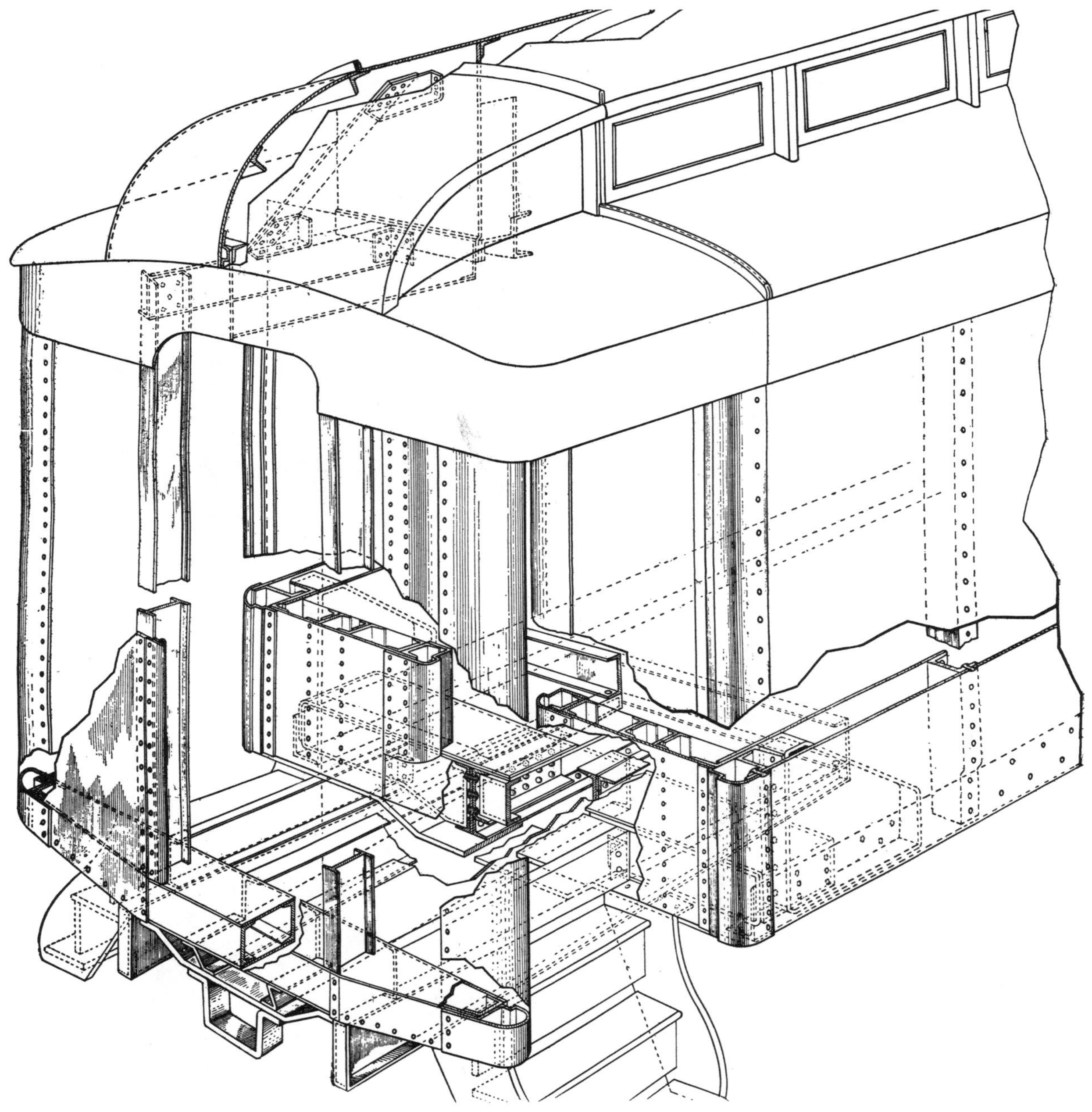

Figure 2.45 In 1913 the Barney and Smith Car Company offered a collapsible car end meant to absorb the impact of rear-end collisions. (American Railroad Journal, *February 1913*)

Figure 2.46 Steel I beams hidden by the exterior plates reinforced the ends of most cars. (Pullman Neg. 27989)

freight train running at 40 miles per hour rammed a passenger train at Mount Union, Pennsylvania (Figure 2.43).[82] The sleeping car *Bellwood* was completely telescoped by the sleeper ahead. The floor of the *Bellwood* passed under the forward car, *Bruceville*. The body of the *Bellwood* was slit open like a tin can; twenty people died. The *Bruceville* sustained little damage beyond a few broken windows, and none of its passengers were injured.

Stronger end roof construction was needed to prevent a recurrence of such horrors, although there was no design that could provide absolute safety while staying within some reasonable weight limit. As early as 1905 William F. Kiesel, Jr., of the Pennsylvania Railroad patented an I- and Z-beam reinforced-end design.[83] The good showing of the Pennsylvania steel cars in the Tyrone, Pennsylvania, wreck of July 1913 was probably attributable to Kiesel's design. In that major accident, which involved a number of cars, there were no passenger fatalities.

Around 1910 Commonwealth Steel offered a massive cast-steel end combined with the end portion of the floor frame (Figure 2.44).[84] Its strength was unquestionable, but weight and cost were against it. Even so, a number of roads used this design, and it was offered throughout the standard steel car era.[85] A less monumental solution to the telescoping danger was offered in 1913 by Barney and Smith (Figure 2.45).[86] A collapsible platform and vestibule were meant to absorb as much of the collision shock as possible. If the entire train were fitted with collapsible platforms, the major impact of a collision could be dissipated through the failure of the platforms. The end bulkhead was braced with heavy steel beams, and the structure was carried up into the roof to prevent a high-level telescope. Whether this plan was widely used is questionable, for Barney and Smith's time as a major builder was nearly at an end.

Stillwell, of course, was advocating a stronger body frame for both light weight and passenger safety during these years. A similar scheme that was presented to the New York Railroad Club in 1921 by F. M. Brinckerhoff suggested adding loops of

Figure 2.47 Heavy blankets of insulation were needed to stabilize car temperatures in all seasons.

wire cable to the ends of cars. The loops, elastic but strong, were to be firmly anchored to the frame so that they would act as giant shock absorbers. It is unknown whether this invention ever received a trial. No complete solution to the telescoping hazard was found. The various partial solutions that emerged were, like so many contributions in the everyday world of engineering, anonymous and largely unnoticed refinements of conventional patterns (Figure 2.46). In general, end construction was reinforced by heavier posts and plates; wide letter boards strengthened the upper body structure.

Good insulation was a basic necessity in transforming the uninviting steel body of the heavyweight car into a comfortable passenger compartment. Because of the high heat conductivity of the steel plates, it was difficult to maintain an agreeable temperature in both hot and cold weather. Here was a new problem for the passenger car builder that did not exist in the wooden car era. Fortunately, designers could draw upon the experience accumulated from the refrigerator car. By then this technology was well advanced, and many effective insulating materials were on the market.

At first builders tried gluing thin sheet insulation materials inside the body panels, leaving a large dead air space. But a completely dead air space was impossible to maintain. In service the body joints loosened, and this together with the movement of the train set up air currents that caused an unwanted heat transfer. Loose insulating materials were equally ineffective, for they tended to pack down and leave large unprotected gaps. The ideal material should have mechanical integrity, should not attract vermin, and should be odorless and nonflammable. To add to the

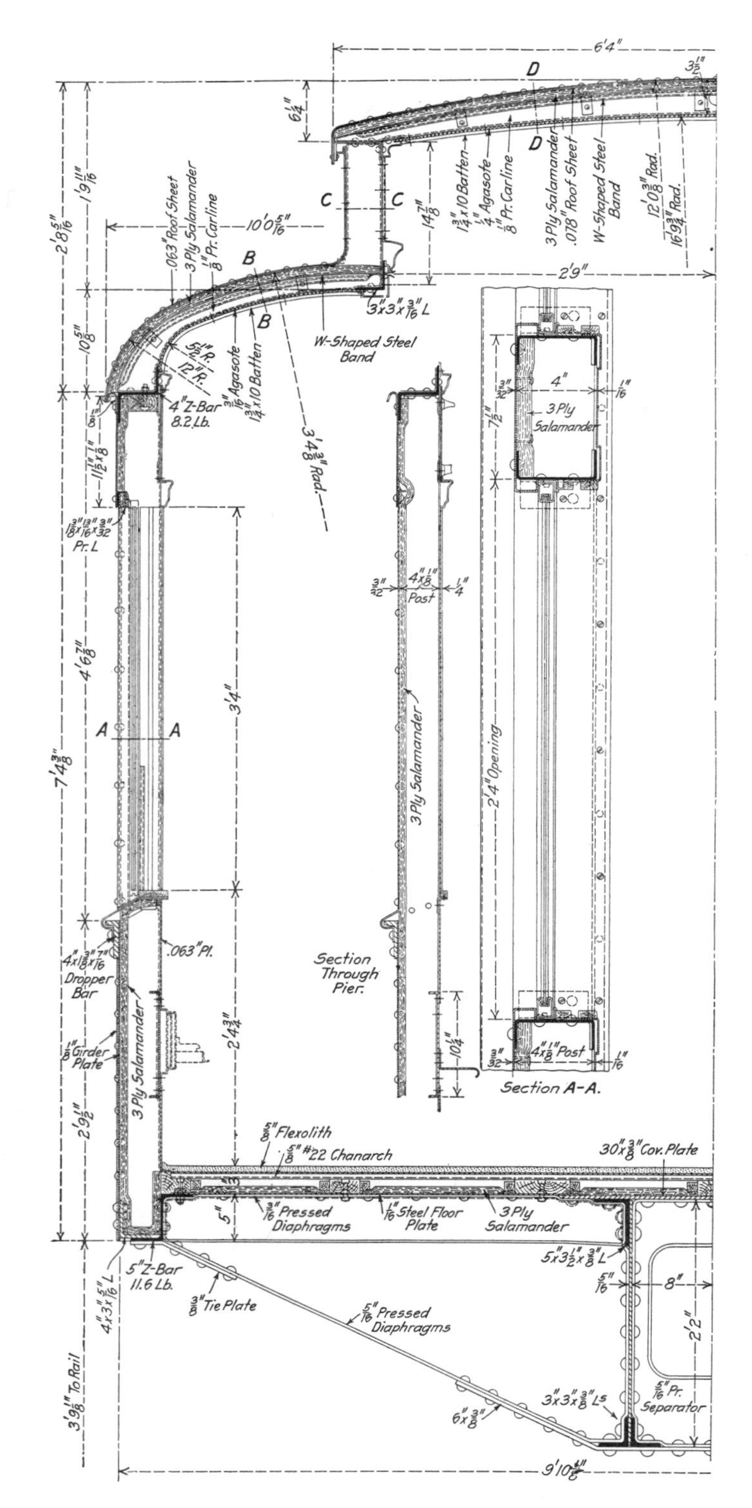

Figure 2.48 The mounting of car insulation is shown in this drawing of about 1925. (Car Builders' Cyclopedia, *1928*)

difficulties, the more fire-resistant materials tended to provide the poorest insulation. One of the best commercial materials was a three-ply, ¾-inch-thick substance developed by the Johns-Manville Manufacturing Company under the trade name of Salamander (Figures 2.47 and 2.48). It was in wide use before 1920.[87] Other hair felts and asbestos board materials were used as well. In extremely cold climates a triple layer of insulation was necessary. The Northern Pacific used an insulated divider panel inside the body walls of its cars.[88] This cut the air space in two, but even so, the temperature differentials experienced along this line (where temperatures sometimes dropped to 50 degrees below zero) caused so much working of the interior and exterior

Figure 2.49 Workmen spread concrete over a pressed-metal subfloor in this 1916 construction scene.

panels that rivets were loosened and tap screws stripped their threads.[89] Wooden interior paneling solved the problem.

Another construction method that improved insulation was introduced in about 1913.[90] Insulating strips were placed between all exterior and interior plates and structural members to retard heat transfer. Double or false flooring was used to provide a large insulating space. But even these techniques did not seem to solve the problem, for there was a continuing demand for better car insulation throughout the 1920s.

Floor construction also underwent a radical change with the coming of the steel car. The wooden floor was largely eliminated, except in baggage cars and Canadian cars. Most American roads felt that wood was a fire and health hazard and adopted various forms of thin concrete floors (Figure 2.49). Such floors had had limited use in toilets before the steel era opened. In constructing them, a corrugated pressed metal or wire mesh was first laid down. The metal was generally galvanized. Over this, composition mortar composed of magnesium or asbestos and cement was poured in a layer ¾ inch thick. An underflooring formed a space both for insulation and for deadening sound. This secondary role of damping down noise was also performed by the car's body insulation.[91] One of the traditional complaints about metal cars was the racket they made. The side-panel sheets produced a drumming sound, while track and undercarriage noises were transmitted through the floor. The fact that the car's heavy insulation filters out noise as well as controlling the temperature is clear to anyone who steps from the passenger compartment to the end platform of a moving train. The din that assails the ears, though some of it is caused by the clatter of the deck plates, draft rigging, and so on, largely represents the contrast between an insulated and an uninsulated area.

The inadequate roof framing of the earliest steel cars, the consequence of efforts to reduce weight, was never really corrected except in the Stillwell car. Part of the problem was the continued use of the clerestory, an inherently weak structure. Here style prevailed over engineering. Designers could argue persuasively in favor of the economy and strength of the arched roof, which they sometimes tried to make more appealing by calling it the semi-

Figure 2.50 Lightweight, arched-roof cars were used by many lines for suburban service. The Number 2601 was built around 1925 by the Standard Steel Car Company.

Figure 2.51 Interior view of the commuter car shown in Figure 2.50.

elliptical or turtleback roof. But to no avail; railroad managements preferred the clerestory. By 1920 it was an even greater anachronism, the "deck lights" (clerestory windows) had been replaced by blank panels, and exhaust ventilation was handled by sheet-metal ventilators mounted at intervals along the blind sides of the clerestory.

A few roads, notably the Harriman Lines, did break away from the clerestory roof. Hundreds of arched-roof cars built to a standard design were operated on the Rock Island, the Union Pacific, the Southern Pacific, and the Illinois Central. But with the breakup of the Harriman empire, most of the associated lines returned to the safe convention of the clerestory.

A new interest in the arched roof appeared in the 1920s, when many roads adopted the design for suburban cars. Among them were the Boston and Albany, the Chicago and North Western, the Rock Island, and the Missouri Pacific (see Figures 2.50 and 2.51). The Delaware and Hudson rebuilt some old wooden cars with turtleback roofs. But what was acceptable for suburban service was not considered good enough for main-line trains. In 1924 Charles E. Barba, chief mechanical engineer of the Bradley Car Company, presented a reasoned argument in favor of the arched roof for all classes of passenger cars.[92] His chief points were not only its strength and cheapness, but also the broader necessity for achieving mass production through standard designs. Although the arched roof did come into common use during the following decade, car building remained a custom business.

During the early years of the steel car, the mechanics of roof construction posed some special problems. The insistence on fireproof cars forced designers into all-metal roof construction. The elaborate clerestory was easily translated from wood to steel, but making it watertight was another matter. The Pennsylvania first tried soft solder joints, but these failed because of flexing of the roof, and around 1910 the road resorted to rivets and welding.[93] Pullman devised a cap joint that proved watertight, though not beautiful.[94] The ends of the short roof plates ($\frac{1}{16}$ inch thick) were flanged and riveted together, and the joint was filled with red lead paint. A pressed-steel cap strip was then placed over the ribs and riveted in place. This roof style was a distinctive feature of Pullman-built equipment until about 1915. The extra initial cost and the maintenance problems connected with an all-metal roof caused many roads to stay with wood-sheathed, canvas-covered roofs. The treated canvas was waterproof and fire-resistant, and the insulation problems were more easily solved.[95]

Among the more interesting developments during the standard steel era was the evolution in the materials used. Some designers of the first-generation heavyweights believed that a flat-sided car body could be made only if precise cold-rolled sheets were used.[96] These plates were indeed true and straight, but they were distorted in the process of fabrication no matter how much care was taken in lining up the frames or how many shims were used. Even if the body left the erecting shop dead true, it soon developed the minor dents, creases, and distortions characteristic of steel cars after a few months' service. The high cost of the cold-rolled plates soon led car builders to substitute common hot-rolled steel sheets. This material was strong, cheap, and readily available, but it had little resistance to corrosion and deteriorated rapidly. By the late 1920s, cars built fifteen or twenty years before were rusting out at an alarming rate.[97] Floors, roofs, sides, and even structural members had rusted through. Regular painting was not enough; a rust-resistant material was needed. In 1916 the American Society for Testing Materials conducted a series of experiments to find a steel alloy that would fill the need.[98] The researchers discovered that even a trace of copper was an effective rust inhibitor. Special copper-alloy steels containing .20 to .25 percent copper were developed for this market. Since the alloys added only about 1½ percent to the cost of a car, they were considered good economy and were widely adopted by the late 1920s.[99] In the following decade U.S. Steel introduced a chromium-copper-silicon steel alloy called Cor-Ten—a low-carbon steel with a durability twice that of the copper-bearing steels.

During the same period there was some talk but little action on the use of high-tensile steel alloys, which would permit the installation of lighter structural members and thus reduce the dead weight. One advocate claimed that passenger car weight could be cut as much as one-third,[100] but other authorities were skeptical and at least one thought that such alloys could be dangerous in practice. Charles E. Barba of the Bradley Car Company feared that shop crews could not be trusted to make repairs with the right grade of steel.[101] What if in their haste or ignorance they replaced a critical structural piece with an ordinary steel plate? Perhaps the alloys actually were tested during this period despite Barba's misgivings, but no evidence of their use can be found. However, alloy steels were incorporated into car construction during the next decade.

A more radical solution to rust and weight problems was the proposal to introduce aluminum. It had received some use since the beginning of the modern steel car era, although it was largely restricted to nonstructural parts such as seat frames, handles, and other bits of hardware. The Pennsylvania Railroad and Pullman tried aluminum sheets for the interior panels of their 1907 experimental cars, but they did not repeat the scheme in the production cars that followed.[102]

The first large-scale use of aluminum was in 1923, when Pullman delivered twenty-five suburban cars to the Illinois Central Railroad.[103] The roof sheets, doors, conduits, junction boxes, and other small fittings were aluminum, with a resultant weight saving of 7,750 pounds. Over the next five years the I C acquired another 235 cars on the same pattern. In 1926 the Pennsylvania Railroad pushed the cause of lightweight aluminum cars even further by building eight 64-foot suburban cars in its own shops. The entire body, including posts, carlines, doors, and interior and exterior sheeting, was aluminum. Heat-treated aluminum was used in the body framing. Over 6 tons of dead weight was saved. It is not known if the experiment was successful enough to lead to additional cars, but six years later Altoona did build a double-deck suburban car for the Long Island Railroad that made extensive use of aluminum. However, it seems that the industry was

Figure 2.52 Large quantities of aluminum were used in many suburban cars to reduce weight. This car was built in 1929 by the Standard Steel Car Company.

Figure 2.53 This coach typical of the heavyweight era was produced in 1921 by the Standard Steel Car Company.

Figure 2.54 A coach built by Pullman for the Central of Georgia in 1925. Steel cars were incredibly durable; many stayed in service for more than forty years. (Pullman Neg. 29243)

not ready for all-aluminum cars, because the Illinois Central's composite model was the example that others followed. The Chicago and North Western acquired 120 semi-aluminum cars in 1927, realizing a saving of 5,700 pounds in each car (Figure 2.52)[104] The Lackawanna acquired a large fleet of semi-aluminum cars for its electric suburban operations in the New York City area. But aluminum made little progress outside of the commuter car; the industry seemed convinced that all-steel construction alone was suitable for main-line service.

The twenty years following the prototype production of steel cars was a period of adjustment. In some ways it was also a time

Figure 2.55 The Chesapeake and Ohio's coach Number 756 was completed in 1929 by the Standard Steel Car Company.

of retreat from the rigid tenets of early steel car construction. The compulsive demand for *all*-steel, fireproof cars was relaxed, and many compromises were accepted. Wooden interiors and roofs reappeared, though not universally. A general softening of the interiors, particularly in first-class cars, was evidenced by a return to carpeting, draperies, and wooden furniture. And the cold, utilitarian, enameled, monochromatic interiors gave way to brighter colors and softer lines toward the end of the 1920s. During the same period the deluxe coach was introduced. Since the first steel coaches had aimed to maximize capacity, seats had been bunched tightly together. Now more thought was given to legroom and comfort. Instead of 80 passengers, a car might seat only 60. Individual chair seats were introduced. Possibly the finest day coaches of the period were the Imperial Salon Cars operated on the Chesapeake and Ohio's *Sportsman*. On one side of the car was a row of single parlor seats; on the other, the usual double seats. The car was thus open and uncongested, with wide aisles and few patrons. David P. Morgan captured the feeling of traveling in one of these luxurious coaches: "It was grand to doze in them over lazy Kentucky miles while the handsome Mountain-type up ahead threw her weight over the ridges. It was travel that savored of good, rich, comfortable tradition."[105] Few coach travelers were treated to such spacious accommodations, of course, but the trend was in this direction, and many older cars were upgraded to meet the new standard.

In general the years between 1910 and 1930 were not notable for substantive changes in passenger car design. It was a period much like the 1870s and 1880s, when orthodox design inspired great contentment. This is reflected in the almost total silence of the railroad trade press on passenger car design. There simply was nothing to report. One order of cars followed another with little or no variation. A slight difference in the arrangement of a new parlor-café car might inspire some notice, but the day coach remained the same. Only in the suburban cars was there some interesting design experimentation in the lightweight aluminum construction, arched roofs, and unusual seating already described.

Figure 2.56 The day coach interior of the 1920s was plain and straightforward, without pretension or unnecessary ornament. (Pullman Neg. 31384)

Figure 2.57 Late in the standard era, some lines began to install more comfortable seating and more decorative interiors as automobiles and intercity motor buses drew away rail traffic. But chair seating was not a new idea even in 1929. (New York Central Neg. 1891)

However, little of this thinking found its way into main-line cars until the following decade.

Age of the Lightweight

The effects of the Great Depression bore heavily on the railroads and their supply industries. Traffic was down, particularly first-class travel; passengers who still had money for luxuries were attracted to the highway and airway, both glamorous new avenues of travel. The car shops that had struggled to complete orders a few years before were now idle.

Some hope during these gloomy years was offered by a new design concept called streamlining. It presented a sleek, modern image of speed and innovation. What had been an obscure technical term in aerodynamics was made into a household word through an astute publicity campaign mounted by several railroad traffic departments. It succeeded in creating a general interest in railroading practically unknown since the opening of the first transcontinental line. In the early and mid-1930s streamlined trains appeared one after another and proved to be effective traffic generators. The attention to styling was wisely buttressed by emphasis on passenger comfort and faster schedules. According to *Railway Age*, "For the first time in many years, the words 'sold out' re-entered the ticket clerk's vocabulary."[106] The new trains were expected to add at least 20 percent in new business, and on certain major runs, such as the Broadway Limited, traffic actually jumped 37 percent. How much of the increase was due to an overall improvement in the economy was never discussed. The reports on "Streamline, Light-Weight, High Speed Passenger Trains" prepared by Coverdale and Colpitts, consulting engineers of New York and for many years high-level advisers to the railroad industry, offer one of the most complete accounts of this subject. The general mechanical, financial, and traffic facts are given in a train-by-train account. By 1950 the firm's eighth report had grown to eighty-nine pages and offered particulars on 260 trains.

From an engineering viewpoint streamlining was little more than "a dramatic gesture" aimed at the public.[107] Its true aerodynamic functions were largely abandoned after the first few automotive trains. These undersized, super-lightweight trains did have scientifically designed front and rear profiles meant to reduce air friction, but they were little more than glamorized rail cars and were not found suitable on heavily traveled runs. For major trains the roads favored full-sized cars capable of interchange service. (The City of Salina, the Zephyr, and other super-lightweights are treated in Chapter 8 on rail cars.)

The more fundamental revolution that occurred in railroad car construction during the 1930s was in fabrication and materials. Many writers have called this period in railroad history the age of the streamliner, but a more accurate title might be the age of the lightweight. Streamlining became a vague word casually applied

Figure 2.58 In 1930 the Chesapeake and Ohio offered even more luxurious and spacious seating with a two-and-one arrangement of reclining chairs. (*Pullman Neg. 36050*)

to any appliance with rounded corners and airflow lines. In reality, railroad cars remained rectangular boxes with only the most superficial vestiges of streamlining in such incidentals as seats, lamps, letter boards, and exterior sheathing. The arched roof—hardly a product of streamlining—was an old form that this book has discussed repeatedly in connection with earlier cars. Possibly the round-end observation car can be fully credited to streamlining, but it was more spectacular than functional.

Streamlining, however, had a prestigious scientific heritage. Its practical application in aerodynamics had been repeatedly demonstrated since the Wright brothers. Now engineers were calling for its application to land vehicles. In 1930 Professor F. W. Pawlowski of the University of Michigan conducted wind tunnel tests for the Brill Company's projected high-speed interurban car. A year later Dr. Oscar Tietjen of the Westinghouse Laboratories carried out similar experiments and concluded that streamlining could save nearly one-third of a train's power at speeds of 75 miles per hour. These and several other wind tunnel tests confirmed the results of experiments by Professor W. F. M. Goss at Purdue University in 1898.[108] Actually, there had been pioneers in railroad streamlining since the 1830s.[109] Prow-fronted locomotives had been tested in this country and in Great Britain during the first decade of the public railway. In 1865 the Reverend Samuel Calthrop (1829–1917) patented a streamlined train of remarkably modern outlines. Some years later a Denver newspaperman, Fredrick U. Adams, patented an equally futuristic design that was given an actual road test in May and June of 1900 (Figure 2.59). The B & O Railroad remodeled several old coaches according to Adams's plan, but they were soon stripped of their fluted shrouding and returned to service as ordinary wooden cars. William R. McKeen of the Union Pacific did his part in perpetuating the wind-splitting idea during the early years of this century with his knife-edge gasoline rail cars. These ancient adventures were forgotten until the 1930s.

The railroads had been slow to capitalize on the potential of streamlining; the pioneers were ignored because they were ahead of their time. But by the thirties the times were indeed favorable, for the railroads were seeking a new image. Enthusiasm for streamlining was clothed in the sacred vestments of science. It seemed a natural—the perfect combination of glamour and utility.

But what was convincing in the laboratory became much less so when tested in the field. For short trains, molded front and rear profiles did have a measurable advantage in speed and fuel economy. But the railroads' express trains were long, not short, and their larger surface areas created a drag or "skin friction" that substantially diminished the promised savings of streamlining.[110] The more zealous advocates of streamlining insisted on flush windows and doors, full-width vestibules, smooth underbody sheathing, and even air-flow shrouds for the trucks. These features did reduce air turbulence, but at a considerable price both in initial cost and in maintenance expense. Most of them became

Figure 2.59 Adams's experimental train, shown here on the B & O in 1900, was visually remarkably advanced. It anticipated the basic designs followed thirty-five years later in the streamline era.

the pet hates of the repair crews. Of the relatively few cars ever treated to full streamlining, most were assigned to the articulated motor trains. The only streamlined feature to be adopted as standard practice was the full-width vestibule covering. After the initial enthusiasm, the excesses of streamlining were abandoned. Superficially the style was perpetuated in the new cars, but many of the true elements of streamlining vanished by about 1937.

Lightweight Construction

Streamlining in a corrupt form governed the appearance of the modern passenger car, but its real essence was lightweight construction. Weight, not air friction, was the chief obstacle to economical operation. Lighter construction had obviously been of long-standing concern to the master car builder, yet cars grew heavier with each major advance in comfort or safety. Every improvement seemed to add that many more tons to the train's groaning load. With the coming of the steel car, weights zoomed. Soon coaches were 65 to 80 tons each, sleepers and diners 85 to 90 tons. Train weights in the period 1910–1930 totaled as much as 1,000 to 1,200 tons. There was some talk, to be sure, of the need for lightweight cars, but little was done about it. The only major weight savings were achieved in suburban cars, where arched roofs, lighter framing, a slight reduction in overall size, and aluminum fittings pared away as much as 6 tons per car. These compromises were not thought safe or suitable for main-line cars. But the light-car advocates persisted, and by the end of the 1920s more serious thought was being given to the problem. Besides the obvious economies possible with lighter cars, some roads were facing serious operational difficulties because of their heavy trains.[111] On densely traveled lines, schedules were congested by the slow starting and stopping of the heavyweights. The alternatives were larger locomotives or more tracks or both, and these were expensive solutions. Lightweight cars began to look very attractive.

The car builder could choose among several remedies to the weight problem. He could:

1. Reduce the overall car size.
2. Reduce the auxiliary equipment.
3. Adopt articulated design.
4. Turn to new structural designs.
5. Use new methods of fabrication.
6. Exploit new materials.
7. Reduce truck weights by using fewer axles.

The first three possibilities were rejected by car builders, while the last four became standard practice. To summarize the consensus of railroad opinion on these points: (1) Undersized cars could effect considerable weight savings—as much as 10 tons in cars of otherwise equal construction.[112] But a major attraction of railway travel was the roomy, full-sized car—the most commodious form of land transportation. (2) The trend toward greater luxury in travel was steadily increasing the amount of auxiliary equipment. Electric lighting and air conditioning in particular added considerable weight. Stripped-down models might be acceptable for short-run suburban cars, but they could never satisfy the long-distance traveler. (3) Articulation eliminated the weight of one truck, which was a substantial saving in a train of fourteen cars. But since it fixed the consists and made it necessary to shorten individual body lengths to keep within swing-out clearances, it was not acceptable in American operating practices. A limited number of "married pairs" (two cars semipermanently coupled together) did come into use in this country. (4) New structural designs were accepted that featured the arched roof and side-panel trusses in place of the traditional clerestory roof and fishbelly underframe. Both had already been used in the wooden car era, but the development of strong alloys made them possible in metal construction. (5) New methods of construction, such as welding that displaced riveting, made light-metal fabrication practical. (6) New materials that were exploited included strong aluminum alloys and high-tensile, rust-resistant steel alloys (Cor-Ten and stainless). It was these, combined with new designs and new methods of fabrication (4 and 5), that produced the modern lightweight car. (7) Improved four-wheel trucks (weighing 18 tons a pair) displaced the heavy six-wheel design (24 tons a pair) typical of the heavyweight car.

All seven remedies were based on techniques and materials available long before their adoption in the 1930s. Why weren't they used sooner? Why did the heavyweight car survive for so long? Tradition, of course, was unquestionably a factor. Once a dependable plan is in use, operating men are reluctant to throw it over. Years of costly development go into every workable design, and practical men rightly distrust those who would discard a proved plan for an unproved one no matter how superior the new plan might seem on first examination. Also, as with the wooden car a few generations earlier, plants and personnel were geared to handle the heavyweights. The commercial builders were equally committed to them. The railroads themselves had an enormous investment in the standard steel car. They immediately asked: Will the lightweights work with our existing stock? Even if they are designed to interchange, will it be safe to run light and heavy cars together?

In addition to these general difficulties, the car builder of the standard era might not be able to obtain the special materials needed for lightweight manufacture. The heavyweight cars were made of common carbon steel, which was available in quantity, in a variety of sizes and shapes, and at low cost. The design was tailored to suit this relatively weak, highly corrosive material. High-tensile steels were specialties, supplied in small quantities, in limited sizes, and at high cost. Like carbon steel, they were highly corrosive; thus their lightweight potential was partially negated by the fear that thin-section structural members would quickly rust out. Availability and first cost, however, remained a large factor in the continued production of heavyweight carbon-steel cars. Only when specialized rust-resistant steels and aluminum alloys became cheaper and more widely available did the lightweight car come into its own.

The first lightweight coaches that appeared in the mid-1930s made the old standard cars look gross. Large weight savings had been effected; body shells were 45 percent lighter. The Milwaukee Road built some coaches of ordinary carbon steel that weighed only 48 tons. Pullman's experimental aluminum observation coach of 1933 probably set an all-time record at 37 tons. After the experimental stage was passed, the weight of the production-model lightweight coach was stabilized at about 50 tons.[113] But this total gradually began to creep up as more extras were added. Ironically, some fully outfitted lightweight cars began to *exceed* the weight of the old standard cars (the dome coaches on G.M.'s Train of Tomorrow, for example, weighed 73½ tons). The reason was the ever increasing size and number of auxiliaries; air conditioning alone added 4 to 5 tons. In general, the modern car's increased power demand greatly increased battery and generator size. There was also the added weight of the massive tightlock couplers, folding steps, high-speed brakes, and roller bearings. By the 1940s some savings were made in running-gear redesigns. Hollow axles could save as much as 500 pounds per axle. Mounting brake cylinders on the trucks simplified brake riggings and thus saved 1½ tons per car. Even greater reductions were anticipated because of new truck designs, but outside of the Milwaukee Road's efforts, little was accomplished until composition brake shoes eliminated the need for clasp brake arrangements. Trucks continued to represent one-third of total car weight. The first steel coaches carried only 11,000 pounds of auxiliary systems, while the lightweight car was encumbered by some 48,500 pounds, which equaled 60 percent or more of body weight.[114] Because the heavyweights could seat more people (80), the dead weight per passenger was only about 1,500 pounds. A 60-passenger lightweight car, on the other hand, carried about 2,100 pounds per seat space.

Even after the lightweights were accepted, some railroads continued to buy new heavyweight cars. In the 1930s these purchases might be explained as inertia or the preference of an old-line mechanical officer. Among the last of the heavyweights built completely on traditional lines (with a clerestory roof, fishbelly underframe, and six-wheel trucks) was a group of twenty-five cars completed in 1938 for the Missouri, Kansas, Texas Railroad by American Car and Foundry. Most were scrapped before 1964.

However, it is more difficult to explain the acquisition of such equipment during the following decade. In 1942 the Chesapeake and Ohio took delivery of twenty 70-ton heavyweights from A.C.F.[115] Several concessions to modern design were made, such as air conditioning and an arched roof, but otherwise all the old features were present, including six-wheel trucks and double vestibules. At about the same time the Atlantic Coast Line received some similar cars from Bethlehem Steel (Figure 2.60). Weighing just under 78 tons, they were among the heaviest passenger coaches ever used by an American railroad and hence the heaviest ever used anywhere in the world. Even more improbable, A.C.F. delivered eleven coaches and four parlor cars to the Alton in 1948. They were built to the outline dimensions of lightweight cars, but much of the car body material was ordinary copper-

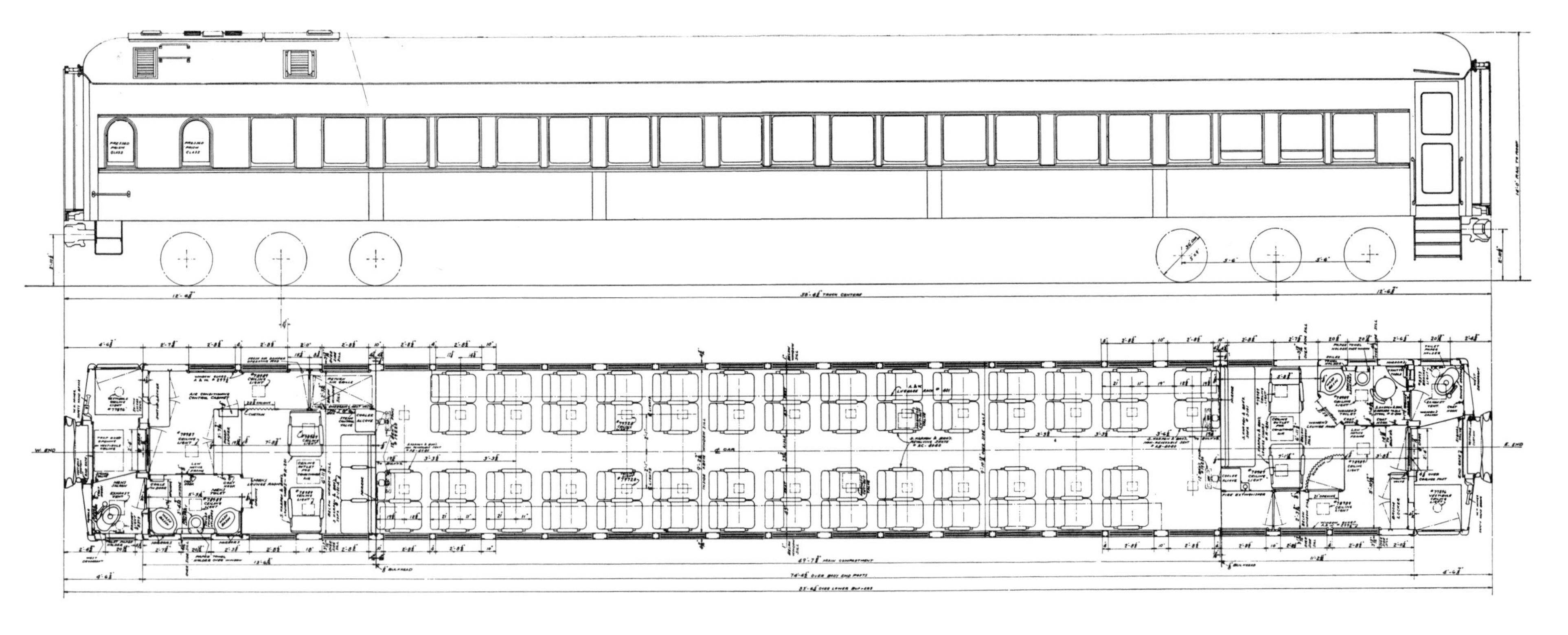

Figure 2.60 *The heavyweight era officially ended in 1934, but a few standard steel cars were still being produced more than a decade later. Bethlehem Steel's Harlan division built several heavyweights from the plan shown here for the Atlantic Coast Line in 1943.* (Car Builders' Cyclopedia, *1943*)

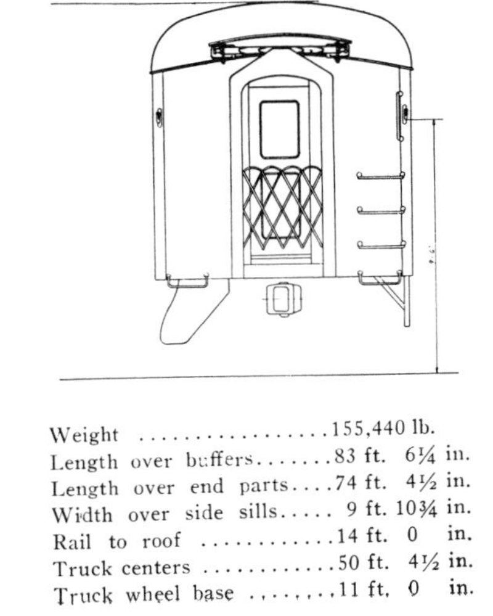

Weight	155,440 lb.
Length over buffers	83 ft. 6¼ in.
Length over end parts	74 ft. 4½ in.
Width over side sills	9 ft. 10¾ in.
Rail to roof	14 ft. 0 in.
Truck centers	50 ft. 4½ in.
Truck wheel base	11 ft. 0 in.

bearing steel. The rivets and six-wheel trucks were a bold testament to their antique design.

The first streamlined motor trains, the U.P. M-10000 and the Burlington Zephyr, naturally monopolized the public's attention, but more lasting developments were underway at the very time that these spectacular trains were stealing the headlines in their 1934 cross-country tours. In 1933 Pullman completed two experimental, all-aluminum observation cars—full-sized vehicles meant to run with conventional equipment. In May 1934 the Milwaukee Road shops turned out their first lightweight coach, and large-scale production immediately followed. Early in the following year, the New Haven began receiving the first of its lightweight coaches from Pullman's Worcester, Massachusetts, plant. A.C.F. entered the lightweight field at midyear with two full-sized trains for the B & O, both of which soon moved to the Alton. Budd, which was initially committed to motor trains, completed its first full-sized coach early in 1936. By this date all the major car builders were producing lightweight cars. By late 1946, 6,500 lightweights were in service, but this figure represented only about 15 percent of all passenger cars running on American railroads.[116] In all, an estimated 9,000 lightweights were built before the passenger car market temporarily vanished in the late 1950s.

ALUMINUM

Of all the new materials available at the outset of the lightweight era, aluminum was the most attractive. It was, in fact, called the light metal. It already had an established reputation among railroad men because of its wide and successful use in suburban cars since the mid-1920s. The automotive industry had used it in vehicle construction even earlier. It was light, corrosion-resistant, plentiful, and reasonably cheap. It weighed only one-third as much as steel. It seemed the ideal material for car construction.

The aluminum producers were anxious to develop a large market for their product and sought every opportunity to promote it to the railroads.[117] Their continuous selling effort is responsible in large measure for the periodic "revivals" in aluminum car construction.

However, the inherent shortcomings of aluminum far outweighed its advantages. Common aluminum was, of course, much too weak for car construction. The wrought or heat-treated alloys that were developed did equal carbon steel in strength (17 ST, a popular alloy, was equivalent to 58,000 pounds of high-tensile steel), but they were somewhat short of Cor-Ten steel alloys (60,000 to 75,000 pounds) and woefully short of stainless (110,000 to 175,000 pounds).[118] As aluminum alloys became more refined, they grew not only heavier and more expensive but also more difficult to work. In the 1930s they cost 35 cents a pound, or ten times the price of Cor-Ten steel. (Part of their cost could be recovered in salvage, however.) The overall cost of an aluminum car was nearly 25 percent above that of a steel one, though aluminum cars were slightly cheaper than those made of stainless steel.

Despite the softness of common aluminum, the alloys first offered for car building were surprisingly hard to work with.[119] They tended to crack when worked cold, and the inconvenience of heating each sheet and member was a major drawback. Because riveting was the only practical method of fabrication, there were countless holes to be punched. Welding was suitable only for nonstructural fastenings. Hot riveting was impossible, because aluminum's coefficient of expansion was so great that without extreme care, the assembly might buckle. Alloy aluminum rivets could not readily be worked cold, common aluminum rivets were too weak, and steel rivets reacted and rusted out quickly. In addition, welding had become the modern metalcrafter's principal tool, and a metal that could not be fabricated by this method was thought seriously deficient.

Another less severe objection to aluminum was that of appearance. Aluminum does not weather well, and it was found necessary to paint the exterior surface to maintain the spruce look expected of passenger equipment. Only when it was anodized could car aluminum be kept bright without regular polishing. In 1946 the Louisville and Nashville received some anodized cars from A.C.F. that had polished, fluted sides which seemed to the unknowing eye like stainless-steel-sheathed cars.[120] These polished panels were fitted in order to hide the unsightly body rivets. An alternate method, used as early as 1941 in a group of Union Pacific cars manufactured by Pullman, was to countersink and putty over the rivets before painting.

The most serious failing of aluminum was that corrosion played havoc with the structural parts and plates of aluminum cars, particularly where they were joined to steel members.[121] These connections were often at such crucial junctions as the center sills and the bolster. Deep pitting from electrolytic action resulted in structural failures. It most often developed on the car's underside, well hidden from view. The usual accumulation of roadway grime prevented detection until after the damage was well advanced. The Union Pacific undercoated its aluminum-framed cars to retard this action.[122] The aluminum industry assured the railroads that it had developed a new alloy to correct the problem. The original copper-bearing 17 ST alloy was replaced by 61 ST 6, a magnesium-aluminum alloy that was a more workable, less corrosive metal.[123] In the future better alloys than the ones that exist today will doubtless be developed, but there is skepticism among some mechanical men that the problem will ever be completely solved. Whether their pessimism is justified will not be known until the latest aluminum freight cars have completed a longer term of service.

Whatever the problems of the aluminum car, nearly 1,000 were built of this material, mostly for the U P, L & N, and K C S. The success of aluminum suburban cars led to planning for a main-line car in 1930. During the following year Pullman received an order for an experimental car, but actual construction was delayed for a year. The upcoming Century of Progress Exposition in Chicago rekindled interest in the project, and Pullman decided to

Figure 2.61 Pullman's ultra-light aluminum coach-observation car, completed in 1933, was one of the first streamline cars built in the United States.

Figure 2.62 The art deco-inspired interior of Pullman's 1933 coach-observation car.

produce two all-aluminum cars. The more famous, an observation sleeper named *George M. Pullman*, is described in Chapter 3 on sleeping cars. The other car, reportedly named the *City of Cheyenne*, was an observation coach (Figures 2.61 and 2.62). This 78-foot, 10-inch car seated 50 passengers.[124] Except for wheels, axles, springs, and couplers, it was completely aluminum. The truck side frames were pressed sheet aluminum. Even the body insulation was of crumpled aluminum foil. The center sill was formed from two shallow U-shaped pressings. It was capable of withstanding the Railway Mail Service buffering load, which was then 400,000 pounds. The most remarkable single feature of the car was its total weight of only 73,880 pounds. Of this, the air conditioning accounted for 6,880 pounds. After the close of the Century of Progress Exposition, the car was briefly in service on the Illinois Central and the Union Pacific railroads. It reportedly rode hard and was apparently returned to Pullman, where its ultimate disposition cannot be determined.

Pullman, undiscouraged by the generally cool reception given its aluminum twins, became something of a champion of this style of construction. W. H. Mussey, Pullman's research engineer, voiced his preference for aluminum cars in a talk before the Western Society of Engineers.[125] He noted his employer's long-standing interest in that material; the Illinois Central's semi-aluminum cars of 1923 were Pullman products. In addition, the City of Salina and other Union Pacific motor trains, all made in the same shops, were light-metal fabrications. Pullman was, of course, encouraged in its loyalty to aluminum by the Union Pacific, a steady customer. After the motor train craze had subsided in 1937, the U P began to purchase full-sized aluminum cars. Over the next few years it became the largest operator of such cars in the country. By 1955 it had a fleet of 450 aluminum cars, most of them built by Pullman.

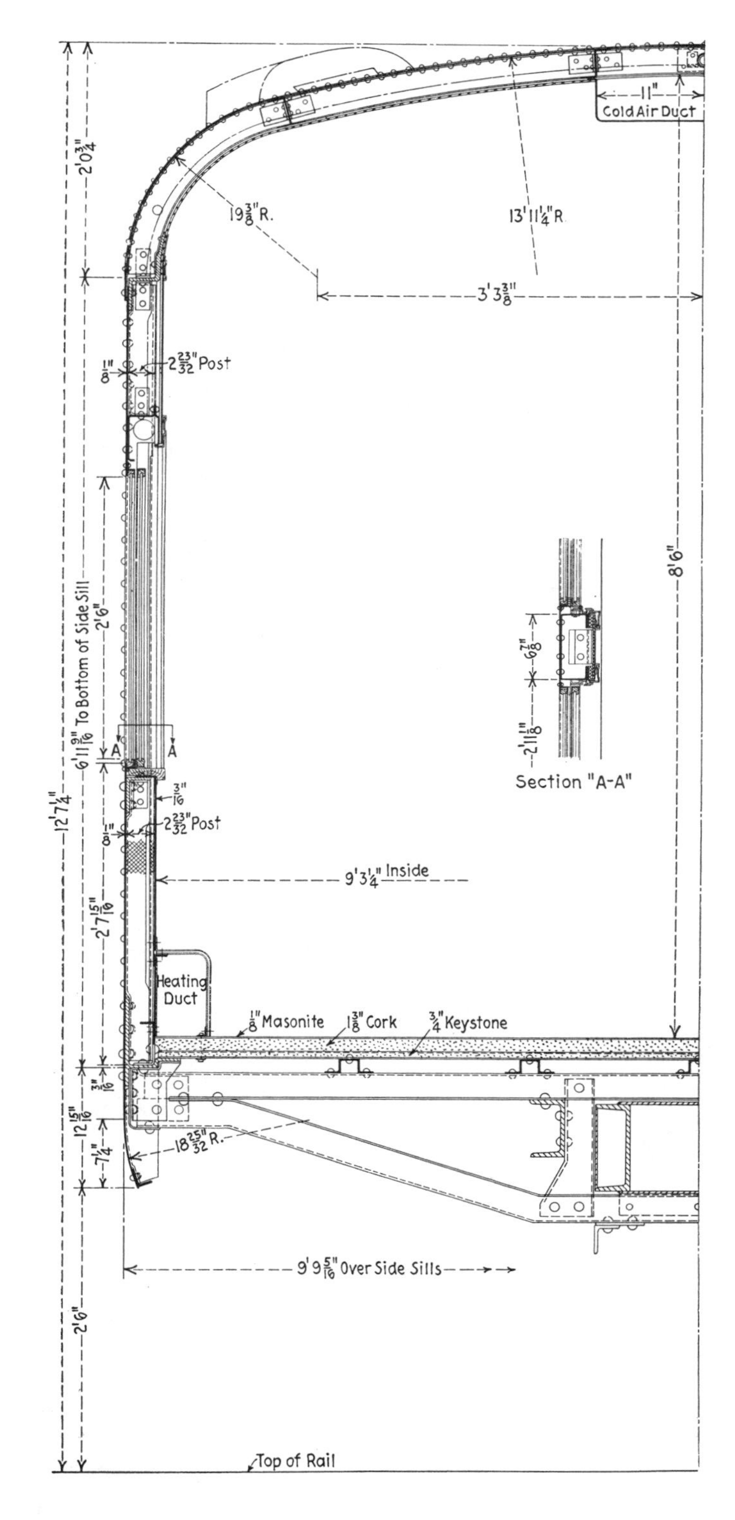

Figure 2.63 Among the earliest production lightweight cars in the United States were the aluminum cars built in 1935 for the Alton by the American Car and Foundry Company. (Railway Mechanical Engineer, *May 1935*)

The American Car and Foundry Company also found itself in the aluminum car business, though perhaps with less outspoken enthusiasm. It supplied aluminum cars to the Kansas City Southern, the Louisville and Nashville, the Union Pacific, and the Missouri Pacific. Its first experience was in making the cars for the Alton's Abraham Lincoln.[126] The Alton's owner, the Baltimore and Ohio, ordered a nearly identical train from A.C.F. to be built of Cor-Ten steel for service on the Royal Blue.[127] The two trains provided a rare opportunity to compare the same equipment made from competing materials. The aluminum train entered service in June 1935, the steel train a month later. The cars of both trains were slightly undersized in all dimensions: they were 70 feet long, 12 feet, 7 inches high, and 9 feet, 9 inches wide. This is an important reason for their good showing as lightweights, and comparison with full-sized equipment is unfair. Comparison between the cars of the two trains, however, revealed that the aluminum coaches were 5 tons lighter. The table below gives other details.

	ALUMINUM COACH	COR-TEN COACH
Body weight	62,300 pounds	72,500 pounds
Trucks	24,600 pounds	24,600 pounds
Total	86,900 pounds	97,100 pounds

The complete eight-car steel train weighed 780,800 pounds, the aluminum train 699,540 pounds—a difference of 81,260 pounds in favor of the Abraham Lincoln.

There were steel center sills in the aluminum cars. Other framing members were aluminum, but steel rivets were used in the assembly (Figure 2.63). Although the body and roof sheets were riveted on, the plate joints were butt-welded and ground smooth (Figures 2.64 and 2.65).

Some attempt was made by both A.C.F. and Pullman over the next few years to use even more aluminum in the underframe structure. Extruded side sills of a complex cross section best described as a combination U and L were a feature of the Abraham Lincoln's frame. Z bars were employed for cross framing, and in later years the side posts were given the same shape. These became rather common structural elements in aluminum cars. A.C.F. favored an all-aluminum U-shaped center sill built up of plates and angles. The top plate of this giant 16- by 16-inch channel was ⅜ inch thick. An example of the plan is shown in the cross section of a car from the Missouri Pacific's Eagle (Figure 2.66). However, even in these cars the bolsters, cross bearers, and end sills were of steel.[128] It was also usual to make the body end posts of steel as a precaution against telescoping. After the A.A.R. raised its yield strength requirements for passenger car underframes in 1946, production was suspended on aluminum-frame cars. Steel frames then became a necessity, and some roads specified cast-steel end frames.

If aluminum was not entirely successful for railway car construction, it did continue as an important secondary material. It was widely used in nearly all modern cars for small fittings, accessories, and interior decoration. And the aluminum manufacturers are not yet willing to admit that an end has come to their railroad adventure. Several thousand aluminum-bodied freight and rapid-transit cars have been built in recent years. United Aircraft's sensational Turbo Train of 1967 was aluminum. And the L.R.C. (Light, Rapid, Comfortable), which was projected in 1967 as the very latest development in railroad passenger cars, was to be of the same construction.[129] A prototype based on the Canadian patent was completed in 1971 by the Montreal Locomotive Works, and the design was said to offer a comfortable and safe ride over existing tracks at speeds as high as 120 miles per hour. But although development work continues in Britain on a similar project, the A.P.T. (Advanced Passenger Train), the L.R.C. appears to be dead in North America. Amtrak's most recent orders have been for stainless or lightweight steel equipment. The aluminum passenger car appears to have no future on American railways.

Figure 2.64 *A.C.F.'s 1935 riveted-aluminum cars for the Alton were somewhat smaller than conventional equipment of this period.*

Figure 2.65 *The 1935 Alton cars were traditionally furnished, in keeping with the views of the line's conservative management.*

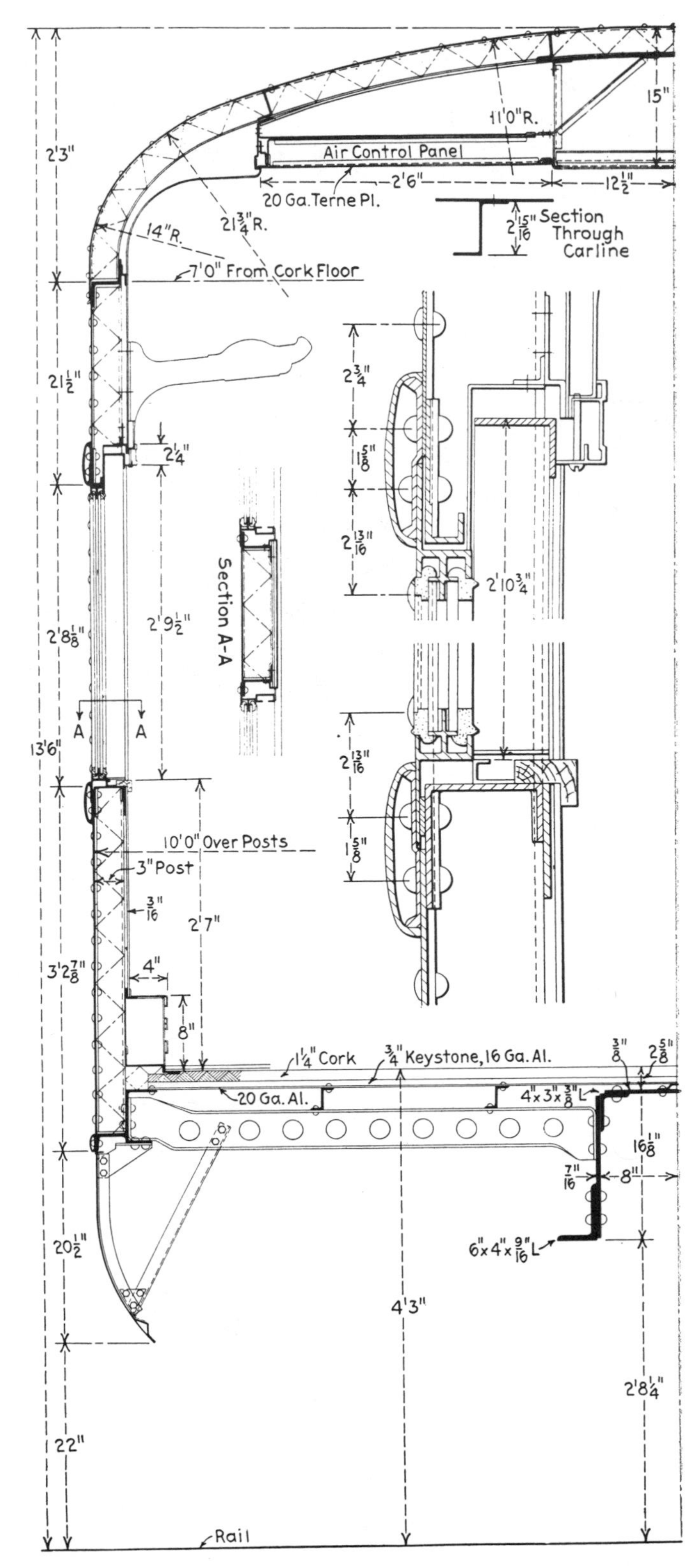

Figure 2.66 A.C.F. built this aluminum-bodied car in 1940 for the Missouri Pacific train, the Eagle. (Railway Age, *February 24, 1940)*

STAINLESS STEEL

At the beginning of the lightweight age the revolution in car building was advanced by an entirely new material: stainless steel. As a miracle metal it outclassed aluminum on several counts. It was incredibly strong, yet ductile. It did not rust, and its bright silver-grey appearance was most appealing. It had been introduced in 1912 by Krupp, but manufacturers had thought it suitable for little more than cutlery and decorative novelties. The metal, a combination of 18 percent chrome, 8 percent nickel, and low-carbon steel, seemed destined for only limited use.

Then the noble properties of stainless steel came to the attention of a Philadelphia automotive parts manufacturer, Edward G. Budd.[130] Budd's specialty was auto body stampings, but earlier he had been in the railway supply field and was looking for an opportunity to reenter it. He first heard about stainless steel in 1928 from a German contact. It did not sound particularly promising, but he investigated. With his knowledge of the sheet-metal trade, Budd soon realized that stainless steel had great potential for lightweight structures. A surprisingly strong body could be fabricated from thin sheet metal, but because of the rapid corrosion of common steel, it was not considered suitable for railway car construction. In addition, the higher-strength steel alloys needed for such fabrication required heat-treating, which would result in warpage in so complex a structure. With a strong, rustproof material, however, extremely light structures would be possible. A beam of the correct proportions made of 30-gauge stainless-steel sheets was equal in strength to a solid I beam nearly three times as heavy. The stainless beam was a closed box, but there was no need for access to it in order to paint; it would never rust out. Budd also discovered that stainless steel was ductile: it could be pressed, drawn, and rolled without difficulty. A deep cup equal in length to its diameter could be formed with no metal failure. It was like working copper, yet this metal was nine times stronger than carbon steel and three times stronger than Cor-Ten. Even subzero temperatures have little effect on its strength or ductility. Cold-rolling resulted in a significant increase of tensile strength (up to 180,000 pounds), but again without sacrificing its easy-forming qualities.

Despite these positive findings, the matter of fabrication presented a serious difficulty. In thin-wall construction, rivets or sheet-metal screws were not considered satisfactory because of the large number of fastenings necessary for an adequate bearing surface. The matter of appearance was a second objection. Conventional gas or electric welding, even spot welding, created too much heat and ruined the metal, reducing it to little more than an inferior mild steel. A special welding technique was needed. The problem was given to Colonel Earl J. W. Ragsdale (1885–1946), Budd's chief engineer. His assignment was to devise a method of quick, extremely high-temperature welding followed by instant cooling, meanwhile preventing the weld from coming to the surface of the pieces being joined. By 1933 Ragsdale and his staff had perfected a method of electric, controlled-energy welding that is known today as the Shotweld process. Precision control is essential; the current, applied for 1/60 to 10/60 second (at 2600 degrees), melts the metal of the two sheets to be joined, allowing them to intermingle and thus bond. The surrounding cold metal chills the weld. In the union of two 1/8-inch sheets, the diameter of the weld is about 1/4 inch.

The combination of stainless steel and Shotwelding added a new dimension to railroad car construction. In Budd's words: "The designer is therefore free to place this metal just where it is in the line of strain, and in just a sufficient amount to withstand the stresses." Andrew Carnegie was fond of saying, "Pioneering don't pay." Budd set out to prove that it can.

Even before the Shotweld process was ready, Budd wanted to demonstrate the possibilities of lightweight stainless fabrication.

Figure 2.67 The Union Pacific was the largest operator of aluminum passenger cars in this country. An unfinished body being fabricated for the U P is shown at A.C.F.'s plant in 1949. (American Car and Foundry Company)

Figure 2.68 Interior of a 1949 Union Pacific aluminum car before insulation was installed. (American Car and Foundry Company)

Figure 2.69 A Union Pacific aluminum car of 1949 after insulation was added. (*American Car and Foundry Company*)

Figure 2.70 An aluminum car delivered to the Union Pacific in 1942 by Pullman. (*Pullman Neg. 45837*)

Figure 2.71 Santa Fe coach Number 3070 was the first full-sized stainless-steel car built in the United States. It was manufactured by the Budd Company in 1936. (Robert Wayner)

In 1931 the firm built an airplane named the "Pioneer" that it claimed was the first all-steel aircraft. The plane was retired to the Franklin Institute after its demonstration days ended. In the same year Budd constructed a small gasoline-powered rail car, and over the next two years he produced several more of this type. Unfortunately, most of them had a rubber-tire drive, which placed them in the freak category and did little to advance the cause of stainless-steel rolling stock among practical railroaders. But in 1934 Budd's construction of the Burlington Zephyr quieted these misgivings, for here was a major motor train purchased by a class-one railroad.

Next Budd ventured into the regular passenger car market, a move which the established car builders regarded as an outrageous intrusion. More detached observers must have thought Budd mad for entering a market already overstocked with producers who had seen few orders since the 1929 crash. Nevertheless, his first full-sized car was delivered to the Santa Fe early in 1936. It was a 52-passenger coach, numbered 3070 (Figures 2.71 to 2.73). The 79-foot, 8-inch car weighed just under 42 tons—the lightweight record for an all-steel car.[131] The general design was a joint effort of the Budd and Santa Fe engineering departments.

Except for the side framing, the Number 3070 followed Budd's normal method of construction: All structural members were made of thin-gauge stainless steel, rolled or folded to shape and Shotwelded. The outer sheathing was formed by fluted, snap-on stainless strips. The center sills were 12- by 1½- by 1/16-inch channels; the cross bearers measured 8½ by 1½ by 1/16 inches and were set at 27-inch intervals. In place of the side-panel truss were large, upright channels (10½ by 3 inches) formed of 1/16-inch stainless-steel sheet. (This method simplified fabrication, but it was apparently not stiff enough; in the next group of Santa Fe cars Budd returned to the Pratt truss side frame.[132]) As was typical of the lightweight cars, the roof served as part of the structure. Like a bridge truss, it was the upper chord or compression member, and was meant to take one-third of the total bending moment of the body. Besides the carlines, no heavy supports were used other than two stringers running the full length of the car and attached to the collision posts at both ends. The corrugated roof sheeting itself was the main roof stiffener. Being the largest unbroken surface area of the car, this great panel provided some strength for the body structure even though it was made of light-gauge sheet. In the words of Colonel Ragsdale: "It was not merely an umbrella to keep the weather out."[133] As an extra precaution against high-level telescoping, a collision bulkhead was formed by a 4-foot sheet welded to the carlines at each end of the roof.

The floor was corrugated stainless, but in place of concrete, a cork composition was laid in. Linoleum was used for the top

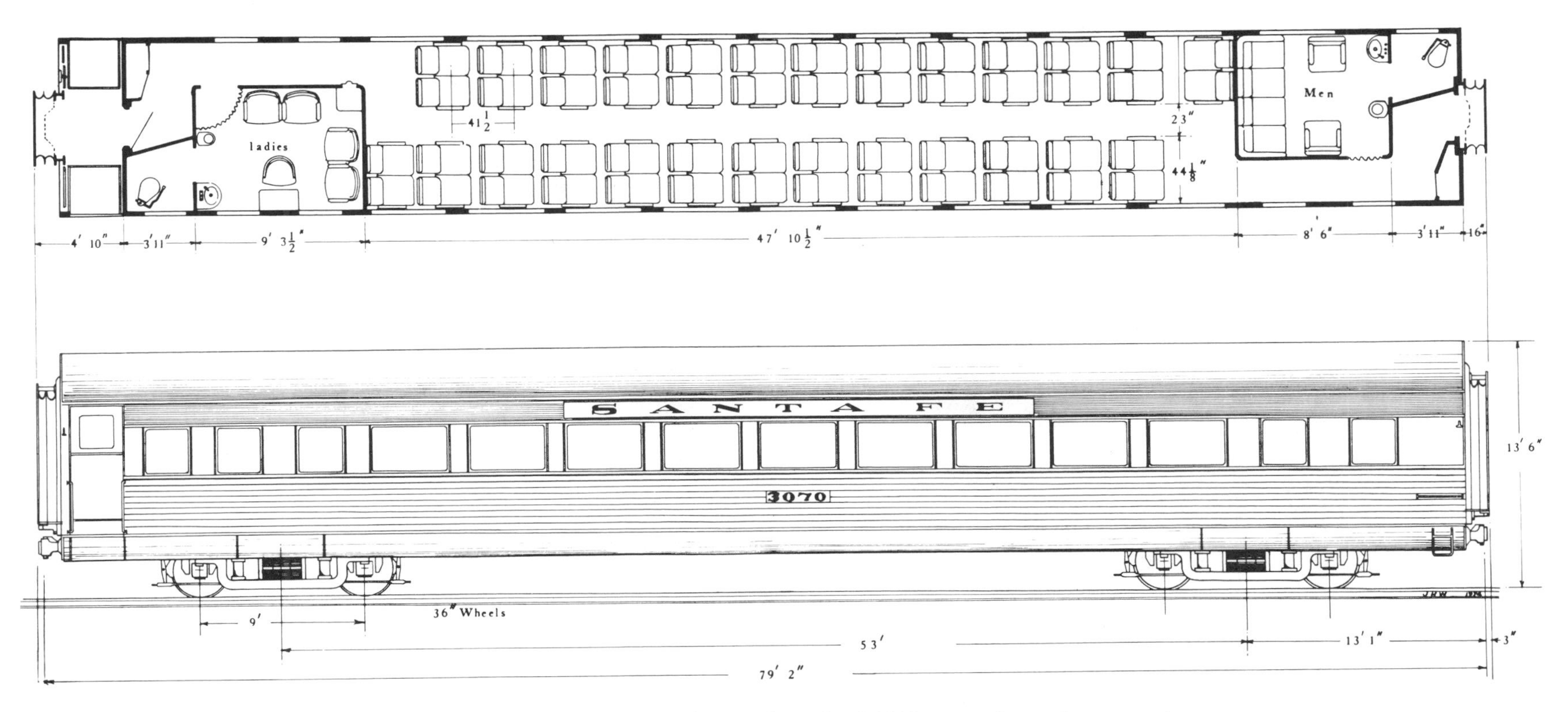

Figure 2.72 This drawing of a stainless-steel coach of 1936 was redrawn from an original tracing too badly soiled for reproduction. (Traced by John H. White, Jr.)

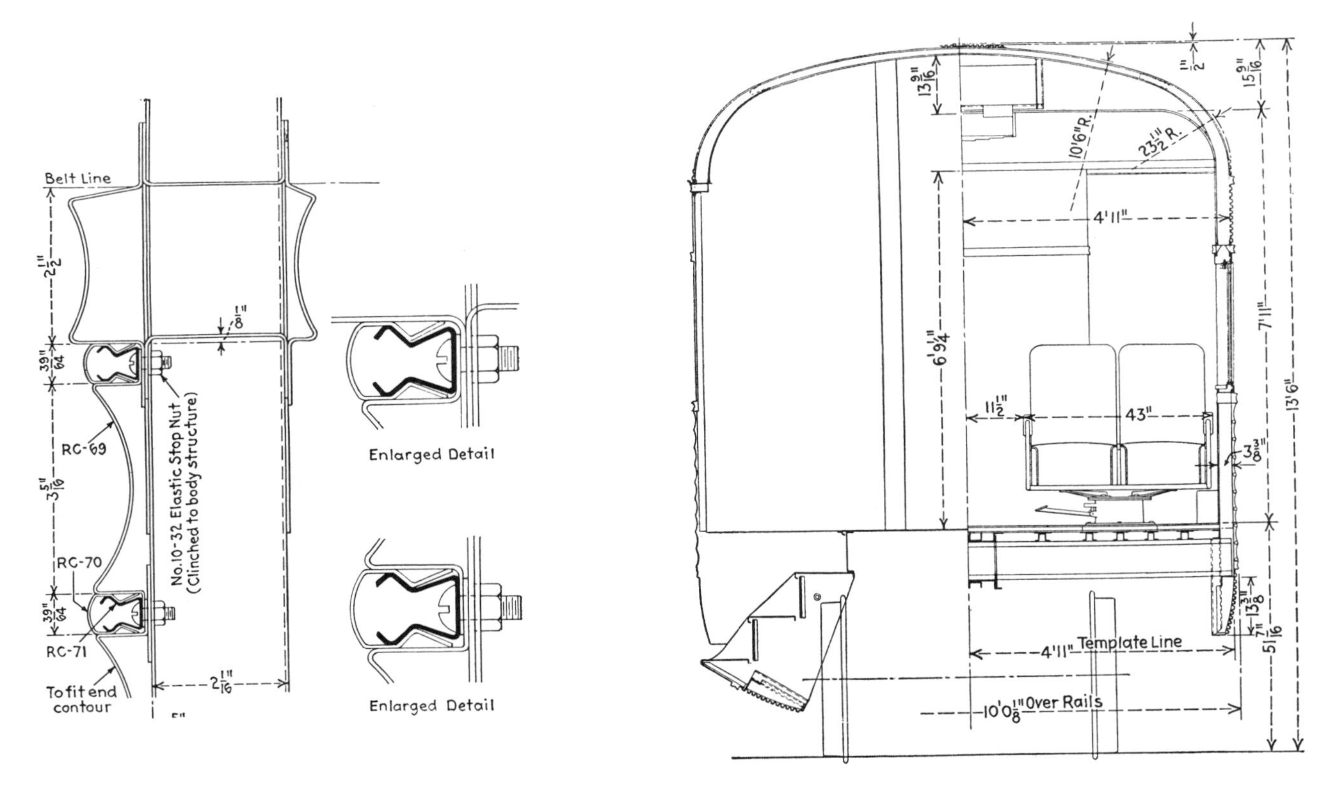

Figure 2.73 The snap-on fastening devised by Budd for fluted car sides. (Railway Mechanical Engineer, *March 1936*)

surface. The underbody was insulated with 3-inch Dry-Zero airplane blanket. As a gesture to earlier times, the interior was paneled in wood, but the paneling was only veneer mounted on presswood sheets.

The Number 3070 was an experiment, and its extra-light framing was found inadequate. Even so, the car continued on the Santa Fe until 1969, when it was sold to the Erie Lackawanna for commuter service. Budd did decide, however, to return to the modified Pratt side-frame trussing and a more substantial center sill. The side truss was taken directly from freight car service and could be traced by the historian (but not the modern car builder) to the dark ages of the wooden car era. If the public could have been sold on triangular windows, a simple Pratt truss would have done the job nicely. But the rectangular form was enshrined in

Figure 2.74 A 1938 construction scene in Budd's shop showing the modified Pratt side-framing truss. (*The Budd Company*)

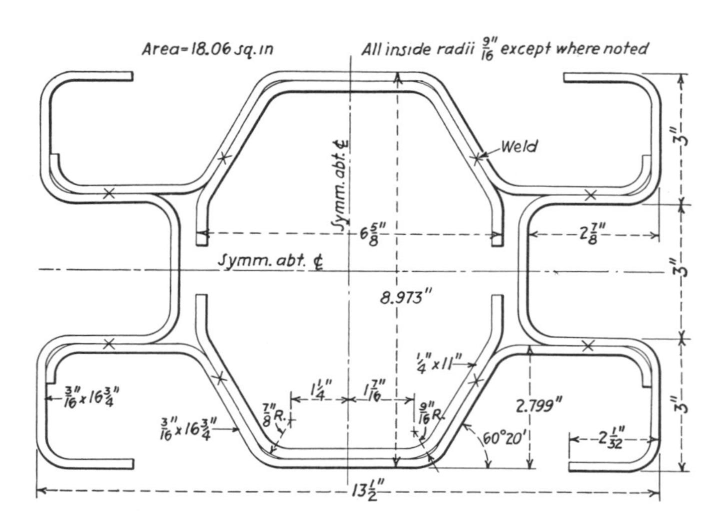

Figure 2.75 The fabricated Shotweld center sill devised by Budd around 1938. (Railway Mechanical Engineer, *April 1939*)

public taste, and it was necessary to use the modified Pratt or Vierendeel variety (Figure 2.74). The old-fashioned belt rail was retained to further ensure the squareness of the window openings. Budd had employed the side-frame truss since the time of the firm's earliest rail cars, but it introduced a new style of center sill in 1938 or 1939. The new sill was a complex weldment formed of six $\frac{3}{16}$- and $\frac{1}{4}$-inch stainless-steel plates drawn to shape specifically for this purpose (Figures 2.75 and 2.76). It was capable of withstanding a 900,000-pound buffering load. These changes materially improved the strength of the Budd cars, but they inevitably increased weight. Coaches built for the Seaboard Air Line in 1939 with the new center sill weighed 51 tons—up nearly 9 tons from the Santa Fe car.[134] Part of the additional weight came from the greater length of the S A L cars (84 feet, 8 inches versus 79 feet, 8 inches), but the sturdier structure accounted for much of the extra burden.

By the late 1930's it was clear that Budd had made a success of the stainless-steel passenger car. Silver cars of its manufacture were rolling on the Reading, the Santa Fe, the New York Central, and other major railroads. By 1944 the company had produced nearly 500 cars. The established car builders were understandably annoyed by the success of this upstart. Pullman's pique was fanned into outright hatred in 1937, when Budd began producing sleeping cars for the Santa Fe; this was sacred territory reserved exclusively for Pullman. Dark rumors were circulated deploring the unreliability of Shotwelding and Budd's general lack of experience, but they appear to have carried little weight outside Pullman's own sales office.[135] Pullman then decided to meet the competition head-on; it would produce its own version of the stainless-steel car.

Shotwelding, however, was a patented process held by Budd, and no satisfactory alternate method of fabrication had been devised. Therefore Pullman decided to produce an ersatz stainless-steel car, in which the structure and major fittings would be of Cor-Ten steel and only the side body panels and other bits of decorative trim would be stainless. A side-panel truss helped support the body; the outer corrugated side-panel covering was "loose-sheathed." In Pullman's lightweight steel cars the side-panel sheets were skin-stressed and thus served as part of the structure. The Southern Pacific was an early customer for stain-

Figure 2.76 *A stainless-steel underframe being assembled upside down at Budd's factory in 1938. The entire car—not just the body or exterior surface—was of stainless steel, which made it extremely durable but expensive.* (*The Budd Company*)

Figure 2.77 *Pullman's answer to the competition from Budd was to produce a cheaper stainless-steel–sheathed car. Shown here is an articulated coach built for the Southern Pacific by Pullman in 1937.* (*Pullman Neg. 40558*)

Figure 2.79 After the weakness of the stainless-sheathed car was exposed, Pullman began to produce all-stainless cars, such as this model completed in 1954 for the Missouri, Kansas & Texas. (Pullman Neg. 63160)

Figure 2.78 Interior of Pullman's 1937 articulated coach. (Pullman Neg. 40638)

less-sided cars, which were painted only for color. They were attractive and obviously cheaper than all-stainless cars. The first order came in 1936 for the Daylight trains (Figures 2.77 and 2.78).[136] In later years the Southern Pacific acquired several hundred similar cars from Pullman. Stainless-sided sleepers were built for the Pullman Company's own use on various railroads; by 1944 about 280 cars of this type were in service on the Southern, the Santa Fe, and the Southern Pacific.[137]

After a few years, however, the cars exhibited a disastrous flaw. To avoid unsightly fastenings, Pullman had copied Budd's snap-on method of applying the stainless-steel side-panel bands. But there was no effective way to seal the cracks, so that water and cleaning compounds gradually seeped inside the cars' lower body cavities. High-pressure car washers only aggravated the situation. The bodies rusted away out of sight, and it was not until major repairs were underway years later that the damage was discovered. The Southern Pacific noticed the condition fairly early but could find no way to seal the sides. The road stopped buying these cars after 1946.[138] Starting in 1958, it was necessary to rebuild the cars with flat sides. Other roads such as the B & O–C & O were too late in discovering the internal-rusting characteristic of stainless-sided cars and were forced to scrap a number of relatively modern ones.[139] Despite this basic defect, however, the cars continued to be built. The New Haven was a late and enthusiastic purchaser, acquiring 200 from Pullman in the late 1940s. The general decline in the passenger car market may have hastened the end of the stainless-sheathed car, but its reputation was badly tarnished by then. Pullman itself gave up on what had at first seemed such an admirable compromise and in the end built some all-stainless cars. These were produced in the 1950s for the Frisco and the M K T. (Figure 2.79).[140] Naturally it was still unthinkable to pay Budd's licensing fee; spot welding and rivets were used instead.

All-stainless cars were not without critics. High cost was the main objection; stainless steel was the most expensive material employed for railway car construction—40 cents a pound even in the 1930s. Budd's argument, however, was that the equipment would last for all time. After thirteen years and 4.5 million miles, the Denver Zephyr was overhauled in 1949 and showed no sign of deterioration in the stainless body or framing. In contrast, significant decay could be expected in an ordinary lightweight car after such extended running. But was this longevity necessarily a desirable characteristic? The useful life of a coach was generally set at twenty-five years. Even if it was not decrepit by this time, it

Figure 2.80 *A production-model stainless coach built by Budd in 1938 for the New York Central. (New York Central)*

Figure 2.81 *Budd produced the Seaboard Airline Railroad's Number 6218 in 1947. The 52-seat coach was used on the Silver Meteor trains. (The Budd Company)*

Figure 2.82 Interior of Seaboard coach Number 6218, completed by Budd in 1947. (The Budd Company)

Figure 2.83 An exterior end view of Seaboard coach 6218. (The Budd Company)

Figure 2.84 In 1956 Budd produced an ultra-lightweight demonstration model, the Pioneer III. *The car was intended for quantity production, but no orders were received except for electric commuter cars. (L. S. Williams, Inc.)*

had become obsolete and needed to be replaced by a more advanced design. Overly durable cars only encouraged the railroads to make do with existing equipment.

Other critics challenged Budd's claim that stainless-steel bodies were maintenance-free. It was true that painting was not required, though the Pennsylvania and the Norfolk and Western both insisted on Tuscan red exteriors. However, keeping stainless cars bright was sometimes almost as troublesome and costly as cleaning painted cars.[141] More serious was the problem of major body repairs. Few railroad shops were equipped to deal with stainless-steel fabrication or Shotwelding, and after a smash-up it was often necessary to ship the body back to Philadelphia. Foreign purchasers found this particularly discouraging; indeed, the Portuguese National Railways gave it as their major reason for declining to accept deliveries on additional stainless cars.[142]

Despite these minor complaints, Budd made steady progress in obtaining orders and advancing the cause of the stainless-steel car. After World War II the company took over a war production plant in Red Lion, Pennsylvania, for a new car shop. Five assembly lines and 4,000 employees were soon at work, and by 1949 they had delivered more than 1,000 cars. Budd developed a good export trade and granted several licenses to foreign builders for the manufacture of stainless cars. The firm even successfully revived the rail car in 1949, calling it the RDC (Rail-Diesel Car). Seven years later Budd developed a cheap, extra-lightweight coach that it hoped to mass-produce.[143]

Budd saw the new coach as a bench-mark development in its company history and, recalling the "Pioneer" airplane and the Pioneer Zephyr train, called the prototype car the *Pioneer III* (Figure 2.84). A full-sized (85 feet, 2 inches long by 11 feet, 6 inches high), full-strength car, it weighed an incredible 26½ tons, or about one-half of a standard lightweight car. This saving produced by a spartan interior, head-end power, and a radical inside-bearing truck design. The weight per passenger was down to 600 pounds, a figure not approached since the days of wooden construction. In some ways the car was a regression, since passenger comfort was sacrificed somewhat for economy and the elimination of dead weight. Seating was provided for 88, with a noticeable reduction of legroom. Small, prefabricated toilets, patterned after the one-piece plastic units made for airliners, replaced the spacious lounges to which passengers had become accustomed. The interior was assembled from precast plastic units. There were no reading lights, footrests, or ashtrays. The *Pioneer III* was a stripped-down economy model which even the manufacturer conceded was not suitable for long-distance travel. Still, the cost was only $90,000, and more luxurious versions were possible if desired. But the *Pioneer III* returned from a two-year, 120,000-mile tour with no customers. The year was 1958, and the railroads

Figure 2.85 The material most often used for lightweight cars was copper-bearing, rust-resistant steel known by the trade name Cor-Ten. The Seaboard's Number 830, built by Pullman in 1936, was an early example of Cor-Ten construction.

were in no mood to acquire more passenger equipment as the big push to abandon all passenger service was beginning.[144] Budd had developed the new design in the twilight of the long-haul passenger train. The company received several orders for electric suburban units but none for conventional Pioneer IIIs, and Budd's car-building division was soon in trouble. Suburban and rapid-transit orders kept the shop open long enough to land the Metroliner contract in the mid-1960s. But this prestigious job was not a profitable one, nor was an earlier order from the Long Island. After a season of steady losses, Budd decided to close its rail division in 1971. Yet only two years later, Amtrak placed an order for more cars and the division was kept open.

COR-TEN STEEL

If aluminum and stainless steel were the glamour metals of the lightweight era, Cor-Ten steel was the workhorse.[145] Developed by United States Steel, it was ready for commercial production in 1934. This combination of carbon steel with traces of copper, chromium, silicon, and phosphrous produced a strong, cheap, rust-resistant steel alloy. With a tensile strength of 70,000 pounds, it also possessed a yield strength double that of common steel and a fatigue resistance 80 percent greater. It could be worked as easily as conventional steel and, most important to the modern car builder, it could be welded. In addition, Cor-Ten was four to six times as corrosion-resistant as carbon steel and two to three times as resistant as copper-bearing steel. U.S. Steel regarded freight cars as the principal market for Cor-Ten, but car builders were quick to apply it to passenger car construction. (In recent years, architects have also found decorative as well as structural applications for Cor-Ten.)

Its low price (3.4 cents a pound) was a major recommendation for a structure as large as a railway passenger car.[146] Cor-Ten represented a cost advantage roughly ten times that of aluminum or stainless steel. Much as Pullman preferred aluminum, it recognized that the price of the new steel alloy was not only an effective argument against Budd's stainless cars but also avenue for the production of low-cost lightweight cars. This conclusion proved correct; more Cor-Ten cars were built than any other type. It was used for more than two-thirds, or about 6,500, lightweights.[147]

Among the earliest passenger cars of Cor-Ten were a set of fifty arched roof cars built in the period 1934–1935 for the New Haven at Pullman's Worcester, Massachusetts, plant.[148] A tubular cross section was partially achieved by the arched roof and convex body sides. The smooth exterior was marred by rivets, but the heads were kept as small as possible. In deference to the high fashion of streamlining, the windows and doors were made flush with the body panels, and deep underbody skirts reached down to within 26 inches of the rails. The large 29- by 60-inch windows were trimmed with aluminum. The seat spacing was not generous; 84 seats were crowded into the 84-foot, 6-inch car. The interior space was even less roomy than that of most light cars, since vestibules were fitted to both ends. This arrangement was a necessity on the New Haven, where stops were frequent and double facilities for loading and unloading were essential. The center sills were formed from two 14-inch channels, which were reinforced and tied together by numerous pressed-steel shapes. The framing members were made of another U.S. Steel alloy called Man-Ten. Somewhat cheaper and less durable than Cor-Ten, it was considered durable enough for the thick cross-section frame pieces. The body was of Cor-Ten. The side posts were box-shaped; the side plates were $\frac{3}{32}$-inch thick, and the roof sheets measured $\frac{1}{16}$ inch.

Weighing $54\frac{1}{2}$ tons, the New Haven cars were comparable to the best lightweights of aluminum or stainless steel. More cars on the same plan were produced during the next three years, and by 1938 the New Haven had 200 coaches in service. The neighboring Boston and Maine and Bangor and Aroostook also placed small orders. Nearly identical cars appeared on the Seaboard Air Line and the Cotton Belt (St. Louis Southwestern). Those built for Seaboard are shown in Figures 2.85 and 2.86.

While the industry was willing to employ Cor-Ten at an early date, the first groups of cars were riveted rather than welded. Like Pullman, American Car and Foundry stayed with the old method of fabrication in its first lot of Cor-Ten cars, which were built in 1935 for the B & O's Royal Blue. Within a year or two after their first orders, however, both Pullman and A.C.F. decided to capitalize on steel's great advantage by producing welded cars. On the other hand, Pressed Steel Car Company stayed with rivets. Once the sole producer of all-steel railroad cars in this country, the firm had lost most of its passenger car orders after about 1915,

Figure 2.86 Interior of Seaboard coach Number 830, completed in 1936 at Pullman's Osgood Bradley car plant.

Figure 2.87 Some New York Central Cor-Ten steel coaches under construction at Pullman's main passenger car shop in Chicago in 1941. (Pullman Neg. 45298)

when the older companies reequipped to produce steel rolling stock. In 1940 Pressed Steel made an effort to recapture a portion of that market with a new design using Cor-Ten.[149] The Pittsburgh builder turned out a pilot car for the New York World's Fair, but the firm had failed to keep up with the most recent methods in car building. The ungainly demonstration model won few admirers with its riveted body and frame. Only the center sill was welded; a molding covered the belt-rail rivets. The Pennsylvania bought this pilot, and twenty-five cars were produced for the New York Central, but no other roads seemed impressed by the design. It was too far outclassed by the all-welded cars then in production.

Pullman not only began to perfect its all-welded plan in the mid-1930s but also developed two styles of framing built of thin-section Cor-Ten steel that were suitable for lightweight cars.[150] First the company tried a self-supporting, welded-truss side fram-

Figure 2.88 A Cor-Ten steel underframe at the Pullman car works in 1946. (Pullman Neg. 48169)

Figure 2.89 An arched-roof forming and welding jig at the Pullman car works, 1946. (Pullman Neg. 48482)

Figure 2.90 This Chicago and North Western coach was built from Cor-Ten steel in 1946 by Pullman. It was one of forty-five lightweight coaches purchased by the North Western to upgrade its postwar car fleet. (*Pullman Neg. 48485*)

ing, with outer sheathing that was "loosely attached" and had no structural function. This scheme became the permanent design for stainless-steel-sheathed cars. In about 1937 Pullman settled on a girder-sided car, where a simple rectangular framing was stiffened by spot welding a stress skin of .078-inch Cor-Ten sheet. The framing members were thin-walled Cor-Ten channels, boxes, and Z bars (Figures 2.87 and 2.88). The underframe was a similar fabrication. The side, roof, and floor frames were welded in jigs (Figure 2.89). A vertical welding machine spot welded the sheets to side frames. These subassemblies were then riveted together. There was some difficulty in producing smooth side panels, as might be expected because of the very lightweight sheets used. Cupping and waves were inevitable, but this visual defect was somewhat overcome by welding ¾-inch channels at close intervals.[151]

WELDING

Since the beginning of the steel era, car builders had been seeking a quick, strong, and cheap method of fabricating railroad rolling stock. For structural and most major body construction, hot riveting was the only technique available. It required drilling or punching, careful alignment of the holes, and maintenance of a steady supply of hot rivets along the entire assembly line. It was tedious and expensive, and strong though the finished product was, the rivet heads were an ugly blemish. The rippled bead of a heavy welded joint might not be beautiful, but it was less noticeable and could always be ground level with the surrounding surface if desired. Gas and electric welding, however, was in its infancy during the early years of steel car building, and rivet construction became established with little competition. Light roof plates were welded at an early date and a welded freight car was built in 1911, but it was clearly experimental and attracted little attention. It was not until the lightweight era that welding came into large-scale use in the car-building industry.

In 1933 the Lincoln Electric Company, a major producer of electric welding gear, sponsored a contest for an all-welded passenger car.[152] One suggested design was for a conventional-looking heavyweight with a clerestory roof, a fishbelly underframe, and the other traditional features; yet a weight saving of over 30 percent (17 tons) was claimed for it. During the following year the Frisco built two lounge cars on this general plan in its Springfield repair shops.[153] Both realized a weight saving of 15 tons, thus vindicating Lincoln Electric's claim. Three more such cars were constructed in 1936 and 1937 by the Frisco.

Yet the real potential of welded fabrication was hardly exploited in these cars. A far more imaginative application was being tried at the Milwaukee Road's shops. The practical genius behind these efforts was Karl F. Nystrom, a Swedish mining engineer who had been attracted to car building soon after coming

Figure 2.91 The Illinois Central's Number 2639 was another Pullman Cor-Ten product of 1947. (Pullman Neg. 48674)

Figure 2.92 American Car and Foundry also produced cars from Cor-Ten steel. The Great Northern 60-seat coach shown here was delivered in May 1950. (American Car and Foundry Company)

to the United States as a young man in 1905.[154] As an employee of the major builders he was schooled in the traditional methods of car design. In 1922 he joined the Milwaukee, where he spent the rest of his engineering career. Nystrom's long service in the railroad industry did not stifle his willingness to try new methods. Past fifty and a seasoned, shop-oriented mechanical expert, he came to champion the then-radical scheme for all-welded freight and passenger cars.

Nystrom's first effort can hardly be called radical, however; he remodeled an old heavyweight sleeper as a smooth-sided coach. It was a slow beginning and to a degree a false start, because the car actually weighed more than many heavyweights of the day. But it proved that the technique of welding was practical, and Nystrom was ready to produce a lightweight prototype. Numbered 4400, it was finished in May 1934.[155] As one of the earliest streamlined passenger cars in this country, it was considered worthy of display at the Century of Progress Exposition. It is unlikely that many visitors were interested in the 4400's construction, but the designer was satisfied and production began at the Milwaukee shops during the same year. The 80-foot car weighed only 48 tons—a remarkable achievement, considering that no specialty steels such as Cor-Ten were used. The framing was made up of 12-inch center sills and 4-inch Z bars. Large rest rooms and wide seat spacing limited the seating to 54 passengers. The 1934 production cars weighed 56.5 tons because of the addition of air conditioning and double-sash windows.[156]

In a group of cars produced in 1936, a 9-ton weight saving was achieved by using Cor-Ten steel and reducing the depth of the center sills by 2 inches.[157] Otto Kuhler was brought in to help with the appearance of the cars, but Nystrom's basic design was not

Figure 2.93 *Interior of the Great Northern car shown in Figure 2.92. (American Car and Foundry Company)*

Figure 2.94 *This Kansas City Southern coach represents the end of conventional lightweight car construction. The ten cars delivered in 1965 were the last of the breed. (Arthur D. Dubin)*

Figure 2.95 An interior view of the 1965 Kansas City Southern coach. (Arthur D. Dubin)

altered. The rib-sided, flat, arched-roof bodies remained the Milwaukee's distinctive feature in the United States, although there are remarkably similar cars in service on Russian railways.

While Nystrom made good progress in paring down weight, refining the general design, and holding the body cross section slightly under size, he encountered a notable setback at the beginning of World War II. Construction was underway on thirty-one cars in 1942, and permission had been granted by the government to complete the lot. As industry began to gear up for the war, however, tight restrictions were placed on critical materials.[158] The aluminum used for conduits and air ducts was unavailable, only small quantities of Cor-Ten could be found, and standard fixtures were unobtainable. These shortages forced Nystrom to make a series of compromises, each apparently small, but all contributing a total increase of 15 tons in car weight. Here again was proof that lightweight car design was a matter of subtleties.

An interesting feature of the first few groups of Milwaukee lightweights was the enclosed underbody.[159] Streamline advocates favored smooth car bottoms for aerodynamic reasons, but this was not the Milwaukee's motive. Because of the extreme winter conditions on its line, the Milwaukee wanted to protect the underbody equipment from ice and snow and avoid the added weight of accumulating ice. A smooth surface offered fewer spots where snow and ice could lodge.

The car's balance and center of gravity was also improved by placing all apparatus under the center sills inside the sheet-metal "gondola." What servicing problems this arrangement caused are not recorded, but apparently they were offset by the advantages.

Between 1934 and 1948 about 300 new lightweight passenger cars were built at the Milwaukee Road shops. Not only had this line pioneered in the construction of welded equipment, but it also produced the cars. Few other roads attempted lightweight car construction, and certainly no other, with the possible exception of the Pennsylvania, built in such quantities. Like a commercial builder, Nystrom had an assembly line. Subelements of the sides, floor, and roof were assembled in horizontal jigs, and a special automatic welding machine with a range of 50 feet was worked three shifts to speed manufacture.[160] The long service life of these cars is a testament to their careful manufacture. In 1961, for example, a group of the 1934 cars was sold to Mexico, where they are still running.

The Revolution that Never Happened

Despite the success of the lightweight cars, they never displaced the existing heavyweight fleet (or at least, not until the near-abandonment of U.S. railway passenger service in recent years). While the conversion from wood to steel seemed slow, the change from heavy to light cars never really happened.

Once the name trains were reequipped, the railroads seemed to lose interest in purchasing more lightweights. The best trains were periodically upgraded with a set of new cars, the older ones being reassigned to lesser trains. In this way slightly worn lightweight cars filtered down to some secondary trains, and this happened with greater frequency as the number of passenger runs declined. New lightweight cars were rarely purchased at the rate of more than 300 cars per year by the entire industry in the period 1935–1955. After this time, of course, almost no purchases were made. The high cost of new cars—$100,000 and up—discouraged the cash-poor railroads from new purchases almost as much as the declining traffic did. Hence the lightweights never represented more than 15 percent of the total passenger car fleet.

The lightweights appeared during the Great Depression of the 1930s at a time when railroads could least afford a major investment in new equipment. The existing car fleet was adequate for the remaining traffic. Some new trains were put on to maintain the roads' public image, but once this cosmetic need was fulfilled, the acquisitions generally stopped. By the time the economic situation improved enough to encourage more buying, the war was on; production of such "nonessentials" as railroad passenger cars was halted. The big buying splurge that began after the war ended by 1950, again after the top-ranking name trains were reequipped. The older cars that were taken off the top trains satisfied most other needs. Robert Young, a passionate believer in modernizing the passenger car fleet, complained in the New York Central's 1955 annual report that at the present rate of acquisition, it would take 115 years to replace the old cars.

But perhaps the main obstacle to a lightweight takeover was the existing heavyweight fleet. It was simply too good to abandon. Many of these cars were relatively new in the 1930s, and few were much over twenty years old. The fears about rapid corrosion of steel cars were unfounded; in fact they proved to be so durable that with reasonable care they were practically indestructible. In addition, the great steel fleet represented a major investment. It was obvious that a more economical alternative to a large-scale lightweight acquisition was to upgrade the existing cars. The average passenger could not distinguish a lightweight from a modernized heavyweight. The degree of refurbishing varied from one railroad to another, as it did from one car class to another. A complete job included an arched roof, new trucks, air conditioning, reclining seats that were widely spaced, enlarged toilets, double-sash windows, new lighting fixtures and baggage racks, and of course, modern decorating, inside and out. Some roads added body skirts and fanciful paint schemes, so that the exterior was convincingly disguised as a streamlined car. But even if the interior alone was redone, most passengers were satisfied that they were riding in streamlined cars. Probably very few noticed or cared that their train was a World War I–vintage battleship thinly veneered in the modern style.

One reason why the railroads were forced to improve their passenger fleets was that their own publicity had raised the expectations of the traveling public. It was not enough to treat the name-train passenger to luxury travel; everyone (except perhaps the lowly commuter) expected greater comfort. The fast, cheap way to meet this demand was to upgrade the heavyweights.

The New York Central was early and prominent in the drive to remodel its old cars, as might be expected from one of the largest passenger operations in the world. Between 1934 and 1937 the line remodeled 150 coaches, according to a statement issued by its engineering department in January 1938. The most extensive work was done on a group of cars for the Cleveland-Detroit train, the Mercury. Seven lightweight suburban cars were thoroughly rebuilt at the Beech Grove, Indiana, shops in 1936.[161] Over and above the usual improvements, the following features were added: single vestibules, tightlock couplers, air conditioning, fluorescent lighting, and special noise-damping vestibule diaphragms. One car was rebuilt with a round observation end. Its interior was redesigned by Henry Dreyfuss, one of the finest designers in the early streamline period. Three years later the Central remodeled a more prosaic set of cars for its low-fare New York–Chicago Pacemaker. They were not given the opulent treatment of the Mercury cars, but the basic amenities—air conditioning and good seating—were provided.[162]

The B & O shops were upgrading old cars during the same period, but here the intent was to use the stock on the road's best trains. Reduced circumstances required this approach; the company simply could not afford to spend freely. Nor was its traditional-minded President, Daniel Willard, convinced that light eight-wheel cars could ever match the ride of a twelve-wheel heavyweight. When the lightweight Royal Blue cars (built by A.C.F. in 1935) were sent to the Alton for the Ann Rutledge, the Mt. Clare shops modernized a set of elderly twelve-wheel heavyweights rather than spend the million or so dollars required for new equipment.[163] Other conversions were carried out at the same time for the National Limited, the Columbian, and the Capitol Limited (Figures 2.96 and 2.97). In 1946 the B & O scooped the C & O by introducing a steamlined daylight train, the Cincinnatian, with modernized cars.[164] While the C & O awaited its new equipment from commercial builders, the B & O worked over some existing thirty-year-old stock and had a new train ready in months. Here was another point in favor of remodeling. The Cincinnatian was more than a paper-and-paste job; the Mt. Clare shops could be proud of their homemade train. Indeed,

Figure 2.96 Many railroads hoped to resuscitate their serviceable but dated heavyweights by remodeling them. An example is this coach modernized by the B & O in 1940.

their official records claim that the cars were built new (Figure 2.98). The B & O was apparently well satisfied with its remodeled heavyweights, and except for the two-year operation of the A.C.F. cars, the Royal Blue never acquired lightweight equipment.

The Pennsylvania set out to modernize its existing fleet in a more analytical way that was appropriate to a road dominated by engineers. In 1936 it remodeled three P-70s on alternate plans.[165] Scheme 1 involved the fewest modifications consistent with a modernized car. Seating was reduced from 88 to 60. New seats and wider spacing offered greater comfort. Air conditioning was provided by the ice system—the cheapest plan available. No changes were made to the body or roof, but aluminum window sashes were installed. Scheme 2 was identical to scheme 1 except that the windows were larger. Scheme 3, the most radical, involved extensive changes to the existing body to incorporate an arched roof, wide windows, and rest rooms at the center of the car (Figure 2.99). Seating in this car, which was intended for long-distance travel, was reduced to 42. Scheme 3 seems to have been applied to only one car. In the entire program, however, about 900 P-70s were modernized between 1937 and 1952. Some were extensively remodeled, with arched roofs, electromechanical air conditioning, single end vestibules, reclining seats, and wide metal-sash windows. Others received more modest upgrading that consisted of little more than an ice-cooling system and new window sashes. The program is summarized in Table 2.2.

By 1963 the Pennsylvania was again actively rebuilding and modernizing. The average age of its coach fleet was thirty-seven years, far worse than the national average of twenty-nine years.[166] To improve this condition the Altoona shops remodeled 186 cars, while the line returned fifty modern sleepers to Budd for conversion to coaches at a cost of $70,000 each. A few years later it was reported that the Pennsy was buying up second-hand heavyweights for renovation.[167]

Western roads also upgraded their heavyweight stock. In general they tended to buy more new equipment, at least after 1950, than the Eastern lines, but most of them also had active remodeling programs. The Santa Fe, obviously concerned about the "mixed consist" appearance of its trains, developed a cheap method of harmonizing the old with the new. In 1947 the road came out with a shadow-line paint scheme that effectively simulated the fluted sides of a modern stainless-steel car.[168] The old car was painted silver, and the shadow lines were air-brushed on by a skilled painter. It was said to be completely convincing at 50 feet. The Denver and Rio Grande Western and other lines copied this paint scheme (Figure 2.100).

By 1946, 17,000 modernized cars were in service. They represented about 30 percent of the entire passenger car fleet.

Safety

Throughout the history of the railroad passenger car, conflicting requirements for light weight on the one hand and safety on the other have challenged the best talent in this area of engineering design. The solution has always been a compromise. An abso-

Figure 2.97 The number 3556 was remodeled by the B & O in 1940 from the standard heavyweight car Number 5287. It had been built in 1924 by Pullman.

lutely safe railroad car is almost as unattainable as a crash-proof airplane. It cannot be achieved within practical weight or cost limitations; consequently the car designer has always aimed for a crash-resistant plan. The goal has been to perfect a design that, with economical use of materials, will produce a car body capable of withstanding any reasonable impact or buffering. The wooden car builder developed his plans empirically, going by what had been done before and what "looked right" to the chief draftsman. The early steel car builder based his designs on a few general formulas, but again one suspects that much of the design was empirical. With the coming of the lightweight car there was a greater scientific input in car design, because each part was expected to play a critical role in the structure. In addition, structural safety standards were refined and enlarged during the years of the lightweight, as will be discussed later.

Just how to measure the effect of car construction on the safety of railway travel is a difficult problem and one that has never been solved. Too many factors are involved, because at the same time that car design was changing, so were operating practices, signaling, traffic density, track construction, and other matters which would affect railway safety. That the coming of the steel car reduced the hazards of travel is obvious, but by just what percentage cannot be determined. Although steel cars cut down the number of casualties in minor mishaps, they certainly did not provide impenetrable shelters in major collisions. Telescoping remained a grim possibility for the railway traveler.

Naturally there were some misgivings about the safety of the new lightweight cars. In May 1936 the I.C.C. said in a preliminary report: "Questions have been raised as to the capacity of these light weight trains to withstand shocks, colli-

Table 2.2 P R R Class P-70 Coach Modifications

CLASS	PASSENGER CAPACITY	WEIGHT LBS.	ROOF HEIGHT	A-C	SEATS	WIN-DOWS	VESTI-BULE
P-70 Original	88	122,000	?	None[a]	Walkover	Narrow	2
P-70R	76	142,000	?	Ice	Walkover	Narrow	2
P-70R "Deluxe"	formerly 39[b] later 80	142,000	?	Ice	Indiv.	Narrow	2
P-70R (Schemes 1 & 2)	60		14'0"C[c]	Ice	?	Narrow	2
P-70ER (Scheme 5)	76	141,200	13'2"A[d]	EM	Rot.	Narrow	1
P-70ER (Scheme 3)	42[e]	141,000	13'2"A	Ice	Rot. Recl.	Wide	2
P-70ER (Scheme 4)	72	141,000	13'2"A	EM	Rot. Recl.	Narrow	1
P-70FR	84	143,800	13'2"A	EM	?	Narrow	1
P-70FR (Scheme 6)	84	144,900	14'0"C	EM	Walkover	Narrow[f]	1
P-70FAR	80	144,200	13'6"A	EM	Walkover	Wide	2
P-70FBR	72[g]	155,900	14'0"C	?	Walkover	Narrow	2
P-70GR (Scheme 4)	68	143,700	13'2"A	?	Rot. Recl.	Narrow	1
P-70GSR (Scheme 4)[h]	68	145,700	13'6"A	EM	Rot. Recl.	Narrow	1
P-70GSR (Scheme 4)[h]	68	146,500	13'6"A	EM	Rot. Recl.	Wide	1
P-70HR	80	131,300	13'6"A	Ice (9 EM)	Walkover	Narrow	2
P-70HR	80	133,700	13'2"A	Ice	Walkover	Narrow	2
P-70KR[j]	56[i]	146,100	13'6"A	EM	Rot. Recl.	Wide	1
P-85GR	84[g]	161,750	13'6"A	EM	Walkover	Narrow	1

[a] An exception was the No. 3558, which had electromechanical air conditioning.
[b] Two dressing rooms.
[c] C—clerestory.
[d] A—arched roof.
[e] Center washrooms.
[f] Exceptions were the Nos. 3906–3935, which had wide windows.
[g] Cars had 8-foot, 6-inch trucks.
[h] Streamlined.
[i] Lounge rooms.
Source: This table was compiled by W. D. Edson, former car superintendent for Amtrak, from material in his files. It may not be a complete record.

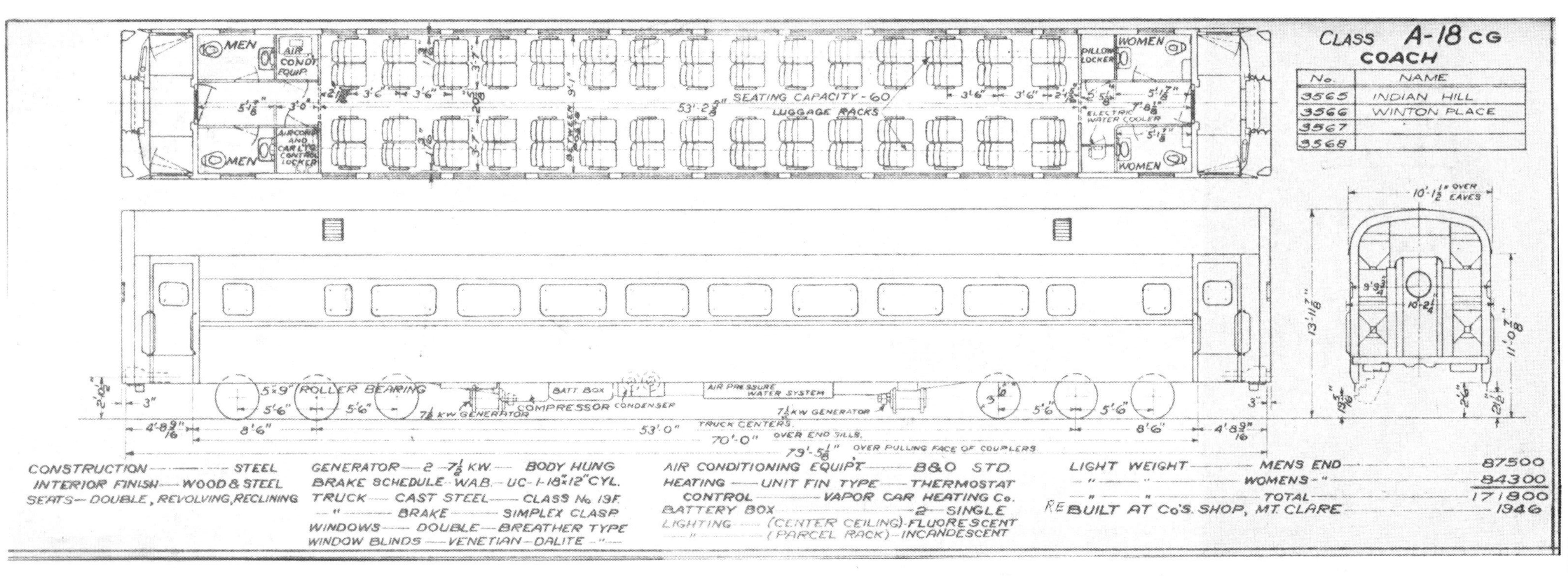

Figure 2.98 In 1947 the B & O manufactured a new deluxe train, the Cincinnatian, *using old heavyweight steel cars. At 86 tons each, they were lightweight in appearance only.*

Figure 2.99 In 1936 the Pennsylvania considered a radical scheme for remodeling its vast old-fashioned P-70 fleet. The road decided that this sample car was too extreme and costly, but it did rebuild many P-70s along more moderate lines.

sions and other accidents."[169] Apparently this mild comment created a stir in the industry, because it was struck out of the final report. It was also deleted from the printed versions reproduced in *Railway Age* several months later. In the same report the Commission said that it was too soon for a definite appraisal of the new cars, that only their record in regular performance could provide the answers. At the time of the report, accidents that involved lightweight trains were relatively minor ones connected with running-gear failures or road-crossing mishaps that proved nothing about the integrity of the cars' structure.

One of the first major accidents to a lightweight train occurred in August 1939 at the Humbolt River bridge in an isolated part of Nevada.[170] The motor train City of San Francisco, running at high speed, was derailed just before it reached the bridge. Several cars fell down the embankment; others went ahead to sideswipe the bridge structure, causing it to collapse. The remainder of the train piled up on the wreckage. Twenty-four passengers and crew members died. Several of the cars were gravely damaged, but it is impossible to say how well heavyweights would have withstood this disaster. Certainly the lightweights were substantial enough to bring down a steel bridge.

A few months later the Pioneer Zephyr collided head-on with a steam locomotive at 45 to 60 miles per hour.[171] The front 28 feet of the Zephyr were demolished and its engine was driven back into the mail room, killing two clerks. There were 38 injuries, but no passengers died.

Neither of these two wrecks furnished a strong case for or against the lightweight car. A more significant test was an accident in January 1948, when the second section of a Missouri Pacific train rammed into the leading train near Syracuse, Missouri.[172] At the rear of the first train was the lightweight Pullman sleeper *Golden Cloud*. The car was totally demolished, twelve people were killed, and thirty-seven others were injured elsewhere on the trains. The locomotive ran 53 feet into the *Golden Cloud*; the front end of the sleeper was telescoped 12 feet by the preceding car, thus effecting an almost complete telescope. The frame of the car was nearly destroyed by the collision. This wreck rekindled the I.C.C.'s suspicions about lightweight construction, and the commission duly noted the inability of a car to withstand "heavy buffering stresses." As the last car of the first train, the *Golden Cloud* received the full impact of the rear-end collision. One can only wonder how well a standard-weight car would have

Figure 2.100 The Louisville and Nashville attempted a faux paint scheme in a rather desperate effort to duplicate the glistening silver of modern stainless-steel equipment.

held up against it. Pullman maintained that the seven-year-old car was built on a design tested to withstand the end buffering standards then existing.

A more direct comparison of heavyweight and lightweight construction was provided by a head-on collision between two cars of an electric interurban line known as Speed Rail which operated out of Milwaukee. In 1950 two excursion cars plowed into each other at high speed, killing ten persons. The lightweight car was demolished. But what might appear to be damning evidence against lightweight construction must be tempered by the fact that the heavyweight car's main frame was a foot or so higher than the other's. This is not the case in regular steam railroad cars, where the floor levels are all built to one standard. (Exceptions, of course, are the motor trains and the more recent experimental underslung units of the Talgo variety.)

Increasing public concern over passenger safety was reflected in the ever tougher standards adopted by the railroads. Standard setting began outside of the industry in 1912, when the Railway Mail Service issued a set of general strength requirements for Railway Post Office cars. the The Post Office hoped to force the early adoption of steel or steel underframe cars. The main requirement was the ability to absorb a buffering impact of 400,000 pounds. For years this figure was unofficially recognized as a basic minimum for all cars in interchange or main-line service. The Post Office, usually at Congressional prodding, periodically revised and strengthened its requirements. In July 1938 a major revision was put into effect covering lightweight cars. The following year the Mechanical Division of the Association of American Railroads decided to adopt the R.M.S. specifications as recommended practice.[173] The required static end load was doubled to 800,000 pounds, and body deflection was not to exceed 1 inch. In 1945 the A.A.R. revised the specifications for new passenger car construction and advanced them to standard practice. All railroads and car builders were now required to follow the specifications. The most important of the new rules were Section 2, item f, and Section 18, item b. The first stipulated that no material could be used in which the yield strength exceeded 80 percent of the tensile strength. This eliminated the choice of aluminum for framing members. The second rule provided for two antitelescoping end posts, each to have a section modulus of not less than 24.375 and an ultimate shear value of not less than 300,000 pounds at a point even with the top of the underframe member to which it was attached. The specifications—composed of twenty-two sections and over 100 rules—covered everything from couplers to insulation.

It was said that the Budd Company actively promoted the tougher strength requirements in these standards. Presumably the firm's motive was to confound its rivals and undercut the position of the aluminum car.[174] It is obvious from the printed record that Budd made a great deal of noise about the superior strength of its stainless-steel cars. Shortly after the 1945 specifications were adopted, Budd dismissed them as a "distinct minimum."[175] Ragsdale, Budd's chief engineer, never failed to mention Budd's superior strength, pointing out that it was in "multiples" of the industry's requirements. Actually, Budd's interest in the matter was substantial. The company installed the first test machine for a full-sized railroad car at the old north Philadelphia plant in 1938 or 1939.[176] Built by the Baldwin Locomotive Works, it could develop 2 million pounds of compression. Its 500 strain gauges could show stresses in every body member of the car. This machine put car design and testing beyond paper calculation or in-service road tests. It was not until 1954 that the Association of American Railroads acquired a testing machine at its Chicago Research Center—ironically, just in time for the virtual end of new passenger car construction in this country.[177]

Standardization

As this book has shown, there was general agreement on car design almost from the beginning of the railroad age in the United States. Details might vary, but materials, methods of construction, overall size, configuration, and interior arrangement were generally in accord from one car to the next during any given period. In the broad sense this can be called standardization, and the word is used here in that sense. Standardization, however, is often misinterpreted. A recent conversation with a prominent author in the field of economic history revealed that he had misunderstood the 1880 Census Report on standardization in the locomotive industry, which stated that parts were made to gauge or pattern so that they might be interchangeable. The historian had taken this to mean that all American locomotives were identical, or that only a few classes were offered, and that parts could be freely interchanged between them. In fact the great diversity of locomotives during this and later periods, and the existence of fifteen or so locomotive manufacturers, each dedicated to its own plan, make the concept of interchangeable parts seem naïve. Some people are misled by looking at illustrations of railroad rolling stock, which do have the same general arrangement or basic design. But there were no Model T Fords in railway equipment, except perhaps for freight cars. A single, identical class of locomotives or passenger cars that numbers even several hundred is notable; any class counted in the thousands is very rare.

A second misconception concerning standardization is the idea that it involves the manufacture of identical duplicates. Such a practice may be possible for certain products, but it does not seem to have been seriously considered for the American railroad passenger car. There was a correct resistance to the cookie-cutter method of production because it would stifle the innovation necessary for the continuous improvement of this complex machine. A more desirable goal, and the one eventually achieved by the Association of American Railroads, was the establishment of certain general standards that all cars should conform to. Under the A.A.R. specifications both the purchaser and the designer had many options. These rules seemed the best compromise between the extremes of custom design and regimented mass production.

There were advocates of standard car design long before the lightweight period. At the beginning of the steel era at least one authority, F. Gutbrod, saw the coming conversion as a grand opportunity to adopt a standard design. A major obstacle was the greater cost of the steel cars, but Gutbrod maintained that a cheap steel car would become possible through the mass production of a few standard models.[178] The lower unit price would then permit a speedy retirement of the existing wooden fleet. However, no one seemed ready to adopt the idea. Each railroad and car builder held to its own plan, claiming special needs or superior requirements. The taste and influence of individual officials and designers was surely a powerful factor contributing to the variety of patterns produced.

The main deterrent to production-line standardization, however, was the rivalry of competing roads for passenger traffic. Speed, comfort, safety, and luxury were the drawing cards. The very last thing competitors wished to admit or practice was the duplication of the service and equipment offered by the opposition. Stylish cars of novel interior arrangement with many special (that is, nonstandard) features were the goal of every major railroad's passenger department. The cars of rivals might have the same basic construction, but there was a conscious effort to make them as individual as possible. One of the most powerful forces in the anti-standardization camp was the realpolitiks of travel promotion.

Government participation in setting passenger car standards began in 1918, when the U.S. Railroad Administration took up the question. The agency had inspired excellent designs by outside designers for locomotives and freight cars, but there seemed less urgency about a similar program for passenger equipment. Relatively few such cars had been built during World War I; the existing fleet was considered adequate, and priorities were given to more necessary products. Even so, the agency prepared a standard design by early 1919.[179] No cars were built on this plan as far as can be determined. Some head-end cars were produced for the U.S.R.A., but within a year the agency's control of U.S. railroad operations was terminated.[180] The railroads were not prodded again by a Federal body until 1935, when the Federal Coordinator of Transportation issued its *Report of the Mechanical Advisory Committee*. It was suggested that cars built to the specifications it outlined (pp. 602–612) might be purchased for as little as $40,000 if they were bought in reasonable quantities. But the Federal Coordinator could only hope to persuade; although the railroads listened politely, they were not bound to act.

Standardization might seem to be in the best interests of railroad car manufacturers, yet they showed little enthusiasm for it. Cars were built to order; they were not off-the-shelf goods, and the builders were in no position to force a ready-made product on their patrons. Like it or not, theirs was a custom trade, and as *Railway Age* said, "so long as the market is not standardized," there was little hope of standardizing the product.[181]

In many ways the custom nature of the business was a nightmare to builders. Between 1945 and 1950, for example, 3,781 cars were produced to 424 different floor plans.[182] In this period A.C.F.'s St. Charles, Missouri, plant received orders from nineteen railroads for a total of 98 lots, or 460 cars. They represented twenty-two different types of car and eight methods of construction, from riveted carbon steel to aluminum. Each lot of cars required 500 to 2,500 drawings and 500 to 1,000 requisition sheets.[183] The effect on cost and delivery time is apparent.

Over the years the car builders had gently pressed the railroads to consider a production model. In the mid-1920s Charles E. Barba, Bradley's chief engineer, suggested a production-line arched-roof car. Twenty-five years later A.C.F.'s mechanical head, Allen W. Clarke, recommended the adoption of standardized floor plans and components. Budd's *Pioneer III* prototype of 1956, described earlier in this chapter, was a singular example of a car builder willing to translate its belief in standardization into an operable prototype.

The larger railroads usually tried to maintain a certain degree of on-line standardization. The Pennsylvania was probably most successful in doing so between 1907 and 1929, when it built or bought passenger stock to a relatively uniform design. From the 1870s forward, in fact, it had followed a rational policy of standard acquisition for all its rolling stock. But statistics have not been found to show exactly how well the policy was carried out in the nineteenth century. The acquisition of existing roads and the semi-independence of the Lines West undoubtedly made it difficult for Altoona to maintain a single common standard. Central authority was stabilized, however, by the time of the all-steel car, and systemwide standardization was imposed. This excellent program was abandoned during the lightweight period, when an increasingly mongrelized fleet of cars was assembled from various makers.

The New Haven adopted a standard body design in 1948 after a five-year study.[184] The plan was capable of accommodating nine different interior arrangements. How far the road was able to follow the scheme before financial difficulties overtook it is not certain; new car purchases stopped before the plan was fully developed.

A large-scale program of standardization was also undertaken by the Harriman Lines in the early years of the twentieth century. Plans for a standard steel car were agreed upon by 1908, and soon arched-roof cars of an unmistakable pattern were running on the Southern Pacific, Union Pacific, Illinois Central, and other roads controlled by E. H. Harriman. But after the system broke up several years later, only the Southern Pacific remained loyal to the common standard design.

The Pullman Company, the largest single passenger car operator in the United States, also struggled to standardize its fleet. Yet at all periods of its history it had a bewildering array of cars with countless floor plans and window groupings. Its acquisition of competing operators and of new and rebuilt equipment added to the variety of stock in service. Basic frame and structural designs were kept uniform as far as possible, and a serious effort was made to standardize mechanical equipment, trucks, and hardware. Even so, Pullman's 1945 *Catalog of Materials and Supplies*, a massive printed document of 600 to 700 pages, furnishes clear evidence of the variety of parts necessary to maintain its 8,600 cars.

While standardization may have been an impossible dream, it is unfair to portray railroad mechanical officers as irrational on the subject. They devoted considerable effort to standardizing heating systems, seats, wheels, journals, and other systems as far as possible for all cars on the line. While floor plans or manufacturers might differ, basic components were all of one style on the majority of cars in service. Consequently a degree of standardization was achieved.

One of the strongest points in favor of standardization was cost. Yet how effective would standardization actually have been in lowering the cost of passenger cars? Not very, according to Paul Kiefer, chief mechanical engineer for the New York Central.[185] Because relatively few new passenger cars were built each year, the savings on standard cars might be insignificant. Kiefer argued that the biggest price break was given to large single orders, and that even bigger savings were available to roads which carefully drafted their specifications and resisted the luxury of requesting changes once the work was underway.

A strong case for standardization can be made for freight cars, and in this area the Master Car Builders Association and its successor, the A.A.R.'s Mechanical Division, had an excellent record. Major components were standardized at an early date, and standard overall designs were generally adopted by the 1930s. But the chief concern of both organizations was freight cars. Passenger cars (except Pullman cars) were rarely involved in unrestricted interchange service and hence were within the jurisdiction of the owning road.

The official reports of either association rarely touch upon the passenger car problem. A few standards were offered and one or two reports were issued on construction, but in general the subject was ignored before 1900. Finally in 1910 a letter classification was adopted to designate various styles of passenger cars: PA was the symbol for short-haul passenger cars, PB for standard coaches, PS for sleepers, and so on. During the next years, standards for handrails and steps were also adopted.

Over time the coupler, axles, and wheel specifications were upgraded and modified, but general design considerations were left to the judgment of the individual railroad. Finally in 1940 it was decided to invade this private territory with overall standards. They were rather mild and general, however, and involved only exterior contour and roof-clearance lines. The purpose was little more than the creation of a more uniform exterior appearance. The following dimensions were established in consultation with Pullman, A.C.F., and Budd:

Overall height from railhead	13 feet,	6 inches
Roof arch radius from center of coupler	10 feet,	9 inches
Width over side posts	10 feet,	0 inches
Height to lower edge of window glass	6 feet,	11½ inches

Other than these and a few minor measurements, car builders were free to construct the car as they wished. Even the contour profile could not be enforced, but most roads conformed voluntarily, including the Milwaukee, which gave in to the standard body by 1943.[186]

It was another ten years before agreement was reached on basic interior arrangements. In 1950 the A.A.R.'s Mechanical Division adopted 19 basic floor plans: 4 coaches, 3 combines, 3 parlors, 1 diner, 1 diner lounge, and 7 varieties of baggage and combinations thereof. (No satisfactory scheme for sleepers could be agreed upon.) The plans were the result of a study conducted by the American Railway Car Institute, a New York–based trade association funded by the car manufacturers.

The A.A.R. Mechanical Division considered strengthening the standards for the floor plans, but it decided to stop short of more specific rules. Interior decoration, flooring, fixtures, seating, and other equipment were left to the discretion of the buyer. It was felt that any attempt to standardize car details more rigidly would disqualify some suppliers and in no way aid the railroads. Unlike the specifications for freight cars, no concrete standard drawings were ever adopted. This step might have come had passenger travel continued as part of normal railroad operations. In 1971 it seemed possible that Amtrak might take on the job the railroads never finished, but this did not happen. Amtrak has purchased both high- and single-level cars in recent years.

Late and Unusual Developments

Streamlining was credited with "saving" the railway passenger business in the 1930s. Twenty years later, railroad managers hoped for another miracle through the introduction of attention-getting passenger equipment. But by then the competition was stronger and the public was more highway-minded—in fact, the rules of the game had changed. Also, what the railroads had to offer in the way of innovations was unimpressive. Most of the schemes were simply old wine in new bottles. The extra-lightweight trains of the 1950s (for example, Train X) were a final and, in the opinion of many observers, a poorly conceived attempt to save the passenger train (their dismal story is covered in Chapter 8). Apparently a recurring theme in industrial history is that radical design changes proliferate both during the development stage of a technology and during its decline. Certainly it seems true that desperate measures are common to the last days of every dying trade.

During the 1950s a few alternatives to conventional designs were offered that were more plausible and successful than the extra-lightweight trains. These were the articulated, bilevel, and dome cars. Articulation is actually an ancient solution, traditionally put forward when all else fails. As in the case of the monorail, an impressive list of hypothetical advantages can be assembled:

1. A great weight and cost saving by the elimination of one truck.
2. An increase in passenger space by the elimination of one vestibule.
3. A shorter train (an advantage at short station platforms).
4. No body overhang at one end of the car.
5. An improved ride, because all cars are joined more solidly together.
6. Simplified end and vestibule construction.
7. A more streamlined exterior, because all cars are joined in one long, smooth tube.

The weakness of the case lies in the fact that these points, all so plausible-sounding, are mostly fallacious. In reality the motion of the cars is amplified through the central truck's center plate. Much of the movement of a conventional set of cars is damped by friction of the draft gears and vestibules.[187] The eminent railway historian and engineer Brian Reed has spoken of the disagreeable ride characteristic of articulated trains in Britain. As for the other real or fancied advantages of articulation, they are more than canceled by the fixed consist necessary with fully articulated trains. This feature has been wholly unacceptable in the operating routine of American railroads, where pickups and setouts are the normal practice.

A convenient though little used compromise was found in the "marriage" of two or three cars as an articulated set. So far as can be determined, the scheme was first tried in Great Britain during the first decade of this century by N. H. Gresley of the Great Northern Railway.[188] Two- and five-car sets were operated, and by 1924 over 200 sets were in service. During the same year the London and North Eastern placed a three-car diner-kitchen-diner unit in service. Gresley's 1907 patent was controlled by the Leeds Forge Company, which in turn gave Commonwealth Steel an American license. There was considerable experimentation with articulation by American streetcar lines, and one of the Brooklyn subways obtained some three-car trains from Pressed Steel. So far as is known, however, no main-line units were tried by United States railroads until the following decade.

In 1936 Pullman produced an articulated pair of sleeping cars named *Advance* and *Progress*, and beginning in 1937, both the Union Pacific and the Southern Pacific acquired a number of articulated diners. But only the Southern Pacific championed articulated passenger cars: it ran a sizable articulated fleet that included many twin-unit coaches (see Figure 2.78).

While articulation was to remain an oddity in this country, bilevel and dome cars were more widely accepted. Somehow the split-level car has always been appealing, since it offers opportunities for greater architectural expression as well as added space. Morgan's land barge of 1828, the *Columbus* of 1831, and the Gothic cars of the 1830s were described in Chapter 1. In the 1890s the East Indian Railway built two triple-deckers—eight-wheel, iron-frame cars that seated 100 third-class passengers.[189] According to a contemporary report they were found unwieldy, and presumably no more were built.

The general scheme was not revived in this country until the 1928 patent of Albert E. Hutt.[190] In 1932 a sample car was

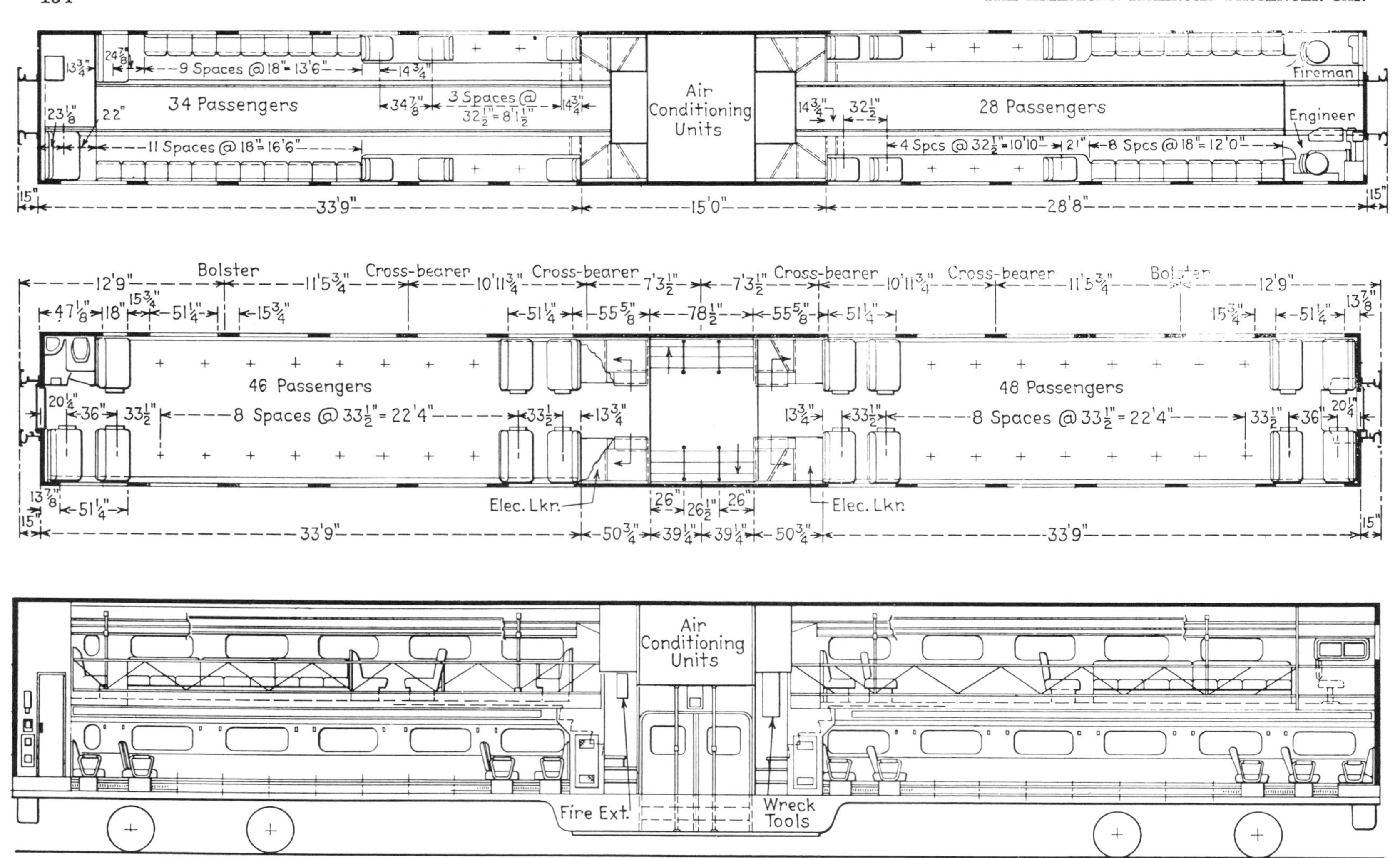

Figure 2.101 The Milwaukee Road was late in introducing gallery cars for commuter service when it acquired a double-decker from Budd in 1965. But this set of cars was among the most expensive of their type to be built. (Car Builders' Cyclopedia, *1965*)

Figure 2.102 An interior view of a gallery car on the Chicago and North Western Railway. Several lines in the Chicago area are now operating similar cars. (Chicago and North Western)

fabricated to this design at the Altoona shops for the Long Island Railroad. Hutt's design was aimed at double seating within the general limitations of standard clearance. This was an important consideration on the Long Island, where tunnel clearances allowed only a slight increase in overall height. With an ingenious staggered seating plan and only 5½ extra inches of roof structure, the 72-foot Long Island car (Number 200) offered seating for 120 passengers. Extensive use of aluminum held the total weight down to only 36 tons. Sixty-two more cars on this general design were built for the Long Island's electrified suburban service (see Chapter 8).

Although the Long Island cars had an impressive seating capacity, they were not well liked by passengers. In addition, they could not take full advantage of the bilevel's space potential because of the restrictions on height. The railroads offering commuter service in the Chicago area, however, were unencumbered by overhead clearances and decided to capitalize fully on the double-deckers (Figures 2.101 and 2.102). Undoubtedly the idea was suggested by the success of similar suburban cars on several European railways.[191] In 1950 the Burlington received thirty 85-foot double-decked "gallery cars" from Budd.[192] They towered 15 feet, 8 inches over the rails and weighed 61 tons. The

Figure 2.103 In 1954 the Santa Fe introduced a luxury bilevel car that has come to be accepted as the best form for long-distance travel. New equipment ordered by Amtrak in 1976 follows this general arrangement. (The Budd Company)

upper seats and overhead floor were supported by brackets attached to the side walls and ceiling of the car, thus leaving the main floor open and free from posts and other encumbrances. Passengers mounted narrow, curving stairways placed near the center doors. End vestibule entrances were entirely eliminated by the central door so that more space was available for seating, which totaled a whopping 148. In the five cars that had toilets, seating was still an impressive 145. For short runs the facilities provided were satisfactory and the cramped, swaying ride was tolerable.

The Burlington acquired more cars of this type during the following year. With 124 gallery cars in service or on order by 1965, the road had converted its entire Chicago commuter fleet to bilevels.

It was obvious that the Burlington had come upon a good thing, and imitators quickly appeared. The Rock Island, the Milwaukee, and the North Western began acquiring similar cars for their Chicago suburban trains. The first group of North Western cars, which were made by St. Louis, had the largest seating capacity (169) of any car on this continent.[193] Within a few years the road had fifty of the cars and today uses equipment of the same design for almost its entire passenger service. Gallery cars were introduced on the West Coast by the Southern Pacific and did much to reduce operating costs on its San Francisco commuter trains, but none have been used on Northeastern railroads because of clearance problems.

The success of these cars suggested their application to main-line service—an idea which illustrates the tendency toward desperate innovation in the 1950s. The Santa Fe was not encumbered by a major suburban service, but its managers found the bilevel cars intriguing and decided that they might stimulate long-haul travel. More commodious accommodation would be necessary, of course, though the gallery plan was not really suited to this end. The road envisioned a double-floor or bilevel car with ample space and generous seating. Placing the daytime passenger compartment on the upper level, some 8 feet above the rails, would give travelers a better view of the countryside and remove them from the track noises. The lower level between the trucks could be used for crews' quarters, lounge areas, equipment rooms, or storage. The lower floor could be dropped between the trucks to reduce the overall height of the car and lower the center of gravity, which was held to 65 inches, or about the same as that of a diesel locomotive. The crawl space above the trucks at each end below the passenger-floor level would be filled with air-conditioning equipment, air brake apparatus, and the diesel generator.

Two experimental high-level coaches were made on this scheme and began service on the Santa Fe's deluxe coach train, El Capitan, in 1954.[194] The El Capitan was announced during the same week that Train X and the Talgos were ordered. In terms of weight per square foot of usable space, the Santa Fe decision looks remarkably wise. The results were so good that the road bought forty-seven more cars from Budd in 1956 at a cost of about $275,000 each. Thirty-five were center-door coaches seating 72 passengers and measuring 85 feet in length and 15 feet, 6 inches in height. They weighed 79½ tons (Figure 2.103). Six were double-deck lounge cars, which were heavier. Six were diners with kitchens on the lower deck, and these cars were so heavy

Figure 2.104 Interest in the underslung or possum-belly style of car has persisted since the 1830s. The Pennsylvania Railroad's Keystone of 1956 was a late example of a long series of failures to perfect a workable design.

that they required six-wheel trucks. All these cars ride as well as conventional equipment does, even at 90 miles per hour. Despite their great height, they have only a slight noticeable lean on curves.[195] The Santa Fe purchased twenty-four more of the coaches from Budd in 1964.

The only road to try gallery cars for long-distance travel was the North Western.[196] More workaday than the Santa Fe high-levels, these cars were put into service in the fall of 1958 between Chicago and Green Bay. Thirteen cars were produced at Pullman's Worcester, Massachusetts, plant: ten coaches, a parlor car, a parlor-lounge, and a coach-lounge. The North Western remodeled four existing cars, two diners, and two Railway Post Office baggage-lounge cars with false high roofs to conform to the new bilevel equipment. The coaches were 85 feet long and 15 feet, 10 inches high. They seated 96 passengers. The ride was described as rather bouncy, with a tendency to noticeable rocking on tight curves. The one parlor car was eventually converted to a coach and pressed into commuter service, but the other twelve cars remained unchanged. They were sold to Amtrak after the Green Bay train was discontinued.

It seems as if no idea ever becomes so discredited that someone will not give it another trial. Certainly the phenomenon of radical innovation in a dying industry is starkly demonstrated by the late revival of the old possum-belly car. This scheme, described in Chapter 1, originated in the early 1830s and was tried by both the B & O and Richard Imlay. It appealed to amateur inventors and was repeatedly patented. During the 1870s the New York Elevated operated a number of drop-center cars, and some thirty years later the London subway acquired some steel cars in the style. Its great advantage was a low center of gravity, but this necessitated a complex and expensive drop-center floor frame. The short set of steps required to connect the two levels was probably even more hazardous than the stairways of the gallery or high-level cars, because it was less noticeable. Unsteady on their feet from the motion of the train, passengers often stumbled on them.

The misguided revival of the super-lightweight articulated motor trains in the 1950s undoubtedly caused designers to dust off the drop-center plan. It had one important advantage over the low-level lightweights: the ends of the car could be made the same level as that of standard-sized cars, so that drop-center cars could be integrated with the existing fleet. In addition, the danger of telescoping was minimized. The scheme was a compromise between the radical and the conventional.

Deciding to give the drop-center plan another trial, the Pennsylvania purchased eight cars from Budd for its New York–Washington service (Figure 2.104). The train, named the Keystone, began running in June 1956.[197] Budd had hoped to produce the cars for about $80,000 each, but the cost proved to be nearly double this figure. The coaches were standard in length (85 feet) and were carried on two four-wheel trucks. Couplers and end platforms were the regulation height. However, the

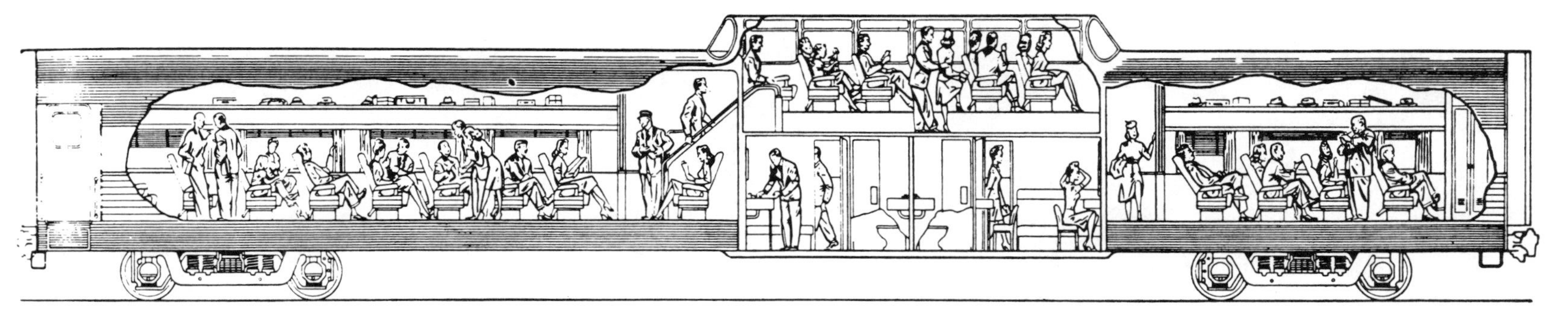

Figure 2.105 The dome car was introduced in 1945 to give passengers a good look at the scenery. Such cars became extremely popular in the West. (Railway Age, *October 11, 1947*)

overall height was 11 feet, 9 inches over the rails, and the coach floor was only 24 inches above the rails, or some 28 inches below standard floor level. Undersized 34-inch wheels further lowered the cars. The center of gravity was said to be 45 inches, or some 9 inches lower than the standard. The overall weight was held to 46½ tons. The interior was divided into three sections. One of the elevated ends over the trucks was outfitted as a smoking section, and in two cars this area was arranged as a buffet. The depressed center section had coach seating, as did the other elevated end section. It should be noted that even the elevated end sections were slightly below the regular floor level. The floor was connected to the vestibule level by a ramp on a 7-inch incline.

In actual service, the Keystone did not measure up to expectations. The ride was not smoother, or the noise level lower, than that of conventional cars, and the steps, as already noted, were a nuisance. The one innovation was the head-end power system, which eliminated generators, batteries, and considerable weight. It was necessary to run all the cars together as a unit, coupled to a conventional diner and parlor cars at the end opposite the power car. Budd received no more low-level orders, but the Keystone piled up a fantastic mileage record before it was retired in 1969 and sent to Altoona for disposition. Amtrak has since purchased it for possible rebuilding.

At the close of World War II there had been expansive talk of what the public might expect in the way of innovative postwar passenger equipment. Most of the prophecies had been promotional hot air, and most of the new trains featured cars not unlike their predecessors. The one spectacular exception that did see service on many U.S. lines after 1945 was the dome car. A glass enclosure and high-level seating offered a thrilling rooftop view of the passing countryside. Vista Dome, Astra Dome, blister, bubble, glass top—whatever name was applied in praise or criticism, these cars were effective traffic generators. Despite their enormous weight and cost, they were probably the most successful of the new designs put forward to save the passenger train. Unfortunately they were really only suitable for use on Western lines, where clearances were not a problem.

Perhaps the caboose cupola could be regarded as the forerunner of the dome, but there seems to be no direct thread of development between the two. Evidence does exist that a passenger car with a high-level observation compartment was in service on the Burlington in the early 1880s.[198] J. M. Forbes requested use of the "birdcage car" for a Western sightseeing trip in 1882. The president of the C B & Q was willing to lend the car, but he warned that it would be unable to clear the tunnels or snowsheds on the Central Pacific. No description-of the car has been uncovered; it may have been a special model used for track inspections, or perhaps it was some form of private business car. The 1891 patent of a Canadian, T. J. McBride, might be cited as a more direct ancestor of the modern dome car. It was called an observation-sleeper by *Scientific American* and was pictured as a luxurious twelve-wheel car with three graceful observation domes.[199] No cars appear to have been built to the patent specifications. The Canadian Pacific's Montreal shops produced a boxy version of McBride's plan in 1902 for sightseeing service in the Rockies.[200] Three more were built in 1906. All were outfitted as parlor cars (described in Chapter 4).

These early efforts appear to have been forgotten by the railroad industry—as well they might, since only a handful were ever constructed. The idea seems to have been reinvented in July 1944. At least, there is a monument along the tracks of the D & R G W which states that the scheme occurred to Cy R. Osborn at this exact spot while riding in the cab of a diesel locomotive. The elevated cab windows offered a magnificent view of the Colorado Rockies, and Osborn, an official of General Motors Corporation, thought that it should be available to more people than the engine crew. The designers at General Motors agreed. They put drawings and models for the projected "Astra Liner" dome car on display in a spring 1945 exhibit on the future of railroad equipment that was especially created for the chief railroad executives. Apparently the only executive much taken with the idea was Ralph Budd, president of the Burlington. Because there was no hope of obtaining a new car, Budd ordered the road's Aurora shops to rebuild a modern stainless-steel coach into a dome car.[201] Certain compromises were necessary because of the existing car's design and the difficulties of obtaining materials. The plan called for a depressed floor below the dome section, but this was not possible without a major reconstruction of the frame. Since the ceiling height below the dome was sufficient for seated passengers, seats were installed. They were placed lengthwise—an awkward arrangement, but the best solution that the Aurora shops could devise. The dome itself was also a makeshift, being built up from flat sheets of glass rather than from the more graceful curved panes called for in the General Motors design (curved panes were not available in an industry geared for war production). Whatever the dome's aesthetic shortcomings, the Burlington aimed to make it as comfortable as possible. Double panes with an air space between them were used for insulation. The dome seated 24, while the lower level carried another 34 passengers. The remodeling was accomplished quickly, and the car, renamed the *Silver Dome*, entered service in July 1945.

The public was delighted with the *Silver Dome*; people felt that it was a preview of the next generation of railroad passenger cars. Within six months the Burlington had placed orders for forty new dome cars (Figure 2.105).

Not everyone shared the Burlington's enthusiasm, however.

Figure 2.106 General Motors' Train of Tomorrow, completed in 1947, featured several dome cars. Here enthusiastic models pose as passengers enjoying the sights. (General Motors Corp.)

During a symposium on postwar car design held just after the *Silver Dome* entered service, some doubts were expressed regarding the merits of the dome car and the desirability of exciting expectations for a revolution in passenger cars.[202] There was much talk of providing plain, comfortable new cars and not wasting energy and time on fanciful innovations. It was also obvious that only roads with generous clearances (up to 16 feet) could handle domes. This excluded virtually all the Northeastern roads. In addition, much of the dome car's seating was "nonrevenue." Passengers could not be expected to travel long distances in the dome; other space would have to be available. Domes might be traffic generators, but there was a price to pay—especially since the drop-center floor, the stairway, the extra air conditioning that was required, and the dome itself made the first cost well above that of an ordinary coach.

But the pressures for innovation were strong. A dramatic response to the competition was thought necessary, and improvements in equipment and service were the only course available. It was hoped that the dome car and similar improvements could hold a substantial share of the postwar passenger traffic market. And so other roads began to follow the Burlington's lead. Of course, the Western roads were the big buyers, not only because of their ample clearances but also because of their longer haul and the glorious scenery along the lines.

In 1947 the first domes came off the assembly line. That spring Pullman completed four dome cars for G.M.'s Train of Tomorrow.[203] All were 85 feet long and stainless-sheathed. The 30-foot domes were squared off, much like the Burlington's *Silver Dome*. Only one of the cars was a coach; it seated 72 and weighed 73½ tons (Figure 2.106). After making numerous press runs and being exhibited at the Chicago Railroad Fair in 1948 and 1949, the train was sold to the Union Pacific for service between Portland and Seattle.

The Burlington received its first commercially built domes from Budd in the fall of 1947.[204] These cars were used on the Twin Zephyrs running from Chicago to St. Paul–Minneapolis. They were the first fully streamlined domes with curved glass. The dome seated 24, the lower level 54. The washrooms were located below the dome (Figures 2.105 and 2.107). A narrow passageway to one side of these compartments allowed transit between the passenger sections on the main floor level. The floor was dropped at this section 19¾ inches for overhead clearance. The possum-belly floor frame and the opening in the roof for the dome required a heavier, more complex structure than a normal coach, but Budd's testing machine calibrated its strength as well above the A.A.R. minimum requirements.

Next came dome cars for the California Zephyr, a Chicago-to-Oakland train sponsored by the Burlington, D & R G W, and

Figure 2.107 The Silver Terrace *was built for the Burlington in 1952 by Budd. (The Budd Company)*

Figure 2.108 The Milwaukee Road's superdome illustrates a good idea that was developed too far. Ten of these 112-ton giants were produced by Pullman in 1952 and 1953. (Pullman Neg. 62835)

Western Pacific railroads. The initial train sets consisted of sixty-six cars, thirty of which were domes.[205] The Missouri Pacific also began to assemble a smaller fleet of domes for its Eagle trains.

The Pere Marquette introduced the dome east of the Mississippi River in 1948. To the surprise of many, the B & O became the next (and for many years the only) dome car operator in the East. Adjustments were necessary on a line notorious for its limited clearances, but the B & O was eager to maintain its pioneer image. In 1949 it added two brand-new dome coaches to the Baltimore–Chicago Columbian. A year later it purchased three dome sleepers from the C & O, which was having second thoughts about its massive postwar equipment orders. About the only other Eastern line to attempt dome service was the N & W, which transferred its three Wabash Bluebird dome coaches to the Norfolk–Cincinnati run in the late 1960s. Perhaps the dome car category should be expanded to include the three sun-roof lounge sleepers built for the Seaboard Air Line by Pullman in 1956.[206] The roof lines of these cars were kept to the standard 13-foot, 6-inch clearance. Seating was at the same level. A unique effect was created by large side and roof windows for about one-third of the

Figure 2.109 Interior of a Milwaukee superdome car of 1952. (*Pullman Neg. 62840*)

car's length. But the high-level view so important to the dome car's success was not available, and no other cars of this type were built.

It was inevitable that the success of the dome concept would lead to the baroque extravagance of the superdomes. The largest land passenger vehicles built for everyday use, they are either the greatest achievement or the most flagrant excess of railway travel, depending on the point of view. Anyone standing beside the track as one of these ponderous cars rolls past is tempted to step back a few paces. The swaying "grand saloons" give the impression that they are ready to fall on some hapless bystander. The first superdome appeared late in 1952, a swollen, glass-topped 112-ton giant.[207] It was superlative in all ways except for its overall height—15 feet, 6 inches, compared with the *Silver Dome*'s 16 feet, 2 inches. Pullman built ten Sky Top domes for the Milwaukee Road at the staggering cost of $320,000 each (Figures 2.108 to 2.110). Seating was provided for 68 in the dome and 28 in the lower lounge. The top-heavy cars required special six-wheel trucks and huge outboard springs to support and stabilize them. The large glass area of the full-length blister (625 square feet) required an air-conditioning plant of 16 tons' capacity with on-board diesel generator sets. Once in service, the Sky Tops were found less satisfactory than their smaller predecessors. The forward view was hampered by low seating and high end bulkheads, and the cars rode hard. Six of the superdomes ran on the Milwaukee's Olympian Hiawatha until the road abandoned its Chicago–West Coast service in 1961. These six were subsequently sold to the Canadian National. The other four remained in Chicago–Twin Cities service and were purchased by Amtrak.

In spite of its disadvantages, the superdome was attractive enough to be copied by other lines. The Santa Fe purchased fourteen similar cars from Budd in 1954, while the Great Northern bought six identical full-length domes in 1955. Stainless-steel construction held the weight to 96 tons. The standard or short dome was generally regarded as providing the best view. All the Santa Fe cars were sold to Auto Train in 1971.

By 1958, 234 dome cars were in service on American railroads.[208] Although they were not able to reverse the downward trend of the American passenger train, they remain the most outstanding of the postwar cars.

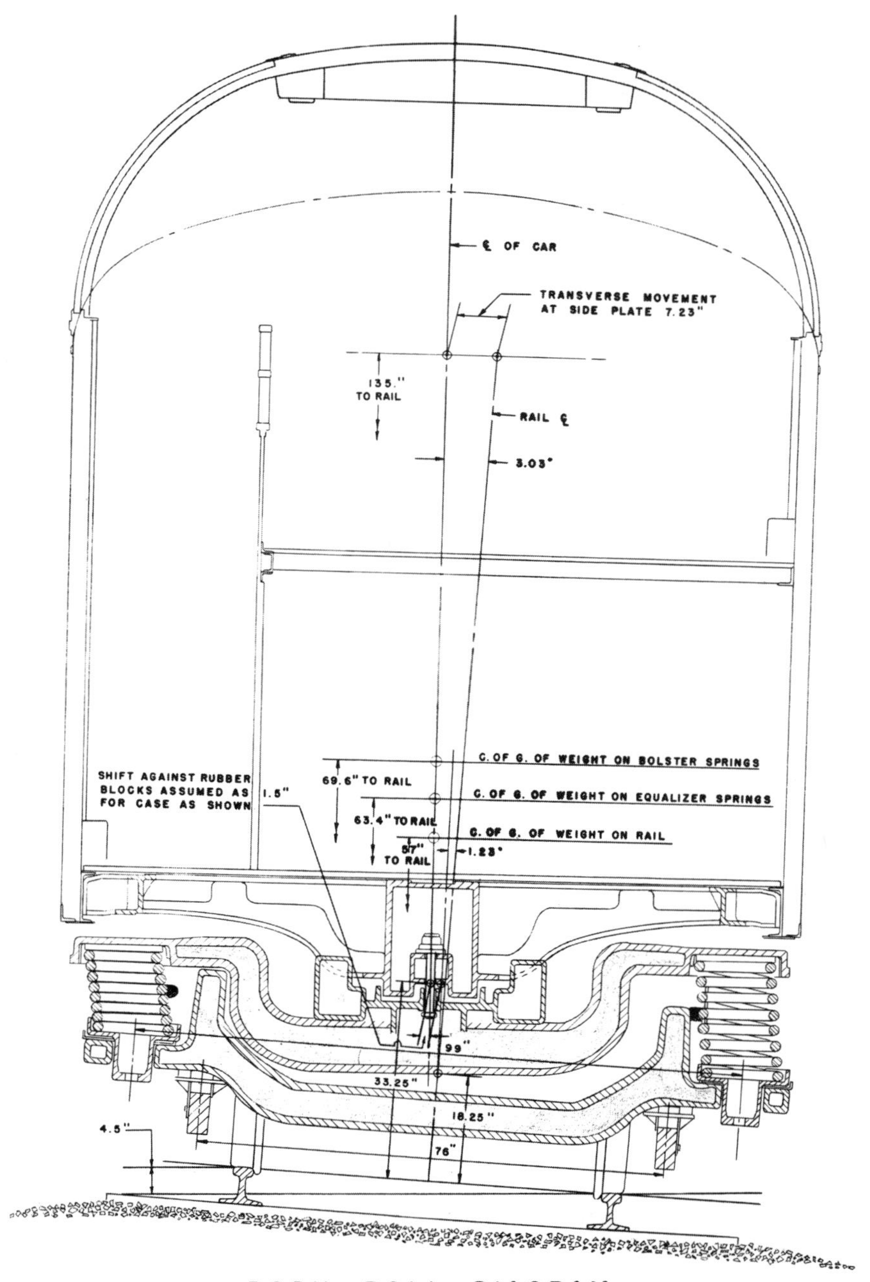

Figure 2.110 *This diagram illustrates the calculated stability of the superdomes. A hard ride and poor visibility were their chief defects.*

CHAPTER THREE

First-Class Travel

SLEEPING CARS

IN THE EARLY DAYS of the republic, Americans warmly denied that any public institution would promote any form of privilege. All citizens belonged to one class, and all would enjoy (or endure) identical treatment. The October 2, 1852, *Scientific American* said proudly, "we have no second class cars for the inferior classes, because all our citizens rank as gentlemen and every man has his own coat of arms." A year later a trial lawyer at the Winans eight-wheel car case eulogized the single-compartment American car as "republican in its character because it allows persons to mingle freely," while he disparaged the European compartment cars that encouraged exclusiveness.[1]

In his *American Notes* of 1842 Charles Dickens mocked this boastful Yankee egalitarianism. He agreed that there were indeed no first- and second-class railway cars as in England, but said, "There is a gentlemen's car and a ladies' car; the main distinction between which is that in the first, everyone smokes; and in the second nobody does."

The philosophical satisfactions inherent in a single class of cars were bolstered by practical considerations. Using one simple style of car for all travelers, the capital-poor American railroads of the early nineteenth century were able to stock their lines cheaply and easily. A long, open-compartment eight-wheel day coach became the standard American car.

Yet from the very beginning of railway travel there were strong pressures for better accommodations than those of the standard coach. And comfort was not the only reason; in 1835, for example, a testy traveler on the Boston and Providence Railroad complained: "The rich and the poor, the educated and the ignorant, the polite and the vulgar, [are] all herded together in this modern improvement in travelling."[2] That passenger would surely have paid double fare for a parlor car seat had one been available, and within a few years special cars did begin to appear on American railways. In his railway inspection tour of this country in 1838 to 1840, the great Austrian engineer F.A. Ritter Von Gerstner reported on the surprising variety of passenger cars in service on American lines.[3] While details in his and other accounts are sparse, there is enough data to conclude that the much-glorified single-class ideal began to break down at an early date. Von Gerstner said, for example, that the Utica and Schenectady inaugurated second-class passenger cars in 1836 but gave up the service three years later because only emigrants would ride in them. The Mohawk and Hudson sold cheap tickets to people willing to ride in freight cars. And according to Von Gerstner, the three major lines running out of Boston were providing second-class cars. Several years later another Austrian railway engineer, Karl Ghega, reported that the Boston and Providence was operating second-class cars and that the New York and Harlem had mounted roof seats for economy-minded passengers.[4]

Moreover, Von Gerstner said that Negroes were often obliged to travel in the baggage car. In many cases head-end cars had a compartment fitted up with seats for Negroes, and half fare was generally charged. The Petersburg Railroad of Virginia, however, permitted Negroes to travel in the coaches at full fare. By 1850 Dionysius Lardner described the segregation of Negro travelers, including freedmen, as a general practice and said that they were obliged to ride in baggage cars.[5] Even after the Civil War, separation of the races continued in the South, where it was often enforced by state law. Cars that were politely called "partition coaches" were built as late as the 1950s. The emigrant car that was furnished for a poorer class of traveler had inferior accommodations, but this was not always true of the Jim Crow car. Particularly in later years, the doctrine of "separate but equal" facilities was obeyed.

Not only second-class accommodations but also first-class cars were in existence by the 1830s. Thus even during the pioneer period of railroads in this country, the American self-portrait of single-class travel was more mythology than fact. Strong demand both for cheap and for luxurious conveyances forced the railroads to develop specialized cars.

Early Sleeping Cars

Sleepers formed the largest group of the specialized first-class cars. They appear to have been the earliest luxury vehicles envisioned for railway use. It has been claimed that Napoleon's traveling coach suggested the railway sleeping car, but more direct antecedents can be found. In 1819 Benjamin Dearborn of Boston speculated that railway cars not only could provide a safe retreat from highway robbers, but also might be outfitted like an ocean packet, with facilities for dining and sleeping. Richard Morgan's proposed double-deck railway car of 1827 and its numerous "births" on the first level was pictured in Figure 1.7. These and other prerailroad projections cannot be dismissed as idle visions, for the traveling public had already experienced comfortable travel on the sea if not on the land. A year before Dearborn's suggestion, scheduled packets began to cross the Atlantic. Some offered the refinements of a good hotel, with bedchambers, dining halls, decorated saloons, and before long, bathing facilities. Canal packets began to emulate their oceangoing counterparts, so that by the time the railway movement began, prototypes for sleeping and feeding the traveler during the journey were well established.

But as Chapters 1 and 2 have pointed out, the great bulk of railway travel was in coaches, even after the development of the sleeping car. That the coach was also called the day car is apt. A trip of half a day in a coach could be a pleasure, but later the passing scenery began to lose its appeal. It seemed as if no comfortable position could be found by turning or shifting in the upright seat. The sway and bounce, the clatter and roar, and the dust, cinders, and wind pouring in through the windows began to seem intolerable. A few miles ahead there might be a transfer point where the travelers, festooned with luggage, rushed for seats and the empty victory of another foul coach, its floor a litter of crumpled papers, boot scrapings, and saliva. Naturally the car was already filled with coarse-faced occupants whose rude conversation was punctuated by the squall of teething babies. The prospect of the night's journey was grim for everyone.

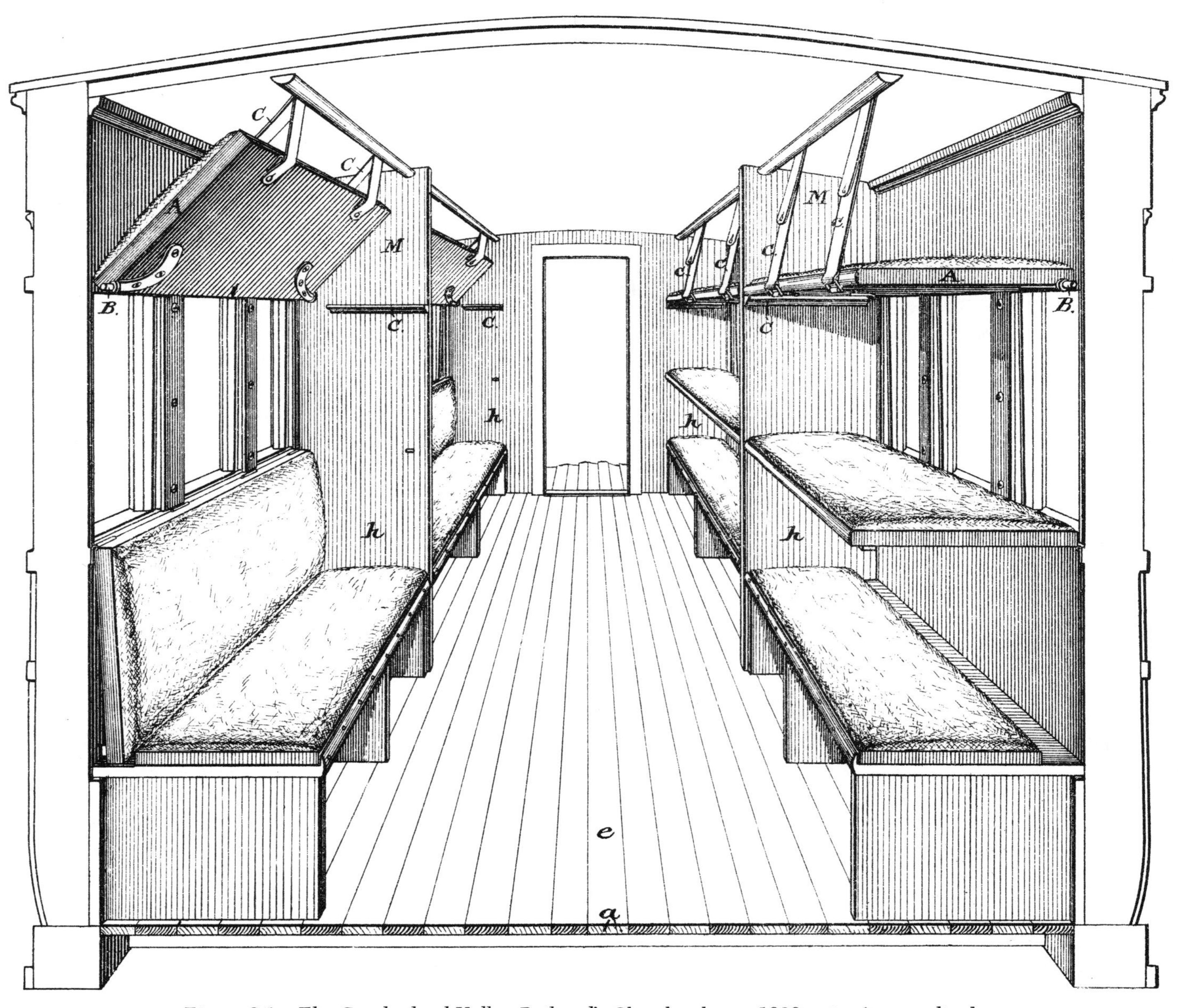

Figure 3.1 The Cumberland Valley Railroad's Chambersburg, *1838c., is often credited as the first railway sleeping car. This drawing was prepared around 1881 for a patent suit.* (Pullman vs. Wagner, *1881*)

The compounded aggravations of a transcontinental trip were summarized by a contemporary traveler:[6]

You will be worn out with fatigue. You will be cramped and stiff with the confinement. You will turn blacker than the Ethiop with tan and cinders and be rasped like a nutmeg grater with alkali dust. You can never sleep a wink for the jarring and noise of the train, and never will be able to dress and undress and bathe yourselves like Christians. Above all, your nearest and dearest, under the influence of the fatigue and the monotony and the discomfort, will be ready to turn and rend you before you get down into the Sacramento Valley—and *you* will desire nothing better than to make a burnt offering of them and every one else insane enough to shut himself up seven days and nights in a railway car!

Travel is a necessity and it was the duty of the railway manager to make it as bearable as possible. The development of more comfortable cars is among the acts of public charity that make life more tolerable in small but deeply appreciated ways. The short-distance traveler and the poor were left to the coaches, but the wealthy and privileged demanded and would pay for something better. This demand led first to the introduction of sleeping cars and later to parlor, lounge, and private cars.

Tradition credits the Cumberland Valley Railroad with operating the first sleeping car, called the *Chambersburg*. Popular accounts say that it entered service as early as 1836 or 1837, but circumstantial evidence indicates that it began running sometime in 1838.[7] Although no contemporary accounts have been uncovered to establish the exact date, it is known that regular train service did not commence on the line between Chambersburg and Harrisburg until February 1838. In later years several old officials of the line said that the sleeping car was not placed in service until some time after the inauguration of the scheduled trains.[8] Stage passengers from Pittsburgh used the Cumberland Valley as a connection with Harrisburg, where trains of the Pennsylvania State Works took them on to Philadelphia. Many travelers complained about the late-evening connection; most were already exhausted by the long trans-Allegheny trip. Could not the railroad, as the marvel of the new age, offer something better than a low-backed coach seat?

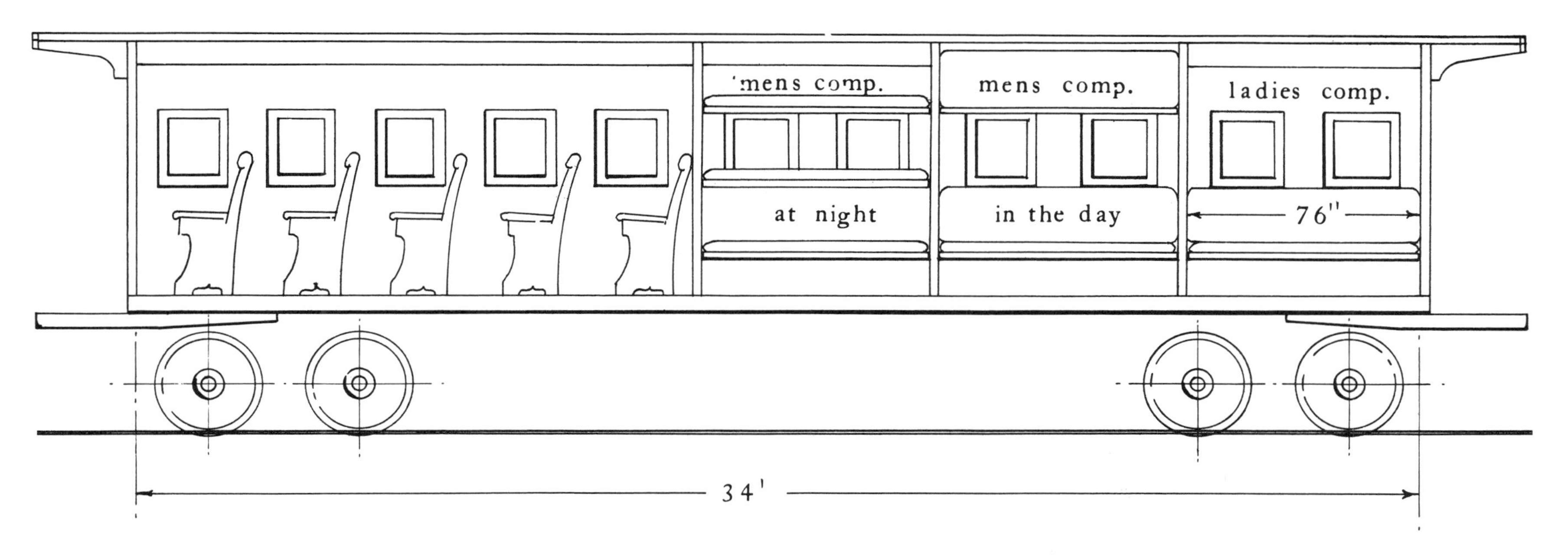

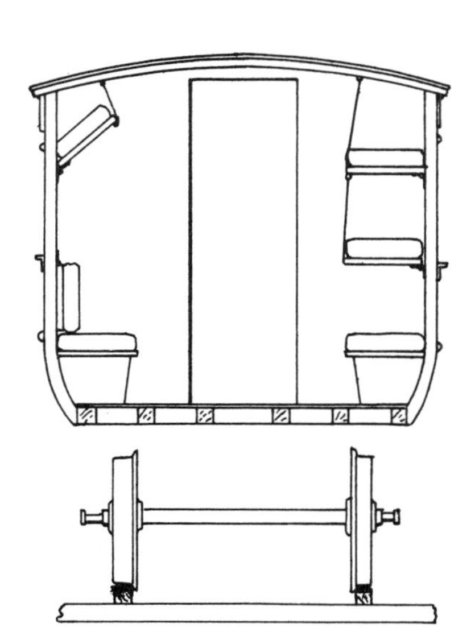

Figure 3.2 A reconstruction of the Chambersburg*'s general plan, based on a sketch by an employee of the Cumberland Valley Railroad in 1871. (Traced by John H. White, Jr.)*

The railroad responded by ordering a special car from Richard Imlay of Philadelphia. Whether he supplied an existing car or built one to order is not recorded, but it may have been an existing car, since Imlay is known to have furnished sleepers for other Eastern lines at this time. A fairly complete description of the *Chambersburg* can be reconstructed from testimony given in two early sleeping-car patent suits.[9] Although the car itself had long since been dismantled by the time of the legal contests, the roof, a berth, and certain other parts were stored in one of the Cumberland Valley repair shops. Measurements were made by the master car builder, Jacob Shaffer, who had worked at the car shops since 1835.

The *Chambersburg* was an eight-wheel car with end platforms and an arched roof. The body measured 34 by 8 feet, with an inside clearance height of 6 feet, 5 inches.[10] The roof arch was given as 3⅝ inches. The exterior was painted brown and was picked out in shaded bronze or gold-leaf striping inside the body panels. The lower body panel was convex.[11] The interior was divided into four compartments, with a center aisle extending the length of the car (see Figures 3.1 and 3.2). At one end was a ladies' compartment with two upholstered benches running parallel to the central aisle. At night hinged edge flaps were raised to make a wide double berth. No upper berths were placed in this section of the car, but the two center sections had three levels of berths on each side of the aisle. They were convertible into side bench seats for daytime travel. The other end section was outfitted with ordinary cross seats. Total sleeping capacity was 16 persons.

The berths in the men's sleeping compartments measured 19 inches wide by 6 feet, 4 inches long.[12] Bedding consisted of some form of coverlet and round or bolster-type pillows covered in figured cotton calico; sheets were not offered. The mechanics of convertible berths were described by Jacob Shaffer:[13]

> The first was a stationary seat with a loose cushion on it, the second one, which formed the back for the first one, was hung with hinges to the side of the car. In day travel this formed the back for the seat or bottom berth, and at night it was raised up horizontally and formed a middle berth. The top berth was hung to the partition at each end by a peculiar kind of an iron; as near as I can describe it it was formed in the shape of a sickle. The one end had a round boss on which part when screwed up to the bottom of this berth held the berth. There was a plate connected with each iron. These plates were let into the partition and the ends of these irons ran into a hole in the middle of the plate. These irons were on the back end of the berth. That berth, in day time, could be swung up towards the top of the car. It was held up by two girths. Those girths were fastened to the carlines or rafters of the roof by a wood strip running along the head-lining. The other ends of the girths had a bottom hole which hooked onto a button on the face of the berth and held it up. In the night it was unhooked from those girths and let down to a horizontal plane, where it was held by two sliding bolts, one at each end of the berth near the front lower edge. There was an opening or slot cut into the partitions into which those bolts entered. There was also two girths extending from the strip at the roof down. On the bottom of that berth there were two brass plates having hooks at the front end. The long girths had a square loop of brass which hooked onto the hooked plate and made a support for the centre of the berth.

Several witnesses in an early sleeping-car patent case said that the *Chambersburg* had an oilcloth ceiling, candle lighting, a water closet in the ladies' compartment, and a small, tapered iron stove for heating in the winter.

This first car proved so satisfactory that a second sleeper, the *Carlisle*, was produced in 1842 or 1843 at the Cumberland Valley shops by remodeling a day coach. The work was supervised by Jacob Shaffer, who patterned it exactly after the first sleeper.[14] In about 1850, however, the road suspended sleeping-car service. Both cars were too worn for continued first-class use, and most of the overnight traffic had been diverted to the Pennsylvania State Works railway, which by that time was offering through service between Pittsburgh and Philadelphia.

The earliest sleeping car for which a contemporary account has been discovered began running on the Philadelphia, Wilmington, and Baltimore Railroad in the fall of 1838 between Havre de Grace, Maryland, and Philadelphia.[15] Since the 24-passenger car was built by Richard Imlay, the probability is that it was very similar to the *Chambersburg*. The arrangement of the berths sounds nearly identical:[16]

> During the day the bottoms of the upper tier of berths fold downward so as to form the back of the lower seat and render them as comfortable as can be. Whenever an upper berth is wanted it is only necessary to draw up the back to a longitudinal position where it is sustained by an iron catch at each end, take a pillow and blanket from the rack above . . . and there is formed at once a snug bed.

The Washington *Globe* of November 23, 1838, carried an advertisement for the new car that promised "a comfortable night's sleep . . . without the least disturbance of rest," and an early-morning arrival in good time for a train to New York. It also said that ladies could expect "commodious retiring rooms, attended by female servants." The 1838 P W & B annual report announced that the service was a success and that a second car was under construction. And then silence; there was no more comment by the road on sleepers until 1847, when a company report noted that some cars had been fitted with special seats for night travel. T. Wesley Bowers, a Wilmington car builder and onetime employee of the P W & B, claimed that the two Imlay sleepers were rebuilt as day coaches in 1844.[17] Bowers offered no reason for the suspension of sleeper operation; apparently the management of the road decided that chair cars were adequate for the run. The journey between Baltimore and Philadelphia was broken by the ferry crossing at the Susquehanna River, so that a through sleeper was not possible. In 1854 the road did make a slight improvement by installing reclining night seats.[18] No more sleepers appeared on the line until 1859, when two coaches were rebuilt for through service between Washington and Philadelphia.[19] By this time the Susquehanna ferry had apparently been remodeled to carry railroad cars. These company-operated sleepers were soon followed by Knight's patent cars, which are discussed in a later section.

At about the time that the P W & B's sleepers first appeared, Von Gerstner reported seeing sleeping cars on other Eastern railways. The Philadelphia and Columbia had one or more eight-wheel sleepers built by Imlay at a cost of $2,350.[20] Water closets and ladies' compartments were provided. The cars were said to accommodate 42 passengers. During an inspection of the Richmond, Fredericksburg, and Potomac Railroad in 1839, Von Gerstner saw an eight-wheel car that was outfitted for night travel.[21]

Stephen W. Worden worked in the Richmond, Virginia, shop of James Bosher when some of the first cars were built for the R F & P in 1835.[22] Worden testified years later in a patent case that two of these cars were long eight-wheelers meant for through service. Four years later they were fitted as sleepers, with folding berths in three tiers. The lower berth could be folded out to form a double bed. A green curtain separated the ladies' and men's compartments. Hair mattresses and pillows covered in maroon were furnished for each passenger. The exterior was distinguished by a small three-paned Gothic sash over each window. Both cars were remodeled as coaches in 1849.

Von Gerstner mentioned one more Southern sleeping car dating from the first decade of American railroading. However, it was not meant for revenue passengers; it was a six-wheel track-workers' bunk car that the Austrian engineer saw on the New Orleans and Nashville Railroad.[23]

Clearly Richard Imlay played a large and perhaps a central part in the early introduction of sleeping cars. Three railroads had sleepers of his construction in service before 1840, and it is possible that there were others for which no records survive. Another scrap of evidence on Imlay's pioneering work in this area comes from accounts that were published in the 1890s of an antique car model owned by the Baldwin Locomotive Works.[24] It was a large wooden model of about ¾-inch scale that the firm had acquired, together with a lot of patterns, years before from Imlay, possibly at the time that his business was closed. His shop had been adjacent to Baldwin's in the Bush Hill section of Philadelphia. A label attached to the model read: "Made by Richard Imbrey [sic] and Jacob Dash, car builders, Bush Hill, Philadelphia, built in 1840."

Whether this model depicted a car that was actually built or not is unknown, but it does contain features Imlay is known to have employed. Note in Figure 3.3 the clerestory roof and longitudinal berths, which are arranged in the same way as those of the *Chambersburg*. Had a full-sized car been built from the plan of the model, it would have been unusually large for the period. Assuming that each berth space was equal to at least 6 feet, the car would have measured 42 feet over the body. If all spaces on both sides were fitted with berths, it could have slept 42 passengers. But according to the drawing in Figure 3.3, one side had only one level of berths, equaling some 28 berths. It is also likely that the blank end-panel section was set aside for a water closet. Perhaps the model builder simply did not bother to install the folding berths on the left side. Unfortunately the disposition of this extraordinary model is not known.

There were a few other sleepers in existence before 1850. In the summer of 1842 the New York and Erie began running two large eight-wheel cars, the *Erie* and the *Ontario*.[25] They were built by John Stephenson on his lattice or Diamond frame pattern. Because of the Erie's 6-foot gauge, it was possible to make the bodies 11 feet wide and to provide extra-large cross seats capable of accommodating three passengers abreast. The seats were made

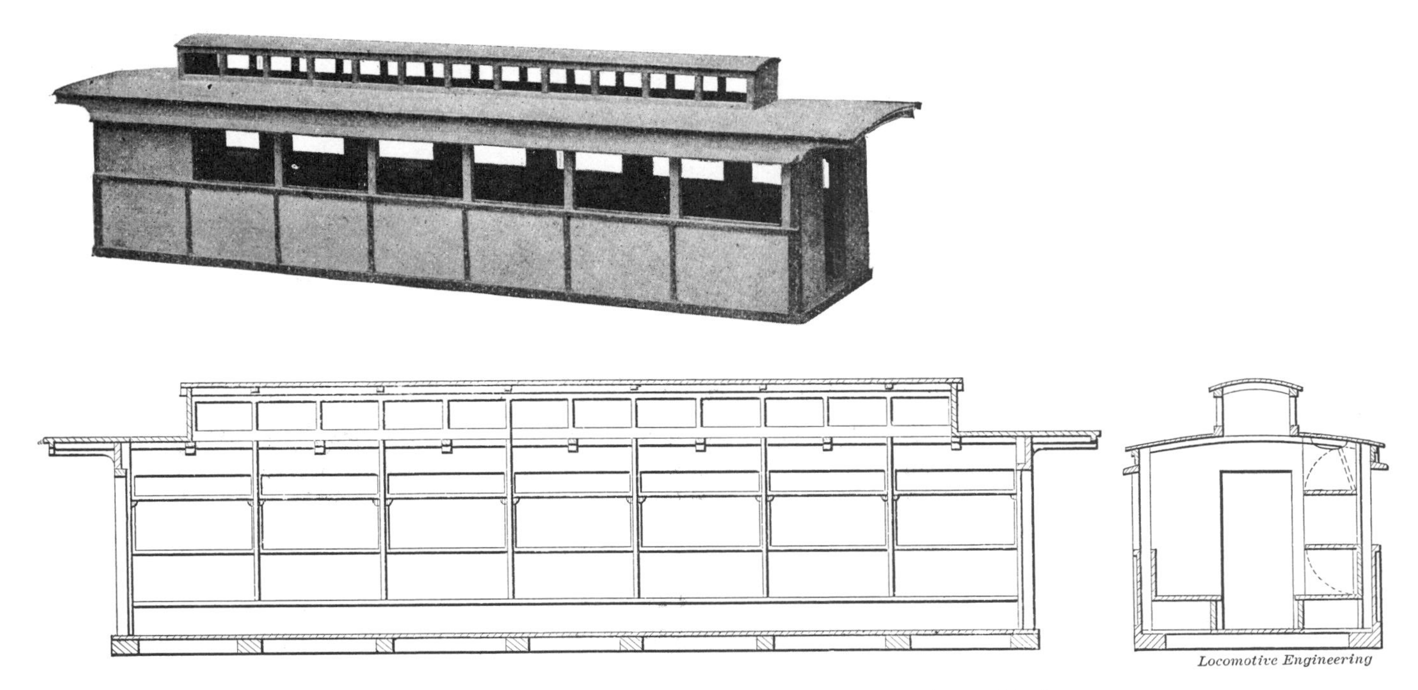

Figure 3.3 A model made around 1840 of an eight-wheel sleeping car built by Richard Imlay. The model was once in the possession of the Baldwin Locomotive Works, but it has since disappeared. (Locomotive Engineering, *January 1898*)

from a plan patented by Nicholas McGraw on December 10, 1838 (No. 1030), for a convertible sofa bed. McGraw's design was intended for household use, but Stephenson adapted it for railway service. Even so, the basic concept of a convertible sofa bed was retained, and it is remarkable how similar the design is to that used generations later by Pullman (Figure 3.4). The likeness was so strong that Wagner and Pullman are reported to have settled one of their patent suits out of court to avoid publicizing the early origins of the convertible seat berth.[26] Apparently neither wished to emphasize that others had anticipated the basic idea of the American sleeper—a convertible day-night car—so many years before their own patents. Such a revelation would undercut their carefully constructed patent deterrents and possibly encourage other firms to enter the field.

The berths on the Erie cars were made up by pulling out two support bars fitted to the underside of the seat frame, thus forming a bridge between the seats. The loose back cushions were laid across the supports to complete the berth. Pillows, blankets, and a berth enclosure were apparently not offered, nor was any provision made for upper berths. In fact, the Erie had so little track when these cars were placed in service that there was no possibility of a long overnight trip. By the time more of the line was opened, the sleepers had been downgraded to slow mixed-train service. They remained in use until sometime in the late 1850s, when they were further downgraded as track-workers' bunk cars.

Competition with a steamship line that ran a parallel route compelled the Petersburg and Roanoke Railroad to build two sleeping cars at its own shops in 1846 and 1847.[27] By offering a good night's sleep, the steamers were taking all the traffic, and the railroad president was determined to recover some of it by providing equivalent service. The car bodies were 48 feet long. The roofs had a pronounced arch of about a foot and provided an inside clearance of 7 feet, 6 inches. Ample space was consequently available for three berth levels. Each car contained eight sections, or twenty-four berths. A ladies' compartment with a water closet was placed at one end of the car. At first the upper bunks were fixed and only the lower berth was convertible, but the lack of headroom made it necessary to refit the upper berths so that they could be folded aside in daytime. For unexplained reasons sleeping service was suspended in 1853–1854, and the cars were remodeled as coaches.

During the same year that the Petersburg car was placed in operation, the Baltimore and Ohio Railroad decided to try a similar experiment. What inspired this move is not known; perhaps it was memories of Imlay cars on neighboring lines, or perhaps some now forgotten official wanted to improve the traveler's lot. Were it not for a French traveler's account, nothing more might be known of this car than the simple fact of its existence. Although it is listed in the B & O's 1846 annual report, no details are offered. However, the July 22, 1848, *L'Illustration* contained an article on a Frenchman's visit to North America in 1847–1848. One convenience of the journey which he particularly praised was the sleeping car. An illustration of the car's interior accompanied the article, but its barn-like proportions and the general lack of detail suggest that it was not a sketch from life but a dubious reconstruction based on the traveler's verbal description. The Frenchman wrote:

> I wish to speak of the interior arrangement of these carriages (sleeping cars) destined to receive travelers for the night. They are actually houses where nothing, absolutely nothing is lacking for the necessity of life. They are divided into several compartments or sleeping rooms, some for men and some for women alone. Each of these rooms holds six beds, or rather couches, placed in three tiers along the sides. During the day the lower couches make excellent sofas. When the sleeping hour arrives, they take the trouble to raise the backs of the couches, and when they reach the horizontal position necessary for their use, strong iron braces moved by some interior mechanism, seize and hold them firmly. Three perpendicular straps guarantee the sleeper from falling. To tell you that these beds are perfectly comfortable would be to lie, but one is very thankful to find them, such as they are, and to be able, thanks to this precaution, to pass a pretty tolerable night. These six-bedded rooms are all connecting, so that one can walk, if necessary, from one end of the car to the other. Lanterns suspended from the ceiling light the interior, which presents a picturesque and novel spectacle.

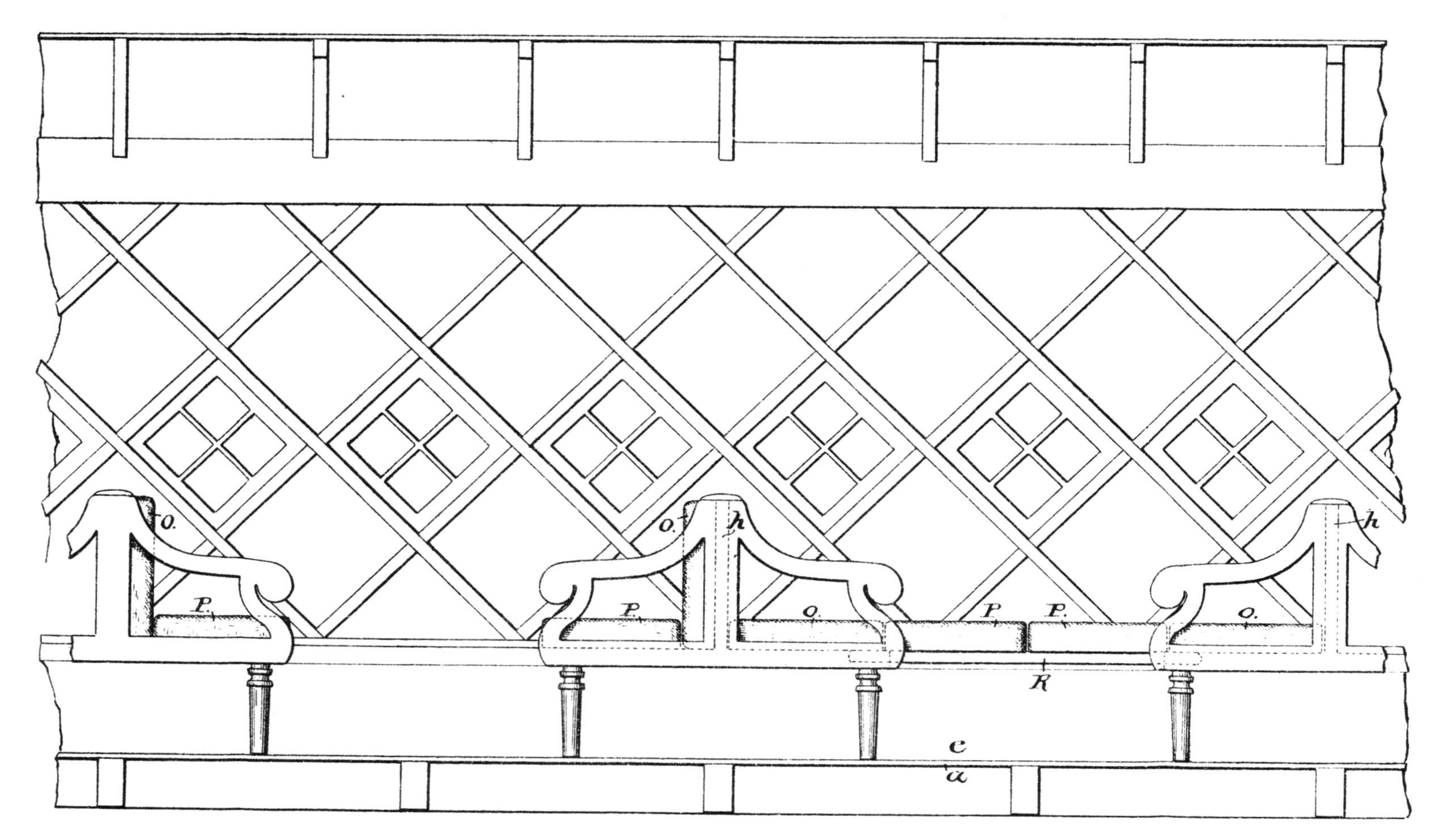

Figure 3.4 A New York and Erie Railway sleeping car of 1842, built by John Stephenson of New York City. The drawing was prepared around 1881 for a patent suit. (Pullman vs. Wagner, *1881*)

Like the other pioneers already mentioned, the B & O sleeper appears to have been withdrawn after only a few years of service. It disappeared from the official roster and was probably returned to work as a day coach. A hint that no more sleepers were operated on the B & O for another decade can be found in the published account of the opening of the line's Western connection to St. Louis in 1857.[28] On the first night out of Baltimore, it notes, the special train stopped overnight at Grafton, Virginia. The adult members of the party were put up in a hotel, while the youngsters spent the night in "comfortable sleeping-cars upon the side tracks of the Road." Presumably these were bunk cars of some variety.

A final sleeper for which there is a record before the middle of the century was reported in operation on the Georgia Railroad by *Disturnell's Railroad Guide*. In the scattered copies of the guide available for 1848 and 1849, stateroom cars for night travel were listed as running on the road between Augusta and Atlanta. The Georgia Railroad's annual reports do not mention these cars, and no other details or sources on them have so far been located.

To summarize, the available evidence, sleeping cars were operating on at least eight railroads before 1850. These sleepers were relatively sophisticated and embodied all the basic features of the convertible day-night car that was to become a resounding success after the Civil War. Why did the idea fail to take hold earlier? Certainly not from lack of technology; the mechanical details had been fairly well worked out before 1840. A principal reason for the railroads' early lack of enthusiasm was doubtless the small demand: few roads offered through travel for much over 100 miles. Even at 20 or 25 miles per hour, the trip was over in a few hours. Of even greater importance, most trains operated in daylight. The few that traveled at night could offer a form of makeshift sleeper. Ladies' cars were relatively common by the mid-1840s, and they usually had a comfortable sofa where a woman could stretch out for a nap.[29] Since husbands could accompany their wives in the ladies' car, a few men enjoyed the same privilege. For the coach traveler, high-backed seats were in use on some roads by the 1830s. Reclining seats were not introduced until the middle of the century, but even standard coach seats were convertible into a poor sort of bed by means of hammocks, foot straps, and headrests fastened to the seat back or frame. Some of these devices were homemade gadgets brought along by the passengers; others were manufactured; and to some degree they satisfied the cheap overnight market.

In addition, sleeping cars were more expensive and heavier than regular coaches, and they carried fewer passengers. The bedding and internal fittings required more maintenance. The fairly limited demand and modest income potential of sleepers were probably all the reason any practical railroad manager needed to discontinue them.

It was during the 1850s that long-distance travel became a dramatic reality and precipitated demand for more luxurious accommodations. The Eastern lines were being rapidly extended into the heart of the continent. By 1851 the Erie reached Lake Erie; two years later the B & O touched the Ohio River; four years later a through rail connection opened between Baltimore and St. Louis, except for the two major river crossings. Passage between New York, Cleveland, and Chicago was soon available. But who

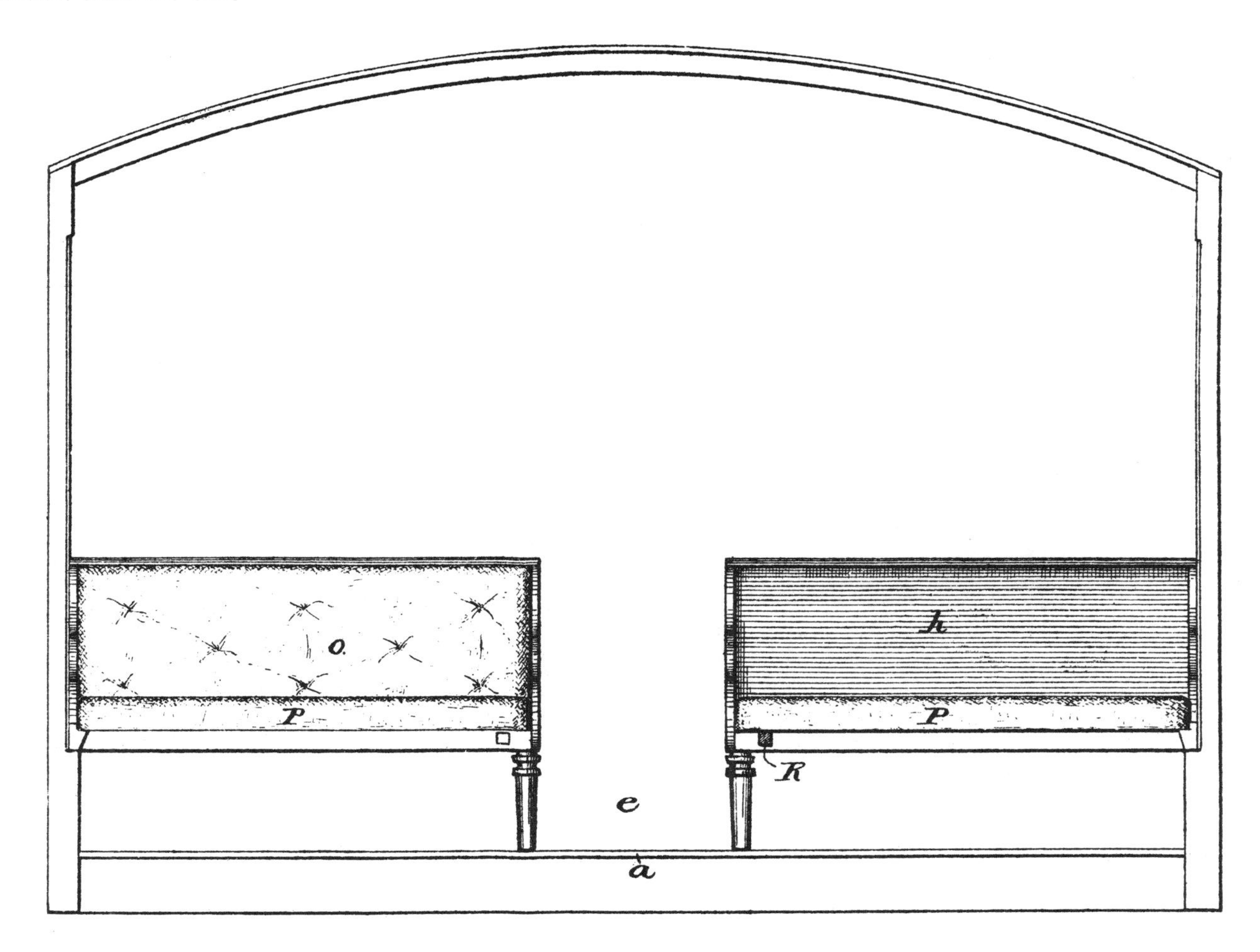

could endure forty hours in a day car? Even though through cars were not run until several years later, and overnight stays at cities along the way were traditional, night travel became more common. For the commercial traveler anxious to get on with his business, the loss of time for rest stops was an intolerable waste. The most valuable part of the day was squandered by sitting on a rattling railway car. To travel at night would save time for the workday hours, and time was money.

All these pressures combined to bring about the sleeping car's renaissance. Some railroads produced their own cars. The Philadelphia, Wilmington, and Baltimore, which had been in and out of the sleeping-car business since 1838, reentered the field in 1859 with two coaches remodeled as sleepers. The road offered through service between Philadelphia and Washington; the cars were moved on the southern leg of the trip by the B & O. During the following year the P W & B remodeled two more cars, and in 1863 it converted a fifth coach more elaborately, giving it a clerestory roof and an "improved interior."[30] But in the same year, the road decided to quit sleeper operation and contracted with the Central Transportation Company to take over the service.

Some years earlier the Illinois Central had made a serious bid to enter the sleeping-car business. It did not dabble in makeshift converted coaches but set out to provide luxurious railway travel long before Pullman contemplated the idea. In 1855 it placed an order with the Buffalo Car Company for some Gothic stateroom cars which were to measure 50 by 10 feet.[31] The first six cars were delivered in June 1856; a Detroit newspaper described their splendor:

> LUXURIOUS RAILROAD CARS—The *Detroit Advertiser* says the cars of the Illinois Central Railroad, for comfort and convenience, excel those of any other road in the West. One of them contains six staterooms, each room having two seats with cushioned backs long enough for a person to lie upon. The backs of the seats are hung with hinges at the upper edge, so that they may be turned up at pleasure, thus forming two single berths, one over the other, where persons may sleep with all the comfort imaginable. In one end of the car is a small wash room, with marble wash bowls, looking glasses, etc. On the opposite side of the car from the stateroom is a row of seats with revolving backs, similar to barbers' chairs, so arranged that the occupant may sit straight or recline in an easy attitude at pleasure. The other five cars each have two or three similar staterooms.

More sleepers were added to the fleet and the Illinois Central continued to operate its own cars until 1878, when a contract with Pullman finally ended the road's sleeping-car operations.[32]

Two years after the Illinois Central introduced its Gothic cars, the neighboring Burlington line altered two coaches for sleeping service.[33] In 1861 it converted a third car, but a few years later this road also decided to contract with Pullman for the specialized service.

As one railroad after another began to fall into line, their master car builders showed a new interest in the sleeping car. At the same time that inventors and private contractors were entering the field, several railroads encouraged their car superintendents to perfect their own designs. S. C. Case of the Michigan Central introduced a sleeper on his own plan in the late summer of 1858.[34] In June of that year Case received a patent (No. 20622) for his "Excelsior Plan for combining seats and couches." His three tier, folding-bunk arrangement offered 56 berths, which the inventor claimed could be converted by one attendant in three minutes. A lady traveler rhapsodized over Case's plan:[35]

> Ah! how pleasant it looked, after leaving the other dusty cars—I could almost fancy I was at home, there was such a parlor-like air about it, and in less time than I can tell it, my seat was converted into a low spring bedstead, and a double one at that, and I and my weary

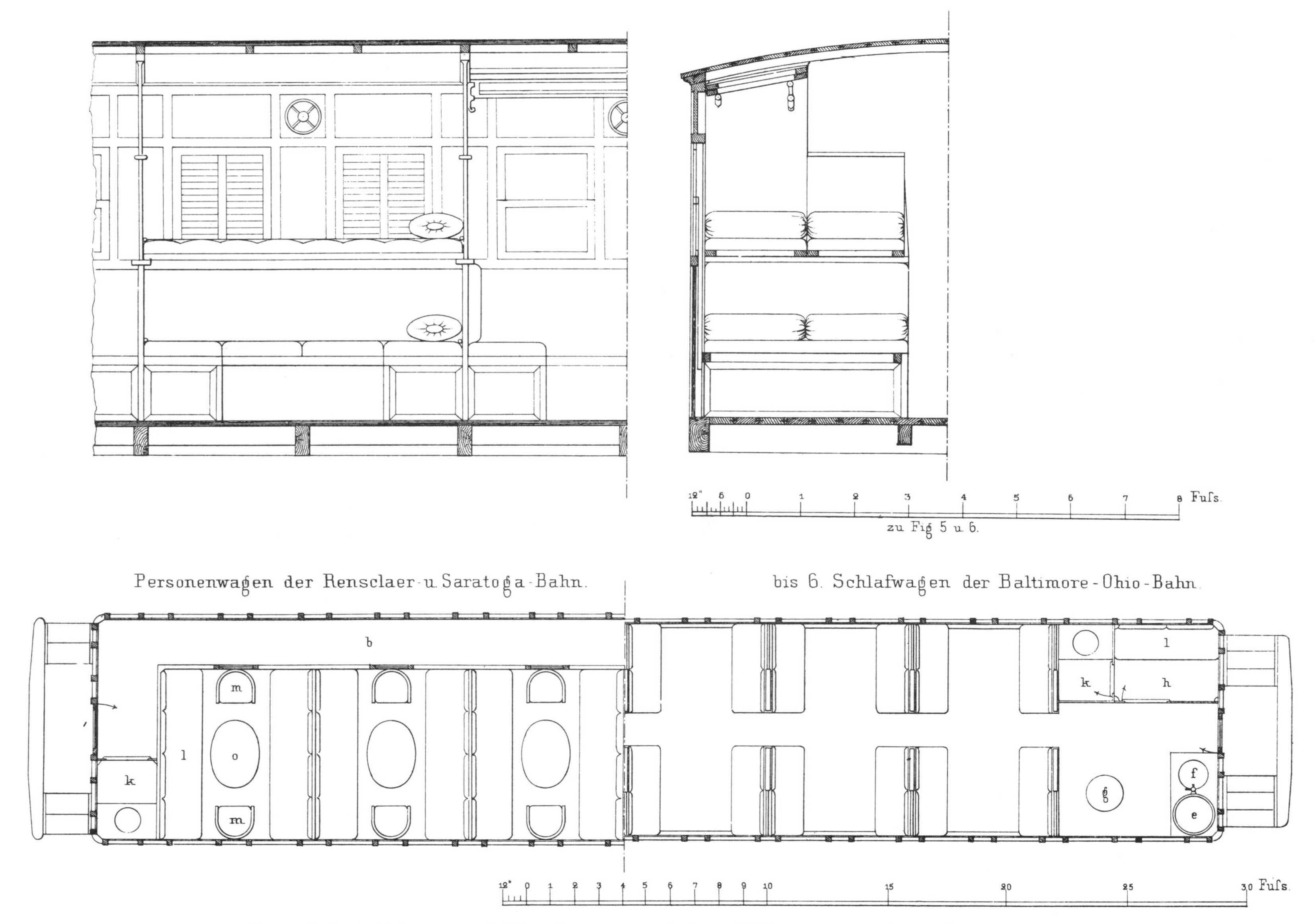

Figure 3.5 A Baltimore and Ohio sleeping car of about 1859 was pictured in a German text on American railways. The berth arrangement does not appear to be one of the many schemes patented during this period. (A. Bendel, Aufsätze Eisenbahnwesen in Nord-Amerika, *Berlin, 1862, plate E)*

little pet were soon wrapt in quiet slumber. At dawn of day I awoke very much refreshed, and on going the length of the apartment, I discovered a neat little dressing room, with all the necessary conveniences attached. I really disliked to leave this huge cradle which had so gently lulled us through the night, and I did so with a feeling that I will not try to suppress: it was gratitude to the inventor of so comfortable a conveyance, and I thank him in the name of mothers, who have hitherto found it so fatiguing to travel by night with little children.

Case's successor, J. B. Sutherland, continued to improve on the Michigan Central's sleepers. His patent (No. 29635) of August 14, 1860, for larger cars superseded Case's design. Using Sutherland's plan, the company's Detroit repair shops produced eight new twelve-wheel cars, which were described by the April 25, 1861, *American Railway Review* as the ultimate in luxury and convenience. The interiors were paneled in native woods; the clerestory roof had stained-glass windows; dressing rooms were located at each end of the car; Ruttan's heating and ventilating system was used. No less than $4,500 had been spent on each car. The operations of the Excelsior Line extended to the Michigan Southern, an arrangement probably facilitated by Vanderbilt control of both roads. The *Elkhart* was one of the many splendid cars built at the Adrian, Michigan, shops by John Kirby. That and a compartment car, the *City of Toledo*, were both meant for the Excelsior Line's Chicago-Buffalo run.[36] The beautiful mahogany, maple, and gilt interior was lighted by gas. The Wilton carpet was in cherry and drab, berth curtains of brown terry were bordered with cherry velvet, and a flowered ceiling completed the interior's rich decor. Fully $20,000 had been poured into the *Elkhart*. By this time Pullman cars were operating on the two Michigan lines, and very soon Wagner would displace both Pullman and the railroad-operated Excelsior Line.

The Great Western Railway of Canada produced a sleeper at its Hamilton, Ontario, shops late in 1858.[37] Most early sleeping cars attempted to seat and sleep an equal number of passengers, but the Great Western car was designed to carry only 36 passengers and hence offered more space for each. Cross seats were provided at one end for day travel. The fact that the car had triple-tiered berths and a single washroom, however, indicates that it may not have been much superior to the maximum-capacity sleepers of the time. The berths were equipped with springs and carpet-covered horsehair mattresses. Pillows, quilts, and silk damask curtains were provided. The car had a black walnut interior, which suggests that it was nicely trimmed inside. This sleeper was built for the Albany-Chicago run via the suspension bridge.

In 1859 Draper Stone produced a sleeper for the La Crosse and Milwaukee Railroad. Meanwhile the Baltimore and Ohio had again become intrigued with sleeping cars, perhaps through memories of its 1847 sleeper, or perhaps because its exchange of through cars with the P W & B reintroduced sleepers to its tracks. The B & O remodeled four coaches for sleeper service on the main line to the west and began to operate other night cars on the north-south line.[38]

Again it was a foreign traveler who has left a record describing one of these cars.[39] The report of a German government official, A. Bendel, offers not only a detailed word picture (based on a spring 1859 journey) of a second-generation American sleeper, but also an excellent mechanical drawing (Figure 3.5). According to the scale on these sketches, the car's length was 54 feet over the end platform; its width was 9½ feet outside. The inside ceiling clearance was 7 feet. The berth and seat arrangement does not seem to follow any of the convertible car patents and was probably the idea of the B & O's own shop personnel. The lower berth was formed by a frame bridging the leg space and the loose seat-cushion backs. The upper berth was supported by two open wooden frames covered with woven cane. The frames were made in two sections for more compact storage against the roof during the day. The bolster pillows and upper-berth mattresses were rolled up for storage in the seat boxes. Sheets for both berths were rolled up like a spring-loaded window blind in the daytime. Stationary wooden partitions separated the berths. Passengers were expected to provide their own blankets.

To sleep 48 in a relatively small car, it was necessary to put two passengers in each berth—a 45- by 60-inch space. It was a cramped, low-ceilinged car like this one that justified Pullman's criticism of the early sleepers as being jingling old arks with macadamized mattresses and a greasy haircloth bolster for a headrest. Actually, the B & O car offered some comforts. A water closet with washbasin and a tank to one side was available at each end of the car. And Bendel reports that the interiors were nicely furnished with oilcloth or carpet floor coverings and wall paintings outlined in gilt and upholstery on the panels between the windows.

Halfway through the Civil War, the B & O shops were producing far more luxurious sleeping cars for the new through Washington–New York service, made possible by the cooperation of the B & O, the P W & B, and the Camden and Amboy railroads.[40] The cars offered one or two staterooms each, with berths like those aboard the best coastal steamers. The B & O soon turned its sleeper service over to contract operators, but in 1880 it once again began operating its own line of night cars.

Because of the soaring interest in sleepers just before and after the Civil War, sleeping cars were soon being operated in all sections of the country, with equipment ranging from the barely adequate to the luxurious. In perhaps a slight exaggeration, A. Bendel claimed that by the time of his 1859 visit, *all* U.S. railroads operating night trains were providing sleeping cars. In 1862 four major lines running out of Chicago advertised their sleeping cars as luxurious and elegant, while the Erie proclaimed its own design a "Splendid Ventilated Palace Sleeping Car!"[41] Contemporary evidence does demonstrate that the cars of this date were more than the "old arks" that popular history has called all pre-Pullman sleepers. While statistics for the period are not available, *Appleton's Railway Guide* gives a general idea of the extent of sleeping-car service. The 1866 *Guide* lists twenty-three railroads, operating in all regions from New England to points west of the Mississippi River, that furnished sleepers.[42] The Chicago and North Western had sleeping cars running to five cities from Chicago. The Illinois Central's night service from Chicago to New Orleans was probably the longest sleeper run in the country.

Another indication of the accelerating interest in sleeping cars was the growing number of mechanics who were attracted by the idea. The usual battalion of threadbare opportunists appeared to patent and reinvent countless old or obvious designs. Even the most talented seem to have offered only modest alternatives to the basic scheme worked out generations earlier by Richard Imlay. All had one goal—the perfection of a combination car that could be easily transformed from a day coach to a sleeping dormitory. The obvious advantage of the dual-purpose car was that it eliminated the need for two separate cars for the same number of passengers. The convertible car was invariably achieved by some form of folding coach seats and a hinged overhead bunk. The more closely the car resembled an ordinary coach during the day, the more skillful the design was considered. It was also desirable to seat and sleep as many passengers as a regular coach could carry, but no plan was ever devised that successfully achieved this goal within the limits of what came to be an acceptable berth size. Some of the earliest three-tier bunk cars did so, but with considerable discomfort for the passengers.

In the late 1850s mechanical trade journals like *Scientific American* were filled with descriptions and engravings of ideas for sleeping cars. The Patent Office suddenly faced a boom in applications in a nearly dormant subclass that had seen only a few additions since the first sleeper patent (No. 11699) was issued on September 19, 1854, to Henry B. Myer.[43] The rush began in 1858, and by the end of the following year over thirty patents had been issued. This record was never approached in later years, but patents continued to be granted even into the 1960s.[44]

Close behind the inventors came the entrepreneurs. Their enthusiasm for sleepers was sharpened by the understanding that the greater profit was in operating them, not in perfecting an ideal sleeping car suited to the needs of the railroads. The majority of the lines were soon weary of the troublesome special cars, with their bedding, porters, and all the fussy details of first-class travel. Since the investment was enormous and the management problems endless, it seemed best to turn sleeping-car operations over to an outside contractor. The railroad would then be free to concentrate on its regular affairs, while sharing in the profits of night cars. A pattern for contracting business out was available in the express and mail lines already in operation. A few railroads, notably the New Haven, the Great Northern, and the

Table 3.1 Railroad-Operated Sleeping Cars

DATE BEGUN	RAILROAD	DETAILS
1838?	Cumberland Valley	Service suspended 1848
1838	Philadelphia, Wilmington, & Baltimore	Suspended 1844, reestablished 1858c.
1838	Richmond, Fredericksburg, & Potomac	Service suspended 1849; Woodruff cars after 1866
1839c.	Philadelphia & Columbia	Reported in service by Von Gerstner
1842	New York & Erie	Two cars retired about 1850
1846	Petersburg & Roanoke	Service suspended 1853–1854
1846	Baltimore & Ohio (and see entries for 1859, 1880)	Service short-lived
1848c.	Georgia Railroad	Stateroom cars
1856	Illinois Central	1878 contract given to Pullman
1858	Michigan Central	First used Woodruff, Case, later Sutherland patents; Pullman 1865, Wagner 1875
1858	Chicago, Burlington, & Quincy	1865c. contracts with Pullman
1858	Chicago, Rock Island, & Pacific	Myer & Furniss patents
1859	La Crosse Railroad	D. Stone's patent used
1859c.	Rensselaer & Saratoga	Compartment Car
1859	Baltimore & Ohio	A second attempt to offer sleeping cars: its own, Knight's, and Central Transportation's
1860c.	Michigan Southern & Northern Indiana	Excelsior Line
1860c.	New York, New Haven, & Hartford	Several private lines as well as its own cars: Mann, Monarch, and Wagner at various times; contract given to Pullman in 1913
1865c.	Boston & Albany	Its own cars and Wagner's
1869c.	Central Pacific (Silver Palace)	Operated its own cars; 1883 contract given to Pullman
1873	Wisconsin Central	Five cars by 1882. Possibly a Woodruff operation.
1880	Baltimore & Ohio	Ended Pullman service, ran its own cars and those of Mann and Monarch; returned to Pullman 1888
1890s	Florida East Coast	Cars by Jackson & Sharp; Pullman contract 1909
1890	Chicago, Milwaukee, St. Paul, & Pacific	Contract with Pullman until 1890; Returned to Pullman 1927c.
1926	Gulf, Mobile, & Northern	Pullman cars were used before this time.
*	Central Georgia	Pullman contracts 1906 for some but not all service.
*	Flint & Pere Marquette	Pullman contract 1904
*	Detroit & Mackinac	Pullman contract 1911
1890	Great Northern	Contracts with Pullman 1922; by 1925 all sleepers operated by Pullman
*	Soo Line (M St. P & S St. M)	Pullman contract on some through lines, 1934; railroad-operated sleepers until about 1953

* Dates service began cannot be supplied.

Note: Railway Age, Sept. 13, 1895, summarized railroad-operated sleeping-car service in the United States and Canada. It reported that of 134 individual lines and 154,224 route miles, 19 railroads representing 22,144 miles were independent. About one-half of these were Canadian.

Milwaukee, continued to operate their own sleepers into the twentieth century. But the real story of the sleeping car in America is that of a concession business (see Tables 3.1 and 3.2).

It was a confused tangle during the first decade as swarms of petty capitalists struggled to gain new sleeping-car contracts. Some were small proprietary operators who personally manned the one or two cars under their contract and owned nothing more than the furniture and bedding. In return for the fares collected, the operator might pay the railroad a fixed amount or a portion of the fares. In addition to this tithe, the concessionaire was usually obliged to pay a license fee to one of the patentees. Arrangements of this type were standard on the Rock Island and Wabash railroads.[45] Adding to the confusion was the fact that some railroads operated their own cars in addition to giving contracts to private concessionaires. The New York Central had Woodruff, Wagner, and Wheeler cars on its eastern lines and company-sponsored cars on its western subsidiaries. In addition, the New York Central was carrying Great Western sleepers over a portion of its road. The B & O permitted Knight, Central Transportation, and Woodruff cars to run among its own cars, apparently without restrictions. The chaos of these mixed operations bothered some far-sighted men, who saw a great need and opportunity to consolidate the sleeping-car business. The potential profit of such a monopoly was obvious. Late in life Andrew Carnegie readily conceded that his fortune was started by an early investment in sleeping-car operation. There was genuine conviction in his epigram: "Blessed be the man who invented Sleep."

Woodruff's Sleeping-Car Enterprise

Theodore T. Woodruff (1811–1892) was the first major figure to emerge from the early concession operators. At the start he seemed destined to dominate the business and technical development of the sleeping car. He was among the first to operate cars on a concession basis on several lines, and before 1860 he had obtained such major contracts as the Pennsylvania Railroad's. His patents covered basic features of the convertible car. The historian of technology Siegfried Gideon thought that Woodruff's plans were the most inventive and clear-headed of all those among the pioneer patentees in this field.[46] But history has largely ignored his accomplishments; he is generally treated as just another pioneer who built a few cars and was then eclipsed by Pullman. George Pullman had reason to play Woodruff down, but it is harder to explain why Carnegie dismissed the inventor as little more than a country rustic, and why such scholars as Gideon, who studied the matter in some detail, did not give more prominence to Woodruff's contributions. Of course, Woodruff's early withdrawal from the sleeping-car business and his subsequent life, which reads like a Frank Norris novel, tend to place him in undeserved obscurity. Had it not been for his brother Jonah, who perpetuated the Woodruff name with his own line of sleeping cars, Theodore might have been an even shadowier figure than he is today.

Like Pullman and Wagner, the other two giants of the sleeping-car trade, Woodruff was born in New York State in the early years of the nineteenth century. He began life as a simple tradesman apprenticed to a Watertown wagonmaker.[47] Later he worked in a pattern shop and eventually became engaged in railroad car work. There are several conflicting stories about how he became interested in sleeping cars. He is said to have considered it as a very young man when the first sleepers began to appear in the 1830s, but he was too busy at the time to pursue the scheme.[48] Another version is that he was inspired to develop a more humane means of night travel after a grueling journey with his chronically ill wife.[49] Or according to an equally romantic story, Woodruff was literally hit on the head by the idea.[50] As a trainman riding atop a freight car in 1854, he was knocked off by a low bridge and while in a dazed condition began work on a sleeping-car model. He made this model in the office of James Tillinghast, superintendent of the Rome, Watertown, and Ogdensburg Railroad. Woodruff asked Tillinghast to help him finance the scheme and secure a patent, but the superintendent declined, seeing little future in the idea. Ironically, Tillinghast

Table 3.2 Sleeping-Car Companies

DATE FOUNDED	NAME	DETAILS
1857	T. T. Woodruff & Co.	Joined Central Trans. 1862
1858	Wagner Palace Car Co., also called New York Central Sleeping Car Co.	Taken over by Pullman 1899
1858	Case Sleeping Car Co. (Mich. Cent. R.R.)	Succeeded by Pullman 1865c.
1858	Eli Wheeler (N.Y.C.)	Patents taken over by Central Trans.?
1859	Knight Sleeping Car Co.	Joined Central Trans. 1862
1859	Gates Sleeping Car Co. (L S & M S)	Succeeded 1873 by Wagner
1859	Pullman	(See Pullman Palace Car Co.)
1860c.	Paine-Harris Sleeping Car Co.	Succeeded by Anderson & Co.
1861	Hapgood Sleeping Car Co.	New York to Boston via Springfield; succeeded by Wagner? 1868?
1862	Indianapolis & St. Louis Sleeping Coach Co.	Woodruff licensee; bought by Pullman 1871
1862	Central Transportation Co.	Successor to Woodruff, Knight, etc.; leased 1870 to Pullman
1864	John B. Anderson & Co. (L & N)	Purchased 1872 by Pullman Southern
1865	Southern Transportation Co.	Controlled by Central Trans.
1865?	Crescent City Sleeping Car Co.	(See Paine-Harris)
1865?	Rip Van Winkle Line	(See Paine-Harris)
1866	Silver Palace Car Co.	Associated with Central Trans.
1867	Pullman Palace Car Co.	Succeeded Field & Pullman
1868	Pullman Pacific Car Co.	Formed to operate cars over Central & Union Pacific, but C P withdrew in 1870c.
1870	Erie & Atlantic Sleeping Car Co.	Used Woodruff patents; succeeded by Pullman 1875c.
1870	Pullman Southern	Consolidated with Pullman Palace Car Co.
1871	Woodruff Sleeping & Parlor Coach Co.	Formed by Jonah Woodruff; joined Union Palace 1888
1875c.	Leighton Sleeping Car Co.	Operated over New Haven; succeeded by Wagner?
1875	Lucas Sleeping & Parlor Car Co.	Cent. of Georgia and other Southern lines; sold to Woodruff 1878
1882	Flowers Sleeping Car Co.	No operation; office Bangor, Maine. William Flowers Patent, 1874.
1883	Mann Boudoir Car Co.	Began U.S. operation 1883; joined Union Palace 1888
1885	Monarch Parlor-Sleeping Car Co.	Operated 30 cars; gone by 1894
1887	Crescent Safety Parlor & Sleeping Car Co.	Jersey City—no operation
1888	Union Palace Car Co.	Combine of Woodruff and Mann; purchased by Pullman 1889
1889	Harris Palatial Car Co.	One car built; in receivership by 1892
1887	W. J. Brashears	Baltimore—no operation
1889 or 1891	Jones Vestibule Sleeping Car Co.	Office in Denver; not certain if any cars built
1890	Allen Hotel Car	Two cars—New Haven
1892	Krehbiel Palace Sleeping Car Co.	Operated briefly on Milwaukee Road
1895	Williams Palace	Kansas City—no operation
1895	L. F. Ruth	Pittsburgh—no operation
1896	Lockwood Stateroom	Cincinnati—no operation
1903	Owensboro Sleeping Car Co.	Maine—no operation
1905	American Palace Car Co.	Several cars on Canadian Northern
1907c.	National Sleeping Car Co.	Texas Central Railroad

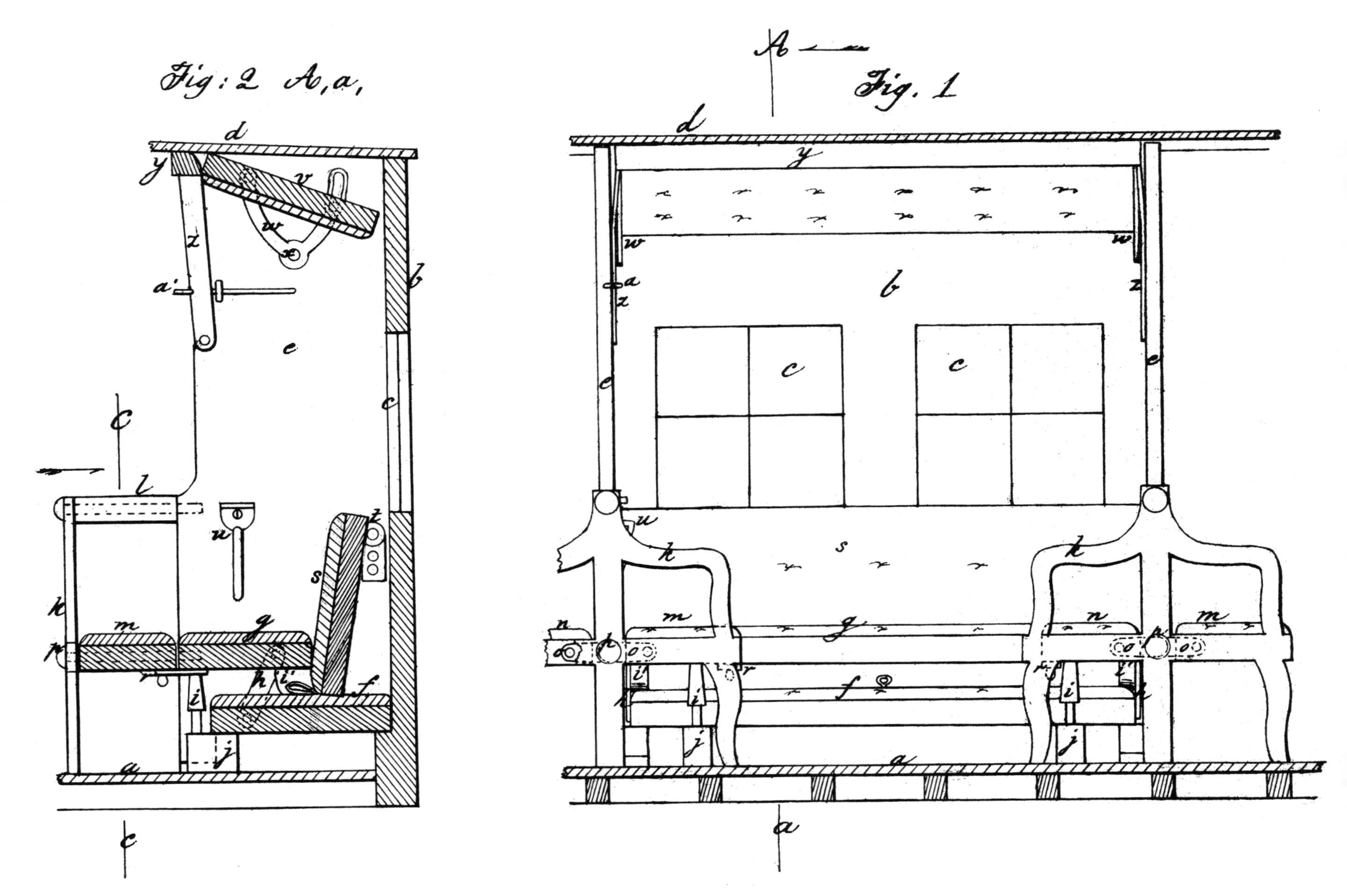

Figure 3.6 The first patent received by Theodore T. Woodruff, the most important of the early sleeping-car pioneers, in 1856. (U.S. Patent Office)

was to become president of the Wagner Sleeping Car Company in later years. Woodruff then packed up his model in an old-fashioned bandanna handkerchief and took it to Thomas W. Wason, the car builder of Springfield, Massachusetts. As the story goes, Wason agreed to build a test car early in 1857, though his employees advised him not to become involved in it.

There appears to be some truth in the third story, except that Woodruff already had several substantial supporters before his visit to Wason. Around 1855 he was employed by the Terre Haute, Alton, and St. Louis Railroad as master car builder. The chief engineer of the line was Orville W. Childs (1803–1870), a fellow upstate New Yorker with a distinguished career in canal building. Childs liked Woodruff's scheme and encouraged the railroad's president, E. C. Litchfield, to draw up patent papers for Woodruff and advance him the money for the filing fees.[51] Childs encouraged the inventor to form T. T. Woodruff & Company, induced two local businessmen to invest in the concern, took a quarter interest for himself, and gave Woodruff a quarter share for his invention.[52] Woodruff received two patents (Nos. 16159 and 16160) on December 2, 1856 (Figure 3.6). A contemporary engraving shows their general plan (Figure 3.7).

Hence when Woodruff approached Wason early in 1857, he had two patents and the backing of three investors. The Springfield *Daily Republican* of October 7, 1857, describes the car:

> We have been much interested in examining a new seat and couch car nearly perfected at the car factory of Wason & Co., in this city, the invention of Mr. Woodruff of Alton, Ill. The car is of the average length, containing upon each side 28 good seats, the whole car containing 56, or four less than the usual number.
>
> These seats face each other in pairs, that is, two persons face two other persons always when the car is full. Everything is elegant, and one would not suspect at the first glance that the whole car could be thrown into couches in which every one of the whole 56 passengers could lay down at full length, yet such is the case, and it is so arranged that each one in a compartment may go to bed singly without depriving the others of their seats, and for the last one two seats will be left. In the first place there is an upper berth six feet long which can be swung down into location and be above the heads of all. Into this one may retire, drop the curtain and go to sleep; when the second wishes to retire, the backs of the seats next to the wall of the car are swung up and united to a stationary section of the berth that stands as a table between the two outside passengers, and thus the second couch is constructed, and over this the curtain drops.

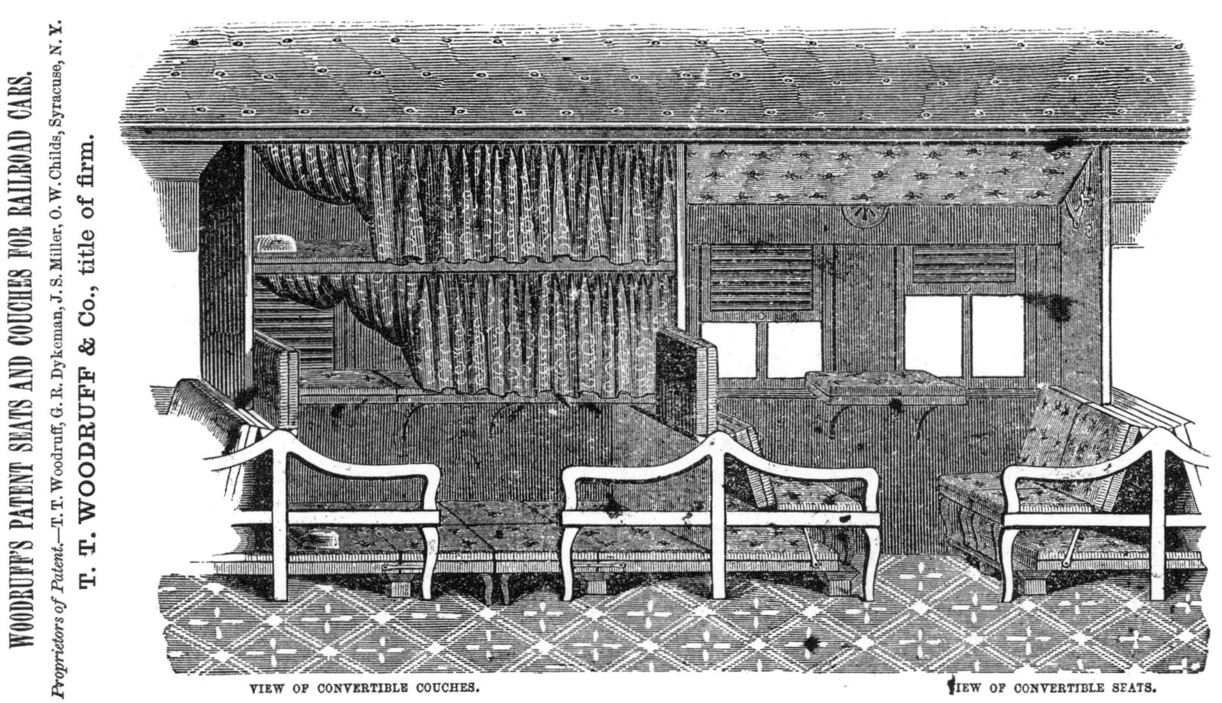

Figure 3.7 The interior of a Woodruff sleeping car of 1859 is shown in this engraving from a railroad guidebook of the period. The section on the right is arranged for daytime travel, while that on the left is made up for sleeping.

There are two passengers left sitting face to face next the passage-way; when the third wishes to retire, the two seats (which are double) under the berths and hung together by an adaptable hinge, are unfolded and just fill the space between the seats, thus making the third couch 6 feet long. There are then two unoccupied seats left, and in these, two men can sit, thus making five in each compartment, or the fourth can make his couch in the same manner the third did. We do not see but the thing is perfect. It certainly needs nothing but to understand it to make it practicable everywhere, and it is a very great improvement; these long night trips are made comfortable by it; it must at once attract the attention of railroad men, and passengers on long routes will give it their universal blessing whenever it may be brought into use.

The car is finished in Wason & Co.'s best style, and will be deemed at the West, whither it is bound, a good specimen of Yankee skill.

The owners, having no service contract, took the car on a tour in hopes of securing a contract for the use of their sleeper. The first trials were reportedly held on the New York Central between New York and Buffalo. The Central's management had little faith in the merits of the car and insisted that Woodruff, who personally acted as the conductor, pay his own fare—which absorbed nearly 70 percent of his earnings.[53] The cheese-paring parsimony of the Central was discouraging, but the other lines were more receptive. Woodruff was allowed to carry out free trial runs on the Michigan Central, the Chicago and Alton, Woodruff's own Terre Haute, Alton, and St. Louis, and other roads. Friendly as these lines were, few seemed willing to build or operate cars under licenses with Woodruff. It was probably at this juncture that Woodruff and his associates realized they must do more than offer designs and advice: they must take over the provision and operation of a full service package. Their investment of $5,000 to $6,000 in the first car was just the beginning.

One way to avoid direct investment was to induce subcontractors to operate sleepers, and in 1858 Woodruff was able to make several such arrangements. Some of these contracts, printed in *Bulletin* 59 of the Railway and Locomotive Historical Society, included agreements with Benjamin Field and Associates (April 5, 1858) to supply service on the Galena and Chicago Union and the Illinois Central Railroads; with T. N. Parmelee, G. B. Gates, Morgan Gardner, and Webster Wagner (May 8, 1858) to operate Albany-Buffalo sleepers for the New York Central and with J. D. Morton (May 12, 1858) to run sleepers on the Michigan Central (with 20 percent of the net profits to go to the railroad). In the same year two large sixteen-wheel Woodruff sleepers built by Wason were placed on the Michigan Southern and the Northern Indiana.[54] They had fourteen sections and small bay windows. They were said to be owned by a Mr. Bates of Utica, presumably a Woodruff contractor who could not place his cars locally. Between 1858 and about 1865 the Wabash contracted with several small concessionaires to operate Woodruff cars; the cars belonged to the railroad, but the operators owned the interiors.[55]

By the fall of 1858 Woodruff claimed to have twenty-one cars in service, and the 1857 sample car continued its tour in search of more customers.[56] Then T. T. Woodruff & Company secured a rich plum in the form of a contract with the Pennsylvania Railroad on September 15, 1858, to operate sleepers between Philadelphia and Pittsburgh.[57] The terms of the contract were very much like those of a twentieth-century Pullman agreement: the railroad was to maintain the cars except for the interiors; it was to transport the attendants (porters) without charge; it was to be free from patent claims; and it would receive coach fare for each passenger. Woodruff was to collect a fee not exceeding 50 cents for each berth and must provide enough sleepers by January 1859 to service all the night trains operated over the road. Four new sleepers, numbered 22 to 25, were ordered from Murphy and Allison, car builders of Philadelphia. The 1857 Wason car was not good enough for this prestigious run; it was sent to end its days on the Ohio and Mississippi Railroad. The results of the Pennsylvania contract were described in September 1859 by Jonah Woodruff, who served as conductor aboard car C.[58] He reported that between 10 and 50 passengers were carried daily, producing a total monthly revenue of $550. Deducting wages and supplies, the sleeping-car company netted a handsome return of $483.09 a month.

A fanciful story about the Pennsylvania contract is often quoted from Andrew Carnegie's *Triumphant Democracy*. The great steelmaker claimed that he was approached on a train one day by a "tall, spare farmer-looking kind of man [who] wished me to look at an invention he had made. With that he drew from a green bag . . . a small model of the sleeping berth for railway cars. He had not spoken a minute before like a flash the whole range of the discovery burst upon me. Yes, I said, that is something which this continent needs." At the time Carnegie was private secretary to the Pennsylvania's general superintendent, Thomas Scott. He claims to have pressed Woodruff's invention on Scott and thus brought "sleeping cars into the world."

The nonsense of this tale is dispelled in Joseph Wall's exhaustive biography of Carnegie, which quotes a letter written by Woodruff himself in 1886 stating the events as they happened. By 1858, Woodruff notes, he was already an established sleeping-car operator, and he was introduced to Scott by the road's president, J. Edgar Thomson. He was hardly the forlorn, "farmer-looking" person wandering about with a model in a green bag, as described in *Triumphant Democracy*. Carnegie knew better, but he liked the story. Without directly denying Woodruff's corrections, he refused to acknowledge his fabrication publicly. He did frankly admit his debt to Woodruff and in later years gave the old inventor a monthly pension. It was a well-deserved subsidy, for Carnegie acquired an eighth interest in the sleeping-car company for a cash outlay of just over $200. By 1860 he was receiving $5,000 a year in dividends.

THE CENTRAL TRANSPORTATION COMPANY

It was clear by 1862 that Woodruff's operation could no longer be effectively managed as a partnership. The firm had expanded wonderfully in five years. The partners were no longer riding the cars; they could afford to hire porters. But if they were to grow and meet the challenge of the new sleeping-car companies then emerging, they needed additional capital to buy out the competition and purchase more and better cars. T. T. Woodruff, who is described as having a rather poor head for business, did not see the need for such radical action, but his older brother Jonah saw that they must expand or expire.[59] Some of their licensees, like Gates and Wagner, had grown increasingly independent and were expanding quickly. Edward C. Knight, the fabulously wealthy Philadelphia sugar merchant, was introducing his cars on many lines and had the capital to develop a serious rival enterprise. It seemed vital that steps be taken to deal with the situation.

At last a compromise was negotiated: the consolidation of the Woodruff and Knight companies, together with the acquisition of the more important sleeping-car patents. The deal was engineered by Jonah Woodruff and O. W. Childs, one of the original partners, who had come to Philadelphia around 1860. The new firm, chartered on December 30, 1862, took the ambiguous corporate title of The Central Transportation Company. Capitalization was set at $200,000; the charter expired in 1891. Into the pot went the contracts, equipment, and patents of T. T. Woodruff & Company—the largest contributor to the assets of the new joint stock company. Of the 4,000 shares first issued, Woodruff received a controlling interest of 2,141.[60] The other patents acquired were those of E. C. Knight, H. B. Myer, W. A. Brown, P. B. Green, Eli Wheeler, J. B. Creighton, and Edward Burke. In this group only Knight and Wheeler appear to have operated any cars. The others, so far as is known, were paper tigers; but they held registered patents, and even if the designs were never used, they were effective tools for discouraging potential competitors. The deterrent effect of these twenty-five patents was emphasized in the company's advertisements and statements. Even the cars carried the patent message on their letter boards.

The inclusion of Knight in the Central Transportation Company was essential to its success. His three patents, issued between 1858 and 1860, were a minor menace, but Knight's wealth and influence (he was an officer and shareholder in several railroads) made him a real threat to Woodruff. It is reported that his interest in perfecting a night car was the result of a sleepless rail journey to New Orleans. Knight's first car was not produced until 1859 or 1860. One of his cars was engaged for the Prince of Wales's trip between Washington and Philadelphia in 1860.[61] This vehicle, built by Murphy and Allison, was described as substantial and elegant, with twelve sections and a total capacity of 60 passengers. It had a double floor insulated with cork chips. Within a year Knight cars were running on the Baltimore and Ohio, the Philadelphia, Wilmington and Baltimore, the Camden and Amboy, the Virginia and Tennessee, and the Cincinnati, Hamilton, and Dayton railroads. Had it not been for his merger with Central Transportation only two years after entering the industry, Knight might have become a major name in the sleeping-car world.

The Central Transportation Company was the strongest sleeping-car operator in the country. It controlled the principal Eastern routes and the basic sleeping-car patents. Its cars rolled over the B & O, the Pennsylvania, the Northern Central, the Central of New Jersey, and many other lines. Childs, the new president, was eager to expand. In 1863 the following advertisement appeared in the *American Railroad Journal* and the *U.S. Railroad and Mining Register*:

CENTRAL TRANSPORTATION COMPANY.

THIS COMPANY, a Corporation organized in pursuance of the provisions of a General Act of the Legislature of the State of Pennsylvania, having, by purchase, recently become the sole owner of

WOODRUFF'S, KNIGHT'S, MYERS'
AND OTHER PATENTS FOR
SEATS AND COUCHES IN SLEEPING CARS,

would respectfully give notice to all RAILROAD COMPANIES IN THE UNITED STATES, that may desire Sleeping Cars on their Roads, that this Company are now prepared to negotiate for placing wholly at its own expense, on such Railroads as may require them, their

S L E E P I N G C A R S,

and operate them upon terms at once liberal and satisfactory to RAILROAD COMPANIES.

The Cars of this Company are constructed of great strength, and contain their late improved and patented plans of Seats and Couches, with STATE ROOMS AND BERTHS, finished in a style the most elaborate and tasteful, and are furnished with all of the modern conveniences and means of comfort usually found in the SLEEPING APARTMENTS OF OUR FIRST-CLASS HOTELS. The interior is lighted with gas; they are well ventilated, and at seasons of the year requiring it are warmed with pure heated air, regulated in degree at pleasure by registers in each State Room and section.

Where known, these Cars are not only regarded as indispensable on all through NIGHT TRAINS, but have become desirable, above all others, as DAY CARS.

Communications from the officers of Railway Companies desiring Sleeping Cars on their Roads, addressed to the Central Transportation Company at their office, No. 1347 BROWN STREET, PHILADELPHIA, will receive immediate attention.

Railway Companies using, or permitting to be used on their respective lines, Sleeping Cars or Couches, that infringe upon the patents owned by this Company, are respectfully notified that satisfactory arrangements for such infringement will be expected. Address as above, O. W. CHILDS, President.

New contracts, of course, created a need for more cars. Again the company advertised in the railroad press, this time for five cars by February 1864 and an additional twenty before the end of the year.[62] Builders were told that the cars must be patterned after the existing cars lettered Q, R, S, M, N, O, P, and T which were running on the various lines serviced by Central Transportation. It is likely that some of the new cars were built by Kimball and Gorton of Philadelphia. A few years before becoming president of the sleeping car company, Childs had been head of the Philadelphia car works. In 1862 Kimball and Gorton were succeeded by J. R. Bolton, and two years later George Dykeman, another of Woodruff's original partners, took a major interest in that firm. It is probable that a good number of Central Transportation cars were produced at the shop before it closed in 1867. This interesting liaison between the operating company and a car manufacturer, with overlapping officers and interests, was a forerunner of the gigantic combine assembled by Pullman in later years.

Two of the new cars appeared on the Pittsburgh, Fort Wayne, and Chicago Railway in June 1864. The *Chicago Tribune* described them as equal in elegance to the few first-class hotels in America.[63] The staterooms were fitted with large French plate mirrors. The cars had hot and cold running water, a hot-air furnace mounted under the rear platform, a perfect ventilation system, and "bedding of excellent quality," all of which guaranteed the traveler a comfortable night's rest. It might be noted that George Pullman was just starting work on his first luxurious sleeper, the *Pioneer*, in a nearby shed. Perhaps some inspiration was offered him by the two Woodruff cars.

Three years after its charter was granted, the company had expanded so much that it was necessary to increase its capitalization tenfold—to 2 million dollars.[64] The business had spread into the Midwest during the same period, and now that the Civil War was settled it was time to move into the South. The Southern Transportation Company was formed in the fall of 1865 by Knight, Childs, and their associates. The controlling interest of 2,000 shares was assigned to the Central Transportation Company. The remaining 1,100 shares were held by Childs and other investors, the most prominent being Henry Plant, the Connecticut Yankee turned Southern railroad promoter. The parent company granted the use of its patents in return for a 15 percent share of the Southern's revenue, in addition to the controlling interest already owned by Central Transportation. During the following year the Southern negotiated contracts with seven Southeastern railroads, including the Virginia and Tennessee, the Alabama and Florida, and the Richmond, Fredericksburg, and Potomac. However, the enterprise did not flourish as well as had been hoped. Southern traffic was not comparable to that in the North, and other independent competitors operating out of Atlanta and Louisville took over many lines that Southern Transportation had hoped to acquire. The concern managed to pay an average dividend of 7.5 percent before being acquired by Pullman Southern in 1870.

Elsewhere, Central Transportation was doing well. In 1866 its eighty-eight cars were valued at $8,000 each.[65] During the summer of that year the company's vice president and general manager, Jonah Woodruff, inaugurated the first Silver Palace cars. A special train pulling twelve Silver Palaces was run from New York to Chicago via Philadelphia and Pittsburgh to promote the new luxury sleepers. The excursion party, made up of railroad men, the press, and their families, was said to have enjoyed a week of purest pleasure. A musician was commissioned to compose a march dedicated to the Silver Palace car.[66] A colored engraving of the Silver Palace car *Chicago* (Number 76) was printed on the cover of the sheet music. *Scientific American* heaped praise on the Silver Palace cars and their promoter, calling Jonah Woodruff "King of the Sleeping Cars" and declaring that he deserved the golden harvest he was now reaping.[67] It said that the cars were the "most brilliant vehicles that ever rolled on iron wheels." It described richly carved black-walnut interiors, enhanced by velvet window hangings, Brussels carpets, and silver-plated lamps and hardware. Similar cars are shown in Figures 3.8 and 3.9. There were stained-glass clerestory windows that "shed a flood of blue-tints upon the glittering silver pillars and fret work below." The builder of these cars was not named. A few years later a York, Pennsylvania, newspaper reported that two Silver Palace cars, Numbers 112 and 113, had just been finished at the Northern Central shops. The bodies were yellow, the platforms a salmon color, the trucks green. The window sash was varnished cherry. Each car cost $10,000. Late in 1866 the *Amercan Railway Times* mentioned that the New Jersey Central had finished three Silver Palace cars in its shops and had three more underway that would be named *New York, Chicago,* and *Pittsburgh.*

In addition to the extravagant decorative treatment, the Silver Palace cars featured an improved berth design patented by Woodruff on November 19, 1867 (No. 71,258). The upper berth was double-hinged and built so that the outer leaf could be folded perpendicularly to the main inner leaf. The whole structure was

Figure 3.8 Woodruff began to introduce palace cars in 1864, at the very time that Pullman was starting work on what is erroneously believed to be the first of the breed, the Pioneer. *The Woodruff cars here were built around 1869 by Jackson and Sharp. (Hall of Records, Delaware Archives)*

Figure 3.9 The exterior of the Woodruff Silver Palace car, the Empire *(Number 85), shown in Figure 3.8. (Hall of Records, Delaware Archives)*

L-shaped when closed and was fastened to the ceiling at a 45-degree angle (Figure 3.10). This allowed a double-wide berth to be folded up against the ceiling. Woodruff also included removable berth-divider panels and folding tables in his patented design. The silver-plated columns were more than decorative; they helped to support the roof and upper berth structure.

Silver Palace cars placed Central Transportation well ahead of its competition and gave the firm invaluable opportunities for further expansion. In the late 1860s the most prestigious upcoming contract was for the transcontinental railroad. Negotiations began in 1867, with both Pullman and Central determined to win the prize. Actually, two contracts were involved because of the division of the line between the Union Pacific and the Central Pacific. It appears that Woodruff was successful in landing the Central Pacific contract. At least one builder's picture survives showing a Central Pacific car ascribed to Woodruff (Figure 3.11). However, some Western railroad historians believe that sleeping cars were never operated by Woodruff on the Central Pacific. All the early sleepers on this line carried Silver Palace lettering and seemed to feature the interior arrangement and decor of Woodruff's cars, but it is possible that nothing was signed except a licensing agreement for the name and design. It is also possible

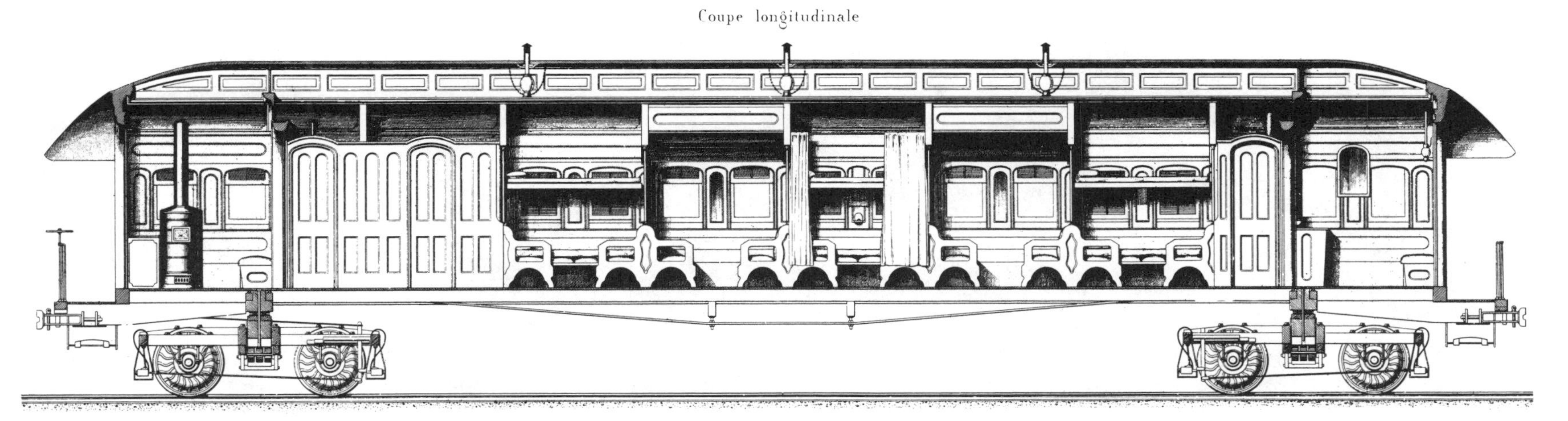

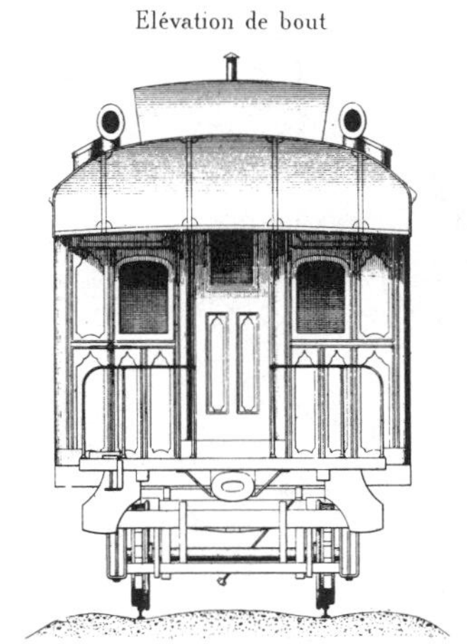

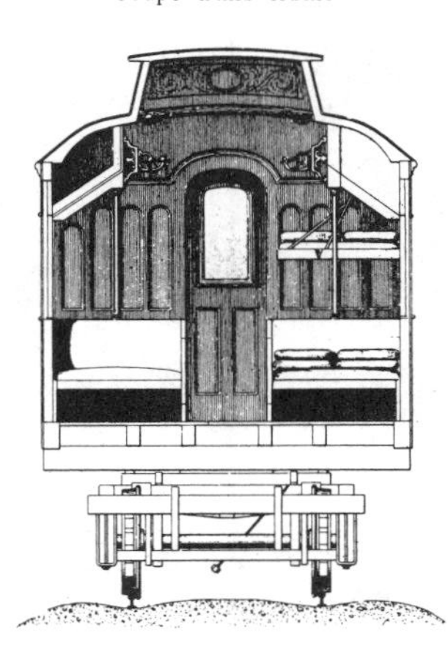

Figure 3.10 A Silver Palace car built in 1869 for the Central Pacific by Harlan and Hollingsworth. This cutaway drawing appeared in a French work on American transportation. (E. Malezieux, Public Transport in the United States, *1873)*

that the operating contract, if one did exist, was canceled at the time that the service was to begin because of Pullman's lease of Central Transportation early in 1870.

Whatever the facts about the contract, the Central Pacific's sleepers were operated under the name of Silver Palace. The first car was received in July 1869 from Harlan and Hollingsworth.[68] Other cars were furnished by Jackson and Sharp. By 1870 twenty-one were in operation; Pullman sleepers were not permitted on the line, and passengers were obliged to change cars. In 1875 Barney and Smith added eight Silver Palace cars to the fleet.[69] They had 55-foot-long bodies and six-wheel trucks, and they were painted a light canary yellow. Before it abandoned independent sleeper operations in 1883, the Central Pacific's fleet had grown to forty-one cars. The cars and service were not considered equal to Pullman's best, however; one waggish British traveler said: "I met with no silver whatever in the Silver Palace sleeping-cars. The fittings, lamps, . . . door handles, etc., are of the white metal called pinchbeck or Britannia metal."[70] Thus within a decade of its appearance, Woodruff's triumph of elegance was being called a cheap sham.

If Central Transportation could not secure a hold on the Far West, it was doing magnificently in the East. In 1869 it had 119 cars in service on sixteen major railroads and was by far the strongest sleeping-car company in the nation.[71] During the fiscal year 1869 it earned over $256,000—equal to a 17.8 percent return on invested capital.[72] A 20 percent profit was confidently expected for the following year. It seemed likely that the firm would enjoy a long and prosperous life.

The seeds of its unexpected demise were sown in the course of competition for the Union Pacific contract. It was during these negotiations early in 1867 that Andrew Carnegie met George M. Pullman. Carnegie, immediately impressed by Pullman's ambition and energy, saw that he was more than "a lion in the path"; he was a business genius likely to outrun his competitors. Carnegie decided to abandon his old associates and join Pullman. He devised a "compromise" that squeezed out the Central Transportation Company and gave the Union Pacific contract to a new company, the Pullman Pacific Car Company, with shares owned by Pullman, Carnegie, Thomas A. Scott, and the Union Pacific Railroad.

Both Carnegie and Scott were shareholders in Central Transportation. The officers of that company, particularly Childs, were annoyed by Carnegie's double-dealing and did not think that the $20,000 patent fee arranged by Carnegie for the Pullman Pacific was adequate. Carnegie finally quieted their complaints by selling his shares in Pullman Pacific to his Central Transportation associates—pocketing a $20 profit per share on the transaction. Although Childs believed that Pullman could be beaten in court with their arsenal of patents, he was finally persuaded to accept the compromise plan. But Pullman, hard-pressed for cash, was putting everything he could raise into new cars needed for the transcontinental service. His delays in paying the patent fee hardly sweetened relations with Childs; yet somehow Carnegie

Figure 3.11 The Central Pacific Railroad operated its own sleeping cars, but its first sleeper, the 1869 car shown here, was patterned on the patented designs of Woodruff. Note the Creamer spring brake on the left side of the end platform railing. (*University of California, Bancroft Library*)

managed an even more startling agreement between the two firms.

After months of quiet parley, Pullman leased the entire property of the Central Transportation Company. The lease, retroactively effective to January 1, 1870, was signed on February 17 and conveyed the rolling stock, patents, and contracts for 4,400 route-miles on sixteen railroads to Pullman.[73] An annual rental of $264,000 was to be paid during the 99-year life of the contract. At the same time, Pullman leased the Southern Transportation Company for a yearly rental of $20,000. Its property consisted of seventeen cars and five contracts covering 1,500 miles of line. Pullman had overtaken his chief rival and was on the highroad to realizing his dream of a national sleeping-car monopoly—all because of Andrew Carnegie. Since the early years of the Woodruff firm, Carnegie had had little regard for "the cautious old men in Philadelphia" who ran its affairs. At the time of the lease Childs was sixty-eight and may have realized that his life was coming to an end; he died in September of that year. Knight was busy with other affairs. Theodore Woodruff had been out of the company since 1864. Only Jonah Woodruff protested, but he was unable to stop the Carnegie steamroller.

The Central Transportation management retreated to a small office and for the next several years did little more than record Pullman's rental payments. A dispute that arose in 1877 over Pullman's failure to pay the Southern Transportation rental marked the beginning of a long legal battle.[74] Central won the first suit and even threatened to reenter the sleeping-car business. A settlement was reached in 1878, when Pullman purchased the Southern Transportation Company outright for $90,000. A more serious battle developed in 1885 over the renewal of the Pennsylvania contract. The railroad demanded a more favorable agreement than the original one that Pullman had inherited from Central Transportation. Because this was one of the principal assets Pullman had taken over from the old company, he contended that the annual rental of $264,000 which he was paying to Central Transportation must be reduced to $66,000.[75] Pullman also maintained that most of the Central cars had been retired and many of its patents had expired. Central countered that Pullman's earnings were never less than $800,000 per year, or nearly three times the rental.[76] Pullman protested that his company's average income resulting from the lease was only $300,000. Central won a judgment, but it was thrown out by the U.S. Circuit Court, which ruled that the firm was *ultra vires*: acting beyond its corporate powers in transferring its operating contracts to another party. The bickering continued through a decade, benefitting only the lawyers.

At last the case reached the U.S. Supreme Court, where a lower court's judgment against Pullman for 4.2 million dollars was set aside in 1898.[77] Speculators who had paid around $40 a share in hopes of a fat settlement with Pullman faced the prospect of clearing little more than $29 a share. They now recognized that their failure to accept Pullman's 1885 offer to exchange Central shares at 4 to 1 was a golden opportunity gone by. The affair was finally settled in 1899, when a special liquidating dividend of $470,000 put to rest the tortuous life of the Central Transportation Company.[78]

WOODRUFF SLEEPING AND PARLOR COACH COMPANY

At the time the Pullman lease was negotiated, Jonah Woodruff was sixty years of age. Most men would have retired, but Woodruff thought that the lease was all wrong, that a sleeping-car monopoly was not a good thing, and that a Pullman monopoly was particularly objectionable. What is more, he thought that a better sleeping car could be built and set out immediately to do it. He had money, and he found a friendly workshop in one bay of the Harlan and Hollingsworth car works. His former patents, assigned to the Central Transportation Company, were now Pullman's property. But no matter; Woodruff had a new scheme in mind and obtained a patent in July, 1870 (No. 105,288), covering some novel ideas.

More striking was the general plan of what Woodruff came to call his rotunda car. It was the most original, if not the most practical, sleeping-car design since the commercial beginnings of sleepers. The upper berths were completely dismantled and stored away during the day. Inside, the car looked like an ordinary parlor car because it had none of the bulky upper-berth structure and paneling. All this gear, including the bedding, was stored in the recesses between and under the seats. Woodruff regarded the openness of his design as a decided improvement over the "burial casket aspect of the old sleepers."[79] He pointed out that it permitted better circulation of air and heat. The car's center of gravity was lowered, he said, and the possibility of being accidentally shut inside the upper berth in the event of a wreck was eliminated.

Woodruff also claimed a weight saving of 4 to 6 tons, which was considerable for the chronically overweight sleeping car. But the greatest novelty of the plan was the oval lounges at both ends of the car. These lounges contained chairs and the water closets. One was usually assigned to ladies; the other was reserved as a gentlemen's smoker. Since the lounges were over the trucks, no berths had to be placed in this undesirable location. The lounges also served as pleasant observation or sitting rooms. The unique domed ceiling carried out the rotunda treatment of the oval rooms, which added a nice architectural touch and, Woodruff said, assisted in the ventilation.[80]

Most of these advantages that Woodruff announced to the press were fabrications, partial truths, and wishful thinking. But the design did have one merit: it was strikingly different, and as such it attracted the attention and support Woodruff needed in order to form a new sleeping-car company. By May 1871 he completed two sample cars, *Lady Washington* and *Lady Franklin*, at the Harlan and Hollingsworth plant. Both went on tour to woo contracts and investors. A few months later a charter was obtained from the state of Pennsylvania incorporating the Woodruff Sleeping and Parlor Coach Company, with Jonah Woodruff as president, August Trump as secretary and treasurer, and James Irvin as vice president.[81] A paid-in capital of $150,000 was reported. Pullman instituted a patent suit not long afterwards, but was forced to withdraw when it became clear that Woodruff's design was unique and could not be suppressed by the 100 car patents in Pullman's control.[82]

Freed of this entanglement late in 1871, Woodruff put his first cars in service on the Indianapolis, Peru, and Chicago Railroad. By mid-1873 eight sleepers and four parlor cars were running on the Chicago, Danville, and Vincennes, the New York, and Oswego Midland, the Camden and Atlantic, and other lines. In 1872 Woodruff took over the sleeping-car contract of Charles Doubleday on the important Cleveland, Columbus, Cincinnati, and Indianapolis. The operation was small but thriving. Woodruff's reputation and old contacts were undoubtedly helpful. Railroad travel and mileage were expanding, and the number of small and medium-sized lines made it difficult for Pullman to corner all the available contracts. In these circumstances small operators like Woodruff could prosper. By the third year his company had twenty-four cars in service (Figures 3.12 to 3.15).

Unfortunately, Woodruff's health appears to have declined in 1874 or 1875. He tried to recover it by going into semiretirement on his New Jersey farm, but to no avail. After his death in March 1876, the affairs of the company were taken over by several Pittsburgh businessmen headed by Frank Rahm. The new management continued to seek new lines. It found the South receptive and was soon providing service between Richmond and Atlanta.[83] It achieved a major advance in 1878 by purchasing the Lucas Sleeping Car Company of Atlanta, which had been formed three years before by Christian E. Lucas with the backing of W. M. Wadley, president of the Central Railroad of Georgia.[84] It

Figure 3.13 The Flora *was a Woodruff rotunda car built around 1875 by Jackson and Sharp.*

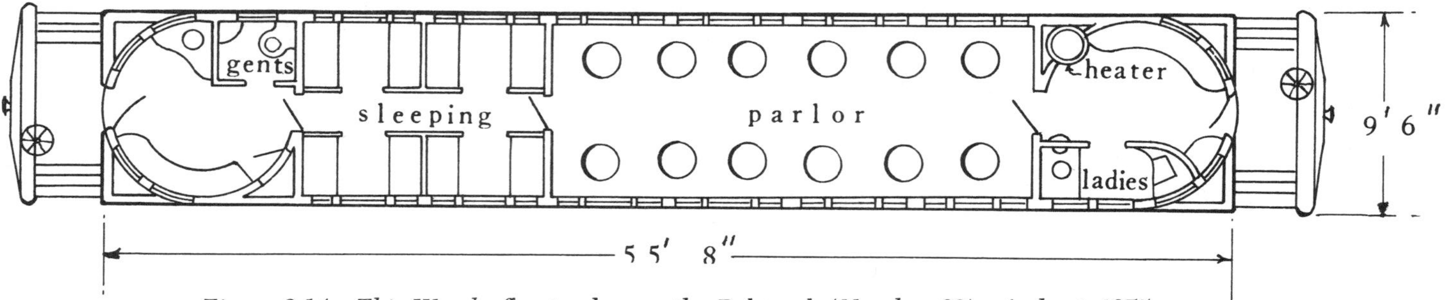

Figure 3.14 This Woodruff rotunda car, the Roberval *(Number 23) of about 1875, combined parlor and sleeping facilities. Not all Woodruff cars devoted so much floor space to parlor chairs.*

Figure 3.12 After Pullman's lease of the Central Transportation Company, Jonah Woodruff organized his own firm. In 1871 he introduced a unique style of car with round ends. The example shown here dates from about 1875 and appears to be named the Lottie.

is said that the line had been founded as a protest against Pullman, who began to admit Negroes as passengers on the formerly segregated sleepers. Lucas had only five cars, but he was a practical builder with several valuable sleeping-car patents, and he had gained the good will of other Southern lines (see Figure 3.16). Lucas was made head of Woodruff's Atlanta office. The Chesapeake and Ohio and the Virginia Midland quickly joined the Woodruff clients. In 1879 it was reported that Lucas patent cars were being completed for the Jacksonville, Pensacola, and Mobile. At the same time new Woodruff cars were being built for the Chicago and Eastern Illinois. Like most of the other early Woodruff cars, they were manufactured by Harlan and Hollingsworth and were painted a rich wine-maroon color with green, blue, and gold striping.[85]

By 1880 Woodruff had 64 cars in service—a puny fleet compared with Pullman's 800-car armada. And the disparity in strength was even greater because the Woodruff operation was so scattered. Most of its lines were isolated, often with only two or three cars running between relatively minor cities. An example was the contract with the St. Louis, Keokuk, and North Western, which involved two cars that began service in 1877. By 1882 the cars needed heavy repairs, but the revenue they brought in barely covered operating expenses.[86] Meanwhile the railroad had come under the Burlington's control, and the new owner considered turning the operation over to Pullman rather than repairing the cars in accordance with the contract. It also had the option to buy them outright for $12,000. The Burlington decided to keep the Woodruff contract alive because its president saw an advantage in having two sleeping-car companies on the string. No matter how poor or weak Woodruff might be, it could be used as a threat against Pullman. A few years later when the contract came up for renewal, Woodruff offered to run the cars at 2 cents a mile and give 25 percent of the net earnings to the railroad. Pullman charged 3 cents a mile, and the Burlington wanted to pay only half that.

Marginal as many of Woodruff's operations might be, the company controlled several good lines. As early as 1877 it was running a through car between New York and St. Louis via the New York Central's Bee Line. This plum was taken away in 1881 by Wagner, but Woodruff received a payment of $1 per passenger for giving up the contract. At that time all New York Central

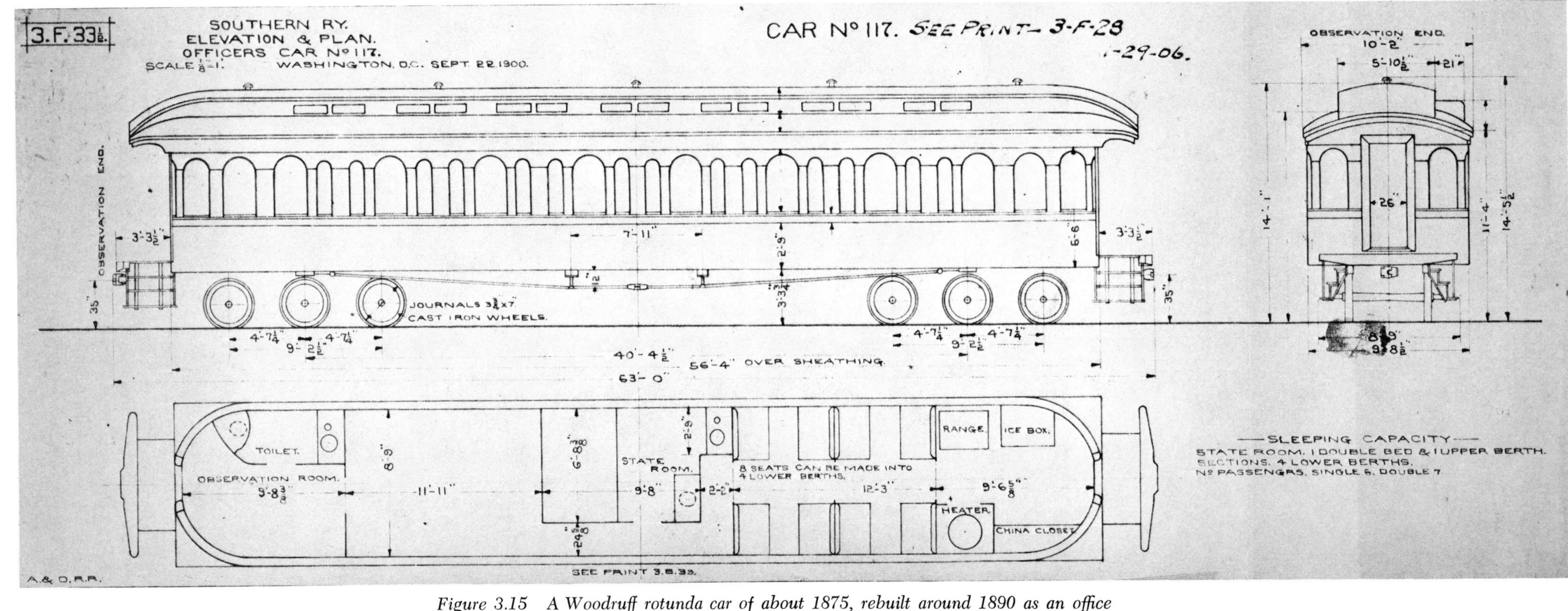

Figure 3.15 *A Woodruff rotunda car of about 1875, rebuilt around 1890 as an office car for the Southern Railway.* (Southern Railway)

Figure 3.16 Christian E. Lucas received six patents for sleeping cars between 1875 and 1880. Some cars built from his designs were used in the South. All became the property of Woodruff in 1878. (Hall of Records)

Figure 3.17 Interior photograph of a Woodruff sleeper of about 1875, showing three upper berths made up.

subsidiaries were under extreme pressure to deliver their sleeping-car contracts to the Vanderbilt-controlled Wagner company. Pullman was also busy trying to grab Woodruff's better contracts. Soon after the Wabash took over the Indianapolis, Peru, and Chicago line, Pullman disputed the I P & C contract (which had been Woodruff's first business deal). Pullman contended that its own contract with the Wabash superseded those made with the now-defunct Indiana line.[87] But for every contract that Woodruff lost, the company found a new one to take its place. In addition to the routes already mentioned, Woodruff eventually secured the Wisconsin Central, the Evansville and Terre Haute, the Grand Rapids and Indiana, and the New York and Manhattan Beach.

During these years the rotunda car design was the mainstay of Woodruff's rolling stock. Jonah Woodruff had improved the stowaway upper berths by mechanisms covered in two patents issued in February 1874 (Nos. 147,538 and 147,539). Neat metal

Figure 3.18 This interior of a Woodruff sleeper displays an upper berth framework patented in 1874. The rotunda parlor can be seen through the end door. Note the hot-water heating pipes visible under the seat end frames.

Figure 3.19 When the Monongahela *was delivered by Jackson and Sharp in 1885, the old-fashioned rotunda ends had long been abandoned. After 1889 the car was operated by Pullman under the name* Melissa. *(Lancaster County Historical Society, Pennsylvania)*

frames, apparently of cast brass, took the place of the 1870 design. Two excellent interior photographs made at the Jackson and Sharp plant sometime in the 1870s clearly show the apparatus (Figures 3.17 and 3.18). One of the rotunda rooms can be seen through the doorway. By the early 1880s, however, the rotunda cars were considered out of date. The design had lost its novelty, and the extra expense of the rounded car ends made it desirable to develop a new plan. In addition, the round ends were weak and not likely to withstand a telescoping accident. The complicated tinwork of the dome roof must also have been difficult to keep watertight. More conventional-appearing cars became the standard. Woodruff's collapsible berth was replaced by those of the folding type, and eventually only an expert could distinguish one of the late Woodruff sleepers from a Pullman (Figure 3.19).

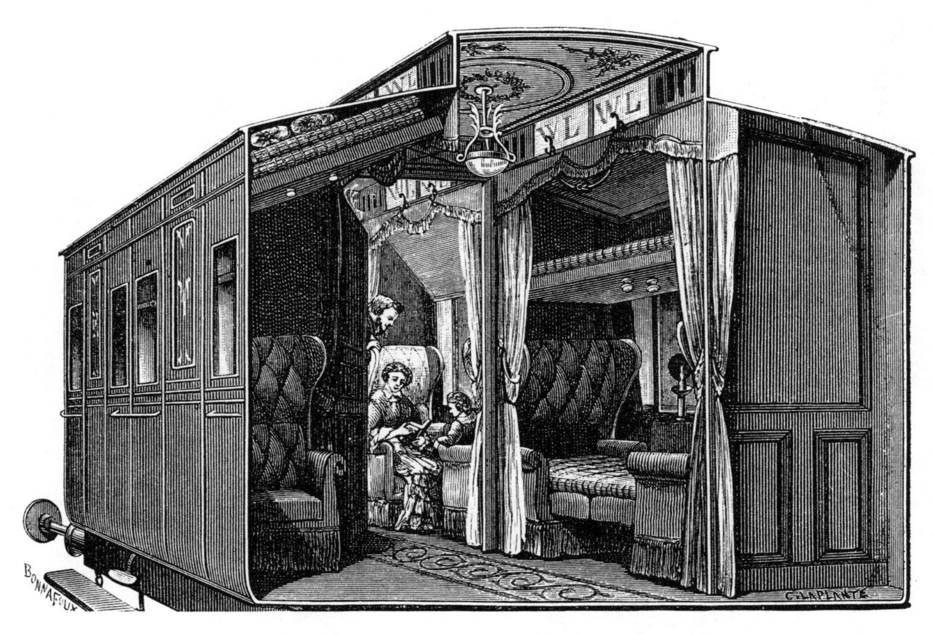

Figure 3.20 W. D. Mann, who helped found Pullman's European counterpart, Wagon Lits, attempted to introduce compartment-style sleepers in America. An 1878 Wagon Lits car is shown here.

With few exceptions, the Woodruff cars were built by Jackson and Sharp or Harlan and Hollingsworth. Both companies had accepted stock in at least partial payment for car orders, and over the years both had become large shareholders in the firm. It is likely that their power brought about the downfall of Woodruff's president, Frank Rahm. Early in 1882 he took over an idle car plant at Swissvale, Pennsylvania, with the idea that Woodruff, like Pullman, could build its own rolling stock.[88] The shop's extra capacity could be used to produce equipment for outside buyers. Whether this venture was a success or failure has not been discovered, but within the year Rahm resigned and Job H. Jackson, president of Jackson and Sharp, took over the management.[89] Afterwards all Woodruff cars were produced at Jackson's establishment; the Swissvale plant closed and was sold a few years later to Westinghouse.

A second management shakeup occurred in January 1884, when Woodruff's longtime vice president and general manager, J. H. Irvin, died. The offices were shifted back to Philadelphia. A year or two later Jackson stepped aside as president in favor of Daniel C. Corbin, a New York financier.

Before 1877, Woodruff paid a steady 12 percent dividend, had no mortgages, and in general was doing rather well in the bad time following the Cooke financial panic.[90] The firm remained profitable after this date, but dividends were suspended for several years and all cash was put into new equipment. In 1881 it was necessary to take out a $300,000 mortage. To win contracts and hold the existing clientele, more new cars were needed. Between 1884 and 1888 fifty new cars were purchased at a cost of over $594,000. All this buying was highly profitable for Jackson and Sharp but proved a severe strain on Woodruff's increasingly shaky finances. Capitalization was set at $5 million, but less than $1½ million was paid in; the bonded debt was up to $122,000 and other notes were held privately. Bonds sold at a 25 percent discount, and little stock had been bought since the 1870s. Some shares had been traded at 75 percent below their par value. The company was not broke, but its obligations so exceeded its earnings and modest cash reserves that one bad year would surely precipitate a failure. Short of selling out to Pullman, the only solution appeared to be a merger with the Mann sleeping-car company.

The Mann Boudoir Car Company

William D'Alton Mann (1839–1920), an American who built a sleeping car that became popular in Europe, had a great ambition to repeat his success in his native land. The life of Mann reads like a caricature of a hero in one of Theodore Dreiser's novels. He was flamboyant, greedy, pursued by scandal, and one of the fascinating characters of American business history. The details of Colonel Mann's nefarious life have been recorded elsewhere.[91] It is enough to say here that after a few years of war profiteering, oil well speculation, carpetbagging, and railroad promotion in Mobile, Alabama, he turned his thoughts to perfecting a sleeping car.

In 1871 Mann settled on a not very original plan for a compartment sleeper in which each compartment had its own side door, like a European coach. It was not possible to pass through the car on the inside—an arrangement totally unacceptable on American lines. Far superior compartment cars with side aisles extending the full length of the car had been running in this country since the 1850s, but they were apparently unknown to Mann and the Patent Office examiners. (A floor plan of a drawing-room car dating from 1859 was shown earlier in Figure 3.5. This chapter has also discussed stateroom sleepers on the Illinois Central in 1856 as well as on other railroads.) Mann received a patent on January 9, 1872 (No. 122,622).

At this stage of his career he left Mobile, apparently for good reasons, and went to England with the idea that his sleeping car might be well received there. However, it turned out that the relatively short runs characteristic of British railway travel were not favorable to sleeping-car service. Later in 1872 Mann went to Brussels and built a prototype car.[92] During the following year he helped to form what was to become the International Sleeping Car Company, better known as Wagons-Lits. He also established a line between Munich and Vienna, and Mann cars were said to

Mann Boudoir Car.

PRIVACY! COMFORT! SAFETY! LUXURY! CLEANLINESS!

Mann's Boudoir Car Company is now prepared to furnish Railways with the service of these MAGNIFICENT CARS for NIGHT or DAY use on highly favorable terms.

The VAST SUPERIORITY of the System and the BEAUTY of the CARS are recognized by every one who has seen or used them.

PRIVATE CARS.

Private cars of the type of the "Adelina Patti," "Etelka Gerster" and "Janisch," well known and undoubtedly the most magnificent private cars in the world, are To Let for long or short trips. They are provided with attendants, kitchen and dining-room, silver, table and bed linen complete.

Full particulars and descriptive pamphlets will be furnished on application to

MANN'S BOUDOIR CAR COMPANY,

Duncan Building, 11 PINE ST., NEW YORK.

☞ The Cars now built in this country are vastly more luxurious than those of same system in use all over the Continent of Europe.

Figure 3.21 The daytime arrangement of a Mann boudoir car is shown in this advertisement of 1885. (Biographical Directory of Railway Officials, *1885*)

be coming into wide use in Germany by July 1873.[93] Within two years, Mann had fifty-one cars placed on a dozen railways and was transporting 3,000 passengers weekly between Berlin, Vienna, Paris, and Bucharest (Figure 3.20).[94] The London, Chatham, and Dover was also testing a Mann car on its line. In 1873 Mann had gained a valuable partner in Georges Nagelmackers, a member of the Belgian banking family. The company greatly expanded its service in the next several years. It introduced improved cars with side aisles, and Mann managed to obtain a U.S. patent covering the design on January 8, 1878 (No. 198,991).

Just as Wagons-Lits really began to prosper, Mann decided to return to the United States. He sold out to Nagelmackers for 2 million dollars and in March 1883 obtained a New York State charter incorporating the Mann Boudoir Car Company. One of the original investors was the political cartoonist Thomas Nast, an odd associate for an entrepreneur as gaudy as Mann.

Within a few years Mann had made the two worst decisions of his business career. The greatest blunder was to sell his interest in Wagons-Lits, which was about to become Pullman's European counterpart and would have made Mann wealthy beyond even his avaricious dreams. His second miscalculation was to start an independent sleeping-car operation in the face of Pullman's growing monopoly. When Jonah Woodruff had attempted the same thing a decade earlier, Pullman's hold on the industry had been much weaker. Within five years Mann was to lose his fortune in the attempt to invade the North American sleeping-car business. But at the beginning he was fully convinced that the technical superiority of the boudoir car was enough to win a large share of the American market. He plunged half of his savings into new cars, a number of which, produced by the Gilbert Car Company, were placed in service between Boston and New York via the New Haven and the Boston and Albany railroads in October 1883.[95]

The first group of Mann's sleepers included a luxurious private car named after the opera singer Adelina Patti. Modestly describing it as the most magnificent private car in the world, Mann made it available to the celebrated star for a national tour. The press printed countless stories about the $40,000 car, its decor, and its glamorous passenger. Mann also produced two other private cars: the *Etelka Gerstner* and the *Janisch*. All three were rented to private parties and served to publicize Mann's new enterprise.

Some years earlier Mann had learned that associating with celebrities was good publicity. In 1878 he had received much favorable press notice through a tenuous relationship with the Russian royal family, for whom he oversaw the construction of two private cars (presumably incorporating his Boudoir arrangement) at the Ashbury Carriage and Iron Company of Manchester, England. America had its own royalty of sorts—government officials, financial moguls, and stage personalities—and Mann was alert to opportunities for ingratiating himself with this elite whenever possible.

For his regular series of sleeping cars, Mann adopted the Wagon-Lits plan for American use. The overall size was enlarged, and the cars were mounted on four- or six-wheel trucks. The bodies were stretched out to 64 feet, but the compartment or boudoir system was retained.[96] Each room had its convertible seat-bed and overhead berth (Figures 3.21 and 3.22). The oversize seat back was folded upward to form the upper berth; bedding was stored in the seat box. Note that there were single and double compartments which slept two and four passengers, respectively (Figure 3.23). The little fold-up seats in the aisle were for passengers who wished to see the view from the opposite side of the train. The interiors were quite different from the typical car of the period. In place of the usual varnished wood paneling, much of the wall surface was covered with embossed leather, which was laid over hair padding and provided a soft surface when passengers bumped against it. The arched roof and 42-inch, spoked Krupp wheels were other design novelties. Mann had devised his own heating and ventilating system. He claimed

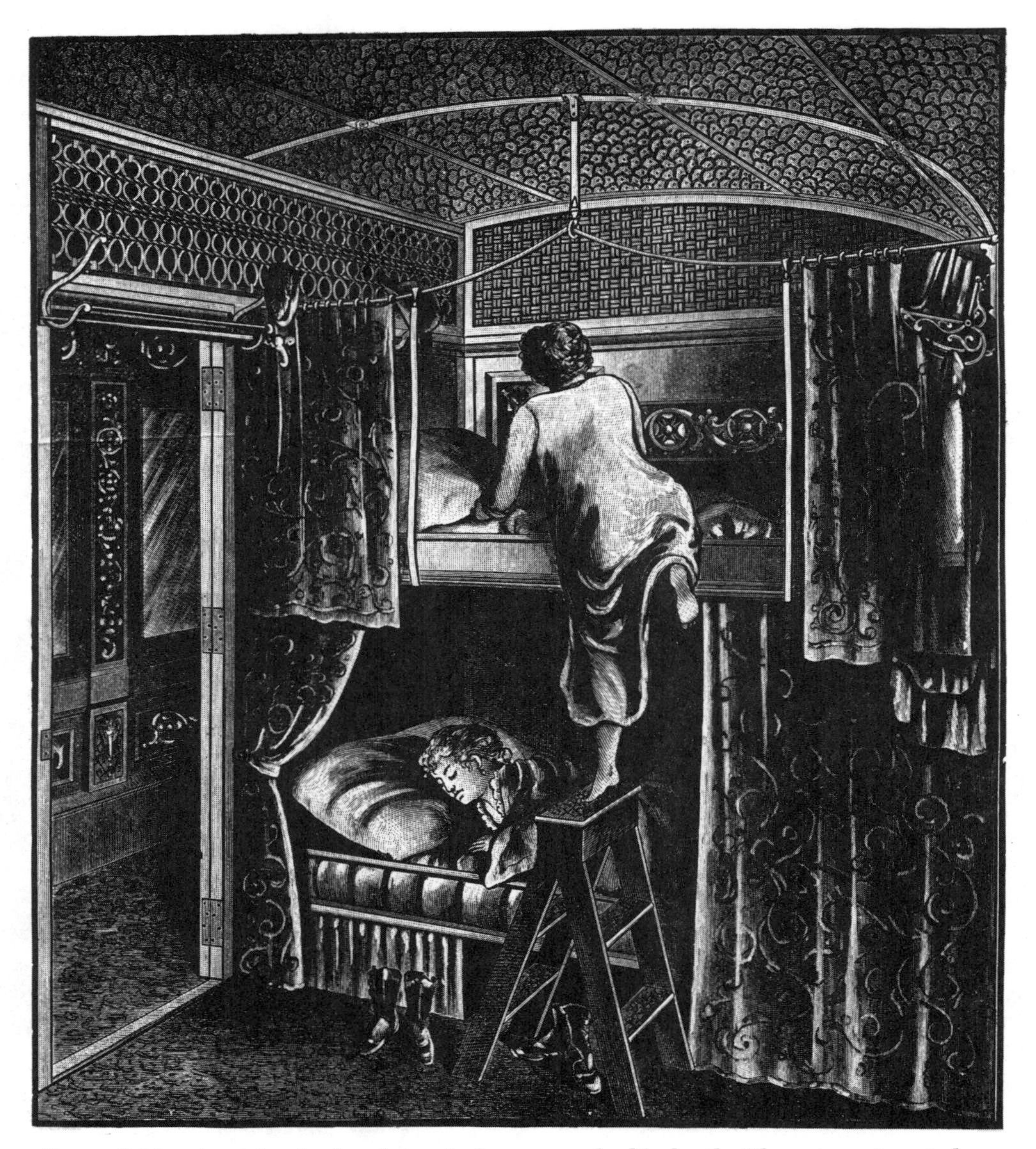

Figure 3.22 At night the boudoir sofa became a double berth. The compartment door stands open to the side-aisle corridor, which ran the length of the car. (*Mann Boudoir Car Company catalog, 1885c.*)

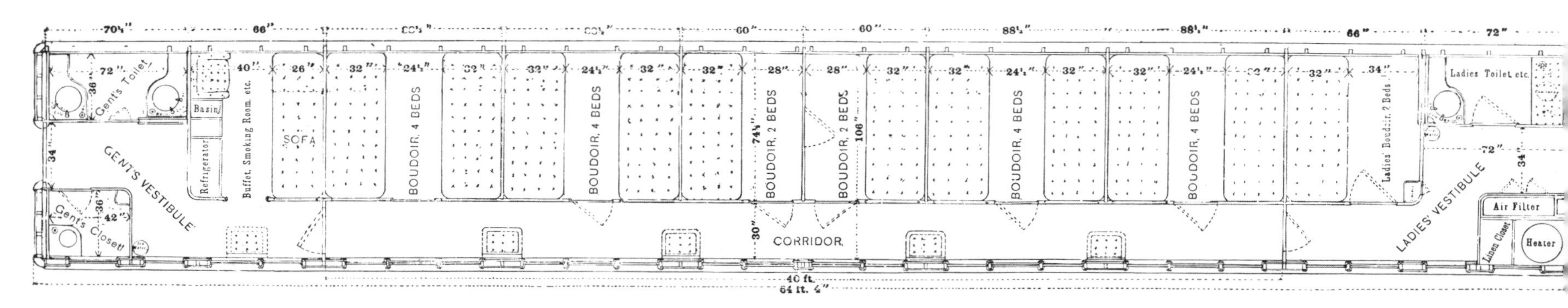

Figure 3.23 Floor plan of the Mann boudoir car La Traviata, *built in 1883 by the Gilbert Car Works.* (*Mann Boudoir Car Company catalog, 1885c.*)

that his fan-ice-water apparatus collected all the dust and cinders and lowered the temperature by as much as 20 degrees. The effectiveness of the device was unquestionably exaggerated, but it was an early move toward air conditioning.

In short, Mann was offering the traveling public a radically different style of sleeper. It answered the most frequent complaint about the Pullman by furnishing greater privacy and more room. Passengers no longer had to begin undressing in the aisle and struggle to complete the task inside the cramped berth behind the aisle curtains. Mann hoped that this basic advantage, together with the other original features of his cars, would create enough demand to displace the traditional open-section sleeper.

Several railroads were attracted by the boudoir cars soon after the first lot went into service. The Cincinnati Southern signed up

Figure 3.24 Exterior of La Traviata, *showing its arched roof. The raised portion of the roof was part of the air-washer ventilating system used by Mann. Note also the Miller couplers. (Mann Boudoir Car Company catalog, 1885c.)*

for six cars in 1884. Then came the Boston and Lowell, the Grand Trunk, the Wabash, and the B & O. The Boston and Albany asked Mann to operate its dining cars as well as some of its through sleepers. So many of his cars were running south out of Cincinnati that Mann opened a repair shop across the river in Ludlow, Kentucky. The Wabash substituted Mann cars for Wagner's on its Chicago–St. Louis trains.[97] Four Mann cars were exported to Australia in 1885 by the Gilbert Car Works; three of these were still in service in 1960, according to the railway historian G. M. Best.

By 1885 forty-three Mann cars were operating over seven railroads, which might be considered good progress in such a competitive field (Figure 3.24). The company was losing money, however, and management decided that acquiring additional cars and contracts was no guarantee of success. A more conservative policy was formulated: rather than striving for blind expansion, the company would seek contracts for more heavily traveled lines.[98]

To implement the new policy, a more experienced and conservative executive than Mann was sought to head the operation. In 1885 Thomas C. Purdy, former general manager of the Mexican National Railways, was named president. Mann was demoted to vice president and general manager. But efforts to consolidate the business's operations seemed impossible. Like Woodruff, Mann was left with the marginal lines that Pullman or Wagner did not control. Also like Woodruff, the company's activities were scattered over the Eastern, Southern, and Middle Western states. New contracts with the Georgia Pacific and other lines did little to aid the sagging fortunes of the Boudoir Company.

Other factors were working against Mann. The smaller capacity of the compartment car has always been the chief economic argument against its use. A Mann car slept 22, while an open-section car of the same size and weight accommodated 30. Because of the smaller capacity, Mann's fares were nearly double those for an open-section car. But Mann maintained that his cars remained popular and that on the New York–Boston trains, where compartment and section cars were both available, passengers always selected the boudoir.[99] His statement was confirmed by a report that travelers made the same choice on the Chicago–St. Paul sleepers, where 75 percent of the passengers preferred the Mann cars.[100] In the very long run Mann's contention was borne out by the success of the roomette, though it took Pullman sixty years to evolve it.

Other objections were made to the compartment car. A major one was that the partitions added weight and expense. Since the earliest days of the American railroad all practical railroad men had agreed on the disadvantages of the compartment arrangement. There was also an undercurrent of disapproval in the Victorian mind concerning the compartment sleeping car. The very name "boudoir" was suspect. The *National Car Builder* thought it too exotic and best "left to thrive in its native soil on the other side of the water, where the people know how to pronounce it properly."[101] There was the suspicion that unspeakable acts took place behind the locked doors of the boudoir compartments. Or some people viewed the compartments as dens harboring villains who were waiting to pounce upon the innocent. No such goings-on were possible in the section sleeper, where all passengers mingled in democratic openness under the watchful eye of the porter. Female passengers might complain about the mixing of the sexes with only curtains for privacy, but somehow this seemed less objectionable than the perfumed privacy of the boudoir. These fears were confirmed by the railroad trade press, which delighted in reporting the latest outrages that had occurred aboard European compartment cars. The *Police Gazette* ran a story called

"Love in a Palace Boudoir" which described how a wife discovered her husband and his sweetheart aboard an infamous boudoir sleeper. Mann's reputation thus suffered a mortal blow.[102] The boudoir was not only sinister and immoral; it was French!

By 1888 Mann was ready to admit defeat. He had lost his fortune trying to keep the company afloat, and now he was heavily in debt. In five years, only the last season had shown a profit, and that was too modest to encourage even the ebullient Mann. It had been a costly experiment; the time had come to end it.

The Union Palace Car Company

By the late 1880s it was clear to the creditors of both the Woodruff and Mann companies that desperate measures were necessary if either firm was to survive. As early as 1884 Woodruff was considering acquiring Mann's patents and was weighing the possibility of taking over his entire operation. But these discussions led nowhere, for Mann still expected to outrun the established sleeping-car firms with his exotic Continental dormitory cars. It took only four years to prove the folly of this dream. Meanwhile Woodruff had overextended itself through purchases of new equipment—a reasonable gamble, but one that did not pay off. It was in no position to buy Mann now that he was eager to sell. An outside agent was needed to salvage the faltering companies.

That the combination of two failures will produce a success seems to be an undying hope in the business world. Such marriages must occasionally work, but the main examples that come to mind seem to be negative, such as the Penn Central and the Erie Lackawanna. Was the Woodruff-Mann merger, which was proposed in 1888, a misguided but sincere effort to save them, or was it a cynical maneuver to extract a large price from Pullman to be rid of these nuisance competitors? On first glance the conspiracy theory might seem the most convincing, but actually the parties who brought about the merger had another motive entirely.

Job Jackson, president of Jackson and Sharp, led the effort to combine the two firms. He was encouraged by another large creditor, Harlan and Hollingsworth. Both of the Wilmington car builders were holding large quantities of Woodruff and Mann paper, and it was very much in their interest to keep Mann and Woodruff alive. If both companies failed, the rolling stock would most likely be sold at a discount, with both car builders taking the loss. More important, there would be no more sleeping-car orders at the Wilmington shops, for both Wagner and Pullman were producing their own stock. The two builders were anxious to maintain an independent sleeping-car operation, no matter how poor that firm's prospects might be.

And the prospects were indeed bleak. Both companies were financially distressed, but Mann was shockingly weak, with an income that rarely paid interest on the bonds. Both had large debts, poor earning records, and too many lightly traveled routes. To gain some contracts they had undercut Pullman to the point where those routes made no profit.

Jackson felt that only a new corporation could save the situation. Mann and Woodruff were both discredited, but fresh capital might be raised to support a new enterprise. Accordingly, in September 1888 the Union Palace Car Company was formed, with a capital of 3 million dollars.[103] Two-thirds of this amount would be needed to buy Woodruff and Mann; the remainder would be used to purchase new cars and expand the business. The scheme was poorly conceived, for even if the stock could be sold at full value, the outstanding liabilities of the old companies, plus the expenditures planned by Union, would result in a shortage of $823,000. From the beginning investors recognized this weakness, and the stock of the Union Palace Car Company sold at a 50 percent discount. But Jackson was happy enough, because he and the Harlan company had transferred their holdings in Union to a group in New York that put up 1.5 million dollars.[104] The sale was managed by Thomas Purdy, the former president of Mann and the new head of Union. The Wilmington car builders were also paid off with an order for thirty-four cars, most of which appear to have been patterned on Mann's design.

At the outset Union had Mann's 54 cars and 2,000 miles of route, together with Woodruff's 127 cars and 3,500 miles of route. Union also claimed to have an additional 4,000 miles through a new contract with the Richmond and West Point Terminal Company. There was talk of running cars on the New Jersey Central and the Reading.[105] During the same year an advertisement appeared in *Official Guide* claiming that the Union Palace Car Company's operations would cover 15,000 miles of railroad by January 1889. Even if Union reached this goal, it was far short of Pullman's 82,000-mile empire.

Shortly after the merger it was already clear that Union was not going to make it. After only three months of corporate life, Pullman purchased the Union Palace in January 1889 for just over 2½ million dollars.[106] *The Commercial and Financial Chronicle* contended that Pullman acquired the stock for 50 cents on the dollar. Yet even if Pullman paid the full price, the sleeping-car giant was satisfied to be freed of this rival at last. Curiously, the corporate identities of Woodruff, Mann, and Union were maintained for a number of years afterwards. Advertisements appeared in *Official Guide* that gave G. M. Pullman or Thomas Wickes (a Pullman official) as head of the firms. Their general offices were located in Pullman's Chicago office. Apparently Pullman saw some advantage in maintaining this scheme, or perhaps old contractual obligations made it necessary.

Monarch Parlor-Sleeping Car Company

A dozen or more insubstantial sleeping-car companies were organized after 1880. All were founded in a futile effort to upset Pullman's monopoly. Most were shadowy paper corporations. A few, like Harris (1889) and American (1905), actually built sample cars, but of these enterprises possibly only Monarch is worth more than passing notice.

The genesis of the Monarch Company appears to be a patent

Figure 3.25 Gustave Leve's 1876 patent for sleeping cars was followed by the Quebec, Montreal, Ottawa, and Occidental Railway for several cars built at its own shops in 1880–1881. Leve's patent was also used by the Monarch and Mann sleeping car companies. (Canadian Illustrated News, *February 19, 1881*)

(No. 181,857) issued September 5, 1876, to Gustave Leve of New York. The design's main objective was to provide a day-night car with parlor seats. The revolving chairs were collapsible, and the bedding was stored in narrow closets along the sides of the car. Curtains formed berth enclosures, as in the ordinary American sleeper. The mechanism and cabinetry to accomplish this transformation was complex and probably not worth the trouble or expense.

Leve's design apparently lay dormant for four or five years; then in 1881 the *Canadian Illustrated News* described some cars of his design on the Quebec, Montreal, Ottawa, and Occidental Railway. The cars were built in the railroad's shops, with certain improvements added by the road's mechanical superintendent. (An engraving accompanying the article is reproduced in Figure 3.25.) In 1883 the Gilbert Car Works constructed two cars using Leve's patent for export to Australia.[107] Early the next year four more cars were in service on the Occidental Railway of Canada and the Florida Transit.[108] Leve had meanwhile acquired a partner named Alden and was engaged in the "excursion business," but little information can be found on the nature of this enterprise. Reports were circulated in July 1885 that the Monarch Parlor-Sleeping Car Company had been formed in New York to exploit Leve's patented car.[109] A. F. Higgs, owner of several refrigerator-car lines, was to head the 5-million-dollar organization. The cars would be named after ancient rulers, and soon sleepers called *Zenobia*, *Cleopatra*, and *Rhodope* were running under Monarch's letter board. Thirteen cars were in service by the following spring.[110] They were 75 feet long, with 66-foot bodies, and offered 24 rotating parlor seats by day and 12 double-berth sections by night. The 35-ton cars were carried on eight or twelve wheels and cost $22,000 each.

In the summer of 1886 the Gilbert Car Works produced an extraordinary parlor observation car for Monarch named *Ymir*. It offered service from New York City to the White Mountains via the Connecticut River Railroad and featured an open side aisle for viewing the scenery. (In keeping with its unusual floor plan, the car was decorated in an original scheme that is described in Chapter 4.)

In addition to its Canadian, New England, and Florida contracts, Monarch operated cars for a time between Baltimore and Pittsburgh, Indianapolis and Evansville, and Fort Smith and New Orleans. Like Mann, it had entered the field too late; and like Mann and Woodruff, its operations were too scattered and marginal to pay. The company appears to have struggled on until 1893 or 1894 and was said to have built a total of thirty cars.[111] The details of its disposition have not been uncovered, but it does not appear to have been taken over by Pullman. It is possible that the Gilbert Car Company had a large interest in the concern, after the fashion of the Wilmington shops' involvements with Mann and Woodruff. Since Gilbert is known to have built all but one or two of Monarch's cars, Gilbert's unexpected and complete failure in 1893 may explain Monarch's disappearance.

The Wagner Palace Car Company

By the late eighties Pullman had vanquished all its major competitors but one. The semi-independent sleeping-car line trading under the names of its founder, Webster Wagner (1817–1882), continued to flourish. Its cars ran principally over the New York Central and the other Vanderbilt-controlled lines. It carried a large, profitable traffic between the major cities of the industrial Northeastern–Great Lakes region of the country. The massive resources and influence of the Vanderbilts seemed to ensure its continued operation. Much to Pullman's frustration, there appeared little hope that the Wagner Company would topple of its own weight or sell out as smaller competitors had done. Wagner was a formidable rival.

Webster Wagner was one of the earliest Woodruff concessionaires. In partnership with George B. Gates, T. N. Parmelee, and Morgan Gardner, he signed a contract with the New York Central Railroad on May 8, 1858, to operate sleeping cars between Albany and Buffalo.[112] Erastus Corning, president of the railroad, favored the scheme; possibly he saw the concession arrangement as a cheap way to introduce some measure of luxurious travel.[113] The hard times following the 1857 panic had left the road with little excess capital for frills. Four cars costing $3,200 each were said to be in service on this route by September 1858. Woodruff reportedly received a royalty payment of $500. Several months later *Leslie's* magazine published an interior view of one of the cars showing its three-tier berths made up for the evening.[114] The engraving, which has been frequently reproduced, is one of the earliest pictorial representations of an American sleeping car.

Just two years after the cars entered service, a British traveler recorded his experience aboard a sunflower-yellow car which carried the following legend in red lettering: "Albany and Buffalo Sleeping-Car." The charge for a single upper was 25 cents; the double lower cost 50 cents. The passenger wrote:[115]

> When we were between Little Falls and Herkeimer the officer of the sleeping-car entered and called out, "Now then, misters, if you please, get up from your seats and allow me to make up the beds." Two by two we arose and with quick hand the nimble Yankee turned over every other seat so as to reverse the back and make two seats facing each other. He shut the windows and pulled up the shutters, leaving for ventilation the slip of perforated zinc open at the top of each. Then he stripped up the cushions and unfastened from beneath each seat a light cane-bottomed frame which he fastened to the side of the car at suitable heights and covered each with cushions and blankets.
>
> I went out on the balcony to be out of the way and when I came back the whole place was transformed. There was no longer an aisle of double seats but the cabin of a small steamer with curtained berths and closed portholes. The bottom berths were wide and comfortable with room to roll and turn. There were two high berths to choose from; both wicker trays, ledged in, cushioned and rugged. The one was about a foot and a half higher than the other and I chose the top one as being nearer the zinc ventilator. Some of the passengers had turned in and were snoring. Others, like young cows balancing on spring boughs, were swinging their legs from the wicker trays and peeling off their stockings or struggling to get rid of their boots.
>
> I clambered to my perch and found it was like lying on one's back on a narrow plank. If I turned my back to the car wall the motion of the train bumped me off my bed altogether and if I turned my face to the wall I felt a horrible sensation of being likely to roll backward into the aisle, so I lay on my back and settled the question. It was like trying to sleep on the back of a runaway horse. At each place the train stopped there was the clashing of the bell and if I peered through the zinc ventilator into the outer darkness a flying scud of sparks from the engine did not serve to divest my mind of all chances of being burnt alive. Then there were blazes of pine torches as we neared a

station, more bell clamor and jumbling sounds of baggage, slamming doors and itinerant conductors.

At last I fell into a precarious and fragile sleep that lasted until daylight returned.

In later years statements attributed to Wagner about the origin of his company are in conflict with what appears to be the true story. Perhaps he was seriously misquoted or had an exceptionally poor memory, or he may have purposely altered the story to flatter his latter-day patron, Cornelius Vanderbilt. The Commodore, a man of no small ego, doubtless approved of the efforts of his faithful associate to keep the name of Vanderbilt central to the action. Wagner's accounts made no mention of Woodruff, Corning, or the original partners.

Wagner claimed that he was allowed to remodel a Hudson River Railroad coach early in 1858 to demonstrate his ideas for a sleeping car. The car was intended to compete with the palatial steamers on the Hudson River. With the help of William H. Vanderbilt, the Commodore's son, the gruff old capitalist was induced to inspect the car. Wagner expected to hear the worst, but instead Vanderbilt asked: "How many have you got of these things?" According to Wagner, "I told him I had only one." "Go ahead," said Vanderbilt, "and build more. It's a devilish good thing and you can't have too many of them." "My heart dropped back in its place and I knew that my fortune was made."[116]

In another version of their first meeting, the story goes that a sample car was attached to a train taking the Commodore through Albany.[117] A chance inspection led to an immediate partnership, with Vanderbilt providing the capital for a 75 percent interest. The Commodore made it clear that he would have no cars operating over his line which did not pay him a major share of the revenue, and that if Wagner did not like the terms he could take his cars elsewhere. The general drift of the story is probably true enough, but the details and chronology are badly distorted. Vanderbilt had little direct connection with railroads until the mid-1860s. He was a steamboat man at the time that Wagner was involved with his first cars, and would not have looked kindly on new ways to kill the overnight river trade. Wagner operated no cars south of Albany until June 1860. Nor is he known to have done more than simply use Woodruff's designs; at least, no sleeping-car patents were issued in Wagner's name.

The meeting with Vanderbilt most likely took place in 1868 rather than 1858, for it was not until late in 1868 that Vanderbilt gained control of the New York Central. It was probably only then that he gave Wagner notice to accept his terms or find a new client. Wagner wisely consented to Vanderbilt's proposition; it was essential to his survival against Pullman, and it made him a rich man. The Wagner Company was basically a New York Central subsidiary. Though it maintained a facade of autonomy, its officers were New York Central men, Vanderbilt agents, or Vanderbilt in-laws. Although stock was issued, it was a closed corporation. Unlike the Pullman Company, Wagner did not publish public annual reports, and considering the scope of Wagner's operation, the trade press gave its affairs little coverage.

Wagner's prospects for success before the alliance with Vanderbilt appeared modest. He had failed as a carriage maker and in 1843 had taken a job as a station agent for the New York Central.[118] In this work he discovered the need for sleeping cars and, as already discussed, became a small contractor for Woodruff in 1858. In 1860 he left his railroad post for the grain and produce trade. But the sleeping-car business continued to hold his interest; his original fleet of four cars grew to a dozen, and he decided to concentrate on expanding this line of business. In the spring of 1860 he contracted to run sleepers over the Hudson River Railroad.[119] By June Wagner and his partners were running cars from New York City to Buffalo; through cars were not possible, however, until the bridge at Albany was opened in 1866.[120]

Noticeable improvements were introduced at this time. Sheets and pillows were added to the bedding, and far grander cars were placed in service. The new Hudson River cars built by Eaton and Gilbert featured clerestory roofs, double windows, and three lamps with large reflectors.[121] (Wagner liked to claim the clerestory as his innovation and managed to obtain a patent for the raised roof, but as discussed in Chapter 1, the idea can be traced back to 1835.) A description of Wagner's improved cars was given in the May 31, 1860, *American Railway Review*:

> Two new and improved sleeping cars yesterday went into operation on the Hudson River Railroad, which promise to inaugurate a more healthy and convenient mode of traveling than has hitherto been enjoyed by our locomotive community. The improvement consists in the thorough ventilation of the car, complete exclusion of dust, and an arrangement by which every four seats are converted into a stateroom, which renders a family of four almost as complete as if they were under the lock and key of their own stateroom on a steamboat. In fact, the car throughout is simply a steamboat saloon on wheels. The roof is raised in the centre, and in the sides of the elevated part are inserted twenty-eight ventilators, which preserve the air pure and sweet, and keep up a steady and healthful circulation—a desideratum which has long been desired, but which, until now, has never been supplied. The system of seats and berths is not unlike that already in use. By day the mattresses and pillows are tucked away beneath the cushions, the cane-bottomed berths rest on their ledges at the top of the car, and four seats are all that meet the eye. By night these are ingeniously brought together and turned into a double bed, while still another tier of horizontal accommodations mysteriously finds its way into existence three feet above, and offers its downy temptation to the unprovided voyager; the whole being inclosed by a curtain, which throws its protecting folds around the sleeping inmates within. In addition to these staterooms, which are paid for at the rate of fifty cents a berth, there is, at the end of the car, a ladies' dressing-room, where are all the conveniences of the toilet, and every thing necessary as a prelude to a breakfast but the meal. The floor is constructed of two layers, with sawdust between to deaden the sound of the wheels; and for a similar purpose, and to exclude the dust, the windows are likewise duplicated. The cost of each car is about $3,500. They are said to be the finest cars ever built in this country.

A few years later, increasing extravagance was evident in eight new cars that Wagner purchased for $8,000 to $9,000 each.[122] The ceilings were covered with green and gold frescoes. More

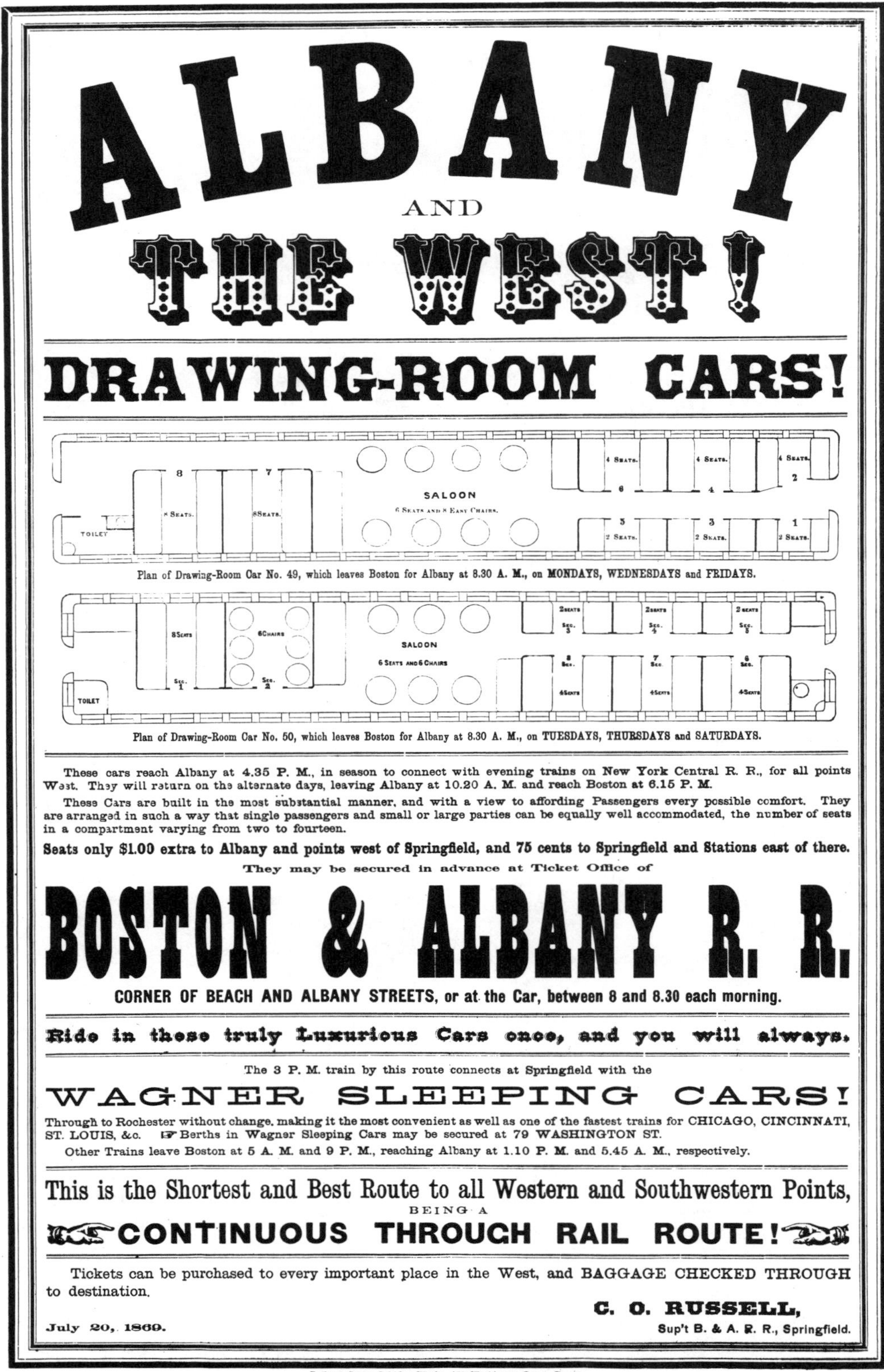

Figure 3.26 The Boston and Albany advertised Wagner sleeping cars in this broadside of July 20, 1869. The parlor car floor plans may also represent Wagner equipment. (Smithsonian Neg. 76336)

important than the costly decorations, however, were the improvements in passenger comfort. The old three-layer Woodruff berth arrangement gave way to the two-berth plan. Here again were spacious, well-appointed sleeping cars in advance of, or at least contemporaneous with, Pullman's *Pioneer*.

At the same time that Wagner was improving his night cars, he decided to introduce parlor cars for day travel. In September 1867 the first Wagner parlors entered service. A heavy first-class daytime traffic could be expected between New York City and the upstate cities. Wagner was eager to capture that traffic, and within a decade one-third of his fleet was made up of parlor cars (Figure 3.26).

By the end of the Civil War, Wagner was solidly in the sleeping-car business and ranked second only to the Central Transportation Company. Since the old partnership arrangement was no longer adequate to the firm's growing business, in January 1866 he formed the New York Central Sleeping Car Company.[123] Its capital was limited to $80,000. George Gates, one of

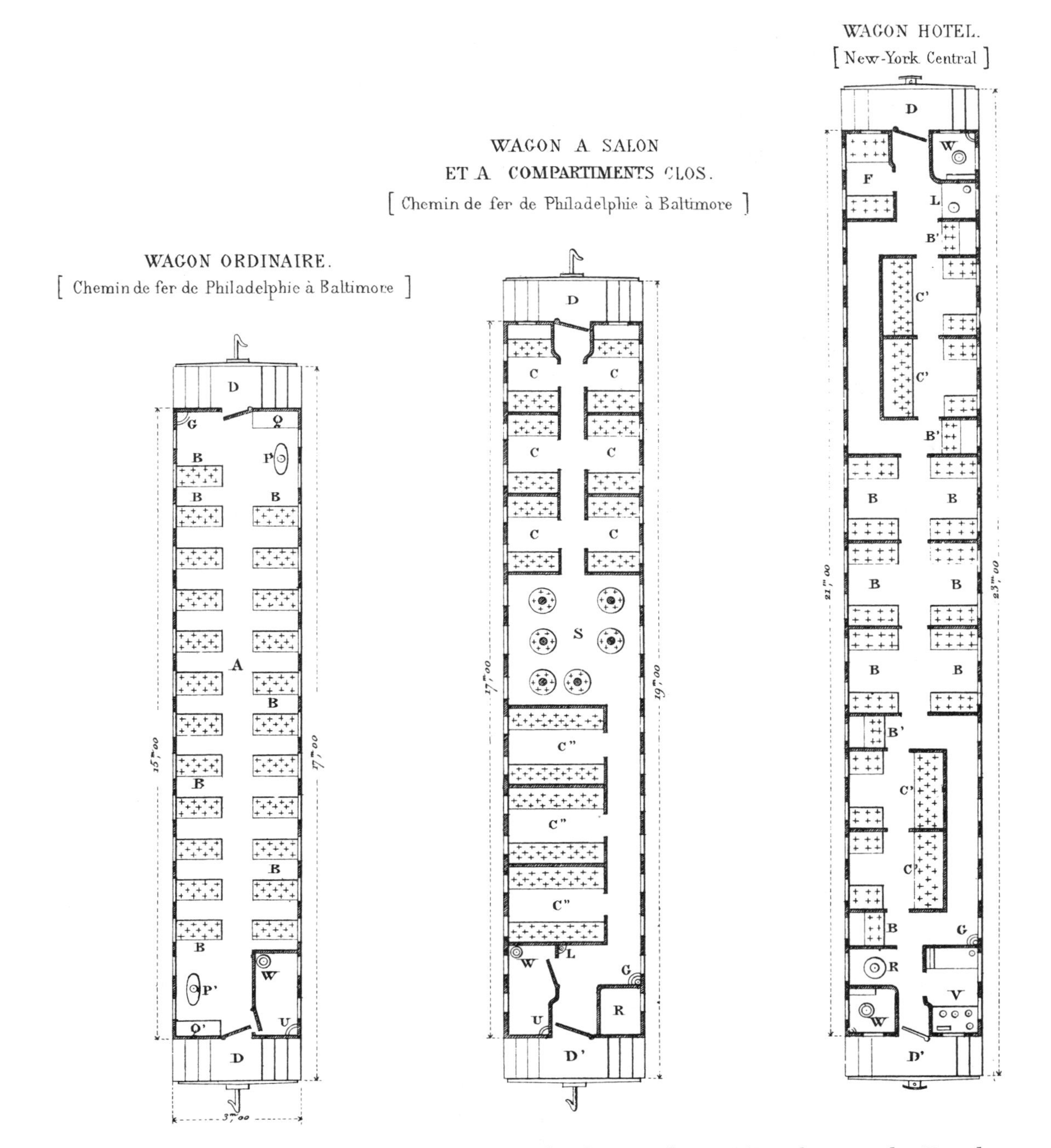

Figure 3.27 A floor plan for a Wagner hotel car of about 1870 is shown in this French engraving published in 1873. The legend reads: A. aisle, B. seats, B′. porter's seat, C. compartment 4 seats, C′. compartment 5 seats, C″. compartment 6 seats, D′. platform, F. smoking compartment, G. ice water fountain, L. wash stand, P. heating stove, Q. coal box, R. heating stove (special type), S. parlor seating area, U. urinal, V. kitchen, W. water closet. (Malezieux, Public Transport in the United States, *1873)*

Wagner's original partners, was made vice president of the new firm. Gates held the Lake Shore and Michigan Southern sleeping-car contract and was also running sleepers between Buffalo and Erie and Rochester and Cleveland. It is likely that Wagner and Gates strengthened their alliance at this time to facilitate through service between New York and the west. Wagner took over the company, entirely succeeding Gates, in 1869 or 1873.[124]

When Vanderbilt consolidated the New York Central and Hudson River railroads, he issued an ultimatum for a more favorable contract. Wagner had little choice but to comply. In the usual sleeping-car agreement, the railroad was paid a percentage of the profits; or the road might even be obliged to pay the contractor a mileage fee if the cars ran at a loss. But in the 1869 Vanderbilt contract, Wagner was obliged to pay a percentage of *gross* revenues to the railroad. Vanderbilt placed a levy of 20 percent on drawing-room car revenues and a twenty-five percent charge on sleepers.[125] Wagner was also responsible for the upkeep of car exteriors and was solely liable in the event of their loss through collision or fire. Typically the sleeper concessionaire was obliged to maintain only the interiors and to provide new cars as required. Moreover, the Commodore demanded a major number of the shares in Wagner's company.

In return Vanderbilt offered capital, contracts, and the prospects of a golden future as his rail empire spread west. He was already buying into the North Western and other lines west. To meet Vanderbilt's demands for more and better cars, Wagner had to sell six stock issues in 1869–1870. Capitalization increased more than sevenfold to $600,000.[126] The business was developing too rapidly for the makeshift methods and the simple corporate structure of the existing firm. It was necessary to form a joint stock

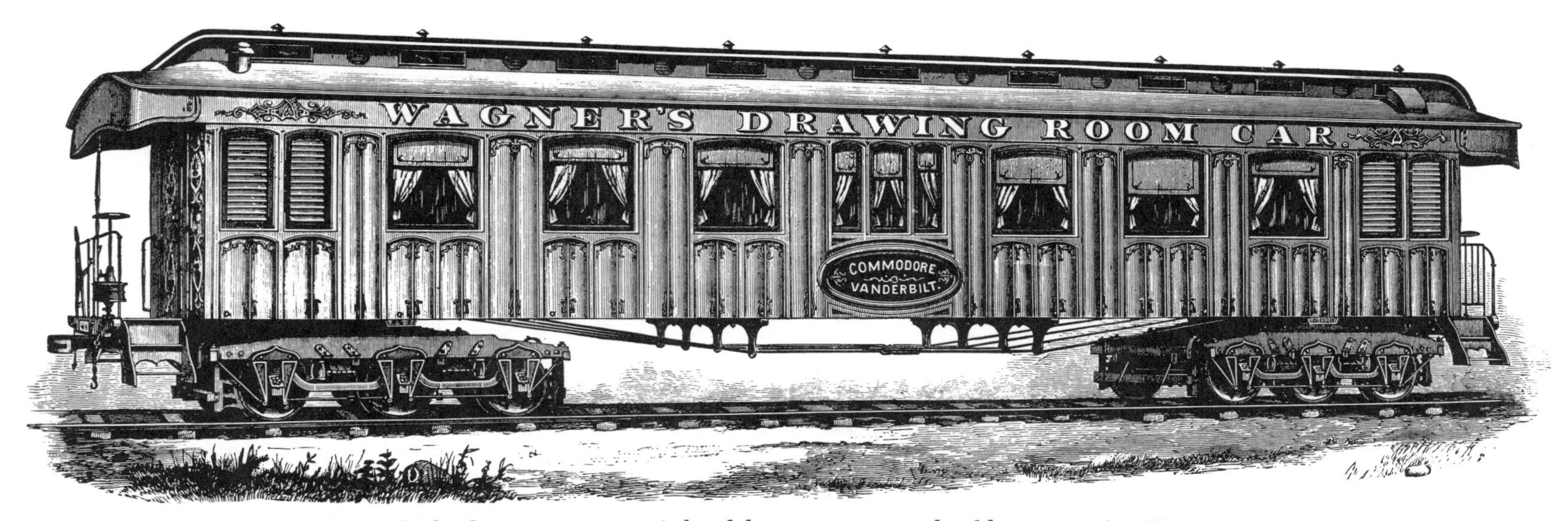

Figure 3.28 Some impression of the elaborate maroon and gold exterior of a Wagner palace car is shown in this engraving made around 1875. (J. R. Asher and G. H. Adams, Railroad Atlas, *1879)*

Figure 3.29 A Wagner sleeper that was operating on the Lake Shore and Michigan Southern Railway around 1875. The interior arrangement differs little from a Pullman car of the same period. (J. R. Asher and G. H. Adams, Railroad Atlas, *1879)*

company in November 1875.[127] Wagner transferred his contracts to the new firm, and Vanderbilt's old gross-revenue tariff was abandoned. The new body retained the old corporate title of New York Central Sleeping Car Company but remained a closed corporation, with all shares held by Wagner, Vanderbilt, and a few trusted associates.[128]

Just months before the reorganization, Wagner cars replaced Pullman's on the Michigan Central. After Vanderbilt gained control of the road in the fall of 1875, one of his first acts was to force a new sleeping-car contract on the directors.[129] Pullman was outraged. It was one of his oldest and best contracts, and he offered new terms: 4 cents a mile, plus the first-class fare. Wagner agreed to provide service for the fare alone. The Michigan Central management would have preferred to do business with Pullman, a

Figure 3.30 New York State Senator Webster Wagner and one of his cars, pictured in the New York Graphic *January 17, 1876.*

Chicago man, but Vanderbilt told them that they could accept Wagner or face an end to through New York–Chicago service. The Commodore embargoed Pullman cars at Buffalo, which would funnel the first-class traffic to the other east-west lines offering through cars. Vanderbilt prevailed; Pullman was banished from the Michigan Central except for the Grand Rapids and Montreal (via the Grand Trunk) trains. Wagner ordered sixteen new cars from the Wilmington car builders. The New York Central–Michigan Central combine offered the smoothest and levelest between New York and Chicago, and it was now all Wagner's.

Pullman, embittered by his loss, became a furious enemy of the Vanderbilts. Since he could find no immediate way to even the score with the powerful family, he turned to their associate as a promising target. Before this incident he and Wagner had been merely business rivals, but now Pullman meant to make trouble. His patents were always a handy weapon for harassing the opposition. After acquiring the assets of the Central Transportation Company in 1870, Pullman had agreed to give Wagner a new contract licensing use of the basic sleeping-car patents now in his control. Wagner used the Woodruff-Pullman designs for the hinged upper berth and the other standard features of the American convertible day-night cars of that day. He had no special designs, and aside from compartment cars, his sleepers were undistinguishable from Pullman's (Figures 3.27 to 3.31). The November 1870 agreement allowed Wagner use of patents for a yearly fee of $5,000 on the New York Central cars and $500 for

Figure 3.31 Another engraving from the 1876 New York Graphic *showed a Wagner interior. The illustration was captioned "The luxuries of Modern Travel."*

the Shore Line cars (Boston–New Haven).[130] It was also agreed that Pullman could run four cars a day over the New York Central and Wagner could run the same number over the Michigan Central. Thus each could have through service between the major cities involved.

That agreement had been broken by the Michigan Central affair, and Pullman meant to prosecute for patent infringements. The first case opened the following spring in New York Federal Circuit Court, but it was withdrawn for some reason. In 1881 Pullman mounted a second attack, issuing a complaint against the New York Central Sleeping Car Company in the U.S. Circuit Court of Chicago. Damages of 1 million dollars were claimed. By this time Pullman controlled over 100 patents and felt that he could humble, if not ruin, his adversaries. Wagner, however, assembled an astonishing number of old car builders whose recollections of ancient sleeping cars cast a deep shadow of doubt on the priority of many of Pullman's patents. Most of the basic ideas had been in use by 1840. The case continued for several years and was then quietly settled out of court for $50,000.[131] However, Pullman took the Wagner company to court again in 1887 over the Sessions vestibule patent and enjoyed another minor triumph.

His attacks on Wagner were not confined to the courtroom. He took pleasure in running down his rival in press interviews, saying that Wagner's success was attributable to his backing by certain "rich men."[132] Pullman neglected to mention the support he himself had received from several Chicago millionaires. He went on to say that Wagner was grossly overcapitalized at 5 million dollars—that his 250 cars would then be worth $30,000 each, when they might not fetch more than $5,000 in a forced sale.

Pullman's effort to punish the Vanderbilts by joining in the West Shore railway scheme was conspicuously less successful. The Hudson River line paralleled the New York Central's east shore main trackage, but it hardly brought ruin to the Vanderbilts. In fact, they acquired it at the 1885 foreclosure sale. Pullman's vendetta had cost him dearly, and when the final victory over Wagner came in 1899, Pullman had been dead for nearly two years.

At the time of the Michigan Central contract, Vanderbilt was also promoting Wagner's fortunes elsewhere. The *New York Graphic* for January 17, 1876, said that Wagner's 125 sleepers and 60 drawing rooms were worth 1.5 million dollars. More of his fleet would soon be operating in the West. After the Vanderbilts had acquired a large interest in the Chicago and North Western, the road was ordered to use Wagner cars (though it was necessary to permit the transit of some Pullman cars to honor an agreement with the Union Pacific). The C C C & I began using Wagner cars in 1875 on its Cincinnati-Sandusky trains for the same reason.[133] In New England similar pressures were brought to bear. The Fitchburg and the Troy and Boston railroads were forced into line.[134] Wagner had had a few New York–Boston cars on the New Haven's Shore Line since 1870 or before, but the schedule was slow because of the New London ferry. Although the Vanderbilts pressed the Boston and Albany to adopt Wagner sleepers for its faster route via Springfield, the company was still independent enough to resist, at least partially, for several years. The road built and operated its own sleepers into the 1880s.[135]

During the seventies and eighties, Wagner attempted to improve his equipment (Figures 3.32 and 3.33). Gilbert produced some drawing-room cars in 1873 with very large windows for the time: 27 by 45 inches.[136] Two years later a few cars were outfitted with panoramic 36- by 54-inch windows. A fancy New York decorating firm was employed to create the finest interiors. More

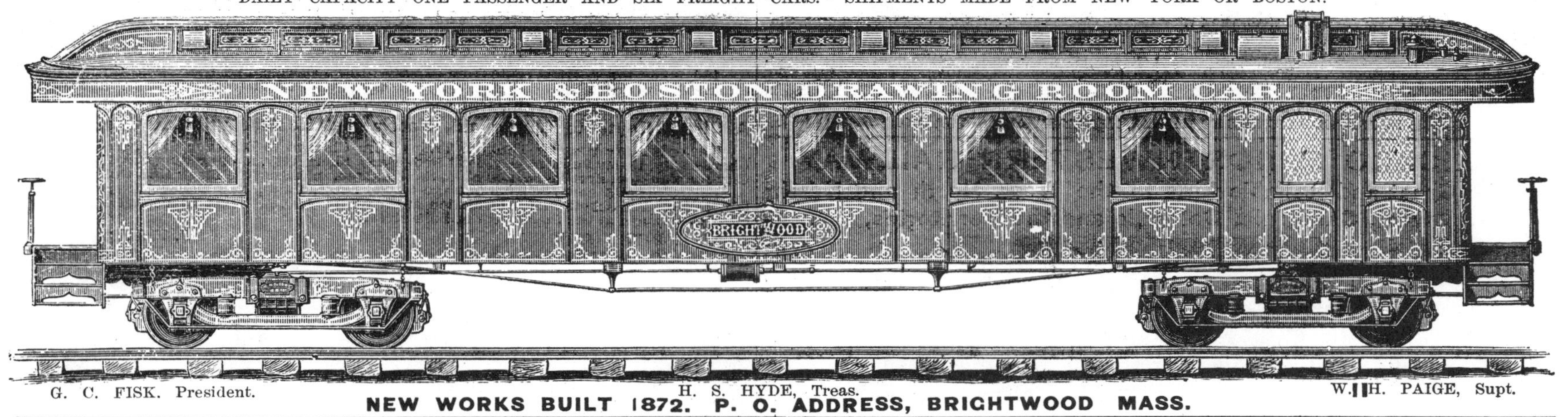

Figure 3.32 The Brightwood *was built for Wagner's New York–Boston service by the Wason Car Works around 1875 to 1879.* (Car Builders' Dictionary, *1879*)

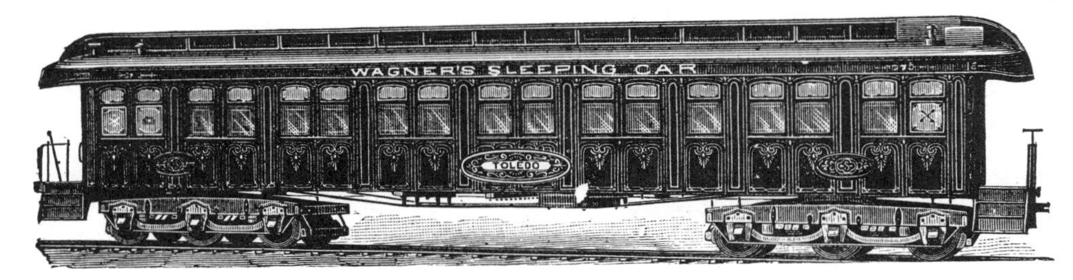

Figure 3.33 The Toledo *was built for Wagner in about 1879 by the Gilbert Car Works.*

important, Wagner made an effort to get around the Pullman-controlled patents. In the late 1870s he tried a patent sleeping car on the plan of James T. Leighton (who died in 1892), the superintendent and part owner of the New Haven Car Company.[137] In 1873 and 1874 Leighton and his brother obtained four patents. Their scheme was similar to other sleeper patents of the time, being an effort to eliminate the hinged upper berth and devise a collapsible bed that could be stored under the seats. Leighton claimed that he had made substantial weight savings, noting that some sleepers weighed as much as 34 tons and that his car was only 19 tons. Seven cars were built early in 1876 for the Atlantic, Mississippi, and Ohio Railroad.[138] Wagner saw enough merit in the design to order several Leighton cars during the following year.[139] How far he pursued the plan is uncertain, but he does not appear to have given it more than passing attention.

Wagner experimented again with the same general idea in 1881, when he purchased three or more cars from the Wason Car Works.[140] The plant's superintendent, William H. Paige (1842–1885), had patented a sleeping-car design in 1876. Like the Leightons', the scheme was for a convertible day-night car free of the hinged upper berth. There seems to have been no large-scale acceptance of either the Leighton or the Paige designs, and Wagner stayed with the conventional Pullman open-section plan.

By the early 1880s Wagner was spending $16,000 to $17,000 a car. Some were built with body lengths of 70 feet, 6 inches and total car weights exceeding 32 tons. The following description offers other details on some Wagner cars produced by Gilbert in 1880.[141]

It is so complete in every way as to leave little or no margin for further improvement. Its entire length, exclusive of platforms, is 69 ft. 5 in. At one end, and occupying a little more than one-third of the length of the car, are a state room, smoking room, ladies' saloon, locker and heater, and at the other end a lavatory and saloon for gentlemen, occupying a much smaller space. In the intermediate space are 12 sleeping sections, and there is also another in the state room. The upper berth frames work with weighted chains, and also have spring beds for the mattresses. The entire finish of the interior is in solid mahogany, relieved with inlay work upon the section panels. The ceilings are decorated oak, and the seats are upholstered in scarlet plush. There are two windows to each section with spring-roller curtains; and also projecting dust-guards on the outside to intercept cinders. These guards are placed vertically on each side of the windows, and work with a hinge and spring catch, so each can be set alternately at a right angle with the car side, according to the direction in which the train is moving. Hitchcock lamps are used for lighting, and the hot-water pipes from the Baker heater run under each of the seats. There are small portable tables provided for each section, in case they are wanted for lunches, card-playing, and the like. Howard's nickel-plate curtain rods, brackets, etc., add to the effect of the interior fitting. As an example of the many little conveniences that are being added to the comforts and luxuries of railway travel, we noticed in the gentlemen's lavatory an elegant nickel-plated receptacle for collars and cuffs, attached to the lower end of the mirror frame. There are the usual supply and exhaust ventilators, and also some of the latter placed in the roof and working with the wind. They may be opened, closed or regulated from the inside.

The cars have 6-wheel trucks, with Allen paper 33-inch wheels, Miller platforms and couplers and Westinghouse automatic brakes. Altogether these beautiful coaches embody about all the luxuries consistent with railway traveling. What their average weight is we do not know, but considering their great length, it must be somewhat ponderous. The average proportion of paying weight hauled, as compared with the non-paying weight, must be something almost frightful to contemplate as it affects railway economy. But unfortunately no skill or cunning of builders can abate the evil without curtailing the luxuries, and these the traveling public will not consent to forego.

Wagner's death in the January 1882 rear-end collision at Spuyten Duyvil prompted rumors that Pullman would take over his firm. But the Vanderbilts moved quickly to silence these speculations. Augustus C. Schell (1812–1884), an old friend of Vanderbilt's and a prominent member of the Tweed ring, was named president. He was a director of several of the New York Central lines, and though familiar with the legal side of railroading, he had little practical knowledge of the sleeping-car business. He was one of the few privileged outsiders with a large interest in the firm. Together with his son-in-law, James D. Taylor, Schell managed the fortunes of the New York Central Sleeping Car Company for just under two years.[142]

After Schell's death in March 1884, James Tillinghast (1822–1898) was appointed president. He was another longtime Vanderbilt associate, but he lacked the financial and political influence of his predecessor. Tillinghast was a practical railroad manager who had been general superintendent of Central since 1869. Ironically, many years earlier he had declined to help Woodruff with his first car. He had, however, been a vice president of the New York Central Sleeping Car Company since 1881 and was Schell's logical replacement. Yet he held the office for less than a year, because the Vanderbilts decided that a younger, more aggressive man was needed to direct the enterprise. William H. Vanderbilt had just the man: his new son-in-law, William Seward Webb, M.D. (1851–1926). Webb had practiced medicine for several years, but he gave it up for Wall Street. He never returned to the care of the sick, but he did have an ailing corporation to treat and he was eager to prove his ability to Vanderbilt.

The Wagner line had fallen into eclipse since the death of its founder. Complaints of "antediluvian sleepers" and "cheese box" smokers were reaching the trade press.[143] A letter appeared in the 1885 New York State Railroad Commission Report protesting the poor condition of Wagner cars. According to one passenger, cold air poured through loose windows, making the car's name, *Arctic*, all too appropriate. The interim management had not concentrated on keeping up the property, but Webb meant to bring about an immediate and full recovery. New cars were purchased and old ones rebuilt. Six months after Webb took charge, the improvement in service was noticeable.[144] Webb had long-range plans for the company that included expansion outside of the Vanderbilt empire and direct competition with Pullman. But first it was necessary to improve the physical plant and consolidate the contracts that could easily be acquired. Webb secured full rights to the Bee Line's sleeping-car operation in 1885. The following spring a contract was signed with the Central Vermont.[145] There was even talk of a Woodruff takeover. Of course there were inevitable setbacks: the Wabash, for example, banished Wagner in favor of Mann, but Webb had the pleasure of driving out the last Pullmans from the Big Four (the C C C & I) a few years later.[146]

Webb felt that the firm's old name, the New York Central Sleeping Car Company, was too parochial. For railroads outside the Vanderbilt empire that Webb hoped to entice into leaving Pullman, a title featuring the Central might prove offensive. In December 1886 the directors adopted a new corporate name: the Wagner Palace Car Company.[147]

During the same year Webb decided to rationalize the entire process of repairing and procuring cars. For years Wagner had acquired his stock from a variety of builders. Most of the orders seemed to go to Gilbert. But Jones; Jackson and Sharp; Harlan and Hollingsworth; Osgood Bradley; and others were also represented. Repairs were contracted out to the builders or handled at the New York Central's Albany and Niagara Falls shops. In 1879 Wagner was said to have taken a large interest in the Gilbert Car Works, which may explain why so many new car orders went there.[148] In 1883 an effort was made to consolidate the repair facilities at Buffalo, but only cosmetic work was attempted at first; mechanical repairs were scattered around as before.

As a stopgap measure Webb leased the Jones Schenectady plant, but his long-term aim was a major upgrading of the Buffalo shops. He did not aspire to a plant like Pullman's, but to an industrial establishment that could construct and repair a large number of cars. Several large buildings were added to the 40-acre site in 1886.[149] Within a few years almost every square foot of the plot was covered with buildings. The idea was that 1,600 to 2,000 men would be employed in fabricating 80 and repairing 500 cars a year. One of the smartest car builders in the business was hired to head the plant: Thomas Bissell, superintendent of Barney and Smith and at one time head of Pullman's Detroit shops.

The new plant was not ready in time to meet Wagner's immediate needs. New cars were required on the West Shore and the Bee Line. Commercial car shops were apparently too crowded to take the work, because Webb turned to Pullman—an unlikely move if there had been any alternative. Between 1884 and 1886 Pullman's Chicago plant received eleven orders for a total of sixty-five cars.[150] During 1887 Bissell concentrated on overhauling the Wagner fleet; 500 cars were run through the shop, but only a dozen new ones were built.[151] During the following year at the Buffalo shops, a large new-car program was initiated in which 1 million dollars was spent on seventy cars (Figures 3.34 to 3.37).[152] About twenty-five of these were for three luxurious vestibule trains scheduled for the Michigan Central. Several private cars were produced for Webb, the Vanderbilts, and various other New York Central officials. Three new private cars for rental parties were built at about the same time. Once Wagner's needs were met, the shops would seek outside orders on a contract basis.

Bissell's work went on public display December 7, 1887, when the new Chicago express entered service.[153] The six-car consist included a buffet-smoker-library, two parlor cars, two sleepers, and a diner. Steam heating and vestibules were special features of what was described as "the most substantial and handsome lot of railway carriages ever constructed."[154] The interiors were designed by Louis Tiffany in an elegant but unobtrusive manner. In

Figure 3.34 The Indiana *featured narrow vestibules, Miller couplers, gaslights, and other improvements typical of the palace cars. It was built in 1888 by Wagner's Buffalo shops. (William Voss,* Railway Car Construction, *New York, 1892)*

Figure 3.35 An interior view of the Indiana, *1888. (William Voss,* Railway Car Construction, *1892)*

addition to the library, the first car had a barber shop and bathroom. Hot and cold water were drawn from a 400-gallon tank. Electric lights were installed in some of the cars a few months later—such a novelty that the *National Car Builder* suggested there might be a bowling alley and a billiard table next. The new train was to compete with the luxurious special put on the Pennsylvania a few months before by Pullman. Like the arms race of modern times, the competition for first-class travel fostered an early version of the industrial-military complex. In both cases the purchaser, fueled by fear, bought more and more elaborate hardware to counter the competition. The contest was an enormous benefit to the supplier as new trains rolled from the production line.

In another effort to combat Pullman, Webb decided to promote the compartment-style sleeping car. Pullman was opposed to the idea and, except for a few experiments, seemed determined to stay with the open-section car. Yet the public's interest had been aroused by Mann's boudoir design, and there was no question that the room car was far more comfortable than the berth car. Wagner had been running such cars for many years, and now Bissell was directed to produce a more modern version. In 1886 some combined berth-compartment cars were put into service.[155] Berths and center aisles were placed at the ends, with six staterooms and a side-corridor passageway in the center. Two ladies' and three men's washrooms were provided. Three years later, four more compartment cars began operation on the New York–Chicago run.[156] The 45-ton, twelve-wheel cars were named *Lorraine, Magenta, Barcelona,* and *Normandy*. The 70-foot body of each was divided into ten rooms and had a staggered side aisle (Figure 3.38). It slept 24 passengers. The small end compartments for the use of all passengers were set aside as a men's and a ladies' lounge, respectively. A small library and a

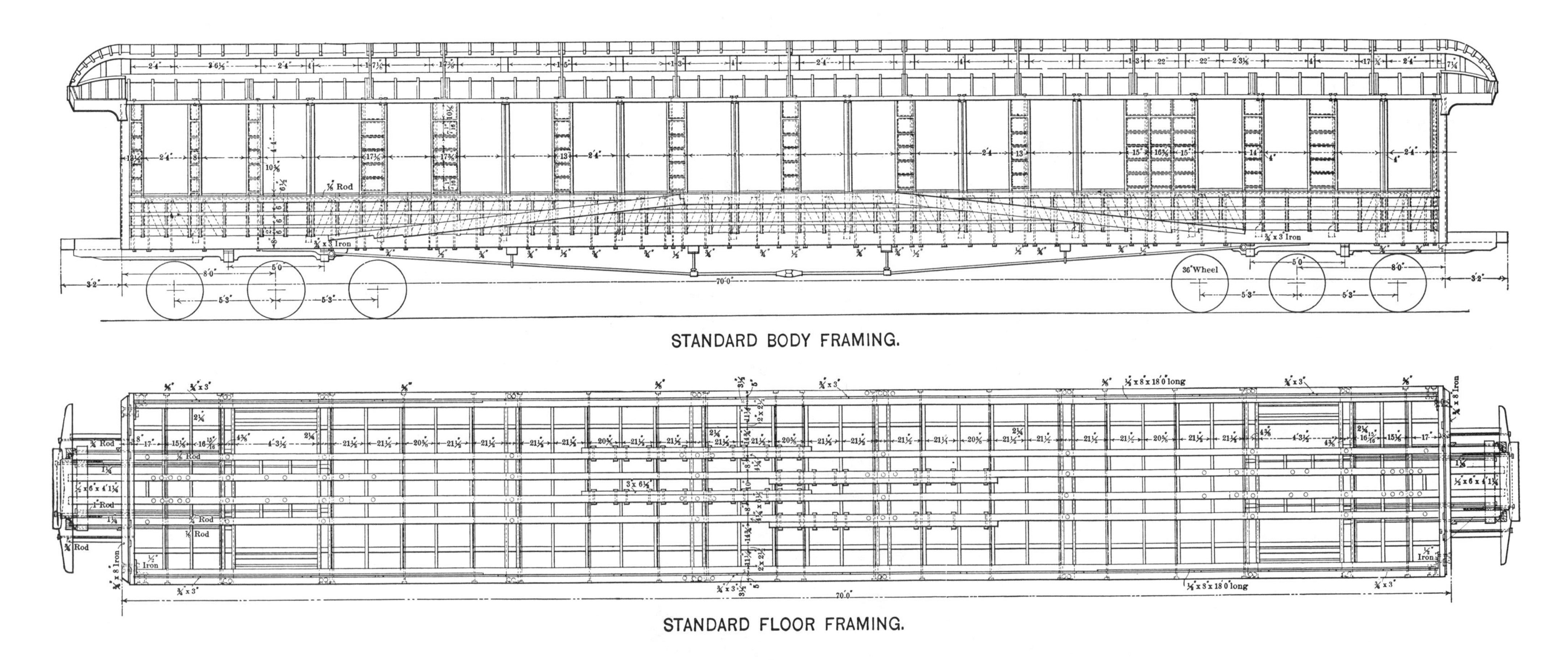

Figure 3.36 Floor and body framing of a 76-foot-long Wagner sleeping car. (*William Voss,* Railway Car Construction, *1892*)

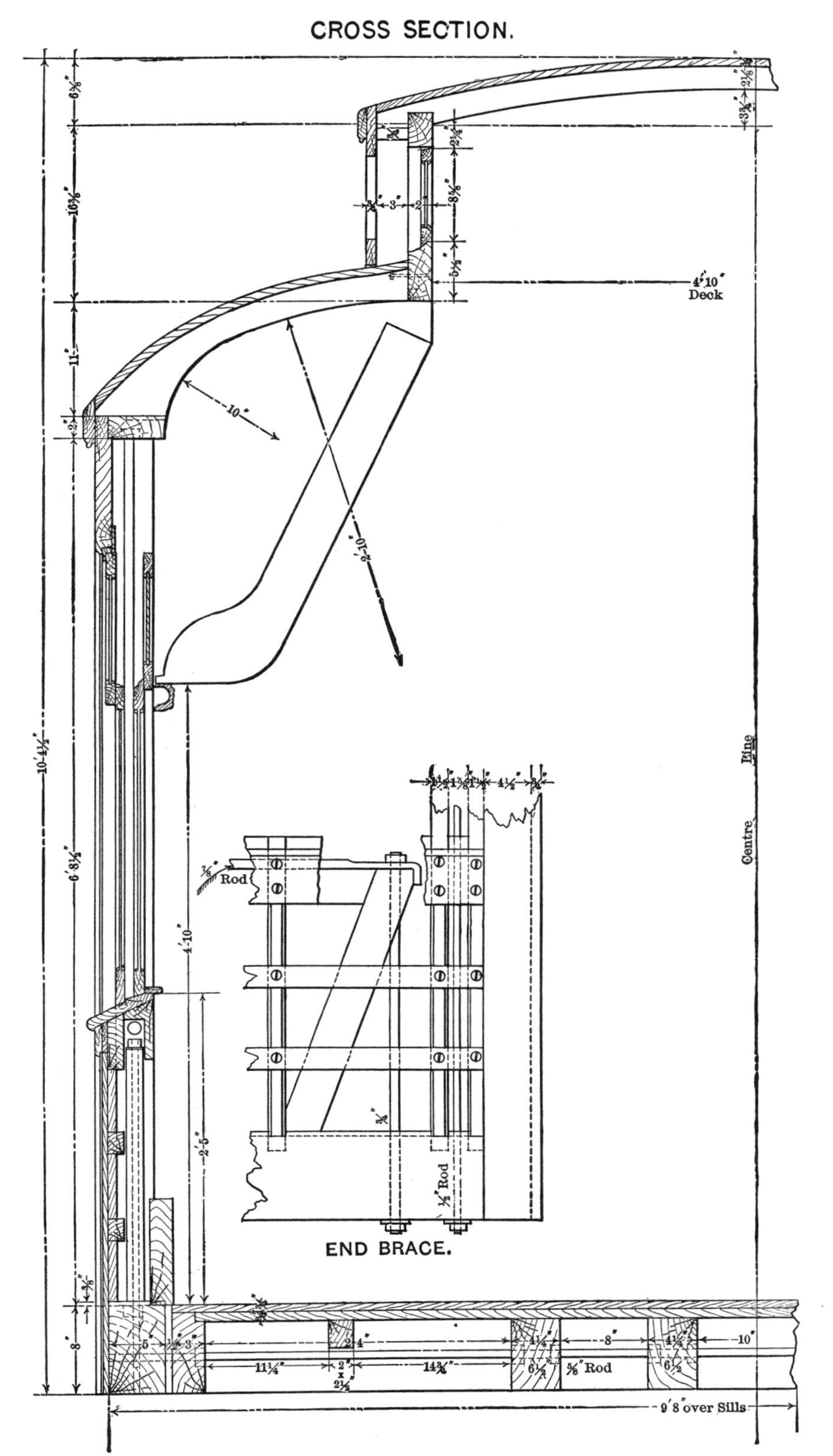

Figure 3.37 A cross-section drawing of a 76-foot-long Wagner sleeping car of about 1890. (William Voss, Railway Car Construction, *1892)*

buffet were somehow worked into the floor plan. The library consisted of several triangular bookshelves in the center cross aisle.

The most interesting Wagner-built room cars were produced in 1890 on the plan of Elbridge G. Allen, superintendent of the New Haven, for the New York–Boston Shore Line trains. The cars had eight open, double-berth sections and eight single-berth staterooms laid out in a curious zigzag pattern (Figure 3.39). They slept 24 passengers and seemed to have no particular advantage aside from their novel floor plan. In 1893 Jackson and Sharp built several more Allen cars for New Haven's Old Colony division.[157] Apparently they were not Wagner-operated and were eventually rebuilt as parlor cars. They cost $15,000 to $20,000 each, which was not excessive for an elaborately finished car, but the limited capacity and great weight prevented their wider use.

At first Webb's strategy in expanding the Wagner line seemed successful. By the early 1890s Wagner was operating cars on 20,000 miles in eighteen states and Canada.[158] It owned about 600 cars worth 10.5 million dollars and employed 3,000 people. By the end of the decade its 725 cars were running on thirty railroads, with routes going as far west as California and as far south as the Gulf of Mexico.[159] Pullman was operating some 2,400 cars over 120,000 route miles; Wagner was carrying one-third the number of passengers on its much smaller system because of the densely populated territory it served.

Webb drove hard to take even more business away from Pullman. In 1896 he won a contract with the Chicago, St. Paul, Minneapolis, and Omaha Railroad.[160] He was unsuccessful in negotiations with the Northern Pacific and the D L & W, however, and competition with Pullman was proving very expensive. The cost of new cars to compete with Pullman's best was staggering. Wagner put together an extravagant train for the 1893 Columbian Exposition and produced new cars to handle the anticipated crowds (Figure 3.40).[161] In 1897 the company spent even more money on costly stock to reequip the Lake Shore Limited.[162]

Since the beginning of his administration, Webb had spent lavishly. The Vanderbilt resources were adequate, and the family had not been timid about investing in a good thing. When Webb took over in 1885, Wagner's capital worth was fixed at 6 million dollars and it was netting a comfortable $137,000 on a $537,000 gross revenue.[163] Three years later, the capital stock was doubled.[164] In 1891 it was boosted to 20 million dollars. At the same time, profits fell to $628,000 on a $3,297,000 gross—a disappointing record compared with that of six years before.[165] The regular 8 percent dividend (1.6 million dollars) continued to be paid out of the treasury's surplus, though actual earnings were rarely above 5 percent. And so Webb's presidency was a mixture of triumph and failure. He had built up the company and competed with Pullman's best trains. But he had won few new customers outside of the Vanderbilt circle, and the company was showing a poor return on so large an investment.

Because the firm remained a closed corporation to the end, few records are available on its internal workings. Speculation based on what evidence exists leads to a few conjectures about why Wagner was sold to Pullman. In addition to Wagner's mediocre financial picture, it must have been obvious that Pullman had won the war, if not the battle. By 1890 it was clear that his monopoly was complete and that the other independents must eventually fall into line. That Wagner did not take over the Union Palace Car Company is one indication of Pullman's growing mastery. Possibly Webb did not even try for the Union Palace, but considering his earlier interest in the Woodruff merger, it seems logical that he would attempt to persuade the Vanderbilts to move before Pullman. It may be that they were already disenchanted with the sleeping-car business, or that they thought Union a poor investment—which indeed it was. Wagner really had nowhere to

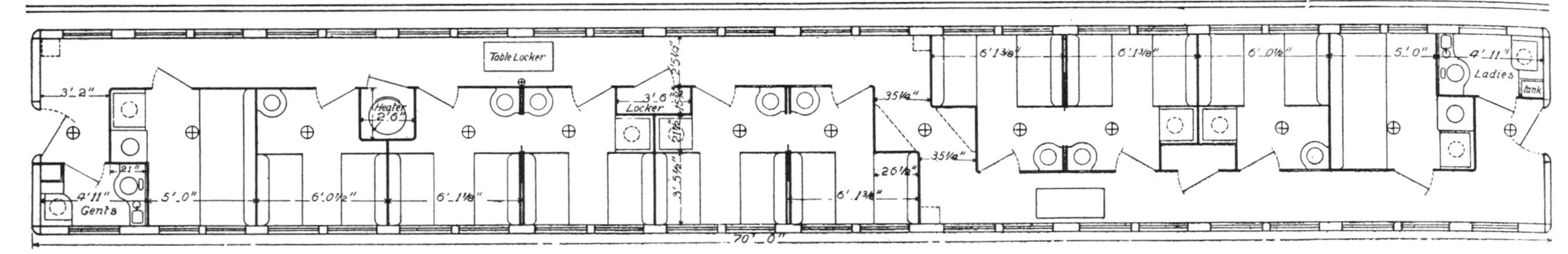

WAGNER STATEROOM SLEEPING CAR "LORRAINE."

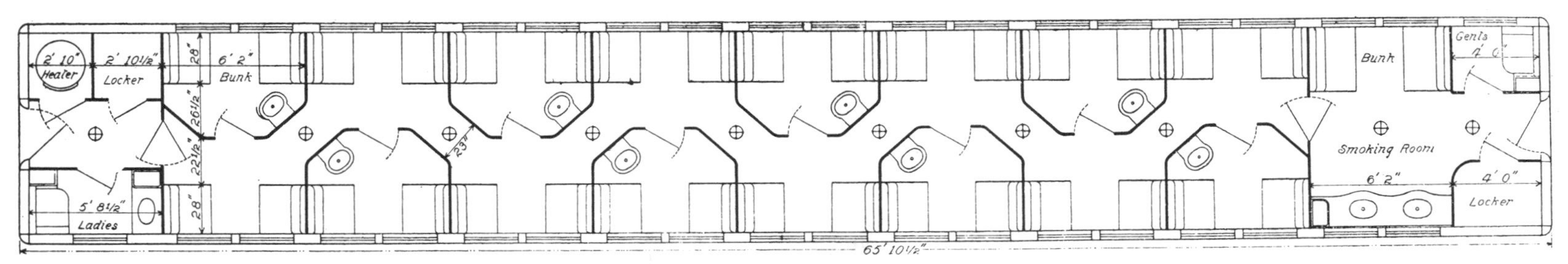

WAGNER SINGLE-BERTH SLEEPING CAR—BOSTON AND NEW YORK SHORE LINE.

Figure 3.38 Wagner was more receptive than Pullman to the idea of compartment-style sleeping cars. The Lorraine, *illustrated in the upper floor plan, was built in 1889 at a cost of $15,000. It weighed 45 tons and looked very much like the* Indiana, *shown in Figure 3.34.* (Railroad Gazette, *October 31, 1890*)

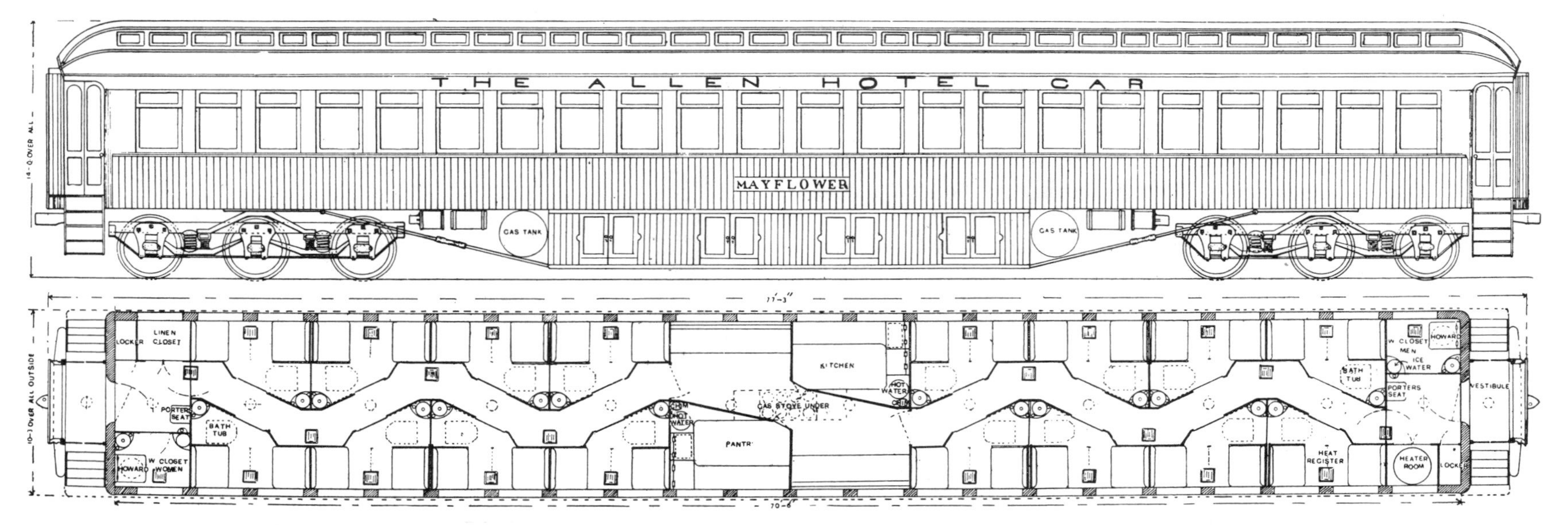

Figure 3.39 E. G. Allen (1850–1899) designed a bedroom-style car, the Mayflower, *that was in use on the Old Colony Railroad. Jackson and Sharp built the* Mayflower *in 1893. Wagner operated similars cars; see Figure 3.38.* (Railroad Car Journal, *October 1893*)

go; Pullman had the contract lines, and the few railroads that were running their own cars seemed content to remain independent. One sure sign of skepticism about Wagner's future was the sharp decline in new-car acquisition after 1893. Before then an average of fifty-five new cars was added annually, but between 1894 and 1899 the average fell to only seven cars a year.[166]

It is also possible that Webb himself was ready to quit. He had labored fifteen years to upgrade Wagner and could well have been weary of that uphill battle. He would soon be fifty, and possibly he felt that it was time to relax a bit and enjoy his fortune. Visitors to the Shelburne Museum, Webb's Vermont estate, can understand the good life that awaited him after retirement from the sleeping-car business.

When Wagner's earnings fell to a dismal $356,000 in fiscal 1898–1899, the shareholders saw another reason to sell.[167] The firm had no debts, but the earnings trend and the need for new cars meant that capital could be raised only through a mortgage. The company's prospects were just not that promising; it might be wise to sell before the present owners were saddled with a fixed debt of major proportions. The Vanderbilts were ready to unload.

The final and perhaps most convincing reason for the sale was the generous price that Pullman was ready to pay. Wagner's physical plant was valued at 8.1 million dollars. Its stock was said to be worth 20 million. Yet the purchase price was $36,667,000—a rather large premium, it would seem, to pay for good will.[168]

Figure 3.40 The Isabella *was part of the elaborate train that Wagner had specially built for exhibit at the Columbian Exposition of 1893. (Pullman Neg. 2269)*

According to an article published in *Pearson's Magazine* during the height of the muckraking era, the sale involved a conspiracy between the giants.[169] The article claimed that Pullman and Wagner had agreed to respect each other's territory and "went forward in harmony to pluck the public." Thus when the managers of Wagner saw that a sale was in the offing, they began to pump out new stock issues to inflate the price. Pullman was also accused of watering its stock to balance the exchange. The author of the piece failed to mention the bitter competition between the firms and the large investments in new equipment that justified their periodic increases in capitalization.

Another rumor circulated at the time was that the Vanderbilts were drawn into the sale by an offer of 5,000 shares of Boston and Maine stock from the Pullman estate.[170] That they coveted a larger share in the B & M was not questioned, but the story of the transfer was disproved by the news that the executors had sold the stock some time before the merger was consummated.

The negotiations were apparently prolonged, perhaps for several years. The first public notice that something was definitely underway was Wagner's move in July 1899 to register its entire stock holdings with the New York Stock Exchange's Unlisted Department. Yet before Wagner could go public, an agreement with Pullman was announced in October. The directors of both concerns agreed to a one-for-one exchange of shares. For 200,000 Pullman shares, Wagner would transfer its rolling stock, Buffalo car plant, contracts, and other assets. Because of its large cash surplus, Pullman had to raise its capital structure by only 20 million dollars to complete the sale. William K. Vanderbilt and Dr. William Seward Webb were given seats on the Pullman board. The plan was ratified by the shareholders in December 1899, and on the first day of the new century, the Wagner Palace Car Company was no more.

The Pullman Company

Until recent years the Pullman Company figured large in the everyday life of Americans. It was a singularly public business. Except for the very poor, most travelers had first-hand experience with the sleeping car; it was the normal conveyance for long-distance travel. And to the average citizen a sleeping car was simply a Pullman. The name was part of the language, and it was the only name that the public readily associated with railroad car history. Pullman was popularly accepted as the inventor of the sleeping car. Overseas, where few true Pullman sleepers ever operated, the word was known to the educated classes, who associated it with luxury and excellence in travel.

The term is rapidly losing its meaning, however; many young people have never ridden on a train, much less spent the night behind the swaying curtains of an open-section sleeper. To such an audience one must explain Pullman as if it were some ancient kingdom—a mysterious, vast business empire once proud and rich, but like Byzantium, gone and nearly forgotten. And like Byzantium, it has a complex and romantic history. As shown statistically in Table 3.3, its rise and decline was epic. The life of its founder was the classic American success story—the poor cabinetmaker who became an industrial titan. Pullman formed a corporate giant that systematically took over all sleeping-car operations in North America, established branches in Great Britain and Europe, and constructed the largest model community and railroad car plant on earth. The resulting multi-million-dollar corporation was remarkably stable. It was celebrated for its profits, comfortable cash surplus, and absence of fixed debt. Its conservative, well-managed progress seemed imperious to the vicissitudes of time.

But the Pullman empire, like so many of its historic predecessors, was brought down unexpectedly. Internal decay may have played a small part, but it was the general decline in railroad passenger travel after 1945 that hurt Pullman irrevocably, for first-class travelers turned to the airways and highways. The separation of Pullman's manufacturing and sleeping-car divisions, which occurred at the same time, weakened the already troubled giant. The end came with merciful speed; on December 31, 1968, in its one hundred and first year, the Pullman Company suspended operations.

George Mortimer Pullman (1831–1897) did not set out to monopolize the sleeping-car business, or even to be a railroad man. His first ambition was to rise above the poverty that his family endured. The Pullmans were New York State farmers and mechanics who had never enjoyed much in the way of comfort, luxury, or social distinction. George seemed destined to follow the same honest but colorless path. He had learned woodworking in his brother's cabinet shop, and at the age of twenty-two he became a contractor who moved buildings near the Erie Canal, which was being widened. He did well, but the work gave out after a few years. Around 1856 Pullman went to Chicago, where the general level of the city was being raised above Lake Michigan. Again Pullman's skill in house moving paid off, but again the work was temporary.

Pullman was soon casting about for a more stable career or some larger enterprise to occupy his considerable energy and ambition. Around 1858 he returned to Albion, New York, to see his old friends, among whom was Benjamin C. Field (1816–1876). Field, a man of influence and money, was active in state politics and railroad building.[171] He had helped Pullman in his first years as a contractor, and their association remained a friendly one. Field mentioned that he had received the rights for Woodruff's sleeping-car patent on certain Western railroads—presumably in return for his help in pushing favorable sleeping-car legislation through the state assembly. He needed a Western partner to help develop the franchise. Chicago was rapidly becoming a railroad center, Pullman knew the local men of affairs, he had some cash to invest, and he was looking to the future. Why not join Field and his brother in the scheme? The older man was persuasive. Many years later, Pullman also claimed that a ride in a sleeper out of Buffalo (presumably aboard one of Wagner's first Woodruff cars) convinced him of the need for better sleeping cars.[172]

In April 1858 Field signed contracts with the Illinois Central

Table 3.3 The Pullman Company, 1868–1968

YEAR	NO. CARS	ROUTE MILES	ASSETS	GROSS	PROFIT, LOSS	PASSENGERS
1868	50*		$ 1,470,800	$ 258,000	$ 169,700	
1870	300*		3,312,800	746,500	331,800	
1875	600*	30,000*	11,257,000	2,568,600	1,260,800	
1880	700*	60,000*	13,280,000	2,635,000	1,416,400	2,000,000*
1885	1,195	71,400	28,466,000	5,613,600	2,793,400	
1890	2,135	120,680	43,013,000	8,860,900	4,563,700	5,023,000
1895	2,556	126,660	62,792,000	8,547,600	4,290,300	4,788,000
1900	3,258	158,500	78,895,900	15,022,800	6,623,400	7,752,000
1905	4,138	184,100	96,151,900	26,922,000	13,038,000	14,969,000
1910	5,285	124,100	135,989,600	38,880,800	18,050,300	20,203,000
1915	7,287	215,800		41,512,800	20,580,100	24,252,000
1920	7,752	119,700	143,678,000	14,519,700	12,913,500	39,255,000
1925	8,776	130,300	329,148,900	81,490,300	16,779,000	35,526,000
1930	9,801	133,800	352,276,400	19,061,600	7,404,000	29,360,000
1935	8,007	115,400	276,275,300	50,063,400	*Loss* – 502,000	15,479,000
1940	6,910	109,500	304,469,800	60,095,500	2,411,800	14,765,000
1945	8,590	95,700	327,446,500	147,855,700	26,150,924	31,484,000
1950	6,226	102,700	83,131,400	103,756,600	1,325,200	15,606,000
1955	4,776	89,100	44,129,000	94,506,500	282,317	11,438,000
1960	2,650	67,400	28,000,000	55,876,900	286,046	4,484,000
1965	1,494	51,000	24,162,000	35,805,900	*Loss* – 21,513,000	2,507,000
1968	765	33,400	20,126,500	17,778,100	*Loss* – 22,000,800	1,073,000

* Estimates
Note: Figures before 1950 represent both sleeping-car and manufacturing divisions.
Source: Poor's Manual of Railroads and *I.C.C. Reports.*

and the Galena and Chicago Union to provide Woodruff-style sleepers.[173] Since Field signed for himself and his associates, it is not known whether Pullman was a partner at this time; it appears, however, that the deal somehow went wrong. Possibly Pullman was brought into Field's enterprise as the local agent after the failure of these first contracts, in order to prevent future fiascos. During the next year Field and Pullman convinced the Chicago and Alton to let them remodel two coaches for sleeping-car service.[174] The exact details of the sleeping furniture are not known, but if the arrangement of an 1897 replica can be believed, it differed from Woodruff's standard three-tier scheme.

The replica, presumably a fair representation of Pullman's first sleeping car, the Number 9, was built during his lifetime. The recollections of old employees were apparently not taken into account. The original conductor recalled eight sections, and one of the car builders involved in the remodeling says that oil lamps were used.[175] The replica, however, was made with twelve double-berth sections. It was 50 feet long over the platforms and 12 feet, 9 inches high, and it weighed 17.5 tons. The ceiling clearance was only 7 feet. Three candle center lamps lit the car. The upper berths were arranged with ropes and counterweights, much on the order of the Eli Wheeler and C. M. Mann 1858 patents. If this was the actual arrangement of the Number 9, it is possible that Pullman was attempting to avoid Woodruff's patents. According to the few records available, the original Number 9 was destroyed by fire.[176] The replica, made in 1897 for display at the Tennessee Centennial Exposition, was produced by rebuilding a Santa Fe tourist sleeper, Number 402—an 1870 product of Barney and Smith. A year or two later one end of the replica was removed, and the car was used to carry tour parties around the Pullman plant. In 1948 it was restored for the Railroad Fair in Chicago. After several years of outdoor storage it rotted badly and was destroyed in 1953 or 1954 at the Calumet shops. Countless interior and exterior photographs of the replica have been reproduced in school textbooks and popular histories, but it is almost never correctly identified as a replica.

The two cars, Numbers 9 and 19, were remodeled at the Chicago and Alton's Bloomington repair shops and ready for service in the fall of 1859.[177] The sleeping-car business was thus underway, but it was not promising enough to hold Pullman's full attention. Two sleeping cars connecting several frontier settlements did not add up to the exciting future that a young man with his ambition had envisioned. Traffic on most Western lines was relatively light, and the more heavily traveled routes were already taken. The Illinois Central chose to run its own cars. Woodruff, Wagner, and their associates were signing up the eastward lines. Field, however, was presumably happy with their progress and began buying more sleepers for the Alton. Barney and Smith produced a new car for him in 1860 or 1861.[178] Operations were extended to the Galena and Chicago Union Railroad. By July 1863 the Alton had four Field-Pullman cars in service and had contracted for four more.

Meanwhile, Pullman left Chicago for the Colorado gold fields. This move might be explained as the act of a restless youth seeking adventure, except that even as a young man Pullman was not impetuous. He was strong-willed, deliberate in his judgments, and brave, but not idly daring. It seems more likely that he felt disappointed in the limited prospects of his Chicago business. The Field brothers might be happy with the sleeping-car investment, but Pullman at twenty-nine was seeking a fortune. He was impatient with the dowdy bunk cars; they failed to pay large profits or excite his interest. Perhaps he decided that they had no future, or perhaps he recognized the need for more capital to build a better line of sleeping cars. It has also been suggested that during the early years of the Civil War, Federal government interference with operations on the Western lines damaged Pullman's business and caused him to invest his time elsewhere. But it appears more likely that Pullman had simply not yet decided to concentrate on the sleeping-car business. He had other investments that seemed more attractive at the moment.

As early as 1860 Pullman was buying land in the Pikes Peak area, and though he retained his interest in sleeping cars, he began to spend more time in Colorado. Stanley Buder claims that he stayed there from the spring of 1862 to the spring of 1863.[179] During that year he was involved in mining, storekeeping, and operating a wagon line. The idea that it was a pastoral interlude spent mooning about in a mountain cabin, dreaming of the perfect plan for sleeping cars, has been largely discredited.[180] Pullman was making money in a business-like way; provisions and transportation were necessities, and he was supplying them. He acquired an interest in a silver mine. He brought his partner in the still active Chicago-based house-moving firm of Pullman and Moore out to Colorado to help expand his growing business.[181] His visits to Chicago were frequent enough to keep him active in the sleeping-car business—and to permit eventual expansion into a shirt factory and into the Third National Bank of Chicago. In the spring of 1863, he returned to Chicago a rich man.

Possibly it was at this time that Pullman decided to concentrate on developing the sleeping-car business. Up to now the partnership with Field may have been just another side venture, like buying an interest in a hotel or a lumberyard. But things were changing; the prospects of the business were now quite different, or at least Pullman was seeing them differently. The possibility of building something grand was taking shape, and Pullman sensed an opportunity that few others saw or were bold enough to attempt. The East and South were blanketed with railways; the war had been decided; the Union would prevail. Work was about to start on a Pacific railroad, so that a transcontinental rail network would be a reality within a few years. Long-distance rail travel would flourish, and the sleeping-car trade with it. Whoever controlled that trade would become rich and famous. Perhaps Woodruff realized this too, but he was moving slowly. As Carnegie said, the Philadelphia-based company was run by "a few cautious old gentlemen." Pullman's vision may not have come until sometime after 1863, but his actions indicate that his course was set by that time.

Evidence of his renewed commitment was the splendid new sixteen-wheel palace car purchased by Field and Pullman in July

1863 for service on the Alton. Originally numbered simply the 40, it was renamed the *Springfield* in honor of the location of its builder, the Wason Car Company. A description of the car was given in the *Western Railroad Gazette* for July 18, 1863:

> The car is 58 feet long and of the usual width. The interior is of black walnut finish, highly polished. Instead of the ordinary cramped seats, the berths are so arranged that in the daytime the car is really a pleasant *salon*, the sofas extending along the sides, with camp chairs for those who desire. The car contains fourteen sections, capable of accommodating four persons to a section, or fifty-six passengers in all. The curtains are of [mohair] plush, with curtains of volin finished lasting. To convert this into a sleeping car at night, all that is necessary is to slide out the sofas, and let down the cushion back, and the couch is complete. When the beds are made up, about three feet of space is left in the aisle, and the berths are screened by costly damask curtains, extending from the roof to the floor. The car is provided with double doors and windows to deaden the sound, and it is said the passengers can converse with as much relief from noise as though they were in their own parlors. In the raised ceiling are eight of Westlake's patent ventilators in the roof. They render ventilation most complete, while dust and cinders are entirely excluded. At either end of the car, by a very ingenious arrangement, is a pleasant and convenient state room, with doors, windows, a table for writing, &c., for the accommodation of a party of three or four persons; the whole stateroom so arranged that in the day time it slides back into the side of the car, and does not occupy more than a foot of room. At night the stateroom is drawn out, the bed is let down and the weary passenger has a soft bed upon which to pass the night, secluded from "the rest of mankind." The car is also provided with every convenience usually found in first-class hotels. It is brilliantly lighted with six large Kerosene lamps, suspended from the ceiling. One very desirable feature, not found in ordinary sleeping cars, the mattresses are of the very best material, and the berths provided with clean, white counterpanes, pillow cases and sheets, that are daily changed.
>
> The body of the cars rests upon two double trucks, having in all sixteen wheels, hence the irregularities of the track are reduced by a compound lever lessening the motion, which, by passing through several sets of springs reduces it practically to nothing. The swing-beam is placed in the frame connecting each pair of short trucks. The cost of this railroad palace is $6,500.

Pullman and Field promised the Alton to provide even more luxurious cars than the *Springfield*. The two men were joining the race to supply the best in sleepers; no more provincial rebuilds or bare-bones bunk cars. Whatever the Central Transportation Company offered, they would attempt to match or beat. Pullman wanted to produce the ultimate palace car and decided to build it himself. His plan was not definite enough to give to a car builder; he would have to work it out in a shop. The Alton rented him a shed not far from his office, and he began work in the winter of 1864 under the direction of a practical car builder.

Pullman's car, completed in the spring of 1865, was called simply A; it is better known by its later name, the *Pioneer*. Over the years its importance became greatly exaggerated. It was credited with revolutionizing the car-building art and as exceeding anything before imagined "even in the wildest flights of fancy."[182] All previous sleepers were dismissed as primitive "rattlers." Pullman himself appears to have started the myth. In a statement before the U.S. Senate in 1874, he claimed that the early sleepers were "crude and unsatisfactory." In later years his remarks became more extravagant, and soon sleeping-car history was divided into two great epics: before and after the *Pioneer*.

The *Pioneer* was Pullman's natural favorite, not because it was his first car but because it was the first that he could call his own. Its predecessors were rebuilt products or the handiwork of a contractor, while the *Pioneer* was like a love child: he had seen each plank and cushion put in place. Its success was a great personal triumph, and in the passing years sentiment and frailties of memory colored its victories ever brighter. It was not just an important advance in the art; it was the greatest milestone since the beginnings of the railway carriage; it was the biggest, the most advanced, and the most expensive car ever constructed to that time. The exaggerations began to feed upon themselves, and what had been slight overstatements grew in the retelling until fantasy overwhelmed fact. Pullman officials, perhaps with the best intentions, began to parrot and embellish the story. Repetition has made it a fixture of technical history. The old truism that history belongs to the victor is borne out in the story of the sleeping car. It is likely that if Wagner or Woodruff had surmounted their competition, we would hear little of Pullman or the *Pioneer*.

In fact, the *Pioneer* does not seem spectacular when compared with other first-class sleepers of its time. Both Wagner and Woodruff were operating magnificent cars before the *Pioneer* had left its shed. Pullman himself had purchased a comparable car two years before; according to the description of the *Springfield* quoted earlier, it was in no way inferior to the *Pioneer*. All the *Pioneer's* supposedly remarkable features were already in use: sixteen-wheel trucks, clerestory roofs, under-the-floor heaters, and fanciful interior decoration.

In overall size the *Pioneer* was said to be a foot wider and 2½ feet higher than any other car ever built. The legend that station platforms had to be reconstructed and bridges raised for it may be true, as Joseph Husband and other writers on Pullman have stated, but it seems unlikely that so practical a man as Pullman would have intentionally built an oversized vehicle.[183] This overconstruction has been explained by Pullman's supposed need to gain space for a better convertible day-night car—especially space for overhead capacity to accommodate the folding upper berth. The berth design was described in a patent issued to Field and Pullman in April 1864 and in a second patent granted in September of the following year. Both involved a hinged upper berth, but neither were original inventions. Imlay's sleepers of the 1830s used it, and both Woodruff and Green managed to patent it some twenty years later. Yet the idea is commonly credited to Pullman. The arrangement does not require any more space than the other clerestory roof cars then in service. Any increase in inside dimensions was made more for show than utility. And the *Pioneer* was unquestionably showy; its interior, outfitted like a gaudy ocean steamer, helped drive the total cost up to $20,000. This was said to be four or five times that of an ordinary coach,

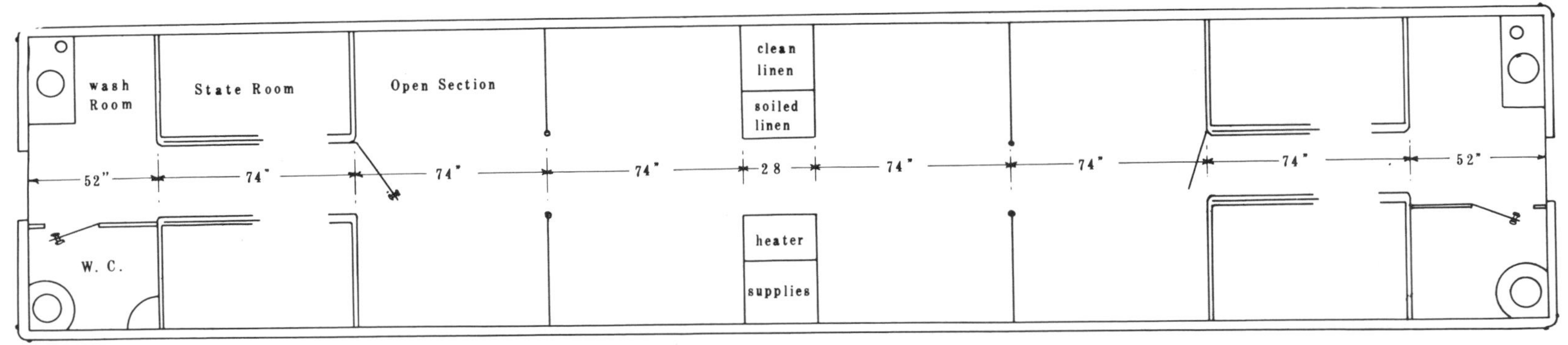

Figure 3.41 Reconstruction floor plan, based on early newspaper accounts, of Pullman's palace car the Pioneer. *Completed in 1865, the car was remodeled several times before its demolition in 1903. (Traced by John H. White, Jr.)*

Figure 3.42 The Pullman hotel car Viceroy *was completed at the Great Western Railway's Hamilton, Ontario, shops in 1865. It seated 56, weighed 37½ tons, and cost $20,000.* (Engineering, *June 12, 1868*)

and equal to the price of two locomotives. But the inflation of the Civil War had doubled and even tripled railroad equipment prices; at prewar figures the *Pioneer* would have cost little more than an ordinary first-class sleeper.

Pullman's invention of the hinged upper berth appears to be a fiction which developed after his lifetime, when the basis of his success was somehow ascribed to a fertile creativity rather than an extraordinary business sense. Pullman himself was frank to disclaim any talents as an inventor or even particular skill as a mechanic. He rightly considered himself a capitalist and was listed as such by contemporary biographical sketches. Skilled mechanics could be hired, and Pullman might generally oversee their work. But he left the details to his artisans; his time was better spent in directing the business.

The actual history of the *Pioneer* cannot be traced before 1893, but legend has it that the Chicago railroads agreed to rebuild platforms and raise bridges so that the car could be used in 1865 on Lincoln's funeral train. The inclusion of the *Pioneer* was a double coup for its owner. The first—the reconstruction undertaken by the Chicago lines—is obvious; the second was the prestige gained from its use by the Presidential party. No photographs of the *Pioneer* as built are known to exist, but it is possible that the car appeared in one of the Lincoln funeral train photographs. In particular, a picture taken at Michigan City shows an unusually large clerestory roof car under the memorial archway.

Little information can be offered about the *Pioneer* beyond its legend. By sifting through the surviving newspaper stories, and by discounting such obvious errors as that it slept 100 passengers, the following details can be assembled.[184] The 54-foot body had twelve open sections. The end sections were fitted with sofas that faced the center aisle. The original floor plan was not satisfactory, and the dressing rooms were enlarged after the car had been in service a short time (Figure 3.41). Over the years it went through several remodelings, including a change to six-wheel and then to four-wheel trucks. It was retired in 1889, and photographs made two years later show it as a well-used relic in the boneyard. Its labors were not at an end, however; Pullman was hard-pressed for equipment during the Spanish-American War, and the *Pioneer* carried troops to San Francisco. The car was held in storage until August 1903, when the scrappers put an end to this famous, if overrated, railway veteran.

While the details of the *Pioneer* legend can be questioned, the essence of the story is true enough. Pullman recognized a widespread public desire for luxurious railway travel and a willingness to pay for it. He saw the sleeping-car trade as a public-service enterprise where taste, fashion, and publicity were essential in-

Figure 3.43 The Pullman sleeper Omaha *was built in 1866 at the Galena, Illinois, shops of the Chicago and North Western Railway.* (Engineering, *February 7, 1868*)

gredients for success. A first-rate restaurant must have more than nourishing food; it must offer a sophisticated atmosphere, excellent service, elegant tableware, and exotic dishes. The affluent classes have always demanded these accessories in the institutions they patronize. Pullman intended to put the premium hotel on wheels. If it meant fanciful cars with stained glass, polished brass, and veneer panels picked out in gilt, he would oblige. If that was the taste of the times, the eye would never lack for some new detail, carving, or rich hanging to gaze upon. Pullman carried his belief in the merits of luxury into his personal life and spent lavishly on his family. Their great stone mansion on Chicago's Prairie Avenue was that of a merchant prince. In the mid-1870s a very special private car was produced for Pullman that was meant to impress, and it created this effect on all who saw it. Pullman believed in a grand scale of living, and so did the patrons of his sleeping cars.

Thus the *Springfield* and the *Pioneer* were only the beginning of a long line of overstuffed palace cars (Figures 3.42 to 3.45). The fleet expanded rapidly during the late sixties as Pullman acquired more contracts. Some idea of their appearance can be gained from a description in the *American Railway Times* of June 9, 1866:

> The wood of the interior is of black walnut so worked as to bring the beautiful grain of the wood out to the best advantage. Exquisite carving adorns every place where it can be tastefully placed. Some of the principal pieces show the designers and executors masters of their art. In the day time, all portions of the furniture of the car, used for sleeping purposes, are carefully concealed from view by richly painted screens, which are carried to and kept in their places by a neatly arranged system of pulleys. The upholstery is of rich Brussels carpet, with curtains of heavy damask. The metal work is all silver plated—the mirrors (one, surrounded with rich carvings, to each section) of heavy French glass. The windows are double, and so arranged as to assist the ventilation; each sash has a newly invented fastening, working simply and effectively. Each car is heated by Westlake's patent heaters and ventilators, with registers in each section. The washing apparatus is beautiful and convenient. Each car has one state-room, accommodating four persons, so arranged as to be, at will, completely separated by sliding doors, from the remainder of the car. The seats are commodious and luxurious; each car is also provided with four sofas.
>
> The exterior is as splendidly finished as the interior, having received the best effort of gilder, landscape painter and varnisher. They are throughout, most perfect specimens of car architecture. Each car is set upon two double eight-wheeled trucks, built upon principles peculiar to the Aurora shops, and so constructed that, when in motion, and even in turning curves, but the slightest jar is perceptible; in fact, when in rapid motion, one can write easily and legibly upon the tables in the car.
>
> The "Omaha" car, built at the Chicago and Northwestern shops in this city, differs somewhat from the Aurora built cars, the seats being all sofas, leaving during the day, the central aisles wider, and the general appearance of the car lighter, and possiby more home-like. Extra seat room is provided by the introduction of camp chairs. The "Omaha" is further provided with an organ, set in the centre of the car, so that the passengers may indulge in music and singing. The "Omaha" cost $28,000.

The need for solid comfort was attended to as well. Pullman wanted to offer amenities beyond that of a good bed. Travel was boring, and a meal or drink would enliven the trip and make it a more social affair. The idea was an old one, but Pullman felt that it had been neglected on the railroads. In 1867 he introduced the hotel car to the American public, and in 1868 he inaugurated the full dining car.

Pullman also proved a master at devising celebrations and excursions that gained newspaper space. The *Pioneer* was attached to the first Alton train between Chicago and St. Louis just weeks after its Lincoln funeral-train journey. The following spring a special train of new Pullman cars carried five hundred Chicago notables to what was described as a "country dinner" in a picnic grove near Aurora. In reality, the meal was catered by one of Chicago's best restaurants. Printed silk menus indicated the sim-

Figure 3.44 A drawing room and a portion of the main compartment of the Pullman hotel car City of New York *(1866) are shown in these two engravings. The drawing room was available for an extra fare beyond the regular sleeping-car tariff.* (Illustrated London News, *October 2, 1869*)

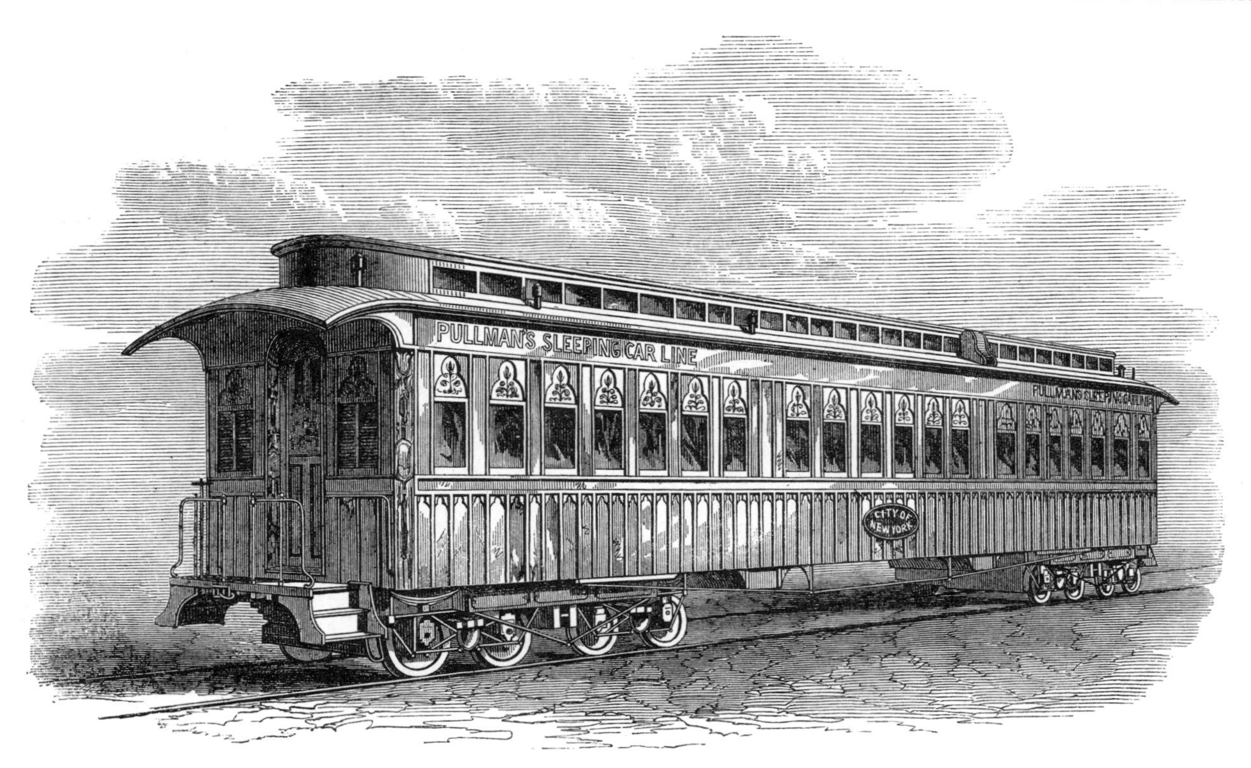

Figure 3.45 The 1866 City of New York *was a product of the Burlington's Aurora shops.* (Illustrated London News, *October 2, 1869*)

plicity of the affair, and the Chicago Opera Company presented the evening's rustic entertainment.[185]

Other excursions followed, usually to celebrate additional Pullman service or new Pullman cars. The most elaborate of all was the Boston Board of Trade's transcontinental trip in 1870. The mayor of Boston, Alvah Crocker, and other wealthy men were Pullman's guests. With his usual attention to detail, Pullman put a small printshop aboard one car so that a newspaper could be published. The trip produced reams of copy and a resolution of thanks from the passengers that included a prayer for "no delay in placing the elegant and homelike carriages upon the principal routes in the New England States." On the other hand, it did little to promote large-scale travel to the Pacific coast, nor was it effective in preserving Pullman's hold on the Central Pacific's sleeping-car trade. Certainly the excursion was useful in building a favorable corporate image and introducing Pullman's name to a wider audience. And now that he was a public figure, excursions were less necessary; Pullman's normal activities became newsworthy in themselves.

BUILDING THE PULLMAN EMPIRE

New cars, investment, and publicity brought new customers. Within a year after the *Pioneer's* inaugural trip, Pullman had signed up every major line operating out of Chicago. The Michigan Central joined his system in October 1865; the Burlington came in a month later. At the end of 1866 Field and Pullman were running over forty cars on seven lines.[186] A dozen more cars were under construction, according to the credit reporter R. G. Dunn. And once the Great Western's third rail was laid down, Pullman cars would be running along the southern boundary of Canada.

But if the firm was to expand beyond the Middle West, Pullman knew that it must attract more capital and influential friends. Incorporation was essential. Apparently Field did not agree; perhaps he was frightened by Pullman's drive to expand, or perhaps he was simply weary of the partnership. In all events, he retired sometime in 1866. In the fall of that year Pullman decided to incorporate. A charter for the Pullman Palace Car Company was granted on February 22, 1867, with capital set at an even 1 million dollars. By this time Pullman was an insider with the Chicago money barons, and John Crerar and Marshall Field were pleased to promote the first issue of stock. Their admiration for Pullman was genuine; he was respected as a clear-headed industrialist with a genius for organization and finance. Carnegie thought that he was one of the ablest men of affairs he had ever known.

Pullman's subsequent career confirmed these opinions. A practical visionary, he began to map out a plan to implement his dream of a national sleeping-car network. At the time of incorporation he had tied up much of the Midwest and established a Southern subsidiary in Altanta. An opportunity lay in the West, for the Union Pacific contract would soon be tendered. The greatest obstacles were in the East, where the Central Transportation Company and Wagner lines exercised control. There seemed little hope of upsetting these giants, and for the moment Pullman decided to press on into the South and West.

The last battle of the Civil War was hardly over before Central Transportation turned toward the South, hoping to take over its sleeping-car business. To counter this move Pullman formed a partnership in 1866 with Hannibal I. Kimball and Robert H. Ramsey of Atlanta.[187] A Southern banker, George Rice, was made head of the firm, but Kimball (1832–1895) appears to have been its driving force. A New Englander who went broke in the carriage business, Kimball moved to the South after the war, like many other bankrupts. He recouped his fortunes with wonderful speed through bank, hotel, and street-railway promotion. The very model of a successful carpetbagger, he became a good man to know in Atlanta. After two years Pullman, Kimball, and Ramsey had cars on ten Southern lines.

Pullman, however, remained dissatisfied with this share of the

Figure 3.46 Barney and Smith of Dayton, Ohio, built the Dayton *in the 1870s for Pullman. Dimensions for a similar car are given in the diagram in Figure 3.49. (Arthur D. Dubin)*

traffic. After acquiring the Southern Transportation Company early in 1870 through the Central lease, he became determined to consolidate the remaining independents of the region. He had the Southeast in hand, but to the west were the so-called Paine Lines. Enoch H. Paine of Louisville had begun assembling a small sleeping-car operation after the war.[188] By 1870 he controlled four companies, including the Crescent City and Rip Van Winkle lines. Negotations for some form of consolidation with the Pullman firm were underway late in the same year, although Kimball had soured on his Chicago partner by then. He wrote to Paine: "You know his [Pullman's] way of dealing: if he could buy us out at half what we were worth he might do it but don't think he would [do so] otherwise until he thought some one else wanted us."[189] An agreement was reached despite these misgivings, and in February 1871 the Pullman Southern Car Company was formed to consolidate the Pullman-controlled lines in the Cotton States. Paine and Kimball both were told to join up or face patent litigation.[190]

The consolidation promised greatly improved service, including through New Orleans–New York cars outfitted with changeable-gauge trucks.[191] The projected 58-hour schedule was declared "one of the grandest railroad arrangements of this fast age." But the new era that Pullman Southern promised was slow in coming. Traffic was light and service not of the best, and Paine managed to keep a bitter legal battle going with Pullman for many years. He accused Pullman of milking the operation, while Pullman claimed that he was subsidizing it. By the early 1880s, however, the affairs of Pullman Southern were much improved. The debt was down to $88,000 and a profit was visible for the first time in years.[192] By 1882 the subsidiary had served its purpose, and it was absorbed by the Pullman Palace Car Company after an exchange of stock.

If Pullman Southern was a mixed success, the Western auxiliary company must be considered a failure. Just months after the Palace Car Company was incorporated, the Pullman Pacific Car Company was also organized. As described in the earlier section on Woodruff and the Central Transportation Company, this new firm was formed at the suggestion of Andrew Carnegie, who had just met Pullman and became so impressed with him that he abandoned the Woodruff enterprises. Joining forces with Pullman, Carnegie helped the Pullman Pacific to beat the Central Transportation Company out of the coveted Union Pacific sleeping-car contract. To sweeten the deal various officials of the railroad were taken in as stockholders, as was Carnegie for his help in securing the contract. Pullman was made president and general manager, but he controlled only 1,200 of the 5,000 shares. The state of New York issued the Pacific Car Company a charter in April 1868, with a capital limit of ½ million dollars.[193]

Pullman was delighted with his triumph over Central Transportation. His dream of a national sleeping-car company seemed ever more real, and he plunged into an extravagant car-building program. Even Carnegie, the supreme optimist, was taken aback by Pullman's enthusiasm and wondered if so many palace cars were needed to traverse 1,200 miles of desert.[194] After the transcontinental opening in May 1869, the future steelmaker's question was answered in the negative. Traffic in general was disappointing once the initial novelty of rail service to the West had faded. Both the Central Pacific and the Union Pacific seriously undercut Pullman's position. After July 1870 the Central Pacific refused to handle through Pullman cars, and passengers were forced to change at Ogden.[195] The C P had built up its own fleet of Silver Palace cars and wanted no competitors on its line. Meanwhile the Union Pacific cut into Pullman's revenues by offering tourist sleeper service. The road acquired twenty-three such cars at about the time that through service began. The glamour of the Pullman Pacific plummeted, and the Chicago headquarters began to buy up its shares in 1870.[196] In October of the following year, a new 12½-year contract was signed with the Union Pacific that ended operations of the Pullman Pacific Car Company. An Association agreement (a joint ownership scheme sometimes used in the sleeping-car business) was made giving equal shares to both Pullman and the railroad. When the agreement was renewed in 1884, the old Pullman Pacific corporation was officially liquidated.

During the first years of the Pullman Palace Car Company, George Pullman's general strategy did not seem to be going well. The Southern and Western ventures were costly disappointments. Then by a fortuitous accident, the largest sleeping-car company in the country came into Pullman's possession. Even he must have been surprised at his good fortune and the relative ease with which he took over the Central Transportation Company. What appeared to be growing into a decade-long battle was settled in a few months of negotiations. After February 1870 Pullman con-

DOUBLE TRACK.

RATES OF FARE ALWAYS AS LOW AS BY ANY OTHER ROUTE.

FEWER CHANGES OF CARS THAN BY ANY OTHER ROUTE.

PULLMAN'S PALACE DAY AND SLEEPING CARS,

Via "Mantua Junction," run as follows,

FROM NEW JERSEY RAIL ROAD DEPOT, FOOT OF COURTLANDT STREET:

LEAVE NEW YORK DAILY, except SUNDAY, at 9.30 A. M.	ARRIVE CHICAGO, 3.00 P. M.
LEAVE NEW YORK DAILY at 7.00 P. M.	ARRIVE CHICAGO, 6.00 A. M,
LEAVE NEW YORK DAILY, except SUNDAY, at 9.30 A. M.	ARRIVE CINCINNATI, 12.45 P. M.
LEAVE NEW YORK DAILY at 7.00 P. M.	ARRIVE CINCINNATI, 10.30 P. M.
LEAVE NEW YORK DAILY, except SATURDAY and SUNDAY, at 9.30 A. M.	ARRIVE LOUISVILLE, 5.45 P. M.
LEAVE NEW YORK DAILY at 7.00 P. M.	ARRIVE LOUISVILLE, 7.40 A. M.
LEAVE NEW YORK DAILY, except SUNDAY, at 9.30 A. M.	ARRIVE ST. LOUIS, 6.45 A. M.
LEAVE NEW YORK DAILY, except FRIDAY, at 7.00 P. M.	ARRIVE ST. LOUIS, 1.30 P. M.

VIEW OF PULLMAN'S PALACE DAY AND SLEEPING CAR AS ARRANGED FOR NIGHT SERVICE.

PULLMAN'S PALACE DAY AND SLEEPING CARS,

Via "Allentown Line," run as follows,

FROM DEPOT OF CENTRAL R. R. OF NEW JERSEY, FOOT OF LIBERTY ST.:

Leave NEW YORK DAILY AT 5.00 P. M. Arrive CHICAGO, 6.00 A. M

COURTEOUS EMPLOYEES.

Figure 3.47 An Erie Railway timetable of 1870 included this engraving of a Pullman interior.

trolled the great majority of sleeping cars. His system was nationwide, and he held the basic patents.

Most men would have been satisfied to consolidate their position and enjoy their new fortune. Pullman did so, but he also pressed for more sleeping-car contracts and continued to acquire such desirable properties as the Erie and Atlantic Sleeping Car Company in 1873. Three years before, he had decided that he needed a car works of his own. He took over an old plant in Detroit and in less than two years transformed it into one of the nation's finest car shops.

By the early 1870s the Pullman organization had become incredibly complex. The Pullman Palace and its two subsidiaries had 400 cars, employed 2,000 people, and controlled some 30,000 route miles (Figures 3.46 to 3.49). Another 400 cars were jointly owned by Pullman and the individual railroads having Association contracts. On top of this was the car plant and Pullman's efforts to break into the general area of commercial car building.

But these activities were not enough to occupy his energy or satisfy his ambition. Now that the United States was conquered, Pullman looked overseas. His new goal was an international

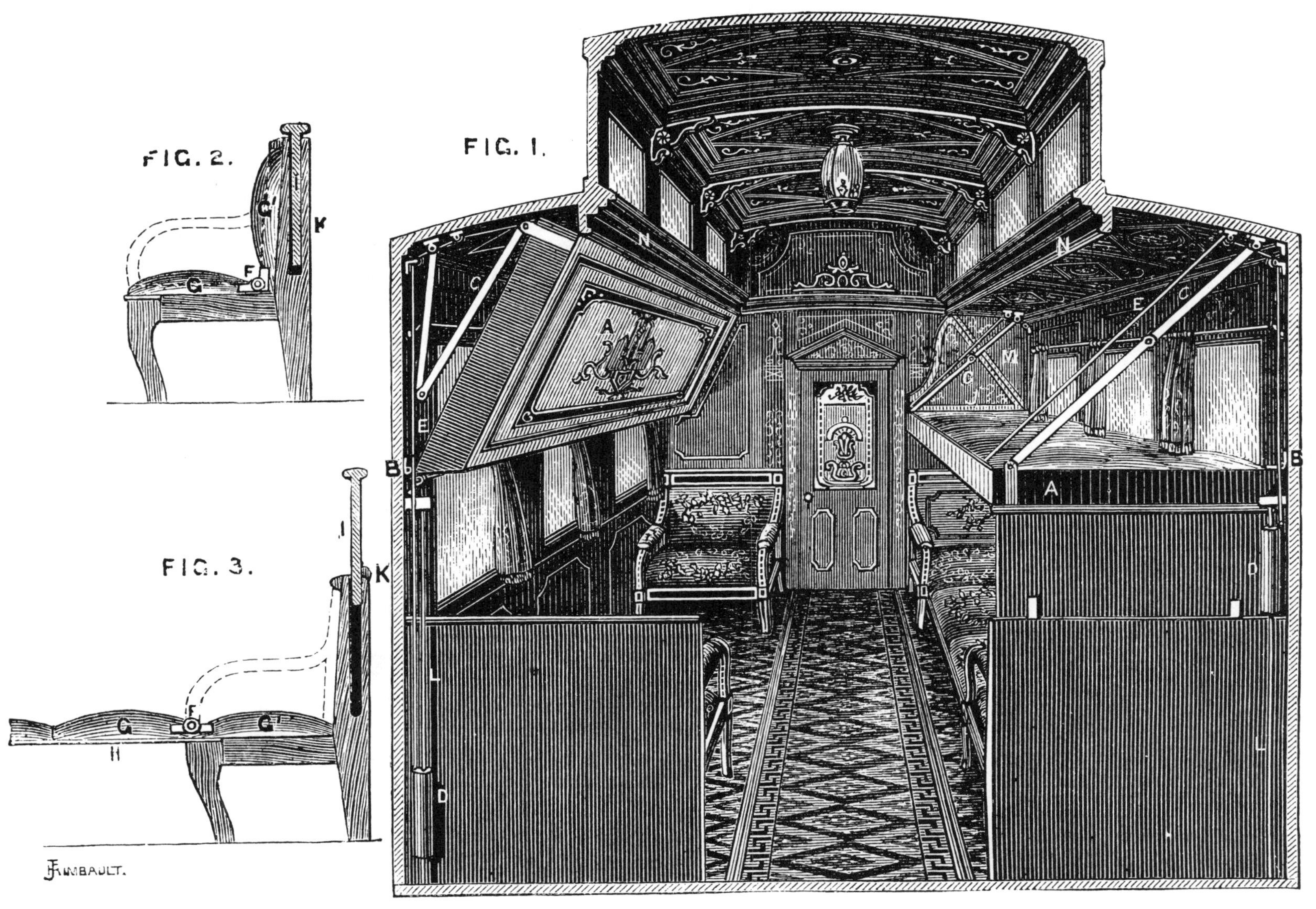

Figure 3.48 A berth and headboard divider plan patented by Pullman and Field in 1865. The patent covered only the details of the hardware, not the idea of the folding upper berth itself. (Engineering, *January 26, 1872*)

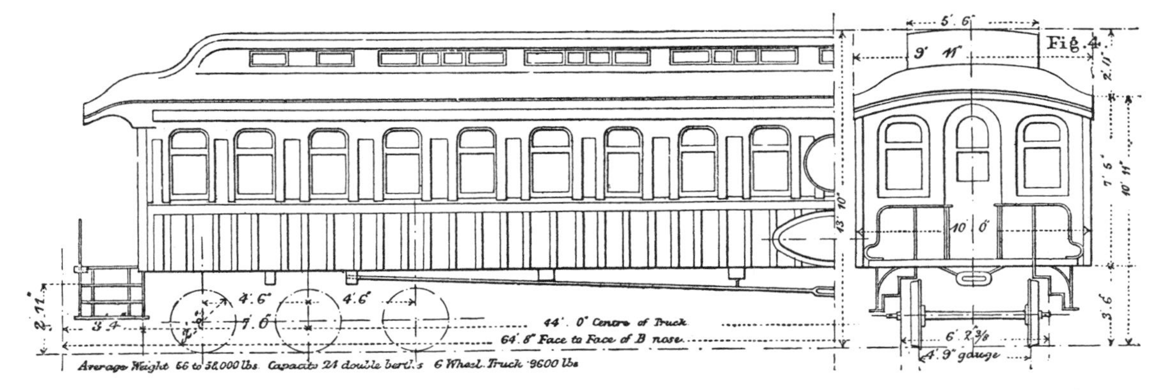

Figure 3.49 This drawing, published in 1877 by the British journal Engineering, *is believed to represent a design of about 1870. (See Figure 3.46). The car slept 24.* (Engineering, *November 2, 1877*)

sleeping-car cartel. In 1872 James Allport, general manager of the Midland Railway of England, made a tour of American railways. Impressed by the superior accommodations of Pullman's equipment, Allport recommended their use on the Midland.[197] Pullman visited England early in the following year and returned to Chicago with an order for six cars and an operating contract. Not long afterwards, a group of English investors organized a company to promote the use of Pullman cars in Great Britain, Ireland, Europe, and India.[198] Calling itself the Car Syndicate Limited, the firm tried to interest foreign railroads in the merits of American-style sleeping and parlor cars. But it encountered a general rejection of the idea: just as North Americans seemed to feel a revulsion toward compartments, Europeans saw little merit in the open American plan.

Pullman was not about to compromise his basic plan, and the cars he sent abroad were faithful copies of their American coun-

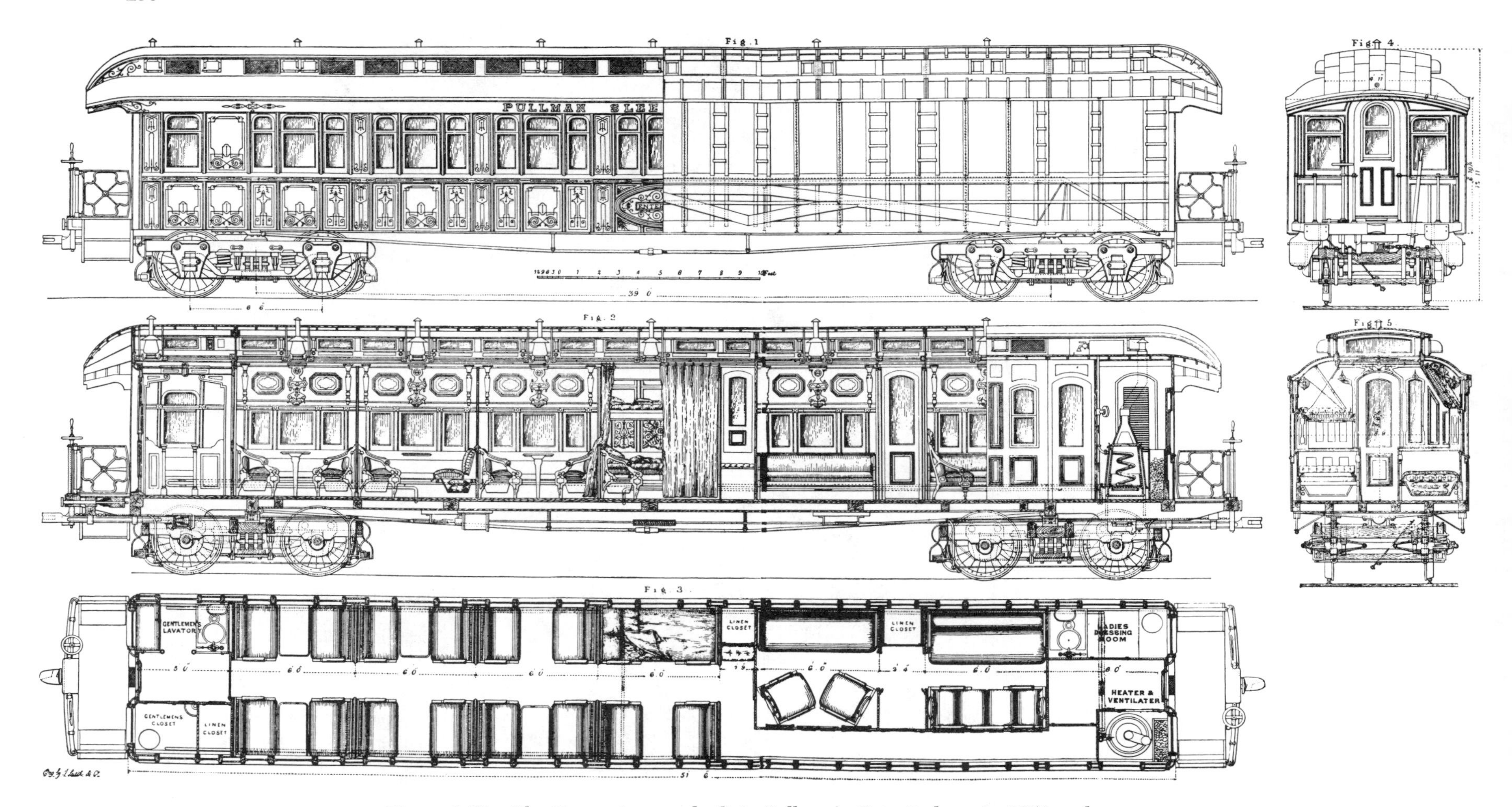

Figure 3.50 The Enterprise *was built in Pullman's Detroit shops in 1873 and was exported to England for service on the Midland Railway. Except for the wooden-center wheels, it appears to be a purely American design in all details.* (Engineering, *April 2, 1875*)

terparts. They were manufactured at the Detroit shops, knocked down, and shipped to Derby for reerection at the Midland's repair sheds. One of the sleepers was ready for service in January 1874. The *Midland* measured 58 feet long by 9 feet wide by 13 feet, 2 inches high. It weighed about 25.5 tons and slept 30. Within a few months *Leo, Enterprise, Excelsior, Britannia,* and *Victoria* were all in service. The last two were parlor cars. In all particulars except for their narrow width (a foot less than the usual 10 feet) and their wooden-centered wheels, these cars were standard Pullman sleepers of the time. Excellent drawings of the *Enterprise* were published by *Engineering* magazine in April 2, 1875 (see Figure 3.50).

Public reaction to the cars was favorable, and Pullman secured the good will of the press by running well-provisioned excursions. The resulting clippings were reprinted by the Pullman Company in a small booklet, *Official Correspondence with Reference to the Relations of the Pullman Palace Car Co. to Railway Companies* (Chicago, 1874). In its pages timid ladies were assured that their safety was more secure under the watchful eye of a Pullman porter than it would be if trusted to the confines of a lonely compartment. Pullman stock had all the modern improvements, and each car was a "museum of Patented appliances." Another newsman proclaimed that "the Pullman car is an absolute civilizer. It encourages the social amenities, . . . and, by its elegance of surrounding, and beauty of adornment, encourages aesthetic tastes and the refinements of life."

These and similar accolades failed to produce a mass conversion to American-style cars; in fact, other British lines were rather slow to sign on. The need for sleeping cars was limited in the British Isles, because most trips were short and there were many excellent railway hotels. Pullman's open parlor cars had little to recommend them over the first-class British carriage. By 1882 only four lines were using Pullmans; the fleet was a meager fifty. In 1888 the Midland itself withdrew to operate its own cars. In 1907 British Pullman was sold to a group headed by Lord Dalziel. Under native management its affairs prospered greatly, perhaps a sign not only of the benefits of local control but also of the universal resentment felt toward foreign ownership.

Pullman had even less success on the Continent. In 1874 he sent the *Midland* on a three-year tour of European railways, an expedition which landed a contract in Italy but nowhere else.[199] Sleeping cars were definitely appropriate to European travel, but the local talent was already engaged in perfecting such a system. Seventy years before the Common Market, Europe was characterized by intense nationalism. Nations were jealous of their borders and disinclined to permit free entry of foreigners. Pullman faced no such problem in the United States, but in the Old World extraordinary skill and patience were needed to arrange for the interstate movement of railway cars. Georges Nagelmackers, who was Pullman's equal in talents for persuasion, finance, and management, had succeeded in winning such concessions for the Wagons-Lits. It was futile to challenge so powerful a competitor, and after Nagelmackers declined a suggested merger, Pullman had nowhere to go but home.

There was enough to keep him occupied in Chicago, since the day-to-day direction of the multilevel Palace Car Company was demanding even for a man of Pullman's ability. The business decline following the 1873 panic made it essential to devote more attention to the firm's management. Independent sleeping-car contractors such as Woodruff and Wagner remained a threat, and the competition for contracts as new railroads opened would continue briskly for another twenty years. Pullman also faced a new problem: his success and far-reaching efforts to "run everything" had earned many enemies, who lost no opportunity to abuse and ridicule Pullman and his business practices. In the first year of the Pullman Company, complaints against the sleeping-car monopoly were voiced by the *American Railway Times*, which said that the contract arrangement was against the best interests of both the traveling public and the railroads.[200] The cars were "tawdry, tapestried, cushioned and over-ornamented, receptacles of filth and fever." Could not the railroads run plain, economical cars, to the benefit of the public and their own stockholders? According to the *Times* article, the sleeping-car companies were a ring of parasites that feed and fatten because railway managers are too inept or lazy to run their own affairs.

Protests against the alleged abuses of the sleeping-car monopoly came to a head in 1874. Maine's State Railroad Commission issued an angry statement about the gross excesses of the palace cars, declaring that they benefited only the privileged class of traveler and were subsidized by the general public.[201] Pullman could ignore the state of Maine, but not the U.S. Senate. That year Senator William Windom asked Pullman to report on its affairs. Rather than appear in person, Pullman prepared a written statement that was published in the Windom Committee's report.[202] Pullman gave a calm, reasoned defense of the sleeping-car monopoly. Its main justification was one of public convenience, safety, and comfort. Because individual railroads were not offering through service for long-distance travel, Pullman was acting as the public's advocate in perfecting a nationwide system. The railroads were relieved of a burdensome, specialized service. And a national system like Pullman's could afford to run cars on poorly patronized lines because of the revenues received from its better-paying runs. Only a national system could offer a uniform standard of service.

Pullman only hinted at the benefits of the car pool—a point that in later years became a favorite justification for a sleeping-car monopoly. According to this argument, only the Pullman Company could afford to maintain a reserve of cars for seasonal emergencies or special charter service. Most of the cars could be in constant use when deployed on a nationwide basis through a central organization. Cars for the Florida trade, for example, could be shifted from the Western lines at the end of the summer holidays. Cars used one week for a Shriners' convention could be used the next for delegates to a political conference.

In his final arguments Pullman noted that his company had developed a specialized conveyance that "by a number of ingenious devices, very greatly improved the comfort, safety and healthfulness of railway passenger-cars, lessening the fatigue of travel and making night-journeys convenient and easy." He pointed out that while the weight and even the decoration of these cars was questioned in some quarters, the standard sleeper weighed 28 tons, or about 2.5 tons more than a coach. He also said that the specialized expertise of the Pullman organization was evident in the discipline and training of its porters and conductors. In general the Windom Committee seemed to accept Pullman's reasoning, for no Federal control was imposed until the early years of the twentieth century.

While the Senate investigation was underway, the stockholders of the Pennsylvania Railroad conducted an investigation of the road's management in which the Pullman contract was a matter of particular concern.[203] The road's officers defended the contract, maintaining that operation of its sleepers was best handled by outside contractors and that no larger share of the revenues could be expected. In addition, Pullman had invested $2,550,000 in 170 cars on their lines, freeing this large amount of capital for other uses and thus effecting a saving to the shareholders. Neither side mentioned possible collusion between certain officers of the railroad who held an interest in the Pullman Company.

While Pullman seemed to enjoy general public approval, there were occasional complaints. It would be impossible for any organization to satisfy everyone, particularly when the majority of its customers were of the fussy, demanding middle and upper-middle classes. In the nineteenth and early twentieth centuries these groups had raised fault-finding and fastidiousness to a neurotic preoccupation. Criticism of toilet facilities and cramped dressing rooms was frequent. In 1870 the *National Car Builder* said that some cars had no washbasins and that the toilet was nothing more than a "dry closet."[204] During the same decade the journal remarked that the term palace car must be a whimsical stroke of American irony, for most sleepers were more like dungeons than palaces.[205] Their closed-in spaces, poor ventilation, and overheating were inexcusable. Worst of all, the evil-smelling oil lamps provided so little light that the passenger could not read and was forced to retire early, suffering the pangs of claustrophobia behind the curtains of a narrow berth that rolled like a cork in a tempest. After viewing Pullman's Centennial exhibit, the *Railroad Gazette* could not resist grumbling: "We always admire the Pullman decoration with a protest and regret that the money is not expended in ways which would give so much more pleasure to passengers."[206]

More substantial accusations were made about rates charged. In 1875 the Illinois Central Railroad contended that Pullman's $2 fee was too much and that a fair profit was possible at $1.50.[207] At the present rate, the road said, cars paid for themselves in only fifteen months. Some years later a complaint in the *Railroad Car Journal*, while admitting that a sleeper might not pay for itself in less than three years, took up the same theme: berth fees were too high and profits were excessive.[208] It said that a typical car might travel 300 miles a day, and even allowing for only 50 percent occupancy, it would earn $3,000 a year.

The various complaints about the Pullman company prompted more than an occasional editorial. Efforts to regulate and tax its operations were attempted by several states. Ohio considered taxing Pullman as early as 1875. The Illinois legislature conducted an investigation of the firm's affairs with an eye toward rate regulation.[209] Robert T. Lincoln, acting in Pullman's defense, contended that the company's profits were very small, that no one was compelled to use sleeping cars, and that it was foolish to go into the regulating business, for where would it all end? The classic statement made during these hearings was by another friend of Pullman, who said simply that "the company was not running cars for the fun of the thing." In the end Pullman's legal department seems to have been successful in fending off state regulation by arguing that the firm operated an interstate business.

Carping critics and suspicious state legislators were so many gnats compared with the economic problems that Pullman faced in the 1870s. During this decade the national economy suffered one of its most severe and protracted financial collapses. Thousands of businesses failed and others barely survived. But it was not the general depression that troubled Pullman; the company weathered the storm with relatively little trouble. It was, however, shaken by the loss of the Michigan Central contract. Its stock fell $20 at the news, but soon recovered to nearly its full par value.[210] Yet a certain doubt about the company's future was abroad, and whereas no shares had previously been available, few buyers could now be found.

Even so, the next year's earnings were down only $33,200, representing a decline of about 1.5 percent.[211] The Centennial traffic admittedly helped cushion the blow, but actual earnings on investment were an impressive 18.75 percent. The corporate structure was wonderfully solid as well, with assets exceeding obligations by 42 percent. A cash surplus of 2 million dollars was on tap to tide the company through difficulties. Pullman might be accused of overexpansion, of a foolhardy overseas adventure, and of extravagance in his personal life, but he was always the practical visionary who took care not to overextend.

Just before the Centennial year began, Pullman decided to retire part of the bonded debt by reducing the annual dividend from 12 to 8 percent.[212] He knew when to retrench. Sleeping-car revenues seemed to hold up well enough, but the Detroit car shops were running at a loss. Wages were cut back to stop this drain. The shop's deficit was too minor, however, to have any real effect on overall earnings. Another economic setback occurred in 1878, when the firm's secretary, Charles W. Angell, absconded with $115,000.[213] It was an embarrassing scandal, but the stockholders were assured that even so large a theft could not adversely affect the company's financial position. The hard times bottomed out in the last year of the decade. In 1879 revenues were down a solid $300,000 compared with average earnings of the previous five years.[214] Even so, profits were substantial enough to cover all expenses, pay the usual dividend, and leave a surplus of $337,000.

The next ten years were an era of growth and prosperity. In 1881 earnings were just short of 3 million dollars.[215] Profits were so large that an extra stock dividend was declared. Some idea of Pullman's expansion in this decade is given by the dramatic increase in capitalization. In 1880 the capital stock was valued at 6 million dollars. This figure increased several times over the next few years, and by 1889 it stood at 25 million.[216] During the same time, route mileage and the number of cars doubled, while assets tripled.

Pullman's progress was the result of more than good times and the general increase in railway mileage and travel. It was the reward of a many-sided effort to obtain new contracts and hold existing ones. One obvious tactic was to drive out or take over the competition, as Pullman did in the Union Palace and Wagner purchases. Another was to seek new markets. Pullman never entirely forgot about European possibilities; he reportedly even considered starting a transatlantic steamship line in the early 1880s. But he was realistic enough to leave this to the Cunards, just as he had prudently left the continental sleeper trade to Wagons-Lits. Mexican railways, however, were too tempting to be ignored. There were not many roads operating in that country when Pullman made a contract early in 1884 for service between El Paso and Mexico City.[217] Except for an interval during the revolution of 1914–1920, Pullman cars continued to run over Mexican lines until the company itself was dissolved.

But the techniques of takeovers and new-market development were less important than the simple maintenance and acquisition of contracts. Pullman saw that good contracts were more vital to his success than the most perfect equipment or the best-trained porters. Prudent contracts were essential to the company's profits and its relations with its landlord railroads. As a tenant he must satisfy a railroad's desires, yet somehow extract terms favorable for a profit. Just the right balance of concession and insistence was needed to forge such an arrangement.

The standard contract required each car to gross $7,500 a year (later this was raised to $9,000). Pullman agreed to furnish suitable cars in the required number; to maintain interiors, supply fresh linen, provide porters, and supervise their conduct; and to hold the railroad safe from patent claims. The railroad was to haul the cars, handle routine running-gear repairs, and provide heat and light. It was generally understood that the road would receive the coach-fare portion of the ticket while Pullman took the first-class portion. In the early years Pullman accepted a simple $2 berth fee. If revenue fell below the $7,500 base, the railroad would pay Pullman a mileage fee, usually 3 cents a mile. Contracts generally ran for fifteen years, though a few were for thirty.

Some lines demanded and received additional concessions. The Burlington required Pullman to give Burlington employees free passes and to pay all repair costs if revenues reached $9,000.[218] The Pere Marquette was granted a "scaled mileage" contract that provided for a graduated schedule of charges: 2 cents a mile if annual revenues were below $5,000; 1 cent a mile if revenues were between $5,000 and $6,000; no charge if revenues averaged $6,000 or over on specific lines; no charge on the entire system if revenues averaged $7,500.[219] The B & O received a special rate of 1 cent a mile for its tourist cars but no mileage fee for its regular sleepers. Another form of contract variation was the split-concession agreement with the Chicago and Northwestern that permitted the railroad to operate its own parlor cars. All such variations in terms aroused contention, jealousy, and suspicion among the roads. In an attempt to avoid quarrels, Pullman kept the contracts confidential until regulation by the I.C.C. made them a matter of public record.

When necessary, Pullman tried various means to placate difficult customers. In his earliest years he had new cars built in the shops of the contracting railroads. This gave the railroad a bit of extra revenue and won the loyalty of the presiding master car builder, who had the opportunity of showing off his skills. But the practice was costly, and it quickly disappeared after the Detroit shops opened in 1870.

A more successful method of winning railroad enthusiasm for Pullman service was the joint ownership scheme called an Association line or contract. This device had been used since the beginning of the sleeping car. In it the railroad and Pullman held a quarter, a half, or a three-quarter share in cars and their maintenance. A fifty-fifty split was the most common Association agreement. Revenues were shared accordingly. In this way the railroad was at least part master in its own house. At the same time, Pullman was relieved of providing capital for that portion of the cars and was thus free to divide his efforts among a larger number of operators. In effect, it gave him control of that much more mileage. Even if his ownership was not total, no competing firm could put its cars on the line. In some cases an Association contract might be the only type that a line run by an independent management would accept. It was a wedge, however, and Pullman could always hope to take over the service in time.

Several roads were drawn into Pullman's web through Association contracts. The Rock Island ran its own sleepers for nearly twenty years before joining Pullman in an Association agreement in 1880. Within eighteen years Pullman had full ownership of the line's sleeper service. The Northern Pacific wanted to run its own affairs, but was forced into Pullman's camp by its weak financial condition. Though it was operating seven cars at a good profit in 1880, it was being hard-pressed to finance its Western extension. It gave in to Pullman, knowing that funds would not be available for a full sleeper fleet once through service began. In 1881 it signed a fifty-fifty Association contract calling for sixty cars to be furnished within two years. To assure his hold on the Northern Pacific, Pullman purchased a large number of its shares.[220] An even more dramatic example of this technique was evident in the case of the Boston and Maine. In 1892 Pullman bought 5,000 of its shares; in the following year a twenty-five-year contract was signed.[221]

During the 1880s roughly one-third to one-half of Pullman's agreements were Association contracts. They included such major lines as the Pennsylvania, the Illinois Central, and the Santa Fe. A breakdown of the number of cars running under the two types of contract shows that Pullman gradually gained control:[222]

1881	502 Pullman	232 Association =	734 total cars
1885	606 Pullman	509 Association =	1195 total cars
1889	1066 Pullman	344 Association =	1420 total cars

By 1890 Pullman and Association cars were almost equally divided—probably the result of the Union Palace takeover. By the late 1880s Pullman was clearly disenthralled with the Association concept. In its early years the company did not have the capital to provide all the cars needed and was content to share investment and revenues with a partner. By 1890, however, its surplus was over 16 million dollars and its credit was unlimited. The monopoly was nearly complete; few lines seemed tempted to join Wagner, and even fewer were venturesome enough to run their own sleepers. There was a shift in Pullman policy toward discouraging new Association contracts and buying out the existing ones wherever possible. The chronically insolvent Erie was one of the first Associations to go in 1887. In the next year the Santa Fe and its subsidiary, the Atlantic and Pacific, gave way. In 1898 the Union Pacific sold its Association interest for 1½ million dollars. That nearly half of this figure was for "nonphysical assets," or good will, shows how eager Pullman was to end all its fourteen Association contracts.[223] In the early years of the new century, such major Association lines as the Southern Pacific, the Missouri Pacific, and the Atlantic Coast Line sold out. The final holdout, the Spokane, Portland, and Seattle, was a very small joint ownership involving only three cars. It was terminated in 1922.[224]

In general, the Association plan had been an effective provisional measure for holding service contracts until a complete takeover was possible. There were a few notable failures, however. The most dramatic breakaway was that of the Baltimore and Ohio. In 1880 the road's president, John W. Garrett, decided to follow a new policy of controlling all services connected with B & O operations. Beginning in the early 1870s he began to take over telegraph, rail supply, and printing services. He hoped to develop on-line passenger travel by providing resort hotels, and he intended to assume full control of the sleeping and parlor car operations once the Association contract with Pullman expired. The agreement itself was a hangover from the days of the Central Transportation Company. Before the contract ended, the road ordered new cars, and on the date of expiration, October 1, 1880, Garrett placed his stock in service and forbade the movement of Pullman cars over the B & O.[225]

Pullman immediately began legal action, asking for an injunction against operation of the railroad's sleeping cars and claiming patent infringements. He initiated a similar suit against Barney and Smith, who were supplying some of the cars to the B & O. Pullman further contended that the road's management was endeavoring to persuade other railroads to take up independent sleeper and parlor operations and was thus guilty of malicious conspiracy to ruin Pullman's business. The Baltimore and Ohio countered that it had every right to operate its own cars now that the contract had expired, and that it had decided to do so because Pullman had furnished it with inferior cars. It also alleged that because of the unfavorable terms in the old contract, interest and repair expenses had resulted in a loss to the railroad. The courts sided with the B & O on all issues except that of the patent claim.[226] The thirty-three cars owned by the Association were placed in the hands of a court-appointed receiver. At a public sale held in 1882 the equipment was sold for $5,000; Pullman claimed they were worth $18,000 each, and so a third legal battle began.[227] Litigation was pursued vigorously over the next few years. To Pullman the B & O represented not only a valuable contract, but a maverick who might tempt others to stray outside the syndicate he had so carefully constructed. If the other Association lines saw that malcontents could go unpunished, they might all develop wanderlust.

In the end the B & O returned to the fold, not because of the courtroom maneuverings but owing to a change in its management. The B & O's new president, Samuel Spencer, found the line in financial trouble and sought partial relief by eliminating its peripheral activities. He signed a new twenty-five year Pullman contract in June 1888.[228] An immediate settlement was made out of court, and the B & O remained forevermore a loyal Pullman subject.

No other Pullman defectors can be directly linked to the B & O's unsuccessful try for freedom, though at least one major Western line decided to go it alone during the same period. The Milwaukee Road had reluctantly formed an Association with Pullman in 1882.[229] They were so unwilling that Pullman was forced to agree to very liberal terms to win a contract. The Milwaukee would give Pullman only a 25 percent interest in the Association, and the contract could be terminated at five-, eight-, or eleven-year intervals, with a maximum lease of fifteen years. Pullman was to furnish forty-five sleeping cars and manage the road's diners. But the road did not consider even this weak Association agreement desirable, and late in 1890 it canceled the lease.[230] The rumors were that Wagner would take over the contract, but this was not so; the Milwaukee intended to run the 150-car sleeper, parlor, and diner fleet on its own. A suit developed over compensation for the cars furnished, but this litigation and other harassments did not persude the Milwaukee to give up its independent course. Only when the road was near financial collapse in the 1920s did Pullman again secure a hold on its first-class car service.

These cases of contract disagreements and legal struggles should not be taken to mean that Pullman built and maintained its business through bullying. It resorted to open conflict only when no other measures were possible. The company recognized its subservient position with the railroads and the need to maintain harmonious relations with them. This was achieved by making contract concessions and by providing good service. Service was the long-term basis of Pullman's success. Throughout its corporate life there was a paramount emphasis on good equipment, polite employees, meticulous housekeeping, and gracious and well-controlled food service. A uniform and uncompromisingly high level of service could be expected not just on the Broadway Limited but on the meanest plug run to Muncie, Indiana. The car to Muncie might be older but it would be well maintained, and its crew would be every bit as professional as those assigned to the fanciest name trains. That so uniform a standard could be upheld in an operation as large and diverse as the Pullman Company speaks well for its management and its employees. Even in its years of decline, when all hope of reversing the slump in railway travel was painfully evident, the spirit of perfect service continued.

At the same time that the Associations were being discouraged, Pullman decided to deal with another menace to its affairs. The tourist sleeper was the logical outgrowth of the old emigrant car. These primitive boarding cars handled a class of traffic that Pullman had no hope or ambition to attract. Emigrant cars had been used since the 1840s and were often little more than converted boxcars. But the Central Pacific began a new trend in the 1870s by introducing a more civilized emigrant car. It followed the general arrangement of the standard sleeping car, though it had no bedding or other frills. Within a few years, other Western roads were running improved emigrant cars that began to rival the comforts, if not the ostentatious luxury, of Pullman's finest.[231] The addition of bedding and porters upgraded the emigrant car to the tourist sleeper, which amounted to a very acceptable cut-rate overnight excursion car. The difference in fares was astonishing: a trip to the West Coast was $13 by Pullman, $3 by tourist car.[232] In the mid-1880s thousands of middle-class passengers were taking advantage of these low-priced sleepers.

Pullman became alarmed; if the scheme spread, it could ruin his business. The Western roads were not interested in abandoning the popular people's sleeper. In the late eighties, however, Pullman succeeded in taking over the operation of these cars. In 1890 he acquired the Northern Pacific's tourist cars and thereby gained control of a potentially serious rival. Now he had just over 250 second-class sleepers in service.[233] Much as Pullman may have disliked the whole idea, the tourist car was established, and many would continue in operation until after World War II.

Thus during the eighties Pullman was busy with the Associations, the rebellious B & O, the tourist cars, and the Union Palace merger. In general it was a decade of consolidation, growth, and enormous prosperity for the company. The car plant and model community south of Chicago was built in the early 1880s. The town of Pullman was to be its founder's fondest creation and the part of his enterprises in which he seemed to take inordinate joy. Ironically it proved his undoing, for the bitter 1894 strike at the car plant exposed Pullman to nationwide censure. Never before had he been subjected to ridicule on such a scale. How greatly it affected him can never be gauged, but it is thought to have contributed to his death in 1897. He died confused and embittered, under the ugly shadow of the strike. Perhaps fate's cruelest trick was that his death came just months before the company's takeover of Wagner. He was cheated of witnessing the ultimate victory, the fulfillment of a thirty-year struggle to create the great American sleeping-car cartel.

EXPANSION AFTER PULLMAN'S DEATH

Pullman's death left a vacancy that the firm's directors seemingly would not or could not bring themselves to fill. The company attorney and son of a former President, Robert Todd Lincoln (1843–1926), was named acting president. But even such a prestigious figure was not thought capable of assuming Pullman's office. Lincoln was not fully elevated to the presidency until after the Wagner merger was consummated.

Lincoln found himself in charge of a vast corporation, rich in property, rolling stock, contracts, and cash. It controlled the great bulk—nearly 90 percent—of the sleeping-car service in North

America, and operated the largest railroad car plant in the world. Its cash surplus was so gigantic that one of Lincoln's first acts was to pay out a 50 percent stock dividend amounting to 18 million dollars. And there were still ample resources to pay an inflated price for Wagner the following year. With the Wagner merger, the company decided to reorganize in January 1900. Capital was jumped to 74 million, and a new corporate title was adopted. The Pullman Palace Car Company seemed fussy and out of date even in the first year of the twentieth century. The firm became simply the Pullman Company; nothing more was required to describe this giant among American businesses.

Pullman flourished during the fifteen-year Lincoln administration, owing either to his stewardship or to the momentum of the company itself. However, it does seem evident that Lincoln was never ardently enthusiastic about his job. He seemed to prefer the study of law or the quiet pleasures of his hobby, which was astronomy.[234] But times were good, passenger traffic was up, and railways had no competition for long-distance land travel. Between 1900 and 1910 Pullman's traffic tripled, while its assets doubled. Net profits peaked in 1910 with a return of 14 percent, which was equal to over 19 million dollars.[235] During these years the stockholders received two spectacular extra stock dividends: a 36 percent dividend in 1906 and a 20 percent bonus in 1910. If the 1898 split is included, the shareholders were awarded extra dividends equal to 106 percent, or 99 million dollars.

John S. Runnells (1844–1929) succeeded Lincoln in 1911. An attorney with the firm since 1887, he had been Pullman's legal confidant for years. He was the last Pullman president to descend from the original organization, and like his predecessor, he inherited a healthy company. Some vexing problems were gathering, however. The Populist movement was adding momentum to the campaigns of the antimonopoly activists, and Pullman was an obvious target. The more radical Populist spokesmen charged that the company operated unsafe (that is, wooden), uncomfortable, unsanitary, and positively indecent cars.[236] They said that Pullman won the favor of judges, legislators, and other influential men by the distribution of free passes. They accused Pullman of reaping exorbitant profits. And finally, they declared that the cynical tolerance of those in power, coupled with public indifference, had permitted the growth of this monstrous swindler among corporations.

The cries of outrage were somewhat amplified by Pullman's takeover of one of the last major independent sleeping-car fleets. By 1912 the New Haven's president, Charles S. Mellen, had pushed the road into a desperate position through a reckless expansion program. In a last attempt to save his crumbling rail-steamer-trolley empire, Mellen agreed to sell the road's sleeping-parlor car business to Pullman for 3.3 million dollars.[237] The cash was used to shore up the tottering company. The transfer also relieved the New Haven of the enormous expense of reequipping with steel cars. Its safety record had been shocking in recent years, and a series of wrecks not only evoked a public outcry but convinced the New York Central that nothing but steel cars should be permitted to run into its Grand Central Terminal. The terminal was to serve as the New Haven's New York City station under a rental agreement. The possibility of establishing a convenient separate terminal was out of the question for the nearly bankrupt New Haven. The opening of Grand Central in 1913 made action by the New Haven imperative. Since Pullman had the resources to provide steel cars, on January 1, 1913, the great sleeping-car monopoly became greater by 252 sleeping, parlor, and composite New Haven cars. After the New Haven sellout, the Milwaukee and Great Northern were the only large-scale non-Pullman operators left in this country.

The real and imagined abuses of Pullman were now so much a part of public debate that in 1906 the company was subjected to regulation by the Interstate Commerce Commission. This official gesture by Congress helped quiet the editorials, but it did not silence the Pullman legal staff, who argued that the Pullman Company was an "inn-keeping" operation and therefore not liable to I.C.C. control. For years Pullman lawyers had warded off state regulation by demonstrating that the business was interstate in nature. Now their own argument had been used to put Pullman under Federal control. While no one could agree on a precise definition of Pullman's operation, it was clearly interstate. In 1910 Pullman at last admitted defeat and began to report on its affairs to the I.C.C. However, the regulation was more in form than in substance; the Commission required uniform accounting procedures, but did little at first to govern Pullman's operations. Its corporate structure was in no way altered; it was not required to shed any part of its property, or to modify its contracts, or in general to do much more than report on its financial situation.

One exception to the I.C.C.'s permissiveness occurred during the early years of regulation. A St. Paul traveling man, George S. Loftus, had tried for years to force Pullman to reduce its upper-berth rate. Finally he took the case to the I.C.C., which decided that uppers were indeed inferior accommodation and that the charge for the top berth should come down by 20 percent.[238] This was hardly a crushing defeat for Pullman, although it did represent an annoying interference. Its cars generally ran half full, and most patrons preferred the lower berth and were willing to pay a small premium to avoid the ladder necessary to mount to the top bunk. In this decision a dissenting opinion was expressed by I.C.C. Chairman Martin A. Knapp. A true believer in Pullman as a public benefactor, Knapp's thinking may have represented the silent majority's view of that day. Knapp believed that the Pullman charge above regular coach fare was a bargain for a ride in an uncrowded, luxurious car, whose "occupants are usually persons of good appearance and unobjectionable manners."

In its own defense Pullman could contend that it was working for the public welfare by its crash conversion to steel cars. This program began in earnest in 1910. It involved the early retirement or downgrading of many first-class wooden cars—an action that increased the depreciation account to 6.8 million dollars, or double its normal rate.[239] The extraordinary expenditure for new steel cars cut Pullman's profits substantially. Millions were spent on retooling and building new shops. Repair facilities had to be remodeled as well. Pullman was genuinely devoted to a fast changeover to steel cars and achieved it long before the class-one railroads did. However, the benefits of the conversion far exceeded Pullman's investment. For its money Pullman acquired a durable, safe fleet, and its manufacturing division enjoyed its most profitable years during the wood-to-steel changeover.

In a more direct effort to polish its image, Pullman turned to the popular press for support. Apologists could be hired, while journalists sympathetic to the cause could be well treated. One of the industry's most faithful defenders was the journalist-historian Edward Hungerford. His 1911 book *The Modern Railroad* contained a flattering chapter on Pullman. Articles by many writers began to appear in national magazines, such as *Harper's Weekly*, explaining the necessity for Pullman's monopoly. The propaganda effort reached its high point in 1917 with Joseph Husband's *The Story of the Pullman Car*. Whether or not it was company-sponsored is unclear, but the author seems to have had the company's wholehearted cooperation and he produced a work filled with praise of Pullman. Despite its factual errors and partisan viewpoint, it became a standard reference and was for many years the only volume available on American railroad car history. A company-sponsored history was written in the late twenties, but did not progress beyond proof sheets.[240] Like Husband's work, it was a lavish tribute but a shallow and unsatisfactory business history. Afterwards Pullman seemed content to tell its story through giveaway booklets, exposition leaflets, and general news releases. Its best public relations series was published late in the twenties as "Pullman Facts." Subjects from the history of the sleeping car to the operation of the company laundries were covered in these informative booklets.

The Populist menace had no sooner faded than nationalization was visited on Pullman in the form of the U.S. Railroad Administration. This was a Federal agency organized to manage the railroads during the World War I transportation emergency. Pullman, now the twenty-fifth largest corporation in the land, lost its sleeping-car operation to the U.S.R.A. in July 1918. The car works remained in Pullman's control. The government paid a fixed annual compensation of $11,750,000 for the sleeping-car concession.[241] Pullman personnel operated the cars as before, and record-breaking numbers of passengers were moved. Federal operation was in reality quite short, ending in March 1920. But to Pullman's management, the entire episode was as unthinkable as it was inequitable. The company's claim against the government was one of the last to be settled.[242]

The postwar era opened the second period of growth for Pullman in the twentieth century. The trauma of Federal ownership was soon forgotten as the nation's prosperity seemed to promise an unending boom. Travel expanded with the good times. Salesmen seeking orders, vacationers exploring the Far West, speculators rushing to Florida—all were filling berth, bedroom, and drawing-room space. The leasing of private cars flourished as new millionaires sprang up. Everyone seemed on the move, and everyone seemed rich enough to go Pullman.

The excursion business was doing well too, so that the pool cars earned their keep. The fleet of extra cars was one of Pullman's greatest assets, and one of the major reasons why the railroads were amenable to the existence of the sleeping-car monopoly. Pullman had about 1,500 cars in its active pool by the late 1920s. Some individual cars ran more than 150,000 miles in a year. Extra cars were kept at each of the company's forty-seven district yards, and they were transferred from one district to another to meet demand. A political or business convention, a major sports event, or some other special event might create a sudden need for several hundred extra cars, all on short notice and all for a limited engagement. No individual railroad could afford to keep such equipment on hand for occasional use, much less to maintain the cars in top condition for instant service. But Pullman could and did, because as a national system it could keep its pool constantly employed. In the railroads' view, the antimonopolists be damned; Pullman was a positive necessity.

The pool cars followed the tide of travel. In winter they made the second and third sections necessary to handle the holiday rush, or they ran south with the Florida traffic. In summer many of the same cars carried vacationers to Maine, Michigan, and the Western mountain and lake resorts. During the summer peak, for example, the Boston and Maine required 115 extra cars. Conversely, the Florida trains ran comfortably with 300 cars in normal times, but in the winter season they needed nearly three times that number.

Special events required skillful deployment of the pool cars, with all assignments coordinated by the Chicago office. A Knights Templar conclave in Denver might require 450 cars; the next week the same cars might be rushed east for anticipated World Series crowds or for a major political event (845 cars were used for F.D.R.'s first inaugural). The greatest single pool-car job was the 1926 Eucharistic Congress, which took 1,199 cars to carry the delegates to Chicago.

New cars were rushed through production during the 1920s to handle the crowds of travelers. In 1925, 25 million dollars was spent on 933 cars. During most of the decade, 300 to 400 new cars were put into service yearly. In this busy period the fleet grew to 9,800 cars; passengers numbered 30 to 35 million yearly (or roughly 55,000 a night); some 2,800 conductors and 12,000 porters were employed.[243] Except for the freak traffic surge created by World War II, the twenties were Pullman's busiest and best years.

It was also a time for expanding the monopoly. Washington was favorable to business, and Federal antitrust actions were rare. Pullman was free to pick up the few stray sleeping-car operations still running. In 1922 the Great Northern joined Pullman and within three years ended its sleeper operations. The sleeping-car historian Arthur Dubin was told by old Pullman officials that this concession was arranged because of the general indifference of James J. Hill's son to the affairs of the railroad. Whereas the father had been determined to control all the operations of the line, his son was less inclined to do so and permitted, if not welcomed, Pullman's management of the pesky sleeping-car trade.

Next to enter the Pullman system was the comparatively minor Central of Georgia.[244] Since 1880 the little Southern road had run its own sleeper and parlor car service, but in May 1925 it came to see the advantages of a national interchange via Pullman. In this way through traffic might be routed over its tracks. Independent operations limited first-class travel to on-line customers.

The final and most satisfying take-over was accomplished in 1927. For nearly forty years the Milwaukee had snubbed Pullman—owing, it has been said, to a personal feud between the road's president and George Pullman.[245] But now the Milwaukee was the weakest of the Western transcontinentals, with a financial structure that was sagging badly. It could not afford to reequip its name trains, and it needed new cars in order to hold even a small share of the traffic. Pullman, ever alert to ailing railroads, promised the Milwaukee to deliver 127 new or rebuilt cars ready for service by May 21, 1927.[246] Even so the road did not give in on a systemwide basis. Pullman moved in slowly, and as late as 1940 the Milwaukee was still running a few of its own cars. However, Pullman now controlled 96 or 97 percent of the sleeping-car operations in the United States.[247] Only 120 independent sleepers were in service, and these were mostly on unimportant runs. The Canadian Pacific–controlled Soo Line and a pitiful operation on the Gulf, Mobile, and Ohio represented the last of the independents. On a dollar revenue basis, Pullman must have held a 99.9 percent monopoly—an achievement rare in American business.

The consolidation of the sleeping-car industry was surely satisfying to Pullman's management, but it was more a filling in of details than a major advance. The company had, after all, dominated the trade since the turn of the century. Some greater challenge was needed, and expansion into car building seemed a logical avenue for investment. The Chicago shop showed excellent profits in good times—far better than the sleeping-car operation, and surely better than the government bonds in which the company had invested heavily. In hard times when car orders fell off, the sleeping-car division could be counted upon for a steady cash flow. Its business was stable, and depressions appeared to have little effect on it. Like a smooth-running flywheel, it always carried Pullman over the economic dead spots that the country seemed to experience every twenty years or so.

Thus Pullman decided to expand its car-building operations—not in the passenger car area, since the company had about all that business it could hope to win from American Car and Foundry Company, but in freight cars, which represented an even larger market. Pullman's own shop was obsolete, however. Purchasing an existing plant was simpler and faster than building a new set of shops. In nearby Michigan City was the nation's largest freight car manufacturer, Haskell and Barker. At its head was Edward F. Cary (1867–1929), one of the most dynamic railroad car builders in the nation. It seemed a perfect match.

For Pullman, there was more to the merger than an expansion into car building. Pullman's president, well into his seventies by 1920, was eager to retire, but finding a suitable replacement was not easy. Cary was asked to take the presidency, but he declined to do so except on his own terms. Finally a deal was made in 1922 that has been called a Haskell and Barker take-over of Pullman.[248] Pullman bought the Michigan City car works for 15 million dollars. Cary then became president of Pullman. His Haskell and Barker associates filled major Pullman positions: David A. Crawford and Charles A. Liddle were made vice presidents, Ellis Test became chief engineer, and Harold Dudley was named comptroller.

The new men were eager to revitalize the wealthy but stodgy old firm. In the summer of 1924 a major reorganization separated the manufacturing division from the Pullman Company. The parent corporation was called the Pullman Car and Manufacturing Company. The firm would now have a free hand in expanding its manufacturing interests and at the same time could remove this important area of its business from the scrutiny of the I.C.C. Cary engineered a second and more fundamental reorganization in June 1927.[249] A parent or holding company called Pullman Incorporated was chartered in Delaware. The charter was perpetual and liberal, allowing Pullman Inc. to hold the capital stock of both the Pullman Company and the Pullman Car and Manufacturing Company. In the process another major stock dividend was engineered. The reorganization enabled Cary and his associates to acquire new manufacturing properties for the holding company. Yet despite this freedom, they moved cautiously and stayed within the traditional Pullman borders. Negotiations were opened with several firms, including the General American Tank Car Company.

The most serious bargaining was with the Standard Steel Car Company, which had plants in Butler, Pennsylvania; Hammond, Indiana; Sagamore, Massachusetts; La Rochelle, France; and Rio de Janeiro. It also held a large interest in the Osgood Bradley car works.[250] The firm was controlled by the Mellons, a wealthy Pittsburgh banking family. Standard's founder and longtime manager, John Hansen, had retired in 1923 and the Mellons now found the car works a troublesome though profitable burden. They were interested in combining the property with an established concern whose management would direct the day-to-day operations. While the merger was under discussion in 1929, Cary was constructing a new freight car plant in Bessemer, Alabama. He was determined to capture a larger share of the car market by any means available. He may have been pushing himself too hard, for he died unexpectedly in April 1929. Daniel Crawford succeeded his old associate and pushed through the Standard Steel merger late in the same year. The price was 50 million dollars and a directorship for Richard K. Mellon. Pullman's board now included representatives of the largest financial interests in the country: the Morgans, the Vanderbilts, and the Mellons.

PULLMAN IN DECLINE

The great stock market crash had occurred that fall, but Pullman had weathered many a panic, and of course no one knew that a worldwide depression would follow. Nor were there many who foresaw how prolonged or catastrophic the business collapse would be. Pullman soon realized its huge blunder; new-car orders evaporated and its newly acquired workshops stood silent. As one railroad after another slipped into bankruptcy, Pullman's greatly expanded manufacturing branch became a colossal liability. In 1930 it showed a respectable profit, but in 1931 it lost over 2 million dollars.[251] The debt mounted alarmingly in 1932 and 1933, averaging nearly twice the loss of 1931.

The sleeping-car side of the business, believed to be depression-proof, was demolished by the Great Depression. The old dependable flywheel stopped revolving. For the first time in its history the sleeping-car operation showed a loss: 1.2 million dollars in 1932, 56,000 dollars in 1933, up again in 1935 to 1.64 million. Only its giant surplus saved the company, and some 50 million dollars from this fat kitty was paid out to cover losses and dividends during the thirties. Even at that, no dividends were paid in 1931–1933, and in 1934 Pullman reduced its capital stock for the first time in history from 120 to 108 million dollars.[252]

To survive, drastic economies were necessary. Personnel in the car plants that had once employed 16,000 were cut to 2,000 men. New-car orders were slashed; old sleepers were scrapped for what salvage money they would bring in. It seemed pointless to go after new outside car orders, since there were none to fight for. Pullman's only real hope was to rebuild as much of the sleeper trade as possible. By 1933 that traffic was down by 58 percent relative to 1929. Parlor car trade had nearly vanished; no one seemed too proud now to ride up front in the coaches. But it did appear possible to increase sleeper patronage, and rate cutting appeared to be the most direct means of doing so.

In the spring of 1930 rates for berths were cut by 22 percent.[253] A novel arrangement, worked out with the Pennsylvania to attract budget-minded travelers offered a combination rail-highway ticket between New York and St. Louis. During the day passengers traveled by bus; at night they boarded a connecting train and took an upper berth. The fare was equal to that of a coach ticket. In the following year, upper-berth rates were cut a full 50 percent on certain lines. To stimulate lower-berth patronage, the S.O.S.—single occupancy section—plan was adopted. For a premium of 40 percent over the normal lower rate, a passenger could enjoy the luxury of an entire berth section for himself. This eliminated the possibility of a noisy overhead companion. If he wished, the S.O.S. passenger could have a double mattress by using the one reserved for the upper berth. The scheme was well received, a fact that illustrated the growing disenchantment most Americans felt for the open-section sleeping car.

A more basic fare revision was made in 1932 when the 50 percent surcharge was removed, thus reducing Pullman fare to just 0.7 cents a mile above coach fare.[254] *Railway Age* over-optimistically claimed that the reduced fare would double Pullman's business; in reality the company did well to maintain the current levels of traffic and had little success in adding to them. A major deficit in 1935 shook Pullman's faith in reduced fares, and in 1936 the company petitioned the I.C.C. for a 10 percent hike. A weak recovery in the following years encouraged Pullman to install some new cars and upgrade the standard car fleet, but the "little" depression of 1938 revived business pessimism, at least temporarily.

By 1940 things had come right again. The manufacturing divisions were busy with lightweight car orders, and the sleeping-car trade stabilized as the economy began moving again. But the Depression had badly shaken the company, and it was only a prelude to the graver problems that lay ahead. The history of Pullman took a downward turn after 1930, and the decades ahead would not be good ones.

The euphoria resulting from the economic recovery ended abruptly for Pullman on July 12, 1940, when the U.S. Justice Department filed a complaint in the Federal District Court at Philadelphia against Pullman Inc., its subsidiaries, and its officers. It was a civil rather than a criminal complaint, because Pullman affairs had been a matter of public record since the time of I.C.C. regulation and hence no secret conspiracy could be alleged. Yet the charges in Civil Action 994 were serious enough. Principally Pullman was accused of creating an unlawful monopoly in violation of the Sherman and Clayton antitrust acts.

The result of that monopoly was, according to the Justice Department, inferior sleeping-car service given the American public by a self-serving, miserly corporation which ran old-fashioned cars of its own manufacture. By its policies and restrictive contracts, Pullman had systematically stifled, thwarted, and restrained the railroads and other independent car builders from introducing modern sleeping cars. So long as Pullman enjoyed its monopoly, there would be no incentive to improve service; and so long as Pullman was at once the operator and builder of sleeping cars, little innovation could be expected. Through its operating contracts, some of them extending into perpetuity, the production of sleeping cars by non-Pullman subsidiaries had been blocked. In an eighty-page printed statement, the Justice Department outlined Pullman's minor and major sins. A separation between the sleeping-car and manufacturing divisions was sought in the public interest. And so began a legal battle that plodded through the courts for seven years.

Why had the Justice Department waited until 1940 to move against Pullman? As a very public corporation its monopoly had been obvious for years. True, the climate in Washington was no longer probusiness, and the White House was less inclined to discourage the Justice Department from implementing the antitrust laws. But the true source of the suit became evident as the case progressed. It appears that the Budd Company had promoted it because of Pullman's success in suppressing the purchase of sleeping cars from other builders, such as Budd. It seems more than coincidental that the case opened in Philadelphia, the home of Budd's car plant, rather than in Pullman's headquarters town of Chicago.

The quarrel with Budd had begun in 1935.[255] In January of that year Pullman's contract with the Burlington expired, and the railroad was not anxious to renew the agreement. Pullman continued to provide service for over a year while new terms were under negotiation. Meanwhile the Burlington decided to purchase some sleeping cars from the Budd Company to match its latest streamlined equipment. This was heresy to Pullman, which refused to consider operating sleepers that were not of its manufacture. In the ensuing argument Pullman claimed that the Budd cars were not up to standard and insisted on innumerable petty changes in bedding and accessories after the cars had been delivered. Pullman also stipulated that a "hold harmless" clause be added to the contract in the event of a structural failure—thus casting doubt on the soundness of Budd's Shotweld construction. It further insisted that the Burlington buy no more than ten sleeping cars from Budd. If the Burlington would not meet these demands, it could establish an independent sleeping-car operation. The railroad gave in to Pullman. A similar battle between Pullman and the Santa Fe developed two years later and again Pullman won, permitting Budd to furnish no more than five cars. After the antitrust suit opened, Budd claimed that it was impossible to secure sleeping-car orders because of Pullman's control.[256]

Pullman denied that Budd had been prevented from gaining a legitimate portion of the passenger car business and that Budd offered a product superior to that of the Pullman shops. To the contrary, Budd had openly threatened to instigate government proceedings if Pullman failed to purchase a license for Shotwelding and buy a percentage of new Budd cars. Thus Pullman accused Budd of blackmail, and to the various charges of the Justice Department, Pullman heatedly denied that it was a sedentary monopoly, indifferent to progress or the public's comfort. Pullman took particular umbrage to the criticisms of its rolling stock. Admittedly the bulk of the fleet was composed of heavyweight steel cars, but many of these were under fifteen years of age and all were well maintained and comfortable. Indeed, the company had spent some 30 million dollars on air-conditioning the standard cars during the past decade. While its 388 lightweight sleepers represented only 5.6 percent of its fleet, this figure compared rather well with the general passenger car stock of all the nation's railroads, in which only 3.7 percent were streamliners. Pullman contended that it was moving with all reasonable speed to improve its equipment: 71 lightweights were on order, and many old cars were being remodeled with room-style accommodations. Its lawyers also argued that Pullman was a benevolent monopoly, even if it was not a natural one like the postal and telephone systems, and should not be disbanded merely because it was a nationwide enterprise.

The final judgment, announced early in May 1944, ordered Pullman Inc. to separate its manufacturing and operating divisions. One or the other must be sold within a year; the company had ninety days to decide which one to keep. Meanwhile it must respond to any reasonable demand for sleeping-car service, whether covered by a contract or not, and it must seek competitive bids for all new sleepers—thus ending the in-house purchase of equipment.

The choice of which division of the business to retain was apparently not difficult. Even in the mid-1940s the declining fortunes of railway passenger travel were clear enough. Pullman decided that the operating division would go on the block. At the same time, management resolved not to appeal the case. Perhaps Pullman was secretly happy to have an excuse for unloading the sleeper business before it slipped into a major decline. The court-ordered sale price was based on an assessed value of 75 million dollars. At first no one seemed interested; the anticipated sale to the user railroads did not materialize.

As a device to lower the asking price, Pullman then agreed to sell its 600 streamlined cars to the railroads, thus trimming 35 million dollars from the valuation. By 1945 several customers were interested. The most active was Robert R. Young, (1897–1958), the controversial financier who had alienated the railroad industry through a publicity campaign about the poor quality of passenger service. His advertisements—"A hog can ride across the country without changing cars but you can't"—received wide attention. Young's simplistic solution for reversing the decline in railroad passenger travel was to install new equipment and offer coast-to-coast service. Through the investment bankers Otis & Company, he petitioned the courts for permission to buy Pullman on the same terms offered to the railroads.[257] He promised a grandiose modernization program costing 500 million dollars.

Suddenly the railroads were keen to negotiate with Pullman. The prospect of Young managing all sleeping-car service, or in any way having a wider field of influence, was anathema to many railroad managers. Pullman then sought and won a court order in January 1946 requiring that the sale be made to a consortium of railroads.[258] Possibly Pullman took this action to retain the favor of its future customers. In March the Justice Department appealed the sale to the Supreme Court, contending that railroad ownership would perpetuate the abuses of the old monopoly because the owner and lessor would be one and the same. Meanwhile the railroads sought I.C.C. approval of the sale, hoping to influence the Court's decision. Young threatened to organize his own independent sleeping-car company and thereby give Pullman a taste of competition it had not experienced since Wagner's demise. Nothing came of Young's plan; like his promise to reequip the Chesapeake and Ohio (a subsidiary of the Alleghany Corporation, which he headed), it just faded away.

Late in March 1947, almost three years after the sale had been ordered and seven years after the case had opened, the Supreme Court approved the sale of Pullman's sleeping-car property to the railroads. The purchase was completed on June 30, 1947, when 731,350 shares of stock were transferred to fifty-seven cooperating railroads. They had established the new Pullman Company for $40,202,482.[259] Carroll R. Harding (1888–1963), a civil engineer and former Southern Pacific executive, was named president of this major enterprise owning 6,000 heavyweight sleepers and employing 30,000 people. While passenger miles had declined 45 percent and revenues were down 39 percent since the wartime peak, its performance was still well above the average traffic of the Depression years.

The anticipated growth in the travel market gave Harding hope that Pullman could survive and prosper. There was some danger that the railroads might undercut Pullman by installing superior coach service, except that the roads could now expect to benefit directly from any profits Pullman might earn, and maintenance of the pool cars was as important as ever. Another advantage was the fact that the railroads were still a major means of intercity travel, and on the long hauls, passengers were willing to pay for sleeping accommodations. Americans wanted to travel and enjoy spending the money they had saved during the war, and Pullman was likely to share in that substantial cash flow.

By eliminating jobs and consolidating operations, Harding trimmed 10 million dollars from the 1948 budget. Railroad purchases of the new lightweight sleepers relieved Pullman of this major investment and built up the stock of attractive new cars. By 1949, 1,000 streamlined sleepers were running. Between 1947 and 1950 about 100 open-section standard cars were remodeled as room-style cars. It was hoped that all the heavyweights could soon be eliminated from regular trains. By retaining the traffic it already had, cutting costs, and upgrading equipment, Pullman seemed to have a future.

Yet the results for 1948 were not encouraging. Traffic continued to slip—not gravely, but enough to cause concern. Profits were a meager 1.7 percent. During the following year the forty-hour week drove wages up by 9.8 million dollars.[260] In the spring of 1949 air passenger travel exceeded Pullman's for the first time. Pullman cut service by 16 percent, eliminating most of the marginal runs. The streamlined cars, once regarded as the saviors of railway travel, were proving more costly to maintain than standard steel cars because of the added wiring, plumbing, and gadgetry. They also earned less because of their smaller payload.

In the early 1950s the decline accelerated. Traffic was below Depression levels; the airlines now moved twice as many passengers. In 1955 Pullman had to borrow 1.2 million dollars to meet the regular dividend payment. A major retrenchment program was necessary to cut losses. In 1953, 82 more lines were suspended. During the next few years all repair shops except the St. Louis and Chicago works were closed or sold.

A greater threat to Pullman's corporate life came in the late fifties, when some of its major users began to pull out. In 1957 the Pennsylvania took over its own parlor car operations. In July of the next year Pullman lost one of its major customers when the New York Central withdrew completely.[261] It was paying Pullman mileage fees of $814,000 in excess of the sleeper revenues. Few cars were in interchange service, and extra train movements were now so small that the pool cars were little used. The Pennsy threatened to follow the Central, but somehow Pullman managed to stall off wholesale withdrawals. Major cuts were necessary; nearly 3,000 employees were let go, so that the Pullman work force shrank to one-third of its 1947 size. Most of the heavyweights were retired or sent to government storage yards. Traffic had fallen off so badly that there was now even a surplus of lightweights.

The long postwar Indian summer that Pullman had enjoyed was at an end by 1960. There was no more talk of rebuilding traffic; all efforts were directed toward cutting costs and somehow keeping the company alive as long as possible. Barely solvent, superannuated like some ancient Civil War veteran, its only ambition now seemed to be to reach its centenary. Pullman's decline was irrevocably linked with the general collapse of railway passenger travel. By 1960 the great mass of intercity travelers were moving by private automobile. The 10 percent or so who still used commercial carriers could choose between buses, railways, or airlines. Bus schedules on new interstate roads now rivaled rail schedules, and buses were far cheaper. The airlines had the great advantage of speed. Businessmen, salesmen, field engineers—Pullman's bread and butter—abandoned overnight travel when they found that they could attend to their business in a single day by flying. Air fares also offered a considerable saving when compared with first-class rail tickets. Pullman began to price itself out of the market: as patronage declined, it raised fares. In 1956 and 1957 two increases jumped fares by 15 percent. Soon it cost twice as much to go by Pullman as it did to fly.

By 1965 the Pullman Company was heavily in debt. Its losses exceeded the value of its stock by 300 percent that year. Its traffic dropped, not by 5 or 10 percent compared with the previous year, but by nearly 50 percent. Its cars were now all railroad-owned except for 135 heavyweights in storage and 300 standard cars leased to Mexico. The Pennsylvania pullout in August 1967 was another blow to Pullman's sagging fortunes. By 1968 Pullman was operating on just twenty-seven railroads, and while every cost had been trimmed, its losses were a runaway 22 million dollars. After twenty-one years of railroad ownership, Pullman decided to quit. Late in 1968 it was announced that all operations except some minor housekeeping and maintenance service would cease on January 1, 1969.[262]

In its one hundred and first year, the sleeping-car company expired. But the Pullman name lives on with the manufacturing division, which has prospered since the 1947 separation.

Mechanical Development of Sleeping Cars

Through most of its history the development of sleeping car design and manufacture was dominated by Pullman. The first generation of Pullman cars began in 1865 with the *Pioneer*. Then came a series of twenty nearly identical sixteen-wheel giants.[263] Floor plans and details varied, but all appeared to follow a standard plan of berth arrangement. Next, many different builders and shops produced a largely mongrel assortment of cars. At the same time, Pullman was experimenting with car types, trying to decide whether to offer a straight sleeping car or some composite form which would include parlor, dining, and sleeping facilities. The diner car immediately proved to be a financial disaster, and Pullman hurriedly ended that service unless the railroad agreed to subsidize it. The hotel car was a more feasible scheme, and various combinations of sleeping-buffets and parlor-buffets became a part of Pullman's regular fleet. Light food service proved both popular and profitable. Luxury cars for daytime travel began to appear in Pullman consists during the early 1870s.

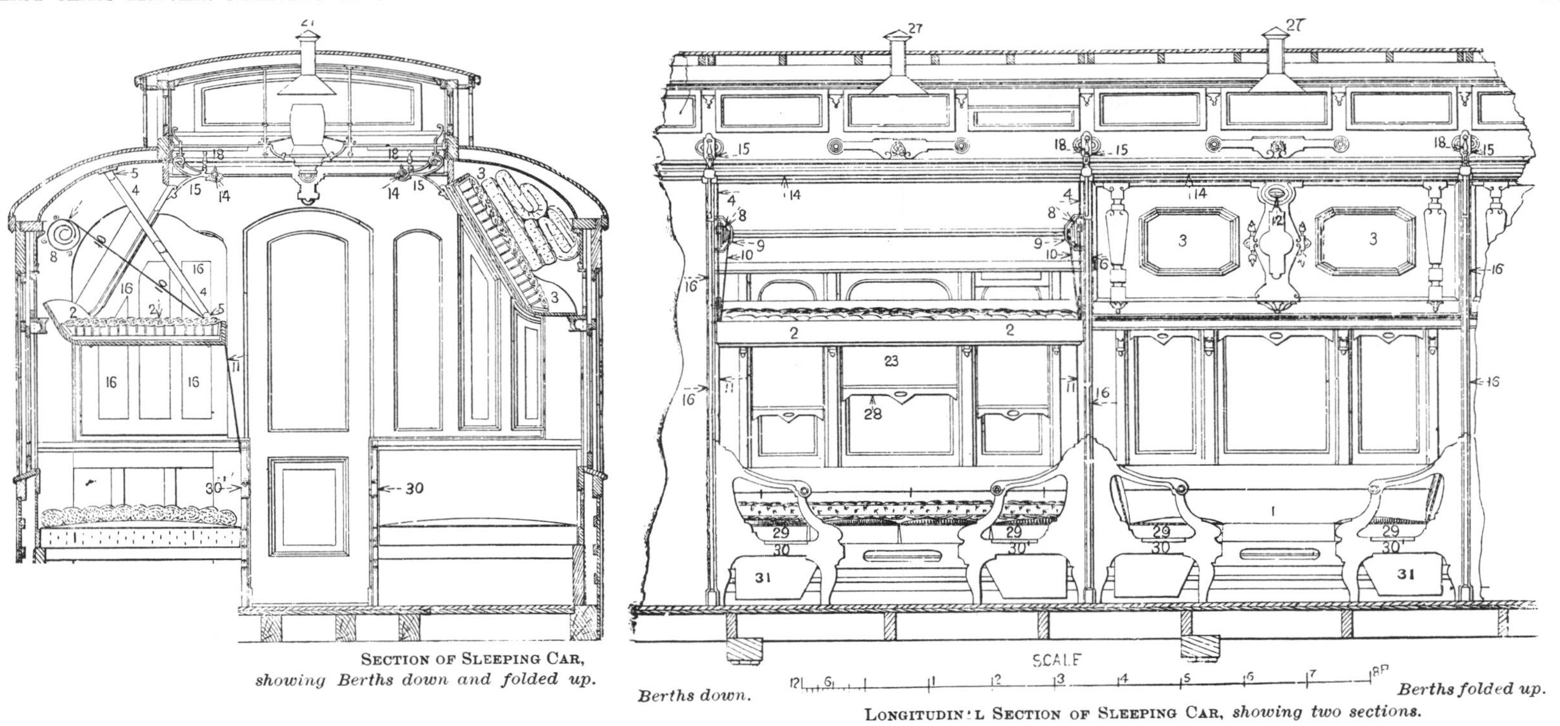

Figure 3.51 The mechanics of the convertible lower and the folding upper berth are shown in two drawings first published in 1879. Note the bedding stored in the right upper berth. (Car Builders' Dictionary, *1879*)

Pullman did not invent the parlor car, as is sometimes claimed. However, he recognized it as a natural adjunct to the luxury night trade and provided and ran such cars under contract for participating railroads.

One of Pullman's most important acts was his early decision to concentrate on the open-section sleeper. It made up the backbone of his fleet and remained the standard American sleeping car for three-quarters of a century. The open-section car was the clear favorite by 1870. Compartment sleepers were rejected again and again during the following decades, and while room cars were occasionally produced, Pullman remained firmly committed to the open-section sleeper until the late 1930s. No matter what its defects in passenger privacy or comfort, it was the cheapest, lightest form of sleeping car.

Coupled with Pullman's loyalty to the open berth was a strong allegiance to a single floor plan. The twelve-section (twenty-four berths), single drawing-room car was the favorite. Again so far as can be determined, Pullman settled upon this plan by the early 1870s. The construction and appearance of a standard Pullman sleeper is shown in many illustrations in this chapter (Figures 3.51 to 3.54, for example). The following description appeared in the July 1879 *National Car Builder*:

The "Keystone" and "Empire" are specimens of the present standard patterns of sleeping cars used by this company. Their length outside is 57 feet 3 inches. The interiors of the cars are finished with walnut near the floors with panels of prima vera, a California wood as above. The raised part of the panels is Circassian walnut, and all the mouldings are of amaranth, a beautiful dark-red wood. The panels on the under sides of the upper berths are decorated with inlaid work of prima vera and ebony. The ceilings are finished in oak panels with amaranth mouldings, and decorated with painted floral designs. The seats are covered with ponceau plush; the window curtains are made of Scotch tapestry of two shades of tan. The berth curtains are of entirely new design, and made in patterns expressly for these cars. The cars contain 24 berths. The ladies' dressing-room is at one end and the gentlemen's at the other. The latter is arranged with a state-room between it and the main saloon of the car, so that it is entirely shut off from the latter. This is the best arrangement for sleeping cars that has yet been devised.

Figure 3.52 The upper berth is down, ready for its nocturnal occupant, while the lower seats remain as they would appear during the day. The center table was removed and stored in a closet when the lower berth was made up. (Car Builders' Dictionary, *1879*)

Pullman followed the general practices of wooden car construction described in Chapter 1. Because sleepers tended to be the longest and heaviest cars in service, the more elaborate methods of side framing were employed. The earliest form of Pullman trussing for which drawings can be found is that of the *Enterprise* (Figure 3.50), presumably a typical Pullman design for the 1870s. The growth in complexity and length over the next

Figure 3.53 A berth is made up in the sleeping car Anton *of about 1890. The curtains are not in place on the second section in this view. The section in the immediate foreground is shown ready for daytime traveling.* (*Pullman Neg. 2827*)

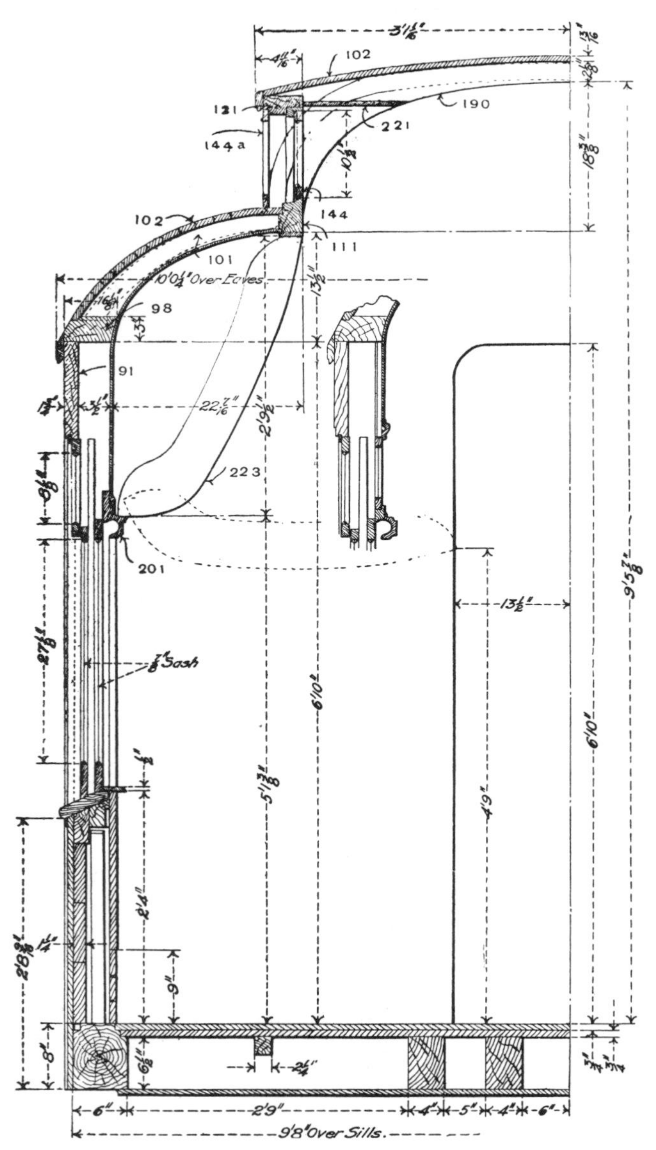

Figure 3.54 A cross section of a Pullman sleeping car from around 1895 shows the upper berth in the closed position. (Car Builders' Dictionary, *1906*)

twenty to thirty years is shown by the wooden body frame drawings in Figure 3.55.

Pullman fought the same losing battle against increasing car weight that faced all American car builders. The trend toward larger, more luxurious cars forced tonnage up, perhaps even more noticeably with sleepers because of their extra burden of bedding. In his 1874 statement to the Windom Committee, Pullman tried to dismiss his car weights of 28 tons as being only 2.5 tons heavier than a twelve-wheel coach. But a more representative coach weight for the period was 20 or 22 tons, making Pullman's cars closer to 6 or 8 tons overweight. And weight became an ever worse problem. By the mid-1880s a 70-foot Pullman car weighed 39 tons (Figures 3.56 and 3.57). By the first years of the next decade 50-ton Pullmans were common, and by the end of the wooden era sleepers had reached 60 tons.[264]

There seemed to be no way to avoid raising weight as overall size increased, as end construction became heavier, and as extra appliances were added. Eight-wheel trucks were abandoned by the late 1860s, but the six- and four-wheel trucks quickly gained in mass and weight as larger wheels, axles, and frames were required to support longer cars and sustain higher speeds. Lighting and heating methods were constantly improved. Gas lighting called for tanks and piping, electric lighting for generators and batteries. Hot-water heating required a massive stove and a system of piping; steam heating did not reduce weight by much. Improved toilet facilities, demanded by the Pullman first-class clientele, were very heavy. Water tanks and pumps were common by the seventies. By the end of the following decade, air-activated water systems were installed. Vestibules were introduced during the same years, adding several tons to each end of the car. But Pullman had to feature the latest improvements, no matter what effect they had on total weight.

Adopting the newest auxiliaries was one way to upgrade the fleet; remodeling older cars was another. Such veterans as the *Pioneer* were kept acceptably modern by progressive alterations in the lighting and heating systems and by adding air brakes and automatic couplers. Roof lines, trucks, and exterior paint schemes were changed to follow the style. Older cars were reassigned to less important runs (Figure 3.58). But it was also desirable to retire obsolete cars, so that while the size of the Pullman fleet might appear relatively stable from one year to the next, the cars themselves were continually changing as new ones took the place of retired stock. In 1873, 120 new cars were placed in service.[265] Most of the old Central Transportation Company's cars were

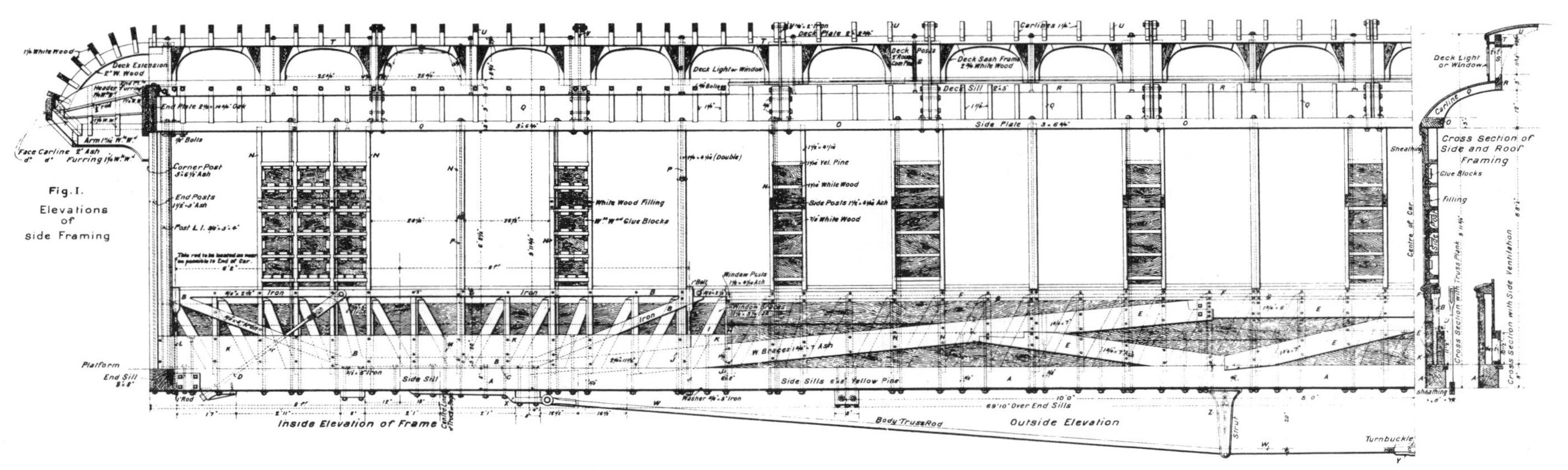

Figure 3.55 Pullman sleeping-car wooden-framing plan as evolved by about 1895. (Engineer, *January 22, 1897*)

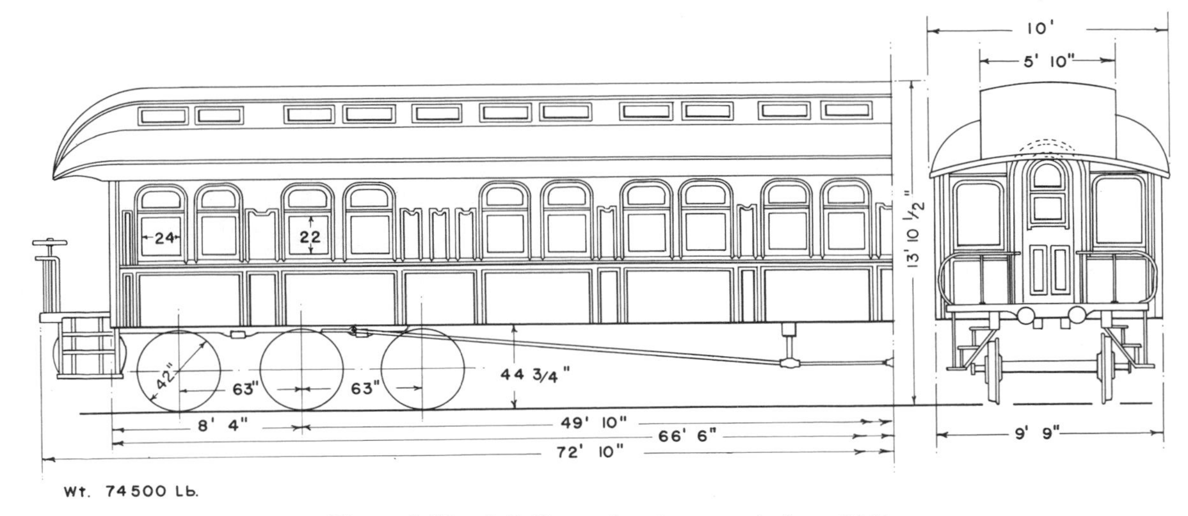

Figure 3.56 A Pullman sleeping car of about 1880.

Figure 3.57 The Vancouver *was built for service on the Rock Island line in 1889.* (*Pullman Neg. 164*)

Figure 3.58 Older cars were often remodeled with new interiors, but the Oconee *was downgraded (see the rattan seat coverings) for tourist service. The berths have been made up on the right side of the car except for the front section.* (*Pullman Neg. 2749*)

Figure 3.59 The Laconia *of 1898 had a wide vestibule end.* (*Pullman Neg. 4254*)

retired between 1879 and 1881, according to Pullman's annual reports. The reports note all additions and retirements—1883, for example, was a very busy year, with 121 new cars added and 50 old ones sold or scrapped.

To maintain the fleet on a nationwide basis, Pullman established regional repair shops between 1873 and 1889 in Wilmington, Delaware, Elmira, New York, St. Louis, Denver, and Buffalo. In later years repair facilities were opened in Atlanta, Georgia, Richmond, California, and Calumet, Illinois (the Calumet shop was adjacent to the main Pullman car plant south of Chicago). New-car construction was undertaken in Pullman facilities at Detroit (1870) and Chicago (1881). Repair work was done at both plants.

Since Pullman's maintenance of the everyday sleeper fleet was not particularly newsworthy, the press emphasized the premium cars that the company furnished for the name trains. There was only so much to be said about a standard twelve-and-one sleeper, but the special cars prompted extended comment. Two luxuriously furnished cars, the *Pilgrim* and the *Puritan*, were turned out for Boston–Montreal service in 1883. The *Railroad Gazette* of February 2 described them with admiration:

> The woodwork is mainly dark mahogany, finely carved and highly polished. The upper berth panels and partitions are inlaid with satin, ebony and other fancy woods elaborately carved. The upper deck and roof are lined with whitewood veneering, which is artistically hand-painted. The upholstery is in crimson plush, and the window and berth curtains are of raw silk. There are ten large centre lamps of beautiful design. These and the other trimmings in the car are of burnished brass or silver. The carpet is of a rich pattern, and the glass in the doors and partitions is of French plate, neatly embossed. The car is luxuriously finished and furnished. A spacious wash-room is located at one end of the car, fitted with bowls, mirrors and toilet utensils of the best make. Adjoining this is a cosy compartment luxuriously furnished for a smoking room. A passageway around this smokers' salon leads to a handsome buffet which faces the main compartment of the car. This little room is attractively fitted up with a side-board of fanciful design, which contains a set of handsome breakfast table-ware for serving out tea or coffee, oyster stews and other light refreshments at a nominal price to the passengers. A small safety stove and other utensils are arranged in a separate compartment, the whole being in charge of a competent steward. At the other end of the car is a handsome parlor or drawing room for ladies, with a commodious lavatory adjoining. The car is warmed by hot-air pipes under each seat. The exterior of the car is painted a dark olive green, and is richly ornamented in gold and colors. The 'Pilgrim' is similar in all respects to the 'Puritan,' and cost some $22,000.

By the end of the eighties whole trains were being produced to capture the luxury trade. One of the first was the Pennsylvania Special.[266] The five cars—the first Pullmans to be fitted with

Figure 3.60 Elaborate exterior gilt decorations were abandoned long before the Lucerne *entered service around 1900.* (*Smithsonian Neg. 48376*)

vestibules—were ready for public display by the spring of 1887, but they did not enter regular service until the following summer. Three of the cars were sleepers: the *America*, the *England*, and the *France*. The interiors were paneled in mahogany and satinwood. The berth seats were covered with a pale-blue glacé plush; the bedrooms were upholstered in a terra-cotta-colored plush.

Other railroads began to acquire whole trains of new stock. In 1888 the Golden Gate Special which began running in the West featured electric lights, steam heat, and an observation car with a recessed end platform and floor-length end windows.[267] Next year came the Montezuma Special, a splendid new train for the New Orleans–Mexico City run. Two of the cars, exceptions to Pullman's open-section rule, were on the compartment plan, with six drawing rooms each.[268]

In keeping with the eclectic taste of the times, Pullman interiors became more diverse in style and even more elaborately decorated as the final decade of the nineteenth century opened. Or rather, ornamentation increased in interior appointments; curiously, the exteriors became more chaste during the same time. The heavy overlay of gilt striping, panels, and baroque scrolls on the outsides of cars slowly gave way to a relatively simple gold border lining (Figures 3.59 and 3.60). The focusing of attention on interior furnishings appears to have been motivated by the desire to create something special for the coming Columbian Exposition. Pullman was a sponsor of the 1893 World's Fair and naturally wanted his own firm to be well represented. Wooden car technology was near its final limits; sleepers with 70-foot bodies were common. But the subtle improvements now possible in car construction would be lost on a general audience, whereas a major change in interior decor would be clear to everyone who had traveled in a sleeping car. Accordingly Pullman's chief draftsman, August Rapp, produced a design in 1892 that skillfully wove the rich Empire style into a form suitable for sleeping-car architecture.[269] The ceiling and berth fronts took on full curves like a Roman vault (Figure 3.61). The seat ends looked like an oversized copy of a Napoleonic bedroom set. In the flashy Empire manner, the moldings, torches, winged lions, and rosettes were finished in gold ormolu. The sleeper-observation *Isabella*

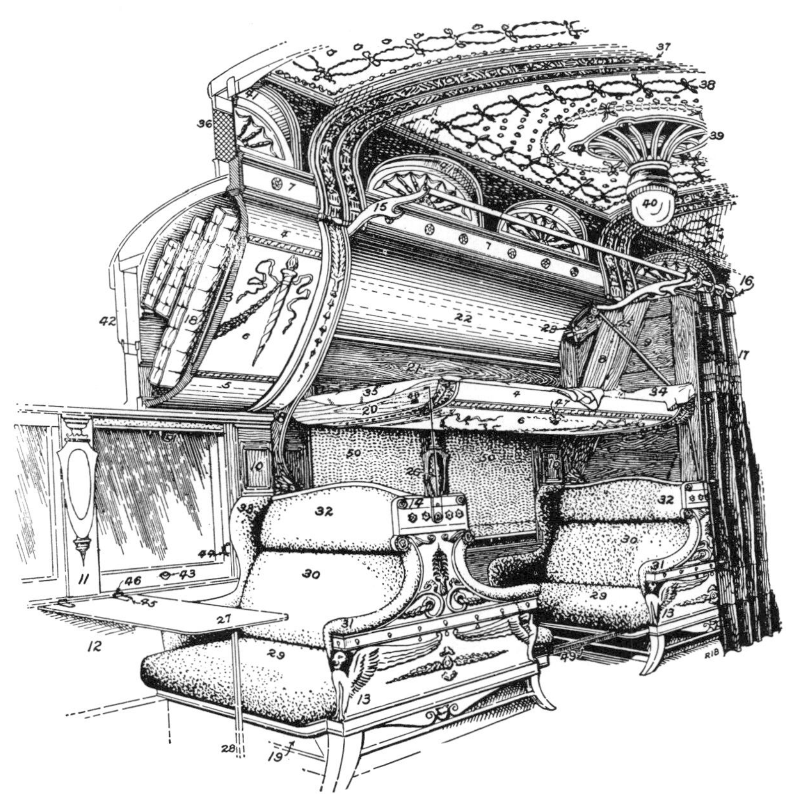

Figure 3.61 While exteriors became more chaste, Pullman designers put greater emphasis on interior decor. The Empire-inspired Columbian style of interior was introduced in 1892. (Car Builders' Dictionary, *1906*)

was one of the special cars built for the 1893 Columbian Exposition. It set the style for the top-grade Pullman sleepers for nearly a decade. No basic improvement or change in general arrangement or comfort was incorporated in what came to be called the Columbian style of interior. The berth and seating mechanisms were identical with those of the 1870s.

With the coming of the twentieth century Pullman continued as one of the great forces in passenger car architecture. In 1904

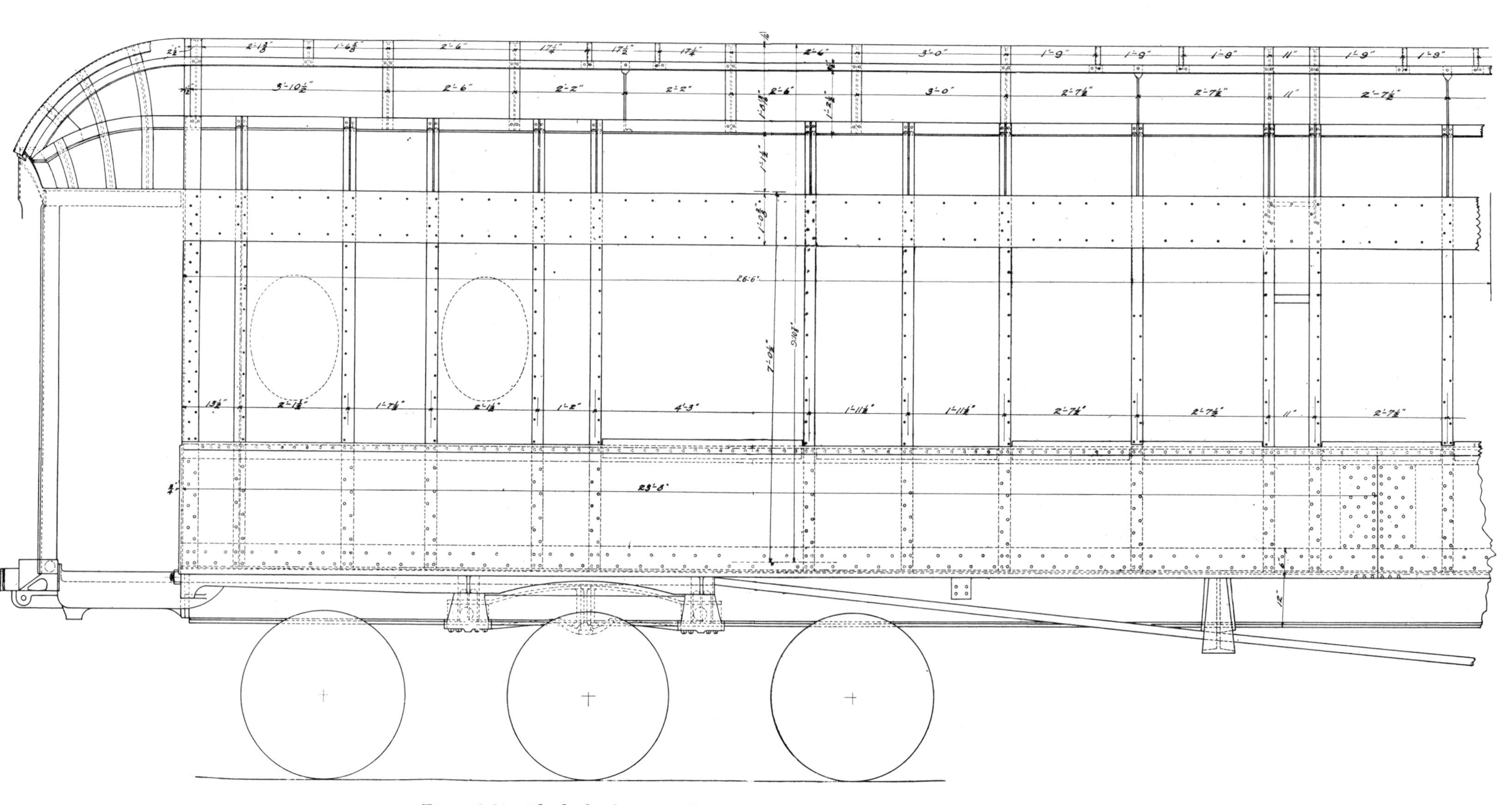

Figure 3.62 The body plan for Pullman's experimental all-steel sleeper, the Jamestown *of 1907. The car was considered too heavy and the design was not repeated.* (Master Car Builders' Report, *1908*)

Figure 3.63 The Carnegie *of 1910 was the first production model all-steel sleeper to enter service. The 68½-ton car was converted for tourist service in 1936. (Pullman Neg. 12112)*

the railroads abruptly dropped the amazing carvings and the North German Lloyd grandeur of the Columbian style for the plain lines of the "sanitary" interiors. The rustic mission style overthrew the luxuriant Moorish. Pullman was quick to accept the new style. It showed considerably less enthusiasm, however, for the shift to steel cars. The company was much more comfortable with the traditional materials of construction. Only after some major customers showed their clear intention of entering the steel era did Pullman begin serious investigation of metal car design. Sometime in 1906 the directors, including J. P. Morgan, met to consider the advisability of such a move. Richmond Dean, the chief engineer, described the historic meeting:[270]

The directors considered and disposed of some unimportant routine matters, and then Mr. Lincoln—Robert T. Lincoln, then president of the Company—raised the question of steel construction. Before he had uttered three sentences of his preliminary statement, Mr. Morgan interrupted:

"Can we build all-steel cars?"

Mr. Lincoln turned to me, passing the question along, and indicating that it was my time to talk. It had come more suddenly than I had expected. But I replied:

"Certainly we can;" and while I was pulling myself together for the detailed exposition, Mr. Morgan went on:

"Then I move that we proceed to do it."

The motion was adopted; nobody wanted to see my plans and estimates; nobody asked any more questions, and we just went back to the works and proceeded to build steel cars.

In truth mass production of steel cars was nearly four years away. Pullman did, however, begin work on an all-steel twelve-section sleeping car, which was completed in March 1907 for exhibition at the Master Car Builders' Convention. It was named the *Jamestown* in honor of the tricentenary of the first English settlement in North America. The *Jamestown* was not mechanically a notable success, although in general arrangement and appearance the car was acceptable enough (Figure 3.62). It was comfortable, pleasing (even if a bit severe in its internal appointments), and as good-looking as any standard sleeper of the day. It was a skillful steel reproduction of its wooden prototype, with rivets that were countersunk and puttied over, and queen-post truss rods.

Pullman apparently attempted to build the car from commercially available steel shapes, using as little special fabrication as possible. After all, the company had little metalworking equipment and was probably trying to see how good a car could be produced without extensive retooling. The center sill was formed from two standard 15-inch I beams joined by a ½-inch bottom plate. The side sills were 6- by 6-inch angles; the belt rail was a flat 3-inch bar. The cast-steel cross sill, platform, and bolster end piece was one of the few special items used in building the *Jamestown*.

The *Jamestown* was said to ride smoothly and quietly; none of the expected battleship rattle developed.[271] But the car was grossly overweight, so much so that its tonnage was kept something of a secret. Of the detailed reports that appeared in the American technical press, none mentioned total weight. The 1908 Master Car Builders' report on steel car construction estimated the weight at 80 tons. The Gutbrod Report of 1913 gave a figure of 81 tons. Presumably the latter was correct; probably the initial embarrassment of the *Jamestown*'s poundage had passed sufficiently by 1913 for the company to release full specifications.

Few engineers expected that a lightweight design was possible in view of the safety, comfort, size, and cost requirements. Nevertheless, 81 tons seemed excessive. Surely a compromise plan could produce a lighter car. A second steel sleeper was reportedly designed in 1907–1908 for service on the Union Pacific, but no record of it can be found except for incidental mention in the 1908 M.C.B. Report. Possibly the car was an improvement on the *Jamestown*. In 1908 Pullman built five steel combine cars for its own service. Again not enough information exists to determine whether they contained notable design advances. During the next year or two, nearly ninety steel passenger cars of various types were produced at the Pullman car works for the Harriman Lines and the New York Central. A new steel car plant was outfitted at the Chicago works in 1909 to facilitate the quantity production of such cars.[272]

Pullman designers were developing a feeling for steel construction. The new plant allowed greater freedom in the manufacture of special shapes, so that plans could be more sophisticated and reductions in weight more substantial. Pullman's experience in building steel cars commercially was of great benefit. In addition,

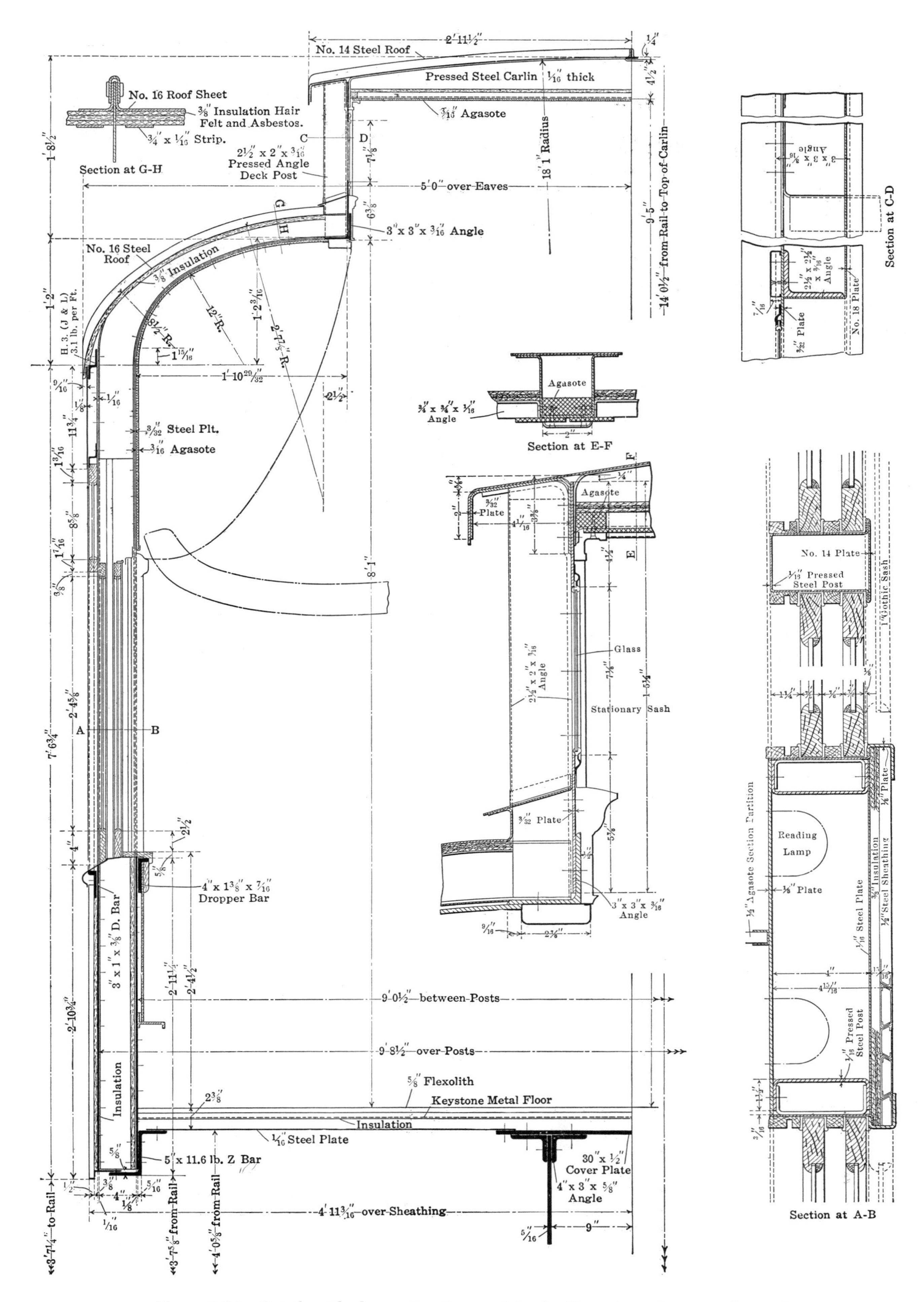

Figure 3.64 Details of body construction used in the Carnegie *and many other early steel Pullmans were published in 1910.* (American Railroad Journal, *October 1910*)

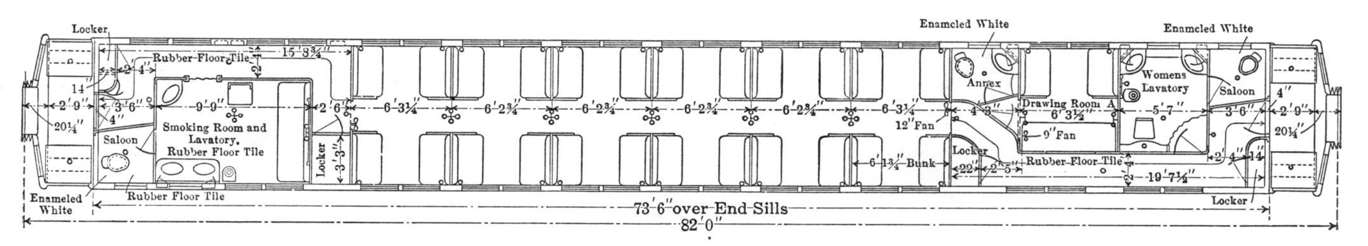

Figure 3.65 The Carnegie*'s twelve open sections and single drawing room repeated Pullman's most favored floor plan.* (American Railroad Journal, *October 1910*)

Pullman as well as most other car builders profited from the achievement of the Pennsylvania Railroad designers who had perfected an improved design for passenger cars. It came to be recognized as the key to practical steel car design.

In 1909 Pullman began work on another prototype all-steel sleeper that would incorporate everything that the company had learned since producing the *Jamestown*. The new car was intended for service on the Pennsylvania. Completed in January 1910, it was named the *Carnegie* (Figures 3.63 to 3.65). The 73-foot, 6-inch body held twelve open sections and one drawing room.[273] The metal interior was grained to imitate mahogany paneling. The total weight was 68½ tons, a saving of nearly 12 tons over the *Jamestown*.

There was no question about the practicality of the *Carnegie*; it was a car that would serve long and well. Within a month Pullman was at work on an order for 500 steel cars. The company made a voluntary commitment to convert its sleeping cars to steel equipment on a nationwide basis as soon as possible. No new wooden sleepers were ordered after 1910. The better wooden cars were sent back to the shop and steel-underframed; by 1913 over 600 had been processed.[274] In the same year, 2,100 all-steel sleeping cars were running on American lines. Pullman claimed that one-third of its fleet was steel; 857 cars at a cost of $22,000 each were installed in 1913 alone. It was a remarkable record, a conversion program in which Pullman outpaced the railroads by decades. In 1926 three-quarters of Pullman cars were all-steel, and all but 4 percent of the remainder were steel-underframed. By the middle 1930s the last wooden-bodied cars were retired.[275]

The coming of the steel car gave Pullman an opportunity to standardize. Because the conversion to new stock was made quickly, it was possible to stay with a single basic design that was not likely to become obsolete within the period of the change-over. Of the 8,000 cars built for Pullman use in the quarter century of the heavyweight era (1910–1935), one-half were of a single uniform pattern.[276] The old standard floor plan of twelve open sections and one drawing room continued into the steel era. Based on the *Carnegie*, one lot after another of twelve-and-ones was manufactured. A new design was introduced in 1923 (referred to as plan 3410), but only an expert could see much difference in it. This is not to say that Pullman did not operate a great variety of cars. Sleepers were offered in all combinations of open-section and room arrangements. Some all-section cars were built; others were parlor-lounge, parlor-buffet, sleeper-lounge, parlor, observations, and all-room cars. Eventually there were 100 different floor plans and, if subclasses are counted, 457 additional styles.[277] It may not sound like such an effective job of standardization, but the cars themselves were manufactured from standard elements. The basic cars were identical shells. The frame, trucks, roof and body parts, and mechanical equipment were the same. Berths, seats, lighting fixtures, hardware, window frames, and doors were standard elements that could be combined to produce a great variety of car types. Hence standardization was realized to a very large extent in the heavyweight era. In the faster-changing lightweight period, such uniformity was not possible.

Standardization, so beneficial to economical operations, also discouraged innovation. One can argue about what is useful innovation and what is merely change for its own sake, but except for air conditioning, Pullman could hardly name any major improvements made during the standard era. A list published in 1932 enumerates items such as legible bulkhead signs, bottle holders, and new window curtain fixtures.[278] Pullman's critics pointed to lists like this as signs of neglect and reaction, as the Justice Department did during the 1940 antitrust suit. The fairness of such criticism might be challenged, for Pullman was offering comfortable, safe accommodations and sincerely believed that little could be gained from any radical design changes.

The most unfortunate consequence of Pullman's conservatism was the decision to stay with the open-section sleeper (Figures 3.66 to 3.69). Criticism of this plan became more pronounced with the passing years, and Pullman came to regret that it had not shifted to room-style cars at the beginning of the steel era when such a change would have been possible. As a convertible day-night car, the open-section sleeper was a compromise, and like many compromises it did poorly in satisfying either requirement. It was an inherently makeshift arrangement. In the daytime the rigid seat was not very comfortable. It could not recline, and because of the face-to-face arrangement, half of the passengers were forced to ride backwards relative to the train's movement. The windows were necessarily small owing to the upper berth. The difficulties of undressing and of clambering into the upper berth have already been mentioned. One magazine spoke indignantly of the degrading spectacle of disrobing in a Pullman aisle: "It is indecent, primitive, rural and jay."[279] The limited toilet facilities were a second failing of the section cars. Ladies' and men's washrooms were located at opposite ends of the car. Going to the toilet in the middle of the night meant a stumbling journey down a narrow, curtained aisle to the far end of the car. In the morning it meant standing in line awaiting a turn at one of the community washbasins. As indoor plumbing became a common feature, the limited facilities offered by Pullman seemed ever more inadequate.

Bedtime entailed minor pandemonium as the porter began to assemble the berths. Passengers had to find seats elsewhere while the beds were unfolded; they could take a seat with a neighbor or retreat to another car. Some passengers wished to retire early, others late. As a result the car was partially made up for much of the evening and thus had a confused appearance that seemed to underscore its makeshift plan. In making up the berths the porter opened the upper berth, where the bedding, curtains, and mattresses were stored. The upper was already made up, but the lower had to be assembled from scratch (Figure 3.70). The seat cushions were first rearranged to form the bed, the mattress was

Figure 3.66 A standard Pullman open-section sleeper as it appeared in daytime, with all the bedding hidden inside the closed upper berths. The seats, unlike those in a coach, could not be reversed. The photograph was made in 1916. (Pullman Neg. 20672)

Figure 3.67 This cutaway drawing shows berths both open and closed. Notice the extra-large men's washroom, which also served as a smoking compartment.

Figure 3.68 The four-square appearance of a standard steel Pullman was well represented by the Lake Garfield *of 1924. It had ten open sections, two compartments, and one drawing room. (Pullman Neg. 27713)*

Figure 3.69 This fourteen open-section sleeper was built in 1927 for the Milwaukee Road's luxury train, the Pioneer Limited. *The built-in headboards and arches formed semiprivate spaces that helped eliminate the open coach-like appearance of most sleeping cars. (Pullman Neg. 31047)*

then put in place, and the bed was made up with sheets, blanket, and pillows. Next the curtains were hung. An experienced porter could do the entire job in just three minutes.

Basically the section cars did not offer the privacy or comforts that American passengers now demanded. And as more Americans traveled overseas, more were exposed to the conveniences of the all-room cars operated elsewhere in the world. Why was this country so far behind the times? Why did Pullman cling to its obsolete cars? *Scientific American* accused the Pullman management of "a blind unreasoning conservatism."[280]

Pullman's answer was the same as it had always been: the open-section sleeper offered the cheapest style of sleeping accommodation possible. Section cars carried 27 passengers; room cars could accommodate only 18.[281] By eliminating the bulkheads of the room car, section cars saved 7,000 pounds per passenger. Because they were lighter and could carry more people, section cars cost less to operate and thus could offer the most economical sleeper space. Pullman also claimed that compartment cars were a proven failure in America. The Mann Company had gone out of business, Pullman implied, because it room cars had been unpopular with the public.

In time, however, changing market conditions forced Pullman to accept the compartment car. Competing forms of transport, coupled with a growing demand for greater luxury, brought about a basic shift in policy, perhaps the greatest since the organization of the Pullman Company.

By the middle twenties Pullman was concentrating more on room- or combination-style sleeping cars that offered several rooms and a lounge or club section. The shift to this style of car in preference to the standard twelve-and-one sleeper was obvious by 1935, when 77 percent of the new cars were of the composite type.[282] Another indication of the open berth's declining popularity was passengers' growing reluctance to travel in the upper berth. By 1936, 80 percent of the riders were choosing the lower berth despite the 20 percent higher fare. Economy was no longer enough to attract passengers. If the uppers were to run empty, there was little point in keeping them. If passengers would accept only single-occupancy space, it was necessary to offer room cars. The space would be more costly, but it seemed to be what the public wanted, and satisfying that want would surely produce more revenue. And so began the search for an inexpensive, single-occupancy design of maximum capacity. These cars would

Figure 3.70 A skilled porter could assemble both the upper and the lower berths in three minutes. In this view the porter is hanging the curtains. (*New York Central*)

Figure 3.71 The Eventide *was an experimental all-room car on the duplex plan. The 16-room model was rebuilt from a library-baggage car in 1933.* (*Pullman Neg. 37692*)

be aimed at the traveling salesman and other business and professional men. Couples and wealthy riders would, as before, choose bedrooms or drawing rooms; what was needed was some ingenious scheme for a single-room sleeper.

Meanwhile the open section did not completely die. It is true that few section cars were built after 1930, but Budd did produce six 16-section sleepers for the California Zephyr in 1948. Other roads purchased a few combination open-section and room cars during the lightweight era. As late as 1955 the New Haven bought some combination cars with six open sections to accom-

Figure 3.72 The Major Brooks *of 1938 was a later form of the duplex single-room scheme. The* Brooks *had several bedrooms as well. It operated on the Pennsylvania Railroad. (Pullman Neg. 41262)*

modate government employees, who were required by law to travel on the cheapest sleeper accommodation. This Federal regulation perpetuated the antique open section into the modern age. The cars remained on a Washington-Boston night train until sometime in 1968.

Room cars were not new to the Pullman Company. From the beginning the standard open-section sleeper contained at least one bedroom for patrons willing to pay for the luxury. By the eighties some all-room cars were being operated on trains catering to the affluent traveler. In 1927 came the first single-occupancy room cars, aimed at the lone traveler who wanted superior overnight accommodations enough to pay a rate set at 25 percent above rail fare plus two lower-berth charges.[283] It was not economy class; for his money the traveler received the privacy of a tiny but fully enclosed compartment, a small but stationary bed (32 by 78 inches), a toilet and folding washstand, and individual heat control. It occupied just over 33 square feet of space, but compact as the rooms might seem, only fourteen could be fitted into an 80-foot car. It is not likely that they were profitable, but they were popular enough so that forty-five were eventually placed in service.

These room cars, which Pullman titled the Night series, were hardly the model that the company was seeking. Something more imaginative than slicing a car up into fourteen narrow rooms was needed to produce a single-room car with enough capacity to replace the open section. Pullman designers experimented with the Bradish plan for remodeling open sections with stationary headboards and semipermanent curtains. But the scheme offered few tangible improvements over the existing plan.

Pullman was looking for a more complete departure from orthodox sleeping-car design. Few more radical plans had been devised than that patented by Albert E. Hutt on April 30, 1929 (No. 711,317). The inventor offered a split-level arrangement for day-night service, but it was initially dismissed by Pullman as too visionary. As Chapter 2 mentioned, Hutt's scheme was tried out by the Long Island for dual-level commuter cars. Pullman began to think better of the general idea, if not of Hutt's specific plan, which was essentially a variation on the old open-section plan. Surely the best use of a given space was to deck it, and Pullman meant to cram as many single-room spaces as possible into the confines of a 10- by 75-foot space. Every other room was raised just a few feet above the car's main floor level. This provided enough free space for a sofa bed and, as the scheme was eventually developed, almost doubled the number of rooms over the one-level arrangement. The design was called the duplex.

To test the duplex idea, in 1931 two heavyweight steel cars were rebuilt at one end.[284] That part of each car was fitted with four duplex rooms and given an arched roof for greater headroom. Small, ungainly windows for the upper duplex rooms were punched into the high side of the body. The rooms were all on one side of the car, with a side aisle on the other. After running these two hybrids, which had been renamed *Wanderer* and *Voyager*, for a year on the Pennsylvania Railroad, Pullman decided to give the duplex a full-dress trial. Two heavyweight baggage-buffets were run through the shop in 1933. Rechristened *Eventide* and *Nocturne*, they emerged with arched roofs and sixteen duplex rooms each, again all on one side. Viewed from the aisle side, the cars appeared reasonably conventional, perhaps lacking the usual number of windows, but in general not conspicuous. But the room side of the car, with its awkward jumble of staggered windows, had the grace of an armor-plated hen house (Figure 3.71). The car's poor looks were not its greatest disadvantage. Pullman did not consider sixteen spaces sufficient for an economical single-room sleeper. The duplex, like all bilevel cars, also had an inherent safety defect: passengers had to mount three steps to reach the upper rooms. Step lights and handrails could not completely eliminate the hazard.

As the Depression worsened, Pullman dropped any plans it might have had for a large-scale car-building program. Duplex rooms were incorporated in a few new cars over the next years, and despite its obvious defects, it was an attractive enough room plan to remain in at least partial favor so long as Pullman stayed in business (Figures 3.72 and 3.73). It had something of a resur-

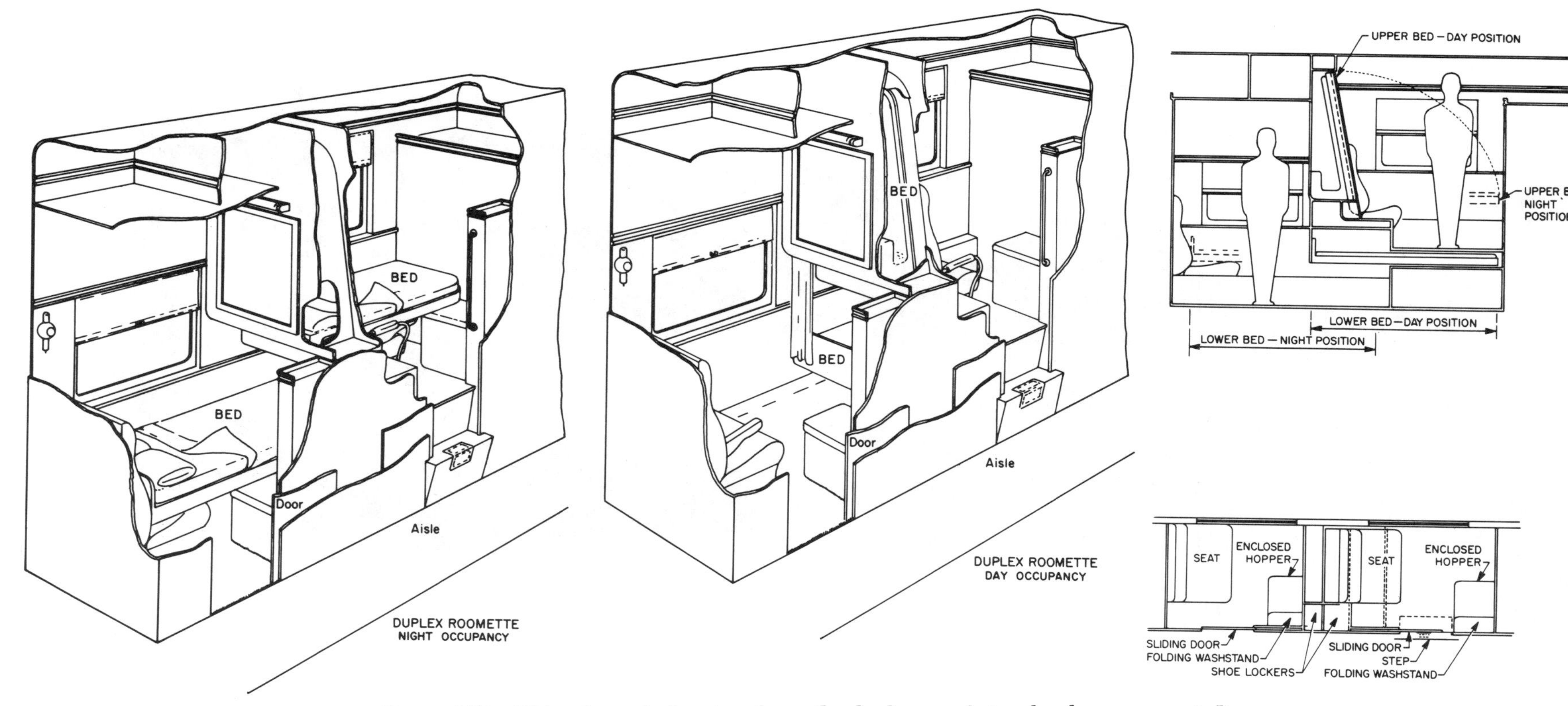

Figure 3.73 This schematic drawing shows the duplex room's two-level arrangement. It was designed to replace the old-fashioned open-section plan. (Traced by E. Tone)

gence in 1942 when a blending of duplex and roomette plans produced a car with twenty-four single rooms—a record at that time.[285] The floor space per unit was squeezed down to only 20.9 square feet, yet somehow each contained a toilet, a sink, and a rollaway bed. The payload was now large enough that a profit seemed possible even if the fare was only 10 percent above that of a lower berth. Some experts claim that the duplex cars built for the Pennsylvania in 1949 were the most comfortable single-room units ever placed in service. These cars, named the Creek series, had twelve duplex roomettes and four bedrooms. In the more usual arrangement there was another bedroom, but by dividing this space among the duplex rooms, enough space was gained for permanent rather than folding beds.

Meanwhile Pullman designers developed a better floor plan, which became known as the roomette. The rooms were arranged on one level on both sides of a center aisle, and they were placed longitudinally rather than transversely. Much of the hardware perfected for the duplex, such as the folding sink, was salvaged for the roomette. The roomette's advantages were that it was on one floor level and that two more rooms could be fitted into a standard 80-foot car. During the day the passenger occupied a wide seat with armrests. The bed was folded upright into a niche at the head of the compartment (Figures 3.74 to 3.77). When the bed was pulled down at night, by either the traveler himself or the porter, the seat and the toilet were covered. The tiny room was in fact almost completely occupied by the bed in its down position. A thin passenger could stand in the narrow space between the doorway and bed. A bit more space could be gained by opening the door, which slid from view between the hollow aisleway bulkhead. Privacy was possible even with the door open, because there were full-length doorway curtains that could be zipped shut. The roomette was a self-contained living unit with its own temperature control, ventilator, fan, sink, mirror, luggage rack, toilet, coat closet, and even such refinements as a night light. The room was small—suffocatingly so, for some passengers—but a great many found it a cozy and pleasant retreat.

The first roomette car was completed in June 1937. The 67-ton, eighteen-room streamliner seemed destined to succeed the old twelve-and-one as the new standard sleeping car. It was hailed as a miniature hotel and as Pullman's best weapon in its battle for overnight travel. More roomettes were built; by 1941 sixty-eight were in service. During the same years (1935–1941) Pullman rebuilt over 300 heavyweights with room-style accommodations. The company anticipated that after the war more new roomettes would join the fleet and that the open section would vanish from the American scene. Indeed, another fifty-five roomettes were built, but the taste of the traveling public took an unexpected turn. Economy seemed less important than comfort, and passengers preferred bedrooms and drawing rooms to the cramped roomette. Fewer full-roomette cars were ordered. Instead the railroads settled on the ten-and-six (ten roomettes, six bedrooms) as the standard sleeping car. Thus after some sixty years the argument over the best form of sleeper was resolved: history had sided with W. D. Mann.

At the same time that Pullman was trying to devise a replacement for the open section, it became involved in a program to develop a lightweight sleeping car. Except for the diner, the sleeper was the heaviest of all passenger cars. Pullman's interest in the problem was more dispassionate than that of the railroads, but the company could not afford to show indifference if it wished to maintain its reputation as a leading car builder. The manufacturing division's investigations into the art of lightweight construction drew Pullman into the new age of aluminum, Cor-Ten, and stainless steel.

In 1931 Pullman produced some aluminum-bodied, high-speed interurban cars for the Indiana Railroad. The same year Pullman

Figure 3.74 The Roomette, introduced in 1937, offered a single-occupancy room on one floor level. This became Pullman's replacement for the open section. The front wall of the Roomette mock-up has been cut away to make the interior visible. (*New York Central*)

Figure 3.75 An interior view of a Roomette taken from the seat. The toilet seat is closed and the sink folded back into the wall when the berth folds down from the opposite wall behind the seat. The oblong door above the toilet is a clothes closet. (*American Car and Foundry Company Neg. 61245-Z*)

decided to build a lightweight sleeping car. The project languished, and work did not begin until 1932. Then the car was rushed through so that the experiment might be ready in time for the 1933 World's Fair.[286] Named the *G. M. Pullman*, the sleeper came as close to being an all-aluminum railway car as any vehicle built (Figure 3.78). The underframe, body, roof, and most of the hardware were of aluminum. The center sill, 20 inches deep, was on the fishbelly pattern and was fabricated from thick aluminum plates. More remarkable, the truck side frames were cast aluminum. Even the wheels had aluminum centers; steel tires provided the running surface. Structurally, the 84-foot-long car was a radical departure from the standard steel sleeper of its day. It featured a round, enclosed observation room, a buffet, and four bedrooms.

The total weight was proudly reported as 48.5 tons, or about one-half that of a steel sleeper. Yet the *G. M. Pullman* outweighed its sister aluminum observation coach, described in Chapter 2, by over 11 tons (Figure 2.61). This demonstrates how much extra weight was added by the bedding, partitions, and other gear necessary for a sleeping room car. The *Pullman* created the hoped-for notoriety during its exhibition. Afterwards it ran on the Florida Arrow and the Super Chief. It proved to be a serviceable car, and after leaving the Pullman fleet in 1952, it was used for another twelve years on the Chicago Great Western. The first lightweight, streamlined sleeping car created little excitement within the ranks of the Pullman management. No more lightweights joined the sleeper fleet for another three years, except for the Union Pacific motor trains.

Again, novelty seemed uppermost in the minds of the designers who laid out the next streamlined sleepers. Aluminum was forsaken for Cor-Ten steel, and the established eight-wheel arrangement gave way to articulation. The two units of the resulting 150-foot, fourteen-wheel mammoth were named the *Advance* and the *Progress*. The rear unit, the *Progress*, was divided into an observation lounge, a buffet, and three bedrooms. The *Advance* was cut up into sixteen duplex rooms. The 112-ton pair hardly proved to be the wave of the future. They had not been long out of the shop when a more conventional lightweight sleeper appeared. This car, named the *Forward*, was ready late in 1936 (Figure 3.79). It was the first to incorporate Pullman's truss side frame and fluted stainless outer skin. The 81-foot, 6-inch, 55-ton car was divided into eight open sections, two com-

Figure 3.77 Some modern sleeping cars were a mixture of drawing rooms, bedrooms, and Roomettes. The car shown here was part of the 1949 California Zephyr.

Figure 3.76 Roomettes are placed on both sides of this narrow center aisle. Metal sliding doors, not visible, can be left open and the curtains closed (or opened), depending on the wishes of the occupant. (American Car and Foundry Company Neg. 61245-FF)

partments, and two double bedrooms. Structurally it proved a satisfactory model for a production lightweight sleeper (Figures 3.80 and 3.81). But Pullman was not ready to make any commitment for a wholesale conversion to streamline equipment.

The reasons were more economic than technical. After all, the existing fleet of heavyweights was adequate for Depression-level traffic. The cars themselves were solid, safe, and comfortable, albeit old-fashioned (Figure 3.82). The cost of a new car was calculated at $75,000 and up, whereas a standard sleeper could be built for $30,000.[287] But the railroads began to join the streamliner bandwagon, and streamliners proved to be profitable traffic generators. Then, as discussed in the previous section, Pullman too began to buy more lightweight cars.

In the same years that Pullman was working to perfect a practical room-style car and lightweight construction, it was also attempting to design a cheap sleeper. The tourist sleeper was still used in the West, but it never gained much of a hold elsewhere in the country. Room cars were driving sleeping-car rates up. Based on lower-berth rates, a roomette cost 140 percent; a compartment cost 200 percent, and a drawing room cost 250 percent. Commercial travelers and the affluent did not protest, and Pullman foresaw even greater profits as more open-section cars were retired. But there was still a market for the economy-class overnight traveler: students, retired people, and just plain poor people.

As early as 1925 Pullman tested a three-tier budget sleeper. In 1940 the coach-sleeper scheme wes revived.[288] Ten compartments with face-to-face seating were placed along one side of the car. A narrow aisle ran on the other side. The arrangement was like a European sleeper, except that the compartments had no ends. Curtains were used at night. The seat formed the lower berth; the middle berth was raised up against the top bunk during the day and pulled down at night; the top berth was stationary. Each unit had a washbasin, but no toilet. The first two cars that were converted to this plan could sleep 45 persons: five compartments slept three each, and the other five slept six each. A second pair of coach-sleepers, fitted in 1941, slept only 42. Because of their large capacity, it was necessary to charge only $5 above the regular coach fare on trips as long as Chicago to Seattle. The coach sleepers seemed successful and certainly appeared destined for wide use after the war. The famous troop

Figure 3.78 In 1933 Pullman produced its first lightweight car. The aluminum-bodied sleeper, buffet, and observation car was used mainly on the Santa Fe. It was scrapped in 1964 after standing idle for many years. (Arthur D. Dubin)

Figure 3.79 The Forward, *1936, was the first of what might be called a conventional lightweight sleeping car. Except for its room arrangement, it set a pattern for Pullman's lightweight fleet. The* Forward *was remodeled as a baggage-dormitory car in 1963. (Arthur D. Dubin)*

Figure 3.80 Pullman built the Paria *and six sister cars in 1938 for service on the Santa Fe. The car, which had seventeen Roomettes and a single open section for the porter, was remodeled into a coach in 1960. (Pullman Neg. 41575)*

sleepers of World War II gave the country a large-scale demonstration of what basic overnight service might be like. Pullman constructed 1,200 of these boxcar-like sleepers on an emergency basis. The chief criticism of the coach-sleeper was the lack of privacy, and the troop sleepers had not only this failing but about every other discomfort that the discerning traveler tries to avoid.

The strong trend in passenger car design after the war was toward greater luxury. Even so, the major car builders came forward with fresh proposals for a coach sleeper.[289] Budd offered a 32-passenger design called the "Budgette," a poor man's duplex with double bedrooms and roomettes. The American Car and Foundry Company created a "Slumberliner," while Pullman came forward with a 30-passenger "Railotel." None of these de-

Figure 3.81 The Shenandoah *was originally built for the Chesapeake and Ohio in 1950. The six-bedroom, ten-Roomette sleeper was sold to the B & O in 1957.*

Figure 3.82 Pullman remodeled many of its heavyweight cars after the coming of the lightweight car. Some were remodeled into room cars, and all were air-conditioned. The Clover Bluff *was built in 1910 as the* Green Tree*; it was rebuilt and renamed in 1935. (Pullman Neg. 39275)*

Figure 3.83 The slumbercoach was an economy duplex-Roomette style of sleeping car introduced in 1958 that could accommodate 40 passengers. The Slumberland *was built by Budd in 1958.*

signs were carried beyond the mock-up stage, and the idea remained dormant for several years.

If Budd never controlled the passenger car-building trade, it did become the leading innovator. The company never let any attractive idea disappear without a trial. In 1953 Budd was ready to reintroduce the coach sleeper. The slumbercoach was a stripped-down duplex-roomette with twenty-four single and eight double compartments (Figure 3.83). It could sleep 40 passengers, surely an all-time record for an all-room car, and something of a benchmark in the history of space-saving design. The beds were small (24 by 73 inches), as were the windows. But the same general luxuries offered by the regular roomette were found in the Slumbercoach.

The first cars of this type entered Pullman service under the sponsorship of the Burlington in October 1956.[290] Cars were run on the Denver Zephyr and the City of Denver. A single room cost $7.50 more than coach fare, a double room $13.50. In 1958 and 1959 the Baltimore and Ohio, the Northern Pacific, and the New York Central bought another fourteen Slumbercoaches for Pullman service. These cars represented final efforts both to perfect a cheap sleeper and to introduce a major design reform in sleeping-car arrangement. The technical problems were admirably solved, even to the costly and involved wiring, waste, and water lines needed to service so many individual rooms. It was the general collapse of long-distance railroad travel that cut short the Slumbercoach's wide acceptance.

The sleeping car's future is now a matter to be resolved by Amtrak. Initially the firm was forced to make do with existing equipment, but large capital-improvement funding (over 400 million dollars) was provided by Congress beginning in 1973. Orders were eventually placed with Budd and Pullman for nearly 800 new cars. The majority of these were coaches, but 70 new bilevel sleepers were included. The basic design follows the Santa Fe hi-level cars; however, no single-room accommodations are included. First-class and economy-fare double bedrooms are planned for the first new sleeping cars built for United States service in about twenty years. Appropriately, Pullman will build this equipment; deliveries are expected to begin late in 1977.

CHAPTER FOUR

First-Class Travel

PARLOR, DINING, AND PRIVATE CARS

Parlor Cars

THE LUXURY DAY CAR was an embarrassing phenomenon in democratic America, which claimed to have a single class of railway travel for a single class of citizen. This equalitarian boast became ever more hollow in the nineteenth century. Some apologists attempted to explain the parlor car as not really anti-republican—as merely a more comfortable style of day coach. But its patrons knew better; the parlor was intended for those who were willing to pay for exclusiveness. The timid and the sensitive found the common mixed-class cars filled with a boisterous crew who seemed to "drink whisky and eat onions simultaneously."[1] Besides those who were nervous of their fellow citizens, there were people who simply enjoyed marching into the svelte, club-like interior of a parlor car while a platform full of coach riders looked on. They could thus buy distinction, and in a nation of new-rich, many were buying.

The extra-fare passenger received a single armchair, a window, and the services of a porter. In terms of real comfort this was little more than that provided by a good coach. Certainly there was no comparison with the difference between a coach seat and a sleeping berth. Sitting up all night in a parlor car was much less comfortable than reclining in a bed, yet the price of a parlor seat was nearly equal to that of a berth. In the nineteenth century a parlor car ticket was generally a flat $1 above the coach price, while in more recent times it has been about 50 percent higher than the coach fare. However, many people have considered it a good buy. In the 1870s one British tourist wrote: "By paying an extra dollar for about every 200 miles, you can have a seat in a luxurious saloon, with sofas, armchairs, mirrors and washing rooms—and the inevitable spittoons."[2]

Parlors were not found on every train and every railroad. They were a relatively rare class of car even compared with sleepers or diners. They never represented much more than 2.5 percent of the total passenger car fleet. Most of them operated in what has come to be called the Northeast corridor, where Boston, New Haven, New York, Philadelphia, and Washington are the principal terminals in a chain of cities. Smaller parlor operations linked such close-lying centers as Chicago–Milwaukee and San Francisco–Los Angeles. State capitals with major satellites, like Chicago–Springfield or Albany–New York City, were connected in order to facilitate the comings and goings of politicians and lobbyists. Affluent resort areas could also support parlor operations, at least during the season. Florida's Jacksonville–Miami–Key West and New England's Boston–Bar Harbor are examples. Thus short runs, usually not over 300 miles in length, between major cities were characteristic of nearly all parlor car operations.

Although parlors were not a noticeable feature of American trains before 1875, many examples can be found. As early as 1845 the Eastern Railroad purchased several fancy cars in which "each seat is a separate arm chair, made to turn on a pivot."[3] Clearly this was parlor car seating, but the fact that each car accommodated 70 persons makes it doubtful that the cars were meant for the exclusive traveler, and it is not stated that an extra fare was charged. Late in the summer of 1853 the Hudson River Railroad began operating the first of several saloon cars built by Eaton and Gilbert.[4] The 45-foot-long body was divided into five staterooms, four of which were 8 feet square. Each had a sofa, chairs, a center table, and a splendid looking-glass. The paneling was ornamented with landscape paintings; the ceiling was hung in silk. At one end of the car was a washroom attended by a chambermaid. A side corridor formed a passageway. The sample car was said to be "a most magnificent fixture—combining sociability and comfort to an extent never before approached in railroad arrangements." The road planned to order another twenty saloon cars if the car became popular, but no record can be found of the success or failure of this experiment. Sometime in the same decade the Rensselaer and Saratoga Railroad introduced a similar car, which a German engineer saw during a visit in 1859. A sketch of the floor plan made by the engineer shows that the car had six compartments, face-to-face seating, tables, chairs, and a 48-foot body (Figure 4.1).[5] Several years later the *American Railroad Journal* noted that this car was very comfortable and had proved a great favorite with families journeying to the spa, but said that such cars were not widely patronized because "they had an appearance of exclusiveness."[6]

During the 1860s parlor and drawing-room cars began to appear on many railroads in the Northeastern and Great Lakes states. In November 1863 the B & O, the P W & B, and the New Jersey lines began to offer through service between New York and Washington. The P W & B provided three stateroom cars to inaugurate this service, but no details on them have been uncovered. A long description of the successors to the first lot of cars does exist, however. The last of four new parlors was portrayed by a writer in 1869:[7]

> A description of this car will answer for all. It is fifty-six feet long and nine and half feet wide, the exterior painted and varnished light yellow, the color agreeably relieved by the bright hues of the lettering. The exterior of the car presents a handsome appearance by the graceful outlines of the mouldings on the sides and at the ovals of the windows. On entering the car at either end the passenger steps into an ante-room or vestibule, separated from the main portion by a dark walnut door, the upper half of which is of ground glass, worked in elegant patterns. Each vestibule contains four roomy arm chairs cushioned with dark velvet, marble washing basins with silver faucets, and the private closets. These apartments are intended for the use of smokers, who will appreciate the provision which enables them to smoke a cigar without being driven into what is usually the most uncomfortable and crowded car on a train. The main apartment of the car contains thirty-seven arm-chairs in double rows on either side of the central aisle, and by merely working a wheel and screw, set below the seat of the chair in the framework, the occupant can adjust it at any incline. When the back is adjusted vertically the chair is an ordinary arm-chair, but by turning the wheel back is lowered and from under the chair a rest for the feet is thrown out, it then becoming a cozy and luxurious arrangement for sleeping. The cushions are of crimson velvet, the wood work of dark walnut grained and polished, and the metal work heavily silver plated. The floor of the car is grained and polished in light yellow, the

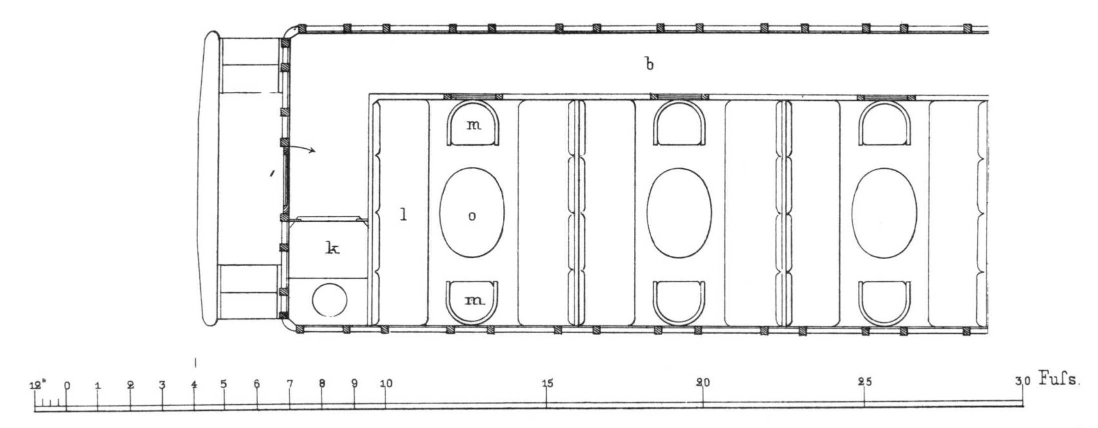

Figure 4.1 The earliest known illustration of an American parlor-stateroom car is this Rensselaer and Saratoga Railroad floor plan of about 1859. (A. Bendel, Aufsätze Eisenbahnwesen in Nord-Amerika, *Berlin, 1862, plate E)*

Figure 4.2 The Boston and Providence operated an English-style parlor car in 1866 on some of its boat trains. Side-door cars remained a rarity in America. (Railway and Locomotive Historical Society)

aisle is carpeted with a neat pattern of velvet carpet, and the interior walls are of dark yellow walnut and light yellow satin wood, each polished in a beautiful manner. The ceiling is of a pure white material, worked out with tracery of blue and gold in graceful designs. The patent ventilators above the roof are of ground glass, and all the metallic finishings are plated with silver. The artistic combination and contrast of colors on the interior of the car, the dark walnut, the brilliant satin wood and crimson velvet, the white, blue and gold of the ceiling and the sparkle of the metal, all form a beautiful and attractive spectacle.

Meanwhile New England railroads were introducing their first parlors. No lusher market could be found than aristocratic Boston and its dignified citizens, who regularly took the boat trains to Stonington or Fall River for excursions to less civilized regions. In 1866 the Old Colony Railroad placed two English-style first-class cars in service on its Fall River line trains.[8] The idea was said to be that of the railroad president's son, who had just returned from Great Britain a confirmed Anglophile. He was convinced that compartment cars were the perfect form for first-class day service. Two were manufactured in the Old Colony shops. They were novel enough in appearance to be featured in the pages of *Leslie's* illustrated newspaper. Eventually two more were placed in service.

The competing Boston and Providence decided that it must have an English saloon car for its Stonington line. The road ordered a 52-foot, six-compartment car from the Salem Car Company, which produced a $14,000 round-end beauty that weighed 14 tons (Figure 4.2).[9] The English cars soon fell into disfavor, however. Passengers disliked being locked into the side-door compartments. The lack of toilets was a nuisance to passengers and crew alike. The conductors feared the hazardous outside running boards when they had to pass from one car to the next. One or more of the Old Colony cars were reduced to a mass of broken timbers in the Wollaston, Massachusetts, wreck of 1878.[10] The B & P car had a less tragic ending: after standing in storage for some years, it was sold for use as a seaside café in 1885.

Parlor car experiments continued in New England. In 1868 the Boston and New York express line introduced two handsome compartment drawing-room cars, the *City of Boston* and the *City of New York*.[11] Built by Wason, they had 45-foot bodies divided into six staterooms, each with accommodations for seven passengers. The seat fee was $1, or an entire compartment could be secured for $6 above the regular fare. The interior was paneled in walnut and cedar and outfitted in the lavish style of the period. Hot-water heating was a new luxury. The exterior was painted in dark green and drab, with maroon panels picked out in blue. The governor of Connecticut, railroad officials, and other prominent men were aboard for the first trip.

An awareness of luxury day cars seemed to spread slowly in some quarters. It is claimed that the New York Central did not consider using them until a vacationing director chanced upon one of the Old Colony English cars during a visit to Newport.[12] Apparently the 1853 Hudson River saloon cars were already for-

Figure 4.3 A conventional parlor interior, dating from about 1875, is shown in this photograph of a Wagner-operated car. (Railway and Locomotive Historical Society)

Figure 4.4 The Garden City, *a progressive style of drawing-room car operated on the Lake Shore and Michigan Southern in 1867. It illustrates the advanced state of the palace cars not associated with Pullman. (New York Central Neg. 357)*

gotten. In all events, operating officials of the Central were summoned to Boston to examine the latest marvel in passenger comfort. They acknowledged the merits of the scheme, but the Central's superintendent, James Tillinghast, objected to the compartment plan and insisted on an open, coach-like style of car. The project was given to Webster Wagner, who introduced his first parlor in August 1867.[13] Among his earliest cars in this style were the *Catskill* and the *Highlander*, which were described as having two staterooms and a large, open central compartment with twenty swivel armchairs. In addition to the usual end doors, they featured a center entrance.[14] Wagner was operating through prime parlor car territory. Albany to New York City was an ideal run for such equipment. Patronage developed quickly, and by 1876 sixty Wagner drawing rooms were in operation. The appearance of a Wagner car of this period was preserved on a stereoptican card, reproduced here as Figure 4.3.

Figure 4.5 A large advertising broadside issued by the Atlantic and Great Western around 1865 included this interior view of a drawing-room car. (Chicago Historical Society)

At the same time that the New York Central was experimenting with parlors, the Michigan Southern began introducing the idea in the West. The Illinois Central had acquired some sleeping cars with revolving parlor seats as early as 1856, but the hybrid scheme was never really intended as a luxury day car. The Lake Shore cars appear to be the earliest true parlors to run out of Chicago. The first two were built in 1867 by John Kirby at the road's Adrian, Michigan, shops.[15] The *Garden City* and the *Forest City* cost $17,500 each—an incredible sum for that day. They had two drawing rooms at each end, with a gaslit central saloon. The interior trappings were furnished by a New York decorator. By chance a photograph of the *Garden City* has survived (Figure 4.4). The picture has been reproduced before, though usually incorrectly dated 1878. By 1870 the Lake Shore was running five parlor–drawing rooms with such names as *Central Park* and *White Pigeon* between Chicago and Cleveland.

The Erie appeared next in the growing number of parlor car operators. Actually, its Western connection, the Atlantic and Great Western, appears to have pioneered parlor service as early as 1865. A broadside of that time (now in the collection of the Chicago Historical Society) illustrates the benefits of travel on the broad gauge. One panel, reproduced here as Figure 4.5, depicts the interior of a compartment-type parlor car. It was not long before the Erie far outstripped its subsidiary in luxury day-time travel. In 1871 the road introduced seven of the most extravagantly decorated parlors ever built. The first four were named *James Fisk Jr., Morning Star, Evening Star,* and *Jay Gould.* The work was done in the Erie's own shops, but even so the cost came to $35,000 each. The *Railroad Gazette* of March 25, 1871, gave the following description:

The Jay Gould is 56 feet by 11 feet; will seat fifty-six passengers; is lighted by Ganster's gas machine, heated with Baker, Smith & Co.'s patent car steam-heating apparatus, with pipes under each seat and sofa; has two water-closets, two wash-rooms in white marble, black walnut, maple, rosewood and gilt; a ladies' lavatory; ice-water tanks and attachments; five reflecting gasaliers suspended from the ceiling, trimmed with silver, by Geo. H. Keitching & Co. The interior of the car is trimmed with silver, wherever metal is used, including arms, hinges, rails, knobs, hat-racks, clothes-hooks, brackets, window fastenings, door-slides, etc., by Newman & Capron, of New York.

The bridal chamber has velvet Brussels carpet; the furniture is covered with drab medalions of crimson scarlet, white and rose colors; drab silk curtains, intermixed with blue and gold, bought of Stewart; sofa beds with easy pillows and bolsters, chairs on revolutionary hinges; beds with improved spiral steel spring mattresses, covered with crimson reps; beds raised by automatic or spiral springs, so complete that a child can raise or lower them. The ceilings were frescoed by Sebastian Brodt.

The car is trimmed with French moquette, and green, drab, and gold crimson are the prevailing colors. The windows are of best French plate glass—one large or center one and two smaller or movable ones—the whole of etched glass by Downing, and in all the panels is the Goddess of Music, with a lyre etched in the glass. The wood used is black walnut, ash-bars, black walnut panels and rosewood mouldings, inlaid with gold, the two former coming from the West. There is a

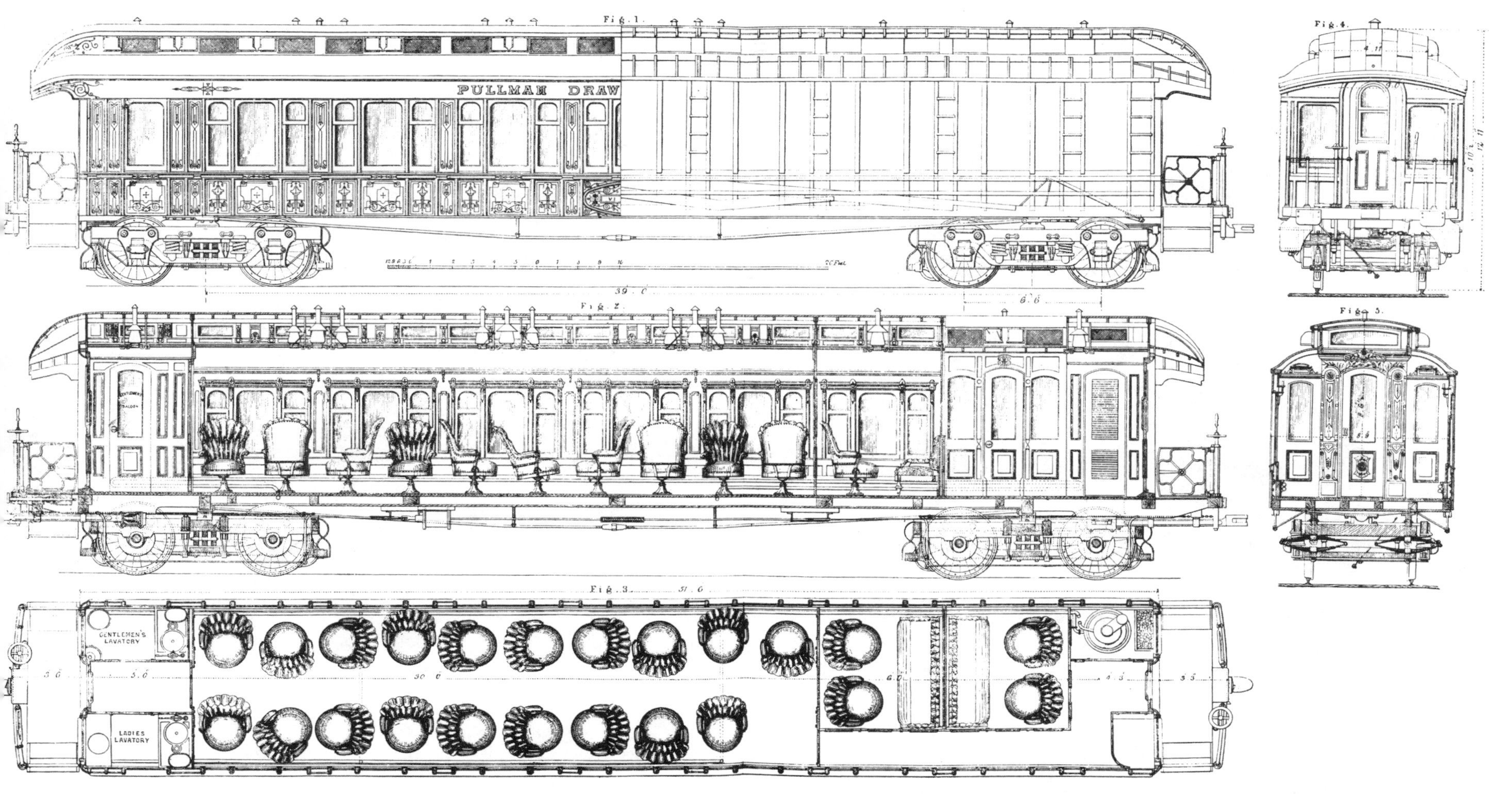

Figure 4.6 In 1874 Pullman exported this American-style parlor car for service in England. The drawing is representative of standard Pullman design except for the wooden-center wheels. (Engineering, *April 2, 1875*)

mirror in each door and panel. The seats are peculiarly adapted for ease and comfort, being upholstered at the ends, back, and extra or lounging back with French moquette of green, drab, crimson and gold. The cushions are on the best of steel spiral springs and are adjustable; arms silvered. The upper berths are self-acting. The busts of Mr. Gould, inside, are carved of white holly, the Erie Company doing all the carving of the car at a cost of $4,000 each. The portraits in oil of Mr. Gould on each side of the outside of the car will attract much attention. The oil paintings in the inside or frescoed panels are representations of scenery along the line of the Erie Railway. Chairs are hung with silver slipjoint hinges. The ventilation is very complete and is formed by wire screens in the roof over each platform, and from this point is conducted through the car, being increased or lessened by the use of beautiful glass-paneled swivel ventilators in the roof of the car. The wire screens prevent cinders from going in. The electric or magnetic brake is used; Miller's platform buffer and bumper; six-wheel trucks with steel tires on wheels, the steel and iron being cast solid together; eight guard chains from trucks to car body, and elliptic and rubber springs combined. The several departments of the company's workshops are so complete that these cars in all their elegance are made therein, embracing cabinet work, upholstering, painting, gilding, body trucks, etc. The Jay Gould weighs but twenty-five and one-half tons on a wide track, whereas a Pullman narrow-gauge car weighs twenty-eight to thirty tons. Commerce, railroad and shipping are represented on the outside corners of the car. None but the best selected timber was used. Mr. Gould gives a personal supervision over the workshops and employes in the construction of these and other cars.

By the early 1870s the parlor car was an established part of the American railway fleet. In addition to the examples already mentioned, small roads such as the North Pennsylvania and the Terre Haute and Indianapolis were reported to be acquiring luxurious vehicles of this type.[16]

Where was Pullman during the upsurge of interest in expanding luxury service? Historians, including Joseph Husband and Stanley Buder, have claimed that Pullman did not introduce his first parlor until 1875. Actually, however, he was running parlors as early as 1871 on the Great Western Railway of Canada.[17] Two cars named *Leo* and *Mars* were produced by the Taunton Car Company for the run between Hamilton, Ontario, and Suspension Bridge. Each 62-foot body was divided into five compartments. The central section seated 20, the two staterooms accommodated parties of 6 each, and the two end rooms were used as smokers. Enthusiastic press reports said that the great 27- by 48-inch windows suggested a crystal palace, and that the large openings would provide easy egress in the event of a smashup. These cars were considered something of a bargain at the advertised rate of 3 cents a mile. Three years later Pullman began operating parlors with reclining seats between New York and Washington. He placed a similar car on the Michigan Central in the summer of 1875.[18] Some months before, Pullman shipped the first American-style open parlor to England for service on the Midland Railway. Drawings of the *Victoria* published not long after the car entered service present the most detailed plans available for a parlor of that time (Figure 4.6).[19] Records uncovered in recent years by Arthur Dubin indicate that the *Victoria* and its companion parlors sent over later by Pullman were not newly constructed; they were existing cars that had been

Figure 4.7 Even some relatively small roads, such as the R W & O, found it profitable to operate parlor car service. The St. Lawrence *began service in 1878. (New York Central Neg. 719-3)*

Figure 4.8 A parlor car interior from an advertising booklet issued in 1876 by the Pennsylvania Railroad to stimulate travel to the U.S. Centennial Exhibition.

refurbished especially for overseas service. The *Victoria* itself had an extraordinary service life, running until 1932.[20]

In 1876 Pullman placed six parlors on the Pennsylvania's best trains out of Philadelphia. Described as "triumphs of elegant comfort and costly ornament," they were named *Hebe, Ariadne, Andromeda, Proserpine, Daphne,* and *Queen.* The *Queen* formed part of Pullman's exhibit at the U.S. Centennial.[21] The major portion of the 48-foot bodies was given over to an open compartment, with a small drawing room at one end and toilets at the other. This became the typical floor plan for an American parlor car. The seating arrangement was not so typical: the chairs were placed in double rows on one side and in a single row on the opposite side. A total seating of 31 was thus achieved, but at the sacrifice of spaciousness, comfort, and exclusiveness—the commodities that parlor car patrons most desired. However, the chairs with reclining seats built to Wood's patent design were intended to mollify the disgruntled first-class traveler. The windows were arranged in the patterns of three that were popular at the time, with a large center glass flanked by two narrow panes.

By the end of the 1870s the parlor car was a customary part of the better trains. Pullman, Wagner, and Woodruff ran a complement of such cars in their concession fleets. If theirs are counted along with the railroad-operated cars, it is probably safe to estimate that between two and three hundred parlors were in service on U.S. railroads by 1879 (see Figures 4.7 and 4.8). Curiously, the term "parlor car" was not widely accepted by this date, for the *Car Builders' Dictionary* of that year continued to list only the "drawing-room car."

As the parlor car proliferated during the next decade, so did the number of its critics. To some, parlors represented an unconscionable swindle that offered doubtful comforts at high prices. Others contended that it was all a conspiracy engineered by rapacious railroad tycoons who ran dingy coaches to force passengers to buy tickets for extra-fare cars. The first school argued that the parlor was nothing more than a common day coach with oversized windows and a single row of chairs.[22] The chairs were uncomfortable, with convex stuffed seats and backs and poorly placed armrests. The big windows were wasted, for half of the seats were aligned with posts. As one writer put it, windows are like toothbrushes; each passenger should have his own.

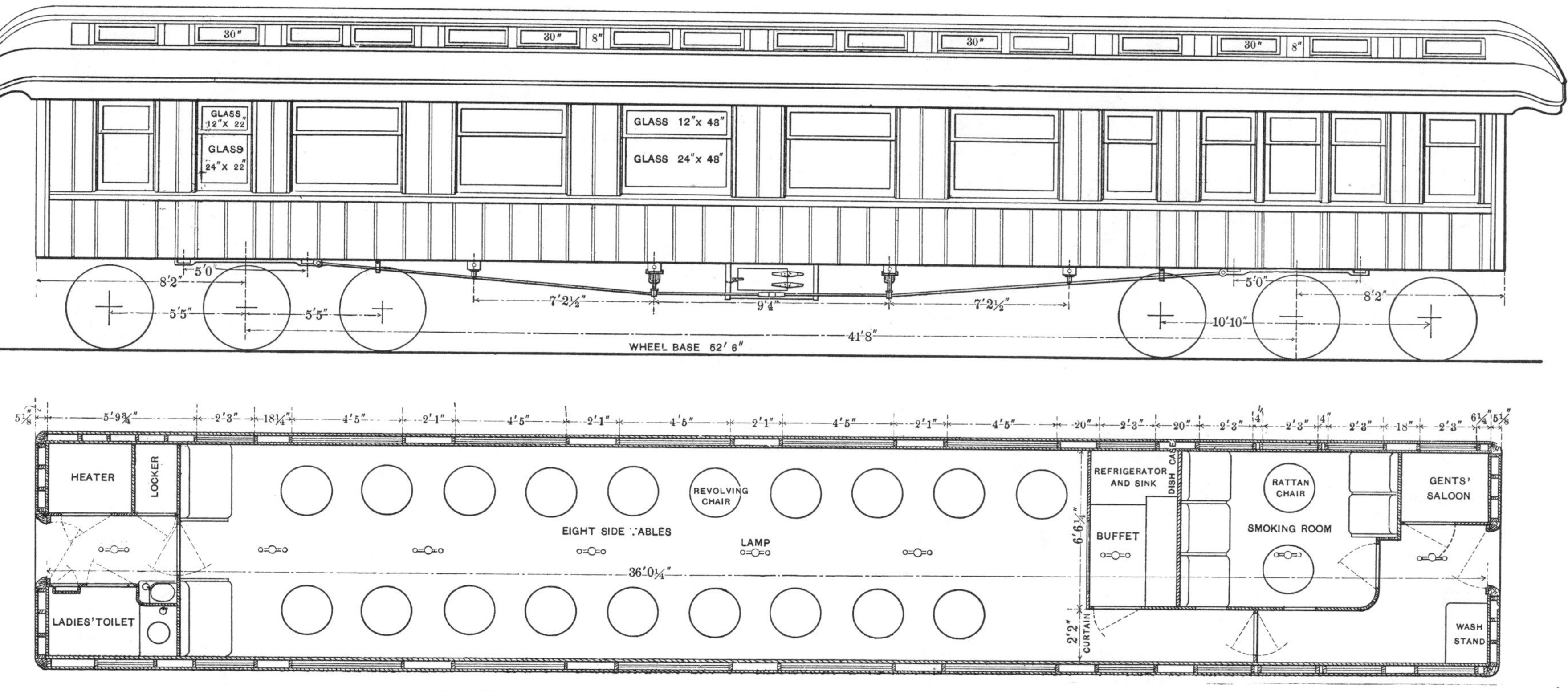

Figure 4.9 The parlor car Brandywine *was built in 1885 by the Baltimore and Ohio Railroad's Mt. Clare Shops in Baltimore.* (National Car Builder, *December 1885*)

Figure 4.10 The Lavinia *was produced for the B & O by Jackson and Sharp in 1886. It is very similar to the* Brandywine *and may have been constructed from drawings furnished by the railroad. (Hall of Records, Delaware Archives)*

This critic was annoyed by the absence of luggage and coat racks and totally enraged by the miserable parlor that he and several other passengers were put aboard during a change of cars in Philadelphia. Contributed by the New Haven for the cooperative through-car fleet, it was a dusty relic in deplorable condition. The window-sash latches were missing, it smelled worse than a fishing wharf, and the stove smoked. Perhaps, the writer suggested, the New Haven thought that the car was good enough for Yankee travelers; or perhaps the road thought it *not* good enough and had therefore sent it south.

Even when the New Haven put some new cars on its routes—and they were very much needed, according to the *Railroad Gazette*—the fault-finding continued. The bodies were duplicates of cars intended for coach service, so that while the big-window humbug was avoided, the seats and window openings did not correspond.[23] The seats, however, had been improved. The backs were not too high, and they were actually shaped to fit the human spine. The top was rolled back, offering a support for the head or neck. The seat cushion was tilted back so that the occupant could settle in. And the back was not too high, a common fault with most parlor chairs. The neighboring Boston and Albany's efforts at parlor car reform were mentioned in the same article, which praised the plainness and lightness of the cars. Car Number 208 seated 30 passengers, yet weighed only 45,275 pounds—some 10 tons less than a Wagner drawing-room car. The article described the rattan seats as a refreshing improvement for summer travel, though it said that the backs were too perpendicular.

During the years when it operated its own sleeping and parlor cars, the B & O made an effort to acquire rolling stock equal to

Figure 4.11 Interior of a parlor car, very likely the Lavinia, *built in 1886 by Jackson and Sharp for the B & O. (Hall of Records)*

Pullman's best. The parlor car *Brandywine* (Number 544), constructed at the Mt. Clare shops in 1885, had nineteen pedestal armchairs in the main saloon that were carefully arranged to correspond with the window openings (Figure 4.9).[24] The 35-ton car measured 65 feet in overall length. A heavy silver-plated handrail was placed in front of each large window. The small buffet was outfitted to serve coffee, tea, and other light refreshments. A very similar car, the *Lavinia*, was produced a year later by Jackson and Sharp. Surviving photographs of this car, coupled with the *Brandywine* drawing, give a good general impression of a parlor car of the time (Figures 4.10, 4.11, and 4.12).

A major attraction of daylight travel is the view. Watching the passing scenery is relaxing, amusing, and free. The designers of parlor cars were deeply concerned with ways to improve the view. By the eighties the big window had been thoroughly exploited, but bay windows, which were then popular in domestic architecture, seemed to offer another approach. In the world of fashion true solutions are not really necessary; novelty is often enough. And so began the bay-window parlor car craze. Actually the term "bay window" was an exaggeration: the windows were set at a slight angle, say 15 or 20 degrees, so that the sash could be made relatively small and would not invade the car's floor space. Even so, the window enabled the passenger to look ahead much more comfortably than he could through a conventional window opening.

In 1884 the Pennsylvania built a sample bay-window parlor at its Altoona shops.[25] Number 901, which seated 37 passengers in its 62-foot body, was successful enough that more bay windows (called class CA) were produced at Altoona (Figures 4.13 and 4.14). The Boston and Albany caught the bay-window fever and ordered Wason to construct the finest car possible on this plan.[26] Bruce Price, the New York designer, and L. C. Hyde of Wason's drafting department took over the design commission. Price claimed the bay-window design in general, but Hyde's lunette window plan was used. The darkly elegant interior was considered remarkable—nothing less than "an art study." The

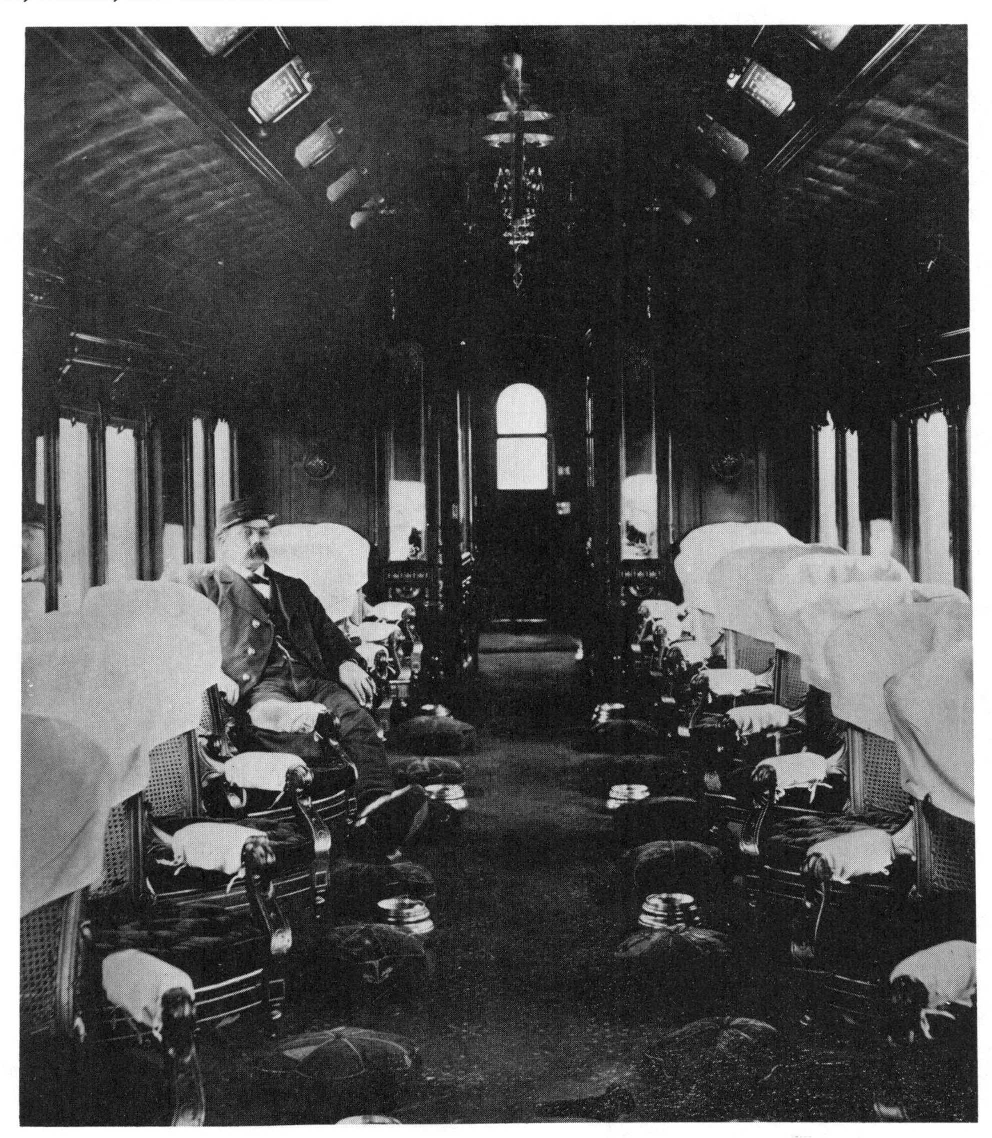

Figure 4.12 *A Baltimore and Ohio parlor car of about 1885. Note the foot cushions and the white cloth arm- and headrest covers.*

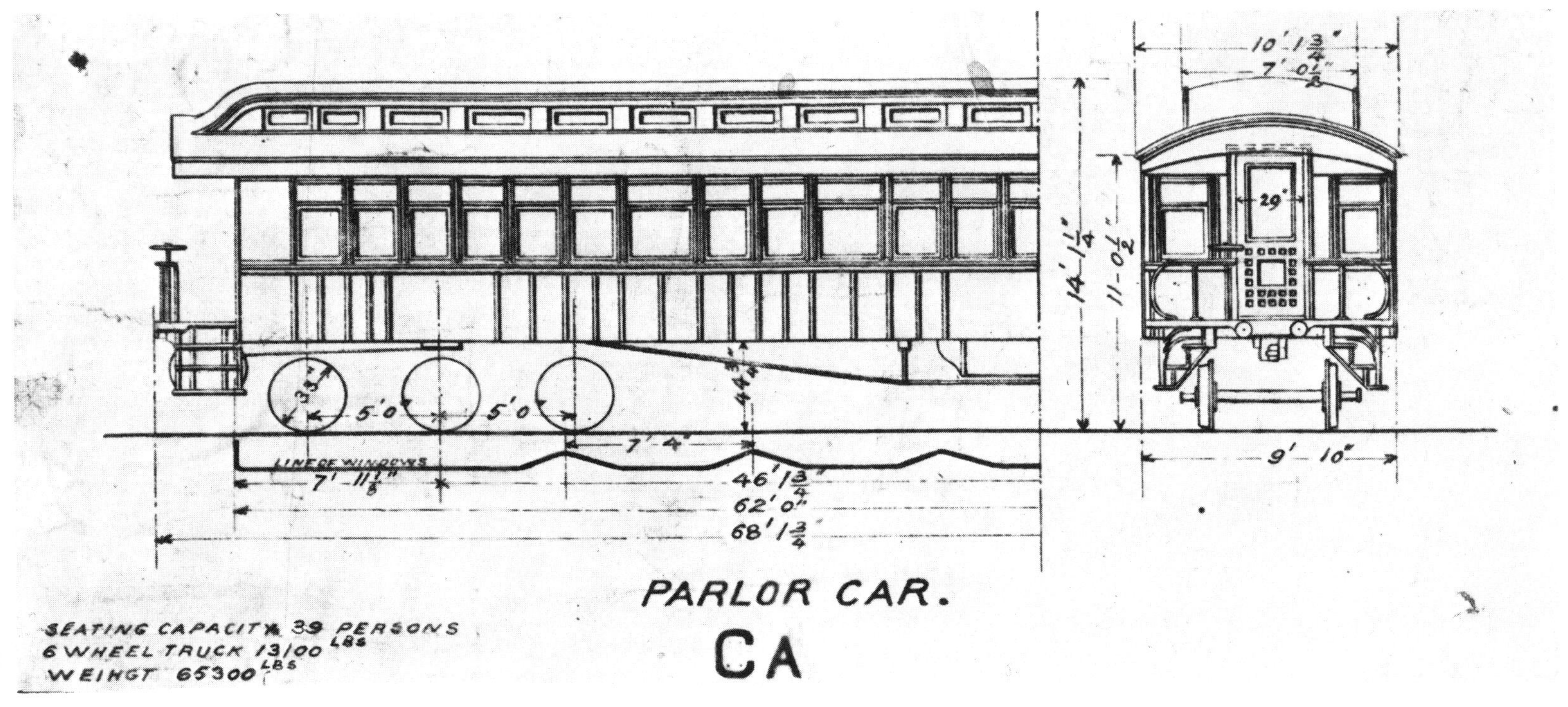

Figure 4.13 *A class CA parlor car built in 1884 by the Pennsylvania Railroad's Altoona shops.*

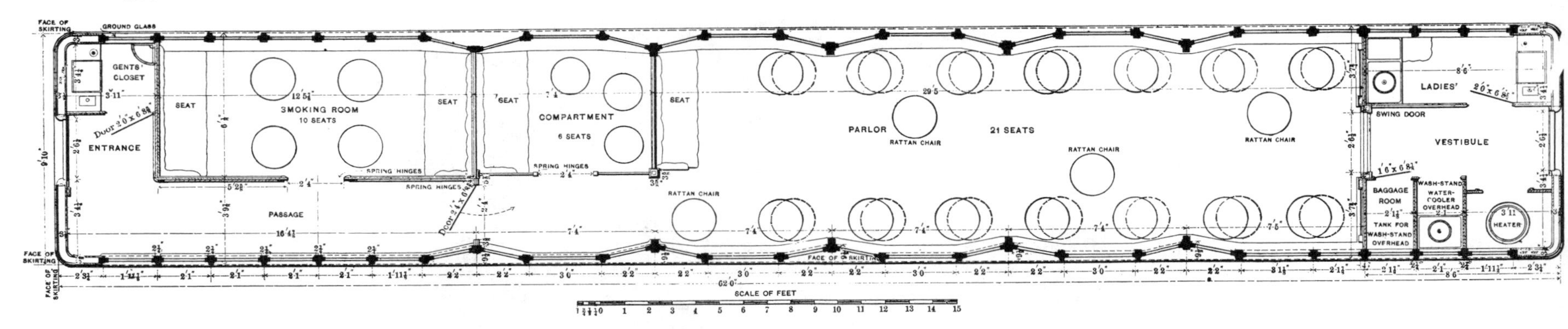

Figure 4.14 A floor plan of a bay-window parlor car constructed by the Pennsylvania in 1884. (Railroad Gazette, *August 29, 1884*)

Railroad Gazette was sufficiently impressed to devote a half page to an engraving of the interior (Figure 4.15). The B & O and Pullman soon followed the fashion. The B & O put a bay-windowed parlor on its plush Royal Limited. Pullman's Golden Gate Special featured a combination observation-parlor-sleeper in which the parlor compartment had bay windows.

In the background of this activity was a patent case doggedly pushed by a poor trainman named W. K. Tubman, who had somehow purchased George S. Roberts's 1877 patent on railway car windows.[27] Tubman claimed that it was the basis for all the fashionable bay windows then in service. During the suit, the defense was able to show a long array of prior examples. Between 1852 and 1866 several patents had been issued that envisioned some form of bay window protruding outside of the car body.[28] The Erie, the New Haven, and a few Western lines were said to have tested one or more of these schemes. Tubman claimed that Roberts's patent was the first to have the bays *inside* a normal car body's outer limits. Roberts's plan, however, resembled the bay window then in use only in the most general sense, for his had a bold zigzag system of interior panels that would have reached 2 or more feet into the interior, whereas the windows in the current designs were set on modest angles that extended only a few inches inside the car. Tubman appears to have had little success with his case against the railroads, and the entire fad seems to have fallen from favor rather quickly. Pullman exhibited a parlor with bay windows at the 1893 Exposition, but that appears to have been the end of it.

In addition to the early examples presented during the course of the Tubman trial, there were some lantern-window cars operated over the New York and Harlem Railroad and the Boston–New York express line that the German engineer Bendel saw during his visit of 1859.[29] He was so impressed by the unique sheet-metal appendages that he included a sketch of them in his published report. They protruded outside the car body from the window openings, but because they were so small there was supposedly no danger of collision with opposing trains. To see out, it was necessary for the passenger to duck his head inside the lantern. The lanterns required constant attention; one or more panes had to be replaced every day. The New York–Boston cars were soon retired, and one of them served long afterwards as a tiny match factory at Saybrook Junction.[30]

During the bay-window era an even more radical parlor car plan was introduced by the Monarch Sleeping Car Company. It featured open galleries at the body's center, with observation rooms at each end. The scheme was reminiscent of the private car built in 1860 for the Caliph of Egypt by Wason. It also resembled the royal train of cars built for Napoleon III at about the same period. Monarch, however, enclosed the midsection so that passengers could choose between the breezy outside gallery and the snug central compartment. The drawings show the arrangement more clearly; note the half round end windows (Figures 4.16 and 4.17). More details on this unique sight-seeing parlor are given in an account from the *Troy Daily Times*, June 15, 1886:

> There is being constructed at the Gilbert car works, Green Island, a novel and magnificent observation car, which is to be used between New York and the White Mountains. The car is constructed on a plan different from that of other observation cars. The coach is about sixty-five feet long, and it is supported upon trucks which have twelve large-sized wheels. The exterior of the car is of a "silver flake" color, trimmed in old gold. It will glitter and sparkle in the sunlight with a dazzling brilliancy. At each end of the car is a drawing-room, about fifteen feet long and the full width of the car. The ends of the car are rounded, and are of French plate glass. The sides are of glass also, which will enable the occupants to have an unobstructed view of the country through which the car will pass. The interior will be fitted up in magnificent style. The floor is of tile. The ceiling is covered with beautiful pearl-tinted Lincrusta-Walton. The small spaces between the windows are also covered with the same material. At the ends of the drawing-rooms are large French mirrors. The rooms will be supplied with easy-chairs. Between the drawing-rooms are three compartments, that will each accommodate four or five persons. These compartments are narrower than the drawing-rooms, which leaves a platform about two feet and a half wide on each side of the car, running from one drawing-room to the other. On the outer edge of this platform will be a silver railing, about four feet high, which alone will cost $1,000. Doors from the smaller rooms open upon the platforms, so that the excursionist can walk out upon the platforms at pleasure. Doors from each side of the drawing-rooms also open upon the platform. The small rooms are fitted with the same elegance as the drawing-rooms. The sides are almost entirely of plate glass. Large mirrors are arranged at each end of the rooms, while from the ceilings are suspended handsome silver chandeliers. There is also a room fitted up for a buffet and there are toilet rooms in the car. A galvanized-iron tube runs along the whole length of the car on top, furnishing ventilation by means of small openings from the ceilings of each room. The railings around the platforms at the ends of the car are all silver plated. The car is supplied with the latest improved platforms and air brakes. On each side of the drawing-rooms appears the name "Ymir" in old gold letters, and along the top of the car the words in old gold, "Monarch parlor-observation car." The coach will cost, when completed, about $16,000, and will be ready for use about July 4.

It is not known whether Gilbert built other cars like the *Ymir*, though it seems improbable. After Monarch suspended operations, the *Ymir* came into the possession of the Jackson and Sharp Car Company, where it was refitted as a demonstration car for

Figure 4.15 The Wason Car Works of Springfield, Massachusetts, produced this magnificent arched-roof parlor-drawing room car for the Boston and Albany in 1887. (Railroad Gazette, *October 21, 1887*)

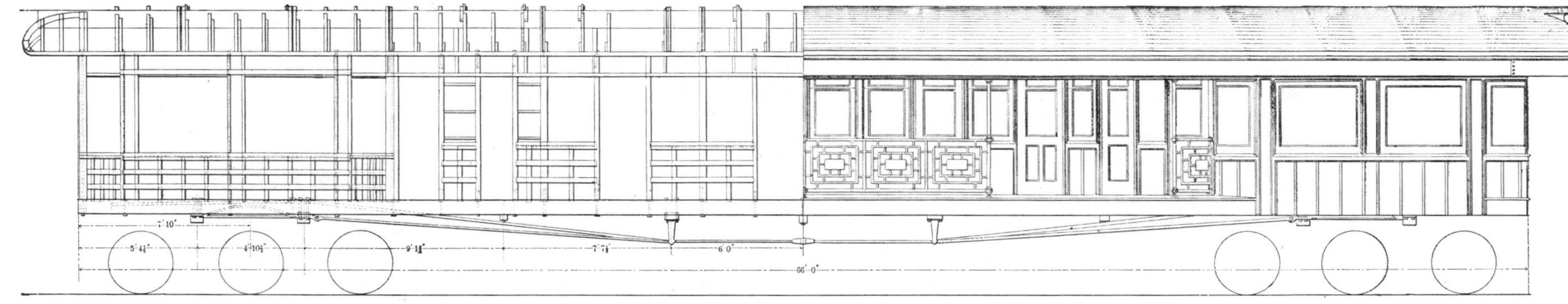

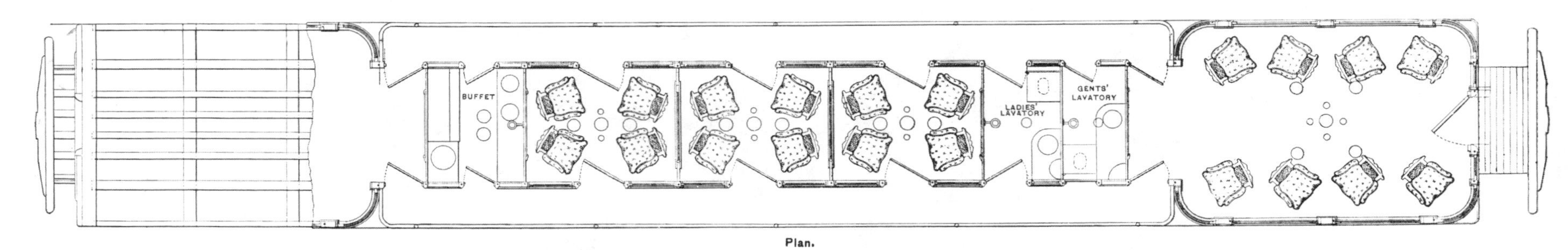

Figure 4.16 The Gilbert Car Company of Troy, New York, built an unusual parlor-observation car for the Monarch sleeping car company in 1886. (National Car Builder, *September 1886*)

Figure 4.17 The Monarch parlor-observation car around 1895, when it was being used to demonstrate a new system of electric lighting. (Lancaster County Historical Society)

the Moskowitz electric car-lighting company and was occasionally used by Job H. Jackson for private excursions.

By the late 1880s the parlor car had developed a multiple personality. It was no longer an extra-fare day coach with swivel seats and a drawing room; it was now expected to serve as a lounge, a buffet, an observation car, a library, and even a gathering place for restless travelers who were weary of their assigned seats. Under the pressure of these demands the parlor took three lines of development. In one direction it evolved into the chair car—a supercoach that often, but not always, had a double row of reclining parlor-car-like seats. In another, the traditional parlor car continued as before. In the third direction the lounge car, whose space was not for sale, emerged as a special luxury car set aside for the free use of first-class passengers.

In the straight parlor, the basic plan remained the same. The car grew in size, changed in decor with changing fashions, and made successive shifts from wood to steel and finally to lightweight construction, but the standard plan of 1875 was identical with that of all later generations (Figure 4.18). The last word in parlor design can be seen in the Metroliner's Metroclub cars. The term "parlor" was considered too old-fashioned to use for them, since it might conjure up a mental picture of a fussy, overstuffed relic of the Victorian age. Yet the Metroclub is a traditional

Figure 4.18 The timeless appearance of the parlor car's interior is illustrated by this B & O streamliner of 1935. (American Car and Foundry Company)

parlor car, and its rows of high-backed armchairs on swivel bases are alternately comforting, disappointing, or amusing, depending on your viewpoint.

The chair car might be called the poor man's parlor. It falls somewhere between the better day coach and the extra-fare parlor. It is analogous to the tourist sleeper as a railroad effort to offer a cheap competitor to the luxury cars run by Pullman and Wagner. It represents an effort by certain railroads to take trade away from the concessionaires or to offer some attractive accommodation not available on a rival line. The chair car was aimed at the parlor car market and at those coach passengers who wanted something better than a stiff-backed double seat. The big attraction was added luxury at no extra charge. Each passenger had a reclining chair that could be turned toward the scenery or the other way to converse with a neighbor across the aisle. At night it could be dropped back like a sofa, and while it was not as good as a berth, it was adequate for a snooze.

To provide this added comfort without extra cost, it was necessary to double up the number of seats. Four rather than two rows were common in the chair car. The legroom and exclusiveness of the parlor was missing, but so was the added tariff. The reclining chair basic to this style of coach can be traced back to the early 1850s. The inventors of that time, however, were preoccupied with perfecting a sleeping chair, and it seems to have been viewed as such over the next twenty years. The shortcomings of the reclining chair when compared with a sleeping-car berth, at least for long-distance travel, are obvious. But for day journeys or short trips it had merit, and in the 1870s the reclining chair was again brought forward. The Chicago and Alton was among the first to push it. In 1877 the road began to run some handsome cars with Horton reclining seats.[31] Regarded as the most scientifically designed of all the new chairs, they were the product of a Kansas City doctor, N. N. Horton. A somewhat stylized engraving of the interior of a Horton car was used in an Alton advertisement of the period (Figure 4.19). Excellent drawings of a later Horton car produced at the road's Bloomington shops are shown in Figures 4.20 and 4.21. In 1882 twelve were in service and two more were under construction.[32] The short bodies carried between 32 and 40 seats. Chair cars gained favor on many railroads in addition to the Alton. They were a compromise between day coach, parlor, and sleeper that seemed to satisfy the class of traveler who would not or could not go first class.

The most exotic of what might be called the parlor's offshoots was the lounge car. Its arrangement and functions were so diverse that it cannot be assigned exclusively to the parlor family, but it fits no other category better. The lounge was the last of the major passenger car types to appear. It developed with the fast luxury trains that were not introduced in this country until the 1880s. A few express trains were operating a decade earlier, and they did include some luxurious sleeping and dining cars. Ac-

Figure 4.19 Reclining chair cars offered parlor car luxury at coach fares. The illustration dates from 1882.

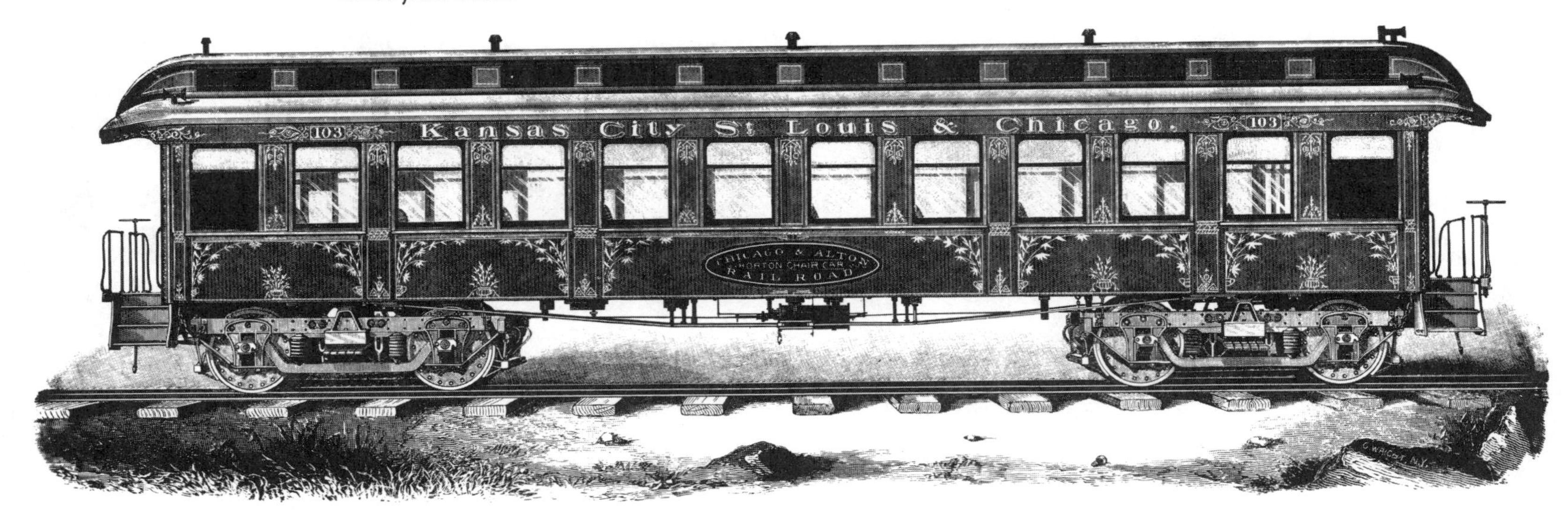

Figure 4.20 Chair car Number 103 was built at the Bloomington, Illinois shops of the Chicago and Alton in 1882 to the design of William Wilson, superintendent of machinery. (National Car Builder, *June 1882*)

cording to the *Railroad Gazette*, as early as 1882 the Pennsylvania was running a parlor-library car on its New York–Chicago express.[33] But such luxuries were rare; most express trains were a mixture of new and old, common and palace. Passengers were essentially isolated in one car. Because of the open platforms, it was dangerous to pass from car to car when the train was in motion. Most roads forbade passengers to do so, and some locked the doors to prevent it.

In 1887 came the luxury train, a unified series of cars designed to serve travelers' needs in the same way that a hotel or a transatlantic liner does.[34] And it was a single establishment, for the cars were linked together with vestibules. Passengers might walk safely through the entire train. If it was stuffy in a sleeper, they could go to another car. If the diner smelled of fish, they could retreat without waiting until the next station stop. The new trains were extra-fast, extra-plush, extra-exclusive, and extra-fare. They were successful as well, and by 1890 more than a dozen Flyers, Limiteds, and Expresses were in service. The name train had arrived, marking a turning point in American railway travel.

It became the function of the lounge car to provide an inviting destination for passengers who were walking around. Strolling through the other cars won only the cold stares of their occupants. The lounge, on the other hand, was a clubroom where anyone might sit (Figures 4.28 and 4.29). Its existence was the result of the vestibule train, and its attractions were the devices of the ages for people with time to kill. Comfortable armchairs, sofas, and side tables were the customary furnishings. If the scenery or a cigar proved dull, there was a small library of popular classics for diversion. A barber shop with a full-sized chair provided another service, which businessmen appreciated as an efficient way to save time that was going to waste anyway (Figure 4.30).

All these features were available aboard what was probably the first true lounge car, the *Esperanza*.[35] It was completed in April 1887 by Pullman for the Pennsylvania Special. Bathtubs became a stock feature the next year, and by the mid-1920s showers were being installed.[36] The Seaboard Air Line went all the way by converting a club car into a traveling gymnasium that

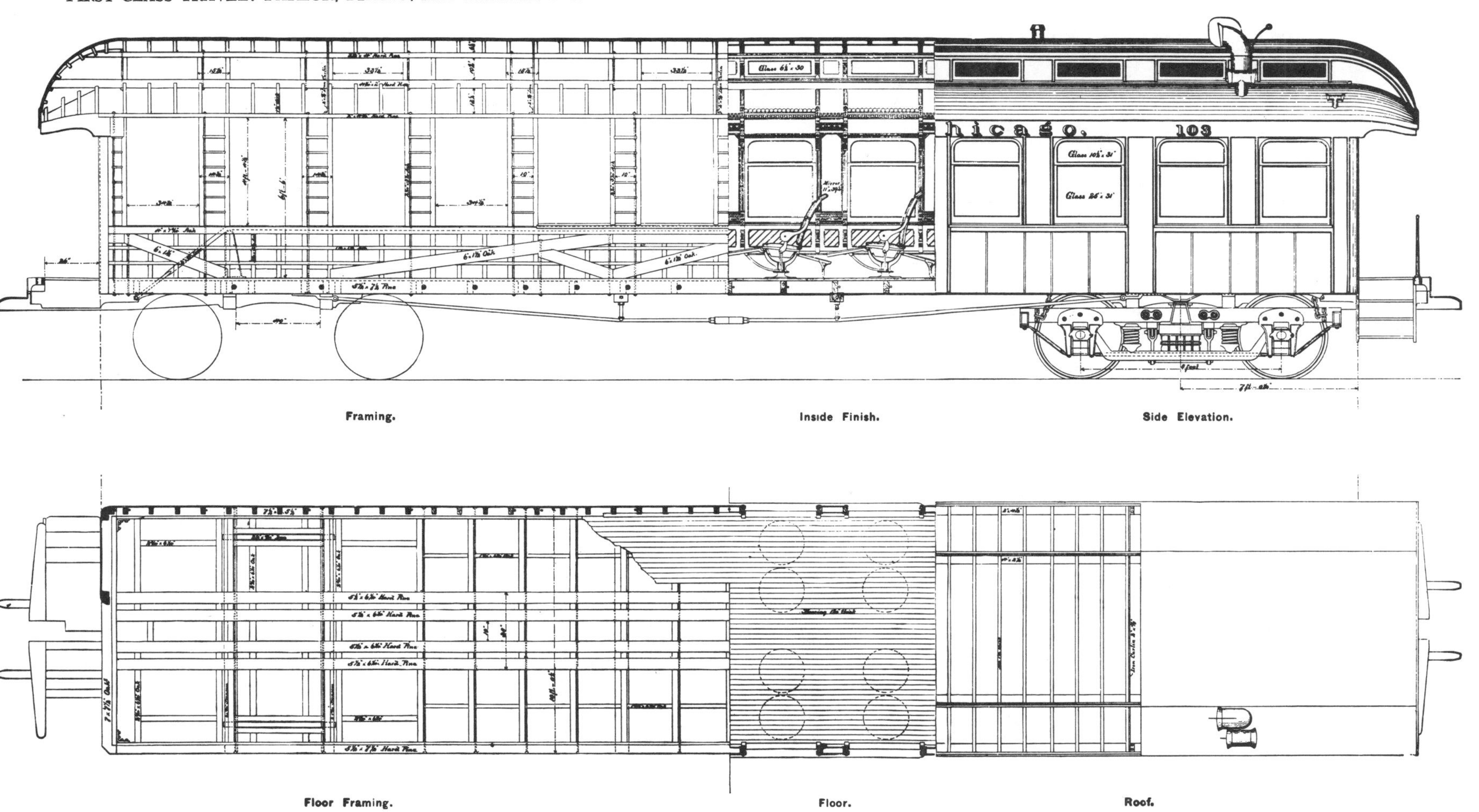

Figure 4.21 *A general arrangement of the chair car illustrated in the previous engraving.* (National Car Builder, *June 1882*)

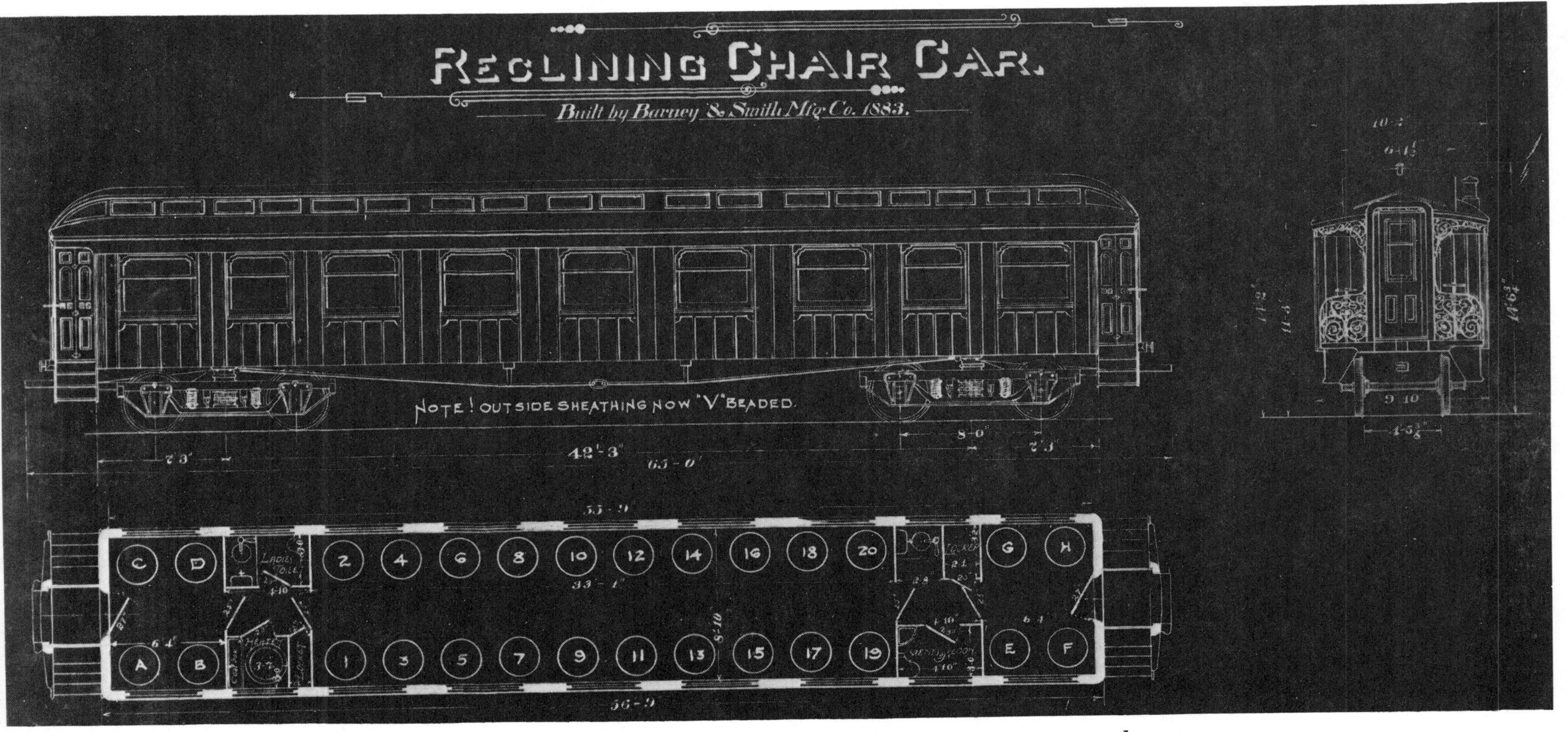

Figure 4.22 *Barney and Smith of Dayton, Ohio, built this parlor car for a predecessor of the Cleveland, Columbus, Cincinnati, and St. Louis Railroad in 1883. The vestibule ends were added around 1890. (Railway and Locomotive Historical Society)*

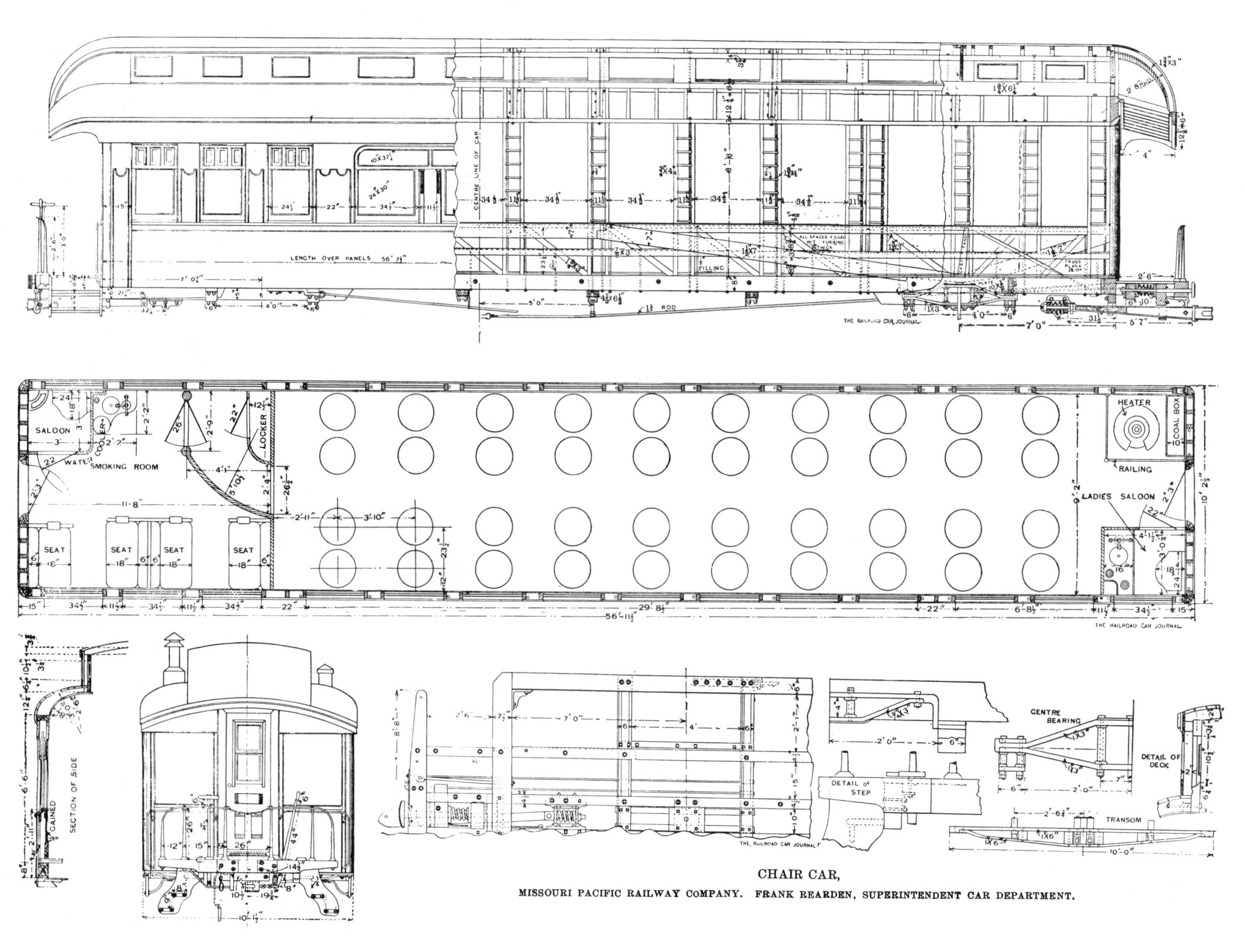

Figure 4.23 *A reclining chair car produced by the Missouri Pacific shops in 1893.* (Railroad Car Journal, *April 1893*)

Figure 4.24 The Santa Maria, *built by Pullman for display at the 1893 Columbian Exposition, was perhaps the most elaborately decorated parlor car ever constructed.* (*Arthur D. Dubin*)

featured mechanical horses, a punching bag, and a small swimming pool.[37] Radios were also installed, and in later years movies and television were available.

Because they were nonrevenue vehicles, full lounge cars were rare. The seats were not sold; they were free to all passengers who had bought space elsewhere on the train. The earliest lounges were usually combined with a baggage area. Later a portion of the car was given over to bedrooms, open berth sections, or buffets. Light food and beverage service was a natural and frequent combination.

The return of alcoholic drinks in 1933 led to the club car, where a bar took the place of the buffet. The library quiet of the original lounge car gave way to a more tavern-like conviviality. It started modestly at first, with little more than an extra shelf for liquor, but full bars were soon installed. Some became flamboyant horseshoe affairs capable of producing both the atmosphere and the full range of drinks offered by a smart Manhattan cocktail lounge (Figure 4.31).

The observation platform was another attraction that became an expected part of the name train's consist. The polished brass railings, drum sign, and scalloped canopy awning added a festive ending to the train (Figure 4.32). Before the coming of the dome car, the observation platform offered the best view on the train. The panorama of the passing countryside was uninterrupted by window posts or curtains. There was something exciting about the ride as well—the clatter and roar, the swirling wind and dust, and the hypnotic sight of the tracks as they disappeared into a subtle triangle at the far horizon. And as the train passed through a small town, passengers could stand on the rear platform to be eyed by farm-bound locals who wished that they too were aboard the glamorous car flying toward Chicago.

The observation platform came to the lounge car after first

Figure 4.26 This arched-window parlor, a Pullman product of 1903, is representative of the last wooden cars built in the United States. (*Pullman Neg. 7394*)

Figure 4.25 The Chicago Great Western lounge parlor car built in 1898 by Pullman had one compartment outfitted as a private library. (*Pullman Neg. 4182*)

appearing on private or office cars as early as the 1850s. The idea was supposedly suggested to Pullman in 1869 by Joseph Becker, an artist on the staff of *Leslie's* magazine.[38] Becker spent many hours on the rear platform of the last car of a transcontinental train. The view was so impressive that he urged Pullman to outfit some cars with proper railings so that his regular patrons could enjoy the scenery in comfort. There is no evidence that such cars were actually built until sometime in the 1880s, perhaps because passengers could not walk through the train until after the introduction of the vestibule.

Apparently no observation-lounge cars were in service before the coming of the luxury trains. If this is true then the first observation car was probably the *Aladdin*, rebuilt in 1888 for the Golden Gate Special, a deluxe train operated over the Union and Central Pacific railroads.[39] In the following year Pullman replaced the *Aladdin* with a new car, the *Sybaris*, which featured a larger open platform that was stepped back a foot or so inside the car body. That is, the rear bulkhead was not flush with the end of the car's sides. This provided more space and shelter for the platform passenger. The end compartment of the car was arranged as a saloon, with easy chairs and large end windows. During bad weather it was still possible to look out from the comfort of this observation room. A few years later the view was improved by moving the door to one side and installing extra-large single-pane windows that extended almost from floor to ceiling. Some windows were 50 by 60 inches. The view was grand but the end framing was rather meager, especially for the rear car, and many roads preferred the less exciting but safer center-door plan.

In its first year the observation platform was only a daytime diversion, but methods were later devised to transform it into a nighttime attraction as well. In 1926 the Northern Pacific acquired ten new 85-ton observation-clubs from Pullman.[40] Each had a 250-watt searchlight mounted in the end of the roof. By means of an overhead wheel, the beam could be directed at

Figure 4.27 Pullman completed the parlor-cafe-observation Number 303 in 1911. (Pullman Neg. 13665)

Figure 4.28 Model passengers enjoy the view in this photograph made around 1925 aboard the B & O's Capitol Limited observation car Mountain Creek.

Figure 4.29 The solidity and dignity of a heavyweight-era lounge car is shown in a view of the B & O's Capitol Bridge. *The photograph dates from 1925.*

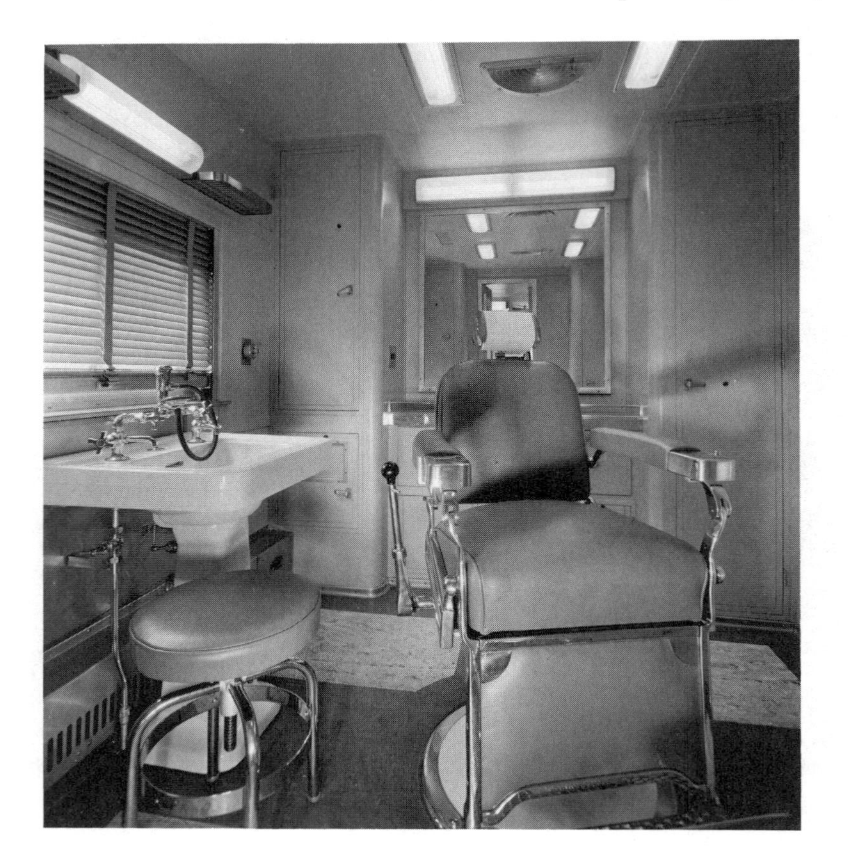

passing mountains and streams. In later years floodlights fixed under the rear platform effectively lighted the passing scene. The B & O also fixed a bank of lights on top of its dome cars for the pleasure of evening viewers.

Open-platform observations continued to be built until about 1930, but their popularity had begun to wane some years earlier. It was argued that they were uncomfortable and unsafe. No recorded instance of a passenger falling overboard can be found, but the hazard was real enough. The exposed platform was comfortable only in good weather, and even then it proved too windy or dusty for many patrons. Because every square inch of a passenger car represents precious floor space, the open platform was considered wasteful since it was so often unusable.

The solution was to enclose the end platform, and in 1909 the sun parlor or solarium car was born. The Burlington built several in its own shops and purchased more from Pullman.[41] The sun parlor did indeed increase passenger comfort and extend the usable floor space, but the view from it was rather limited. The Burlington liked the scheme, however, and acquired more sun-room observations. The Illinois Central became a convert in 1912. Pullman itself operated a number of sun-room cars that

Figure 4.30 Some first-class trains featured barber shops as early as the 1880s. This 1949 photograph shows such a facility incorporated in a Chicago and North Western lounge car. (American Car and Foundry Company)

Figure 4.31 Name-train lounge cars normally showed the latest thinking in interior design. This interior was created by Henry Dreyfuss in 1948 for the 20th Century Limited. (Henry Dreyfuss Associates)

Figure 4.32 The open observation platform offered a splendid vantage point for travelers. This scene was recorded on the Southern Pacific around 1925. (Southern Pacific)

Figure 4.33 A fully enclosed observation end, or sun-room, can be seen at the far end of this Burlington route lounge car. It was built in 1909 by Pullman. (Pullman Neg. 11790)

Figure 4.34 A sun-room observation car used on the B & O's National Limited in the 1920s.

Figure 4.35 Rounded-end observation cars were popular in the streamline era. The tavern-lounge car Franklin Roosevelt *went into service in 1941. (New York Central Photo)*

saw service on railroads in all parts of the country.[42] These boxed-in observations were not particularly attractive cars (Figures 4.33 and 4.34). They formed a blunt, inconclusive finish for a train, though this made them practical for runs where more cars were added en route.

In the style-conscious age of streamlining, the observation car's appearance was to undergo a radical change. The open-platform style was regarded as obsolete, while its blunt-end counterpart, the sun parlor, was discarded as just plain ugly. The observation car was the single form where the designer might readily display his talents. It is true that the car's exterior design was generally a source of frustration, for beyond smoothing its contours there was little opportunity to do much else. The observation end, however, could be molded into many shapes. The basic bullet or swallowtail configuration that came to be adopted after 1935 had been used some thirty-five years before by F. U. Adams. Adams's streamliner was a wooden prototype produced from a series of old B & O cars, but its general appearance was remarkably similar to those turned out by the aeroflow school. Apparently Pullman was the first to resurrect Adams's plan, for use in the two aluminum observation cars which the company built for the 1933 World's Fair. The automotive lightweights produced by various designers a year or two afterward featured many strange ideas for the shape of the train's last car, but the graceful swallowtail plan won the day. The design that had seemed so startling at first soon became a cliché (Figure 4.35).

A notable exception to this prevailing style was the Skytop lounge cars on the Milwaukee Road. From the inside the passenger could not be sure if he were aboard a train or a spaceship. The windows swept not only around but above him: 90 percent of the car's end was transparent (Figures 4.36 and 4.37). Because of their great expanse of glass, the cars became known as "rolling greenhouses."[43] The imaginative design was created by the industrial designer Brooks Stevens. The Milwaukee shops built four parlor cars to his plan in 1948. Pullman built six in 1948 and 1949 which were nearly identical, except that they had bedrooms in addition to the observation lounge. These six were sold to the Canadian National in 1964.

Figure 4.36 The Milwaukee Road's Skytop cars of 1948–1949 were the most daring and original observations built in North America. (*Association of American Railroads*)

Figure 4.37 The observation parlor car Cedar Rapids *was built by the Milwaukee Road's shops in 1948.* (*Association of American Railroads*)

Figure 4.38 Burlington's midtrain or square-end observation parlor car Silver Tower *was built in 1952 by Budd. It is shown here in 1971 leaving Chicago on the Denver* Zephyr. *(Douglas Wornom)*

In time most railroads came to regret their investment in round-end observation cars, since they could function properly only at the end of the train. Cars coupled on were isolated because there was no vestibule; thus an extra switching move was necessary wherever cars were added to the train. For this reason few swallowtails were built after 1950, and those that were came equipped with end vestibules—an addition that did little for their appearance. Square-end observations, much on the order of the old sun-parlor cars, came into favor (Figure 4.38). These so-called mid-train observations could be placed anywhere in the consist. Older swallowtails were retired or rebuilt. The Santa Fe, for example, sent six of the Vista series back to Pullman for remodeling as square-end cars.[44]

In summary, the lounge car fulfilled the railways' promise of luxurious travel—a promise that had been spoken of before the first tracks were laid in this country. Although the common man could not afford its splendors, the lounge car did bring the posh comforts of the private car within the budget of the middle-class traveler.

Dining and Buffet Cars

The diner was one of the last major classes of passenger car to come into general use. Always a great favorite with the traveling public, it was viewed by the railroads as a costly liability. It was the heaviest and most expensive of all the cars in regular passenger service. It rarely made a profit and generally incurred appalling losses. The solution of unloading this service on a concessionaire, such as Pullman, proved vain; its red-ink reputation was too well known. Yet it was a service that first-class travelers expected and that the railroads were obliged to maintain.

At one time there were over 1,700 dining cars in this country.[45] Each was a marvel of compactness and efficiency. There was little unused space; cabinets and lockers occupied every corner, storing an array of tableware and provisions that included 1,000 pieces of crockery and glass, 900 tablecloths, 700 pieces of silverware, and food for 400 customers.[46] Since the car also carried a fully equipped kitchen, dining-room furniture, and heating, lighting, and ventilating fixtures, it is no surprise that it

Figure 4.39 When this engraving appeared in 1871, only a few Western railroads were offering dining-car service. (National Car Builder, *March 1871*)

weighed 80 tons and cost $50,000. The weight and cost increased again with the advent of air conditioning in the 1930s. The crew was equally large and expensive: seven to sixteen men were required, depending on the status of the train. On a top train such as the Broadway Limited, nothing less than a steward, an assistant steward, a chef, three cooks, and ten waiters could do the job. In 1925 10,000 men were employed in dining cars to serve 80,000 meals a day at a net annual loss of 10.5 million dollars.[47]

Like parlor cars, diners were primarily intended for first-class travelers. The great majority of railway travelers never patronized the diner. Among commuters there was little need for it; only long-distance travelers had to eat en route. Even then the thrifty person could bring his own box lunch, buy a sandwich at the station news counter before boarding, or pick up snacks at station stops along the way. At one time food could also be bought through the car windows from platform vendors. It is impossible to say how common this practice was, but it is documented in an engraving from a volume of American pictures published in 1872.[48] The scene, reminiscent of India or Guatemala, shows a crowded platform at Gordonsville, Virginia, where Negro women are selling coffee, hot cakes, and chicken from trays, some of which are balanced on their bandanna-covered heads. The newsbutcher offered another alternative, but his wares were meager, usually consisting more of smokes and cheap publications than food.

As for railroad station restaurants, they could be wonderful or terrible. The depot hotels at Altoona, Springfield, and Poughkeepsie had a national reputation for excellent fare,[49] and the Harvey Houses along the Santa Fe were recognized as model establishments from their beginning in 1876. But most depot eating houses seemed to fall below any conceivable standard of decency and were vigorously denounced in the nineteenth-century press. One account insisted that they were the most infamous of all the "pernicious institutions" in America.[50] It pictured the proprietors as dispensers of dyspepsia or descendants of the Borgias. It described piles of mouldering dishes and crumpled napkins amid swarming flies and an atmosphere ripe with the smell of yesterday's breakfast. It portrayed the help as unkempt, rude, and slow; the coffee as a black, gritty fluid; the fruit as knotty and decayed. It said that everything was dusted with coal soot, and that the prices would cause Delmonico's to blush.

Because of the time factor, meals at even the best depot restaurants were not always a pleasant experience. The train could afford to stop for no more than half an hour; fifteen or twenty minutes was more common. Two or three hundred passengers all wanting to be fed at the same time created an atmosphere that was not conducive to civilized dining. If the train were running late, the scheduled thirty-minute stop might be cut back by a conductor. The fear of being left behind ruined many a stomach.

The contrast between the depot restaurant and the dining car was remarkable. To the good food, good service, and pleasant surroundings of the diner was added the enchantment of the passing landscape. Few other places can induce the same feeling of serenity and well-being than a seat on a railway dining car. It was in many travelers' experience one of the most civilized pleasures created by Western society. J. E. Watkins captured the mood:[51]

> The cry "Wilmington, fifteen minutes for refreshments!" is no longer heard, but instead the passenger enjoys his soup, his fish, joint, Cliquot, and coffee while the train runs fifty miles an hour. In fact, a dinner that is ordered at Newark is scarcely finished by the time Trenton is reached, and the stump of the cigar lighted shortly after completing it is often thrown away to the gamins at Broad Street Station, Philadelphia.

In the nineteenth century, small appetites were as suspect as six-course dinners were admired. A menu of the 1870s would include over eighty dishes.[52] To start, there would be Saddle Rock oysters on the shell or a choice of four soups. Among the entrees there might be broiled trout, stuffed loin of veal, and grilled mutton kidneys. Wild game like venison, quail, and prairie chicken were available as well, and so were a dozen vegetables and twice that many desserts. Such a meal was sold for 75 cents. The price was about equal to that of a good restaurant; it was not cheap, nor was it within the means of the average traveler, for it nearly equaled a day's wages for a workingman.

Only the best big-city restaurants of the time offered such a variety of dishes. In the menu just referred to, nearly everything was served with a sauce or was prepared in some special way. Even though some items were cooked at terminal commissaries, the assembly of such elaborate meals in a rolling, swaying kitchen that had little more than a coal stove seems miraculous. The kitchen of the 1875–1880 period was perhaps only two-thirds the size of modern diner kitchens. Curiously, as meals became simpler the size of the kitchen grew, though it remained a very busy place. In the mid-1920s, for example, the Broadway's galley produced twenty-four different dishes.[53] In its 50 square feet of open floor space (30 inches by 20 feet), four cooks simultaneously broiled steaks, made pies, cooked bacon, and baked potatoes. A dining-car cook had to be agile, well organized, and quick. He needed to know where each of a hundred pots and implements was stored and how to get at it without impeding the other kitchen workers. There was no room for waiters inside this busy area; the finished dishes were placed on a shelf for pickup and delivery.

In his 1819 memorial to the U.S. Congress, Benjamin Dearborn of Boston proposed that a system of railways be built and furnished with carriages offering meals to passengers during their journey. He cited the packet boat as an existing model for such service. A few years later as the railway slowly evolved into a public carrier, the same idea was proposed in England by William Chapman. Early railway managers had enough to do, however, to provide basic transportation. They could not achieve such luxuries as food service. Trips were short; let passengers dine before or afterwards.

Still, the idea of providing meals persisted, and in 1835 it was given an actual test. Imlay's eight-wheel car *Victory*, described in Chapter 1, went into service on the Philadelphia and Columbia Railroad. The fact that a serving counter and shelves occupied one end of the car indicates that food and beverages were available, though apparently no cooking was attempted. This was the earliest known instance of railway food service and represents the beginnings of the buffet car, if not of the diner. But like so many firsts, it seems to have been an isolated incident that led nowhere and was soon forgotten.

Yet food service was such an obvious travel requirement that it was always being discussed. In 1838 one of the Baltimore papers praised the new sleeping car installed on the P W & B and said that the only basic comfort required by the traveler was a dining car. This service, the paper predicted, would soon be offered. Five years later a passenger on the B & O tried to revive interest in the scheme. The B & O tracks reached Cumberland, Maryland, late in 1842, and in the next year Robert McLoed petitioned the directors of the company to place what he called a "refectory" car in service between Baltimore and Cumberland.[54] He suggested that an ordinary day coach be remodeled for the purpose, but apparently nothing came of his proposal.

Two special trains run over the B & O in January 1853 to celebrate the line's opening to Wheeling did contain refreshment cars.[55] Many notables were aboard, including the governors of Virginia and Maryland, state legislators, and judges. Food and drink, catered and served by the United States Hotel, were provided throughout the trip in new cars equipped with long center tables. However, it appears that the cars had been temporarily fitted up for the celebration; they did not offer regular food service.

It is claimed that a public dining car began running between New York and Chicago during the same year.[56] The reliability of this account can be questioned on several points, particularly the route and the date. It is doubtful that a through car could have

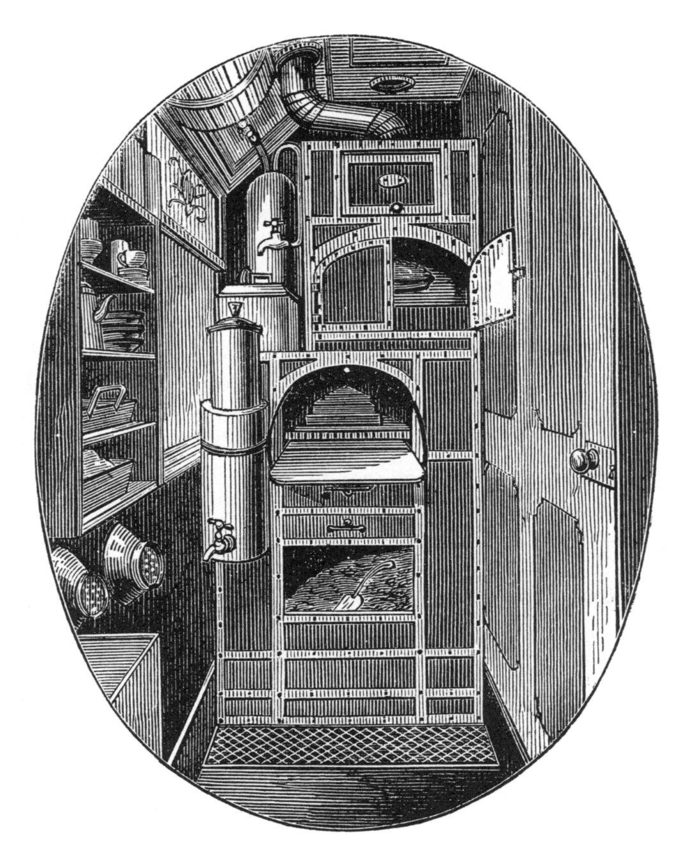

Figure 4.40 The diminutive kitchens of Pullman's hotel cars were capable of furnishing full-course meals. This engraving dates from 1869. (Engineering, *July 17, 1869*)

found its way between the two cities via the Northern Central, the Grand Trunk, and the Michigan Central, as the article claimed. The date of 1853 also seems a trifle early. But even if the story is as much as ten years off, it is early enough to be of interest. The diner was described as a remodeled coach, with tables and stools installed in place of the seats. The kitchen was located in one end of an adjoining baggage car. The waiters had some trouble carrying the food between the cars, but at 50 cents a portion, the service was very popular. Two stops were avoided, saving forty-one minutes. Who owned the car or initiated the service, and how long it remained in operation, are unknown.

During the 1860s, contemporary notices of dining cars began to appear. One journal reported a diner on a railroad in Pennsylvania in the fall of 1860.[57] Another spoke of a restaurant car on the Lehigh Valley Railroad in the same year.[58] That reporter was unimpressed: "We should require to be very hungry before we should desire to see another one. The 'restaurant' was a deserted entry to the baggage room; no one in attendance, and nothing to eat. We did not like its appearance—nor the odor with which it was pervaded." The same reporter, presumably Zerah Colburn, was in fact opposed to the entire idea of the dining car.[59] He questioned the wisdom of running diners because of their expense and weight. Since not everyone could be accommodated in a special car devoted to the purpose, it would be necessary to serve passengers at their seats. An army of waiters would be needed. Gravy, scraps of food, and dirty dishes would be everywhere. He insisted that the solution lay in upgrading the depot restaurants. Some railroads, like the B & O and the Central Pacific, did concentrate on improving their hotel-restaurant facilities, but others were willing to experiment with meals on wheels.

One such road, the P W & B, began diner service in the early 1860s. There are two contradictory accounts of these cars.[60] According to an old employee, they were placed in service in 1863 and ran for three years. They consisted of two 50-foot coaches that had been remodeled with a center partition. One-half of each was used as a smoking compartment; the other half housed an "eating bar." The food, which had been prepared at the depot kitchen, was kept warm in a steam box. On the other hand, an editor of the *New York Sun* described a diner of the same period in a 1924 memoir. In 1862, he said, he was traveling with an uncle to Washington. In the course of the journey they came upon an improvised diner, an old baggage car which showed signs of long service. The interior was bare except for an oblong counter in the middle that was surrounded by high stools. The center of the counter was cut away so that waiters could stand inside. The menu was limited to oyster stew, coffee, and strips of a deep-fried sweet cake known as crullers. Possibly both accounts are correct; the P W & B annual report for 1863 lists three refreshment cars, so that the third may have been the converted baggage car described in the editor's memoirs.

Information is available about several other dining cars of the Civil War period. On his trip to Washington in February 1861, President-elect Lincoln was served dinner between Buffalo and Albany aboard a car specially fitted for the purpose by the New York Central.[61] During a more solemn trip, Lincoln and his party were seated at a table in a baggage car for the noon meal while they traveled to Gettysburg.[62] A special train was run for the purpose, and although an office car was assigned to the President, it either did not have dining facilities or was not large enough to accommodate his entire staff.

During the war years the wounded were served food aboard hospital trains. Injured soldiers were at first carried in boxcars furnished with little more than straw or hammocks, but in January 1863 the P W & B refitted some cars with bunks and kitchens.[63] Hot soup and coffee were provided. In 1863 vastly superior hospital trains were assembled for the Army of the Cumberland by the U.S. Sanitary Commission. Full-scale kitchen cars were outfitted. At one end was the kitchen, with a range, cupboards, and a sink. A storeroom compartment was next to the kitchen. A narrow aisle to one side of the storeroom led to the dining area, which contained a long center table and benches. Although not elegant, these cars were neat and serviceable. They introduced the idea of dining on rails to many young men who probably had never before ridden a train.

Most of the early dining cars that have been described were makeshifts which created little interest outside of their own locale. None precipitated any movement toward the general adoption of dining cars, but they were the prelude to a new era in railway travel. In reality, the custom of eating while in motion had been established by canal packets and extended by river and coastal steamers. Rival lines competed through their cuisine. The great dining hall became the steamer's architectural showpiece. An attraction of travel was the promise of an excellent meal in a sumptuous cabin. Thus a public that was accustomed to eating and traveling considered it a nuisance to detrain every few hours for a snack, and not always a very good one. Yet curiously, the Eastern railroads, which competed most directly with the overnight steamers, did not lead the way in experimenting with din-

Figure 4.41 A British publication, Engineering, *published four interior views of Pullman's hotel car* City of Boston *in July 1869. Portable tables and folding berths converted the car, as required, to function as a coach, sleeper, and diner.*

ers. It was the lines running out of Chicago that introduced the dining car; Chicago was the place and Pullman was the man.

The beginnings of the commercial dining car were intertwined with Pullman's effort to establish a fleet of cars for luxury travel. Sleeping cars eliminated the most basic discomforts of long trips. Pullman thought that some form of food service would prove a popular and profitable auxiliary. A small kitchen added to a palace sleeper resulted in the hotel car. In 1866 and 1867 several new cars were built on this plan for service on the Michigan Central, the Burlington, and the Great Western of Canada. A tiny kitchen was placed in the center of the car (Figure 4.40). It served the two ends, which were outfitted like a standard open-

Figure 4.42 Pullman's first full diner, the Delmonico *of 1868, is shown on an excursion train operated over the Chicago and North Western Railway. (Chaney Neg. 9816)*

section sleeper except that small folding tables were installed during the day between the seats. These could be used as writing or dining tables; at night they were stored in a closet. Hotel cars slept 30, but they could seat and feed as many as 40 passengers. The food was prepared and served by the crew—generally four to five men. The menu was surprisingly varied considering the kitchen size, which in the case of the *Western World* was said to be only 3 by 6 feet.[64] Built-in cabinets adjacent to the kitchen stored the 133 items of food, the liquor, the 1,000 napkins, the 150 tablecloths, and the china and glassware needed for a long journey.[65]

One of the first hotel cars, the *City of Boston*, was illustrated and described by the British technical journal *Engineering* in the year following its emergence from the Burlington's Aurora shops.[66] The body, 60 feet by 10 feet, 8 inches, was divided into four sections. At one end was a small drawing room; next came a larger compartment with open sections; next the central kitchen; and the far end was occupied by a compartment with open-section berths. The illustrations accompanying the article mistakenly suggested that various sections of the car were permanently set up for dining, sleeping, and day travel (Figure 4.41). Only the drawing room, shown in the lower right of the four-part engraving, remained the same day and night. The other three areas shown are actually one of the open-section compartments as it might be converted for dining, sleeping, or day travel.

The hotel car, though hailed as a great advance in railway travel, had its faults. Travelers who suffered from motion sickness found it disagreeable to be cooped up in a car where the smell of food lingered. The price of the meals was high: an entree with potatoes cost 50 to 60 cents, coffee and tea 25 cents apiece. The crew was too large for the number of patrons. The investment required for a kitchen, linens, and the other accessories added to the economic risk of the hotel car.

Pullman hoped to solve all these defects by building a special car that did nothing but prepare and serve meals. A complete diner could provide more and better food for everyone aboard the train. It would be cheaper, since it would leave more revenue space in the sleeping cars and would surely cut labor costs. And so only a year after the first hotel cars had appeared, work began on a dining car.

That Pullman meant to build a magnificent rolling restaurant is implied in the car's name, the *Delmonico* (Figure 4.42). No restaurant in the New World enjoyed a greater reputation for luxury in cuisine, service, and decor. Pullman planned to create a

Figure 4.43 A rare interior view of the Pullman dining car Cosmopolitan. *The sixteen-wheel car was built in 1869 by the Burlington's Aurora, Illinois, shops. (Charles Clegg)*

mobile eating carriage equal to its namesake. It would be no crude feedin' car with a steam table, or anything that could be criticized as greasy, untidy, or smelly.

Again the Aurora shops were assigned the job of construction. The car, which cost $20,000, was the same size as a hotel car—60 by 10 feet.[67] A kitchen 8 feet square was placed at the car's center and flanked by two dining rooms with twelve tables each, providing total seating for 48 persons. A large ice chest was mounted under the floor of the kitchen. Gaslights and a hot-water heating system were part of the up-to-date equipment. Two cooks and four waiters served 250 meals a day. The crew slept in the car at night by converting the Morocco leather seats into berths. The floor under the tables was covered with oilcloth for easy cleaning, while the aisle was carpeted. In the kitchen a side door to the outside permitted easy access for loading supplies. In April 1868 the *Delmonico* was ready for public viewing. After a test on the Illinois Central, the car was sent to New York and in June formed part of a special excursion train to celebrate Henry Keep's appointment as president of the Chicago and North Western. It was during this trip that the only surviving photograph of the historic car was made (Figure 4.42).

At the time of its introduction the press announced that the car would run regularly between New York and Chicago, but the lines involved thought otherwise. Either they were satisfied with the hotel cars already in service or they found that the *Delmonico* did not pay. In the summer of 1868 it was leased to the Chicago and Alton. This and other disappointments led Pullman to move very slowly in the dining car trade; never in the firm's history were more than a handful of diners on its roster (see Figure 4.43). The hotel car remained in favor, however, demonstrating that there was a profit in light meals for first-class passengers. The kitchen was moved to one end of the car, less ambitious menus were attempted, and improved ventilation helped to dispel cooking odors. The railroads were also pleased with the hotel car, because it satisfied customer demand and relieved them of the necessity for providing dining-car service. Pullman managed the hotel cars with no trouble to the railroads, and if he could make a profit while doing so, all the better. The cars remained in vogue until about 1880. One was shown by Pullman at the Centennial, and a new hotel car named the *St. Nicholas* was described in 1879 as having twenty-four berths and a 10- by 6-foot kitchen-pantry at one end, all within its 63-foot, 10-inch body.[68] Within a few years, however, these cars fell from favor as the railroads reluctantly accepted the dining car. Pullman reduced the kitchen space, adopted the designation "buffet" in place of the obsolete term hotel car, and continued operating the cars much as before.

Following the *Delmonico*'s introduction only one railroad

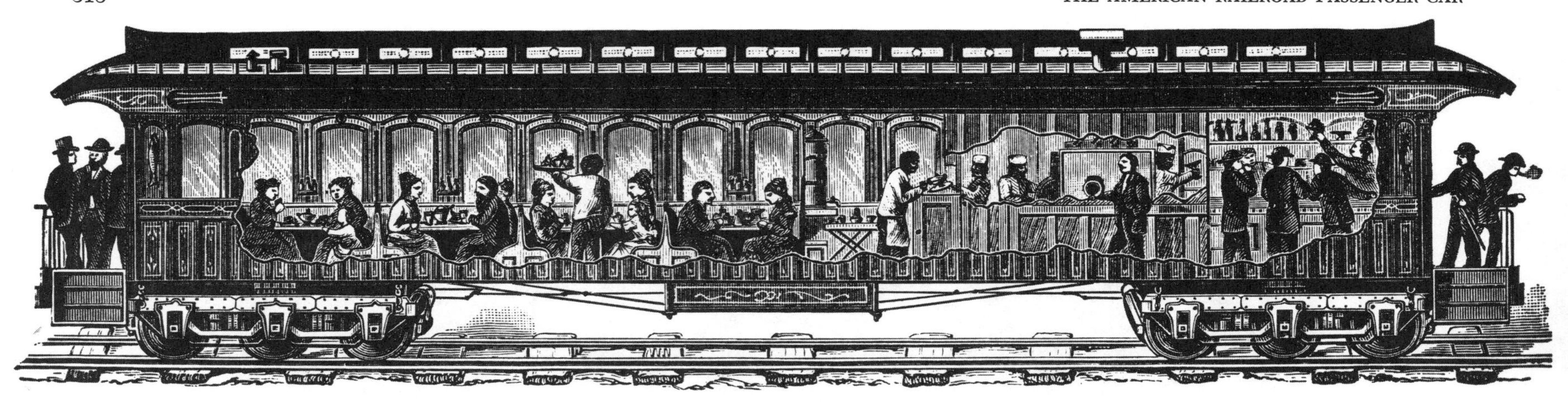

Figure 4.44 A cutaway engraving of a Rock Island dining car of about 1877. Note the men's bar at the right end.

Figure 4.45 A dining-car kitchen with its coal-fired range was depicted by an artist for Leslie's Weekly Magazine *in 1877.*

seemed to have much faith in the full diner. In competing with the other lines connecting Chicago and St. Louis, the Chicago and Alton decided to try improving its food service. It encouraged Pullman to operate diners on its trains, presumably with a cost-plus guarantee. By 1872 the road was using five dining cars, which offered a bountiful meal at the fixed price of $1.[69] Trade picked up, and when the road's rivals countered with diners of their own, the Alton bought out Pullman's interest and reduced the meal price to 75 cents.[70]

By 1875 the dining-car fever had spread to the Michigan Central, and within a year or two the Burlington, the Rock Island, and the Wabash were also running diners.[71] All the Chicago railroads had decided that the dining car was a competitive necessity, no matter how great the losses. Increased patronage was expected to offset the out-of-pocket costs. A Rock Island timetable of this period described the dining car as a democratic experiment which, unlike the hotel cars, was open to all travelers, be they millionaires or miners. The Rock Island also intended to have the best equipment possible, and in 1877 four cars, *Oriental, Overland, Occidental,* and *Australia,* were manufactured at its Chicago shops.[72] Each of these twelve-wheelers had a body 57 feet long. A 27-foot dining room seated 16 at four tables. A 10-foot kitchen with a gleaming copper floor was in the center, and a 7-foot lunch counter occupied the far end of the car. Light meals, snacks, and drinks were served at the counter. Formal meals—*table d'hôte* at 75 cents or *à la carte*—were available in the dining room. Royal Dresden china, elegant silverware, rare wines, and cigars were all part of the service. The car's exterior was painted a wine color. In addition to the usual gilt striping there were some "elaborately executed game pieces indicating the purpose of the car and the nature of the larder." A cutaway engraving of one of the cars is reproduced in Figure 4.44.

By the early 1880s the dining car was an established fact in the Midwest. The last major holdouts in the Chicago area were the North Western and the Milwaukee roads, and they fell into line in 1881 and 1882 (see Figures 4.45 to 4.47). An excellent contemporary account describes one of the C & N W's first cars:[73]

> Each car is 69 ft. in length over all, the bodies being 62½ ft. long by 10 ft. wide. They have 6-wheel trucks, 33-inch Allen paper wheels, and well adjusted springs for easy riding. The main or dining saloon is 32 ft. in length, and is divided into five sections on a side, with a table and seats for four persons in each. Large plate glass mirrors cover the end partitions, and also the wide window panels, the effect of which is to make the size of the room seem much larger than it is. At the end of each double seat is a large window, the panels over the seat backs are narrow and those over the tables wide, with three alcoves for tumblers, salt, spices, etc. The seats are covered with maroon leather, and the floor laid with 3-inch alternate strips of walnut and maple. The seat ends and frames and the inside finish are cherry trimmed with mahogany; five Adams & Westlake 2-light chandeliers with argand burners hang over the aisle—one to each pair of tables. These, and all the metal trimmings of the car are gilt. The head-linings are decorated oak, paneled with mahogany moldings. Those below the clear-story have paintings of game, fruit and flowers. Hartshorn roller shades are used instead of blinds, and roll up out of sight under box-moldings. The windows are all double, the outside ones of the kitchen having game and fish designs ground in the center. The kitchen is 13 × 8 ft., and built of solid walnut, with all requisite utensils and conveniences. Water is carried in a long cylindrical reservoir in the clear-story overhead. Between the kitchen and dining saloon is a pantry 4½ × 8 ft. In the end corner of the car is an ice closet with glass-bottomed drawers opening into the kitchen, also a closet for small stores. At the opposite end is a Baker heater, linen and wine closets,

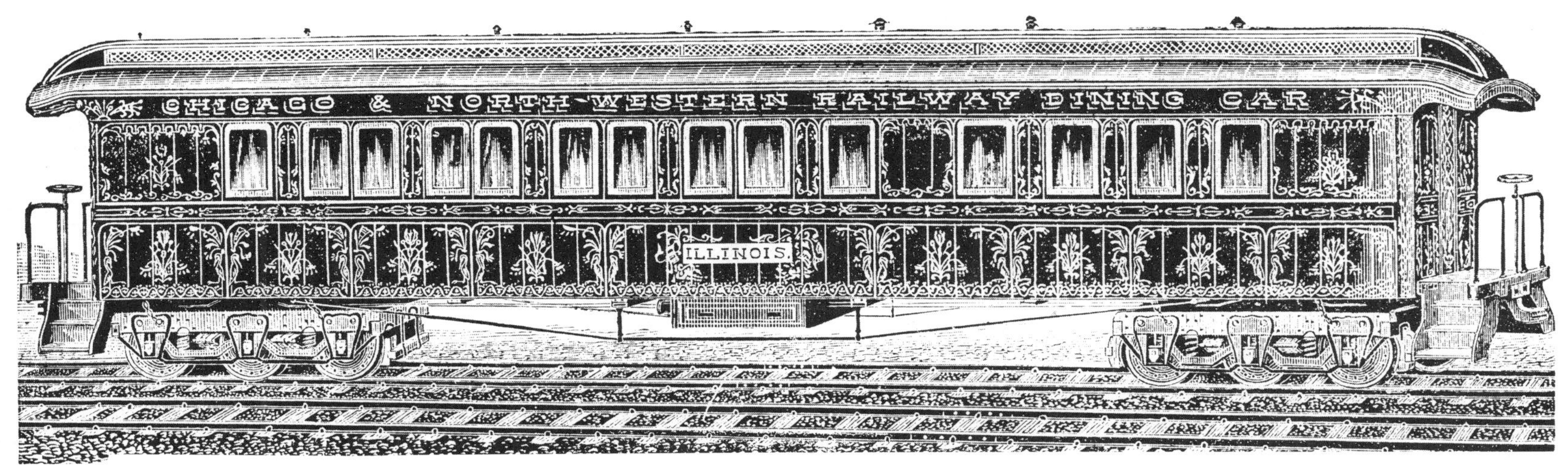

Figure 4.46 The Illinois *was an 1881 product of the Chicago and North Western Railway's West Chicago shops. The car was painted wine color with gilt embellishments.* (National Car Builder, *July 1881*)

Figure 4.47 Harlan and Hollingsworth of Wilmington, Delaware, completed this car in 1882. (Smithsonian Neg. 47984-G)

lavatory and ice chest. Without particularizing further, it may be said that these cars are believed to embody all the improvements pertaining to their class, and are unsurpassed in their luxurious accomodations and handsome finish by any others in the country. The outside is painted a dark wine color relieved with gilt tracery and ornamental designs representing storks, passionflowers, cornucopias, etc.

The demand for dining cars was also mounting in the East, though railroad managers there hoped that the entire diner system would quietly die. Certainly there was talk among the Chicago lines of mutual abandonment of food service because of the continued losses, but this never came to pass.[74] A few lines, believing that diners paid for themselves through increased traffic, would not join the abandonment collusion. The Eastern trunk lines were at last forced to admit that the diner would prevail. To the surprise of many, the conservative B & O was the first to give in. Early in 1881 it placed five new restaurant cars in service.[75] The Pennsylvania, deciding that it must follow suit, became mildly enthusiastic once the change in policy was made, and the Altoona shops were ordered to produce the best dining car possible.[76] The designer was Theodore N. Ely, head of the mechanical department. Four cars, which replaced hotel cars, were ready for the New York–Chicago express in the spring of 1882. A special feature was that each car carried a 125-bottle wine rack.

The New York Central was now forced to act. For years it had been content to run on a leisurely 36-hour schedule between New York and Chicago, providing a breakfast stop at Rochester, lunch at Erie, and dinner at Toledo.[77] Its more experimental Western subsidiaries, the Michigan Central and Canada Southern, had been using diners for some time. Finally in the autumn of 1883 the Vanderbilts decided that it was time to join the ranks of dining-car railroads.

No diners were seen in New England until the following year. The New Haven cautiously put two cars in service between Boston and New York in the summer of 1884.[78] The same pair was expected to double back to New Haven for the noon train to Worcester. Now the New Haven, like other Eastern lines, came to champion the diner, and it was one of the few railroads that realized a profit from food service.

By the mid-eighties the dining car was no longer an unwanted stepchild but the pride of the passenger car family. Every major railroad had at least a few restaurant cars, and they were common on all first-class express trains. Even the Southern railways, usually accused of lagging a decade behind, began to adopt diners almost as early as the Northern roads.[79]

It was the Far Western roads that resisted the dining car longest. The Union Pacific had no diners until almost 1890,[80] nor did the Central Pacific. Both companies had permitted Pullman to operate diners over their lines until the early 1870s, but then they returned to the old-fashioned system of stopping at station restaurants. The Central Pacific was known for the excellence of its eating houses, and it saw little reason to reverse this policy until the demands of faster schedules forced it to do so. For similar reasons the Santa Fe put off running diners for many years. Their contract with Fred Harvey had proved an excellent one. Beginning in 1876, the English immigrant who had turned station restaurateur set out to overcome the notorious reputation of such establishments. Within fifteen years his chain of dining rooms was serving 5,000 meals a day, and its reputation for good food was worldwide. But as other Western lines began to adopt dining cars, the Santa Fe felt obliged to do likewise. Friction developed between Harvey and the railroad, which ignored his protests. Harvey took his patron to court, where in 1891 he obtained an injunction restraining the operation of dining cars on the Santa Fe west of the Missouri River.[81] The railroad was forced to recognize its contract, but within a year Harvey was persuaded that both their interests might best be served if he were to operate the road's dining-car service.

Why did it take nearly a quarter century before passengers could expect to dine aboard a train in any region of the United States? The diner's poor earning performance was to blame, but why was the diner an inveterate loser? Even well-patronized trains could not seem to make money on their restaurant cars. High fixed costs and a limited market were the basic reasons for their fiscal difficulties. A sizable crew, generally ten men, was necessary to serve thirty to forty patrons efficiently. Even in the days of cheap labor, and even though the same crew served several seatings, this was a disproportionate ratio. The crew had to be fed and housed throughout the trip, and at least some of them had to be carried back to their homes as deadhead passengers. Unlike a restaurant, the railroad could not use part-time help. Free meals to other trainmen created an additional new expense. A dishonest crew could also add to the losses through pilferage of food, linens, and silverware, while a dishonest steward could steal supplies in wholesale lots and juggle his accounts to escape with cash as well.

Unused or spoiled food was another large expense. Cars were usually overstocked so that no customer would be disappointed, and what was not used could not always be salvaged. The old-fashioned icebox was not entirely effective in storing perishables.

Moreover, the clientele was limited to those aboard the train. No matter how good or cheap the meal, the dining car could not draw upon the population at large. And many travelers brought their own food; others, being country folk intimidated by the posh diner and its crew, skipped their meals. Only the first-class passengers could be depended on to patronize the diner.

The investment required for a diner could usually be applied more profitably elsewhere. In the 1880s such a car cost $12,000, or nearly three times the amount for a coach.[82] In 1905 the price was up to $17,000, plus $12,000 for dishes, linens, and other necessary gear. By the 1930s the average price was $50,000, and by the 1960s it was nearly four times that. Total investment on roads with large dining-car fleets was substantial. In 1930 the New York Central had 7.8 million dollars tied up in its 156 restaurant cars. Repairs and upkeep were higher than those for a coach or even a sleeper. In addition to renewing the tableware and linen, the kitchen required continual repairs. In 1925 a typical diner needed a major overhaul every eighteen months at a cost of $5,000 to $7,000.[83]

There was no effort to set prices in accordance with costs. No one would pay, say, $10 for a meal, or whatever figure the accountants might estimate was necessary to cover expenses. In the early years, 75 cents in the West and $1 in the East became established traditions. The dollar dinner had gradually disappeared by the time of World War I except on the New Haven, which continued to honor the old standard.[84] The *table d'hôte* vanished during the same period; Pullman and Wagner learned that *à la carte* meals offered the only hope of turning a profit in the food trade.

In 1887 it was estimated that each diner could be expected to lose $100 to $600 a month.[85] Some railroads were losing as much as $21,000 annually. A few, like the Pennsylvania, were just barely in the black—its diners showed a pitiful $29 profit in 1886. The Milwaukee had been losing $16,000 a year, but it succeeded in reversing this to a $4,000 profit by hiring a new manager. It was a temporary victory, however; many years later the same railroad reported an annual loss of $60,000 on just one train.[86]

These losses were justified as a necessary business expense—a cheap, direct form of advertising. Good meals increased traffic and won friends. Railroads became famous not only for the general quality of their food but for individual dishes. The B & O was known for its corn bread and country sausage breakfasts, the Union Pacific for steaks. The Northern Pacific was the great baked-potato route; a giant 20-ounce spud was the minimum size served.[87] This road also offered fresh fish in season. Its homemade fruitcake, which won a grand prize at the 1889 Paris Exposition, was produced in 15-ton lots to satisfy the demand

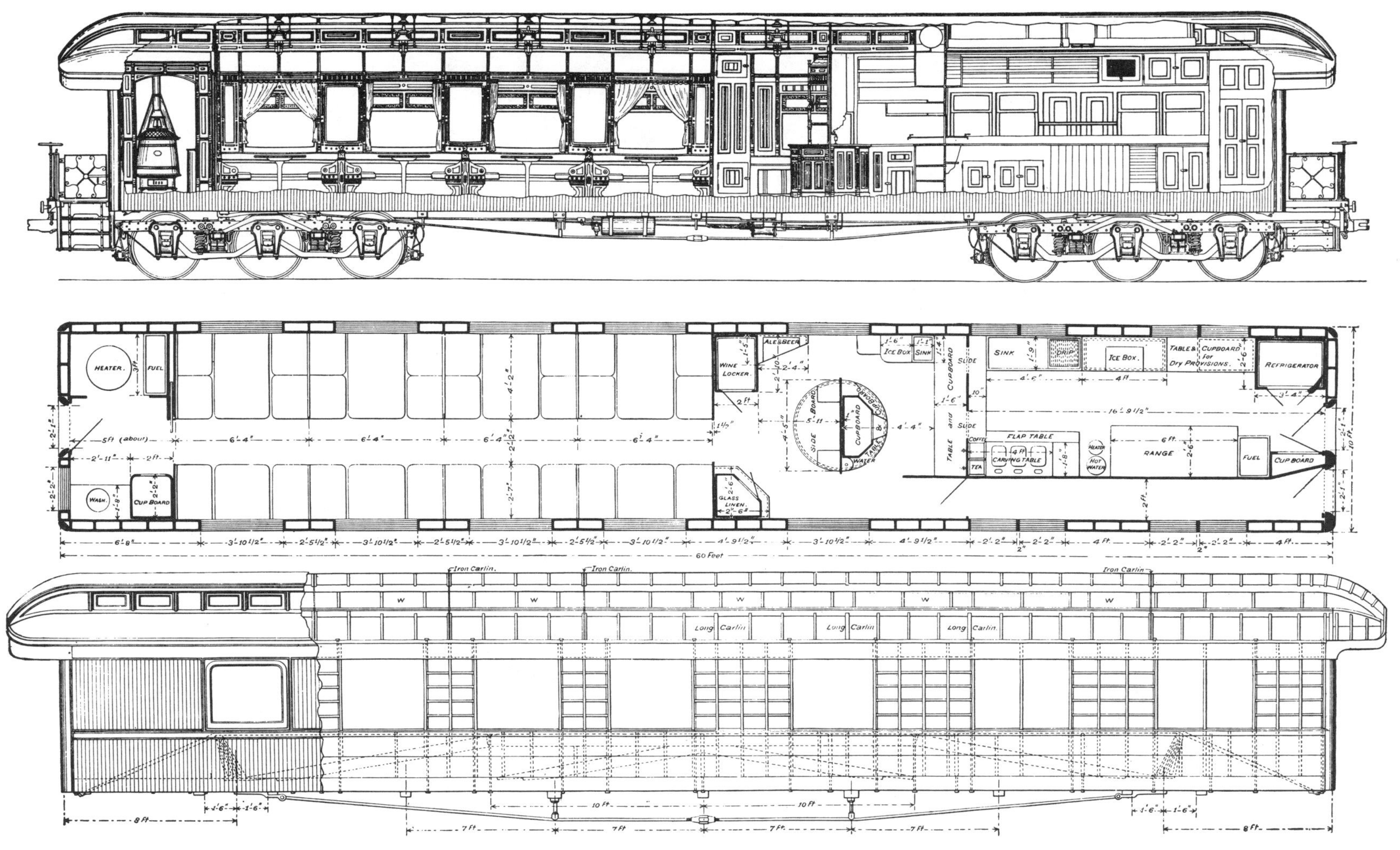

Figure 4.48 Plans for a Michigan Central 37-ton, 24-passenger dining car were published by the Railroad Gazette *in 1887.*

from its dining-car and on-line patrons. As losses mounted and general passenger travel declined, however, the good will created by dining cars was no longer enough to justify their operation.

By the early 1880s the dining car's arrangement was fixed in the general pattern shown in Figures 4.48 to 4.50. The kitchen was placed at one end, with a narrow side aisle connecting the vestibule and the dining compartment. Fixed cross tables and bench seats, much like those used in Pullman open-section sleepers, were at first favored. But sliding in and out of the seats was not easy, particularly for the overweight or elderly, and in the early eighties some lines began to install lift-up seats like those in theaters.[88] Apparently loose chairs were not used for almost another decade. This was, of course, the best form of seating, because it allowed easy entrance to the table and because the chair could be positioned to suit the comfort of a tall or short diner. By the middle eighties the table arrangement was altered so that one side of the car had the regular four-place tables while the other had smaller tables for only two persons.[89] The smaller tables were preferred by couples and single diners. The four-and-two table pattern created a wider aisle, improving the work space for the waiters. It also reduced the car's capacity, a result which was appreciated by both the crew and the diners. Forty seats in a 60-foot car were too many; twenty-four to thirty places came to be recognized as more realistic in view of the kitchen's capabilities and the general space needs of all who inhabited the car.

Even when car sizes grew to 70 and 80 feet in the 1890s there was no effort to crowd more seating into the cars, except on some Eastern lines where patronage was unusually heavy[90] (see Figures 4.51 and 4.52). During these years car designers allotted more space to the kitchen, so that the cramped 10-foot kitchens of the 1870s were expanded to as much as 27 square feet.[91] More space permitted a greater work area and larger ranges. In 1895 a full-sized coal range was 6 feet long, 34 inches deep, and 5 feet, 6 inches high, and it weighed 1,250 pounds. These substantial furnaces were also connected to a heater that supplied hot water for cooking and dishwashing.

During the mid-eighties a distinctive architectural feature emerged that was to remain in favor for many generations. This was the sideboard or breakfront, which was placed at one end of the dining room between the pantry and the rest of the car. It was a decorative focus, usually a handsome piece of cabinetry, and it served to screen off the working section of the car. China or silver plate was exhibited behind its glass doors, and a bowl of flowers was often placed on it.

As in the parlor car, large windows became a feature of the dining car (Figure 4.53). They were used for the same reason: a fine view added to the pleasure of the trip. The kitchen end of the car normally had half windows, which in some cases were made of opaque art glass. Its distinctive fenestration customarily set the diner apart from the other cars in the train. In 1893, however, the Delaware and Hudson broke with this tradition by purchasing a diner with a coach-like body.[92] All the windows

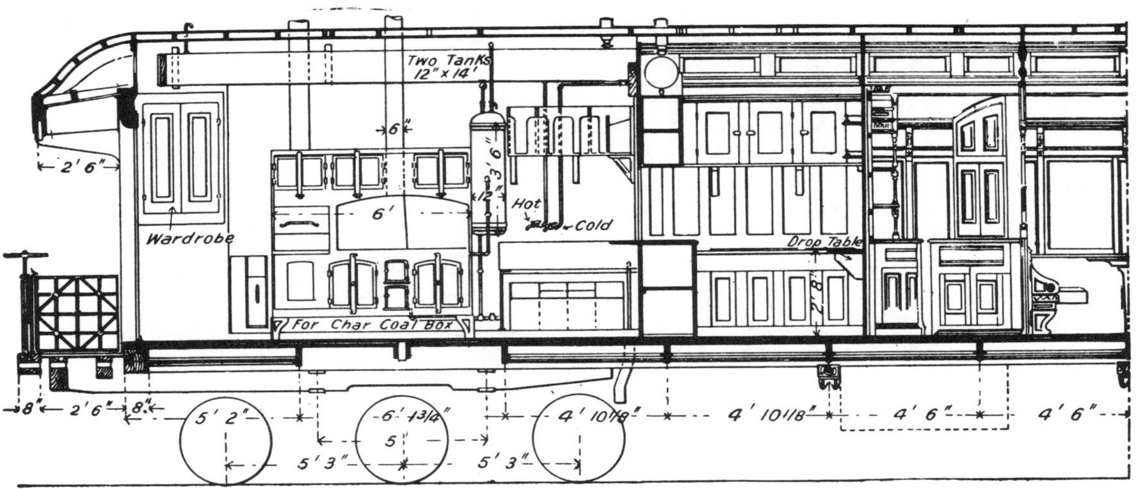

Sectional Half Side Elevation.

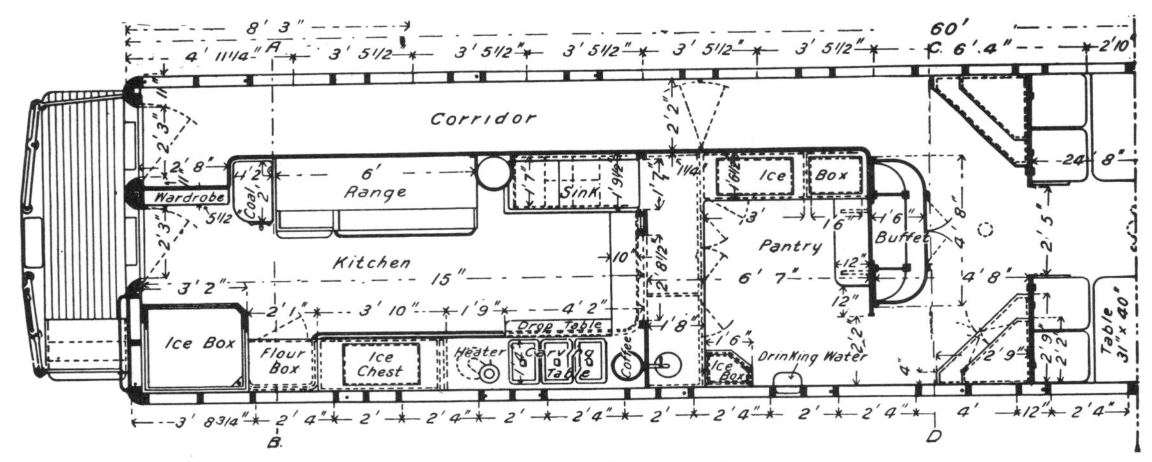

Plan of Kitchen, Pantry and Buffet.

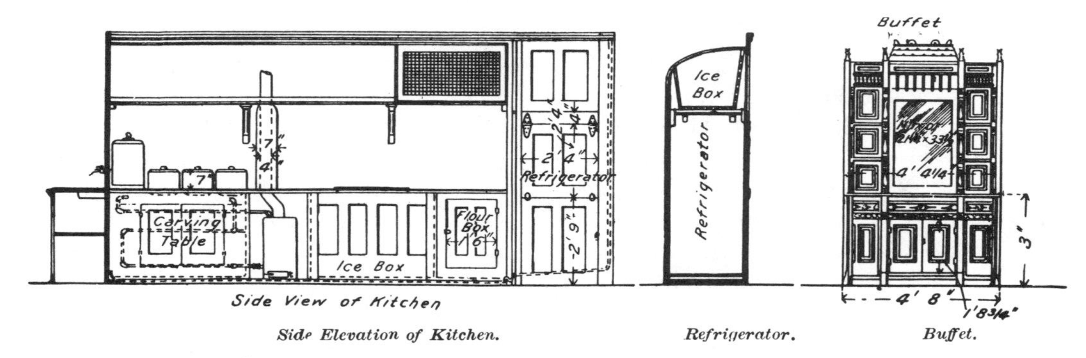

Side Elevation of Kitchen. Refrigerator. Buffet.

Figure 4.49 The general arrangement and furnishings of a Chicago, Burlington, and Quincy dining car of about 1890 are given in this drawing. Other details are shown in Figure 4.50. (Car Builders' Dictionary, *1895*)

were uniform in size, and the table and chairs were loose. The idea behind it was that these furnishings had only to be removed and chairs or seats installed to convert the car into a buffet-parlor or coach. The standard coach body was much cheaper than a regular diner, but sensible as the scheme may have been, it was never widely adopted.

Changes over time in dining-car structure and size followed the same general development of all passenger equipment. In 1886 a good-sized diner had a 64-foot body and weighed 41 tons. By the end of the wooden age, there were some cars with 72-foot bodies and total weights exceeding 76 tons. An example of these large wooden-bodied diners was the New York Central's Number 405, rebuilt at the West Albany shops in 1910 for the Lake Shore Limited.[93] Its overall length was 80 feet, 8 inches. A crew of 13 was required to serve only 30 diners. A diagram of the car is shown in Figure 4.54.

The steel era brought no dramatic growth in size; few cars exceeded 80 feet in overall length, but their weight crept upward

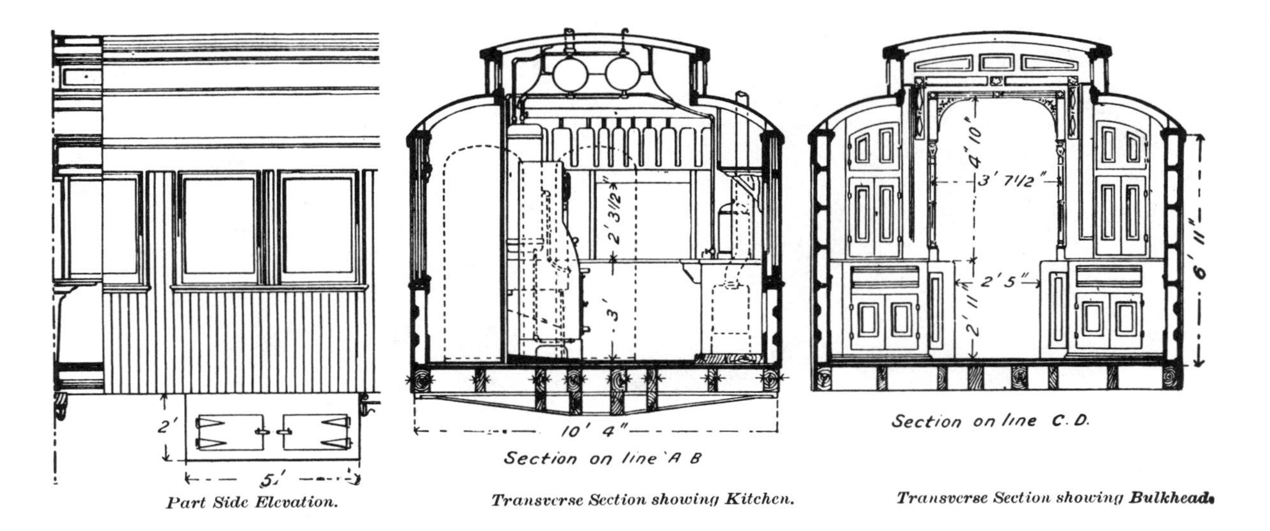

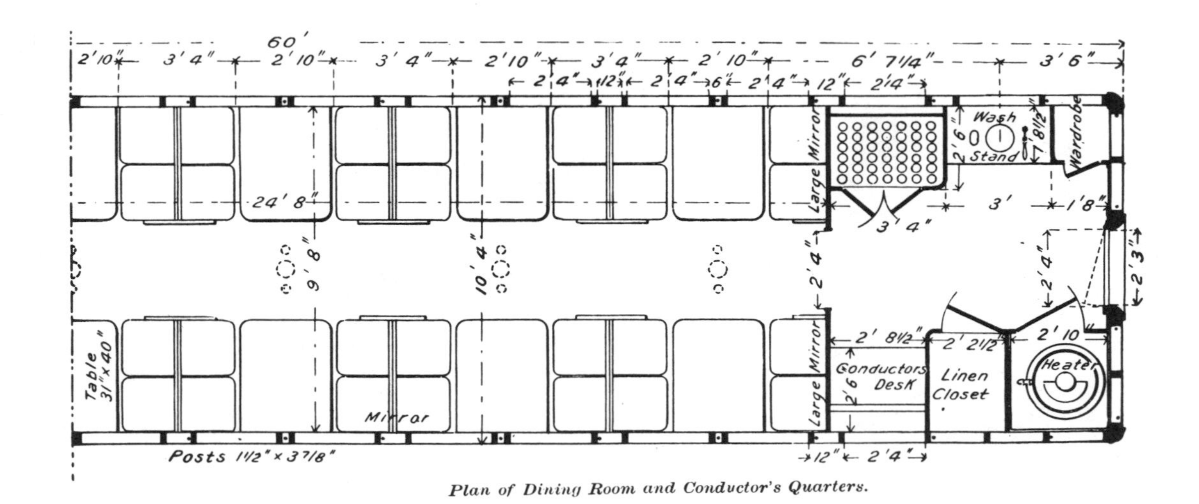

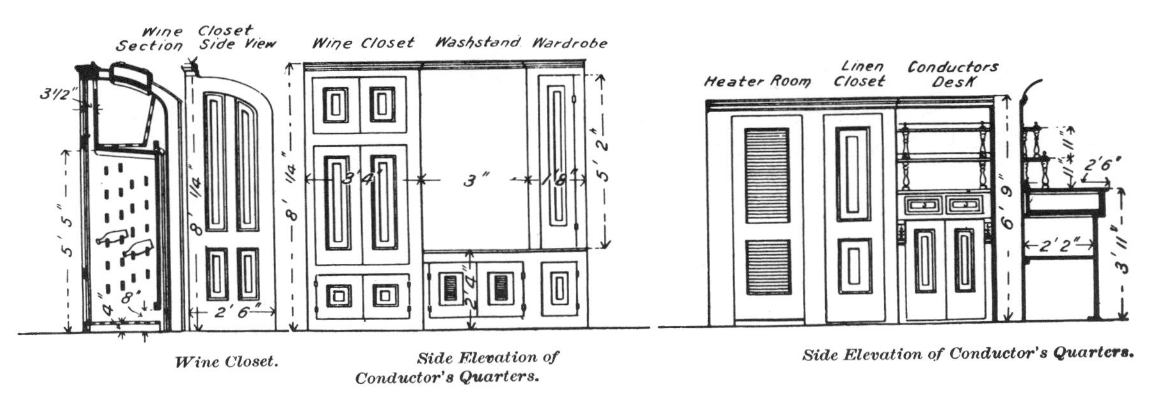

Figure 4.50 A Chicago, Burlington, and Quincy dining car of about 1890. (Car Builders' Dictionary, *1895*)

until it was between 80 and 85 tons (see Figures 4.55 to 4.60). The general floor plan did not change during this period except for a single major alteration: one or both vestibules were eliminated, which increased usable space by as much as 8 feet. An early example was the 1910 Lake Shore Limited car in Figure 4.54, which had no vestibule at the kitchen end. In 1914 the Burlington acquired a steel diner with no vestibules at either end,[94] and by the early 1920s nonvestibule diners were relatively common. These entranceways were considered nonessential, since most passengers did not board the train through the diner in any event. End doors and diaphragms were necessary, of course, for passage to and from the diner.

Beyond eliminating the vestibule platforms, most railroads seemed content with the standard floor plan in which the kitchen was at one end, the dining room occupied the other, and a side passageway ran alongside the kitchen. There were a few interesting exceptions, however. The earliest was a center-table plan tried out by the Erie in 1917 to provide greater privacy.[95]

Figure 4.51 Pullman's diner La Rabida *was an extra-elaborate display piece created for the Columbian Exposition in 1893. (Charles Clegg)*

Shoulder-high semicircular partitions in the car's center created a series of private dining alcoves. Each had a round table with four or six chairs. A center panel split some of these areas into small tables for two. Total seating was for 36. The tables could be reached from aisles on both sides of the car. The kitchen was placed at one end, as in a conventional car. So far as can be determined, the two sample cars built for the Erie by Barney and Smith were the only examples of this style.

In 1925 the Santa Fe, which was forced to operate two diners on certain heavily patronized trains, was seeking ways to cut costs and improve food service.[96] The road purchased a special two-car unit of which one was a full diner and its mate was a combination club car and open-section sleeper. The cars ran as a pair, with the club end of the first car always coupled to the diner end of the second car. The head end of the diner was blind and was attached to the baggage car. Because no one needed to pass beyond this point, there was no side passage, and the kitchen was made full-width. The space gained was used to expand the dining-room seating from the usual 30 or 36 to 42. When the tables were filled, passengers could wait comfortably in the adjacent club lounge rather than stand in the vestibule or aisle, as was necessary with an ordinary diner. The club car was outfitted with a soda fountain which sold cigars, candy, and magazines in addition to ice cream and soft drinks. Service in the diner itself was speeded by an additional employee, who checked the trays against orders and operated a cash register. The scheme worked so well that the road bought a second pair of the cars from Pullman in 1927.

It may have been the Santa Fe's successful experiment that led to the development of twin diners. About thirty-five of these units were built between 1937 and 1953. Some were articulated, but most were fairly conventional cars operated as a married pair. The twin units were large-capacity food-service vehicles, with one car serving as the dining room and the second as a

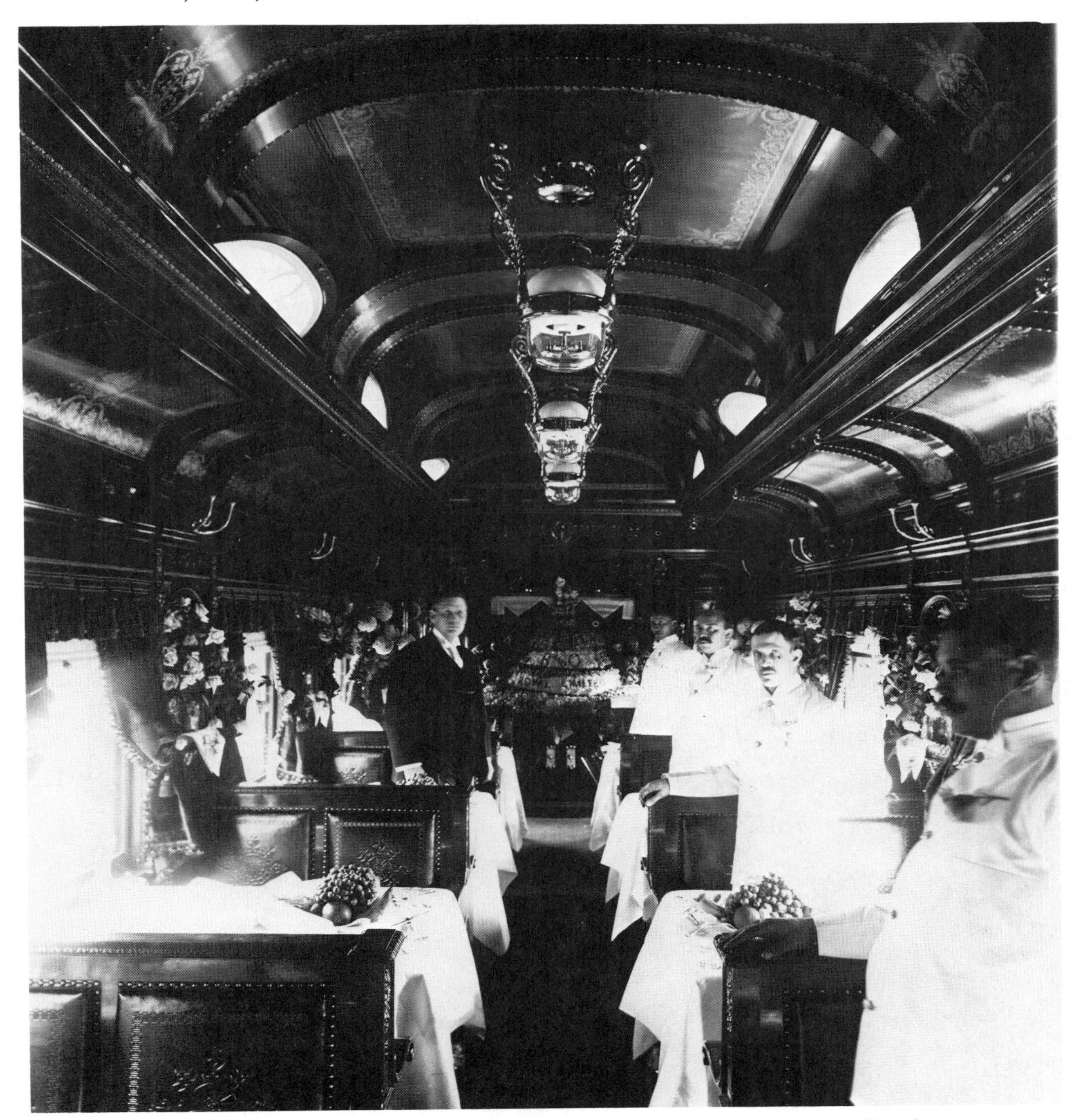

Figure 4.52 *A photograph of the B & O's Royal Limited dining car, presumably taken in 1898 when a new set of cars was placed in service.* (*Smithsonian Neg. 48382*)

Figure 4.53 *The wooden diner Number 452 was outshopped by Pullman in 1909 for the C & O.* (*Pullman Neg. 11559*)

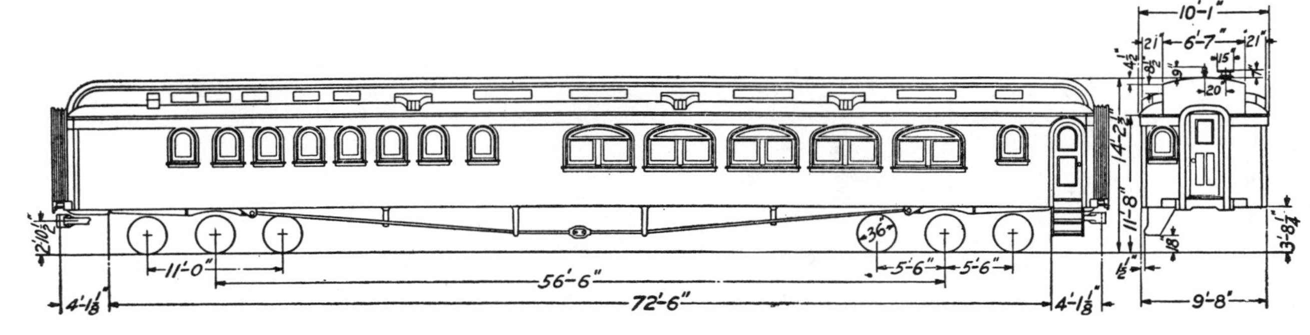

Figure 4.54 A diagram drawing for one of the diners used on the New York Central's train, the Lake Shore Limited, in 1910.

Figure 4.55 Santa Fe's Pullman-built diner of 1914 was equipped with indirect lighting. A battery of fans was required for comfort in the high temperatures of the Southwestern route. (Pullman Neg. 17610)

Figure 4.56 Passengers aboard an arched-roof steel diner on the Southern Pacific around 1925. (Southern Pacific)

kitchen and the crew's quarters. The Union Pacific had been satisfied with the articulated cars on its motor trains and had decided to use this plan of carriage for some full-sized passenger cars. The result was the twin-unit diners built for use on the Challenger trains by Pullman in 1937.[97] Each 140-foot pair had eight-wheel trucks at either end and a single six-wheel truck at the center connection of the two bodies. Aluminum construction held the total weight to just under 104 tons. The diner unit had a pantry at the center end toward the kitchen, while the remainder of the car was devoted to seating and tables for 68 passengers. In the second car a large, fully equipped kitchen occupied more than half the floor space. The rear portion was divided into four rooms for the twenty-man crew. Most of these bedrooms had two three-tier bunk beds. The public was delighted with the twin-unit giants, but the crews showed little enthusiasm. The long walk from the kitchen to the end tables created difficulties and extra work. Apparently the railroad management was not enthusiastic either, for it ordered no more twin diners until 1953. These new cars were nonarticulated, indicating that part of the design had not proved satisfactory.

In 1940 the Pennsylvania transformed eighteen heavyweight diners into nine streamlined, twin-unit ones for service on the Trail Blazer, the Jeffersonian, and the Congressional.[98] The cars were not articulated; four pairs of roller-bearing six-wheel trucks carried the 176-ton couples. The dining-room halves of these units were identical and seated 68 passengers each, but the kitchen cars varied. Those intended for New York–Washington runs had lunch counters, while those meant for overnight service to the West had dormitories built into the same space. Following World War II, the Pennsylvania decided to expand its twin-diner fleet. Seventeen double units were produced in 1948 and 1949 by the Altoona shops, American Car and Foundry, and Budd.[99] They cost $335,000 a pair and measured 165 feet in length. They could run through to Chicago on the most heavily patronized trains without restocking food or supplies. They provided storage for 2,016 pieces of linen and 2,721 other items.

If twin units were good, triple units would be better, or so reasoned the Southern Pacific in 1939–1940 when it acquired two sets of three-car diners for its popular San Francisco–Los Angeles train, the Daylight. Formerly it had operated separate coffee shop and dining cars, each with its own kitchen. A substantial economy could be realized by placing a single kitchen between the two food-service vehicles. To ensure maximum efficiency, a full-sized mock-up of the 201-foot-long triplex was constructed at the Sacramento shops.[100] The fourteen-man crew was invited aboard to collaborate with the design staff, and the interior working space was shifted around until an optimum layout was achieved. More cars of this general design were purchased in 1941 and 1949 for other Southern Pacific trains (Figures 4.61 and 4.62). In the later sets, the kitchen dormitory

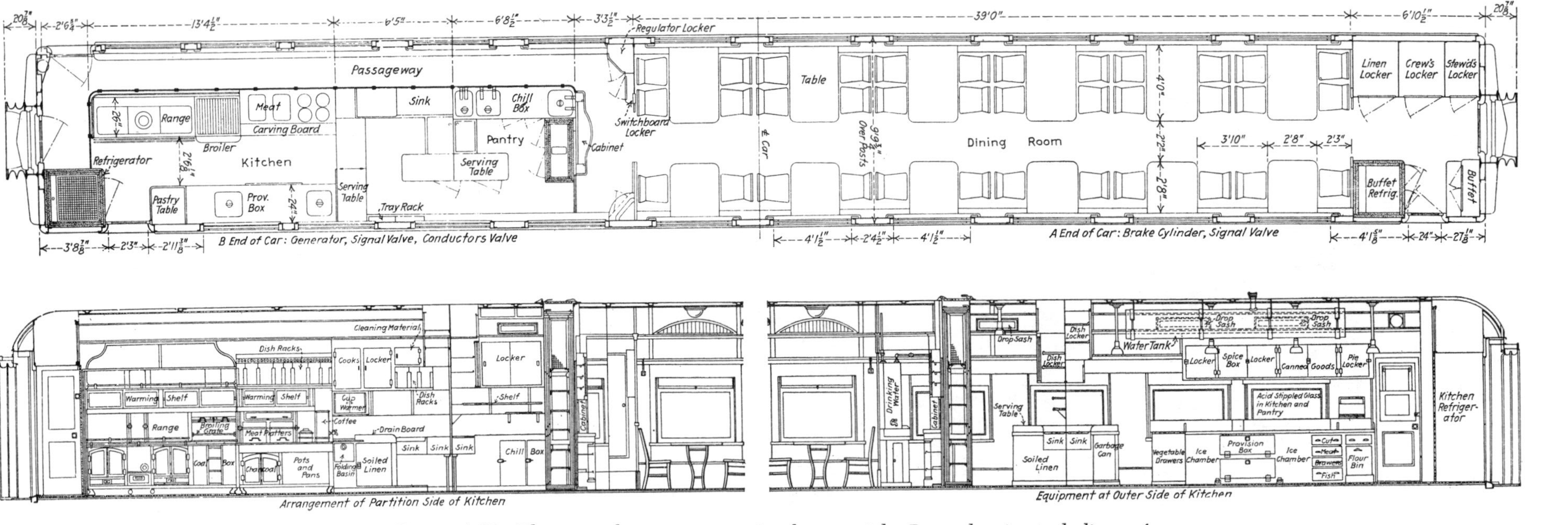

Figure 4.57 The general arrangement of a heavyweight Pennsylvania steel diner of 1922. (Railway Mechanical Engineer, *December 1922*)

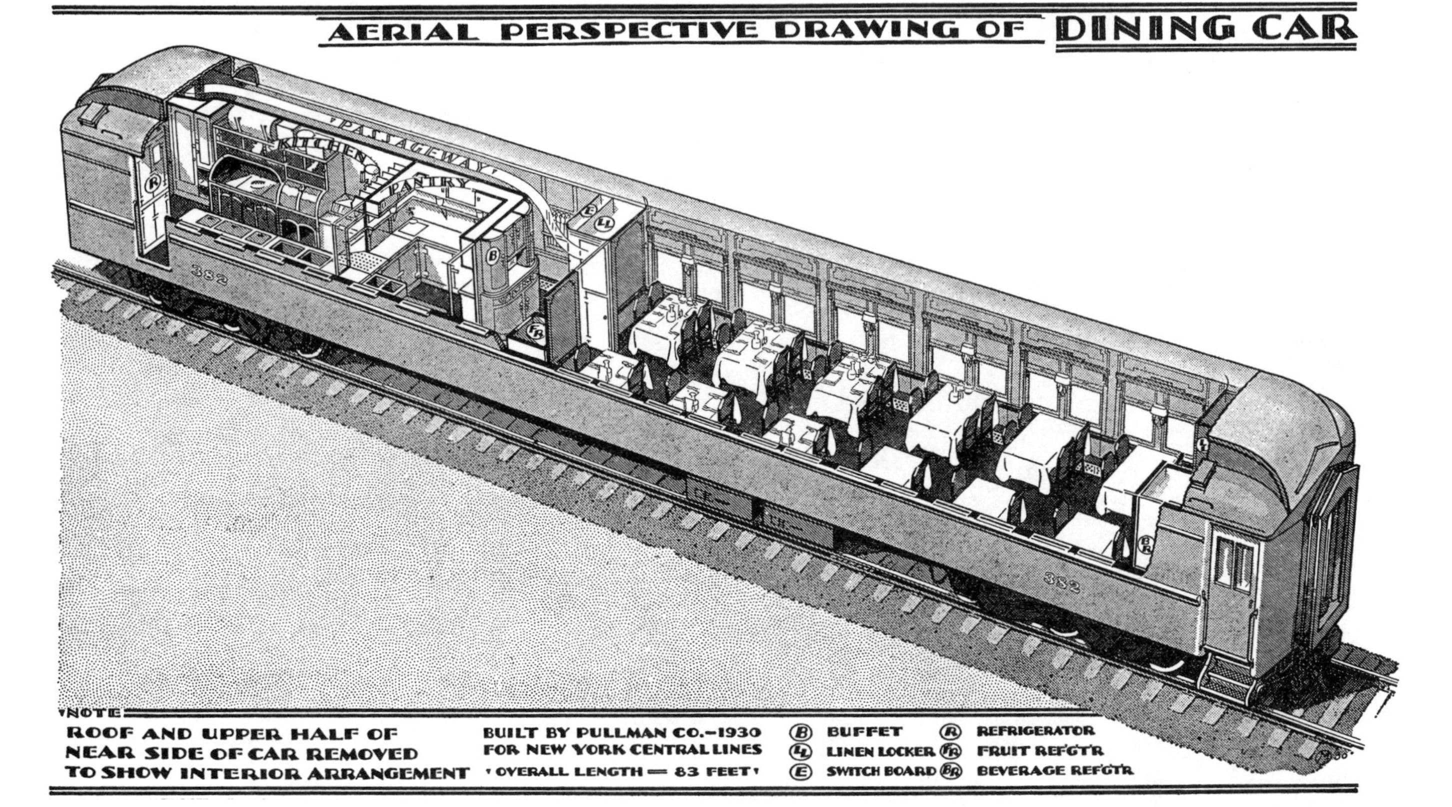

Figure 4.58 A perspective drawing of a heavyweight steel diner used on the New York Central's 20th Century Limited. (Edward Hungerford, Run of the 20th Century, *1930)*

Figure 4.59 The Virginia Dare, *a standard diner delivered in 1927 by Pullman. It was remodeled in 1938 as a streamliner.*

Figure 4.60 Union Pacific dining car Number 383 was delivered by Pullman in November 1923. (Pullman Neg. 27544)

was put at one end. The diner and tavern-lounge sections were coupled at the point of articulation by a "hidden" joint, so that the two units had the appearance of one long room.

More exotic than multiple-unit diners were the dome and hi-level dining cars that appeared in the 1950s. The Santa Fe bought six hi-level diners from Budd in 1956 for its plush El Capitan.[101] The upper-level dining room seated 80; the kitchen was below. These were the largest single-unit diners ever built: 85 feet long and 15 feet, 6 inches high, with a weight of 96.7 tons. The food was sent up to the dining room on a dumbwaiter. The system and general arrangement were so new that service was slow and orders were often confused at first, but the crew soon caught on.

A year before the hi-levels entered service, the Union Pacific offered its passengers the luxury of dining in a dome car. Competition for the 18 dome seats was keen. Another 28 passengers could be served in the lower dining room. Food orders were telephoned to the kitchen and the meals were sent up by dumbwaiter. Ten aluminum cars of this type were built for the U P at A.C.F.'s St. Charles plant in 1955.[102]

The dome cars on the Union Pacific represent the final development of the full diner. Few dining cars were built after that

Figure 4.61 *A Southern Pacific diner publicity photograph made around 1955. (Smithsonian Neg. 73-739)*

Figure 4.62 *A triple-unit dining car built by Pullman in 1949. (Pullman Neg. 60431)*

Figure 4.63 On shorter runs where full meals were not required, trains had café cars, such as this 1905 Pullman product. (Pullman Neg. 7881)

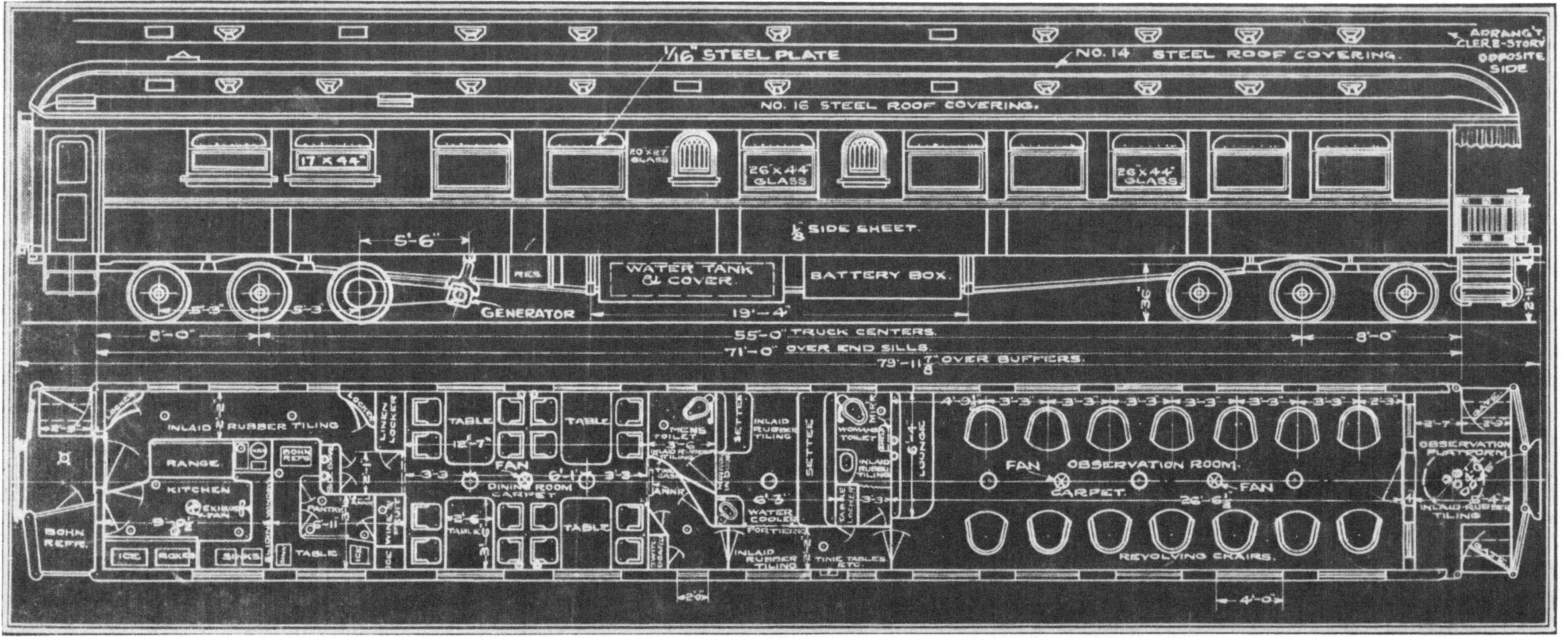

Figure 4.64 The Buffalo, Rochester, and Pittsburgh's café-observation cars Number 206 and 207 offered light food service for first-class passengers. Pullman completed the steel pair in 1913.

date, although a large number of half diners, variously called buffet, snack, grill, club, and café cars, performed an important on-train food service (see Figures 4.63 to 4.65). These cars were in many ways more significant, if less glamorous, than the full diners, since they probably fed more travelers.

The buffet car served drinks and snacks and might consist entirely of a closet-like kitchen and a serving bar. The remainder of the car was devoted to parlor or sleeping space. On the other hand, the café car might be a true half diner—that is, a small kitchen and dining room serving cooked meals, but without the variety or grandeur of a full restaurant car. Cafés were useful both on express trains whose runs were too short to serve many

Figure 4.65 Buffet cars were often combined with baggage cars. The Number 1336 was a Pullman product of 1914. (Pullman Neg. 17713)

***Figure 4.66** The diminutive kitchen of this B & O buffet car, which was produced in 1929, is visible through the doorway.*

Figure 4.67 In many modern buffet cars, the kitchen became more of an open serving bar. The example shown here dates from 1935.

full-course dinners and on less well patronized first-class trains that had to offer proper if somewhat limited food service. And then came a mongrel group of second-class diners that were difficult to catalog. In some the entire car was devoted to food service, but in others part of the floor space was given to alternate uses. Many were rebuilt from old cars that were not necessarily diners. In design and finish they ranged from the professional to the amateur, and they went by many names befitting their uncertain origin and station: grill car, snack car, sandwich car, and lunch-counter car. Still another class of diner that developed in the years before Amtrak took over passenger train operations was the Automat car. Whether it can be accurately described as a third- or fourth-class accommodation is a matter of opinion.

The buffet car has been credited to Pullman; it is said that the idea came to him during a trip through Italy, where the need for a small snack such as a cup of coffee or tea, a sandwich, or some bouillon might "relieve faintness" between meals.[103] On a train a full diner's menu was too heavy for such needs, and the car would probably be closed during off-hours. How nice it would be if a pleasant sideboard were available at any time. In 1883 Pullman ordered some cars equipped in this manner for the West Shore Railway. A buffet measuring 3.5 by 8 feet was placed between the smoking and drawing rooms of a new car. This innovation seems to be more a natural outgrowth of the hotel car, which was then passing out of favor, than a new idea. There was also a precedent in the buffet aboard the *Victory* back in 1835.

Whatever its historical forerunners, the significance of the buffet car is that it permanently reestablished Pullman in the food trade. From that time forward, light food was sold on many Pullman-operated cars. The commissaries, stores, and personnel necessary to offer this service seemed to lead Pullman ever farther into the dining-car trade. Some roads, like the Milwaukee, bought diners from rival manufacturers but asked Pullman to operate them.[104] Full diners were rare in the Pullman fleet; there were usually no more than twenty, and these were in pool service for lease to heavy seasonal or special convention trains. Most Pullman food service cars were of the buffet or café class. In 1933, for example, Pullman's commissary department was operating 64 restaurant cars on nineteen different railroads.[105] It also had 171 buffet cars and 137 other composite sleeper, parlor, or lounge cars that were combined with a soda fountain, a sandwich bar, or a broiler buffet capable of serving simple grills. As the operation of independent food service grew more expensive, more railroads began to contract with Pullman's commissary department. In 1950 Pullman was catering aboard 400 cars.[106] The service ranged from full-course meals to beverages only.

Figure 4.68 The oaken interior of this Burlington café car was paneled with antique timbers imported from England. (*Pullman Neg.* 8255)

Twenty-six commissaries and 1,200 employees were required to run this part of the business.

Buffets were almost always first-class in their appointments (Figures 4.66 and 4.67). They were often combination cars, with a baggage compartment ahead and a parlor or sleeping section behind. Some buffets were little more than a counter with a gas-fired hot plate. Others, like those Pullman produced for the Burlington in 1906, had a kitchen, two small dining compartments for private parties, a public dining room seating 18, and a café smoking section seating 12.[107] The interior of these cars was decorated in high style, like a fine German restaurant, somber and rich in dark, old wood (Figure 4.68). The paneling was in fact sawed from some ancient English oaks that Pullman managed to buy from the Rockingham Park estate. When the planks were cut, a bullet dating back to Cromwell's time was discovered embedded in one of them.

During the same year that the Burlington's elegantly cabineted buffets were put in service, the American Palace Car Company was trying to revive the hotel car in the East. For some years the company had been operating a combination sleeper-parlor-diner built by Wason in 1890 on the patents of L. J. Harris.[108] In 1907 it sold four cars of the same general design to the Canadian Northern. Barney and Smith produced these 20-seat passenger cars. Meals were cooked in a buffet-sized kitchen at one end, and folding tables were set up for the meal. It was ingeniously worked out, but since the plan could not overcome the inherent defects of the old hotel cars, little more was heard from the venture.

Simpler forms of dining cars rather than more complicated ones were now receiving attention. The common man's eating car came into being in the early years of the new century. Fast, cheap food service was the purpose of the lunch-counter car (Figure 4.69). These utilitarian vehicles were often remodeled coaches. Out came the seats; in went a long stand-up counter, a grill, some shelves, and if necessary, a row of stools. The Pere Marquette put such a car in service in 1904 to accommodate the weekend crowds traveling from Chicago to the Michigan resorts.[109] It was designed for 60 people, but 150 to 200 at a time

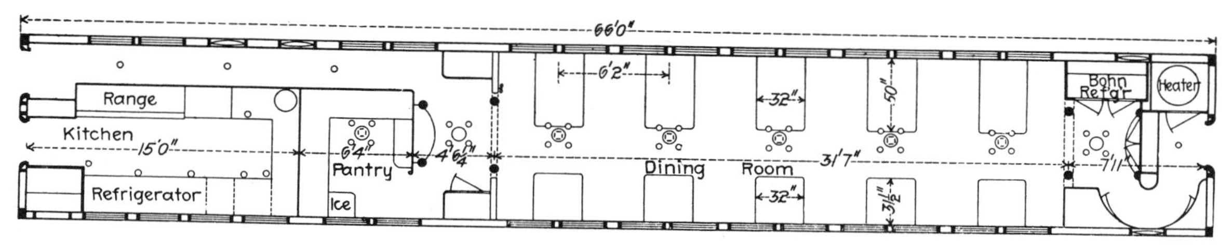

Dining Car, C., B. & Q.

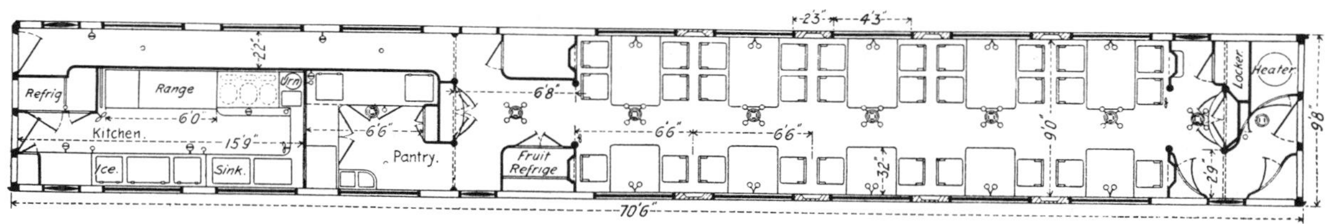

Dining Car. Pullman Co., Builders.

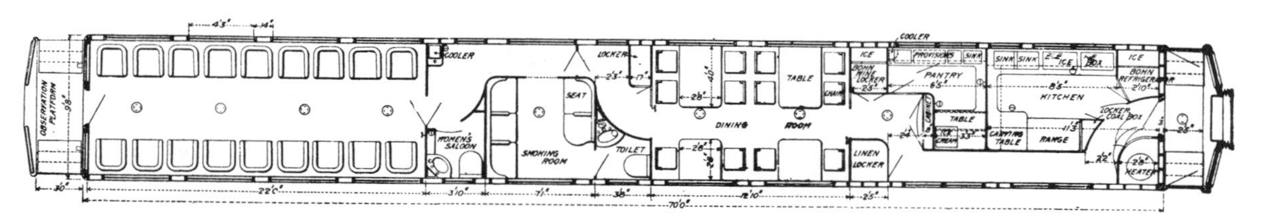

Café Parlor Car, El Paso & Southwestern. Pullman Co., Builders.

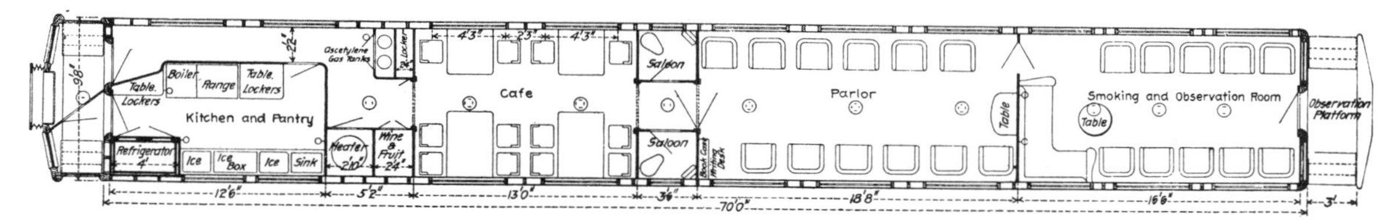

Observation Café Car, Chicago Great Western. Pullman Co., Builders.

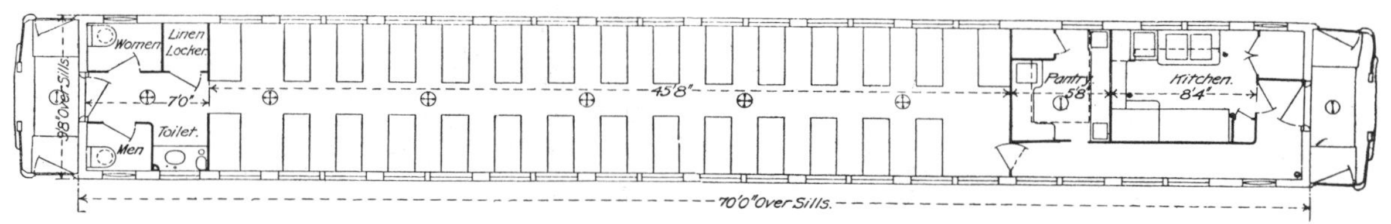

Café Coach. New York Central. Barney & Smith Car Co., Builders.

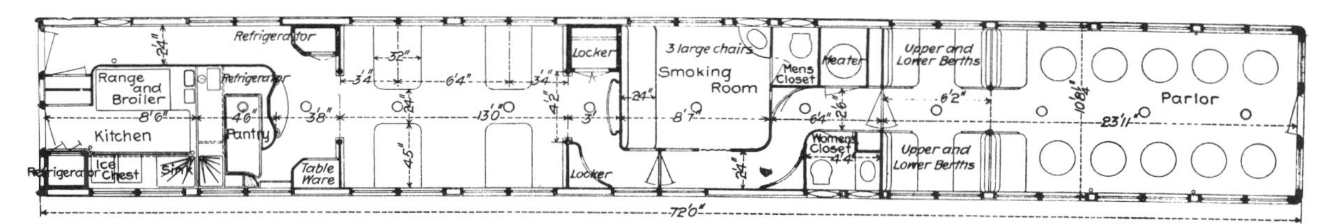

Café Parlor Car, C., B. & Q. Pullman Co., Builders.

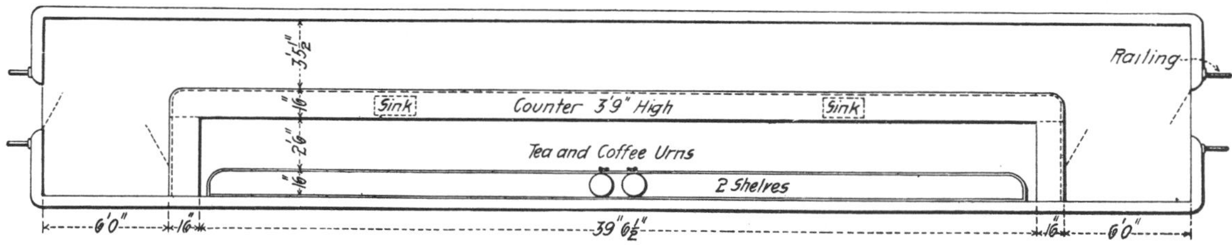

Lunch Counter Car, Père Marquette.

Figure 4.69 The variety of food service cars is illustrated by this series of floor plans published in 1906. (Car Builders' Dictionary, *1906*)

Figure 4.70 Lunch-counter cars offered fast and economical service to passengers who did not want the complete meals offered in the regular diner.

would crowd in on a busy day. The Illinois Central operated a similar car in about the same period; the Pennsylvania produced an updated version in 1914. Gradually lunch-counter cars spread. On the Union Pacific they were called sandwich cars. More elegant versions, known as coffee shop cars, were built new for the Baltimore and Ohio, the Pennsylvania, and the Southern Pacific in the 1930s and 1940s (Figure 4.70).

A direct descendant of the lunch-counter diner was the cafeteria car, which served more elaborate hot meals. After passing the steam table, the passenger would carry his tray to a table. In 1945 Pullman offered to produce cars on this plan, which could serve two or three times as many people as a conventional diner.[110] Self-service would greatly reduce the labor overhead, and because only a small staff was necessary, it would be economical to keep the car open in off-hours. It seemed an attractive plan, but few railroads were much impressed by it. Perhaps passengers, inexperienced in carrying food trays on a moving train, too often found the task beyond their skill. The Union Pacific did, however, remodel four of its lunch-counter cars for cafeteria service in 1960.

The economics of the dining-car business had a direct effect on the design of the rolling stock and the general style of service eral ways: they offered more attractive service to increase vol- offered. Managers tried to reduce their continuing losses in sev- ume and revenue, they installed more efficient equipment to increase productivity, and they experimented with streamlined operating practices to lower out-of-pocket costs. Second-class diners, like the lunch-counter and cafeteria cars, answered the needs of the coach traveler, but more effort was expended on the luxury traveler. Sumptuous decor, food, and service continued to be pushed through the 1950s and is still offered on a few Far Western trains (Figure 4.71). But nothing remains today that is comparable to the New Haven's Yankee Clipper deluxe diner of 1930.[111] The waiters were in smart white uniforms and shoes; the steward wore formal attire (Figure 4.72). Only the choicest cuts were served, and all vegetables and fruits were fresh—no canned foods were used. Patrons might even choose between eggs with white or brown shells. When the C & O's George Washington was inaugurated, a colonial dining car was part of the consist (Figure 4.73). It featured colonial decor as well as

Figure 4.71 *Western lines continued to purchase first class dining cars for the affluent long-distance traveler. This example was produced by A.C.F. in 1949. (American Car and Foundry Company, Neg. 61236-B)*

Figure 4.72 *The fact that dining cars were labor-intensive is amply documented by this B & O photograph of the 1920s.*

Figure 4.73 Beginning in the mid-1920s, both the B & O and the C & O railroads operated cars with colonial interiors on their better trains.

colonial dishes. Fresh-cut flowers were common on all the best diners (Figure 4.74). The railroads' concern for their first-class diner is evident in the fact that the first passenger cars to be air-conditioned were always the restaurant cars.

These efforts to appeal to the luxury traveler undoubtedly attracted more business, but the increased revenues rarely covered costs. Cutting costs seemed to offer a more certain solution to the dining-car dilemma; how to minimize losses became the chief concern of the dining-car department. One of the most successful methods was through wholesale buying. Large railroads required prodigious quantities of food, and they could turn this to the same advantage enjoyed by a grocery store chain. The Southern Pacific, for example, consolidated its dining-car, hotel restaurant, and maintenance-of-way kitchen car purchasing.[112] Every effort was made to buy a one-year supply of basic commodities when they were in season, thus securing the cheapest price and best quality. Since the railroad had its own refrigerator cars and warehouses, transport and storage presented no problem. Vast stores were kept in reserve, since the road's food consumption in a single day included 1.5 tons of potatoes, 8,400 eggs, 500 pounds of butter, 300 to 400 pounds of coffee, and 350 pounds of chicken.

Food preparation was another area in which costs could be cut. Most large railroads and the Pullman Company maintained commissaries at major terminal points—kitchens that made pies, ice cream, puddings, and soups. There potatoes were machine-peeled on a production basis, and meats were cut, trimmed, and wrapped. The commissaries saved time and work for dining-car cooks, as well as ensuring uniform portions and quality. When frozen food became a commercial reality after World War II, railroads quickly adopted it. The Rock Island, in fact, began to serve frozen precooked dinners in the early 1950s.[113]

Idle equipment makes no money, and deadheading is equally profitless. Careful scheduling achieved maximum utilization of each dining car. A diner sent through from New York to Chicago

Figure 4.74 Name-train diners, like first-class restaurants, featured such luxurious accessories as linen tablecloths and cut flowers. (*General Motors*)

would spend long, empty hours rattling through the night. Hence it was common to cut out the diner along the line after the last evening meal was served. The car would be cleaned and restocked so that it was ready to serve breakfast on a train that picked it up the next morning. Cars were worked back and forth over a large system, serving meals on several trains before returning to their home terminal. Like the cars, crews would be picked up and dropped off en route. This method of operation was employed from the earliest days of dining-car service. The New Haven's first diner introduced in 1884, for example, ran on the noon train between Worcester and New Haven. It was then deadheaded to Boston to provide dinner on the evening express to New York.

Technical improvements in the dining-car kitchens themselves could also effect savings (see Figures 4.75 to 4.77). Simple changes like adding a side-door entrance to the kitchen from the outside speeded the loading of supplies. It was also an efficient way of ventilating. When business was slow, the side door provided a breezy view for a lounging cook. Improving the stove was a more complex process. Coal-burning iron ranges were traditional. It was also a tradition that the junior cook had to come aboard hours before the scheduled departure to light the giant black monster. Charcoal was sometimes used, but coal was the standard fuel and remained so until very recent years, when the older cars went into retirement. More convenient fuels were tried out as they were developed, but diners were relatively slow to put them into general use. Steam was widely available after 1900, because vapor heating became common. It was not considered useful for cooking, but it did serve to keep dishes hot. Gas appears to have been the first practical alternative to coal. Nearly all first-class cars had gas lighting by the 1890s, and the large reservoir of this combustible fuel under each car naturally suggested its possibilities as a cooking fuel. Pintsch gas was used aboard buffet cars during the period. To a limited extent, bottled gas was employed to fire full-sized ranges in the 1920s. By the 1940s propane was installed on some of the better new diners,

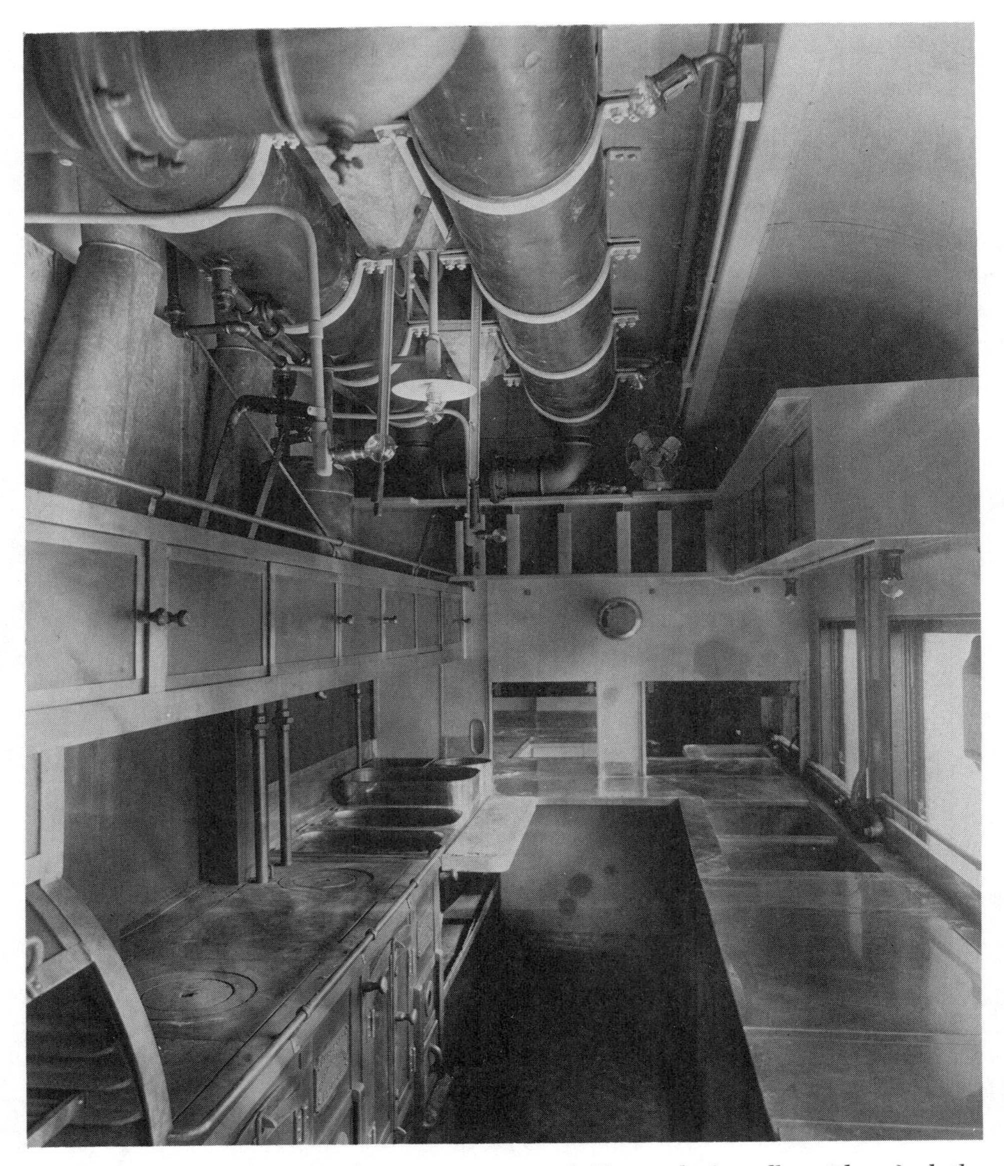

Figure 4.75 Dining-car kitchens were arranged like a ship's galley. They had the utilitarian look of a miniature food-processing factory. (Pullman Neg. 19955)

such as those assigned to the Daylight. Before that, the Southern Pacific had been using Presto-logs.[114]

Electric ranges held the greatest promise for dining-car service. This type of stove was installed in a few diner kitchens in the twenties and thirties, but the limited power supply on board train and the old standard of low-voltage direct current made it impractical. Abundant high-voltage power became common with the advent of air conditioning, thus opening a way for the electric range. The Pennsylvania's twin diners of 1940 had combination cooking: electric for regular dishes and a charcoal grill for broiling.[115] An electric coffee urn was a marvel of efficiency, producing several gallons of fresh brew in just two minutes. In 1946 the Illinois Central began work on an all-electric kitchen at its Burnside shops.[116] Basically the plan was an adaptation of a design perfected for submarine galleys. Postwar shortages prevented completion of the first car (Number 3987) until early in 1949. Besides the range and refrigerators, the car had a 3-gallon coffee maker and an electric deep-fry kettle capable of producing 100 pounds of fried potatoes, chicken, or fish an hour. Mounted on the car's underside were two 50-kilowatt diesel generators that supplied the power. Not many other all-electric kitchens were installed, but electric ranges became relatively common after 1950. During this period experiments with super-fast infrared and microwave ovens were also underway.[117]

Other gadgetry entered the twentieth-century dining-car kitchen. In 1940 came dishwashers, garbage disposals, and automatic door openers that eased the way of tray-laden waiters into and out of the pantry.

The electric refrigerator was adopted before these niceties and proved to be a more basic improvement. A well-insulated icebox could keep food for as long as three days without servicing, but not all boxes were this well constructed, and losses from spoilage grew less tolerable as food prices continued their historical rise. Another disadvantage of the icebox system was the space taken by the great blocks of ice. There seemed little enough room to stow the 1,200 to 1,500 pounds of food needed for a fully stocked car without giving up space to this fast-melting refrigerant. Around 1925 the Milwaukee Road began to experiment with electromechanical refrigerators.[118] The technology was already well advanced for home and commercial use, but it had yet to find acceptance on the dining car. The Milwaukee's installation was a success, and within a few years seven cars had mechanical refrigerated main meat lockers, although their other units remained iceboxes. In September 1927, however, the road con-

Figure 4.76 A chef at work in a B & O dining car in February 1936. Coal-fired stoves and ice-box refrigerators remained standard equipment.

verted all the boxes on diner Number 5138 to mechanical refrigeration. In 1930 the New Haven followed the Milwaukee's lead.[119] Other roads gradually joined them, although the old-fashioned icebox remained a part of the dining-car picture for many years. A few roads, such as the Union Pacific, regarded dry ice as the preferable cooling medium.

Labor was the largest single operating cost, and historically it rose faster than any other. In the twelve years between 1937 and 1949 the wages for cooks doubled, while those for waiters tripled.[120] So far as the full dining car was concerned, wages were the most difficult item to keep under control. The diner was labor-intensive by its very nature. The only way to reduce the work force was to change over to lunch-counter cars or self-service grills, and this was not an acceptable alternative on first-class trains. As trade unions became stronger during the twentieth century, it became more difficult to close out jobs. Thus a dramatic drop in patronage was not necessarily followed by a comparable decline in the number of chefs, stewards, and waiters. In 1925 when railway passenger travel was in full flower, some 10,000 people worked in dining-car service. Yet in 1961 after passenger traffic had dropped by more than 75 percent, 9,600 people were still employed.[121] Some managers hoped that waitresses would offer a way out of the rising labor cost spiral. Women hostesses were hired by the New Haven and the C & O, but most roads stayed with all-male crews. The only effective way found to control labor costs was to eliminate the full diner.

When the railroads were hauling the vast majority of intercity travelers before 1950, they were content to accept dining-car deficits as a necessary business expense. The 8- to 10-million-dollar annual loss was charged off to advertising and good will.[122] For every dollar of revenue in 1950, it was necessary to spend another 40 cents to cover dining-car expenses.[123] The C & O, which had a rather small passenger train operation, reported a $700,000 dining-car loss during the same year. A few years later the industry as a whole claimed a loss of 29 million on diner operations.[124] Individual large systems, like the Pennsylvania, were losing as much as 12 million dollars a year; cutbacks were inevitable. Some lines, like the Cotton Belt and the Long Island, simply abandoned diners altogether. Others, like the Lehigh Valley, cut back on personnel and menus. Cooked-to-order breakfasts gave way to a Continental morning menu of sweet rolls, dry cereal, and coffee. One man did the job formerly requiring three. In 1961 the Pennsylvania pulled off all full diners running between New York and Philadelphia. In their place mobile food carts were pushed through the train by a single vendor, who offered a sad variety of packaged sandwiches, cakes, and soft drinks.

Many of the genteel refinements of the dining car were already gone by the time the food carts took over. The New York Central once spent $2,000 a month for fresh-cut flowers. By the early fifties only passengers on the Twentieth Century Limited could expect to find flowers on the table. On other lines, linen gave way to paper napkins and paper tablecloths. Ever more rigorous economies soon sliced away the last vestiges of the elegance once common to all American diners. Only a few Western lines attempted to maintain the old standards. These lines did not include the Southern Pacific, which set out to cut its losses when dining-car patronage took a nose dive in the late fifties. The first step was not drastic; the triple-unit diners were replaced with single-unit cars.[125] But trade had fallen off too much to support a conventional car, and in 1961 the road decided to further downgrade the food service by substituting Automat cars. Eleven old sleeping cars were remodeled to dispense soft drinks, candy, and hot dishes through coin-operated food machines. Tables and chairs were available at one end. An attendant made change and tidied up. This scheme became the subject of so much criticism that the cars were withdrawn from service after Amtrak assumed passenger operations on the Southern Pacific.

The lesson of the Automat was not lost on passenger train designers. Yet patrons expecting a return to the palmy days of the traditional dining car were disappointed by what the Metroliner and Turbo Train have to offer. On Metroclub (parlor) cars and the Turbo Train, airplane-style meals are served at the seats. The food is prepared in advance and kept warm in a heated tray box at one end of the car. The meals are acceptable and not overly expensive, but the special ambiance that the experienced traveler had come to associate with dining on the rails is not present. Now the passenger must crouch over a mechanized meal which is perched on a low, folding table in front of the seat. The Metroliner coach riders feed in even less comfort at snack bars located in the center of every other car, where pastry, sandwiches, and coffee are served from a stand-up counter. A single attendant dispenses these modest items, which passengers may consume at their seats or at a counter opposite the serving bar.

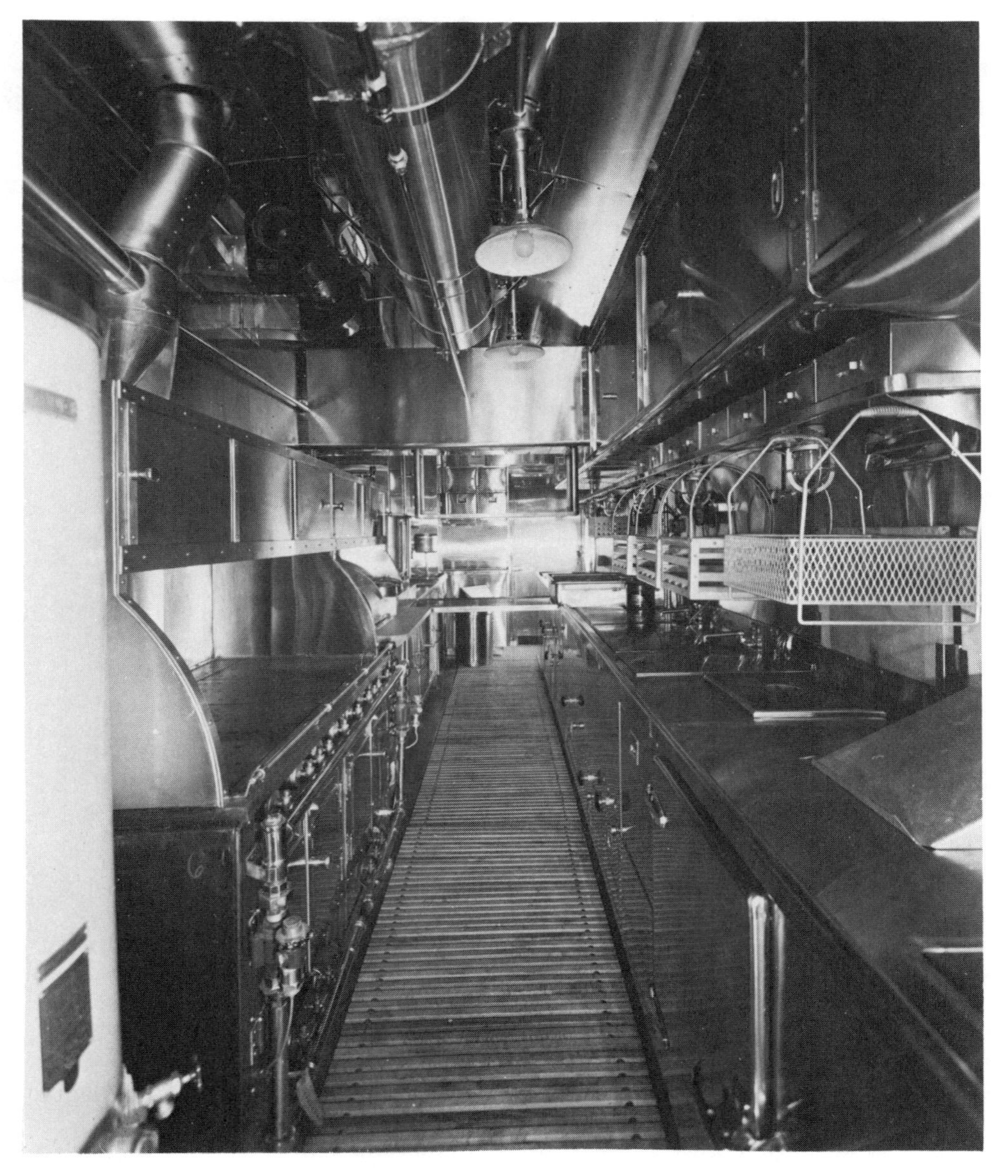

Figure 4.77 American Car and Foundry equipped this Kansas City Southern dining car kitchen (1948) with a gas stove and electric refrigerator. (American Car and Foundry Company, Neg. 61229-R)

Such a dining area is objectionable on at least three counts: standing to eat is essentially a fueling operation, trainmen and passengers block the aisles at station stops, and the absence of windows in this section of the car is oppressive. The passenger who chooses to take the food to his seat must balance it on his lap and ignore the looks of his neighbors as he crumbles and spills his way through the meal. Ironically, the very line on which commercial dining cars first appeared has come full circle back to the stand-up lunch counter of the 1860s.

Private Cars

The private car was the least representative of the general passenger car fleet; the average traveler rarely saw one, much less rode in it. Their numbers were small even when business cars were included in the total. In 1930, which appears to be their peak year, only about 900 were in operation. Wayner lists approximately 750 in his *Car Names and Consists*. In the whole history of American railroads it is doubtful if more than 2,000 private cars were built for North American service, and again this figure is considerably inflated by the inclusion of business cars.

Yet no other form of railroad conveyance has so completely captured public attention. A treatise on the all-important day coach has little popular appeal; the story of the sleeping car interests a small group; but the private car appears to stir enthusiasm among all sorts of readers. Fascination and curiosity about the very rich, their lives and possessions, seem indigenous to all societies and eras. Romance and glamour have developed around the excesses of kings and princes, and in America we look to our merchant princes to do their duty and spend royally on such baubles as private railroad cars. Lucius Beebe insisted that the private car added another dimension to an age of wealth, well-being, and conspicuous consumption (Figure 4.78).

Some private car owners, like A. A. McLoed, went to considerable lengths to obtain the showiest car possible. The more extravagant and costly, the greater the publicity. Like Jim Brady's diamond ring, a flashy railroad car immediately established credit and social position.

Car manufacturers were equally interested in the private car as an image builder. An order for a thousand boxcars or even a

Figure 4.78 The private car was a symbol of luxury and prestige in the nineteenth century. This unidentified group posed with obvious pride beside their private car sometime in the 1880s. (*Bar Harbor Historical Society*)

dozen coaches rarely gained much newspaper space, but a new rail palace for Senator William Sharon was news. The purchaser was a big name; hence any major acquisition was news in itself. The price of $50,000 was news, and the splendors of its interior furnishings made exciting copy for journalists. The handsome paneling, rich hangings, and overstuffed furnishings were standard, but what about the dozen master wood-carvers who worked on it, or the Italian artist imported expressly to do the ceilings and the cluster of golden angels ringing the transom frieze? A dedicated reporter could even find magic in the kitchen: "The fire glowed in the range and shone on polished pots, pans and porcelain. . . . Pudding and jelly molds, skewers, steamers and saucepans as bright as silver hang on the hooks of the dresser. Stores of delicate china are nestled in the snug closets, crystal and silverware crown the oaken buffet."[126] The journalists were encouraged with luncheons in their honor and special predelivery trips in the car. Their descriptions were inspired as much by free champagne as by the car itself, according to *Railway Age*.[127] In an effort to gain free advertising some builders, like Ohio Falls, were claiming prices as great as $30,000 when the car actually sold for $9,000. They would also attempt to spread gaudy rumors —say, that the car would be used by President Garfield.

Figure 4.79 A special eight-wheel car was furnished the Duke of Wellington and his party for the opening of the Liverpool and Manchester Railway in 1830. (Detail of Smithsonian Neg. 8953)

In the strictest definition, a private car was the personal property of an individual owner. Such cars did exist, but the greater number of cars commonly referred to as private were actually railroad-owned. Technically they were the railroad's office or business cars that were intended to transport its officials over the line in the performance of their duties. The Association of American Railroads came to classify both private and business cars by the letters PV. It is often difficult to differentiate between private and official ownership, since many private car owners were directors or major shareholders in railroad corporations. In some cases a railroad-owned car was reserved for their exclusive use, yet it might be carried on the company's books as an official car. Who inside the ranks would argue the point with a Vanderbilt or a Gould? In other instances the car might be purchased by an individual, but as a director of a railroad he could expect free storage, maintenance, and carriage. Interchange privileges with other lines were a welcome extra dividend. Again, who would challenge a financial tycoon to prove that his itinerary was official or personal? In 1930 884 PVs were reported in service. The Pullman Company estimated that less than 50 of these were privately owned, and even in that group many were undoubtedly somewhere in the shadowland between private and railroad ownership. The semantics of the subject are also significant. Railroads became very concerned about emphasizing the utilitarian nature of their PVs through the designation of "business" or "office" car. The intent was to dispel criticism of wanton extravagance and to demonstrate that the stockholders' investment was being used for the affairs of the railroad. Yet in most office cars there was little space devoted to desks, file cabinets, or typewriters.

The abuses associated with private cars came to silence the high-flown reports of their luxuriousness that were once considered such good publicity. To obscure the size of the private car fleet, some major roads began to buy cars under the name of a subsidiary. In 1895 the B & O charged a new PV to the account of the moribund Sandusky, Mansfield, and Newark. In this way it could be made to appear that the parent corporation had only six office cars, when in fact it might own twice that number. Shareholders' complaints about profligate or self-indulgent officers led to suppression of news about private cars. It is conceivable that the paucity of data available for the historian on the earliest PVs is due to this self-imposed censorship.

The proliferation of business cars had reached scandalous proportions by the early 1890s.[128] On the Union Pacific, for example, every employee above a flagman seemed to have his own car. Railroad officers would have better understood the problems of the average traveler if they had taken passage in coaches rather than in the splendid isolation of PVs. Money always seemed available for office cars; small roads, as well as larger ones on the verge of receivership, spent their funds on PVs. Junkets to fashionable watering places were commonplace. On a single day in February 1891, twenty-one private cars were counted on a siding in Jacksonville.

The exchange of privileges was another costly abuse indirectly borne by the shareholders and the public.[129] An official of the smallest line might travel free of charge over the entire national network. For example, during the 1930s the president of a minor Southern line rode a mere 167 miles on his own property and 61,000 miles on other systems. At one point so many private cars were awaiting free carriage east that the New York Central was obliged to run a special train. Eventually the I.C.C. felt obliged to limit these abuses. In 1928 it ended the free carriage of private cars except for on-line movements.

The ancestor of the private car was an elaborate vehicle for dignitaries who attended the opening of the Liverpool and Manchester Railway on September 15, 1830. The Duke of Wellington,

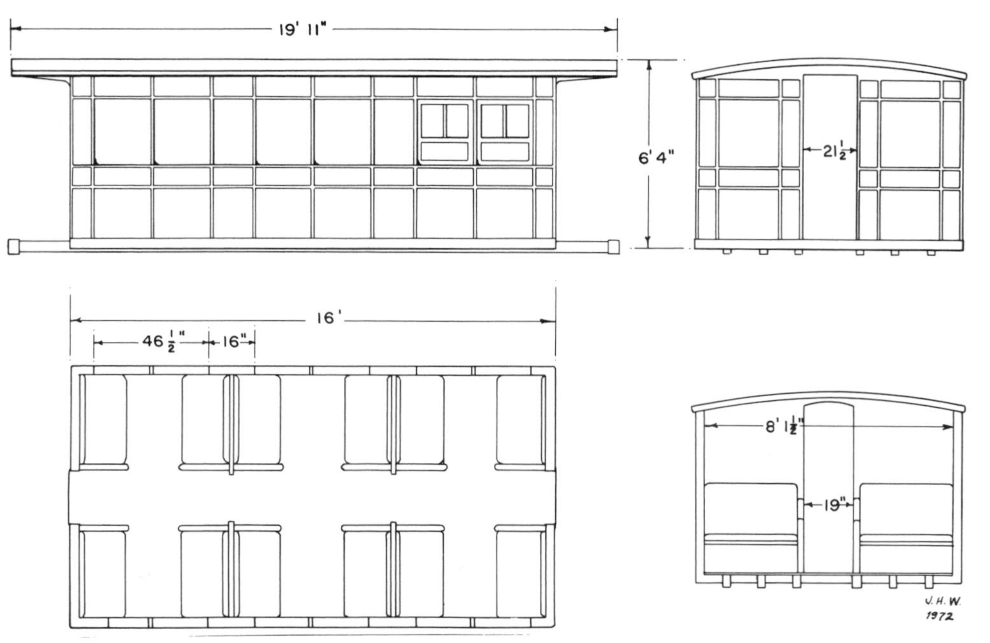

Figure 4.80 The Western Railroad (Massachusetts) produced a private car for its directors in 1839. The drawing is a reconstruction based on early photographs and verbal descriptions by employees of the Western Railroad. (Tracing by John H. White, Jr.)

hero of the Napoleonic wars and now Prime Minister, was the principal guest. He and his party rode in the imposing eight-wheel car shown in Figure 4.79. The following description is taken from a newspaper of that day:[130]

> Built by Messrs Edmundson's of Liverpool, the floor is 32 feet long by 8 wide, and is supported from 8 large iron wheels. The sides are beautifully ornamented, superb Grecian scrolls and balustrades, richly gilt, supporting a massy hand rail all round the carriage, along the whole centre of which an ottoman will be the seat for the company. A grand canopy 24 feet long is placed aloft upon gilded pillars, and is so contrived as to be lowered for passing through the tunnel. The drapery is of rich crimson cloth, and the whole is surmounted by the ducal coronet.

The ducal car was rarely used and appears to have excited no demand for more private cars in Great Britain. The Dowager Queen Adelaide acquired what is the next recorded instance of such a rail vehicle in 1842, and Victoria's first royal car appeared two years after that.

In theory America's First Citizen was expected to sit among the common folk, yet in 1841 when President Harrison was in transit between Baltimore and Washington, the superintendent asked if he should provide a "distinct car." Two years later President Tyler journeyed to Boston for the Bunker Hill Monument dedication. The Camden and Amboy felt obliged to offer a separate car, and since the road had no official car, it provided its coach Number 19 fresh from the shops after a complete rebuilding.[131] The coach had been fitted with six-wheel trucks, a variety of seats including several styles of reclining chairs, and a privy on one platform. It was regarded as very special and hence suitable for so honored a patron, and after the trip it was renamed for Tyler.

After leaving the Camden and Amboy, it is possible that the President continued his journey in an actual private car, for two New England lines did own them. In 1841 or 1842 the New York and Hartford completed a PV for its chief executive, Samuel R. Brooks.[132] And the Western Railroad had acquired a special car for its directors as early as 1839. This is the first American private car for which a record can be found, and because it figured in two Pullman sleeping-car patent suits, some details of its construction have been preserved.[133] It was a boxy, four-wheel car that hardly seems like a grand beginning for the private car in America, but it may have been luxurious in its day. The body plan has been reconstructed from dimensions supplied by old employees during the lawsuits (Figure 4.80). An engraving published by *Scribner's* magazine shows the body as it appeared in the 1880s. The car was divided into three compartments. It had face-to-face seats that could be converted into sofas by placing a plank between them and rearranging the cushions. The car was not a favorite with the directors, however; either it was too small or there was not enough call for it. In addition, the introduction of the double-truck car probably made it almost immediately obsolete. The car was reportedly operated as a regular coach between 1840 and 1845. Later it was used on mixed trains and provided the crew with a deluxe caboose. By the late 1850s it was set off the tracks near West Springfield as a watchman's shed, and a few years later a car inspector moved the body to his garden for a toolhouse. There it stood until rot and neglect obliterated the last vestige of America's first mansion on rails.

So long as railways remained short, there was no real need for private cars. A brief excursion could be tolerated in the meanest car. But as the system was extended, railroad officials, rich people, and prominent people who were obliged to travel yearned for a better class of car. During Jenny Lind's tour of North America in 1850–1852, for example, she rode thousands of miles. Added to the normal fatigue of travel was the adulation of her public, which assembled in oppressive numbers. The problem was magnified by her marriage in Boston toward the end of the tour. So that the honeymoon couple could travel to Pittsfield in peace, a coach was quickly remodeled for them.[134] The seats were pulled out and replaced with luxurious household furnishings. Some months earlier, Miss Lind was reported to have made a portion of her Southern tour in a coach converted into a private drawing room. In later years other stage celebrities found private cars a necessity.

As the B & O reached out toward the Middle West, office cars for its officials were quietly added to its stock. No PVs are listed until the 1854 annual report, which records two cars. One of

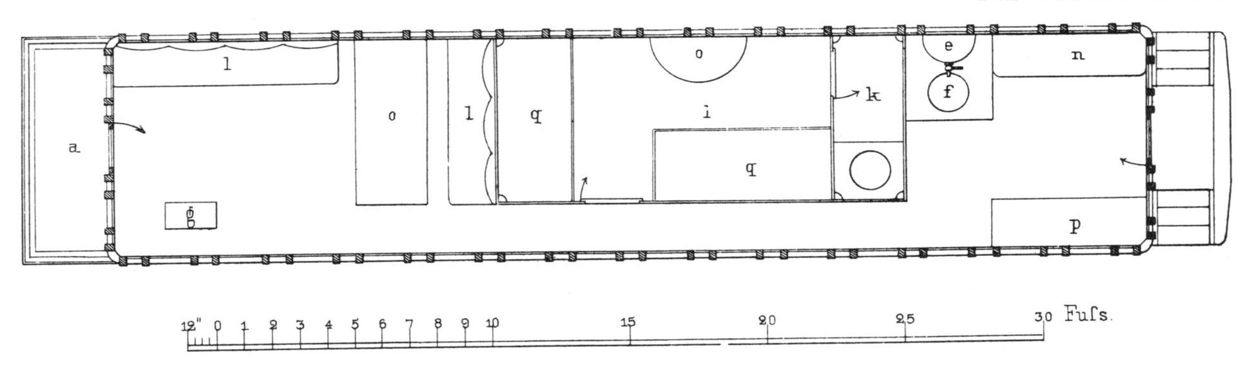

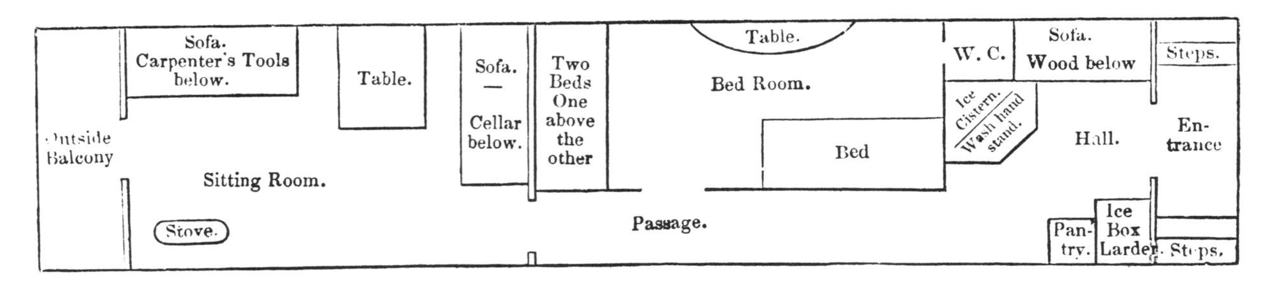

Figure 4.81 The two floor plans of the Baltimore and Ohio Railroad's director's car of about 1855 are from A. Bendel's 1862 book on North American railways and Isabella S. Trotter's travel book of 1859.

them was described as "new," indicating that the other had been placed in service some time before. Three years later the line had three office cars. In 1857 the Ohio and Mississippi built a 64-foot director's car with four compartments and 30-inch-wide passageway along one side.[135] The interior paneling was Gothic in style and painted white with gilt highlights. A Brussels carpet and crimson-velvet-covered furnishings added to the richness of the interior. Two of the sofas could be converted into double berths. The washroom had a "toilet table." A patented ventilating system was effective in cooling and cleaning the air. The car was said to be something "new in the history of railroad traveling," and if this statement is accurate it would indicate that private cars were still a rarity.

The most complete data on the second generation of American business cars is found in the published accounts of two foreign visitors.[136] Both described B & O cars observed in 1858–1859, and both include floor plans. As shown in Figure 4.81, the drawings differ slightly although the general arrangement and the overall size—roughly 8 by 40 feet—agree almost exactly. The three compartments offered an observation-sitting room at one end and a central stateroom, with a side aisle leading to the pantry–crew's quarters. The observation room sofas probably doubled as night roosts for junior members of the party. The absence of a cooking stove indicates that only light meals were served. By far the most interesting feature is the rear observation platform; the drawings are documentary evidence that this basic feature of the private car was in existence before the Civil War. And it may be that the car (or cars) shown in Figure 4.81 was used by Lincoln on his trip to Gettysburg. He was known to have borrowed a PV from the B & O for this historic journey.

In the fall of 1858 a young English gentlewoman, Isabella Stuart Trotter, toured America. One of the floor plan drawings comes from her travel account, which also gives the following description of life aboard a private car in that distant time. Luxurious railway travel, at least for powerful railway men and their favorites, was a reality long before George Pullman's creation of civilized train-board life. Mr. Ralph Greenhill of Toronto, Canada, kindly brought Miss Trotter's memoir to my attention and provided the following excerpt:

> Though Mr. Garrett talked of the directors' car, we presumed it was only a common carriage such as we had been accustomed to, but appropriated to their use; instead we found a beautiful car, forty feet long by eight wide, of which the accompanying diagram shows a plan drawn to scale. Outside: painted maroon, highly varnished with Canada balsam; the panels picked out with dark blue. Inside: painted pure white, also varnished. Ceiling the same, divided into small narrow panels, with excellent ventilators at each end. Round the car there were twenty-two windows, not shown in the plan, and three brilliant lamps in the sitting-room and hall, and one in the bed-room; these were lighted when passing through the tunnels. There were three hooks in the wall serving for hat pegs, and at the same time to support two flags for signals. A large map of the mountain pass from Cumberland to Wheeling hung over the sofa opposite the table. The table was covered with green baize stretched tightly over it. On the table were placed a large blotting-book, ink, and pens, three or four daily newspapers which were changed each day, the yearly report of the railway, a peculiar timetable book, containing rules for the guidance of the station men, times of freight and passenger trains meeting and passing each other, &c. Papa has these. The sofas are covered with a pretty green Brussels carpet (small pattern) quilted like a mattress with green buttons, chairs covered with corded woolen stuff, not a speck or spot of ink or smut on anything. A neat carpet, not a speck or spot on it, a sheet of tin under and all round the stove. Pantry cupboard containing knives and forks, spoons, and mugs. Bed-room berths much higher and wider than in a ship. Red coloured cotton quilts, with a shawl pattern, two pillows to each bed, pillowcases of brilliant whiteness, sofa bed larger and longer than a German bed. White Venetian blinds occupied the places usually filled by the door panels and window shutters. Green brussels carpet like the cover of the sofa; three chairs to match. The windows in the sitting-room had grey holland curtains running on wires with very neat little narrow strips of leather, and a black button to fasten them, and a button and well made buttonhole below to keep them from blowing about when the window is open. Looking-glass in neat gilt frame, hung over a

Figure 4.82 This extraordinary sixteen-wheel car was produced for the directors of the Chicago and North Western Railway in 1867 by the road's Fond du Lac Shops. (Chicago and North Western Railway)

semicircular console in the bed-room, another near the washhandstand, where a towel also hangs. Two drawers for clothes, &c. under berths. Table-cloth for meals, light drab varnished cloth, imitating leather, very clean and pretty, china plates, and two metal plates in case of breakages. Luncheon consisted of excellent cold corned beef, tongue, bread and butter, Bass's ale, beer, whiskey, champagne, all Mr. Tyson's. We supplied cold fowls, bread and claret. The door at the end opens on a sort of platform or balcony, surrounded by a strong high iron railing, with the rails wide enough apart to admit a man to climb up between them into the car, which the workmen always do to speak to Mr. Tyson. Usual step entrance at the other end. The platform can hold three arm chairs easily, and we three sat there yesterday evening, talking and admiring the view.

During the 1860s, accounts of official cars became more common. The *American Railway Review* described four during the first years of that decade.[137] The Michigan Southern and Northern Indiana provided an elegant car for its directors modeled after a special car that the Michigan Central had fitted out for the Prince of Wales. In 1862 the La Crosse and Milwaukee produced a directors' car with such features as a clerestory roof, bay windows, and built-in furnishings. Native Wisconsin woods were used for the interior paneling. During the following year, T. W. Kennard, chief engineer of the Atlantic and Great Western, acquired an "elegant little dwelling house" on wheels. Officials lower on the corporate ladder were now beginning to enjoy the comforts of the private car. Kennard's car, however, represented more than an executive status symbol to be used for an occasional inspection. Construction of the road was in full swing, and the chief engineer was traveling over 1,200 miles a week. His rolling office was well justified and well used. The 48-foot body contained a parlor, a bedroom, a washroom, and a kitchen.

The grandest private car of this early age was built in 1866–1867 at the Fond du Lac shops of the Chicago and Northwestern Railway (Figures 4.82 and 4.83). A contemporary newspaper described the sixteen-wheel giant as more like an "ocean steamer than a car."[138] It might be properly called the *Great Eastern*'s counterpart on rails. The 70-foot-long Gothic structure came complete with a 24-foot sunken central parlor. The parlor floor was dropped nearly 2 feet through the use of a shadbelly frame, making the ceiling 15 feet high. Here was the car with every-

Figure 4.83 A portion of the interior of the Chicago and North Western Railway's directors car of 1867. (*Chicago and North Western Railway*)

thing: mirrors, a frescoed ceiling, Brussels carpets, ottomans, sofas, easy chairs, a piano, crown-glass windows, and oak, cherry, walnut, and ash panels—in short, every trick known to the car builder's trade. At each end were 12-foot-long parlors; bedrooms and closets took up the remaining free space. Since there was no room for a kitchen, a small commissary car was attached for long runs. In later years the once-splendid directors' car was given to a lowly division chief. It ended its days as a station house somewhere in South Dakota, and sometime before 1910 was destroyed by fire.[139]

In 1870 the directors of the Lake Shore and Michigan Southern acquired their own car, the *Northern Crown*.[140] It was a mere 60 feet long, its central drawing room a paltry 17 feet. But it did have extra-large windows for the time (37 by 47 inches), and its chrome-yellow exterior was decorated with "well executed landscapes representing the great industries of the age." The iron handrails had silver mountings. Still, the car was no match for the hulking magnificence of the North Western's.

By 1870 the business car was well established: most trunk lines owned one or more. The opening of the Pacific railroad had provided an even more plausible argument for expanding the PV fleet. Only the most heartless shareholder would begrudge the comforts of a clean bed, a writing table, and good meals to any railway official forced out along the line in the Far West. Thomas Durant had a fine car as early as 1866 for his end-of-the-line inspections on the Union Pacific.[141] During the same year he purchased a special car constructed by the government for President Lincoln. This second car was used by guests or government officials traveling over the line. Meanwhile the Central Pacific shops built a large private car for Leland Stanford that could sleep ten and was paired with a provision car on long journeys.[142]

Eastern managers' need for private cars was less convincing, but there the Directors' car was already a well-established institution and it was not difficult to justify individual cars for the top brass. John W. Garrett accepted the car *Maryland*, produced at the Mt. Clare shops in 1872, as a necessary fixture for the office of the president of the Baltimore and Ohio. This historic car was magnificently engineered and furnished:[143]

> Some months ago, a car was built at these shops designed for the special use of the president of the road, John W. Garrett, Esq. The best skill and resources of the car department were brought into requisition to produce a vehicle creditable to the builders and suited to the service for which it was designed. It has already made several trips to New-

York, and has elicited the admiration of all who have examined its structure and finish. When not in use it is kept in a special house at the Camden station, in Baltimore, where we had an opportunity to inspect it, and for which we are indebted to the politeness of Mr. Schryack, of the Mount Clare works.

The car is named the "Maryland," and is 51 feet long by 10 wide outside the body, runs on six-wheeled trucks with strong check-chains attached, and with Dinsmore springs on the equalizers. Iron body-transoms are used of the kind represented in the foregoing cut. The outside of the car is painted a light yellow, with nothing in the way of external ornamentation to attract special attention or indicate the quality of the interior fitting-up, which is in the best Pullman-coach style. The car is designed to run always in the rear of a train, so as to afford a view of the track from the end windows. A glance at the arrangement of the interior indicates at once that the vehicle is designed for but one principal occupant with his traveling suite. There are four distinct compartments, a porter's room, state-room, a sleeping and toilet-room with side passage and closets, and the parlor or drawing-room. The porter's room occupies about ten feet of the forward end of the car, with a closet for table-ware on one side of the door, and a Baker & Smith heater on the other. Next to this is the state-room with an upper sleeping-berth on each side, and seats for eight persons underneath, and which can be transformed into two lower berths. Next is the central compartment, occupying a space of about 17 feet in length, and containing the principal sleeping and toilet-room, and corridor, with a door and two windows in the partition between. There are also included in this division a water-closet and linen-locker, communicating with the sleeping-room. This room is sumptuously furnished. Across one end is a curtained lounge. The bedstead is an elaborate piece of cabinet-work in French walnut, surrounded with heavy double damask crimson and green curtains. The window-curtains are of the same description, and hung, like the others, on silver-plated rods. Two large mirrors occupy the spaces between the windows. The remaining portion of the car, comprising a space of about 15 feet and extending to the rear end, constitutes the drawing-room, which is entered by a door from the side passage above mentioned. Its furniture consists of a large and elegant sofa-lounge, an oblong black-walnut centre-table with marble top, two easy-chairs of the Pompadour style and two others of a different pattern. This apartment has five windows on each side and three in the end, these last affording a fine view of the track. The spaces between the side windows are occupied with mirrors, and the curtains are of the same kind as those in the sleeping-room. The floors have Brussels carpets. The interior finishing throughout is in solid black-walnut, with elaborate raised paneling of French walnut "burl," with semi circular tops. The general effect is somewhat sombre, but is relieved by light gilt mouldings above the windows. The artificial lighting is done by four of Williams, Page & Co.'s improved plated lamps. Carroll ventilators are used, and all the windows have double sashes. In the central passage-way is a wash-stand with a reservoir underneath holding a barrel of water, which is raised by a small force-pump. The inside door windows are beautiful specimens of embossed glass.

During Garrett's lifetime the car was used by many dignitaries, including several U.S. Presidents.[144] In 1908 it was sold to the Quebec Central and was reported still in operation in 1949.

Thomas Scott of the Pennsylvania Railroad acquired a personal car the year before Garrett's *Maryland* was completed. An experienced operating man, Scott believed in frequent over-the-line inspections. Several years later he was partially paralyzed by a stroke, so that a special car was even more of a necessity. In addition he was attempting to build a second railway empire in the Southwest, a venture that called for regular journeys to Texas. Scott's car was one of the first PVs to be numbered rather than named. The simple designation 120 indicated its utilitarian purpose, and the same number was used by P R R presidents until recent years. The first Number 120 was a product of Altoona. The 63-foot twelve-wheeler was completed in 1871.[145] In addition to its pointed Gothic windows, red body, brown trucks, and green wheels, the 120 is thought to be the first American railway car equipped with a bathtub.

Cornelius Vanderbilt was a greater railroad chieftan than either Scott or Garrett, and as a poor boy become suddenly rich, he was quick in acclimatizing to the luxuries of this world. In the spring of 1871 reports began to appear about a private car nearing completion for the Commodore.[146] It was shorter than a coach but ample enough to accommodate a 25-foot drawing room, a parlor, a bedroom, and a pantry. The triple style of windows was used, with the large center pane measuring 42 by 36 inches. The exterior was painted cream with buff trim, but no gaudy ornamentation was permitted. Nor would its owner allow a public showing; he wanted no more publicity. The car builders wanted to install gaslights, but this scheme was also overruled by the old man, who was content with candles. According to Charles Sweet, the car was not a favorite with its owner.[147] For his personal jaunts Vanderbilt is said to have preferred the Wagner parlor car *Dutchess*, with some of the parlor seats pulled out, other furnishings installed temporarily in their place, and a portable zinc floor and kitchen added at one end. Some years after the Commodore's death, his private car could be seen in service between New York and Chicago. Wagner was running it as a parlor car under the name *Iroquois*.

The early history of the private car revolves almost exclusively around vehicles built to serve railroad officials. It is difficult to identify the first individually owned passenger car, particularly in view of the tangle between railroad and private ownership. For example, the cars just described may actually have been true private cars, since Durant and Vanderbilt may have paid directly for theirs even though they were railway officials. The car built for President Lincoln between 1863 and 1865 would seem to be a clear-cut example of a private car, because it was surely not built for a railroad official. Yet it was never used by Lincoln during his life, and after the funeral trip it was sold to the Union Pacific as a business car. The personal private car does not appear to have come into being until sometime in the 1870s.

Some historians have suggested that P. T. Barnum was the first private citizen to own his own car. The date of his purchase, said to be in the seventies, is uncertain, and like so many facts ascribed to the great showman, remains a source of continuing speculation.[148] According to Lucius Beebe, the Western banking-mining king Darius O. Mills bought a private car in 1870 or 1872. However, a contemporary description puts the date ahead to 1875 or 1876. Early or late, Mills's car set the pattern for other

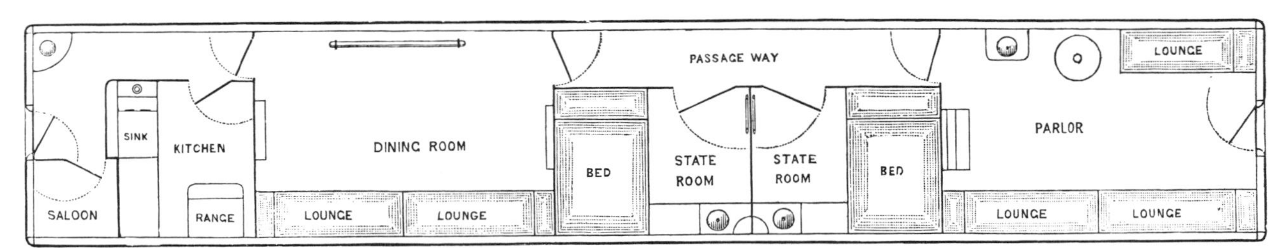

Figure 4.84 Darius O. Mills named his private car the D.O.M. *It was built in 1875 or 1876 by Harlan and Hollingsworth.* (National Car Builder, *January 1876*)

Figure 4.85 Mills's D.O.M. *stands ready for delivery at the factory's transfer table.* (*G. M. Best*)

Western moguls to follow. He commissioned Harlan and Hollingsworth to fabricate a first-class vehicle; there was to be no scrimping on price and no Vanderbilt-like repression of ornament.[149] The exterior was in fact a festival of color. The body was canary yellow with brown, red, and dark-green trim. The trucks were orange with green wheels. The body measured 53 feet, 10 inches by 9 feet, 8.5 inches. The car, named *D.O.M.* after the initials of its owner, cost $15,000 (Figures 4.84 and 4.85). Peter Donahue, a West Coast capitalist, hoped to eclipse the *D.O.M.* by having a PV built locally from native wood by the Kimball car works of San Francisco in 1876–1877.[150] Leland Stanford, realizing that his Sacramento-built car was outclassed by these late arrivals, placed an order with Harlan and Hollingsworth for a car equal to Mills's or Donahue's. According to the Pacific Coast railway authority G. M. Best, the *California* was delivered to Stanford in 1877.

Of all the PVs built during the seventies, George M. Pullman's own car was easily the most famous. The car was called the *P.P.C.* for the Pullman Palace Car Company, and the letters were interwoven to form an elaborate monogram on the car's side. The *P.P.C.* was thus not only a conveyance for Pullman and his family but a subtle rolling advertisement for the sleeping-car business. It was put to other uses as well, for Pullman delighted in lending the car to traveling dignitaries. Presidents, South American emperors, European royalty, great men of industry and finance—all enjoyed complimentary excursions aboard the *P.P.C.* As James W. Holden said, "Its deep luxurious carpets cushioned the tread of a long array of celebrities."

Despite its normally glamorous entourage, the *P.P.C.* was actually built as an excursion car for rental parties. A newspaper account of the day and the order book listing confirm that it was in fact leased out at $85 a day when not otherwise engaged.[151] Pullman thus had the best of all reasons for building a PV: it would be a revenue producer and not a cash-draining parasite

Figure 4.86 George Pullman's own private car, the P.P.C., *was remodeled in 1892, but the interior retains much of the look of the 1870s. (Arthur D. Dubin)*

like other private cars. As late as 1892 the availability of the car on a rental basis was reported in the New York press. Although the *P.P.C.* was technically one of the company's rental cars, it has always been known as Pullman's personal pet. Little else would explain the car's long service record and the special attention it received in later years.

The *P.P.C.* was completed at the Pullman Detroit shops in June 1877. Mounted on twelve wheels, its body housed an observation room, several bedrooms, a large central parlor-dining room, and a kitchen-pantry at the forward end. The parlor-dining room featured a reed organ. The car could sleep 10. After the fashion of the day the interior was finished in the Eastlake style, and the side walls were pierced with triple-window sets laid out with a broad center pane that was flanked by narrow sidelights. Fifteen expert wood-carvers worked on exquisite geometric turnings, panels, and moldings of the most opulent cabinet woods obtainable. The ceilings were painted to represent such scenes as the "bulrushes—with young Moses omitted; another of fuchsias and humming-birds."[152] The lamps and other metal fittings were gold-plated. A lilliputian bathtub completed the car's special equipment. Exaggerated estimates on the cost have been circulated ever since the car was produced; $25,000 was quoted in 1877, while in later years the figure was doubled.[153]

Within a decade the *P.P.C.* was just another veteran private car, yet Pullman exhibited a strong affection for it. Although he could easily have afforded a replacement, and no man on earth was in a better position to obtain one, he would not give the car up. In 1887 electric lights were installed in it, and in 1892 the body was lengthened to 66 feet, 10 inches and remodeled on the outside with a more modern style of clerestory roof.[154] A drawing and a few photographs survive from this period (Figures

Figure 4.87 The P.P.C. *was built in Pullman's Detroit shops in 1877. This drawing shows it as remodeled in 1892. (Pullman, Inc.)*

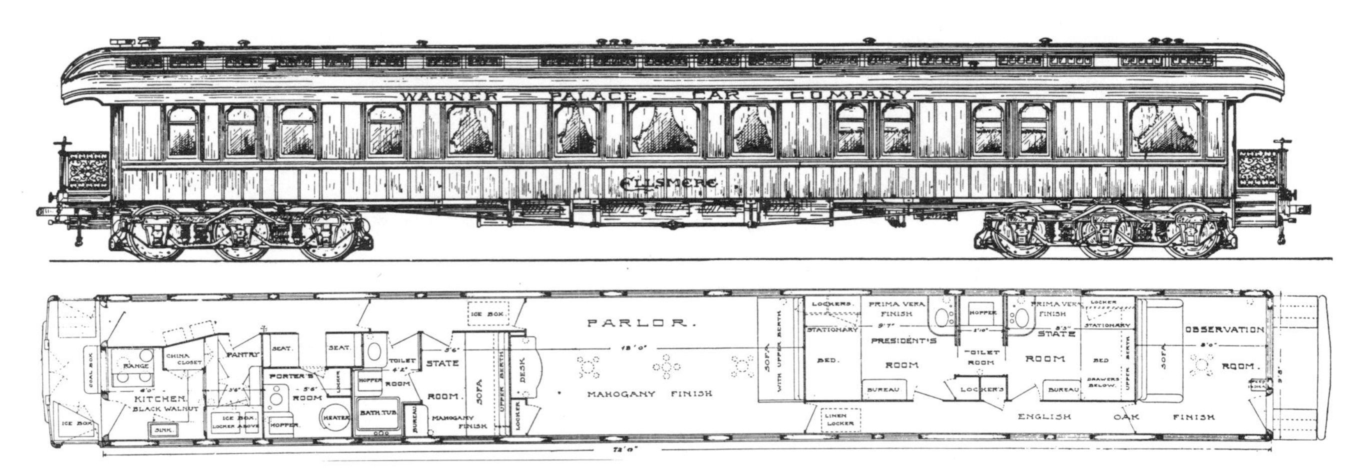

Figure 4.88 The Ellsmere *was used by W. S. Webb, President of the Wagner Palace Car Company. It was built in 1888 at Wagner's Buffalo shops.* (Railroad Car Journal, *August 1897)*

Figure 4.89 The Manhattan *was produced for Austin Corbin, president of the Long Island Railroad, in 1885 by Jackson and Sharp. (Hall of Records)*

Figure 4.90 Observation-room interior of the Manhattan, *1885. (Hall of Records)*

4.86 and 4.87). Mrs. Pullman shared her husband's affection for the old *P.P.C.*, and after his death in 1897 she kept the car, renaming it *Monitor.* When Mrs. Pullman died in 1920 the aging wooden relic, now totally unfit for service among the heavyweight steel equipment of the time, was put in storage at the Calumet shops. Unfortunately no one in the Pullman Company had the sentiment or foresight to preserve this historic memento. It was unceremoniously broken up the next year.

During the decade of the eighties the private car came to its full development. It was no longer a rare plaything to be enjoyed by a few top executives and financial leaders. Its use now spread to a wide circle of lesser officials. In addition, actors, actresses, and opera stars—people who were viewed with some suspicion by the polite strata of society—were now arrogantly flying about the country in fabulous private cars, which were often the gift of a patron or lover. On a less affluent level were circus, medicine-show, and theatrical groups, which found comfort and convenience in private car travel. Frequently these troupes were content to take over some shopworn veteran or remodel an existing car for their use. The Robbins iron passenger car, discussed in Chapter 2, is an example (see Figure 2.12).

During the eighties, with the notable exception of Pullman, the wealthiest car owners developed the habit of trading a relatively recent purchase in on a new one. The Vanderbilt sons switched cars frequently and soon discovered that one or two could not satisfy the needs of the entire family. In 1882 W. K. Vanderbilt had a 74-foot yellow giant built by the New York Central shops.[155] Scenes painted on its sides depicted the major improvements on the line—such as Grand Central Depot and the West Albany bridge—and such scenic spectacles as Niagara Falls. A Wagner sleeping car was connected to the *Vanderbilt*, with some form of vestibule fixed between the two for the safe passage of travelers. Within a few years, however, the railroad king tired of his twin cars and had the Wagner shops fabricate a single-unit successor, the *Idlehour.*[156] His brother Cornelius decided it was time to trade cars as well, and in the following year he acquired a new one from Wagner done up in the Louis XVI style.[157] But it was a brother-in-law, W. S. Webb, who showed the family how to go through cars. In 1885 he obtained a handsome one from Barney and Smith.[158] Two years later a replacement was under construction at the Wagner shops.[159] At his direction a number of private cars were produced for Wagner's private rental fleet and placed at the Vanderbilts' disposal. Of this group Webb favored the *Ellsmere*, with its 18-foot dining room, Italian Renaissance decor, and embossed leather ceiling (Figure 4.88).

Jay Gould and his family were also steady patrons of the PV. Gould's delicate health and secretive personality made a private car a positive necessity. At first he used existing official cars; a photograph shows him standing on the rear platform of the Atlantic and Great Western's Number 200, which was produced at the road's Kent, Ohio, shops in 1876 to replace the old *Atlantic.*[160] After acquiring control of the Missouri Pacific, he was content to use Commodore Garrison's old car, the *Convoy.*[161] In 1886 he is said to have purchased a set of twin cars from Pullman named *Penola* and *Bedford.* The cars were built under lot number 1346, but no other record of their existence can be found. During the following year he acquired a new car named in honor of his yacht, the *Atalanta.*[162] This 70-foot, 78,000-pound, twelve-wheeler was produced at Pullman's Detroit shops at a cost just over $20,000. It was described as rich but not garish. The *Atalanta* served out Gould's remaining years, but both his son and daughter preferred to travel in their own cars. The family could easily afford it.

Other wealthy users of PVs included the Stanfords, husband and wife, who had separate cars, and C. P. Huntington, who

Figure 4.91 The Alexander, *a Pullman creation of 1892, was one of the most lavish private cars ever produced. It was made to the order of A. A. McLeod. (Pullman Neg. 2108)*

roughed it for several years with a single car.[163] Huntington found that his guests, servants, and luggage were always underfoot, however, and bought a second car in 1895 for the overflow.

Other money barons less well known today than the Vanderbilts and Goulds also had sumptuous private cars. Although many were Pullman products, other builders too were busy satisfying this small yet prestigious market. Jackson and Sharp constructed some of the best PVs of the period, but because the cars were ordered by relatively obscure figures they did not attract publicity. In 1885 the New York real estate and railroad enterpreneur Austin Corbin bought from Jackson and Sharp what some authorities regard as one of the best cars ever built up to that time.[164] He called it the *Manhattan*. The Munich-lake-color sides were emblazoned with rich gilt decorations. The windows were glazed with clear French plate and color-leaded glass. Sage-colored carpets, Mexican mahogany, dark brass hardware, and silk tapestry in subtle hues created a somber but rich interior. One unusual decorating note was the sheepskin mats in the master bedroom. Other features included the bathtub that was hidden under a sofa, the five water tanks storing 300 gallons, and the air and vacuum brakes that were part of the mechanical equipage. The builder's photographs in Figures 4.89 and 4.90 give a better idea of the car's appearance.

The financier Angus A. McLeod was at the height of his fortune during the early 1890s, when he managed to assemble several railroad and coal properties, using the unstable Reading for an anchor. During those brief peak years McLeod purchased two private cars from Pullman. The first, the *Alexander*, delivered in 1890, did not satisfy McLeod's patrician taste. In 1892 he ordered Pullman to produce a truly deluxe car, something that would please a Hapsburg prince. According to a Pullman official, the resulting vehicle was "without a doubt, one of the finest ever built."[165] Surely few cars were more elaborately finished, from its polished brass end rails to its Empire dome ceiling. The 70-foot body was literally crammed with costly hangings, carvings, and leaded glass. Because of the car's vulnerable position at the rear of the train, McLeod had insisted on safety as well as splendor, and the second *Alexander* was solidly built. Its frame was in fact unique—formed of a solid row of 4- by 8-inch timbers (Figures 4.91 and 4.92). Canvas and flooring were nailed over this deck. The body was heavily reinforced with steel plates. No tally of the car's total weight has been found, though with such a massive floor frame it must have set a record. Sometime after McLeod's business failure in 1893, it was sold to the Southern Pacific.

McLeod's *Alexander* was in many ways the apex of the private car during its gaudiest period. Other cars imitated and perhaps a few even surpassed the *Alexander*, but most private purchasers backed away from such excesses. Henry M. Flagler, the developer of Florida who had been an early partner of John D. Rockefeller, ordered a new PV that would be rich-looking but in an Empire style less ostentatious than McLeod's. The new car, unobtrusively numbered the 90, was produced by Jackson and Sharp in 1898.[166] The body measured 71 feet, 6 inches, and the

Figure 4.92 Interior of the Alexander's *drawing room, 1892. (Pullman Neg. 2111)*

platforms added another 7 feet, 2 inches. The interior was paneled in white mahogany and satinwood and was lighted with Pintsch gas fixtures. The plan and appearance of the car are shown in Figures 4.93 and 4.94.

Much of the private car's glamour can be traced to the association of stage personalities with railway travel. Although the pattern was established in 1850 by Jenny Lind, it does not appear to have taken hold for another thirty years. And then in the 1880s came three exciting women whose American tours made their every movement a matter of public interest. The first two were opera stars: Etelka Gerster and Adelina Patti. The third was a beautiful socialite turned actress: the legendary Lily Langtry. All three crossed the country aboard Mann-styled private cars.

The singers traveled in cars bearing their own names—both rented from the Mann Company. Patti's car was completed by the Gilbert Car Works late in 1883.[167] Its dark embossed-leather interior was decorated with morning glories picked out in gold and silver leaf. The 55-foot interior was divided into three compartments, with furnishings that included a piano and a pink boudoir. Beebe summarized the decor: "Looped, fringed, frizzled and ferned, its rococco splendors were the tinsel stuff of grand opera itself."[168] Within a few years the *Adelina Patti* was taken over by Pullman and renamed the *Coronet*. Late in 1901 it was sold to a used-equipment dealer who specialized in restocking circus trains, and no more tales of its Byzantine splendors were circulated to the press. In its day Patti's impressario, Colonel Mapleson, had boasted of a solid silver bathtub, 18-carat-gold keys, and other such nonsense for the entertainment page.

The *Etelka Gerster* toured the country several times before Gerster's retirement in 1896. This 1884 product of Jackson and Sharp lingered in Pullman's private car rental fleet for several years and was then sold to the Bangor and Aroostook Railroad.[169] Between 1899 and 1924 it was B & A business car Number 97, after which it was moved to Stockton Springs, Maine, for use as a summer cottage.

Mrs. Langtry rode in rental cars until a wealthy admirer presented her with her own PV, the *Lalee*, built by Harlan and Hollingsworth in 1888 for a reported cost of $65,000. The bright Jersey blue *Lalee* was compared to Cleopatra's barge, but the dazzling car received less attention than it might have because it was overshadowed by the private life of its owner, particularly her rumored dalliance with the Prince of Wales.[170]

The private car was not the exclusive property of bizarre financiers or ladies of flamboyant reputation. Some business cars were actually used for business travel, and after the initial extravagance of the first-generation "directors' cars," their plan and finish seems to have become more practical (Figure 4.95). Although the elaborate cars of the very rich received most of the attention, records do exist that provide details on a few office cars. For example, during the 1870s the general superintendent of the Pennsylvania Railroad traveled over the line in a small, spartan, 46-foot-long car. A floor plan, together with a brief description in James Dredge's 1879 report on the Pennsylvania, supply these facts.[171] The end saloon was outfitted as an office, with long desk-like tables on either side. The central compartment could be used as an auxiliary office by day. A collapsible table converted it into a dining room, and at night, folding berths made it into a sleeping room. At the rear of the car was the general superintendent's tiny stateroom and kitchen (Figure 4.96).

Another office car of the same period was described in a French report published in 1882.[172] Owned by the Louisville and Nashville Railroad, it was a combination pay and track-inspection car. The paymaster's compartment at the forward end contained tables, seats, a cage, a safe, and a signal bell. The right-of-way superintendent occupied the far end of the car. The end bulkhead had extra-large windows for a clear view of the tracks. The stove, linen closets, and toilet were located here. At the center of the car was a bunk room shared by both crews. Double-glass windows were an interesting feature. The framing plan in Figure 4.97 shows other structural arrangements.

The cost of business cars was well below the reported prices of the more widely publicized PVs. In 1880 the Northern Pacific paid the Ohio Falls Car Company only $4,339.13 for its office car

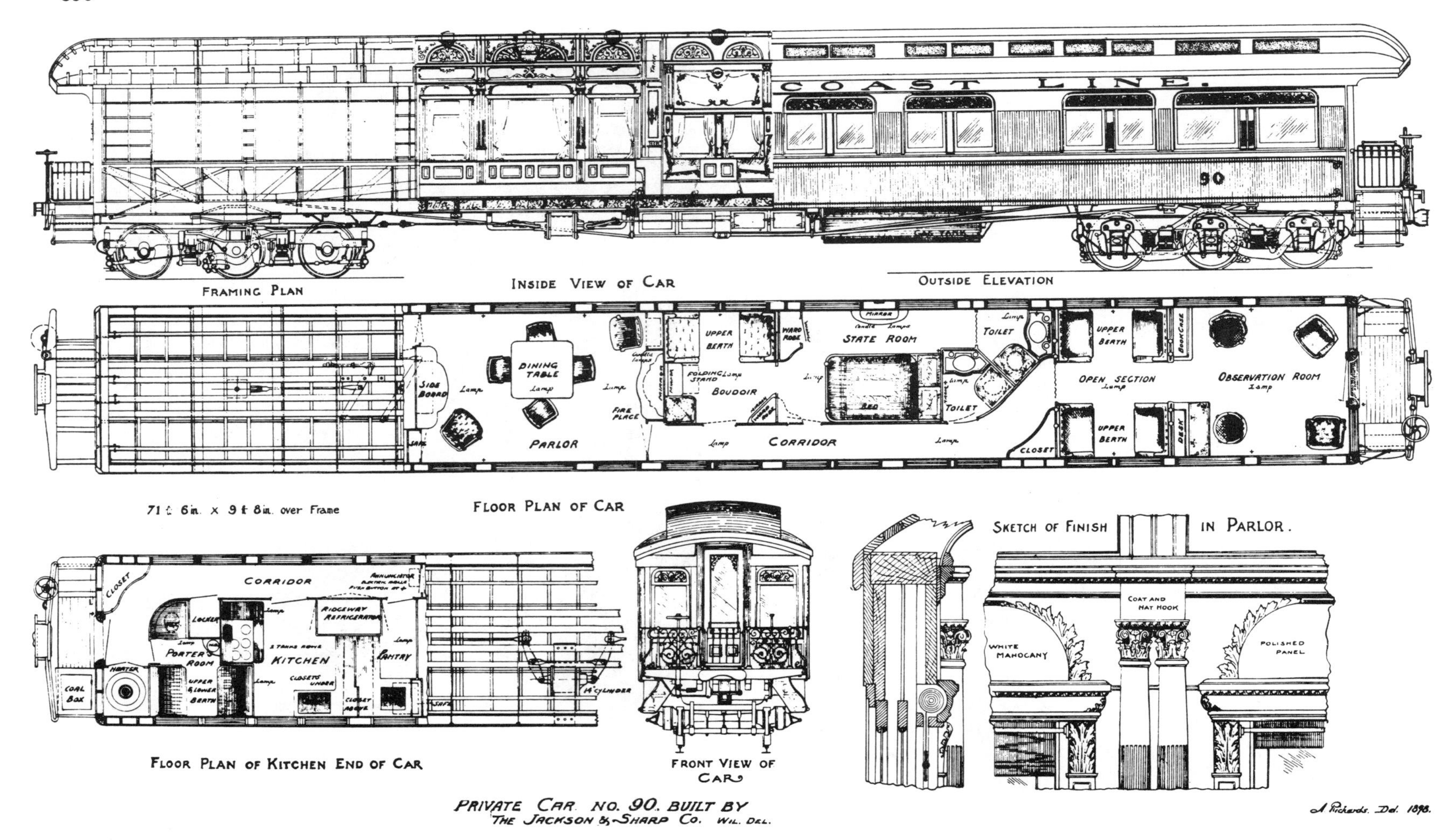

Figure 4.93 Jackson and Sharp produced the Number 90 for Henry M. Flagler, president of the Florida East Coast Railway, in 1898. (Railroad Car Journal, *August 1898*)

A-3.[173] Some years later it spent $2,481.75 to produce the A-8 in its own shops. The figures of $30,000 to $50,000 for the fancy private cars of the time were admittedly exaggerated; but even if the actual prices were only one-third of those amounts, the cars were a silly luxury. The nineteenth-century business car offered the same space and comfort at relatively tiny prices by providing a plain interior and simple furnishings (Figure 4.98).

Because PVs were so often custom-made to suit individual needs, they were the class of car least amenable to standardization. In the first rush of enthusiasm, according to a Pullman official, the customer overwhelms the designer with novel concepts; his car must be radically different and much better than all that have preceded it. "But when he comes face to face with certain blue printed mechanical and engineering limitations that are imposed on the car builder, the enthusiast finds that much of the striking originality in his scheme will have to be eliminated."[174] Yet within the tubular confines of a standard car body, decoration, furnishings, and even floor plans to some degree, were matters of idiosyncratic taste. In 1901 Charles M. Schwab, who spent $40,000 on his *Loretto*, chose an enclosed sunroom in place of the traditional open observation platform. Four years later Louis Hill had his second PV fitted out as a mobile garage so that he could travel with his favorite automobile. Max Fleischmann took an active hand in the design of his *Edgewood*, which is remembered for its heavy underframing, noiseproof construction, and elaborate filtered air system. The Woolworth heiress, Mrs. J. P. Donahue, insisted upon gold fixtures and other extravagances for her *Japauldon*, built by American Car and Foundry. Fairfax Harrison, the aristocratic president of the Southern Railway, required a pair of cars for his travels.

Try as they would, however, designers were confined to what a 10- by 8- by 80-foot box could contain. It was a foolish shape for a "house," since all the rooms were to be private, a hallway was necessary, but then the rooms or the hall had to be made impossibly narrow. The designer was forced to produce a compromise floor plan. Excluding the freaks, there appear to be three basic plans. The most common had an observation platform at one end, a parlor or observation room, and then a series of three or four bedrooms, bypassed by a narrow side aisle which led to a dining room. Both the parlor and dining room were full-width. Behind the dining room were the pantry, kitchen, and crew's quarters (Figure 4.99). This arrangement was in full vogue by 1900. In the next most popular plan the observation platforms were fixed to both ends of the car, the dining room was at the forward end (in place of the kitchen and crew's quarters), and the kitchen was toward the center of the car (Figure 4.98). The third plan, which seems to have gone out of favor by the 1890s, had a large central, full-width room that doubled as parlor and dining room (Pullman's *P.P.C.* was an example).

The first steel PV was Robbins's steel car of 1889. It was not until 1907, however, that another such vehicle was constructed. In that year Pullman's Buffalo (formerly the Wagner) shop produced an all-steel business car for J. M. Schoonmaker, vice president of the Pittsburgh and Lake Erie.[175] Early in 1910 the Al-

Figure 4.94 The dining room of Flagler's Number 90, 1898. (Hall of Records)

toona shops were building a similar car for an executive of the Pennsylvania.[176]

It is unlikely that many new wooden private cars were produced after this date for domestic service (see Figure 4.100). However, steel underframes and steel side plating rejuvenated many an aging wooden PV for long years of use. An example is Austin Corbin's *Oriental*, which Pullman built in 1890. Corbin died in 1896, and the *Oriental* was taken over by the Louisville and Nashville for its board chairman, August Belmont.[177] A steel underframe was subsequently added. Metal sheathing, applied in 1947, so disguised the car that few persons would have guessed its true age. In 1958, after almost seventy years of service, the *Oriental* was retired. It is now exhibited in the Adirondack Museum at Blue Mountain Lake, New York.

Car builders produced new private cars rapidly during the great prosperity of the 1920s. By this time a new PV took up to four years to construct and cost between $70,000 and $100,000.[178] The price was figured at cost plus a 15 percent profit plus an allowance of 175 percent of direct labor for overhead. Furnishings, china, silver, linen, and kitchen utensils added another $7,850. Railroad officials and wealthy private individuals were not the only customers for the cars; corporations such as U.S. Steel, Maryland Oil, Fruit Growers Express, Anaconda Copper, and even the automotive giant General Motors also became PV owners. The crest was reached in 1927, when builders turned out 108 new private cars.[179] During the next year a total of 886 PVs was reported in service, but the Pullman Company estimated that no more than 40 of these were actually privately owned.

The economic chaos following the 1929 collapse abruptly ended the private car era. In 1930 only one new PV was produced. A handful of business cars were constructed after that time, but for all practical purposes the PV was obsolete. Many owners sold or stored their cars during the Depression years. Those who could afford to keep them could not always afford to operate them. *Fortune* magazine published the following table of annual operating charges in its July 1930 issue.

SALARIES *(per year)*		
Steward	$2,400	
Chef	1,800	
Waiter	1,500	
		$5,700
RAILROAD CHARGES		
(Depending, of course, upon the car's use. Here it is figured that the car's habitat is Chicago, that it makes one trip to the Kentucky Derby, one trip to the Yale-Harvard boat races, one trip to Pasadena, three trips between Chicago and New York, two trips to Washington, one trip to Hot Springs, one trip to Quebec, and one trip to Jaspar Park. The remainder of the time it is paying parking charges in the yards.)		
Tickets *(Twenty-five for each trip)*	$20,219	
Surcharge *(10 percent)*	2,021.90	
Parking *($3.60 per day for first seven days; $12 a day thereafter)*	681.60	
		$22,922.50
FOOD		
Passengers *(average five)*	$750	
Allowance for crew *(Depending again upon whether the owner maintains a full crew, or takes on the chef and waiter when needed. In general, however, it will be found more satisfactory to keep a full crew. At $3 per man per day, crew's food will cost . . .)*	3,285	
		$4,035
PARKING CHARGES FOR PERMANENT STORAGE *(Usually at $2.40 per day. After trips are subtracted, total days in Chicago storage are 240.)*		$576
TIPS *(Station masters, yard masters, and their assistants must be tipped.)*		$500
	TOTAL	$33,733.50

By 1940 the private car fleet was down to 572, a one-third decline in a single decade. Furthermore, economic recovery did not stimulate PV construction, for other negative factors were at work. The very rich were beginning to fly or drive; the private railroad car had acquired a stodgy, Victorian aura. And by late 1941, conservative travelers who would have continued using PVs could not because of the war effort. There was no place for luxury pleasure travel; the railroads handled essential business movements, but the PVs were otherwise sidetracked. Many private car owners sold out. Franklyn Hutton's *Curley Hut* was bought by the C & O for use as an office car. Barbara Hutton's car went to the Western Pacific, while Henry Ford's fabled *Fairlane* was sold to the Cotton Belt.[180]

At the war's end the PV did not return to favor (see Figures 4.106 and 4.107). Between 1949 and 1955 only fourteen lightweight business cars were produced.[181] Some roads found it cheaper to rebuild existing cars. The business car fleet soon stabilized at its prewar level, but privately owned cars virtually disappeared. Benjamin Fairless, president of U.S. Steel, was an exception; he commissioned Pullman to produce the *Laurel Ridge* in 1949 at a cost of $200,000.[182] After his retirement in 1955 his company's use of the car declined, and in 1962 the maroon and silver streamliner was quietly sold to the Santa Fe. Except for the Anheuser-Busch Brewing Company's *Adolphus* (1954), the *Laurel Ridge* was the last of the species.

By 1971 there were only 179 business cars in operation. A new generation of private car owners has emerged in recent years, but they are like an exclusive club of antique yachtsmen. The railroads' top management has largely abandoned the office car for the speed and prestige of the executive jet. The general manager or division superintendent still finds the business car useful, but for railroad presidents it is too slow and cumbersome for everyday business or that "fast trip to the Coast." With the coming of Amtrak and a skeleton system of intercity passenger trains, it is no longer possible to travel between many major cities. The inspection trip, once a major justification for the business car, is increasingly made in luxurious automobile-rail cars. In 1954 the New York Central, for example, remodeled a Chrysler four-door sedan for the use of its chief executives.[183] What was once the railway passenger car's most prestigious form has gradually faded away.

Rental Cars

For generations aspiring plutocrats have rented jewelry, seaside mansions, and limousines. In an earlier time they rented private cars. Some of our wealthiest citizens preferred to rent (or borrow) rather than own a PV. J. P. Morgan, who could well afford to buy every private car in North America, declined to own any but was willing to charter an entire train as the occasion arose.[184]

From the beginning of the railroad era it has been possible to charter a private car. Special parties and regular commuter clubs have taken advantage of this service since the 1830s.[185] So far as is known, the early renters made short trips and used ordinary day coaches. Apparently it was that genius of railway travel, George M. Pullman, who first recognized the potential profit in rentals, particularly in luxury rental cars for long-distance travel. After the transcontinental railway was opened in 1869, the scenic wonders of the West might be seen from the comfort of a passenger car. How much better it would be if one's family, friends, or close business associates could enjoy the trip together in the home-like privacy of their own car. In 1870 Pullman offered to

Figure 4.95 A Baltimore and Ohio office car, possibly Number 41. Its plain appearance suggests that it was a division superintendent's car. The photograph dates from 1858.

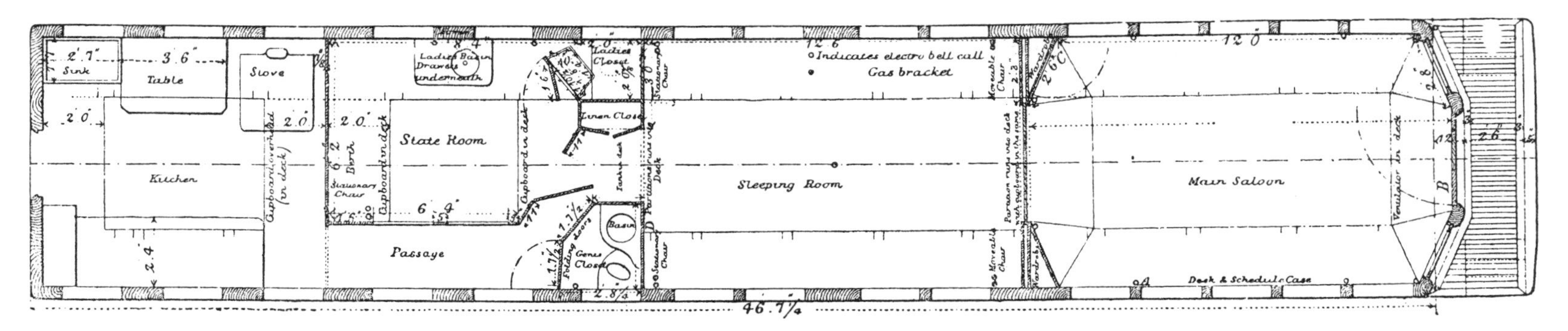

Figure 4.96 G. C. Gardner, superintendent of the Pennsylvania Railroad from 1871 to 1878, used this office car. Such cars were considerably less elaborate than the private cars previously shown. (James Dredge, The Pennsylvania R.R., *London, 1879)*

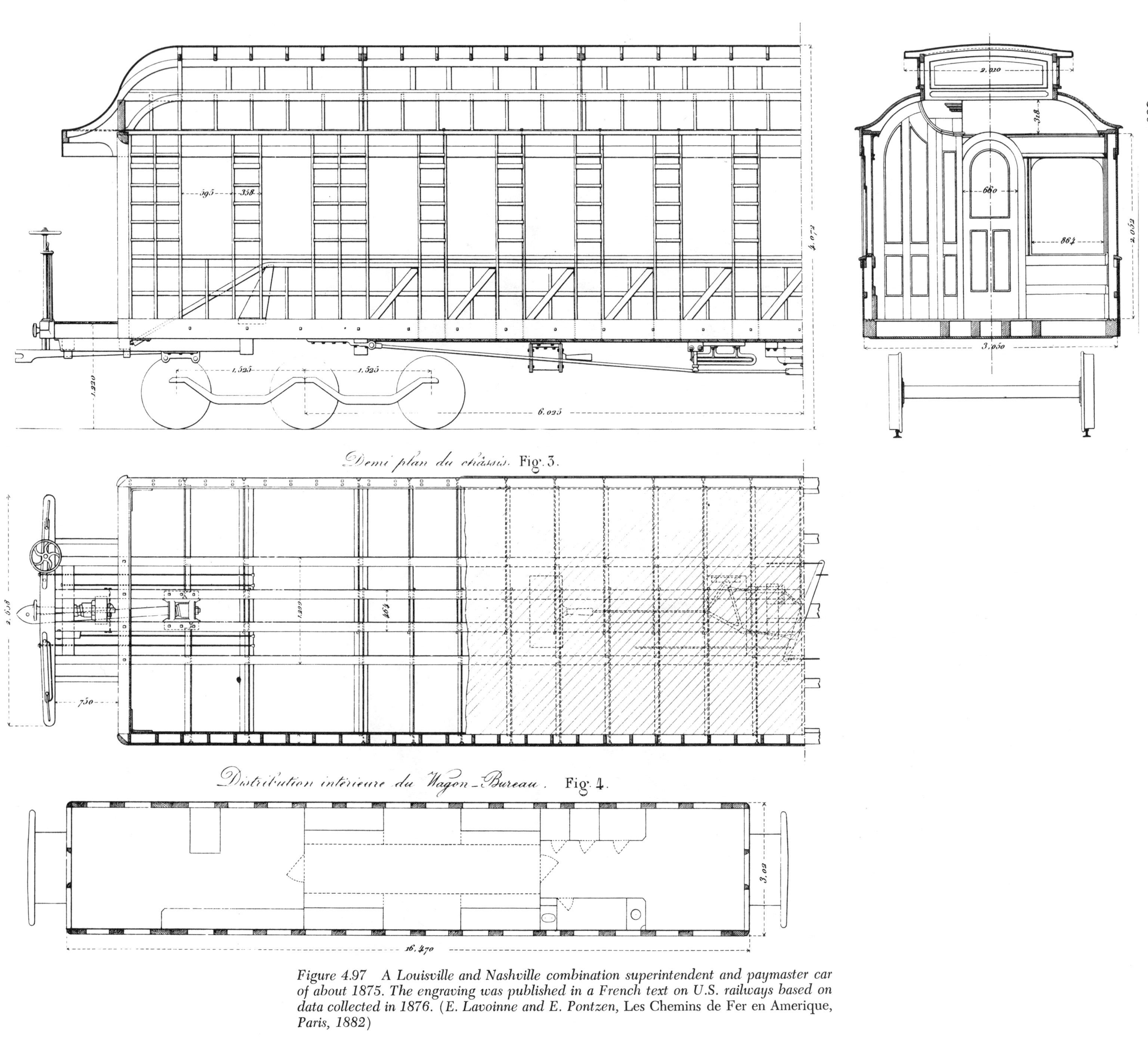

Figure 4.97 *A Louisville and Nashville combination superintendent and paymaster car of about 1875. The engraving was published in a French text on U.S. railways based on data collected in 1876. (E. Lavoinne and E. Pontzen,* Les Chemins de Fer en Amerique, *Paris, 1882)*

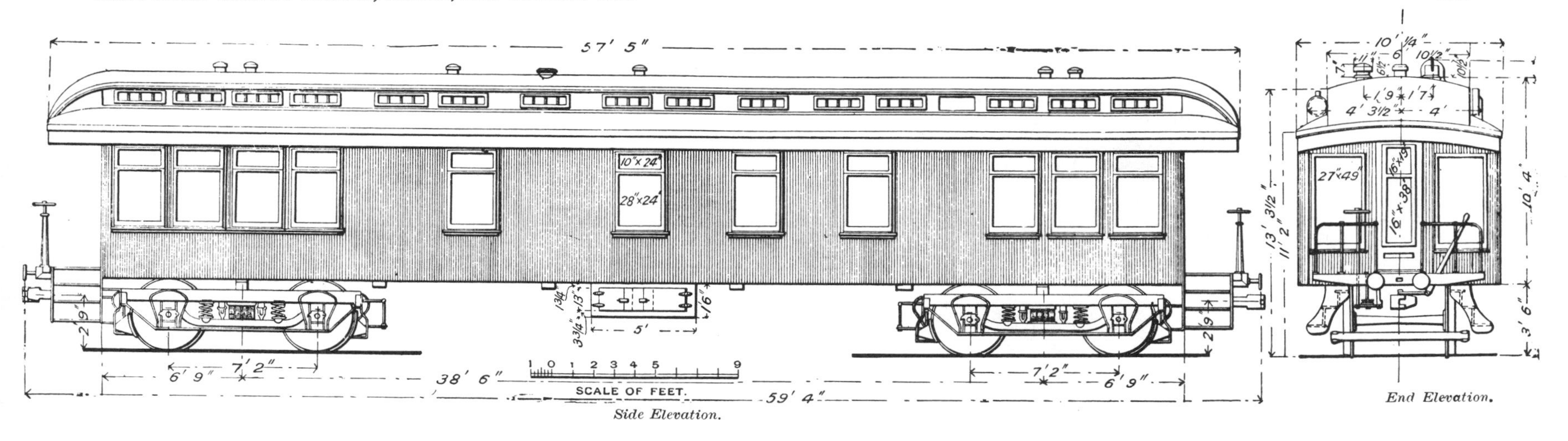

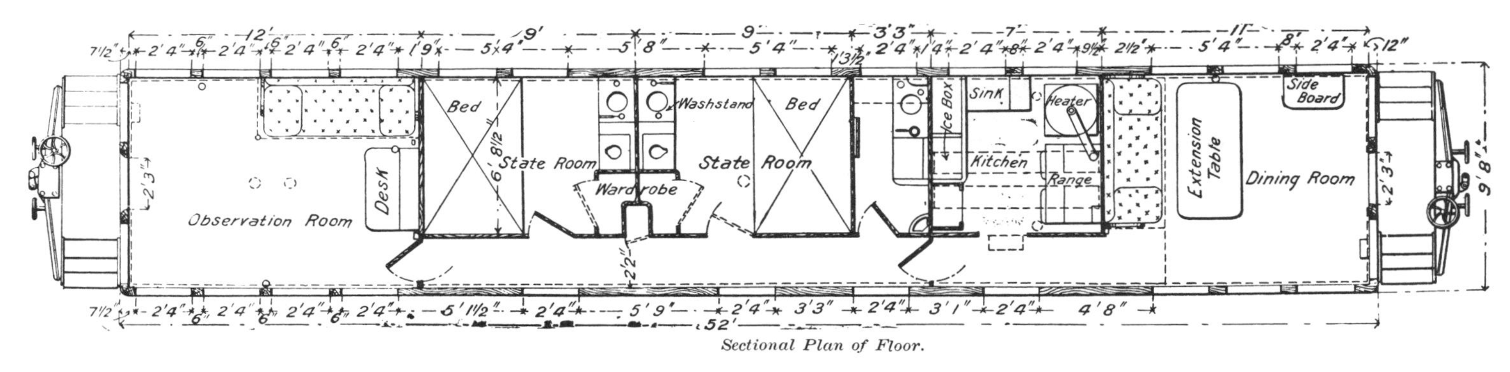

Figure 4.98 *A Chicago, Burlington, and Quincy business car of about 1890.* (Car Builders' Dictionary, *1895*)

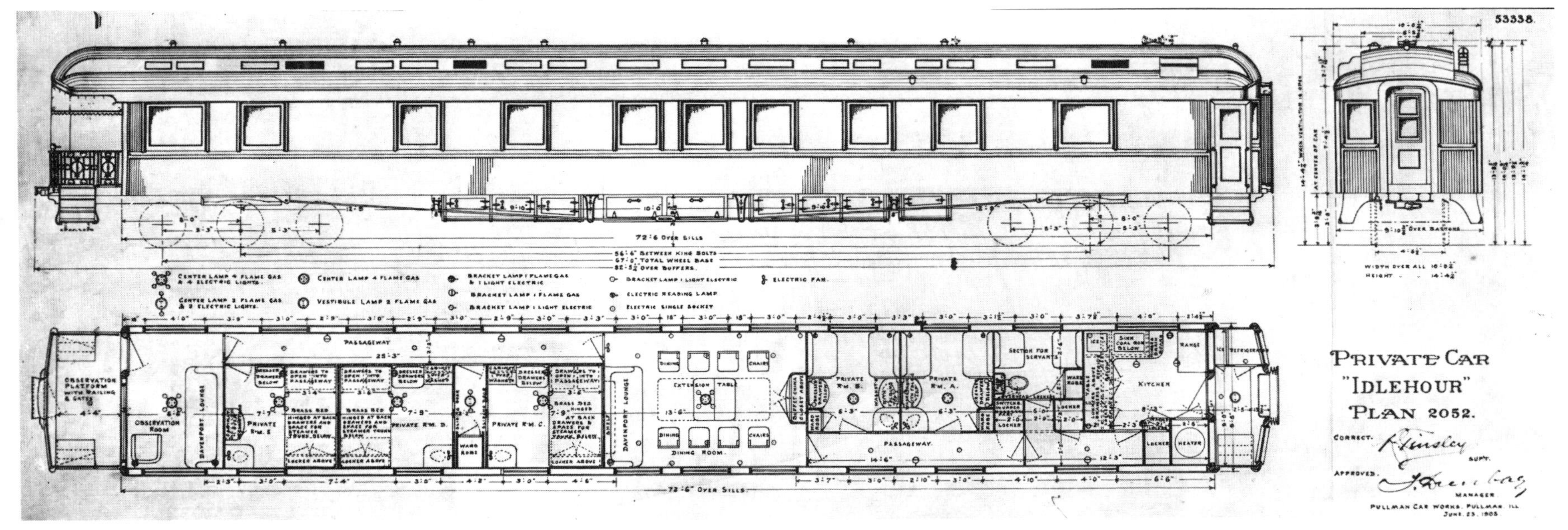

Figure 4.99 *W. K. Vanderbilt's* Idlehour *was a Pullman product of 1905.*

Figure 4.100 *Business cars generally carried low numbers. The Number 100 was completed by Jackson and Sharp in 1911.* (*Hall of Records*)

Figure 4.101 James C. Brady, "Diamond Jim," was a successful railway supply salesman who could afford his own car. This early steel PV dates from 1914. (Pullman Neg. 17014)

Figure 4.102 The Santa Fe purchased some very short 52-foot office cars in 1924 for its division superintendents. (Pullman Neg. 27959)

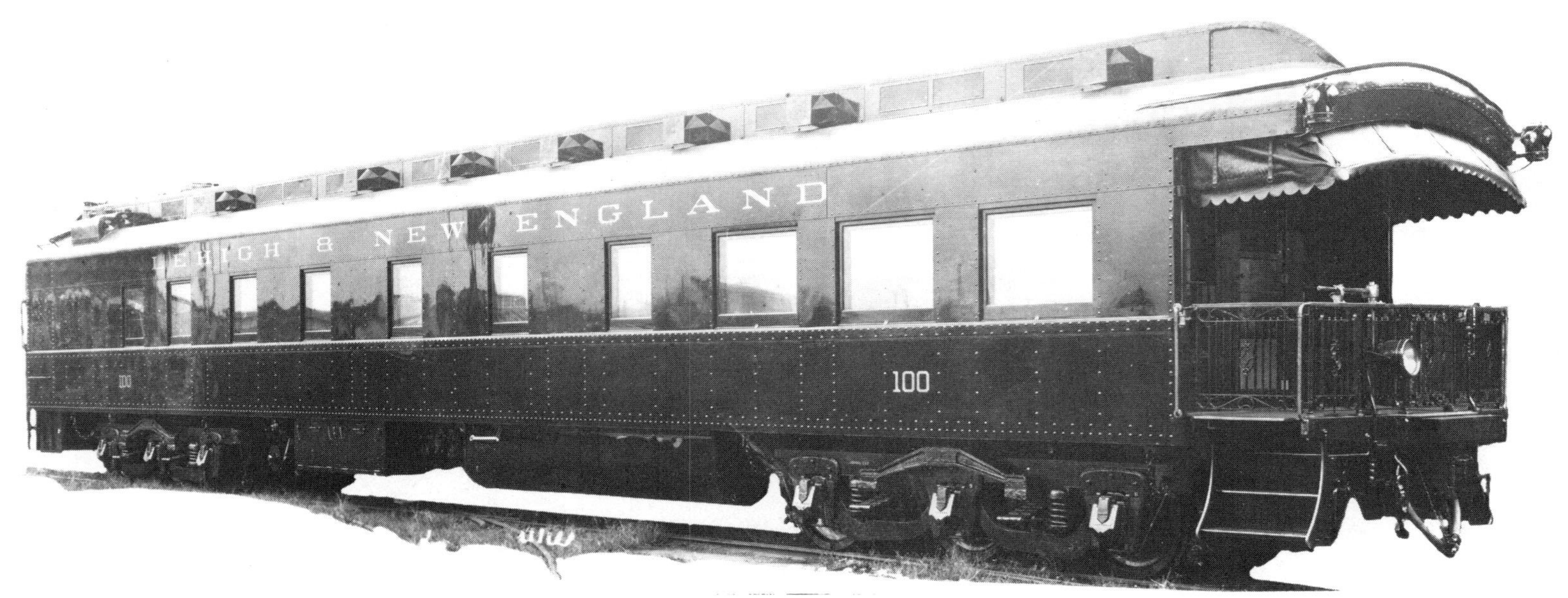

Figure 4.103 Jackson and Sharp was a division of A.C.F. when it produced this office car in September 1925. (Hall of Records)

Figure 4.104 Top officials of major trunk lines suffered few privations even when railroad business kept them away from headquarters.

rent a hotel car, with its crew, for $85 a day.[186] Food would be supplied for an additional $2 a day per passenger. Within two years Pullman was leasing sleeping, drawing-room, and hotel cars. A prospectus dated April 1, 1872, offered the following terms:[187]

For a regular SLEEPING CAR, containing twelve open sections of two double berths each, and two staterooms of two double berths each, (in all twenty-eight berths), with Conductor and Porter, SEVENTY-FIVE DOLLARS per day.

For a DRAWING-ROOM CAR, containing two drawing-rooms, having each a sofa and two large easy chairs by day, and making up at night into two double and two single berths each, three state-rooms, having each two double berths and six open sections of two double berths each, (in all twenty-six berths), with Conductor and Porter, SEVENTY-FIVE DOLLARS per day.

For a HOTEL CAR, containing one drawing-room, as above described, one state-room, having two double berths, and ten open sections of two double berths each, (in all twenty-six berths), and having also, in one end, a kitchen, fully equipped with everything necessary for cooking and serving meals, with Conductor, Cook and two waiters, EIGHTY-FIVE DOLLARS per day.

The Conductor, if desired, will make all arrangements for the excursionists with the railroads for procuring transportation of the car, and in the case of their taking a Hotel Car, will also act as Steward, purchasing for them the requisite provisions for the table.

The number composing the party will, of course, make no difference to us, the car being chartered, with its attendants, at a certain rate PER DAY, from the time it is taken until we receive it back again.

So far as the Railroad Companies are concerned, they require eighteen (18) fares for a car-load. If this number of fares is furnished, there will be no EXTRA charge on the part of the Railroad Companies for hauling the car.

In 1877 Pullman produced the *P.P.C.*, a private car equal to the most deluxe standards of the day, for rental parties. As mentioned earlier, Pullman came to use the *P.P.C.* more for himself and less for rentals as time went on. But it appears to be the first true private car built for the rental market. Around 1880 Pullman added two special cars, the *Davy Crockett* and the *Izaak Walton*, to the fleet for hunting and fishing parties.[188] Sportsmen could hire one of them, plus a cook and a waiter, at the bargain rate of $35 a day. These cars were less highly finished than the typical PV and thus appealed to parties wishing to "rough it" in comfort.

Figure 4.105 The interior of the B & O office car 97 as shown in a photograph of 1930. (The car's dining compartment is shown in Figure 4.104.)

Figure 4.106 Business cars lost their club-like interiors with the coming of the streamline era. This Great Northern office car, the A-28, was completed in 1947. (Pullman Neg. 52052)

They carried eight to ten persons and were outfitted with a kennel for hunting dogs.

By the late 1880s Pullman's rental fleet had grown to nineteen cars—thirteen private cars and the rest hunting and hotel cars.[189] They were available for single trips or extended tours. Pullman offered to stock the cars with provisions and, if desired, to furnish "polite and skilful attendants."

Pullman's rental business was apparently profitable, because the company's competitors began to seek a share of it. Among the first was Jerome Marble (1824–1906) of Worcester, Massachusetts, a banker and man of affairs who returned from a private car hunting trip to the Dakotas convinced that luxury excursion cars would prove a good investment.[190] In 1878 he had Jackson and Sharp remodel the car *Delaware*, which the firm had built two years before for display at the U.S. Centennial Exhibition.[191] The 52-foot body was refitted as a private car, an observation platform was added to one end, and four-wheel trucks were installed in place of the original six-wheel Ashbel Welch–style trucks. Renamed the *City of Worcester*, the transformed coach began making cross-country tours with parties as large as twenty (see Figures 4.108 and 4.109). Marble thought so well of the scheme that he organized the Worcester Excursion Car Company in July 1878.[192] In the following year the *National Car Builder* said that a trip to California could be made cheaper aboard Marble's car than on a first-class train. Fourteen passengers took the *City of Worcester* to the Dakotas—a 78-day, 4,000-mile journey—at a cost of only $203 each.[193]

Marble bought two more cars and in 1883 acquired a fourth

Figure 4.107 As rail executives turned to private airplanes, the business car became something of an anachronism. This late example was produced by Budd in 1957 for the Santa Fe. (Robert Wayner)

Figure 4.108 The Worcester Excursion Car Company rented private cars, mainly for Western travel. The City of Worcester *was originally built in 1876 as a luxury coach but was remodeled as shown here for rental service. (Hall of Records)*

Figure 4.109 The City of Worcester*'s interior as remodeled for rental service. (Arthur D. Dubin)*

car from Jackson and Sharp that he named the *Edwin Forrest*, after the actor.[194] The *Edwin Forrest* featured a central 24-foot-long grand saloon. This car, and perhaps the earlier Worcester cars, used William Paige's patented collapsible lower berths. The metal-framed, canvas-covered beds were put away during the day, leaving the floor clear for easy chairs. Wagner-style folding berths were used overhead. Through these arrangements Worcester hoped to avoid Pullman's patent trust. Another interesting feature of the Worcester cars was a fireproof safe for valuables. The company purchased more cars and by the late 1880s was operating nine, including the luxurious *Railway Age*, which it had leased from the publisher of that journal. Among Worcester's customers were Sarah Bernhardt, Edwin Booth, and Theodore Thomas. Its main business continued to be Western hunting parties. When the American frontier vanished in 1890 the plentiful supply of wild game disappeared with it, and according to Lucius Beebe this was a major reason for the Worcester Excursion Car Company's suspension around 1895.[195]

Worcester was unique in concentrating on private car rentals, but it was not Pullman's sole rental competitor. Mann, Woodruff, and Wagner were all engaged in the rental trade to one degree or another, though only Wagner proved to be much of a rival for Pullman. Wagner was rather late in entering the field; it was after Seward Webb took charge that the firm acquired its first rental car. The *Marquita*, produced at Buffalo in 1887, was soon followed by the *Riva*, the *Grassmere*, and the *Wanderer*. But before Wagner could invade this sphere on a large scale, the Vanderbilts at last accepted Pullman's natural monopoly, and Wagner's rental cars became part of the Pullman fleet after the 1899 consolidation of the two firms.

Pullman was now without rivals, even in the rental field. In 1903 private car leasing was a fast-growing trade; Pullman regularly had twenty-four cars engaged at one time.[196] Wealthy commuters, particularly in the New York and Philadelphia suburbs, often banded together and leased a parlor car. These arrangements were not made exclusively with Pullman. In a few instances the car was jointly owned by the commuters themselves, as when a group of New Orleans cotton brokers purchased the *Club-on-Wheels* in 1886.[197] A similar private parlor-style car was produced in 1893 for a group of Plainfield, New Jersey, businessmen by Harlan and Hollingsworth.[198] The *Plainfield*'s many comforts included lounge chairs, card tables, and a smoking and nonsmoking section.

Pullman rentals rose and fell in concert with the stock market. In 1931, despite the deepening gloom of the Depression, there was enough demand to keep twenty-three cars employed.[199] Daily rates were adjusted for the length of the lease. One- to two-day rentals cost $175 per day, with a graduated scale that declined to $50 a day when ninety or more days were involved. These prices included a cook and two attendants. Food was supplied at cost plus 25 percent. In 1938 Pullman was still operating twenty-three rental cars, but in 1939 the number was cut back to seventeen.[200] It had been a bad year, with financial reverses that prompted many business cutbacks throughout the economy. Pullman's rental division not only reflected the downturn by reducing the number of cars in the fleet, but also added a poor people's private car. This was a restaurant-sleeping car with open-section berths for 16 passengers at one end, a central kitchen, and a parlor-buffet compartment at the opposite end.

World War II put a stop to PV rentals, just as it closed down private car traffic in general. Pullman disbanded this portion of its operations in order to concentrate on more essential classes of passenger transport. Several cars were sold during the war to serve as business cars. The Central of Georgia thus acquired the *Marco Polo*, while the Seaboard Air Line picked up the *Pioneer*. During the same years a car similar to these two was extensively remodeled for President Roosevelt. At the war's end, the Pullman Company was embroiled in its legal contest with the Justice Department (described in Chapter 3) and was apparently too preoccupied to revive its rental car pool.

Presidential Cars

Just how to accommodate the President of the United States was a dilemma in the early days of the republic. In a society self-consciously opposed to privilege, the head of state was in theory no better than any other man, and on a train he might be expected to take his seat among coach passengers. Yet he was a very special citizen holding an office that seemed to call for some mark of distinction when he traveled. Monarchs routinely acquired royal railway carriages with as little fuss as they enlarged their collection of golden state coaches. Pope Pius IX casually accepted an incredibly ornate baroque train from Napoleon III. The nineteenth-century rulers of Mexico showed a special fondness for private rail cars and were inclined to go through one or more sets during their dictatorships. But the U.S President was expected to be more restrained. As temporary head of state, he had the use of a mansion, one or more carriages, and a yacht. Ironically, during the great age of railway travel there was no Presidential private car. If the President wanted one, he had to seek the favor of a railroad executive.

A questionable tradition was thus established that encouraged a measure of obligation toward anyone who provided a car for the President and his party. The practice persisted until 1942, although it seems to have caused little concern among the usually vocal critics of those holding political office. The watchdogs apparently saw nothing amiss when Tom Scott or John Garrett regularly offered the President free use of a private car. Garrett's *Maryland* was probably the most frequently occupied by Chief Executives during the Gilded Age. It was so simple to run the car down from Baltimore, and Garrett was so ready to oblige, that Hayes, Harrison, Cleveland, and McKinley came to think of it as their very own.[201] Cleveland, in fact, borrowed the *Maryland* in 1886 for his honeymoon trip. Pullman, always sensitive to the value of publicity, was delighted to lend the *P.P.C.* to traveling Presidents. Theodore Roosevelt, who came to feel uneasy about this practice and its potential for breeding scandal, was dismayed to learn that during his first term he had unwittingly accepted well over $100,000 in free rail travel.[202] In 1906 Congress added a travel allowance to the White House budget, but it was only for $25,000. Presidents continued to depend upon the generosity of the railroads until the private jet became the Chief Executive's mode of long-distance travel.

Plans for a government-owned private car were made during Lincoln's administration, when the exigencies of the Civil War required him to travel. In May 1862 he went to Fredericksburg, occupying a baggage car for part of the trip.[203] A few weeks later he journeyed to West Point by train. Late in the following year, when it was necessary to make a fast trip to Gettysburg for the battlefield dedication, the B & O lent its directors' car. It seemed clear, however, that a special car should be available to the President. Both his prestige and his security demanded it, and both would best be served by providing a car that was under the control of the War Department.

It is not entirely clear who authorized construction of the Presidential car. The Lincoln scholar Victor Searcher suggests that Secretary of War Edwin M. Stanton initiated the scheme.[204] Stanton was solicitous of the President and took a great interest in his safety and comfort. Another account attributes the idea to B. P. Lamason, superintendent of the U.S. Military Railroad's car shops in Alexandria, Virginia. If it was Lamason's idea, he would surely have needed approval from higher authority before beginning the project. A preliminary search of the U.S.M.R.R. papers at the National Archives failed to yield any information on this subject. Like so many incidents in Lincoln's life, the story of the Presidential car is rich in mythology but short on documented facts. There are, in fact, several foolish tales about the car. The most persistent is that it was armor-plated. In later years, however, two veteran car builders who actually assisted in the car's construction heatedly denied that boiler plate was inserted between the inner walls to form a bulletproof shield. The following narrative is drawn largely from the recollections of these two mechanics, who appear to have recorded the most dependable information available on the Lincoln car.[205]

Construction is said to have begun in November 1863. Lamason based his general design on a Pennsylvania Railroad coach of the period. The car's 42-foot body was divided into three compartments: a parlor and drawing room at either end, with a side aisle and stateroom in the center. The stateroom was planned as the President's compartment. A small washroom was located in the drawing room. The absence of a kitchen and crew's quarters suggests that extended journeys were not envisioned, although the sofas were convertible into berths for overnight travel. In reality the car was outfitted more as an elegant drawing room–parlor than as a self-sufficient private car. The interior woodwork was oak and black walnut. The upper panels were upholstered with a corded crimson silk which had a tufted pattern for added richness. The headcloth, also of crimson silk, was gathered at the center of each panel to form a rosette. The furniture was covered in dark-green plush, and the curtains were light-green silk. The interior of the upper deck, or clerestory, was painted zinc white and decorated with the seals of state in full color.

The exterior was painted a rich chocolate brown, rubbed out to a fine polish with oil, rotten stone, and the bare hand. The national coat of arms was painted on the oval sheet-metal panels. In the window panel directly above, the words "United States" were lettered in gilt. The body and clerestory were highlighted in a restrained manner by fine gold striping. The decorative polished-brass washers for the bolster tie rods are visible below the third window at each end of the car (Figures 4.110 and 4.111).

Lamason's original plan called for conventional four-wheel trucks, but during the course of construction he decided to mount the car on sixteen wheels. He adopted Ambrose Ward's patented bolster. Each truck had a short wheelbase of only 4 feet, 10 inches. There were no equalizers. Broad-tread, 33-inch cast-iron wheels were used so that the car could travel on both standard and Southern gauge tracks. The axle box pedestals were very ornamental; Lamason was said to have devoted weeks to their design.

According to W. H. H. Price, the car was not finished until February 1865 because the Alexandria shops were busy with emergency repairs to keep trains running to the front. Another employee in the U.S.M.R.R. shops claimed that the car was actually finished in May 1864, but that Lincoln then refused to set foot on it. Articles in a hostile New York newspaper had severely criticized the pretentious private car being completed for a head of state who so earnestly proclaimed his homespun, plain-folks origins.[206] The great commoner could not afford to jeopardize his public image, and so the car sat in readiness for its inaugural trip, a trip that was not made by Lincoln during his lifetime.

Figure 4.110 The U.S. Military Railroad shops in Alexandria, Virginia, completed this private car for President Lincoln in 1865. (Library of Congress)

Figure 4.111 Lincoln never rode in the car during his lifetime, but his body was conveyed in it from Washington to Springfield, Illinois. Afterwards it was sold to the Union Pacific Railroad. (Library of Congress)

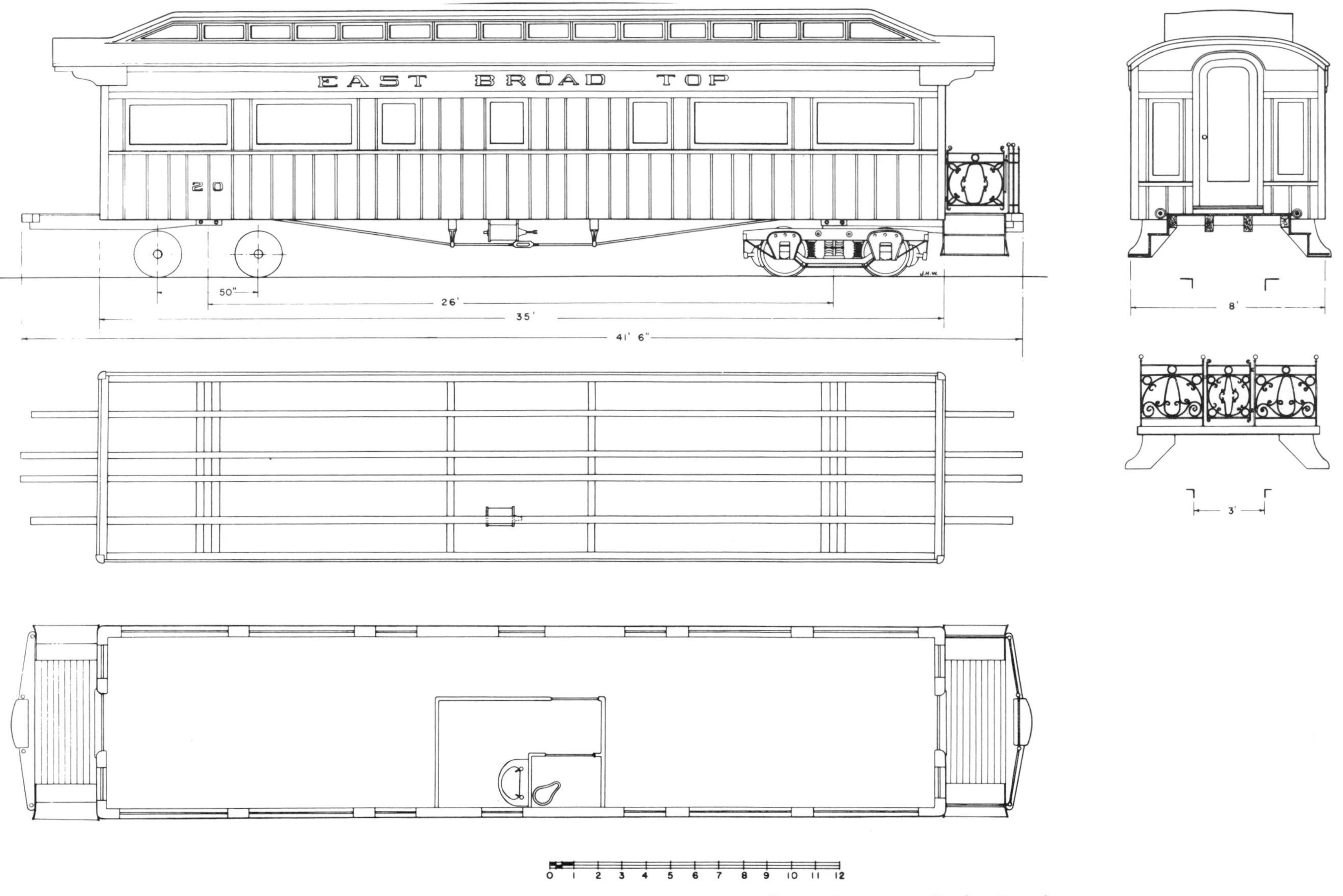

Figure 4.112 This narrow-gauge car was reportedly used occasionally by President Cleveland. It was built in 1880 by Billmeyer and Small of York, Pennsylvania, as a coach for the Bradford, Bordell, and Kinzua Railroad. It was sold in 1907 to the East Broad Top Railroad. (Edmund Collins, Jr.)

After his assassination on April 15, 1865, his body was carried to Springfield, Illinois, aboard this car. The 1,700-mile journey, which meandered among the major Northeastern cities, took thirteen days. After the funeral the car was returned to Washington; late in June it carried Mrs. W. H. Seward's body to Auburn, New York.

Now that the war was over, the military railway began to wind up its affairs. Among the surplus rolling stock put on the block was the Presidential car. Even while the funeral train was on its somber journey an advertisement appeared in a Washington newspaper offering to sell the car to "some enterprising man."[207] A rival paper characterized the scheme as "humiliating," but crass or not, the car was put up for sale. One account claims that Lincoln's old law partner bought it to keep it out of the hands of exhibitors.[208] Others say that it was sold at auction, after spirited bidding, to T. C. Durant, vice president of the Union Pacific.

Whatever the circumstances of the sale, it is certain that the car came into the possession of the Union Pacific sometime in 1866.[209] Durant was proud of his new acquisition and enjoyed carrying excursion parties in it over the U P tracks, which were then being laid. The Lincoln car formed part of the special train that ran in October 1866 to celebrate completion of the line to the one-hundredth meridian. The car's glamour, however, could not really make up for its notoriously rough ride, ascribed by some critics to its cumbersome sixteen-wheel undercarriage by others to its broad-tread wheels. Durant abandoned the Presidential car late in the same year for the greater comfort of a Pullman.[210] During the next three years it served as an office car for government inspectors and various railroad officials. It was then converted into an emigrant car. Some of the elegant interior furnishings were retained by the Union Pacific's president, Sidney Dillon.[211] Later it was sold to the Colorado Central, but ultimately it returned to U P ownership, for the Union Pacific took over the Colorado line.

By 1886 the car was stationed on the Marysville branch of the U P—"a relic of faded gentility."[212] Now it was fitted with rough bunks to shelter section hands; few would have recognized the former palace on rails. And Lincoln's ghost could settle comfortably into the shabby interior with no fear of censure, for it was as humble as the meanest cabin car. It was stored in the North Platte yard by the early 1890s, a weather-beaten hulk with its windows boarded over.[213] There was talk of restoring it for the 1893 Columbian Exposition, but the necessary $5,000 could not be found. After another five years of neglect it was refurbished for exhibit in Omaha at the Trans-Mississippi Exposition, where souvenir hunters picked at its rotting flanks. The remains

Figure 4.113 *The* Ferdinand Magellan *was a private rental car placed in service in 1928. It was remodeled for President Roosevelt in 1942. (Pullman Neg. 33771)*

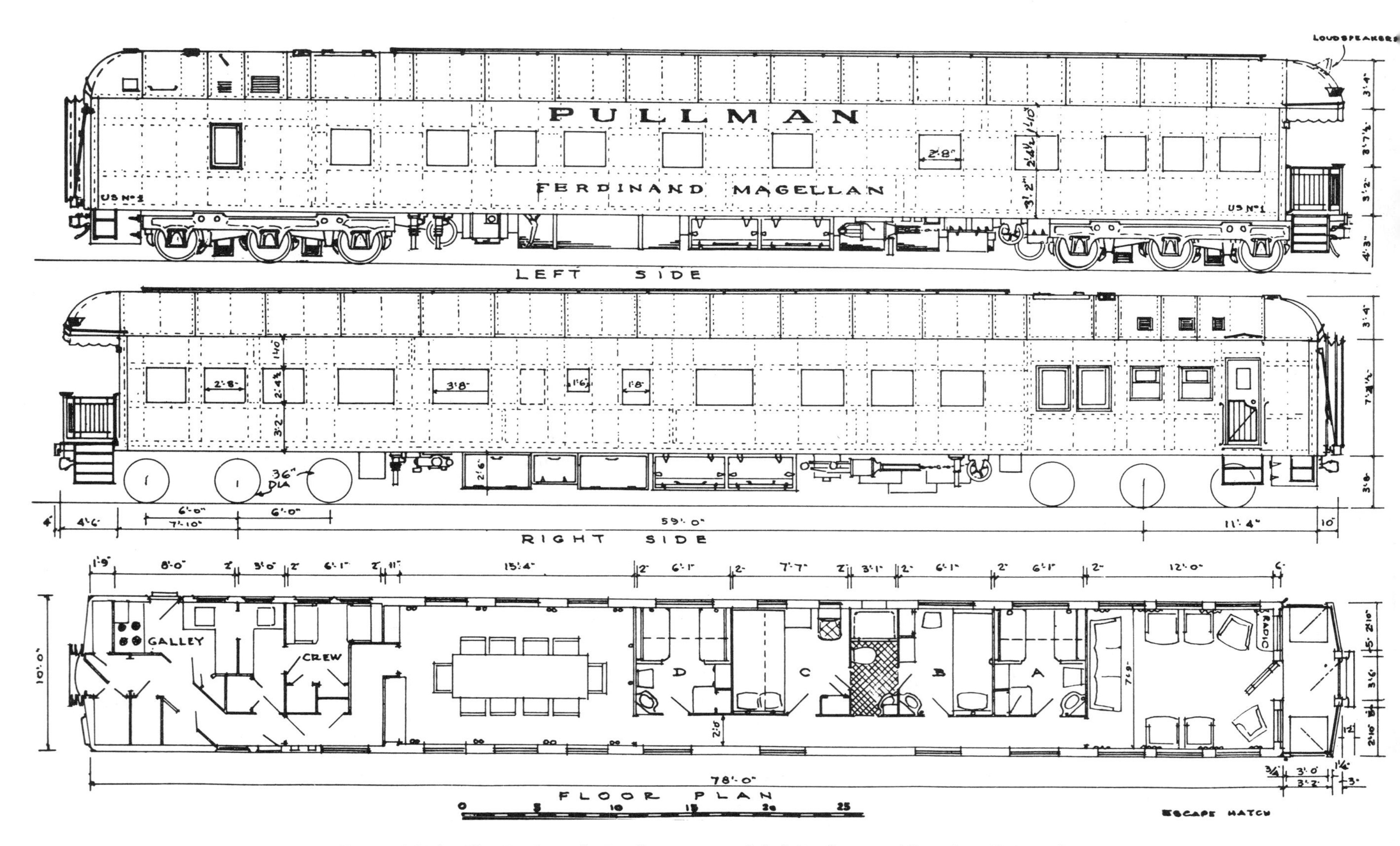

Figure 4.114 *The* Ferdinand Magellan, *as remodeled for the use of President Roosevelt. Presidents Truman and Eisenhower also traveled on it. (National Model Railroad Association)*

were again placed in storage. In 1903 came the final humiliation: a showman named Franklyn B. Snow decided that "the most sacred relic in the United States" should not be hidden from the American public.[214] He persuaded the Union Pacific to sell the car and arranged to exhibit it at the 1904 St. Louis World's Fair. The cross-country carnival following the close of the fair so horrified the growing band of Lincoln admirers that a group was organized to take the relic away from Snow. In the fall of 1905 Thomas Lowry, president of the Soo Line, purchased the car with the intention of placing it in a museum.[215] He presented it to the Minneapolis Park Board, which found that it could not furnish the imposing museum Lowry had envisioned. The board erected a temporary wooden shed for it in a park in the northeastern section of the city. In March 1911 while plans were underway to move the car to a state museum, a group of boys built a bonfire in a grassy field surrounding the shelter.[216] The fresh spring wind spread the flames to the car, and the damage was so severe that the remains were broken up and given away for souvenirs. From its initial rejection to its fiery cremation, the proud creation of the Alexandria shops had suffered a series of degradations. Today a handful of relics preserved at the Union Pacific Museum in Omaha is all that remains of the historic Lincoln car.

The Lincoln car was already a venerated relic before another effort was organized to provide the President with his own car (see Figure 4.112). It galled some railroad men that the Chief Executive should have a purely ceremonial conveyance like a yacht while he lacked the practical benefits of a railroad car. In February 1897 the *Railroad Car Journal* announced a scheme to put matters right.[217] The editor had formed a committee of prominent master car builders headed by Fitch D. Adams and was busy persuading the supply industry to donate the necessary materials. By April the committee had completed its preliminary design for a married pair of eight-wheel cars. The matter was discussed in the editorial pages at some length, but not with such gravity that it did not print several humorous designs incorporating the outlines of the capitol and the White House. The supply trade showed little enthusiasm, however, and the Master Car Builders Association refused to endorse the plan. At this point the project was tabled.

A special car for the President's use became an issue again during World War II.[218] Since the dangers of sabotage and assassination were more acute than an peacetime, there was no opposition in 1942 to the idea that the President should have a secure car to travel in. The Pullman Company had been chauffeuring Franklin Roosevelt even before he became President, for in his years as governor of New York he often used Pullman's rental car, the *Marco Polo*. It was coincidental that the Association of American Railroads, in cooperation with Pullman, selected a sister car, the *Ferdinand Magellan*, for remodeling into a mobile security vault (Figure 4.113). Protection against attack was the first priority; the car was not notably luxurious, and its exterior was far from handsome. The *Magellan* was one of six rental cars built in 1928 at Pullman's Chicago shops. The remodeling was carried out at the nearby Calumet repair shops in 1942. One of the five bedrooms was eliminated so that the observation and dining rooms could be enlarged. The roof and sides were sheathed in ⅝-inch-thick armor plate, and 3-inch-thick bulletproof glass was installed in the windows. The rear platform door was like a bank vault's; it weighed 1,800 pounds. The roof was redone as an ungainly turtleback casement. The undercarriage became a clutter of boxes, pumps, and extra mechanical gear. The total effect was that of a monstrous caterpillar. The swollen giant's weight of 142½ tons made it the heaviest passenger car ever to operate in this country. One engineer said that it "drags like a brickbat on the tail of a kite.[219]

Roosevelt made many journeys in the *Magellan* after its metamorphosis was completed in December 1942 (Figure 4.114). He traveled 50,000 miles aboard it in the following two years. After his death in April 1945, the Presidential car was part of the funeral train. However, it did not carry the President's body, as is often assumed.

In 1946 the A.A.R. sold the car to the U.S. government for the sum of one dollar. No figures have ever been released on the cost of the remodeling. President Truman was a devoted railroad traveler and ran the car throughout the United States during his tumultuous election campaign of 1948. During this time the car lost its identity. In the war years it had been lettered "Pullman" in an effort to avert suspicion that it was something special, but it had carried no name or number. After the war even the Pullman lettering disappeared, and the only marking that distinguished the car was a dark-bronze Presidential seal affixed to the observation railing.

Truman's successor showed less interest in rail travel and rarely used the car. It was Mrs. Eisenhower, in fact, who made the last official trip in it in 1954. Four years later it was declared surplus property. Early in 1959 it was given to the Florida Development Commission, which in turn presented it to the University of Miami.[220] The university in turn passed the car on to a group of railroad enthusiasts who operate the Gold Coast Railroad in Fort Lauderdale, Florida. Today it is tourists and not Presidents who tread its thickly carpeted corridors.

NOTES

CHAPTER ONE

1. Edward Hungerford, *Daniel Willard Rides the Line,* New York: Putnam, 1938, p. 240.

2. The *Pioneer* has been pictured in various accounts as a simple clapboard box-body with three windows. This rendering arose from an artist's erroneous reconstruction, which was based on a sketch in Smile's biography of George Stephenson that supposedly showed the first passenger car on the Stockton and Darlington Railroad. And that sketch is itself considered purely imaginary by British authorities.

3. August Mencken, *The Railroad Passenger Car,* Baltimore: Johns Hopkins Press, 1957, hereafter referred to as Mencken. A notice from the *Baltimore Gazette* of Aug. 14, 1830, is given on p. 10. A nearly identical account from the *Baltimore American* is given in vol. I of Edward Hungerford's *The Story of the Baltimore and Ohio Railroad,* New York: Putnam, 1928.

4. Minutes of Newcastle and Frenchtown Railroad, Nov. 12, 1831. The original book was in the possession of the late C. L. Winey; its present location is unknown to this writer.

5. *American Railroad Journal,* March 3, 1832, p. 149. Hereafter cited as *A.R.R.J.*

6. Ibid., April 28, 1832, p. 288, and May 12, 1832, p. 307. Lithographic views showing Imlay-style cars on the P G & N and the Philadelphia and Columbia Railroad are given in Figs. 238 and 241 of Seymour Dunbar's *A History of Travel in America,* vol. III, Indianapolis: Bobbs-Merrill, 1946.

7. *Scribner's Magazine,* August 1888, p. 197.

8. In his two-volume study of American railroads (*Die innern communication der vereinigten Staaten von Nord-America,* Vienna, 1842–1843), F. A. Ritter Von Gerstner consistently described three-section compartment cars whenever he discussed four-wheel passenger cars in enough detail to indicate the body style.

9. *Poulson's American Advertizer,* Nov. 11, 1831.

10. E. Harper Charlton, *The Street Railways of New Orleans,* vol. 13, no. 1, Los Angeles: Interurbans, 1955, p. 250.

11. *A.R.R.J.,* Nov. 17, 1832, p. 737.

12. This drawing was probably first used in the 1882 annual report of the Lake Shore and Michigan Southern Railroad. The report contained a short history of its predecessor line, the Erie and Kalamazoo.

13. Alvin F. Harlow, *Road of the Century,* New York: Creative Age Press, 1947, p. 17.

14. Frank W. Stevens, *The Beginnings of the New York Central Railroad,* New York: Putnam, 1926, p. 305.

15. Harlow, *Road of the Century,* p. 56.

16. *National Car Builder,* December 1870, p. 5.

17. *Railway Patent Claim Reporter,* 1862, p. 262.

18. Dunbar, *A History of Travel in America,* vol. 3, p. 1006, gives an account of travel on the Boston and Providence Railroad in 1835.

19. *A.R.R.J.,* Sept. 28, 1850, p. 609.

20. L. T. C. Rolt, *The Railway Revolution: George and Robert Stephenson,* London: Longmans, 1960, p. 46.

21. C. F. D. Marshall, *Centenary History of the Liverpool and Manchester Railway,* London: Locomotive Publishing Company, 1930, p. 59.

22. Alvin Harlow, *Steelways of New England,* New York: Creative Age Press, 1946, pp. 56–57, and plate opposite p. 274.

23. William Whiting, *Arguments of William Whiting, Esq., in the case of Ross Winans vs. Orsamus Eaton et al.,* Boston, 1853, p. 129. Hereafter cited as *Whiting, Winans vs. Eaton.*

24. *A.R.R.J.,* Dec. 8, 1855, p. 773. Imlay made his proposal to the directors of the Newcastle and Frenchtown.

25. *Ross Winans vs. The Eastern Railroad Company,* U.S. Circuit Court, Massachusetts District, October 1853 term (Boston 1854). This 1299-page transcript of the eight-wheel-car patent suit is the basic source for the following discussion of the early B & O eight-wheel cars. Hereafter referred to as *Winans vs. Eastern Railroad.*

26. B. & O annual report, October 1834, p. 32, lists the *Winchester, Dromedary,* and *Comet* as having been constructed during that year. But both Gatch and Winans said that they were built a year or two earlier.

27. Von Gerstner, vol. 1, p. 36.

28. B & O annual report, October 1834, p. 11.

29. John T. Scharf, *The History of Philadelphia,* vol. 3, Philadelphia, 1884, pp. 2259–2260.

30. In *Winans vs. Eastern Railroad,* p. 740, the model is credited to Proctor (1833). In *Whiting, Winans vs. Eaton,* p. 82, it is credited to Fultz (1829).

31. Minute Books, Philadelphia, Germantown, and Norristown Railroad, October 14, 1835, and March 29, 1837. These books are in the possession of the Reading Company, Philadelphia.

32. *American Society of Civil Engineers Transactions,* 1878, p. 206. Here W. M. Roberts reminisces about early engineering.

33. Von Gerstner, vol. 1, p. 129.

34. *Locomotive Engineering,* February 1895, p. 92. Walter De Sanno, son of the P & C master mechanic, recalled the use of drop-center cars by Hall & Company, one of the many private operators on the line in the 1830s.

35. M. W. Baldwin Letters, Historical Society of Pennsylvania, Philadelphia.

36. *Winans vs. Eastern R.R.,* pp. 1242–1243.

37. *Railroad Gazette,* April 4, 1874, p. 124. Hereafter cited as *R.R.G.*

38. *A.R.R.J.,* Nov. 11, 1837, p. 629.

39. South Carolina annual reports, 1839–1842. S. M. Derrick, *The Centennial History of the South Carolina Railroad,* Columbia, S.C.: The State Company, 1930, provided information on the hogshead cars.

40. *A.R.R.J.,* April 1, 1840, pp. 209–210.

41. Von Gerstner, vol. 1, p. 265.

42. The information on the Diamond cars is from Stephenson's own testimony in the printed record of *Pullman vs. Wagner,* 1881.

43. *Rochester Daily Democrat,* Aug. 28, 1839. Richard F. Palmer kindly provided this reference.

44. Dunbar, *A History of Travel in America,* vol. 3, p. 858.

45. The two-page 1839 prospectus is reproduced fully in H. T. Gause, *The Semi Centennial Memoir of the Harlan and Hollingsworth Company,* Wilmington, Del., 1886.

46. *A.R.R.J.,* Nov. 15, 1838, pp. 297–299.

47. *Winans vs. Eastern R.R.,* p. 897.

48. Letter from Ross Winans to Hartford and New Haven Railroad, Jan. 30, 1836. Maryland Historical Society, Baltimore.

49. *Winans vs. Eastern R.R.,* p. 1001.

50. Ibid., pp. 897 and 1005.

51. Ibid., p. 996.

52. *A.R.R.J.,* Sept. 28, 1850, p. 609.

53. The information for the eight-wheel car case was taken from the following sources: *Winans vs. Eastern R.R.,* 1854; *Whiting, Winans vs. Eaton,* 1853; William Whiting, *Twenty Years' War against the Railroads,* Boston, 1860; and mss. transcripts of Winans vs. the New York and Harlem R.R. (U.S. District Court, Southern New York, 1855), and Winans vs. the New York and Erie R.R. (U.S. Supreme Court, 1858): both transcripts are in the U.S. National Archives, Washington, D.C.

54. D. B. Eaton, *Ross Winans against the New York and Erie Railroad Co.; General Statement of Facts,* New York, 1856, pp. 6–7.

55. *Railway Review,* June 9, 1883, p. 318.

56. *R.R.G.,* May 19, 1893, p. 369.

57. *A.R.R.J.,* Oct. 9, 1845, pp. 647–648.

58. *Railroad Car Journal,* August 1894, p. 167. Hereafter cited as *R.R.C.J.*

59. Ibid., January 1894, p. 2.

60. Railway and Locomotive Historical Society, *Bulletin 65,* 1945, pp. 61–63.

61. *National Car Builder,* February and June 1883, pp. 16 and 65.

62. *Locomotive Engineering,* June 1904, p. 248.

63. *American Railway Times,* Jan. 19, 1862, p. 25.

64. *A.R.R.J.,* May 14, 1853, p. 307.

65. *Railroad Advocate,* Aug. 29, 1857.

66. A. Bendel, *Aufsatze Eisenbahnwesen in Nord-Amerika,* Berlin, 1862, p. 36. Hereafter cited as Bendel.

67. Railway and Locomotive Historical Society, *Bulletin 6,* 1923, p. 81. Recollections of J. H. French, a former officer of the Old Colony; reprinted from 1900 New England Railroad Club reports.

68. A history of the Reading was published serially in an employees' magazine, *The Pilot,* between 1909 and 1914. The articles, prepared by J. V. Hare, have been reprinted in book form in recent years (published by John H. Stock, Philadelphia, 1970c.).

69. Hamilton Ellis, *Railway Carriages in the British Isles from 1830 to 1914,* London: George Allen & Unwin, 1965, p. 25.

70. *American Railway Review*, May 31, 1860, p. 328.
71. Ibid., Aug. 9, 1860, p. 71.
72. *American Railway Times*, July 20, 1867, p. 230. Hereafter cited as *A.R.T.*
73. Ibid., January 1861, p. 25.
74. *A.R.R.J.*, May 7, 1864, p. 459, and Jan. 7, 1865, p. 28.
75. *R.R.G.*, April 1887, p. 54.
76. Ibid., Sept. 4, 1875, p. 367, shows a diagram for an X-brace clerestory roof. A similar sketch is given in the 1879 *Car Builders' Dictionary*.
77. The literature of the narrow-gauge railway is large and growing. The best general survey is Howard Fleming's pioneer study, *Narrow Gauge Railways in America*, New York, 1875 (reprinted in 1949). A chapter is devoted to cars. A few of the more recent studies, largely pictorial, that have been published by enthusiasts are cited in the following footnotes.
78. *Cincinnati Daily Gazette*, March 16, 1876, in speaking of the College Hill Railroad.
79. *R.R.G.*, July 20, 1871, p. 201.
80. Fleming, p. 52.
81. *N.C.B.*, November 1875, p. 165.
82. A history of several Carter narrow-gauge cars, together with working drawings, is given in Bruce Mac Gregor's *South Pacific Coast*, Berkeley, Calif.: Howell North, 1968.
83. Two book-length studies of 2-foot-gauge railroads have been written: H. T. Crittenden, *Maine Scenic Route*, Parsons, W. Va.: McClain Printing Company, 1966; and Linwood W. Moody, *The Maine Two-Footers*, Berkeley, Calif.: Howell North, 1959. Crittenden's work is the most valuable to those seeking information on cars because of the drawings included in the appendix.
84. *R.R.G.*, Aug. 4, 1882, pp. 472–473, provided a detailed list of the weights of all materials for a B & A coach.
85. *R.R.C.J.*, December 1898, p. 371.
86. *R.R.G.*, Aug. 3, 1872, p. 338.
87. Ibid., Aug. 17, 1872, p. 361.
88. *Bulletin of the International Railway Congress*, 1904, pp. 1067–1070.
89. *Franklin Institute Journal*, January 1879, p. 62.
90. Ezra Miller suggested this reason for dropped platforms in a prospectus of 1873.
91. *Master Car Builders Reports*, 1884, p. 125.
92. *N.C.B.*, July 1872, p. 8.
93. *R.R.C.J.*, May 1893, p. 195.
94. *A.R.R.J.*, August 1901, pp. 242–243, and *M.C.B.*, 1902, pp. 137–140.
95. *M.C.B.*, 1887, p. 224.
96. *Rly. Age*, July 28, 1881, p. 434.
97. "The Forest and Railroad," *New England Railroad Club*, November 1897, pp. 30–58.
98. *Railway Mechanical Engineer*, February 1919, pp. 85–88.
99. *R.R.G.*, Aug. 5, 1881, p. 431.
100. *R.R.C.J.*, January 1899, p. 1.
101. *Engineering News*, Dec. 14, 1893, p. 471.
102. The Chalender truss is described in the following sources: *R.R.G.*, June 22, 1877, p. 282; *M.C.B.*, 1885, pp. 74–75; *R.R.C.J.*, October 1894, p. 207.
103. *R.R.G.*, Feb. 14, 1890, p. 107.
104. *R.R.C.J.*, May 1892, p. 131.
105. Patent No. 562,343, June 16, 1896.
106. *Rly. Rev.*, Dec. 6, 1890, p. 733.
107. *M.C.B.*, 1895, pp. 258–265, and 1896, pp. 277–281.
108. *Rly. Age*, Feb. 5, 1897, p. 112.
109. *R.R.C.J.*, May 1899, p. 118. Drawings for the Bissell platform are reproduced in the 1906 *Car Builders' Dictionary*, figs. 1931–1940.
110. Statistics for car orders are taken from *Railway Age Gazette* statistical reports for the year noted. Figures on cars in service are from the I.C.C. annual reports.
111. *Rly. Rev.*, May 6, 1893, p. 274.
112. Stevens, *Beginnings of the New York Central*, p. 305.
113. Von Gerstner, vol. 1, p. 88.
114. *Railway and Locomotive Engineering*, May 1902, p. 225.
115. Von Gerstner, vol. 2, p. 87.
116. Ibid. Von Gerstner also said that an iron band was "fastened" near the flange to prevent shattering should the wheels crack. This sounds like Davis's patented wheel, where a wrought-iron rod was cast integrally with the thread.
117. *Winans vs. Eastern Railroad*, p. 1005. Griggs made a similar statement in *Whiting, Winans vs. Eaton*.
118. Von Gerstner, vol. 1, p. 36.
119. Karl Von Ghega, *Die Baltimore-Ohio Eisenbahn*, Vienna, 1844.
120. From a letter of Oct. 9, 1843, by Eaton and Gilbert to M. W. Baldwin. The Baldwin papers, Historical Society of Pennsylvania, in Philadelphia.
121. The perspective exterior view is from *A.R.R.J.*, Aug. 8, 1846, p. 511. The same engraving appeared in the Aug. 25, 1845, issue of *Scientific American*. The mechanical drawing is from *A.R.R.J.*, Aug. 7, 1845, p. 368.
122. *A.R.R.J.*, June 15, 1842, pp. 380–382.
123. *Railroad Advocate*, Nov. 15, 1856, p. 2.
124. These drawings are from Douglas Galton, *Report to the Lords of the Committee of the Privy Council for Trade and Foreign Plantations, on the Railways of the United States*, London: Eyre and Spottiswoode, 1857–1858.
125. *Railroad Advocate*, June 2, 1855, p. 4.
126. Ibid., July 19, 1856, p. 1.
127. *N.C.B.*, February 1873, p. 35.
128. The description and drawing are from Gustavus Weissenborn's *American Locomotive Engineering and Railway Mechanism*, New York. The first section of this serialized work was published in 1871. The car description (pp. 189–194) was printed in about 1874.
129. Roy B. White, former president of the B & O, was head of the C R R N J when the old cars were sent to the fair. Around 1954 he confirmed their origins to L. W. Sagle, curator of the B & O Transportation Museum, who was suspicious of their pedigree as original B & O equipment. The cars were formally deeded to the B & O in January 1928.
130. *N.C.B.*, March 1884, p. 28.
131. *N.C.B.*, July 1884, pp. 86–87.
132. *A.R.R.J.*, April 1899, p. 114.
133. *N.C.B.*, August 1881, p. 90; *R.R.G.*, Aug. 4, 1882, p. 472; *N.C.B.*, April 1884, p. 40.
134. *A.R.R.J.*, October 1891, pp. 441–442; a St. Charles standard coach is described in *N.C.B.*, February 1891, p. 22.
135. *R.R.G.*, May 19, 1893, p. 370.
136. The drawings and data are from *Engineering News*, Dec. 14, 1893, pp. 470–471.
137. The drawings and description are taken from *Railroad Gazette*, Aug. 16, 1894, p. 193, and *Railway Master Mechanics Magazine*, March 1894, p. 46.
138. The facts presented here were furnished by the late Charles E. Fisher, longtime president of the Railway and Locomotive Historical Society, and by an article on the wooden passenger cars of the New Haven in *Bulletin 103*, October 1960, of the society.
139. More information on the passenger cars of this company is given in John A. Rehor, *The Nickel Plate Story*, Milwaukee: Kalmbach, 1965.

CHAPTER TWO

1. Hamilton Ellis, *Railway Carriages in the British Isles from 1830 to 1914*, London: Allen & Unwin, 1965, pp. 35 and 37.
2. Ibid. Also *American Railroad Journal*, Oct. 20, 1855, p. 665.
3. *Scientific American*, March 19, 1846. The drawing accompanying this article is reproduced in August Mencken, *The Railroad Passenger Car*, Baltimore: Johns Hopkins Press, 1957, p. 36. An early handbill describing Lewis's car is in the collections of the New Jersey Historical Society, Newark.
4. *Scientific American*, Aug. 23, 1851, p. 388.
5. Information about La Mothe is drawn primarily from Walter A. Lucas, "The First Iron Passenger Cars," *Bulletin*, Passaic County Historical Society, vol. 5, no. 5, October 1961, and from the following leaflets:

La Mothe's Patent Life-Preserving Car, New York, 1855, Landauer Collection, New York Historical Society.

How to Make Railroads Pay . . . , New York, 1856, Cincinnati Historical Society.

La Mothe's Patent Iron Railroad Cars, New York, 1859, Roebling Collection, Rensselaer Polytechnic Institute, Troy, N.Y.

The last prospectus includes reprints from *American Railroad Journal*, *Scientific American*, and various city newspapers. Other sources are noted individually.

6. *A.R.R.J.*, March 31, 1855, p. 201, and Oct. 22, 1853, p. 679.
7. See note 5.
8. *A.R.R.J.*, March 31, 1855, p. 201.
9. *Railroad Advocate*, Sept. 20, 1856, p. 2.
10. *American Railway Times*, April 11, 1868, pp. 118–119.
11. *American Railway Review*, Sept. 8, 1859.

12. *A.R.R.J.*, March 3, 1860, p. 190.
13. *National Car Builder*, August 1870, p. 3.
14. *Railroad Gazette*, Aug. 8, 1879, pp. 464–465.
15. Correspondence between D. C. McCallum and the La Mothe Iron Car Company is among the U.S. Military Railroad papers in the National Archives, Washington, D.C. Refer to record group 92.
16. *A.R.R.J.*, July 4 and 11, 1863, pp. 625 and 645; *Western Railroad Gazette*, Aug. 8, 1863, p. 2.
17. *Western Railroad Gazette*, Oct. 27, 1860, p. 2.
18. Annual report, Pittsburgh, Fort Wayne, and Chicago Railroad, 1862.
19. *Amer. Rly. Rev.*, Nov. 15, 1860, p. 293.
20. Annual report, P F W & C, 1863.
21. *Locomotive Engineering*, Dec. 1898, p. 568.
22. George B. Abdill's *Civil War Railroads*, Seattle: Superior, 1961, is a convenient source for these views.
23. *Amer. Rly. Rev.*, March 28, 1861, p. 182.
24. The Roebling papers are housed at Rensselaer Polytechnic Institute, Troy, N.Y.
25. *Railroad Record*, May 18, 1865, p. 161; *A.R.T.*, June 3, 1865, p. 174.
26. *N.C.B.*, August 1892, p. 127.
27. *R.R.G.*, April 20, 1877, p. 181.
28. *Engineering*, April 22, 1870, p. 275.
29. *N.C.B.*, August 1870, p. 3, and June 1874, p. 88.
30. *N.C.B.*, July 1884, p. 90, and September 1884, p. 115.
31. *N.C.B.*, July 1888, p. 104, and *R.R.G.*, Jan. 16, 1891, p. 16.
32. *Engineering News*, Aug. 25, 1888, p. 139; see also *National Cyclopedia of American Biography*, vol. 3, 1893, pp. 68–69.
33. Ibid., vol. 13, p. 508, and trade papers listed in the following notes. Perky organized the 1882 Denver Mineral Exposition and the St. Joseph Cereal Exposition, and he established the National Food Company. His erratic career continued with the founding of a girls' school.
34. Anderson's career is summarized in *A.R.R.J.*, June 1889, p. 293, and in many biographical dictionaries.
35. *Engineering News*, March 26, 1887, p. 203, and *N.C.B.*, January 1888, p. 15.
36. *N.C.B.*, August 1889, p. 113.
37. *Engineering News*, Aug. 25, 1888, p. 139.
38. C. L. Rutt (Ed.), *History of Buchanan County and St. Joseph*, (Chicago, 1904), pp. 232–233 and 250.
39. *Railway Review*, Jan. 17, 1891, p. 44; also *The Zephyr* of February 1939, a magazine published by the Burlington Railroad, which reported on documents found in St. Joseph pursuant to the demolition of the old Federal Building. The location of these papers is now unknown.
40. This description of the final years of the Robbins car is taken from correspondence in 1948 between J. P. Roberts of the National Museum of Transport (St. Louis) and Leo Blondin's widow. Mrs. Blondin sold the museum a German silver lamp from the car. The *Washington Post* of Jan. 23, 1916, also contains useful information.
41. *Railway Age*, July 12, 1890, p. 493.
42. C. B. Hutchins, Dec. 16, 1890, No. 442,894; G. M. Bird, Jan. 24, 1893, No. 490,469.
43. *N.C.B.*, June 1887, p. 80.
44. Prospectus of The American Fire Proof Steel Car Company, 1890c., in the Bureau of Railway Economics Library, Association of American Railroads, Washington, D.C. Other information on the Green and Murison car is from *R.R.G.*, Jan. 4, 1889, pp. 7–8 and 12; *Rly. Rev.*, Feb. 1, 1890, p. 60.
45. *R.R.G.*, Feb. 28, 1890, p. 148.
46. *Rly. Age*, Sept. 13, 1901, p. 232.
47. *A.R.R.J.*, May 1905, p. 152.
48. Abundant published material is available on Gibbs and the early steel cars. The railroad trade press followed these events closely. A summary review is offered in the *American Society of Mechanical Engineers Transactions*, vol. 35, 1913, pp. 93–96. A detailed personal recollection was given in a May 23, 1929, news release by the Pennsylvania Railroad. See also Gibbs's obituary in the *American Society of Civil Engineers Transactions*, vol. 105, 1940, pp. 1840–1844.
49. *R.R.G.*, Aug. 21, 1903, p. 598.
50. Ibid., Sept. 30, 1904, pp. 383–386, offers a detailed description and drawings of the Gibbs subway cars.
51. Ibid., April 7, 1905, pp. 318–319 and 329.
52. *R.R.G.*, June 26, 1903, p. 452.
53. The trade press was generous to the Illinois Central's side-door cars; see particularly *R.R.G.*, Sept. 4, 1903, p. 630. See also *Master Car Builders Reports*, 1904, p. 207.
54. The three Erie steel cars are described in *M.C.B.*, 1908, pp. 307–308 and 322.
55. *M.C.B.*, 1904, pp. 195–216.
56. *R.R.G.*, Nov. 1, 1907, pp. 530–531.
57. *R.R.G.*, June 22, 1906, p. 686, and *M.C.B.*, 1908, pp. 309–310.
58. *International Railway Congress Bulletin*, vols. 26 and 27, 1912–1913, a five-part series: "The Construction of Iron Passenger Cars on the Railways in the United States of America" by F. Gutbrod. Hereafter referred to as Gutbrod Report.
59. *Western Railroader*, March 1966, pp. 3–9; *A.R.R.J.*, January 1907, pp. 6–8; Gutbrod Report, part 2, pp. 1264–1276.
60. *Railway Age Gazette*, June 12, 1908, pp. 83–84.
61. *A.R.R.J.*, June 1907, p. 232.
62. Information supplied by John S. Fair from Pennsylvania RR mechanical records.
63. *R.R. Age Gazette*, Feb. 18, 1910, p. 367.
64. "The Passing of the Wooden Car from This Railroad," a pamphlet issued by the P R R in June 1928.
65. *A.R.R.J.*, December 1910, p. 489.
66. *Safety on the Railroads: Hearings before a Subcommittee of the Committee on Interstate and Foreign Commerce of the House of Representatives Sixty-third Congress on Bills Relative to Safety on Railroads*, Washington: Government Printing Office, 1914. Hereafter referred to as *Safety on Railroads.*
67. *Railway Mechanical Engineer* (formerly *A.R.R.J.*), February 1922, p. 68; hereafter referred to as *Rly. M.E.*
68. *Safety on Railroads*, pp. 363–366; Gutbrod Report, part 4, pp. 110–114.
69. *Rly. Age Gaz.*, Dec. 15, 1911, pp. 1207–1210.
70. *A.R.R.J.*, November 1909, pp. 447–451; Gutbrod Report, part 4, pp. 114–124 (misidentified as C B & Q).
71. *A.R.R.J.*, June 1911, pp. 208–211.
72. *Rly. M.E.*, September 1928, pp. 506–508.
73. Ibid., January 1928, pp. 23–26.
74. *Rly. Rev.*, Oct. 3, 1903, p. 721.
75. *Rly. Age Gaz.*, 1912, p. 638. Most of the general discussion in this section is drawn from the report *Safety on Railroads* (1914) previously cited.
76. Gutbrod Report, part 3, pp. 36 and 57.
77. *A.R.R.J.*, Jan. 1911, pp. 1–5.
78. A good survey of Stillwell's career is given in *Steam Locomotive and Railroad Tradition*, May 1963, pp. 38–45.
79. *R.R.G.*, June 14, 1907, pp. 831–835.
80. *Rly. M.E.*, July 1917, pp. 387–389.
81. *Rly. Age Gaz.*, Jan. 19, 1912, pp. 84–85. See also R. B. Shaw's *Down Brakes*, London: Macmillan, 1961, for other examples of early steel car wrecks.
82. *Rly. M.E.*, March 1917, p. 187.
83. *A.R.R.J.*, April 1905, p. 112.
84. *Rly. Age Gaz.*, June 16, 1911, p. 1483, and June 15, 1912, p. 1430.
85. *Car Builders' Cyclopedia, 1928*, p. 611.
86. *A.S.M.E. Transactions*, vol. 35, pp. 74–92.
87. Ibid., July 1921, p. 444. See also descriptions of individual cars in various trade journals, 1907–1920.
88. *A.S.M.E. Transactions*, October 1915, pp. 520–521.
89. *Central Railroad Club*, March 1909, pp. 52–67.
90. *Rly. Age Gaz.*, February 1913, p. 58.
91. *International Railway Congress Association Bulletin*, vol. 11, 1929, p. 663. This volume contains an extensive survey of steel passenger cars through the world. Hereafter cited as *Int. Rly. Congress*, vol. 11.
92. *New England Railroad Club*, April 8, 1924, pp. 70–81.
93. Gutbrod Report, part 3, p. 41.
94. Ibid., part 5, p. 325.
95. *Car Builders' Cyclopedia, 1928*, p. 599.
96. *Rly. M.E.*, March 1923, pp. 161–165, contains a summary of a paper read before the Canadian Railway Club in December 1922.
97. Ibid., July 1927, p. 470.
98. *Rly. Age*, Sept. 17, 1921, pp. 518–519.
99. *Int. Rly. Congress*, vol. 11, pp. 602–610 and 664; *Rly. M.E.*, March 1923, p. 164.
100. *Rly. M.E.*, March 1923, pp. 161–165.
101. *New England Railroad Club*, April 8, 1924, pp. 70–81.

102. Gutbrod Report, part 5, p. 331.
103. *Rly. Age*, Feb. 8, 1930, p. 377; Aug. 22, 1931, p. 283; July 22, 1933, p. 147. *Light Metals*, May 1945, p. 223, offers a worldwide survey of aluminum rolling stock; it is clear from the article that the United States was the leader in the early years.
104. Ibid., October 1927, p. 665.
105. *Trains*, August 1951, p. 41.
106. *Rly. Age*, April 22, 1944, p. 762; Nov. 22, 1941, p. 854; Oct. 14, 1939, p. 551. See also *Rly. M.E.*, June 1938, p. 213.
107. *Rly. Age*, Jan. 12, 1946, p. 148.
108. *Rly. M.E.*, October 1934, p. 358.
109. The early and late history of railroad streamlining is treated in Robert C. Reed's *The Streamline Era*, San Marino, Calif.: Golden West Books, 1975. For more detail on Calthrop, see Harold L. Van Doren's *Industrial Design, A Practical Guide*, New York: McGraw-Hill, 1940. For more detail on Adams, see *Dictionary of American Biography*, vol. 1, 1928, p. 58.
110. *Rly. M.E.*, January 1935, p. 23; *Rly. Age*, Nov. 20, 1937, p. 729.
111. *Rly. Age*, Nov. 4, 1939, p. 693.
112. A comparison is made here between the cars of the G M & N Rebel motor train and those of the full-sized Royal Blue. Both were built in 1935 by A.C.F. and both were of Cor-Ten steel.
113. *Mechanical Engineering*, January 1938, p. 29.
114. *Rly. M.E.*, August 1944, p. 356; Dec. 7, 1946, p. 961. *Rly. Age*, Dec. 28, 1946. See also *Int. Rly. Congress*, vol. 24, February 1947, for a worldwide survey of lightweight car construction.
115. Information on late-model heavyweights is from *Rly. M.E.*, March 1938, p. 40, and May 1942, p. 212; *Car Builders' Cyclopedia, 1943*, p. 613; *Rly. Age*, Jan. 10, 1947, p. 47. W. D. Edson of Amtrak reports that as of October 1971, the C & A parlors were still in service on the successor G M & O. The coaches were out of service awaiting a buyer.
116. These figures, like nearly all statistics offered on cars, are guesstimates based on data in *Railway Age*, publications of the American Railway Car Institute, and census reports. None of these sources give clear figures for lightweight cars. See also *Rly. M.E.*, November 1946, p. 614.
117. Articles and advertisements on the merits of aluminum cars appeared in railroad trade journals. See *Rly. Age*, Feb. 8, 1930, Aug. 22, 1931, and July 22, 1933. The author recalls a luncheon in recent years sponsored by a major aluminum maker where the virtues of aluminum freight cars were extravagantly praised.
118. *Rly. Age*, Nov. 20, 1937, p. 729; *Western Society of Engineers Proceedings*, April 1939, p. 69.
119. *Rly. M.E.*, August 1944, p. 358.
120. *Rly. Age*, Dec. 14, 1946, p. 996.
121. The failings of aluminum cars were recorded in a conversation in September 1971 with William D. Edson, chief of equipment for Amtrak and former chief mechanical officer of the New York Central Railroad.
122. *Rly. Age*, Dec. 19, 1955, p. 28.
123. *Rly. M.E.*, December 1950, p. 717.
124. *Rly. M.E.*, June 1933, p. 185; *Light Metals*, May 1945, p. 241.
125. See note 118.
126. *Rly. M.E.*, May 1935, p. 171; Coverdale and Colpitts "Report on Streamline Trains," 1938 ed., p. 59.
127. The Royal Blue cars were sent to the Alton in 1937 for use on the Ann Rutledge. Remodeled heavyweight cars were used in their place on the Royal Blue.
128. *Rly. M.E.*, March 1940, p. 87.
129. *Rly. Age*, June 14, 1971, p. 14, and Nov. 8, 1971, p. 29.
130. Most of the data on the stainless steel car is from the writings of E. G. Budd and his son:
A Lecture . . . Harvard Graduate School of Business, March 6, 1934. No publisher.
Genesis of An Art by E. G. Budd, Newcomen Society, American Branch, Princeton University Press, 1938.
Edward G. Budd, 1870–1946, E. G. Budd, Jr., Newcomen Society, Princeton University Press, 1950.
"Lightweight Champion," *Railroad Magazine*, October 1949, pp. 12–39.
The best technical article on Budd's methods is given in *Heat Treating and Forging*, May 1936, pp. 236–239.
131. *Rly. M.E.*, March 1936, p. 89.
132. Ibid., July 1937, p. 303; *Mech. Eng.*, January 1938, p. 29.
133. Ragsdale gave a good account of his ideas on passenger car construction in a paper before the Society of Automotive Engineers just eleven days before his death. It was printed in *Rly. M.E.*, March 1946, p. 111.
134. *Rly. M.E.*, April 1939, p. 127.
135. Budd-Pullman rivalry was reported to the author by Arthur Dubin during a conversation in May 1971. Dubin referred to interoffice correspondence among Pullman officials and to his own earlier conversations with employees of both firms.
136. *Rly. Age*, March 13, 1937, p. 418.
137. Ibid., April 1944, p. 773.
138. R. K. Wright, *Southern Pacific Daylight Train 98–99*, Thousand Oaks, Calif.: Wright Enterprises, 1970, p. 510.
139. Conversation with William Howes, director of passenger services, B & O–C & O, May 1970.
140. *Railway Locomotives and Cars*, July 1954, p. 39.
141. *Rly. Age*, Jan. 12, 1946, p. 150.
142. *Int. Rly. Congress*, Report on Lightening Rolling Stock, 14th session, Lucerne, 1947.
143. *Trains*, November 1956, pp. 43–47; *Railway Locomotives and Cars*, August 1956, p. 49.
144. *Trains*, September 1958, p. 10.
145. References to Cor-Ten steel are taken largely from advertisements appearing in *Rly. Age*, *Car Builders' Dictionary*, 1943, pp. 357–360, and *Report of the Mechanical Advisory Committee to the Federal Coordinator*, Washington, D.C., 1935. It should be noted that other steelmakers produced similar products.
146. *Western Society of Engineers Proceedings*, April 1939, p. 72.
147. *Rly. Age*, April 1944, p. 774, in a survey of passenger cars built in the period 1934–1944, states that of 2,319 cars, 1,626 were of Cor-Ten construction. A U.S. Steel advertisement in the 1943 *Car Builders' Dictionary* says that 2,300 passenger cars had been built with Cor-Ten. A similar advertisement in the 1955 *Rly. Age* gives a total of 6,388.
148. *Rly. M.E.*, February 1935, p. 43.
149. Ibid., October 1940, p. 382; December 1942, p. 515. Also *Car Builders' Cyclopedia*, 1943, p. 596.
150. *Steel*, Aug. 22, 1938, p. 51.
151. *Rly. M.E.*, November 1941, p. 476.
152. Ibid., April 1933, p. 118.
153. *Rly. Age*, December 1936, p. 825.
154. *Rly. M.E.*, March 1942, p. 93, contains a detailed sketch of Nystrom.
155. *The Hiawatha Story* by Jim Scribbins, Milwaukee: Kalmbach, 1970, offers a fine pictorial record of the Milwaukee Road's lightweight cars. It contains basic historical data but almost no engineering material. Technical data on Number 4400 are given in *Rly. M.E.*, October 1934, p. 361.
156. *Rly. M.E.*, June 1935, p. 266.
157. *Welding Engineer*, September 1936, p. 30; *Rly. M.E.*, November 1936, p. 467; March 1939, p. 95.
158. *Rly. M.E.*, April 1943, p. 157.
159. *Rly. Age*, May 20, 1939, p. 865; October 14, 1939, p. 586.
160. *Rly. M.E.*, December 1934, p. 444; see also note 157.
161. Ibid., August 1936, p. 331.
162. *Rly. Age*, Feb. 1, 1941, p. 246.
163. *Rly. M.E.*, May 1937, p. 241.
164. *Rly. Age*, Feb. 22, 1947, p. 398.
165. Ibid., Sept. 12, 1936, p. 367.
166. *Trains*, June 1963, p. 3.
167. Ibid., June 1965, p. 9.
168. *Rly. M.E.*, October 1948, p. 584.
169. *Report on Light Weight Trains*, Statement 3639, May 1936, Interstate Commerce Commission. A copy of the original mimeographed report is on file in the I.C.C. Library, Washington, D.C.
170. Robert Shaw, *Down Brakes*, p. 327, and *Rly. Age*, Dec. 17, 1939, pp. 878–80.
171. *Rly. M.E.*, March 1940, p. 107.
172. *Rly. M.E.*, April 1948, p. 173.
173. The A.A.R. specifications are published yearly in the *Manual of Standards and Recommended Practices*. A full set may be seen at A.A.R. headquarters in Washington, D.C.
174. This allegation was made to the author by a former railway mechanical official who had no motive for slandering Budd but who wishes to remain anonymous for personal reasons.
175. *Rly. Age*, Jan. 12, 1946, p. 151.
176. *Rly. M.E.*, March 1946, p. 111, and April 1947, p. 212; *Railroad Magazine*, October 1949, p. 26.
177. *Rly. Age*, March 21, 1955, p. 50.
178. Gutbrod Report, part 2, p. 1250.

179. *Rly. M.E.*, February 1919, p. 81.
180. *Car Builders' Dictionary*, 1919, figs. 2877–2878.
181. *Rly. Age*, Jan. 12, 1946, p. 148.
182. *Rly. M.E.*, September 1950, p. 489.
183. Ibid.
184. Ibid., July 1948, p. 59.
185. *Rly. Age*, Jan. 12, 1946, p. 148.
186. *Rly. M.E.*, April 1943, p. 157.
187. *Report of the Mechanical Advisory Committee to the Federal Coordinator*, 1935, p. 599.
188. *New York Railroad Club*, Nov. 21, 1924, p. 7441.
189. *Rly. Rev.*, Sept. 18, 1897, p. 533. *Rly. Gaz.* (British), March 4, 1949, pp. 227–228 claims that a bilevel car was used in India as early as 1862.
190. Information on the Long Island double deckers is from *Rly. Age*, August 1932, p. 221; *The Locomotive Magazine*, Nov. 15, 1933, p. 338, and *Railroad Magazine*, October 1947, p. 80.
191. *Railroad Magazine*, March 1952, p. 58.
192. *Rly. M.E.*, December 1950, p. 713, and Richard C. Overton's *Burlington Route*, New York: Knopf, 1965, pp. 526 and 563.
193. *Rly. Age*, April 5, 1955, p. 10.
194. *Rly. Age*, April 13, 1954, p. 24; April 5, 1955, p. 9.
195. *Trains*, October 1956, p. 21.
196. Ibid., January 1959, p. 16.
197. Ibid., October 1956, p. 25; January 1969, p. 54. R. J. Wayner (Ed.), *Pennsy Car Plans*, New York: Wayner Publications, 1969.
198. Richard C. Overton, *Burlington Route*, p. 200.
199. *Sci. Am.*, May 2, 1891, p. 274.
200. *Rly. Rev.*, Aug. 2, 1902, p. 581.
201. *Rly. M.E.*, September 1945, p. 378.
202. *Rly. Age*, Jan. 12, 1946, p. 14.
203. *Rly. M.E.*, August 1947, p. 395.
204. *Rly. Age*, Oct. 11, 1947, p. 604.
205. *Rly. M.E.*, May 1949, p. 241.
206. *Trains*, August 1956, p. 9.
207. Jim Scribbins, *Hiawatha Story*, p. 114, and *Rly. Locos. & Cars*, February 1953, p. 49.
208. *Trains*, May 1958, p. 19.

CHAPTER THREE

1. Arguments of W. W. Hubbell in William Whiting, *Arguments of William Whiting, Esq., in the case of Ross Winans vs. Orsamus Eaton et al.*, Boston, 1853, p. 3.
2. Seymour Dunbar, *A History of Travel in America*, Indianapolis, 1915, p. 1006.
3. F. A. Ritter Von Gerstner, *Die innern communication der vereinigten Staaten von Nord-America*, Vienna, 1842–1843.
4. Karl Ghega, *Die Baltimore-Ohio Eisenbahn*, Vienna, 1844.
5. Dionysius Lardner, *Railway Economy*, New York and London, 1850, p. 346.
6. *Leslie's Illustrated Newspaper*, Aug. 25, 1877.
7. *Railway Age* (Oct. 20, 1893, p. 773) says that the car *Chambersburg* entered service in the winter of 1836–1837. The most frequently cited date is the winter of 1837–1838, which was given in W. B. Wilson's *History of the Pennsylvania Railroad*, vol. 1, Philadelphia, 1899, p. 395. (The Pennsylvania was the Cumberland Valley's successor.) The son of the Cumberland Valley's pioneer master car builder, Jacob Shaffer, said that the car entered service in the spring of 1838, according to *Locomotive Engineering*, April 1900, p. 150.
8. *George M. Pullman and Pullman's Palace Car Company vs. the New York Central Sleeping Car Company and Webster Wagner. In Equity. Answer as Amended.* U.S. Circuit Court, Northern District of Illinois, 1881. Printed transcript of testimony, 920 pages. Hereafter cited as *Pullman vs. Wagner.*
9. *Pullman vs. Wagner* and *Pullman's Palace Car Company vs. Jonah Woodruff and J. Hervey Jones; Equity Docket No. 28, November term, 1871.* U.S. Circuit Court, Western District of Pennsylvania. Hereafter cited as *Pullman vs. Woodruff*. Data were copied from ms. records housed in U.S. Federal Building at Pittsburgh in September 1969.
10. *Pullman vs. Wagner*, pp. 395–396.
11. Ibid., p. 410.
12. Ibid., pp. 399 and 408.
13. Ibid., pp. 387–388.
14. Ibid., p. 394.
15. *American Railroad Journal*, Nov. 15, 1838, p. 328, copied from the Oct. 31, 1838, *Baltimore American*.
16. Ibid.
17. *Pullman vs. Woodruff*, 1871.
18. *A.R.R.J.*, March 25, 1854, p. 183.
19. P W & B annual report, 1859.
20. Von Gerstner, vol. 1, p. 129.
21. Ibid., vol. 1, p. 238.
22. *Pullman vs. Wagner*, pp. 229–280.
23. Von Gerstner, vol. 2, p. 305.
24. *Locomotive Engineering*, November 1893, p. 503, and January 1898, p. 14.
25. *Pullman vs. Wagner*, pp. 710–743.
26. Charles S. Sweet, "History of the Sleeping Car," Chicago, 1923, a book-length study that was never published. Sweet's uncle was said to be G. M. Pullman's personal secretary. Through this association Sweet apparently gained access to early records of the Pullman Company and its predecessors. The mss. was in the possession of his daughter, Jane Sweet of Winnetka, Ill., in 1972. Hereafter cited as Sweet's History.
27. *Pullman vs. Wagner*, pp. 320–360.
28. William P. Smith, *The Great Railway Celebration of 1857*, New York, 1858, p. 163.
29. As previously noted, the P W & B had ladies' cars as early as 1838. *A.R.R.J.*, June 15, 1842, described them as equipped with "luxurious sofas . . . washstand and other conveniences."
30. P W & B annual reports 1859, 1860, and 1863.
31. C. H. Corliss, *Main Line of Mid America*, New York, Creative Age Press, 1950, pp. 76–77, and *A.R.R.J.*, Aug. 16, 1856, p. 520. The *Detroit Advertiser* description which follows is from Corliss.
32. *50th Anniversary of Illinois Central R.R.*, 1901, no page numbers. (Souvenir booklet published by the railroad)
33. R. C. Overton, *The Burlington Route*, New York: Knopf, 1965, p. 52.
34. A leaflet dating from about 1859 describing Case's Patent Sleeping Car is in the Harvard Business School (Baker Library) business history mss. collection.
35. Ibid.
36. *Western Railroad Gazette*, April 24, 1869.
37. *A.R.R.J.*, Jan. 1, 1859, p. 12.
38. B & O annual report, 1859, p. 151.
39. A. Bendel, *Aufsätze Eisenbahnwesen in Nord-Amerika*, Berlin, 1862, p. 38, plate E.
40. *A.R.R.J.*, Nov. 28, 1863, p. 1132.
41. I. D. Guyer, *History of Chicago*, Chicago, 1862, rear advertising section.
42. Compiled by Arthur D. Dubin from *Appleton's 1866 R.R. Guide*, in his personal collection.
43. Earlier, some sleeping-car–related patents had been issued on folding furniture for shipboard cabins (such as H. King's of 1841). McGraw's 1838 folding sofa bed, used in the Erie sleeping cars of 1842, was mentioned earlier in the chapter.
44. The U.S. Patent Office has assigned sleeping-car patents to class 105, subclasses 314–326. A total of approximately 500 patents are in this group.
45. *Pullman vs. Wagner*, 1881, pp. 580–656 and 684–707.
46. Siegfried Gideon, *Mechanizaion Takes Command*, New York, W. W. Norton, 1948, 1969, p. 461.
47. *Dictionary of American Biography*, vol. 20, p. 497, provides a good summary of Woodruff's life. Other details are offered by the *National Cyclopedia of American Biography*, vol. 14, pp. 203–204.
48. Joseph F. Wall, *Andrew Carnegie*, Fair Lawn, N.J.: Oxford University Press 1970, p. 141. Hereafter cited as Wall.
49. *National Car Builder*, October 1880, p. 166.
50. *Rly. Age*, Sept. 6, 1889, pp. 586–587.
51. W. S. Kennedy, *Wonders and Curiosities of the Railway*, Chicago, 1884, p. 194; *Appleton's Cyclopedia of American Biography*, vol. 1, p. 605.
52. *Manufacturing and Manufacturers of Pennsylvania*, 1875, p. 185.
53. John A. Haddock, *History of Jefferson County New York*, Albany, 1895, pp. 33–35, and *Rly. Age*, April 5, 1889, pp. 221–222.
54. *Railroad Car Journal*, October 1894, p. 208.
55. *Pullman vs. Wagner*, 1881, pp. 580–660.
56. Wall, pp. 139–140.
57. Railway and Locomotive Historical Society, *Bulletin 59*, October 1942, pp. 34–36.
58. *Rly. Age*, April 5, 1889, pp. 221–222.

59. Wall, p. 199.
60. Sweet's History, p. 128.
61. *Engineer* (Philadelphia), Oct. 4, 1860, p. 63. *Dictionary of American Biography*, vol. 10, pp. 463–464, offer a sketch of Knight's career that includes several factual errors.
62. *A.R.R.J.*, Oct. 10, 1863, p. 969.
63. *Chicago Tribune*, June 5, 1864.
64. Sweet's History, p. 130.
65. *Railroad Record*, Jan. 18, 1866, p. 580.
66. *American Heritage*, December 1970, p. 40.
67. *Scientific American*, Aug. 10, 1867, p. 89.
68. G. M. Best, *Iron Horses to Promontory*, San Marino, Calif.: Golden West Books, 1969, pp. 61–69.
69. *N.C.B.*, May 1875, p. 68.
70. August Mencken, *The Railroad Passenger Car*, Baltimore: Johns Hopkins Press, 1957, p. 161.
71. *Manufacturers of Pennsylvania*, 1875, p. 489.
72. Wall, p. 209.
73. Sweet's History, pp. 130 and 284.
74. Wall, p. 211.
75. Ibid., p. 18.
76. *Rly. Age*, April 27, 1888, p. 269.
77. *Financial and Commercial Chronicle*, June 4 and 11, 1898, pp. 1090 and 1141.
78. Acquisition Reports. During the 1940 Federal suit against the Pullman Company, a series of reports was prepared by the firm's legal department giving financial summaries of the companies taken over by Pullman. Copies were made available to me by Arthur D. Dubin of Chicago. Hereafter referred to as Pullman Acquisition Reports.
79. *Railroad Gazette*, Dec. 3, 1871, p. 382.
80. Ibid., May 3, 1878, p. 223.
81. Sweet's History, pp. 238–242; Dun and Bradstreet records (Harvard Business School), Philadelphia and Pittsburgh volumes; also Pullman Acquisition Reports.
82. Records of this suit are on file in the U.S. District Court of Western Pennsylvania.
83. *Railway World*, Nov. 3, 1877, p. 1046.
84. *N.C.B.*, November 1878, p. 167, and Sweet's History, p. 259.
85. *N.C.B.*, November and December 1879, pp. 170 and 180.
86. C.B.&Q. Railroad papers, Newberry Library, Chicago, boxes 8 and 33.
87. *Railway Review*, Sept. 30, 1882.
88. Ibid., June 21, 1882, and April 8, 1882.
89. *Rly. Rev.*, Dec. 29, 1883, p. 776.
90. Financial data on Woodruff is from Dun and Bradstreet and Pullman Acquisition Reports.
91. Andy Logan, *The Man Who Robbed the Robber Barons*, New York: W. W. Norton, 1965.
92. *Rly. Rev.*, May 9, 1885, p. 227.
93. *Engineering*, April 18, 1873, p. 262, and July 11, 1873, p. 33.
94. Ibid., June 11, 1875, p. 493.
95. *Rly. Rev.*, Aug. 25, 1883, p. 504.
96. *N.C.B.*, January 1884, p. 4; *R.R.G.*, Dec. 14, 1883, p. 819.
97. *Rly. Rev.*, May 9, 1885, p. 227.
98. Sweet's History, p. 252, quoted from Mann Boudoir Director's Minute Books.
99. *Rly. Rev.*, Dec. 18, 1886, p. 675.
100. *R.R.G.*, Jan. 3, 1890, p. 10.
101. *N.C.B.*, January 1884, p. 5.
102. L. M. Beebe, *Mr. Pullman's Elegant Palace Car*, Garden City, N.Y.: Doubleday, 1961, p. 254.
103. Pullman Acquisition Reports.
104. *Rly. Rev.*, Oct. 27, 1888, pp. 621 and 633.
105. *R.R.G.*, Nov. 2, 1888, p. 726.
106. Ibid., March 1, 1889, p. 151.
107. *R.R.G.*, March 30, 1883, p. 202.
108. *New York Elevated Railway Journal*, January 26, 1884.
109. *Rly. Age*, July 16, 1885.
110. *Rly. Rev.*, March 6, 1886, p. 117.
111. The April 1893 *R.R. Car Jour.* (p. 155) mentions Monarch, but the article does not clearly state whether it had suspended operations. Advertisements for the firm appeared in the *Official Guide* through October 1894.
112. Railway and Locomotive Historical Society *Bulletin* 59, p. 32.
113. *N.C.B.*, September 1881, p. 104.
114. *Leslie's Illustrated Weekly*, April 30, 1859.
115. Mencken, pp. 135–141, quotes Walter Thornbury's travel account of 1860.
116. *A.R.R.J.*, Jan. 21, 1882, p. 33.
117. Ibid., Sept. 27, 1879, p. 1067, and *American Railway Times*, April 11, 1868, p. 119.
118. Several biographical sketches are available on Wagner. The most accurate account is offered in *R.R.G.*, Jan. 20, 1882, p. 48.
119. Sweet's History, p. 155i.
120. *R.R. Car Jour.*, January 1892, pp. 53–57, offers a short history of Wagner.
121. *West. R.R. Gaz.*, May 26, 1860, p. 2.
122. Alvin F. Harlow, *Road of the Century: The Story of the New York Central*, New York: Creative Age Press, 1947, quotes the *Schenectady Evening Star* for Oct. 21, 1865.
123. *R.R. Car Jour.*, January 1892, pp. 53–57, and Pullman Acquisition Reports.
124. *N.C.B.*, August 1894, p. 118, and Mencken, p. 77.
125. Sweet's History, pp. 155D–155G; *Reports of the New York State Legislature*, Special Commission on Railroads, 1879, vol. 9, pp. 182–185.
126. Pullman Acquisition Reports, p. 10.
127. Sweet's History, p. 155H.
128. *A.R.R.J.*, July 21, 1882, p. 501.
129. *Rly. World*, Sept. 4, 1875, p. 583, and Oct. 30, 1875, p. 704. Similar accounts are given in Harlow, *Road of the Century*, and Edward Hungerford's *Men and Iron: The History of the New York Central*, New York: Crowell, 1938.
130. *Pullman vs. Wagner*, 1881, p. 56-A, and Sweet's History, p. 155H.
131. Sweet's History, p. 155H.
132. *A.R.R.J.*, July 21, 1882, p. 501.
133. *N.C.B.*, May 1875, p. 68.
134. Alvin F. Harlow, *Steelways of New England*, New York: Creative Age Press, 1946, p. 256.
135. *N.C.B.*, March 1881, p. 32.
136. Ibid., September 1873, p. 220.
137. Ibid., October 1892, p. 158.
138. Ibid., January 1876, p. 6.
139. *R.R.G.*, April 20, 1877, p. 181.
140. Railway and Locomotive Historical Society, *Bulletin* 59, pp. 28–29.
141. *N.C.B.*, October 1880, p. 156.
142. *A.R.R.J.*, April 15, 1882, p. 238.
143. *R.R.G.*, June 8, 1883, p. 359.
144. *N.C.B.*, September 1885, p. 113.
145. New York Central Sleeping Car Company Minute Books for 1886–1899 are housed in the Newberry Library, Chicago. Hereafter N.Y. Central Minute Books.
146. *Rly. Age*, Sept. 13, 1889, p. 599.
147. N.Y. Central Minute Books, Dec. 22, 1886.
148. *N.C.B.*, April 1879, p. 57.
149. *Rly. Rev.*, April 10, 1886, p. 175.
150. Pullman order books—from a transcript prepared by Evelyn Danielson of Pullman–Standard Mfg. Co.
151. *Rly. Rev.*, Jan. 28, 1888, p. 44, and *R.R.G.*, Aug. 12, 1887, p. 526.
152. *Rly. Rev.*, Jan. 28, 1888, p. 44, and *R.R.G.*, March 3, 1888, p. 123.
153. *Rly. Rev.*, Nov. 19, 1887, p. 672, and Dec. 17, 1887, p. 720.
154. *N.C.B.*, April 1887, p. 43.
155. *N.C.B.*, March 1886, p. 29.
156. *R.R.G.*, Oct. 31, 1890, p. 750.
157. Railway and Locomotive Historical Society, *Bulletin 105*, p. 65.
158. *R.R. Car Jour.*, January 1892, p. 57, and April 1893, p. 155.
159. *U.S.A. vs. Pullman*, U.S. Justice Department, 1941, *Complaint.*
160. *R.R. Car Jour.*, November 1896, p. 278, and N.Y. Central Minute Books.
161. *Rly. Rev.*, March 19, 1892, p. 206.
162. *Rly. Rev.*, Nov. 13, 1897, p. 650, and *R.R. Car Jour.*, December 1897, p. 368.
163. *New York Railroad Commission Report*, 1885, p. 1548.
164. N.Y. Central Minute Books, April 10, 1888.
165. *Rly. Rev.*, Aug. 29, 1891, p. 573.
166. *Commercial & Financial Chronicle*, Aug. 12, 1899, p. 332.
167. Ibid.

168. Ibid., Oct. 28, 1899, p. 909, and Pullman Acquisition Reports, p. 27.
169. *Pearson's Magazine*, July 1913, p. 19.
170. *R.R.G.*, Nov. 3, 1899, p. 754.
171. F. W. Pierce, *Field Genealogy*, Chicago, 1901, p. 674; and Sweet's History, p. 157.
172. *The World*, Dec. 23, 1892.
173. Railway and Locomotive Historical Society, *Bulletin 59*, Oct. 1942, p. 30.
174. Joseph Husband, *The Story of the Pullman Car*, Chicago, 1917, pp. 28–32; hereafter Husband. This book gives the recollections of two early employees: Leonard Siebert claimed that the cars were remodeled in 1858, while J. L. Barnes claimed that they were ready for service in September 1859. The date cited by G. M. Pullman in several newspaper interviews was 1859.
175. Husband, pp. 28 and 31.
176. Pullman Collection in Newberry Library: file folder marked History of No. 9.
177. *R.R.G.*, April 6, 1877, p. 153.
178. Ibid., and Sweet's History, p. 186.
179. Stanley Buder, *Pullman: An Experiment in Industrial Order and Community Planning, 1880–1930*. New York: Oxford, 1967. Hereafter cited as Buder.
180. *Colorado Magazine*, 1940, p. 113, and *Pullman News*, April 1941, p. 118.
181. Dun and Bradstreet Reports, mss. in Baker Library, Harvard School of Business Administration, Cambridge, Mass. The first entry is dated July 6, 1863.
182. Husband, p. 33.
183. The story of the *Pioneer's* oversized dimensions can be traced to Pullman's lifetime. *Rly. Rev.*, Oct. 20, 1893, p. 774.
184. Sweet's History, p. 196.
185. Buder, p. 13.
186. *Pullman News*, August 1927, p. 114.
187. D.&B. Reports, July 1866.
188. *Courier-Journal* (Louisville, Ky.), Aug. 22, 1877.
189. Sweet's History, p. 209.
190. *U.S. Justice Department Civil Action 994, Stipulation of Facts*, U.S. District Court of Eastern Pennsylvania, 1940, pp. 82–85.
191. *R.R.G.*, April 22, 1871, p. 39.
192. *Rly. Age*, May 19, 1881, p. 277.
193. Sweet's History, p. 211.
194. Wall, p. 208.
195. Best, *Iron Horses to Promontory*, pp. 69 and 141.
196. Pullman Acquisition Reports, 1940.
197. George Behrend, *Pullman in Europe*, London: Ian Allen, 1962, p. 21.
198. Sweet's History, p. 291.
199. Behrend, pp. 27–30.
200. *A.R.T.*, June 29, 1867, p. 206.
201. *N.C.B.*, March 1874, p. 42.
202. *Transportation Routes to the Seaboard*, U.S. Senate Report, 1874, vol. 1, pp. 150–152.
203. *Report of the Investigating Committee . . . on the Pennsylvania Railroad*, Philadelphia, 1874, pp. 125–128. See also George H. Burgess and M. C. Kennedy, *Centennial History of the Pennsylvania Railroad*, Philadelphia, 1949, pp. 327–329.
204. *N.C.B.*, December 1870, p. 1.
205. *N.C.B.*, January 1878, p. 2.
206. *R.R.G.*, July 7, 1876, p. 302.
207. *Rly. World*, Jan. 2, 1875, p. 10.
208. *R.R. Car Jour.*, January 1892, p. 59.
209. Ibid., Feb. 11, 1882, p. 89.
210. D.&B. Reports, August and October 1875.
211. *R.R.G.*, Oct. 5, 1877, p. 446.
212. *A.R.R.J.*, Nov. 20, 1875, p. 1476.
213. *N.C.B.*, October 1878, p. 154.
214. *A.R.R.J.*, Oct. 4, 1879, p. 1118.
215. *Rly. Age*, Sept. 15, 1881, p. 521.
216. *R.R.G.*, Oct. 23, 1891, p. 756.
217. *N.C.B.*, January 1884, p. 1.
218. C B & Q Papers, Newberry Library, box 8C 5331.
219. Pullman Papers, Newberry Library, contracts file, 1913.
220. Sweet's History, p. 303.
221. Ibid., and *R.R.G.*, Oct. 27, 1893, p. 792.
222. Figures are from Pullman annual reports.
223. Pullman Acquisition Reports, 1940c., report on Association contracts.
224. *Trains*, November 1969, p. 26.
225. *Rly. World*, Oct. 2, 1880, p. 943.
226. *Rly. World*, Jan. 8, 1881, p. 39.
227. *Rly. Rev.*, Feb. 11, 1882, p. 89.
228. *Rly. Age*, June 22, 1888, p. 396.
229. Ibid., Oct. 5, 1882, p. 548.
230. Ibid., Dec. 11, 1891, p. 966.
231. *Rly. Rev.*, Sept. 8, 1883, p. 532.
232. Ibid., March 16, 1889, p. 150.
233. *R.R.G.*, Oct. 21, 1891, p. 796.
234. *Fortune*, January 1938, p. 45.
235. Pullman annual reports, and I.C.C. *Sleeping Car Statistics, 1890–1935.* statement 3631, August 1936; hereafter *I.C.C. Statistics, 1936.*
236. *Pearson's Magazine*, July 1913, pp. 12–23.
237. Justice Department, *Stipulation of Facts*, 1940, p. 93.
238. *Rly. Age Gaz.*, April 15, 1910, p. 987, and Nov. 18, 1910, p. 976.
239. *Boston Globe*, May 1, 1913.
240. J. C. Welliver, Pullman Progress. Proof sheets dated 1929, are held by the Chicago Historical Society.
241. *Moody's Industrials*, 1920, p. 1432.
242. *I.C.C. Valuation Docket 1079, Abstract of the Evidence*, June 25, 1928.
243. *Military Engineer*, September–October 1930, p. 426.
244. *Rly. Age*, April 25, 1925, p. 1053.
245. Ibid., June 11, 1927, p. 1914.
246. Ibid., May 28, 1927, p. 1583.
247. Ibid., Aug. 29, 1936, p. 309.
248. *Fortune*, January 1938, p. 94.
249. *New York Evening Post*, Feb. 9, 1929.
250. *Fortune*, January 1938, p. 94, and *New York Times*, Dec. 28, 1929.
251. *Fortune*, January 1938, p. 42.
252. *I.C.C. Statistics, 1936*, pp. 16–18.
253. *Business Week*, April 16, 1930, p. 14.
254. *Barron's* Nov. 28, 1932, p. 6.
255. *Civil Action no. 944 in U.S. District Court, Eastern District of Pennsylvania, U.S.A. vs. Pullman Company, Answer*, Oct. 15, 1940.
256. *Rly. Age*, Nov. 22, 1941, p. 881.
257. *Pullman News*, October 1945, p. 50; *Trains*, October 1945, p. 4.
258. *Federal Supplement*, vol. 64, 1946, p. 108.
259. *Pullman News*, July 1947, p. 3.
260. *Rly. Age*, Nov. 19, 1949, p. 896, and *Pullman News*, January 1950, p. 29.
261. *Trains*, March 1958, p. 10, and *Daily News* (Chicago), April 16, 1958.
262. *Rly. Age*, Dec. 2, 1968, p. 40.
263. *Pullman News*, October 1929, p. 191.
264. *R.R.G.*, Nov. 6, 1908, p. 1317.
265. *N.C.B.*, October 1873, p. 251.
266. *Rly. Rev.*, April 16, 1887, p. 226.
267. *R.R.G.*, Dec. 7, 1888, p. 805.
268. *Rly. Rev.*, Nov. 9, 1889, p. 645.
269. *R.R. Car Jour.*, December 1892, p. 46.
270. *Pullman Progress*, pp. 44–45.
271. *R.R.G.*, April 19, 1907, p. 636.
272. Ibid., Feb. 4, 1910, p. 287.
273. Ibid., April 29, 1910, p. 1084.
274. *Safety on the Railroads: Hearings before a Subcommittee of the Committee on Interstate and Foreign Commerce of the House of Representatives Sixty-third Congress*, Washington, D.C., 1914, part v, p. 354.
275. *I.C.C. Statistics, 1936*, p. 21.
276. R. J. Wayner, *Pullman Panorama*, New York, vol. 1, 1967, p. ii.
277. William Kratville, *Passenger Car Catalog: Pullman Operated Equipment, 1912–1949*, Omaha: Kratville Publications, 1968, p. 30.
278. *Rly. Age*, April 16, 1932, p. 642.
279. *Pearson's Magazine*, July 1913.
280. *Scientific American*, February 1923, p. 87.
281. *Harper's Weekly*, Feb. 7, 1914, p. 22.
282. *Rly. Age*, Aug. 29, 1935, p. 310.
283. Ibid., April 2, 1927, p. 1071.

284. Ibid., April 16, 1932, p. 640.
285. *Pullman News*, April 1942, p. 133.
286. *Railway Mechanical Engineer*, June 1933, p. 186.
287. *Fortune*, February 1938, p. 78.
288. *Rly. M.E.*, June 1940, p. 241, and *Trains*, August 1960, p. 59.
289. *Rly. Age*, Nov. 17, 1945, p. 797, and *Railroad Magazine*, October 1949, p. 15.
290. *Pullman News*, October 1956, p. 17.

CHAPTER FOUR

1. *Railroad Gazette*, April 28, 1882, p. 254.
2. J. G. Medley, *An Autumn Tour in the United States and Canada*, London, 1873, pp. 70–71.
3. *American Railroad Journal*, Oct. 9, 1845, p. 647.
4. *Scientific American*, Aug. 27, 1853, p. 396.
5. A. Bendel, *Aufsätze Eisenbahnwesen in Nord-Amerika*, Berlin, 1862, plate E.
6. *A.R.R.J.*, Sept. 17, 1864, p. 897.
7. Ibid., Sept. 11, 1869, p. 1043.
8. *Old Time New England*, April 1932, p. 197; *Locomotive Engineering*, April 1893, p. 173.
9. *American Railway Times*, June 9, 1866, p. 182.
10. *National Car Builder*, November 1878, p. 169.
11. *A.R.T.*, March 1, 1868, p. 76.
12. *Loco. Eng.*, April 1893, p. 173.
13. *New York Graphic*, Jan. 17, 1876, p. 603.
14. Alvin F. Harlow, *Road of the Century*, New York: Creative Age Press, 1947, p. 287.
15. *Railroad Car Journal*, November 1894, p. 243.
16. *N.C.B.*, January 1871, p. 3, and May 1875, p. 68.
17. *R.R.G.*, April 22, 1871, p. 41.
18. Ibid., June 26, 1875, p. 269.
19. *Engineering*, April 4, 1875, p. 274.
20. Hamilton Ellis, *Railway Carriages in the British Isles from 1830 to 1914*, London: Allen & Unwin, 1965, p. 94.
21. *N.C.B.*, September 1876, p. 130.
22. *R.R.G.*, April 28, 1882, p. 254, and *A.R.R.J.*, Sept. 9, 1882, p. 664.
23. *R.R.G.*, July 7, 1882, p. 415.
24. *N.C.B.*, December 1885, p. 159.
25. *R.R.G.*, May 16 and Aug. 29, 1884, pp. 380 and 630.
26. Ibid., Aug. 21 and Oct. 21, 1887, pp. 530 and 678.
27. *R.R. Car Jour.*, February 1892, pp. 65 and 74.
28. *American Journal of Railway Appliances*, June 1884, p. 202.
29. Bendel, p. 36 and plate 13.
30. *R.R.G.*, Aug. 20, 1886, p. 577.
31. *N.C.B.*, March 1882, p. 31.
32. Ibid., June 1882, p. 66.
33. *R.R.G.*, May 12, 1882, p. 285.
34. Both the Pennsylvania and the New York Central railroads began operating luxury trains in 1887. The first was the Pennsylvania Special, better known by its later name of Broadway Limited. In the same year came New England's first name train: the Mt. Desert Limited. In 1888 the Golden State Special was introduced; in 1889 the Montezuma Special. The history of these trains is given in A. D. Dubin's *Some Classic Trains* (Milwaukee: Kalmbach, 1964) and L. M. Beebe's two-volume work, *Trains We Rode*, Berkeley: Howell-North, 1965 and 1966.
35. *Rly. Rev.*, April 16, 1887, p. 226.
36. *Railway Mechanical Engineer*, April 1926, p. 221.
37. *Popular Mechanics*, March 1933.
38. *Chicago Daily Tribune*, June 15, 1950.
39. Dubin, pp. 170–171.
40. *Rly. M.E.*, April 1926, p. 221.
41. F. Gutbrod, "The Construction of Iron Passenger Cars on the Railways in the United States of America," *International Railway Congress Bulletin*, vols. 26–27, 1912–1913, part 4, p. 121. Hereafter referred to as Gutbrod Report. Also see Dubin, p. 266.
42. W. W. Kratville, *Passenger Car Catalog: Pullman Operated Equipment, 1912–1949*, Omaha: Kratville Publications, 1968, pp. 82–87.
43. Jim Scribbins, *The Hiawatha Story*, Milwaukee: Kalmbach, 1970, p. 109.
44. *Trains*, September 1956, p. 8.
45. The peak year for dining cars was 1930, when 1,742 were reported in service by the I.C.C.
46. *Popular Mechanics*, October 1925.
47. Ibid.
48. W. C. Bryant, *Picturesque America*, New York: Appleton, 1872.
49. Edward Hungerford, *The Modern Railroad*, Chicago, 1918, p. 304.
50. *R.R.G.*, Jan. 25, 1884, p. 72, and *N.C.B.*, November 1883, p. 125.
51. *Harper's Weekly*, Aug. 25, 1888, p. 643.
52. L. M. Beebe, *Hear the Train Blow*, New York: Grosset and Dunlap, 1952, p. 377, reproduces a Burlington menu of the 1875–1880 period.
53. *Popular Mechanics*, October 1925.
54. *B & O Magazine*, September 1942, p. 64.
55. W. P. Smith, *History and Description of the B. & O. R.R.*, Baltimore, 1853, p. 147.
56. *Rly. Rev.*, March 4, 1893, p. 129.
57. *Western Railroad Gazette*, Sept. 8, 1860.
58. *Engineer*, Philadelphia, Dec. 8, 1860, p. 130.
59. Ibid., Aug. 23, 1860, p. 15.
60. *R.R.G.*, Dec. 19, 1884, p. 893, and E. P. Mitchell, *Memories of an Editor*, New York, 1924.
61. *Lincoln Day by Day*, vol. III, Washington, D.C.: Lincoln Sesquecentennial Commission, 1960, p. 17.
62. J. W. Starr, *Lincoln and the Railroads*, New York, 1927, p. 253.
63. *Models of Hospitals Cars*, a pamphlet published by the U.S. Army Medical Museum in Washington, D.C., 1885.
64. Stanley Buder, *Pullman: An Experiment in Industrial Order*, New York: Oxford, 1967, p. 20.
65. Charles Nordhoff, *California: For Health, Pleasure and Residence*, New York, 1872, p. 29.
66. *Engineering*, July 17, 1869, pp. 46 and 50.
67. *A.R.T.*, April 11, 1868, p. 119, and *Rly. Rev.*, May 7, 1868, p. 1.
68. *N.C.B.*, July 1879, p. 102.
69. Ibid., November 1872, p. 13, and Poor's *Manual of Railroads*, New York, 1871–1872.
70. *N.C.B.*, March 1882, p. 31.
71. *R.R.G.*, Feb. 24, 1882, p. 115.
72. *N.C.B.*, May 1877, p. 68.
73. Ibid., July 1881, p. 84.
74. *Rly. Rev.*, Dec. 10, 1892, p. 777.
75. *Rly. Age*, Jan. 13, 1881, p. 17.
76. *R.R.G.*, May 12, 1882, p. 285.
77. Edward Hungerford, *Men and Iron*, New York: Crowell, 1938, pp. 247 and 305.
78. *Rly. Rev.*, July 19, 1884, p. 378.
79. Ibid., May 7, 1881, p. 253.
80. G. M. Best, *Iron Horses to Promontory*, San Marino, Calif.: Golden West Books, 1969, p. 141.
81. *R.R. Car Jour.*, August 1891, p. 120.
82. Cost figures are from the following sources, in the order given: *R.R.G.*, Aug. 19, 1887, p. 536; ibid., May 19, 1905, p. 540; and American Railway Car Institute *Reports*, 1940 through 1960.
83. *Popular Mechanics*, October 1925.
84. Hungerford, *Modern Railroad*, p. 308.
85. *R.R.G.*, Aug. 5, 1887, p. 517.
86. *Trains*, March 1943, p. 6.
87. Ibid., December 1959, p. 40.
88. *Rly. Rev.*, June 9, 1883, p. 321.
89. *N.C.B.*, August 1886, p. 107.
90. *Rly. Rev.*, Sept. 9, 1899, p. 504.
91. *R.R.G.*, Jan. 15, 1886, p. 35.
92. Ibid., Sept. 22, 1893, p. 700.
93. *R.R.G.*, March 4, 1910, p. 447.
94. *Rly. M.E.*, February 1914, p. 77.
95. Ibid., June 1917, p. 346.
96. Ibid., March 1927, p. 163.
97. W. W. Kratville and H. E. Rank, *Union Pacific Equipment*, Omaha: Kratville, 1969, p. 37.
98. *Mutual Magazine*, August 1940.
99. *Rly. M.E.*, May 1950, p. 265, and *Trains*, September 1951, p. 37.
100. R. K. Wright, *Southern Pacific Daylight Train 98–99*, Thousand Oaks, Calif.: Wright Enterprises, 1970, vol. 1, p. 325.
101. *Trains*, October 1956, p. 24.
102. Ibid., May 1958, p. 19.
103. *N.C.B.*, October 1883, p. 110.
104. *Rly. Rev.*, Nov. 4, 1882, p. 626.

105. *Rly. M.E.*, March 1933, p. 109.
106. *Pullman News*, October 1950, p. 16, and April 1953, p. 2.
107. *R.R.G.*, March 2, 1906, p. 207.
108. *R.R.G.*, Dec. 20, 1895, p. 834, and May 3, 1907, p. 623.
109. Ibid., July 29, 1904, p. 190.
110. *Rly. Age*, Nov. 17, 1945, p. 799.
111. W. W. Kratville, *Steam, Steel & Limiteds*, Omaha: Kratville, 1962, p. 48.
112. Pacific Railroad Club *Proceedings*, August 1920, p. 9.
113. *Trains*, February 1953, p. 18.
114. Wright, *Southern Pacific Daylight*, p. 325.
115. *Mutual Magazine*, August 1940.
116. *Railroad Magazine*, September 1946, p. 88, and *Rly. M.E.*, April 1949, p. 187.
117. *Rly. M. E.*, July 1949, p. 376.
118. Ibid., May 1928, p. 273.
119. *Rly. Age*, April 12, 1930, p. 854.
120. John F. Stover, *Life and Decline of the American Railroad*, New York: Oxford, 1970, pp. 211–212.
121. I.C.C., *Transportation Statistics in the U.S.*, 1961, p. 52.
122. *Popular Mechanics*, October 1925, and J. H. Parmelee, *The Modern Railway*, New York, 1940, p. 116.
123. *Trains*, September 1951, p. 36.
124. John F. Stover, *American Railroads*, Chicago: University of Chicago Press, 1961, p. 242.
125. Wright, *Southern Pacific Daylight*, p. 442.
126. James W. Holden, "Mansions on Wheels," *Railroad Magazine*, November 1939, pp. 6–27; hereafter cited as Holden.
127. *Rly. Age*, Feb. 3, 1881, p. 59.
128. *N.C.B.*, April 1891, p. 56.
129. Holden, p. 26.
130. R. E. Carlson, *The Liverpool and Manchester Railway Project*, New York: Augustus Kelley, 1969, p. 232. See also Ellis, *Railway Carriages in the British Isles*, pp. 19 and 44.
131. S. M. Whipple, *Record of Thompson and Bachelder Center Lever or Tanner Brake*, New York, 1867, p. 20.
132. Harlow, *Road of the Century*, p. 129.
133. *Pullman's Palace Car Company vs. Jonah Woodruff and J. Hervey Jones*, U.S. Circuit Court, Western District of Pennsylvania, 1871, pp. 5 and 9. *George M. Pullman and Pullman's Palace Car Company vs. the New York Central Sleeping Car Company and Webster Wagner*, U.S. District Court, Northern District of Illinois, 1881, p. 786.
134. *N.C.B.*, November 1893, p. 185.
135. William P. Smith, *The Great Railway Celebration of 1857*, New York, 1858, pp. 217–218, and *Railroad Advocate*, Aug. 8, 1857, p. 35.
136. I. S. Trotter, *First Impressions of the New World on Two Travellers from the Old in the Autumn of 1858*, London, 1859, pp. 141–144; and Bendel, p. 38 and plate E, fig. 7.
137. *Am. Rly. Rev.*, Feb. 21, 1861, p. 99; March 27, 1862, p. 398; Nov. 22, 1863, p. 375.
138. *Columbian Weekly Register*, Nov. 24, 1866.
139. W. H. Stennett, *Yesterday and Today: a History of the C.&N.W.*, Chicago, 1910, pp. 71–72.
140. *N.C.B.*, August 1872, p. 10.
141. Best, *Iron Horses*, pp. 110–111.
142. Ibid., p. 48.
143. *N.C.B.*, February 1873, p. 35.
144. *Trains*, November 1949, p. 64.
145. *N.C.B.*, October 1881, p. 117, and Holden, p. 15.
146. Ibid., May 1871, p. 5.
147. Charles S. Sweet, "History of the Sleeping Car," Chicago, 1923, p. 156c. (Unpublished mss.)
148. A Dayton, Ohio, newspaper dated Aug. 2, 1878, speaks of a new car produced for Barnum by Barney and Smith. It is not clear whether this was Barnum's first PV.
149. *N.C.B.*, January 1876, p. 5.
150. Lucius Beebe, *Mansions on Rails*, Berkeley, Calif.: Howell-North Books, 1959, p. 195; hereafter Beebe, *Mansions*.
151. Arthur Dubin of Chicago owns a copy of Pullman's Detroit shop order book. *Pullman News*, April 1936, quotes a Rochester newspaper of Nov. 26, 1877.
152. *R.R.G.*, June 22, 1877, p. 282.
153. *Pullman News*, January 1932, p. 277.
154. *R.R.G.*, Sept. 30, 1887, p. 641, and *Pullman News*, April 1936, p. 117.
155. *Rly. Rev.*, June 10, 1882, p. 336.
156. *R.R.G.*, April 6, 1888, p. 227.
157. Ibid., Sept. 13, 1889, p. 599.
158. *Rly. Rev.*, Sept. 5, 1885, p. 431.
159. *R.R.G.*, Aug. 19, 1887, p. 547.
160. *N.C.B.*, October 1876, p. 150.
161. *Rly. Rev.*, June 10, 1882, p. 336. Beebe claims that the remains of the *Convoy* are in a museum at Jefferson, Texas.
162. *Rly. Rev.*, June 2, 1887, p. 393.
163. Beebe, *Mansions*, pp. 102 and 146–151.
164. *R.R.G.*, Sept. 11, 1885, p. 580.
165. *R.R. Car Jour.*, April 1897, p. 87.
166. Ibid., August 1898, p. 234.
167. *Pullman News*, Jan. 1941, p. 91.
168. Beebe, *Mansions*, p. 67.
169. *Christian Science Monitor*, Oct. 11, 1957, p. 15.
170. Beebe, *Mansions*, p. 204.
171. James Dredge, *The Pennsylvania R.R.*, London, 1879, p. 153 and plate 53.
172. E. Lavoinne and E. Pontzen, *Les Chemins de Fer en Amerique*, Paris, 1882.
173. Beebe, *Mansions*, p. 24.
174. *Scientific American*, October 1931, p. 258.
175. *Rly. Rev.*, May 11, 1907, p. 393.
176. *A.R.R.J.*, January 1910, p. 36.
177. Data on the *Oriental* was furnished by the Adirondack Historical Association, Blue Mountain Lake, N.Y.
178. *Fortune*, July 1930, p. 65.
179. Holden, p. 23.
180. Beebe, *Hear the Train Blow*, p. 391.
181. W. D. Randall, *From Zephyr to Amtrak*, Park Forest, Ill.: Randall, 1972, p. 198.
182. *Cincinnati Enquirer*, Sept. 16, 1962.
183. *Popular Science*, May 1955, p. 153.
184. Beebe, *Mansions*, p. 271.
185. Alvin F. Harlow, *Steelways of New England*, New York: Creative Age Press, 1946, p. 108.
186. *Van Nostrand's Engineering Magazine*, June 1870, p. 657.
187. This 1872 Pullman car rental pamphlet is in the collections of the Newberry Library, Chicago, Ill.
188. *Rly. Rev.*, June 10, 1882, p. 336.
189. *Poor's Directory of Railway Officials* 1887, page x.
190. L. M. Beebe, *Mr. Pullman's Elegant Palace Car*, Garden City, N.Y.: Doubleday, 1961, p. 217, is mistaken on the exact dates involving Worcester. It also incorrectly states that the first car was built new in 1875.
191. Data transcribed on an original print of the *City of Worcester* in the author's possession. See also *Rly. Age*, May 22, 1884, p. 329.
192. *Biographical Dictionary of Railway Officials of America*, New York and Chicago, 1885, p. 159.
193. *N.C.B.*, August 1879, p. 114.
194. *R.R.G.*, July 27, 1883, p. 501.
195. Beebe, *Mr. Pullman's Elegant Palace Car*, p. 218.
196. *Rly. Rev.*, Jan. 10, 1903, p. 20.
197. Ibid., June 19, 1886, p. 319.
198. *A.R.R.J.*, June 1893, p. 285.
199. *Scientific American*, October 1931, p. 258.
200. Holden, p. 26, and a 1939 promotional booklet on Pullman's private rental cars, reprinted in 1972 by Robert Wayner of New York City.
201. *Trains*, November 1949, p. 64.
202. *Railroad Magazine*, November 1945, p. 13.
203. *Lincoln Day by Day*, vol. III, pp. 114, 122, 220.
204. Victor Searcher, *Farewell to Lincoln*, New York: Abingdon Press, 1965, p. 87.
205. W. H. Price wrote an account of the Lincoln car for *Locomotive Engineering*, September 1893, p. 415. A similar account by S. D. King appeared in J. W. Starr's *Lincoln and the Railroads*, p. 276.
206. *Lincoln Lore*, May 1957, no. 1431, published by the Lincoln National Life Insurance Company, Fort Wayne, Indiana.
207. *National Republican*, April 25, 1865.
208. *R.R.G.*, Aug. 27, 1886, p. 603.

209. All accounts cited above agree on the U P purchase.
210. E. L. Sabin, *Building the Pacific Railway*, Philadelphia, 1919, p. 282.
211. *R.R. Car Jour.*, April 1897, p. 89.
212. *Denver Tribune Republican*, quoted in *R.R.G.*, Aug. 27, 1886, p. 603.
213. *N.C.B.*, May 1893, p. 80.
214. Advertising materials published by Snow are in the collections of the Chicago Historical Society.
215. *Rly. Age*, Sept. 22, 1905, p. 347.
216. *Minneapolis Journal*, March 19, 1911.
217. *R.R. Car Jour.*, 1897: February, p. 30; March, p. 57; April, p. 95; May, p. 120; July, p. 183.
218. Information leaflet issued around 1969 by the Gold Coast Railroad, Fort Lauderdale, Florida.
219. *R.R.Mag.*, November 1945, p. 24.
220. *New York Times*, April 2, 1961.